D1283611

The Triune Brain in Evolution

Role in Paleocerebral Functions

Library of Congress Cataloging in Publication Data

MacLean, Paul D.
The triune brain in evolution: role in paleocerebral functions / Paul D. MacLean.
p. cm.
Includes bibliographical references.
ISBN 0-306-43168-8
1. Brain—Evolution. 2. Animal behavior. 3. Neuropsychology. I. Title.
QP376.M185 1989 89-22899
596′.0188—dc20 CIP

First Plenum printing 1990

Plenum Press is a division of
Plenum Publishing Corporation
233 Spring Street, New York, N.Y. 10013

Printed in the United States of America

To my wife Alison

Preface

Succinctly put, the brain appears to be a kind of detecting, amplifying, and analyzing device for maintaining us in what we perceive as our internal and external environment. At the same time, one may wonder how an organ so tenuous has the capacity to provide the impression of the dependable solidity of the world in which we live. Quantum mechanics and molecular biology do not provide much help in understanding these matters. In the scientific world the brain has often been regarded as imprecise and untrustworthy. This traditional view is exemplified by a statement made by James Jeans, the well-known physicist and mathematician. He contrasted the difficulties of a philosopher who must rely on his own brain with those of "the modern physicist who can trust his instruments to give absolutely objective and unbiased information." It was as though in scientific matters a way had been found to bypass the brain. Yet it would be just as reasonable to challenge the human brain's ability to design and construct instruments of precision, to record their exact measurements, and to interpret the findings. Given the traditional untrustworthy view of the subjective brain, where does confidence lie in any field? Hence, the question is raised as to how there can be substantive advances in knowledge without trying to learn the workings and limitations of the subjective brain.

One might respond by saying that, thanks to human preoccupation with mastering the external environment, ways and means have been found to make it increasingly worthwhile to investigate the structure, chemistry, and functions of the brain. Anyone who harbors this optimistic attitude might also hope that, in addition to the foregoing question, such investigation might provide the best promise of shedding light on the universal, basic question, "What is the meaning of life?" In view of the foregoing considerations, I call attention in Chapter 1 to the need for concerted effort to comprehend the nature of the subjective brain—a pursuit of knowledge complementary to epistemology and referred to as epistemics. Bertrand Russell contended that introspective data are inappropriate for scientific study because they do not obey physical laws, but we will find evident inconsistencies in such claims.

Granted that the brain is an analyzer, it consists of many subanalyzers such as those associated with the sensory systems. It is of special import, however, that if one takes an evolutionary approach to the study of cerebral function, one finds that the brains of advanced mammals comprise an interconnected amalgamation of three main analyzers that in their structure and chemistry reflect developments identified, respectively, with reptiles, early mammals, and late mammals. This triune organization suggests approaches to neurobehavioral studies that otherwise would not be apparent.

The main substance of the present book concerns comparative neurobehavioral and clinical studies germane to the evolutionary considerations just mentioned. The findings would appear to contribute to an understanding of seemingly idiosyncratic aspects of animal and human behavior and of human subjective experiences. But in the end, one uncovers a most disquieting question when assessing the subjective functions of the human limbic system, a part of the forebrain stemming from early mammals. Here the evidence, along with other considerations, seems to present an insurmountable obstacle to our ever obtaining confidence in scientific or other intellectual beliefs—a confidence that is essential to make it worthwhile to pursue a search for the meaning of life. If the reader were able to contradict or circumvent the interpretations that have been reached in this respect, that would be a contribution of endless value.

Paul D. MacLean

Poolesville, Maryland

Acknowledgments

For appointments and support of the studies referred to in this book, I acknowledge my special indebtedness to Stanley Cobb† of Harvard Medical School; John F. Fulton,† Frederick C. Redlich, and Vernon W. Lippard† of Yale University School of Medicine; and Robert B. Livingston, Wade H. Marshall,† and John C. Eberhart of the National Institute of Mental Health. Among associates I am particularly beholden to Robert S. Schwab†, Mary A. B. Brazier, John A. Abbott, C. N. Hugh Long,† Donald Barron, Seymour Kety, Robert A. Cohen, and David Shakow.† I must count on citations within the book itself as acknowledgement of the friendship and stimulation I have received from collaborators in my own and other laboratories. For administrative help I owe special thanks to Gordon Klovdahl, Hector Ragas, Hazel Rea, and Alice Muth Cohen. During a later period I also acknowledge the scientific administrative support of Frederick K. Goodwin, Dennis L. Murphy, and Thomas R. Insel, and their respective staffs. Beyond the confines of the above institutions, I wish to express deep appreciation for the interest and encouragement of James W. Papez,† Paul I. Yakovlev,† Lawrence S. Kubie,† Leonard Carmichael,† and Robert L. Isaacson.

As in other technical areas, including neuroanatomical reconstructions, Robert E. Gelhard was of invaluable help in preparing many of the illustrations and doing the drawings for some of them. I owe special thanks to Levi Waters and C. Ronald Harbaugh for dedicated animal care; to Martha Carmichael Oliphant for general technical help; to George F. Creswell and Thalia Klosterides Bussard for beautiful histology; to John Lewis† for photography; to Anthony Bak† and Patrick J. Carr† for invaluable contributions in electronics; and to several persons who typed portions of the manuscript, including Helene K. Sacks, Deborah F. Lonsdale, Marcha K. Carroll, Katherine V. Compton, Mary Ann Bandy, Nancy Sawyer, Judith Osler Newman, and, finally, Francie L. Kitzmiller, who finished and assembled the entire manuscript.

Aside from the universities, financial support derived largely from the munificence of the U.S. Public Health Service, the U.S. Veterans Administration, the Foundation Fund for Research in Psychiatry, and the National Science Foundation.

A final acknowledgment requires some explanation. The impetus to write the present book might be attributed to the occasion when I presented the Thomas William Salmon Memorial Lectures on December 7, 1966. Originally it was the stipulation that the lectures be published in a bound volume. At a later time, when the number of lectures was reduced from three to two, it meant that more material had to be added in order to provide sufficient pages for a bound volume. In my case, the intended solution developed as follows: The first slide of my lectures showed a colored representation of Figure 2-1 in the

†Deceased.

present book, illustrating that in evolution the brains of advanced mammals comprise a hierarchical organization of three main neural assemblies that reflect an ancestral relationship to reptiles, early mammals, and late mammals. The focus of my lectures was on the functions of the paleomammalian brain, which corresponds to the limbic system that molds itself around the protoreptilian forebrain. Regarding the protoreptilian brain, I was able to do little more than speculate that it participates in orchestrating basic forms of behavior required for self-preservation and preservation of the species. In this respect, I said, "We are planning experimental brain studies at the new NIMH Field Laboratory to test these and other possibilities. Instead of the traditional procedure . . . we shall go to the field to learn what parts of the brain are required . . . for survival in a natural habitat." Accordingly, it seemed promising at that time to round out a volume based on the Salmon Lectures with research findings on the protoreptilian brain. But beyond delays in architectural plans and construction, the research itself became a greater gobbler of time and interest. Consequently, I am only now able to acknowledge the proffered Salmon gift toward the publication of the present book and to express my esteem for, and gratitude to, the late Francis J. Braceland, who was chairman of the committee at the time of my award, and to his associates on the committee, who were Kenneth E. Appel, Lawrence C. Kolb, David M. Levy, Nolan D. C. Lewis, William T. Lhamon, H. Houston Merritt, and S. Bernard Wortis.

* * *

It remains to express my appreciation to Plenum Publishing Corporation, and to thank especially Mary Phillips Born, Senior Editor, for her skillful guiding hand from start to finish; Kirk Jensen for judicious help in the preliminary phases; Shuli Traub, Production Editor, for attention to every detail; Mary Safford, Senior Production Editor, for splendid supervision; and David Bahr, discerning copyeditor.

In my own office, I am especially indebted to Francie Louise Kitzmiller for her hard work on the indexes.

P.D.M.

Contents

I. INTRODUCTION

II. THE STRIATAL COMPLEX WITH RESPECT TO SPECIES-TYPICAL BEHAVIOR

III. THE LIMBIC SYSTEM WITH RESPECT TO THYMOGENIC FUNCTIONS

IV. NEO-ENCEPHALON WITH REGARD TO PALEOCEREBRAL FUNCTIONS

V. CONCLUSION

I

Introduction

1

Toward a Knowledge of the Subjective Brain (“Epistemics”)

Introductory Considerations

In the human quest for a cosmic view of life, it would seem to be of primary importance to obtain a better understanding of the brain. In both its substance and communicative capacity, the brain is incommensurate with our presumed instruments of precision used to gauge “the world out there.” Might not certain problems be resolved if we were more conversant with the limitational workings of the subjective brain? Take, for example, questions regarding the origin of the universe. Subjectively it is as difficult to think of a state of nothingness as it is to conceive of how space and things within space came into being. Some physicists calculate that there was a moment when there was infinite density at a point in space, whereas others would claim that at time zero “the whole universe, the infinite space, was filled with an infinite density of matter.”[1] In considering elementary particles, superstring theory requires us to go beyond our present measuring devices and imagine in the case of quantum gravity the Planck distance of 10^{-33} cm^2. Perhaps, Chodos comments, some version of superstring will be the ultimate description of nature, but “since the Planck scale is so hopelessly beyond the reach of . . . experiment, how would we ever know?”[2] Or consider the nature of time and space (Kant’s “transcendental aesthetic”[3]), which do not exist *per se,* but are derivatives of the subjective brain, being purely *information* that is of itself neither matter nor energy.

Foremost of all, is it possible that further knowledge of the subjective brain might give insights into the meaning of life and the justification for the perpetuation of life with the untold suffering that afflicts so many forms of life? A theoretical physicist has queried, “Might not life have a more important role in cosmology than is currently envisioned? That is a problem worth thinking about. In fact, it may be the only problem worth thinking about.”[4]

Questions of this kind may seem quite esoteric and removed from the real world until one pauses to reflect that the subjective brain imprisoned in its bony shell is the sole judge of its own existence and the presumed existence of what lies outside. Moreover, because the brain reconstructs the world we live in, it does not have, nor ever can have (because of self-reference[5]), any yardstick of its own by which to measure itself.

Understandably, there is the apprehension that unless one can be assured of an outside existence, there can be, in selfish terms, no reason for continuing an inside

Parts of this and the next chapter appeared in an introductory preview of the present book published as a separate article and subsequently reprinted in a number of compendia (see MacLean, 1975a).

existence. Since Hiroshima there has been universal concern that the capability of thermonuclear war threatens the existence of life on this planet. Yet in this respect, and apropos of one of the introductory questions, it is curious that those who most loudly warn against the possibility of nuclear war and the extinction of life, fail at the same time to articulate reasons for justifying the perpetuation of life either here or elsewhere in the universe.

In recent years there has been a competing concern that humankind and many other forms of life may be on the way to extinction because of scientific developments that have made possible overpopulation, pollution of the environment, and exhaustion of critical resources. A curve showing the growth of the world's population indicates that each successive doubling of the population has taken place in half the time of the previous doubling.[6] At this rate, the present population would be expected to double in 30 to 40 years. In 1969, U Thant, speaking as Secretary of the United Nations, made the pronouncement that there remained only 10 years to find solutions for the exploding population.

All of the foregoing kinds of warnings have tended to focus attention on the external environment. In connection with overpopulation, planning experts have been so preoccupied with problems pertaining to demands for food, water, energy, and the other basic requisites that they seem to have overlooked the lessons of animal experimentation indicating that the psychological stresses of crowding may bring about a collapse of social structure despite an ample provision of the necessities of life.[7] Systems analysts who have attempted to predict the limits of growth with the aid of computer technology either admit to an inability to deal with psychological factors or neglect them altogether.[8]

Objectivity versus Subjectivity

Michael Chance[9] has remarked that the parts of the universe first chosen for formalized study were those furthest removed from the self—meaning, of course, the heavens and the science of astronomy. In modern times the remote intangibles of nuclear physics have become of rival interest.

Objectivity

Achievements of the "exact" sciences have helped to promote the attitude that solutions to most problems can be found by learning to manipulate the external environment. It has been traditional to regard the exact sciences as completely objective. A statement by Einstein illustrates the self-conscious cultivation of the objective approach. C. P. Snow[10] quotes him as saying, "A perception of this world by thought, leaving out everything subjective became . . . my supreme aim." Monod,[11] in an essay on the contributions of molecular biology, is equally insistent on applying the "principle of objectivity" in the life sciences. "The cornerstone of the scientific method," he writes, "is the postulate that nature is objective."[12] Even in the world of fiction one finds a book reviewer saying, "Humanity is likely to be saved, if it is at all, by a search for an objective reality we can all share—for truths like those of science."[13]

Early in this century, John B. Watson and others of the behaviorist school sought to revive the spirit of the Helmholtz tradition and establish psychology as an exact science on an equal footing with physics and the other "hard" sciences.[14] As Watson explained, "In

1912 the behaviorists reached the conclusion'' that they would drop from their ''scientific vocabulary all subjective terms such as sensation, perception, image, desire, purpose, and even thinking and emotion as they were subjectively defined.''[15] In a satirical examination of behaviorism, Koestler commented that Watson's precepts have, unbelievably, continued to take hold in modern times.[16] He quoted one contemporary authority as saying that ''mind'' and ''ideas'' are nonexistent entities, ''invented for the sole purpose of providing spurious explanations. . . . Since mental or psychic events are asserted to lack the dimensions of physical science, we have an additional reason for rejecting them.''[17] This last statement is an echo of Bertrand Russell's[18] claim that psychological phenomena are inappropriate for study because they do not obey physical laws.

Jeans[19] has stated that ''physics gives us exact knowledge because it is based on exact measurements.'' But if the ultimate scientific instrument, the human brain, is for one reason or another predisposed to artifactual interpretations, where does confidence lie in any field?

The irony of the completely objective approach is that every behavior selected for study, every observation, and every interpretation, requires subjective processing by an introspective observer. Logically, there is no way of circumventing this or the other inescapable conclusion that the cold, hard facts of science, like the firm pavement underfoot, are informational transformations by the viscoelastic brain. No measurements obtained by the hardware of the exact sciences are available for comprehension without undergoing subjective transformation by the ''software'' of the brain. The implication of Spencer's[20] statement that objective psychology owes its origins to subjective psychology could apply equally to the whole realm of science. For such reasons, one is obliged to consider how the nondimensional emanations of the subjective brain may account for a dimensional view of the world.

Subjectivity

As will be remarked upon in a later chapter, the neocortex subserving ''intellectual'' functions appears to be primarily oriented toward the external world. This may help to explain why the sciences from the very beginning have focused on the outside world. By contrast, and perhaps for similar reasons, there has been a retarded interest in turning the dissecting lamp of the scientific method onto the inner self and the psychological instrument accounting for the derivation of all scientific knowledge. Until fairly recent times, theologians and philosophers were the recognized authorities on psychological matters, among whom Aristotle, as in so many of his systematic endeavors, would rank high on the list. Although having their modern origins in the 18th century, psychology and psychiatry did not begin to acquire scientific status until the latter part of the 19th century. As explained in one encyclopedic article on psychology written in 1892, ''After having long occupied a doubtful place as a department of metaphysics . . . , its character as a science dealing with a special order of facts . . . may now be said to be established.''[21] According to Kathleen Grange,[22] the term *psychology* was used in titles as early as 1703, while *psychiatry* first appeared on a title page in 1813. The precedent for including psychiatry in the medical curriculum occurred in 1854, when Griesinger at the University of Munich united for the first time the teaching of neurology and psychiatry.[23] Meynert, Gudden, Forel, and others followed this practice and established it as a tradition in Europe. Since the middle of the present century, neurology has tended to follow an independent course, delving into psychological functions only insofar as particular disturbances in cerebration make it possible to diagnose the nature and location of brain disease. Psychoanalysis, which gave new conceptual and methodological dimensions to psychiatry, first began to attract public interest in 1900 with the publication of Freud's[24] *The Interpretation of Dreams.*

The late development of the psychological sciences raises a question of epistemological interest. Why is it that none of the psychological sciences devotes itself specifically to epistemological questions concerning the origin, nature, limits, and validity of knowledge? Except for sensation and perception, it is also curious how little attention philosophers and others have given to the role of the brain in matters of epistemology.

In stating the truism that epistemology exists because of human societies and that human societies exist because of individual persons, it serves to emphasize the incontrovertible centricity of the individual person with respect to public knowledge. In constitutional language, public knowledge, just as society itself, derives authority from individuals. In extending the analogy, it was men and women as individuals who gave John Adams[25] the authority to say we are a "government of laws, *not of men.*"

Central to every individual is a subjective self—a self that Descartes referred to as "this me."[26] A dissection of the subjective self requires that it be laid open not only in terms of its inner workings, but also in relationship to the societal and nonsocietal elements of the external environment. There are two sides to each of these relationships: the side that is intuitively and unsystematically experienced and the side that becomes known through the analytic and synthetic approaches of the sciences. The animate relationships become systematically known through the social and life sciences, while formal knowledge of the inanimate derives from the natural sciences.

"Epistemics"

There exists no branch of science that deals specifically with an explanation of the subjective self and its relation to the internal and external environment. While such a study would draw from every field of knowledge reflecting on the human condition, it would build fundamentally on the psychological and brain-related sciences. In order to have a matching expression for epistemology, as well as an equivalent term for science, one might borrow a word from the Greek, and refer to an "episteme (επιστημη) of the self." Then the body of knowledge or the collective disciplines dealing with this subject could be succinctly referred to as *epistemics.*[27]*

It requires emphasis that the domains of epistemics and epistemology are the same. The difference lies in the point of view. Epistemics represents the subjective view and epistemic approach from the inside out, whereas epistemology represents the public view and scientific approach from the outside in. The two are inseparable insofar as epistemics is nuclear to epistemology, and epistemology embraces epistemics. What is entailed is an obligatory relationship between a private, personal brain and a public, collective, societal brain.

In Plato's[28] Protagoras we are told of two far-famed inscriptions in the temple of Apollo at Delphi—"Know thyself" and "Nothing too much."[21] The latter precept might

*The author first used the word *epistemics* in its present sense in an article published in 1975 (see MacLean, 1975a).

otherwise be stated as "Nothing in excess." If, as Plutarch[29] claims, all other commandments hang on these two precepts, why are they contradictory? Was it the implication that it is possible to have too much self-knowledge? When toward the end of this book we consider the subjective phenomenology associated with limbic function, we uncover what seems to be a subjective manifestation that is not to be confused with the familiar Cartesian perceptual illusions. It presents an epistemic impasse that seems to have been ignored by philosophers and others because of a void in the knowledge of brain mechanisms. Unless some way can be found around it, it presents an impasse that might lead one to ask in the words of William Morris,[30] was it "all for this?"

References

1. Weisskopf, 1983
2. Chodos, 1986
3. Kant, 1899
4. Pagels, 1982
5. Bronowski, 1966
6. Calhoun, 1971; von Foerster *et al.*, 1960
7. Calhoun, 1962; Myers *et al.*, 1971
8. e.g., Meadows *et al.*, 1972
9. Chance, 1969
10. Snow, 1967, p. 90
11. Monod, 1971
12. Monod, 1971, p. 21
13. Weisberger, 1972
14. See Shakow and Rapaport, 1964
15. Watson, 1924, p. 6
16. Koestler, 1968
17. Koestler, 1968, p. 7
18. Russell, 1921
19. Jeans, 1943
20. Spencer, 1896
21. Chambers' Encyclopedia, 1892a
22. Grange, 1961
23. See Ackerknecht, 1959
24. Freud, 1900/1953
25. Adams, 1780
26. Descartes, 1637
27. MacLean, 1975a, 1977
28. Plato, 1937
29. Plutarch, 1962
30. Morris, 1858/1933

2

Specific Indications for Brain Research

Herein too may be felt the powerlessness of mere Logic, the insufficiency of the profoundest knowledge of the laws of the understanding, to resolve these problems which lie nearer to our hearts, as progressive years strip away from our life the illusions of its golden dawn.

George Boole, *An Investigation of the Laws of Thought* (1854), p. 416

Wherever the human brain is described or pictured, it appears to us as a large, global organ completely enveloped by cortex and dominated by cortex. One imagines that John Locke had the outer cortical surface in mind when he compared the newborn brain to a "white paper,"[1] or, as he had earlier referred to it, a "tabula rasa," a clean slate for each individual to record afresh all the impressions that provide the basis of learning and knowledge. Two hundred years later, Pavlov's work on conditioned reflexes, with its main emphasis on the role of the *neo*cortex in learning and memory, seemed to provide scientific verification of Locke's thesis.[2] In psychology, it became the traditional view that the cortex accounts for all human learning and expression. Since verbal communication is a unique human function, many people presume that learning is primarily dependent on language. The emphasis on learning is illustrated by one psychological book in which the first sentence reads, "All human behavior is learned."[3]

Among cultural anthropologists the cerebral cortex also seems to be regarded as a clean slate for transcribing and transmitting the totality of culture from one generation to another. Here again one finds instances of a total emphasis on learning. In a critical essay on this matter, Ardrey quotes one authority as saying, "The evidence indicates quite clearly that everything that human beings do as human beings they have had to learn from other human beings."[4]

The above monolithic view of the neocortex is what one may expect in taking any of the usual philosophical and scientific approaches to the brain. The situation would be somewhat comparable to approaching the Pennine Alps from northern Italy, where the massive Monte Rosa obstructs a view of the Matterhorn and Mont Blanc just beyond. Quite a different view of the brain and its functions derives from a comparative evolutionary approach. A comparison of the brains of existing vertebrates, together with an examination of the fossil record, indicates that the human forebrain has evolved and expanded to its great size while retaining the features of three basic evolutionary formations that reflect an ancestral relationship to reptiles, early mammals, and recent mammals (see Figure 2-1). Radically different in chemistry and structure and in an evolutionary sense

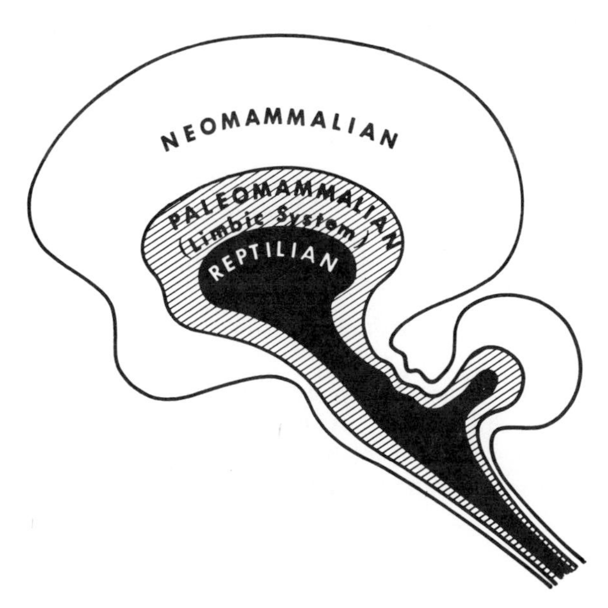

Figure 2-1. The "triune brain." In its evolution the human forebrain expands along the lines of three basic formations that anatomically and biochemically reflect an ancestral relationship, respectively, to reptiles, early mammals, and late mammals. The three formations are labeled at the level of the forebrain that constitutes the cerebral hemispheres comprised of the telencephalon and diencephalon. From MacLean (1968).

countless generations apart, the three neural assemblies constitute a hierarchy of three-brains-in-one, a triune brain.[5] Based on these features alone, it might be surmised that psychological and behavioral functions depend on the interplay of three quite different mentalities. Of further epistemic interest, there is evidence that the two older mentalities lack the necessary neural apparatus for verbal communication. Stated in popular terms, the three evolutionary formations might be imagined as three interconnected biological computers, with each having its own special intelligence, its own subjectivity, its own sense of time and space, and its own memory, motor, and other functions.

Improved anatomical, chemical, and physiological techniques have afforded a clearer definition of the three basic formations than was previously possible. It has also proved that the three formations are capable of operating somewhat independently. My emphasis on these distinctions seems to have led some writers to conclude that higher mammals are under the control of three autonomous brains. It was to guard against such an interpretation that some years ago I began to use the expression "the triune brain."[5] *Triune,* a concise term, derives letter by letter from the Greek. If the three formations "are pictured as intermeshing and functioning together as a *triune* brain, it makes it evident that they cannot be completely autonomous, but does not deny their capacity for operating somewhat independently."[6] Moreover, as diagrammed in Figure 2-2, the triune relationship implies that "the 'whole' is greater than the sum of its parts, because the exchange of information among the three brain types means that each derives a greater amount of information than if it were operating alone."[6]

It is perhaps worth noting at this point that some workers misinterpret the triune concept as implying a consecutive layering of the three main neural assemblies, somewhat analogous to strata of rock. Others, in their criticism, seem to be of the opinion that the concept leads one to believe that the counterpart of reptilian forebrain ganglia in mammals is just the same as in extant reptiles. This would be the equivalent of stating that the modern automobile engine is the same as that of the first car.

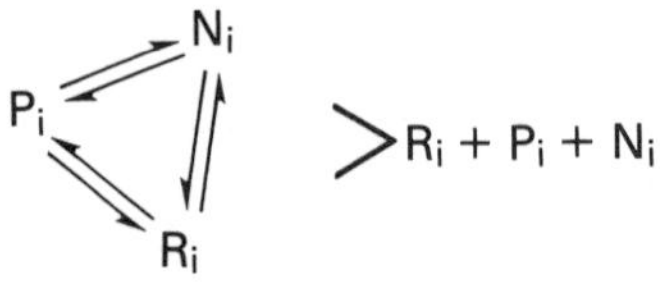

$$P_i \rightleftarrows N_i \rightleftarrows R_i \rightleftarrows P_i \quad > R_i + P_i + N_i$$

Figure 2-2. A scheme for indicating that through their intercommunication the three neural assemblies of the triune brain obtain a greater amount of information (i) than would be the case if each were operating independently. Abbreviations: R, reptilian; P, paleomammalian; N, neomammalian.

Special Focus of the Present Investigation

Operational Terminology

The focus of the present book will be on the protoreptilian and paleomammalian formations and their role in nonverbal aspects of mentation and behavior. Dealing with such a subject requires a brief explanation of operational terminology now to be considered under the headings (1) subjective experience, (2) "facts," and (3) communicative behavior.

Subjective Experience

For each one of us as individuals, there is nothing so vital as subjective experience. Without the essence of subjectivity, there would be no means of realizing our existence. Since subjectivity and its associated emanations of the mind represent forms of information, they have no material substance. As Wiener stated more succinctly than Berkeley or Hume, "Information is information, not matter or energy."[7] If psychological information is without substance, how do we put a handle on it for scientific purposes? Here, we are saved only by the empirical evidence that there can be no communication of information without the agency of what we recognize as physical, *behaving entities,* no matter how large or small. This invariance might be considered a law of communication. Metaphorically, behaving entities and information would compare to particles and waves of quantum mechanics.

"Facts"

Behaving entities provide the means of ascertaining what are known as "facts." Hence, facts are conventionally identified with something substantial, real, and true. In scientific discourse, they compare to the pitons of a mountain climber, being so hard and rigid that they can be nailed or wedged in anywhere for gaining a purchase. A step-by-step analysis shows, however, that "facts" *apply only to those things that can be agreed upon publicly as entities behaving in a certain way* (see Chapter 29). The term *validity* does not apply to the facts themselves, which are neither true nor false *per se,* but rather to what is agreed upon as true by subjective individuals after a public assessment of the facts. What is agreed upon as true by one group may be quite contrary to the conclusions of another group.

Communicative Behavior

Next in importance to our subjective experience is the ability to share what is felt and thought with other beings. Such communication must be accomplished through some form of behavior. Human communicative behavior can be categorized broadly as verbal and nonverbal. In his book, *The Way Things Are,* P. W. Bridgman, the Harvard physicist-philosopher, reaffirms the popular view when he says that "most communication is verbal."[8] Because of the accustomed emphasis on the unique verbal capacity of human beings, relatively little attention has been given to the role of nonverbal factors in communication. This neglect is evidenced by the lack of a specific word for nonverbal communication; the use of a negation to define it has tended to depreciate its significance.

Contrary to the popular view, many behavioral scientists give greater importance to nonverbal than verbal communication in human relations. For example, when a psychologist, a behavioral ecologist, a specialist in environmental design, and an ethologist were asked to draw two squares representative of the weight that they would give verbal and nonverbal communication in everyday human activities, there was a striking similarity in their responses. In each case, the square for nonverbal behavior was drawn about three times larger than that for verbal behavior. When shown the striking similarity of their drawings, each was quick to deny the possibility of achieving any quantitative assessment of the relative influence of verbal and nonverbal communication. To cite some obvious imponderables, how does one sort out and weigh the influence of nonverbal factors affecting one's choice of spouse, friends, associates; a vote for a particular candidate; one's judgments as a member of a committee or jury? All that the four specialists meant to signify, they agreed, was that verbal communication represents but the small, "visible" part of the communicative process, the tip of the iceberg.

Nonverbal ("Prosematic") Communication

Nonverbal communicative behavior mirrors in part what Freud called primary processes.[9] In drawing a distinction between verbal and nonverbal communication, it is easier to see differences than similarities. But in a less strict sense, nonverbal behavior, like verbal behavior, has an orderly arrangement (syntax) and meaning (semantics). It has been the special contribution of ethologists to provide the first scientific insights into the syntax and semantics of animal behavior.[10]

Many forms of human nonverbal communication show a parallel to behavioral patterns of animals. Since it is hardly appropriate to refer to nonverbal communication of animals,[11] there is the need for another term for this kind of communication. The Greek word *sema* (σημα)—a sign, mark, or token—conjoined with the prefix *pro* (προ) in its sense of "rudimentary," forms the word *prosematic,* which is appropriate for referring to any kind of nonverbal signal—vocal, bodily, or chemical.[12]

An analysis of prosematic behavior of animals reveals that analogous to words, sentences, and paragraphs, it becomes meaningful in terms of its components, constructs, and sequences of constructs (see Chapter 9). Since the patterns of behavior involved in self-preservation and survival of the species are generally similar in most terrestrial vertebrates, the customary designation of "species-specific" for such patterns in a given species is hardly suitable. But since various species perform these behaviors in their own typical ways, it is both correct and useful to characterize the distinctive patterns as "species-typical."

Introspectively, we recognize that prosematic communication may be either active or passive. When two or more individuals are within communicative distance, there is the possibility for either active ("intentional") or passive ("unintentional") communication to occur with respect to the "sender" or "receiver." Even when an individual is alone, a sound, utterance, movement, or odor emanating from the self may have self-communicative value as it originates either actively or passively.

Paleopsychic Processes and Paleomentation

My research has been primarily concerned with an attempt to identify forebrain mechanisms underlying prosematic forms of behavior that, on phylogenetic and clinical grounds, might be inferred to represent expressions of "paleopsychic" processes. For immediate purposes, I will use the term *paleomentation* to cover two main kinds of cerebration presumed to be involved in originating and organizing paleopsychic processes. One will be identified as protomentation, which applies to rudimentary cerebration involved in regulating the everyday master routines and subroutines, as well as the expression of four main behavioral patterns (displays) used in prosematic communication. The other will be referred to as emotional mentation—a form of cerebration that appears to influence behavior on the basis of information subjectively manifest as emotional feelings (Chapters 23 and 24). Defined briefly in such terms, *paleomentation* will now serve as a convenient label for forms of cerebration that are to be distinguished from rational mentation (ratiocination) which, in its formal properties, at least, lends itself to verbal description.

In drawing a distinction between emotional and rational mentation, I am aware that many people maintain that it is inadmissible to distinguish between "emotion" and "reason." Raphael Demos, in an introduction to the dialogues of Plato, expresses a traditional philosophical view: ". . . [W]e are apt to separate reason from emotion. Plato does not. Reason is not merely detached understanding; it is conviction, fired with enthusiasm."[13] Piaget, the founder of the Center for Genetic Epistemology, is quite vehement, saying that "nothing could be more false or superficial" than to attempt "to dichotomize the life of the mind into emotions and thoughts." ". . . [A]ffectivity and intelligence," he insists, "are indissociable and constitute the two complementary aspects of all human behavior."[14]

Granted the complementary aspects of "emotion" and "thought," we shall encounter evidence that the two may occur independently because they are products of different cerebral mechanisms. Such a physiological dichotomy has profound implications for epistemics and epistemology.

Why Brain Research?

Why is it necessary to investigate brain mechanisms in order to gain understanding of forms of paleomentation under consideration? After all, the laws of formal thought have been derived without taking the brain apart piece by piece to look at the machinery. It is the unique characteristic of rational mentation that it lends itself, as in the case of logic, to symbolic representations in the form of words or other signs that, when semantically specified and syntactically related according to certain rules, result in inevitable conclusions. A parallel situation applies to numerical procedures in which the steps of calculation are so interlocked as to assure an outcome as predictable as the movements of a geartrain.

In formal rational mentation, we have the advantage of being able to specify the inputs into our brains or the "prosthetic" brains of computers. But the situation is quite different in the case of paleomentation. Here the known input is so obscured by an indefinable input from the person's ancestral past and personal life history, that there is no means of ascertaining what the outcome will be. The successive mentational processes have neither been identified nor shown to obey laws that allow predictable conclusions.

Here it is significant to point out how different the situation is in connection with perceptions, excluding the consideration here of perceptual delusions that were of so much concern to Descartes. As opposed to mental states such as compulsions and emotions that have no specific gateways to the sensorium, perceptions depend on a publicly available information derived via the sensory systems. To be sure, as Livingston[15] has emphasized, what is perceived and how it is perceived may be greatly conditioned by one's cultural heritage and living conditions. The important distinction between perceptions and the other stipulated conditions is that the former depend on signals affecting auditory, visual, and other sensory receptors that can be publicly compared.

Since it is not possible to specify and deal operationally with inputs affecting paleomentation, an indirect approach for gaining an understanding of paleopsychic processes is, first, to identify brain structures providing the underlying mechanisms and, then, to conduct detailed studies on the neural architecture, nature of inputs and outputs, and the neurochemistry and physiology of these structures. The first step of identifying the underlying cerebral mechanisms depends on comparative neurobehavioral studies and on an analysis of clinical case material, including detailed postmortem findings on the brain. In regard to clinical data, we need to develop much more sophistication in obtaining, recording, and reporting histories and to use computer-assisted technology in performing a complete examination of the brain. Eventually, it is to be hoped that a national facility will become available for investigators to build up a "brain library" of unusual, well-studied cases. Such a facility could also be useful as a place to receive collections of comparative material, including, most importantly, the brains of species threatened by extinction.

An Evolutionary Approach to the Study of the Brain

An evolutionary approach to the study of the brain has special appeal because it requires both reductionistic and holistic analysis.[16] As Jacob[17] points out in his essay on "Evolution and Tinkering": "If molecular biology, which presents a strong reductionist attitude, yielded such a successful analysis of heredity, it was mainly because, at every step, the analysis was carried out simultaneously at the level of molecules and at the level of . . . the bacterial cell." Commenting on the value of evolutionary studies, Darlington[18] remarks that although "viruses and bacteria are not the precise ancestors of our nuclear and cellular organisms . . . they show us how such ancestors must have worked." It is now recognized that in all animals there are molecular commonalities with respect to genetic coding, enzymatic reactions, and so on, that carry over into complex cellular assemblies. Nowhere is the uniformity of complex cellular assemblies more striking than in the cerebral evolution of vertebrates, both as it applies to similarities within classes and to certain commonalities across classes.

In my investigations, I have taken a comparative evolutionary approach which also has the special value that it allows one to telescope millions of years into a span that can be seen all at once, and somewhat like a plotted curve, reveals trends that would otherwise not be apparent.[19] Since animal experimentation provides us our only systematic knowledge of brain function, there is need here for a statement about the justification of using findings on animals for drawing inferences about the workings of the human brain. At the molecular and cellular levels, there is general enthusiasm for applying findings on animals to human biology. In the field of neurology and psychiatry, neurochemical and neuropharmacological discoveries have radically changed the medical treatment of certain disorders. But many people believe that neurobehavioral and neurological observations on animals have little or no human relevance. Standing opposed to such a bias is the evidence that in its evolution, the human brain has developed to its great size while retaining the chemical features and patterns of anatomical organization of the three basic formations characterized as reptilian, paleomammalian, and neomammalian. The comparative anatomical studies of such investigators as Jolicoeur *et al.*,[20] Armstrong,[21] and others are providing additional information about variations of development of components of the three formations that may in turn help to direct attention to the examination of certain behavioral functions.

"Fractal Biology" in Relation to "Dynamic Morphology"

Although one could cite several other reasons for conducting comparative studies on animals, I will single out only one, "fractal biology," because it relates to a developing field that promises to renew interest in the role of morphology in evolution, in normal function, and in disease. It is a strength of the comparative approach that it affords insights into how variations of pattern affect these different processes. New directions in the winds of change can also be sensed in a recent redefinition of mathematics as "the science of patterns."[22] And new directions foretell of a reuniting of the study of flora and fauna. The patterns of flora themselves challenge the imagination to harbor the possibility that, given the right cosmic conditions, life itself represents a crystallizing out of organic matter somewhat akin to the formation of crystals of inorganic substances.

The subject of fractal geometry[23] has rekindled interest in D'Arcy Thompson's book *On Growth and Form* first published in 1917.[24] The growth and form of a tree has symbolic applicability not only as it relates macroscopically to various phylogenetic trees and cladograms, but also to the treelike microstructure of parts of plants and organisms.

Thompson[24] felt compelled to establish biology on as firm a mathematical footing as that for the physical sciences. He gave emphasis to scaling, which as illustrated by musical scales, refers to recurring regularities. He drew attention, for example, to observations that many growth processes such as arborization in various forms of vegetation, the spiral patterns of pine cones and of snails,[25] develop according to ratios expressed by the series of numbers named after the discoverer Filius Bonacci, nicknamed Fibonacci and formally known as Leonardo of Pisa. Beginning with one, the Fibonacci series develops as follows: 1, 1, 2, 3, 5, 8, 13, 21, 34, 55, 89. . . . Each succeeding number equals the sum of the two preceding it. The series converges to 1.618 . . . , an irrational number referred to as ϕ. This ratio is identified with the golden mean of the Greeks and the golden rectangle of the Renaissance.

Scaling also applies to grossly irregular formations such as coral and cumulus cloud with its gyrallike accumulations. Anyone who works with the brain cannot help but wonder if it will be shown that its gyral accumulations pile up somewhat like cumulus.

In nature, successive branchings have the function of providing more surface area, as in the case of the bronchial tree of the lung, and more contacts, as in the case of the nerve cells of the brain. In the lung, the bronchial tree and the vascular tree conform as hand in glove. The branching of the bronchial tree adheres roughly to Fibonacci proportions for the first ten generations.[26] Such regularity in branching is referred to as self-similarity. But the greater the number of generations of branches, the less is the similarity because of the increasing number of irregularities.[26] Hofstadter[27] characterized the self-similarity in such complexity as the "sameness-in-differentness." In 1977, Mandelbrot[28] introduced the term *fractal* (from the word *fraction*) for referring to structures of this kind. In the case of the lung, West and Goldberger[26] found that by applying a new scaling principle based on solutions used in the physical sciences, they obtained curves that compensated for the irregularities. In another study involving cardiac decompensation, they suggested that disease in the conduction system of the heart introduces the consideration of "fractal time" as it relates to periodicity.[26]

These various examples of *dynamic morphology* suggest parallels in the structure and function of the nervous system and raise questions, for instance, as to how branching affects axoplasmic flow in development; in health and disease; and in regeneration. And, timewise, how does arborization and the consequent slowing down of nerve impulses propagating into terminals affect chemoreceptor mechanisms at synaptic junctions under conditions of health and disease? How might neural mechanisms influenced by scaling be implicated in the periodicity or aperiodicity in neurological and psychiatric disorders, such as the periodicity of manic–depressive illness or the seemingly aperiodicity of the fluctuating responses of the parkinsonian patient to levodopa therapy?

At the macroscopic level the inspection of phylogenetic trees shows the extinction or perpetuation of species symbolized by the disappearance or continuation of branches at varying distances from the trunk. How does distance of branching from the trunk affect the evolution of species? At the microscopic level, how does distance from the trunk influence disease processes? In Alzheimer's disease, for example, it is the evolutionarily newest areas of the frontal and temporal cortex that are the most severely affected. The extensive involvement of the archicortex might appear to be an exception, but in the human brain there is considerable "newness" of development there also (see, e.g., Chapter 18). In his studies of nerve degeneration, Ramón y Cajal[29] commented on the tendency of abnormalities to occur at the "bifurcations," a word much used by those concerned with catastrophe theory when discussing evolution and related topics.[30]

Synopsis of Neurobehavioral Studies

After a brief account in the next chapter of the functional role of the lower brainstem and spinal cord, I will deal successively with the neurobehavioral studies on the protoreptilian and paleomammalian formations of the forebrain, and then, before a final chapter on implications, consider the neo-encephalon in connection with paleocerebral functions. The following summary provides a perspective of the investigation as a whole.

The Protoreptilian Formation (R-complex)

The protoreptilian formation is represented by a particular group of ganglionic structures located at the base of the forebrain in reptiles, birds, and mammals. As the German neurologist Ludwig Edinger[31] (1855–1918) commented, these ganglia must be of "enormous significance" for otherwise they would not be found as a constant feature in the vertebrate forebrain. Because of previous ambiguities in naming, the entire group of structures in question will be referred to as the striatal complex, or for brevity's sake in a comparative context, as the R-complex (reptilian complex). As explained in Chapter 4, the R-complex is characterized by distinctive anatomical and biochemical features. It is of key significance that more than 150 years of investigation has failed to reveal specific functions of the R-complex. The traditional view that it is *primarily* a major constituent of the motor apparatus under the control of the motor cortex is inconsistent with clinical and experimental findings that large amounts of certain parts of its gray matter may be destroyed without apparent loss of motor function.

It has been a primary purpose of our research to conduct comparative neurobehavioral studies in an attempt to disclose functions of the R-complex. As background for this work, it was desirable to obtain a detailed analysis of reptilian behavior. The first of six chapters dealing with this subject gives a summary of what is known about the mammal-like reptiles (therapsids), which are the presumed antecedents of mammals. Long before the dinosaurs, the therapsids populated the earth (then a single continent now known as Pangaea) in great numbers.

No existing reptiles are in direct line with the therapsids. Of living forms, lizards would probably have the closest resemblance to mammal-like reptiles. An analysis of reptilian behavior reveals more than 25 special forms of behavior, and at least 6 kinds of general "interoperative" forms of behavior, that are also characteristic of mammals. The comparative neurobehavioral work to be described (involving animals ranging from lizards to monkeys) indicates that the R-complex is involved in the regulation of an animal's daily master routine and subroutines, as well as the behavioral manifestations of four main types of displays used in prosematic communication (Chapters 11–14). A review of pertinent clinical findings (Chapter 15) is followed by a concluding chapter on implications of the neurobehavioral work regarding such basic behavior as the struggle for power, adherence to routine, "imitation," obeisance to precedent, and deception.

The Paleomammalian Formation (Limbic System)

In the evolutionary transition from reptiles to mammals, three cardinal behavioral developments were (1) nursing in conjunction with maternal care, (2) audiovocal communication for maintaining maternal–offspring contact, and (3) play. In mammals, the origination of nursing conjoined with maternal care marks the beginning of the evolution of the family and its associated parental responsibility.

Judged by the few available endocranial casts and other considerations (Chapter 17), the mammal-like reptiles probably had only a rudimentary cortex. In the lost transitional forms between reptiles and mammals, the primitive cortex is presumed to have ballooned out and become further differentiated. In mammals, the evolutionarily old cortex is located in a large convolution that Broca[32] called the great limbic lobe because it surrounds the brainstem. Just as the R-complex is a basic part of the forebrain in reptiles, birds, and mammals, so is the limbic lobe a common denominator in the brains of all mammals. The limbic cortex and structures of the brainstem with which it has primary connections have been known since 1952 as the limbic system.[33] The limbic system corresponds to the paleomammalian formation.

In the last 50 years, clinical observations and animal experimentation have demonstrated that the limbic system plays a basic role in thymogenic functions reflected as emotional behavior. The limbic system comprises three main subdivisions subserving different functions. The two evolutionarily older subdivisions closely associated with the olfactory apparatus have proved to be involved, respectively, in oral and genital functions requisite for self-preservation and procreation. The third subdivision, for which there appears to be no rudimentary counterpart in reptiles, has been found to be implicated in parental care, audiovocal communication, and play behavior.

The clinical study of psychomotor epilepsy (complex partial seizures) (Chapters 22–26) provides the best evidence that the limbic system is basically involved in the experience and expression of emotion. If the brain were likened to a detecting, amplifying, and analyzing device, then the limbic system might be imagined as particularly designed to amplify or lower the intensity of feelings involved in guiding behavior required for self-preservation and preservation of the species. At the onset of epileptic discharges involving the limbic cortex, patients may experience one or more of a broad spectrum of vivid emotional feelings that range from intense fear to ecstasy. An analysis of the phenomenology of limbic epilepsy provides a basis for a classification of three categories of affects that is particularly relevant to epistemic questions pertaining to ontology, including a sense of time and space. It is of special epistemic significance that the limbic cortex has the capacity to generate free-floating, affective feelings conveying a sense of what is real, true, and important. In regard to global functions, there is diverse evidence that the limbic system is essential for the interplay of interoceptive and exteroceptive systems required for a sense of personal identity and the memory of ongoing experience (Chapter 27). The phenomenology of psychomotor epilepsy indicates that the limbic system is implicated in dreaming and in certain psychotic manifestations.

The Neomammalian Formation

The term *neomammalian formation* applies to the neocortex and the thalamic structures with which it is primarily connected. Compared with the limbic cortex, the neocortex is like an expanding numerator, ballooning out progressively in evolution and reaching its greatest proportions in the human brain. On the basis of its extensive connections with the visual, auditory, and somatic systems, it appears to be primarily oriented toward the external world. In its evolution the neocortex, together with its brainstem and neocerebellar connections, has afforded a progressive capacity for problem solving, learning, and memory of details. In human beings it provides the neural substrate for the linguistic translation and communication of subjective states accompanying various forms of mentation. Because of its capacity to generate verbal communication, the human neoencephalon is able to promote the procreation and preservation of ideas that, purely as information (i.e., without mass or energy), not only afford the transmission of culture from generation to generation, but may also affect the course of biological evolution.

Because of the focus of the present investigation on paleopsychic processes, consideration of the neo-encephalon will be largely confined to its role in connection with paleocerebral functions. In this respect, both clinical evidence and evidence based on the expansion of the cranium during hominid evolution call particularly for a review of the part played by the granular frontal cortex in the elaboration of thymogenic functions.

Relevant to epistemics, it is significant to conclude this synopsis by pointing out that the phenomenology of psychomotor epilepsy, together with the findings that seizure discharges tend to propagate in, and be confined to, the limbic system, suggests a dichotomy (a "schizophysiology"[34]) in the function of the limbic system and the neo-encephalon. Such a dichotomy helps to explain the dissociation of their respective thymogenic and intellectual functions observed under certain abnormal conditions. At the same time, the potential for dissociation has disturbing epistemic implications that will be discussed.

References

1. Locke, 1690/1894
2. Pavlov, 1928
3. Miller and Dollard, 1941, p. 1
4. Ardrey, 1972, p. 8
5. MacLean, 1970a, 1973a
6. MacLean, 1973a, p. 5
7. Wiener, 1948, p. 155
8. Bridgman, 1959
9. Freud, 1900/1953
10. e.g., Lorenz, 1937; Tinbergen, 1951
11. Hinde, 1972
12. MacLean, 1975a, 1977
13. Demos, 1937, p. 9
14. Piaget, 1967, p. 15
15. Livingston, 1978
16. MacLean, 1986c
17. Jacob, 1977
18. Darlington, 1978, p. 447
19. MacLean, 1967, 1970, 1973a
20. Jolicoeur *et al.*, 1984
21. Armstrong, 1982
22. Steen, 1988
23. Mandelbrot, 1983
24. Thompson, 1917/1952, Vol. 1
25. Thompson, 1917/1952, Vol. 2, p. 923
26. West and Goldberger, 1987
27. Hofstadter, 1979
28. Mandelbrot, 1977
29. Ramón y Cajal, 1928/1959
30. Nicolis, 1975; Prigogine and Lefever, 1975; Nicolis and Protonotarios, 1979
31. Edinger, quoted by Vogt and Vogt, 1919a
32. Broca, 1878
33. MacLean, 1952
34. MacLean, 1954

3

Role of Forebrain Contrasted with That of the Neural Chassis

Since the main focus of the present investigation is on the forebrain, this will be an opportune place to define the anatomical extent of the forebrain and to emphasize an important distinction between its functions and the rest of the neuraxis.

Main Subdivisions and Functions of Forebrain

Definitions

The anatomical terminology is easiest to explain by reference to the early embryologic development of the central nervous system. The "keel" of the brain and the spinal cord is laid down as a thickened band of ectoderm called the neural plate. Growth takes place on each side of the midline and reflects itself by the appearance of two parallel neural ridges with an intervening neural groove. With continued growth, the two ridges meet above the groove and form the so-called neural tube. When the human embryo is 4 weeks old and 4 mm in length, the forward part of the neural tube shows a succession of three swellings that represent the initial formation of the forebrain, midbrain, and hindbrain.[1] Pictured in Figure 3-1a and b, the three swellings (vesicles) are otherwise referred to as the prosencephalon, mesencephalon, and rhombencephalon.[2] By the sixth week, when the embryo is somewhat longer than 6 mm, two additional swellings appear. These two extra vesicles result from a differential expansion of the forebrain into the telencephalon and diencephalon, and of the hindbrain into the metencephalon and myelencephalon (Figure 3-1c). Since the midbrain remains undivided, there appears at this early embryonic stage the foundation for the *five* main subdivisions of the mature brain. The metencephalon of the hindbrain becomes the pons and overlying cerebellum, while the myelencephalon constitutes the medulla. The telencephalic division of the forebrain eventually divides into two hemispheres that encapsulate the diencephalic division. The term *forebrain* will be used here in the usual sense as constituting both the telencephalon and the diencephalon.

Behavior of Representative Animals Deprived of Forebrain

Extensive experimentation has shown that in terrestrial vertebrates the forebrain is essential for *spontaneous, directed* behavior (see below). But the performance of such

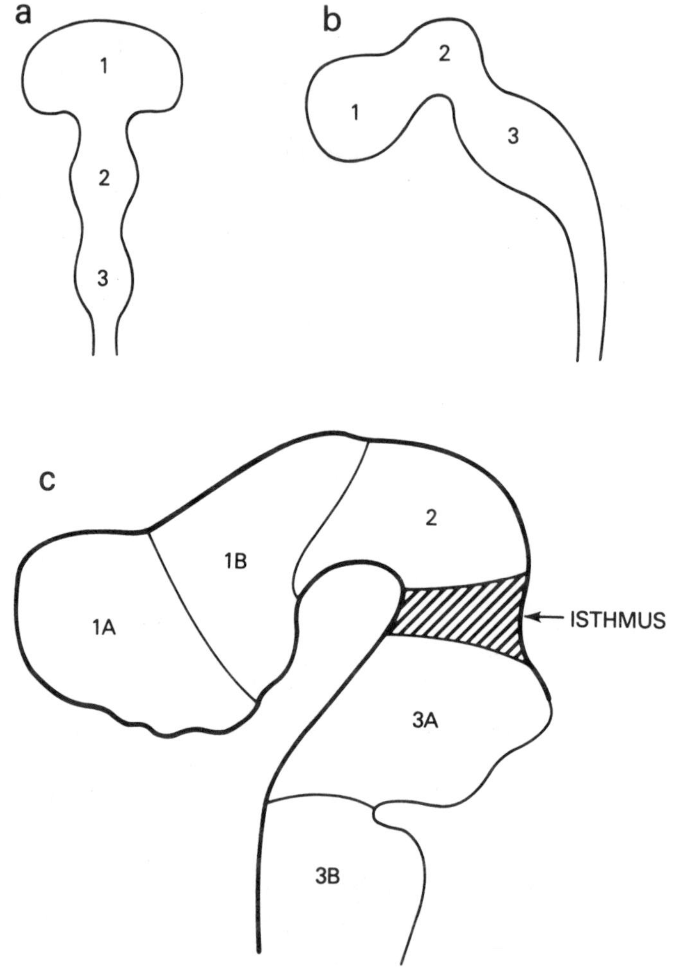

Figure 3-1. Compartmentation of the brain early in ontogeny. Diagrams in (a) and (b) show dorsal and lateral views of the three initial swellings (vesicles) of the brain. The three vesicles are identified as the (1) prosencephalon, (2) mesencephalon, and (3) rhombencephalon. In the human embryo, the swellings are evident at 4 weeks. As shown in (c), shortly thereafter (2 weeks in the case of the human embryo) the prosencephalon subdivides into the telencephalon (1A) and diencephalon (1B), while the rhombencephalon becomes apportioned into the metencephalon (3A) (pons and cerebellum) and myelencephalon (3B) (medulla). The present investigation is primarily concerned with the forebrain (prosencephalon), with special emphasis on the telencephalon. The shaded area represents the isthmus. Its special connections with the forebrain will be described in Chapter 18. Drawings adapted from ones appearing in His (1904) and in Villiger (Figure 3, p. 4, 1931).

behavior depends on the remaining neural chassis contained in the midbrain, hindbrain, and spinal cord. These latter divisions of the neuraxis are essential for the maintenance of posture, locomotion, and the integration of actions involved in self-preservation and procreation.

Fish

It is of evolutionary interest that in animals in which the forebrain is smaller than the midbrain, it is not altogether essential for spontaneous, directed behavior. The left-hand

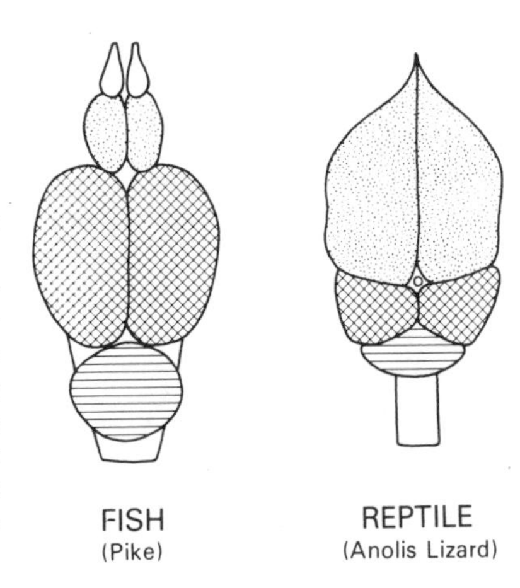

Figure 3-2. Brains of pike and green anolis lizard, illustrating the contrast between the forebrain of a fish and that of a reptile. In the fish the optic tectum (crosshatch) is of much larger proportions than the forebrain (stipple). With the evolution of amphibia the forebrain becomes larger than the midbrain. In reptiles the forebrain/midbrain ratio is almost the reverse of that in the fish. The horizontal shading identifies the cerebellum. The olfactory bulbs (shown in white in the pike's brain) lie too far forward in the anolis lizard to be illustrated. The forebrain of a 2- to 3-foot pike is about 3 mm long, while that of a finger-length lizard is also about 3 mm.

diagram in Figure 3-2 depicts the brain of a teleost fish in which the forebrain is less than half the length of the midbrain. After surgical removal of the forebrain, a teleost fish is constantly on the move, and though tending to swim in a straight line, will avoid obstacles. Some observers attribute the constant swimming to the continuous stimulation of the watery environment.[3] Despite the spontaneous activity and evidence of directed behavior, however, fish deprived of the forebrain show a number of deficits. Noble, for example, found that such preparations fail to school.[4] The fighting *Betta* fish of his experiments were less prone to fight, and, though capable of sexual discrimination, showed deficiencies in mating behavior. Other workers have described impairments in conditional learning.[5]

Amphibia and More Advanced Forms

With the evolution of amphibians the forebrain becomes larger than the midbrain, whereas in reptiles such as the lizard a dorsal view of the brain suggests the picture of a *turned-around* fish brain (Figure 3-2). In the last century, there was much experimentation attempting to show how removal of the cerebral hemispheres affects the behavior of different vertebrates ranging from fish to mammals. In his classical summary of the behavior of decerebrate animals (1876), Sir David Ferrier[6] (English physiologist, 1843–1928) described the effects of removing the cerebral hemispheres in several different kinds of animals, taking into account the observations of such well-known investigators as Flourens, Longet, Vulpian, and Goltz. He emphasized that, above all, the decerebrate animal is characterized by a lack of spontaneity of movement and the absence of any tendency to explore. The frog is representative of an amphibious form transitional between fish and reptiles. "Deprived of its cerebral hemispheres," Ferrier wrote, "the frog will maintain its normal attitude, and resist all attempts to displace its equilibrium. If laid on its back, it will immediately turn on its face, and regain its station on its feet. . . . If its foot be pinched, it will hop away. If it is thrown into the water, it will swim until it reaches the side of the vessel, and then clamber up and sit perfectly quiet. If its back be stroked gently, it will utter loud croaks. . . . Indeed, in many respects, it would be difficult to say that the removal of the hemispheres had caused any alteration in the usual behaviour of the animal."[7]

"But yet," Ferrier continues, "a very remarkable difference is perceptible. The brainless frog, unless disturbed . . . will sit forever quiet in the same spot, and become converted into a mummy. All spontaneous action is annihilated. Its past experience has been blotted out, and it exhibits no fear in circumstances which otherwise would cause it to retire or flee from danger. . . . Surrounded by plenty it will die of starvation; but unlike Tantalus, it has no psychical suffering, no desire, and no will to supply its physical wants.''[8]

After describing the results of similar operations on pigeons, guinea pigs, and rabbits, Ferrier concludes that the behavior is generally similar in all. They are capable of maintaining equilibrium, performing coordinated locomotion, and responding to various forms of sensory stimulation. Most significantly, Ferrier reemphasizes, ''If the animal be left to itself, undisturbed by any form of external stimulus, it remains fixed and immovable on the same spot, and unless artificially fed, dies of starvation. . . . If artificially fed . . . the animal may live an indefinite period.''[9]

Ferrier did not include reptiles, but the behavior of the decerebrate reptile has been described in similar terms. Goldby[10] found that decerebrated lizards (*Lacerta viridis*) ''tended to remain in any position'' in which they were placed and failed to eat spontaneously. When stimulated they could walk or run ''in a perfectly coordinated manner.''[11] He concluded that ''there is no evidence from ablation experiments that any movement is localized in the forebrain in such a way that removal of the forebrain, or any part of it, leads to the loss of that movement.''[12]

Domestic Carnivores

Ferrier was unable to give descriptions of ''higher'' animals such as cats and dogs because they did not survive the shock of the decerebration. Some of the reports of such preparations by subsequent workers are of dubious value because the postmortem examination revealed attached remnants of forebrain. As C. Judson Herrick (noted American neuroanatomist, 1868–1960) has commented, ''In most of the decerebrate dogs described in the literature some portion of the striatum [striatal complex] was preserved.''[13] He cited, for example, the famous case described by Rothmann (1923). This dog ''would seek food, avoid obstacles, and in general behave like a dog deprived of his higher sense organs.'' In contrast, Herrick described the absence of directed behavior of a dog prepared by Dresel (1924) that proved to have no striatal remnants. The dog lived 3 months. In Herrick's words, ''The animal can stand and walk, but it does not do so spontaneously, only under stimulation, external or internal. Generally it lies quiet, but becomes restless if the bladder is full, if hungry, or if stimulated. Then it arises and runs around with depressed head. It runs into every obstacle, avoiding nothing, and gradually comes to rest standing, or it falls down and remains lying until a new stimulus excites it. *It lacks all spontaneity* [author's italics]. In general, the animal until death learned absolutely nothing.''[14]

The best example that I have found of a chronically surviving cat is from a study by Bard and Rioch (1937).[15] Identified as Cat #228, this animal was maintained in a good physical condition for more than a year. All of the cerebral hemispheres had been removed except ''a shred of striatum'' and a small part of the diencephalon. The behavior of this cat, like that of Dresel's dog, was reminiscent of Ferrier's description of the decerebrate frog, bird, and rabbit. It was lacking in spontaneity of movement and failed to investigate its surroundings or to seek nourishment. Except for one instance of ''claw-sharpening behavior'' there was no sign of grooming. There was no evidence of pleasure reactions, but *undirected* angry behavior could be provoked by noxious stimulation.

The decerebrate animal shows evidence of sleeping and waking.[15] It will assume a normal, species-typical posture for urination or defecation, but will not clean itself. Decerebrate female guinea pigs in estrus will not seek courtship but will respond to a male's nuzzling by adopting a typical receptive posture (lordosis and tail to one side) and thereafter engage in copulation.[16]

Generalizing Comment

It is on the basis of the foregoing kind of evidence that one concludes that the forebrain is essential for spontaneous, directed behavior in amphibians, reptiles, birds, and mammals. The remaining brainstem and spinal cord constitute a neural chassis that provides most of the neural machinery required for self-preservation and the preservation of the species. By itself, the neural chassis might be likened to a vehicle without a driver. Significantly, in the more advanced vertebrates the evolutionary process has provided the neural chassis not with a single guiding operator, but rather a combination of three, each markedly different in its evolutionary age and development, and each radically different in structure, chemistry, and organization. In the chapters to follow, it will be my purpose to show that just as the three basic evolutionary formations of the forebrain can be distinguished on the basis of their anatomy and chemistry, they can also be shown to account for certain different psychological and behavioral functions.

For the sake of generalization, no distinction has been made in the present chapter between the functions of the two divisions of the forebrain, namely the diencephalon and telencephalon. For conceptual purposes they might be imagined as complementing one another like the cone and screen of a television set. The telencephalon would represent the screen. In regard to epistemics, the problem of greatest interest is to discover what is portrayed in the telencephalon.

Special Relevance of Peripheral Autonomic Nervous System

Introductory Considerations

The peripheral nervous system may be regarded as an extension of the neural chassis. For the present account, it is not obligatory to go into detail about the physiology of the peripheral somatic nervous system. As to their respective capacities to convey information and execute commands, the sensory and motor nerves may be regarded as an extension of the neural chassis to the periphery of the organism. There is, however, another part of the peripheral nervous system that, as a preliminary, deserves consideration in some detail because of its role in prosematic communication and emotional expression. It was mentioned that the brain may be regarded as a detecting and amplifying device for providing the organism information about the internal and external environment that is requisite for self-survival and survival of the species. It was noted in this respect that the limbic system has the capacity to turn up or down the "volume" (intensity) of feelings that guide behavior required for self-preservation and preservation of the species. Short of inducing physical exercise, emotional mentation represents the only form of psychological information that may provoke marked, and often prolonged, physiological changes *within* the organism. Such changes are in large part mediated by the outflow from the *peripheral autonomic nervous system* now to be considered.

As to be explained, it is only within relatively recent times that the autonomic nervous system would merit consideration as a peripheral extension of the neural chassis. The term *autonomic* applies to something that is self-regulating or functionally independent. Ironically, this name was selected as a substitute for other terms that had become synonymous with an independently functioning nervous system (see below).

There is now paleontological evidence that life in the form of individual cells can be traced back at least three billion years.[17] The life functions of unicellular organisms depend on a continual process of assimilation of nutritive material and elimination of waste products. The same processes are requisite for multicellular organisms. Among vertebrates (see Ref. 18), a special peripheral nervous system develops for the assimilative and eliminative needs not only of individual organs, but also of the entire animal.

Assimilative and eliminative processes, however, are not confined to individuals. We find that in the socialization of animals, the same kind of process is symbolically at work. In other words, in social groups some individuals may be assimilated, while others are eliminated (see Chapters 23 and 24). Curiously enough, for reasons that have not been explained, the housekeeping part of the nervous system involved in assimilation and elimination is called symbolically into play in the prosematic communication involved in social assimilative and eliminative functions. In birds and mammals, for example, respective ruffling of the feathers and horripilation originally developed as a defense against (and hence elimination of) the cold. Somewhere along the evolutionary process, these same thermoregulatory manifestations became symbolically appropriated for making the animal look bigger in size, either under conditions of threat (challenge) or defense. Another autonomic manifestation—namely, dilatation of the pupil—has the effect of making the eye appear larger. Autonomic manifestations that occur in conjunction with acts of elimination and ejection are salivation associated with spitting that familiarly occurs among cats, and defecation upon intruders illustrated by the howler monkey.

Before going into further detail, the autonomic nervous system may be defined as a special neural apparatus for coordinating and integrating the assimilative, metabolic, and eliminative functions of the internal organs of the body on which vertebrate animals depend in the struggle for survival and procreation. In addition to glands and viscera, the autonomic system innervates the blood vessels and sweat glands, as well as piloerector muscles, that account for the elevation of feathers in birds and the hair of mammals (see Figure 3-3).

Historical Unfolding

We find our first description of the autonomics in the anatomical works of Claudius Galen (c. A.D. 129–200).[19] Born in Asia Minor, Galen attended the medical school at Alexandria, and served for a time as physician of Marcus Aurelius in Rome. He recognized two chains of nerves with interspersed little swellings called ganglia, that ran along each side of the vertebral column from the level of the cervical atlas to the coccyx. He observed that the ganglia gave off ramifications of nerve fibers to the viscera. He believed that the two paravertebral chains were connected to the brain by the large nerve that has since become known as the vagus nerve, while there were additional connections with the spinal cord by nerves later named the white communicating branches (*rami communicantes*). Galen believed that nerve fibers were hollow tubes that conducted the so-called animal spirits. He concluded that the viscera receive an ''exquisite'' sensitivity directly from the brain and their motor activity from the spinal cord.

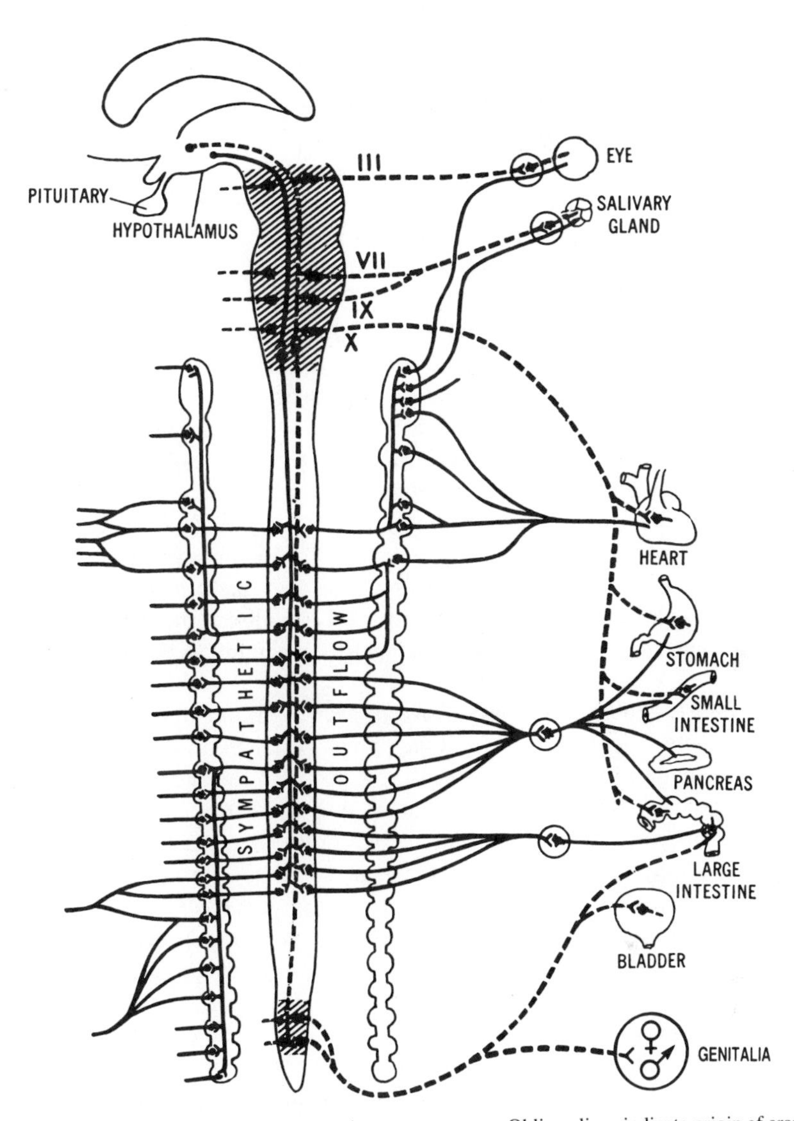

Figure 3-3. A simplified diagram of the autonomic nervous system. Oblique lines indicate origin of craniosacral parasympathetic outflow. The sympathetic outflow (labeled) originates from the 12 thoracic and upper 3 lumbar nerves. The supply to the viscera is shown on the right, while the outflow to the blood vessels, sweat glands, and smooth piloerector muscles is indicated on the left. For simplification, the sympathetic supply to the bladder and genitalia via the inferior mesenteric ganglion is not shown; it originates from the last three thoracic and first lumbar nerves. From Gardner (1959) with one addition.

The English anatomist Willis[20] (1621–1675) referred to the pair of vertebral chains as the intercostal nerves and concluded that they were connected to the cerebellum (literally, little brain) by the same large nerve mentioned above to which he gave the name vagus, meaning wandering nerve. He believed that the cerebellum produced the animal spirits that caused involuntary movements of the viscera, whereas the cerebrum was the source of the animal spirits responsible for voluntary control of the skeletal musculature.

How did the paravertebral chain become known as the sympathetic nerves? One finds the answer in the authoritative 18th century anatomical textbook by the Danish anatomist

Winslow[21] (1669–1760). "These nerves," he explained, "are commonly called intercostals. . . . I believe that the great sympathetic nerves would be more suitable because of their frequent communication with most of the other principal nerves of the whole body."[22] Adhering to a concept handed down from Greek times, he believed that these nerves were ideally suited for conveying "sympathies from one part of the body to another." He suggested that the ganglia interspersed among the sympathetics might be looked upon as "*so many little brains*" (italics added).

In 1800, the young French physiologist Bichat[23] (1771–1802) arrived at an interpretation that was to have a major influence on physiology and psychology for more than 100 years. He proposed that "the little brains" mentioned by Winslow functioned independently of the brain and spinal cord and controlled both the viscera and the passions. Echoing an Aristotelian concept, Bichat conceived of animals as being constituted of "organic" and "animal" life. Organic life controlled the involuntary functions of the viscera and sustained the internal needs of the body, whereas the animal life made possible the voluntary adjustments to the external environment. "One knows," Bichat said, "that the external functions, sensations, locomotion, voice, are all dependent on the cerebral nervous system; that on the contrary most of the organs serving internal functions, derive their nerves from ganglions and with them the principle of their action." Accordingly, he explained, "I shall henceforward . . . divide the nerves into two great systems, one emanating from the brain, the other from ganglions; the first a single center, the second a very great number."[24]

In Section II, Article VI of his book *Recherches physiologiques sur la vie et la mort,* Bichat[23] presented evidence that everything which pertained to the passions (emotions) fell within the realm of the organic life. He pointed out how anger affects the heart and circulation, grief the respiration, resentment the stomach, and so on. It therefore followed that the emotions were generated in the internal organs and the *little brains* controlling them, being entirely independent of the voluntary nervous system. In appraising the influence of Bichat, Claude Bernard said: "The ideas of Bichat produced in physiology and medicine a profound and universal revolution. . . . All the ideas of his contemporaries on life, all the attempts to define it, are in some way only the echo or the paraphrase of his doctrine."[25]

In the physical sciences that grew out of the studies of Galileo, people could hope for increasing emancipation from the vicissitudes of their environment; in the American and French revolutions they could see the possibility of breaking the chains of human enslavement; but owing to the persuasive appeal of Bichat's independently functioning nervous system, it was to be more than 100 years before there could be hope of partial liberation from the ruling passions within.[26] Even Freud[27] was so immersed in Bichat's doctrine that he believed that visceral malfunction and illnesses referred to as psychosomatic today, could not be psychological in origin.

In the latter part of the 19th century, it was the special contribution of two English workers—Gaskell (1847–1914), an anatomist, and Langley (1852–1925), a physiologist—to demonstrate that, contrary to Bichat's teaching, the sympathetic nervous system received its innervation from the spinal cord.[19] In examining the spinal nerves, Gaskell[28] observed that the anterior roots, representing the motor nerves, of the 1st thoracic through the 3rd lumbar nerves, contained many fine medullated fibers (1.8–2.7 μm) that made up the white rami which could be traced to their appropriate ganglia. He wrongly believed that in the ganglion, the medullated fiber lost its myelin and divided into several unmyelinated (gray-appearing) fibers that innervated the viscera. But on the basis of these findings he rightly concluded that there was a flow of nerve impulses from the spinal cord to the viscera and not vice versa. In addition, he detected nerves of a similar caliber in the 3rd, 7th, 9th, 10th, and 11th cranial nerves and in the 2nd and 3rd sacral nerves.

On the basis of these painstaking observations, it was evident that there were three major outflows of fine medullated fibers from the lower brainstem and spinal cord. Gaskell referred to them as the (1) bulbar, (2) thoracolumbar, and (3) sacral outflows, respectively (see Figure 3-3). He prophetically stated that "The evidence is becoming daily stronger that every tissue is innervated by two sets of nerve fibres [(1) thoracolumbar and (2) craniosacral] of opposite characters so that I look forward . . . to the time when the whole nervous system shall be mapped out into two great districts of which the function of one is katabolic, of the other anabolic, to the peripheral tissues. . . ."[29]

Over a period of several years, Langley[30] provided physiological confirmation of Gaskell's anatomical observations. By stimulating the white rami and observing the response, he showed that nerve conduction led from the spinal cord to the viscera. He and Dickinson[31] devised an ingenious means of demonstrating which ganglia were responsible for a particular response. By painting a ganglion with nicotine they blocked the transmission of nerve impulses generated by stimulating the white ramus. However, by stimulating the gray ramus leading from the ganglionic nerve cells to the viscera, the response could still be obtained. In this way, Langley subsequently mapped the autonomic nerve supply to different parts of the body. He called the white rami "preganglionic" and the gray rami "postganglionic" fibers.

By the end of the 19th century, physiologists recognized that stimulation of the thoracolumbar outflow generally resulted in effects that were opposite to those obtained by stimulating the cranial and sacral outflows. For example, stimulation of the nerves associated with the upper part of the sympathetic chain would produce acceleration of the heart beat, whereas stimulation of the vagus nerve resulted in slowing. In 1895, Oliver and Schäfer[32] reported their findings that injection of a glycerin extract of the adrenal glands induced acceleration of the heart rate and respiration, along with a fall in body temperature. After isolation of the active principle called adrenaline, Elliot[33] (1904) observed that the administration of this hormone induced changes similar to those obtained by stimulating the thoracolumbar outflow. Somewhat later, Dale[34] found that an extract of ergot (acetylcholine) produced effects similar to those obtained by stimulation of the cranial and sacral outflows. Langley[35] introduced the term "para-sympathetic" to apply to the system of nerves representing the cranial and sacral outflows, an expression alluding to the location of the two outflows, respectively, "next to" the sympathetic division (see Figure 3-3).

Langley[36] also introduced the term *autonomic* for referring to both the sympathetic and parasympathetic divisions. As he explained:

> It is . . . convenient to have some term to include the whole nervous supply. The words 'organic,' 'vegetative,' 'ganglionic,' and 'involuntary' have all been used, but they have also been used in senses other than that we require. . . . The word 'visceral' . . . is obviously inapplicable to some of the structures brought under it, such as the nerve fibres which run to the skin. I propose, then, following a suggestion of Professor Jebb, to use the word 'autonomic,' including under that term the contractile cells, unstriated muscle, cardiac muscle, and gland cells of the body, together with the nerve cells and fibres in connection with them.[37]

Altogether, experimentation revealed that with the exception of the piloerector muscles and sweat glands, all of the structures mentioned receive a dual innervation from the parasympathetic and sympathetic divisions of the autonomic nervous system. Langley admitted that the term *autonomic* suggests "a much greater degree of independence of the central nervous system than in fact exists," but he thought it was "more important that new words should be used for new ideas than that the words should be accurately descriptive."[38]

Autonomic Functions

If the body were likened to a house, the activities of the parasympathetic and sympathetic nervous system would correspond, respectively, to those of the mother and father of a household in a traditional rural setting. The role of the mother is primarily concerned with activities within the house, whereas the father has the responsibility of looking after the household's relationship to the outside world, serving as both provider and protector. In the following summary of functions of respective divisions of the autonomic nervous system, Figure 3-3 serves as a simplified reference to the dual innervation, with the main supply to the viscera shown on the right and that for the blood vessels, skin, and piloerector muscles on the left.

Parasympathetic Functions

The household duties of the mother are preparing and canning food; cleaning; throwing out garbage and wastes; keeping intruders, such as flies and mice, from passing the threshold of windows and doors; quieting the household for sleep; and serving as partner in the act of procreation. In the preparation and the assimilation of food, the cranial nerves of the parasympathetic division activate salivation and the secretion of the digestive juices, relax the sphincters separating the various compartments of the gut, and induce peristalsis to push along the digestive residue. Following assimilation, parasympathetic activity promotes the storage of sugar, fat, and protein.

The cleaning eliminative functions of the sacral nerves of the parasympathetic division involve activation and contraction of the smooth muscle of the bladder and lower gut for the evacuation of urine and feces. Its protective eliminative actions come into play when noxious agents affect the membranes of the eyes, respiratory passages, and gut. The shedding of tears washes away irritants from the eyes, and the secretion of mucus and constriction of the bronchioles help to prevent dust and other forms of matter from entering and irritating the lungs. Salivation and the reverse peristalsis of vomiting eliminate noxious ingested material from the upper alimentary tract, while increased forward peristalsis with diarrhea serves a like purpose for the lower part of the gut. The parasympathetic constriction of the pupil prevents a harmful amount of light from entering the eye. In preparing the organism for rest and sleep, the parasympathetics slow the heart and respiration and reduce the stimulating effects of light by constricting the pupil of the eye. Parasympathetic participation in the act of procreation results in genital tumescence and the production of lubricative secretions.

Sympathetic Functions

Comparable to a father's role as breadwinner and protector, the sympathetic division comes primarily into play in dealing with exigencies of the external environment. In times of emergency or excessive exertion, the sympathetics call upon all of the body's internal resources required for survival. During combat or vigorous exercise the sympathetics induce changes that increase circulation of the blood and provide a ready supply of energy for the working muscles. Sympathetic dilatation of the upper and lower respiratory passages affords an increase in pulmonary ventilation, and sympathetic effects on the cardiovascular system result in an increased action of the heart, rise in blood pressure, and through local vasodilatation, an augmented flow of blood to the exercising muscles. In addition to supplying more oxygen, the increased blood supply brings with it an augmented supply of energy through the release of the body's store of glucose. Compensatory sympathetic constriction occurs in the blood vessels of the skin and gut where less blood is needed. While the stressful situation continues, sympathetic inhibition results in a cessation of digestion and peristalsis and a closure of the sphincters that act like gates between various compartments of the gut. In the eventuality of blood loss in combat, the blood clotting time is shortened and, in some animals, sympathetic contraction of the spleen ensures an extra supply of red blood cells.

Comparable to the father's role in stoking the fires and putting up the storm blinds to keep out the cold, sympathetic activation results in vasoconstriction, piloerection, and shivering. These defenses against the cold, as has been noted, may also be manifested during times of struggle, having a special symbolic signification as social signals in the display of aggressive and protective forms of behavior. The same applies to other thermoregulatory effects of sympathetic stimulation such as induction of sweating for expelling heat.

Just as the mother and father of the household have their independent responsibilities, so also are there occasions requiring their close cooperation, such as in the case of dire emergency or the act of copulation. In unusually stressful or threatening situations, the eliminative functions of the parasympathetics come into play along with those of the sympathetics, being manifest by secretion of saliva employed for spitting and by urination and defecation. In the act of procreation, parasympathetic activation results in genital tumescence and the production of lubricative secretions, while sympathetic impulses trigger ejaculation.

Humoral Aspects of Autonomic Functions

As amply demonstrated by the well-known American physiologist Walter B. Cannon[39] (1871–1945), sympathetic activity at times of emergency also results in the secretion of adrenal medullary hormones that act on autonomic effector sites in such a way as to mimic, as well as potentiate, the effects of widespread sympathetic discharge. Since these hormones are destroyed relatively slowly, they also have the effect of prolonging sympathetic excitation.

As is well known, the pituitary gland, which lies below the base of the brain, releases hormones that in turn activate the release of hormones from the so-called glands of internal secretion. The term *hormone* derives from the Greek word for messenger, and as some writers have commented, hormones are like messages "to whom it may concern." Particularly to be mentioned in the present context is the pituitary adrenocorticotropic hormone, which releases corticoid hormones from the cortex of the adrenal gland. The corticoids have been shown to play an important role in the body's defenses against shock and infection.

Later, as in the chapter dealing with brain mechanisms underlying maternal behavior (Chapter 21), there will be occasion to allude to other aspects of hormonal function, such as hypothalamic mechanisms regulating the release of pituitary hormones, and feedback mechanisms for terminating the release of pituitary hormones once they have accomplished their effects.

Concluding Comment

Despite the clarification that the central nervous system innervated the autonomic ganglia, the traditional views continued to be pervasive well into the 20th century. Thanks largely to the work of Karplus and Kreidl[40] prior to World War I, and later the physiological studies of Hess,[41] Cannon,[39] Bard,[42] and others, the representation of visceral and emotional functions was recognized to exist at as high a level as the hypothalamus in the diencephalon. But as late as 1936, Cannon authoritatively stated that the cortex was concerned with emotional behavior only insofar as it could inhibit those aspects of emotion under voluntary control. During the same period, Fulton[43] and his colleagues in the Department of Physiology at Yale were beginning to demonstrate the effectiveness of cortical stimulation in eliciting autonomic responses. But it remained for investigations on psychomotor epilepsy and the limbic system to demonstrate how extensively the cortex—and particularly the phylogenetically old cortex of the limbic system—can influence autonomic function, particularly as it comes into play as a result of emotional mentation. In this respect, it deserves emphasis that, short of induced physical activity, emotional mentation represents the only psychological process that may lead to profound, and often prolonged, autonomic activity.

References

1. Patten, 1953
2. His, 1904
3. Ferrier, 1876/1966; Aronson, 1970
4. Noble, 1936
5. Aronson, 1970; de Bruin, 1980
6. Ferrier, 1876/1966
7. Ferrier, 1876/1966, p. 34
8. Ferrier, 1876/1966, p. 35
9. Ferrier, 1876/1966, p. 39
10. Goldby, 1937, p. 348
11. Goldby, 1937, p. 347
12. Goldby, 1937, p. 352
13. Herrick, 1926, p. 117
14. Herrick, 1926, p. 118
15. Bard and Rioch, 1937
16. Dempsey and Rioch, 1939
17. Barghoorn and Schopf, 1967
18. Langley, 1903, pp. 15–16
19. See Sheehan, 1936
20. Willis, 1664
21. Winslow, 1732
22. Winslow, 1732
23. Bichat, 1799/1800
24. Bichat, 1815/1955, p. 73
25. Quoted by Solovine, p. 10, Ref. 24
26. MacLean, 1960, p. 86
27. Freud, 1938
28. Gaskell, 1886
29. Gaskell, 1886, p. 50
30. Langley, 1921
31. Langley and Dickinson, 1889
32. Oliver and Schäfer, 1895
33. Elliot, 1904
34. Dale, 1914
35. Langley, 1900, p. 403
36. Langley, 1898; 1900
37. Langley, 1900, pp. 659–660
38. Langley, 1921, p. 6
39. Cannon, 1929
40. Karplus and Kreidl, 1909, 1910
41. Hess, 1954
42. Bard, 1928
43. Fulton, 1949a

The Striatal Complex with Respect to Species-Typical Behavior

4

The Striatal Complex (R-complex)
Origin, Anatomy, and Question of Function

In 1975 we cosponsored with the Smithsonian Institution a first conference on "The Behavior and Neurology of Lizards" at our Laboratory of Brain Evolution and Behavior. In my opening remarks I said that we were interested in lizards

> because they are distant relatives of the long-extinct mammal-like reptiles that are believed to have been the antecedents of mammals. Ordinarily, there is no apparent causal connection between a current happening and something that took place a few hundred million years ago. It is perfectly correct to say that this Laboratory would not be here and that this Conference would not have been held, had there not been an historical link between mammals and reptiles. When beginning to plan for the present facility over 20 years ago, a primary purpose was to use new behavioral approaches in investigating the functions of a basic part of the forebrain that reflects our reptilian ancestry.[1]

For comparative neurobehavioral studies it is unfortunate that none of the existing reptiles is directly in line with the mammal-like reptiles (therapsids). There have existed approximately 17 orders of reptiles, all of which are presumed to be derived from the so-called stem reptiles (cotylosaurs).[2] The cactus tree in Figure 4-1 shows the different lineages. All but 4 of the 17 orders are extinct. The existing orders are the Chelonia, the Crocodilia, the Rhynchocephalia, and the Squamata. The Chelonia (including turtles and tortoises) are so named because of the carapace (protective boxlike shell) in which they are encased. As indicated in Figure 4-1, the turtles appear to have followed an independent course since branching off from the stem reptiles. The lineage of the three remaining orders points to an origin from the eosuchians ("dawn crocodiles"), which represent a separate branch from the stem reptiles. The Crocodilia (including the crocodiles and alligators) belong to the large subclass Archosauria, the so-called ruling reptiles that include the dinosaurs. They have a number of resemblances to birds. The Rhynchocephalia are represented by a single species, the tuatara (*Sphenodon punctatum*) of New Zealand. Finally, the Squamata, or "scaly" reptiles, comprise two suborders—the Lacertilia (lizards) and Ophidia (snakes).

In summary, except for their derivation from the same original stock (the stem reptiles), all extant reptiles have a lineage entirely separate from the therapsids. For comparative neurobehavioral studies on reptiles I chose to focus on lizards because, as will be explained, they suggest a closer resemblance to early mammal-like reptiles than other existing forms. As background for an analysis of the behavior of lizards, the mammal-like reptiles will be described in the next chapter.

Prior to that it is requisite to characterize the structures of the basal forebrain that represent a common denominator in reptiles, birds, and mammals, and, finally, to com-

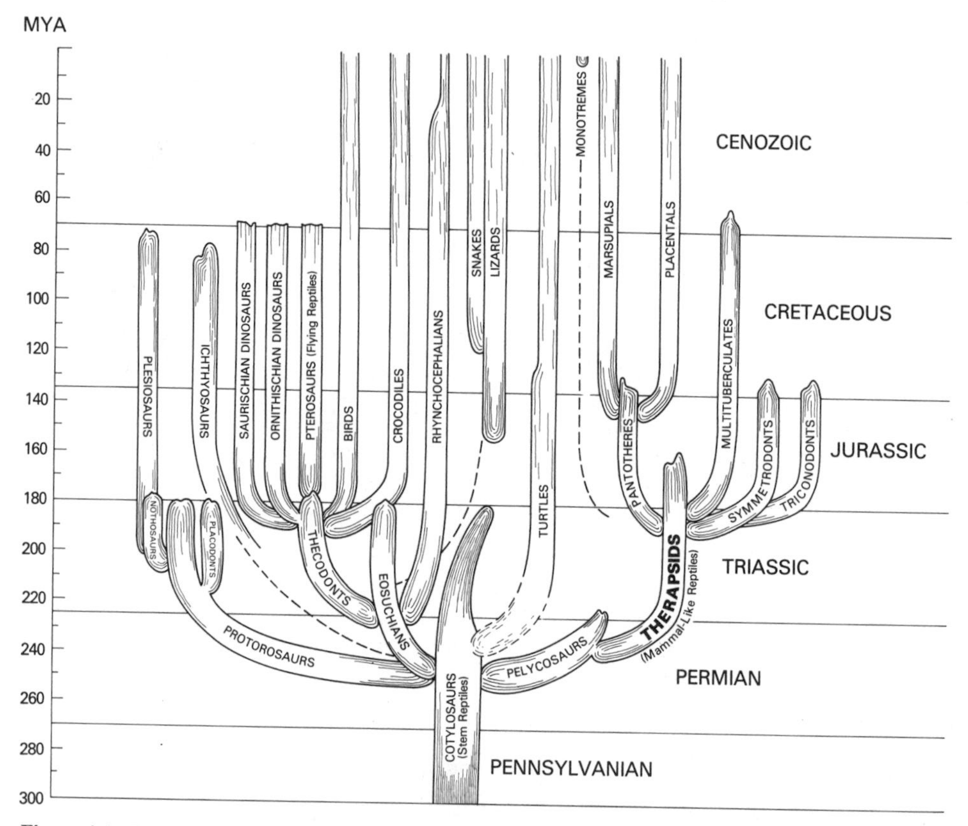

Figure 4-1. Cactus tree depicting the phylogeny of reptiles. The main trunk identified with the stem reptiles (cotylosaurs) is shown arising in Pennsylvanian times. The first right-hand branch represents the line of reptiles leading up to mammals. The therapsids were mammal-like reptiles that are believed to have been the crucial link in the evolution of mammals. Branching off from the pelycosaurs in Permian times, they were almost extinct at the beginning of the Mesozoic when the dinosaurs began to flourish. Note also the temporal relationship of the therapsids to the present-day reptiles—turtles, lizards, snakes, crocodiles, and the rhynchocephalians (represented by a single species, the tuatara of New Zealand). Marine reptiles are identified by the branches on the left (reverse facing labels). Details of the tree are based largely on Romer (1966) and Colbert (1966). The dates for the geological periods are those of Holmes's "final time scale" (see Faul, 1978). MYA, million years ago.

ment on the long-standing enigma regarding their functions. In regard to the question of correspondence of structures in the three classes of animals, the evidence rests on phylogenetic, embryological, neuroanatomical, and neurochemical data.

Introductory Anatomical Considerations

Terminology

The forebrain structures under consideration are masses of gray matter located predominantly in the telencephalon, but including inextricably associated parts of the adjoining diencephalon. In conventional terms, they are defined as belonging to the basal ganglia, a designation applying to "knotlike" masses of nerve tissue located near the base of the brain. In neurological texts, they are further characterized as constituting the cerebral hub of the extrapyramidal "motor" system. In higher primates, the structures in question are identified as the olfactostriatum (olfactory tubercle and nucleus accumbens), corpus striatum (caudate nucleus and putamen),* the globus pallidus, and satellite collections of gray matter known as the substantia innominata and basal nucleus of Meynert. The total mass comprises more than three-fourths of the gray matter at the center of the human cerebrum.[3] Figure 4-2 depicts the two largest ganglia—the caudate nucleus and putamen—as they would appear in a cutaway of the human brain. The putamen was so named because it has the shape of a fluted seashell, while caudate means "having a tail." The two nuclei have essentially the same internal structure and together are called the corpus striatum (the striped body) because upon section, their outgoing bundles of nerve fibers give them a striped appearance.[4] Figure 4-3 illustrates a major striatal projection via the globus pallidus ("pale globe"). The putamen and globus pallidus are so enmeshed that they have been commonly referred to as a single structure called the lenticular nucleus because of its lens-shaped appearance.[4] There is still uncertainty as to what parts of the globus pallidus derive, respectively, from the telencephalon and diencephalon. The medial segment (entopeduncular nucleus of subprimate forms) appears to be of diencephalic origin, representing a dorsolateral part of the hypothalamus (see below). Just rostroventral to the pallidum is the phylogenetically more ancient olfactostriatum. Since there is no accepted term that applies to all of the ganglia in question, I shall refer to them as the *striatal complex*. In a comparative context, I will use alternatively the abbreviated expression *R-complex* (reptilian complex).[5] It should be noted that two other structures often designated as basal ganglia—the claustrum and amygdala—are not included in the present designation (see Chapter 18).

*In the Basel, Jena, and Paris versions of *Nomina Anatomica*, the corpus striatum is defined as including the caudate nucleus and lenticular nucleus (putamen + globus pallidus). In his classic article on the basal ganglia, however, Papez (1942) defined the "corpus striatum" as composed of the caudate nucleus and putamen (p. 25). Other authorities such as F. H. Levy (1942) have suggested the same specification. Crosby *et al.* (1962) have since referred to this definition of the "striatum" as perhaps the "most common clinical usage" (p. 360).

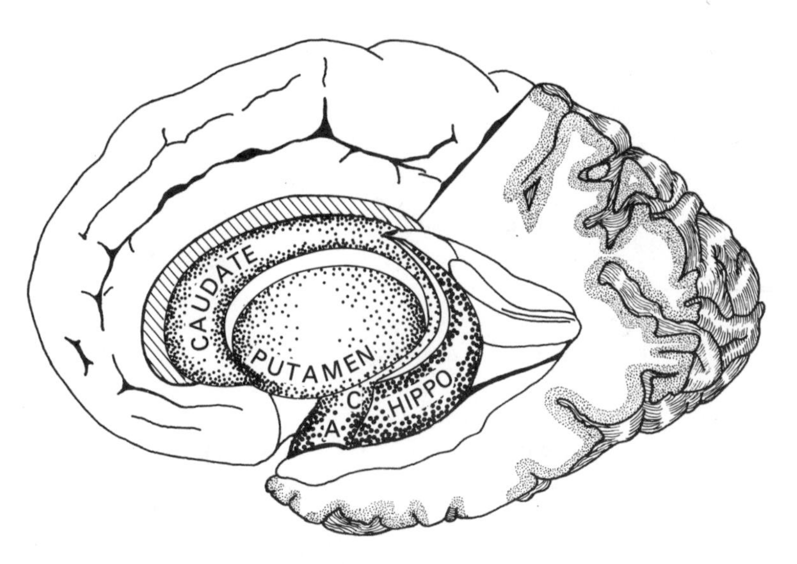

Figure 4-2. A cutaway of the left hemisphere of the human brain, showing location and configuration of the corpus striatum (caudate nucleus and putamen), which constitutes the largest part of the striatal complex. Note how it is encircled by the limbic lobe, including the infolded part of the hippocampus (HIPPO) containing the archicortex. The club on the tail of the caudate (identified by the letter C) is contiguous with the amygdala (A) of the limbic system. Compare with diagram of monkey brain in Figure 4-3. Redrawn after Crosby *et al.* (1962).

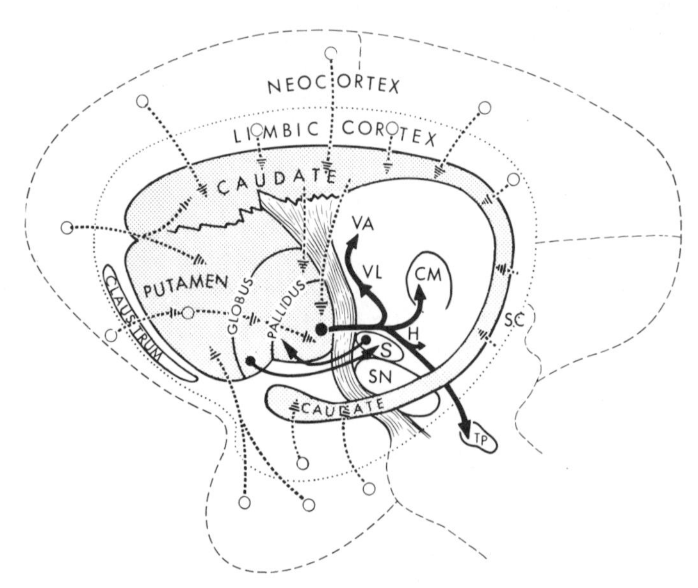

Figure 4-3. Ringlike configuration of striatal complex (stipple) of squirrel monkey, indicating the source and destination of some of its connections. Note origin of some connections from the encircling limbic lobe, as well as from the neocortex. The diagram gives emphasis to the major outflow from the internal segment of the globus pallidus (heavy arrows), projecting, respectively, to the thalamus and tegmentum (see section on connections and also Chapters 13 and 14). The lighter weight arrows show projections from the lateral segment of the globus pallidus to the subthalamic nucleus (S), and from the latter back to the internal segment. The olfactostriatum lies rostroventral to the globus pallidus. Note that, contrary to the opinion of some neurologists, the claustrum is not included as part of the corpus striatum (see legend of Figure 4-5). Abbreviations: CM, N. centri mediani thalami; H, area tegmentalis (Forel); S, corpus subthalamicum; SC, colliculus superior; SN, substantia nigra; TP, N. tegmenti pedunculopontinus; VA, N. ventralis anterior thalami; VL, N. ventralis lateralis thalami. Slightly modified after MacLean (1972a).

The Question of Homology

In addressing the question as to what are similar parts of the striatal complex in reptiles, birds, and mammals, I will use the word *corresponding* rather than the conventional expression *homologous,* a term introduced by Richard Owen (1804–1892) ''about 1848.''[6] Through long and various usage, the meaning of *homologous* has become unclear, being interpreted by some authors to signify ''the same'' or ''identical.'' In dealing with different taxa of animals one can say ''corresponding'' with respect to structures identified by a particular set of attributes, without implying that they are developed to the *same* extent or have the *same* degree of complexity in their organization.

As illustrated in Figure 4-4, a sagittal section through the striatal complex of anolis

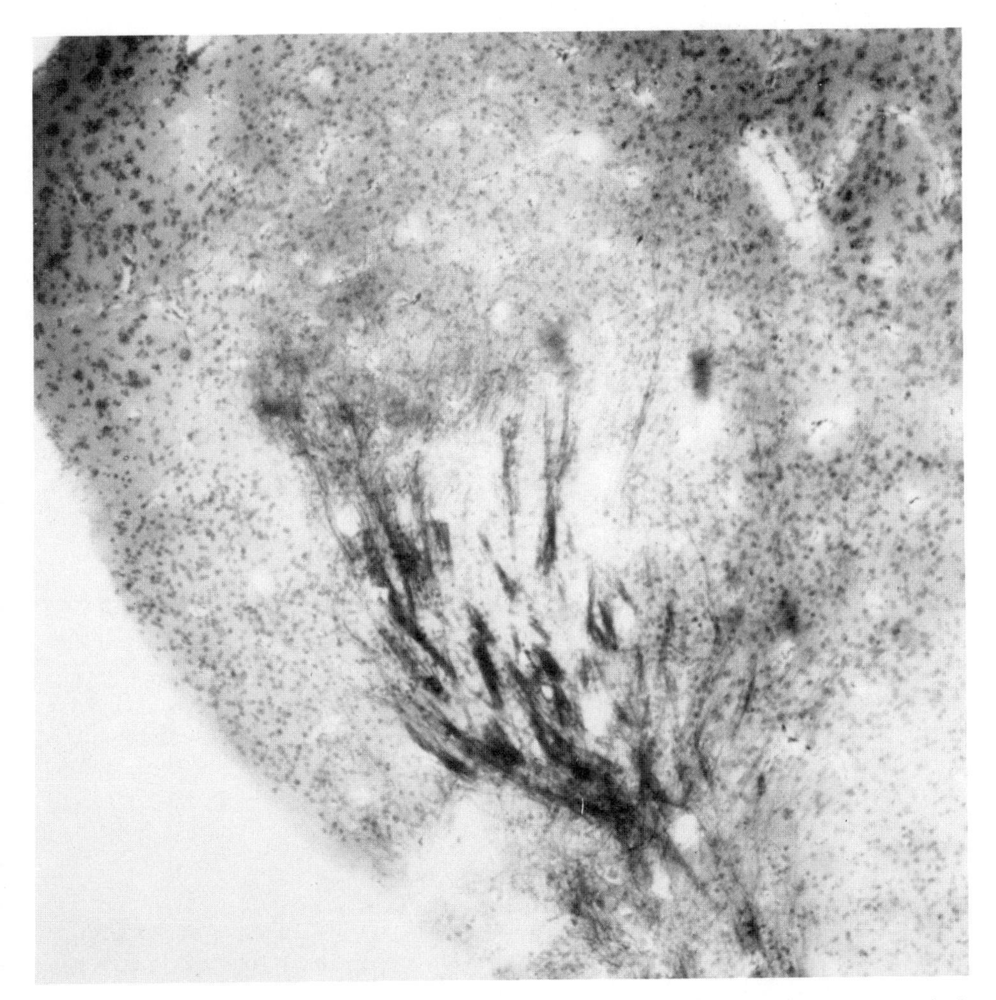

Figure 4-4. Sagittal section of the brain of a green anolis lizard (*Anolis carolinensis*) illustrating the striped appearance of the paleostriatum, which resembles the stripes (sometimes called Wilson's pencils) of the corpus striatum of mammals.

lizards suggests a miniature of the corpus striatum in mammals. But anatomical studies alone have been inadequate for identifying corresponding parts of the striatal complex in reptiles, birds, and mammals. An attempt to answer that question has been greatly helped by recent developments in histochemistry.

Neurochemical Features

In 1925, S. A. Kinnier Wilson (noted English neurologist, 1878–1937) began his Croonian lectures on disorders of motility by summing up the outcome of his own efforts over a period of 20 years to understand the functions of the basal ganglia. To a large extent, he said, "The ganglia at the base of the brain still . . . retain the characteristics of basements—viz., darkness."[7] Until recent times, one could say the same thing with respect to the question of what represents the corresponding basal ganglia in reptiles, birds, and mammals. In reptiles, and to a much greater degree in birds, there is a mass of cerebral gray matter bulging into each ventricle for which there is no apparent counterpart in the mammalian brain (but see below). Having been originally described by John Hunter (1728–1793), the ventricular mass was for a time referred to as Hunter's eminence,[8] but as explained later on, it is now commonly identified as the dorsal ventricular ridge (DVR). The names given to its various subdivisions need not be considered at this time. Suffice it to say, there was argument for many years as to whether or not it represented an elaboration of gray matter corresponding to the corpus striatum of mammals. Clarification occurred after George Koelle, in 1954, reported a histochemical method for revealing the presence of the enzyme cholinesterase.[9] The application of Koelle's method results in a deep copper-colored toning of most of the structures of the striatal complex. Figure 4-5 illustrates the striking demarcation of striatal structures in the monkey's brain. Similarly, as depicted by the comparative picture in Figure 4-6, the method reveals a sharp delineation of structures at the base of the forebrain in reptiles and birds.[10] In each case the structures are located below the dorsal medullary lamina that separates them from the overlying ventricular gray matter. In reptiles the submedullary structures include the olfactostriatum of Cairny[11] and the paleostriatum of Ariëns Kappers,[12] both of which, respectively, bear a resemblance to the olfactostriatum and corpus striatum of mammals. In birds the submedullary structures include the paleostriatum and the counterpart of the olfactostriatum of reptiles and mammals (see below).[13]

Subsequent developments in histochemistry have been a further aid in delineating the R-complex. Among the most significant discoveries of modern times has been the identification of three systems of monoamine-containing neurons that innervate extensive parts of the nervous system—namely, three respective systems with neurons containing dopamine and norepinephrine (both catecholamines) and serotonin (an indoleamine). It is the dopamine system in particular that demarcates the greater mass of the R-complex. These and some other neurochemical features will now be considered.

Dopamine

Falck and Hillarp, two Swedish investigators, were primarily responsible for developing a method by which the three monoamines in question could be made to fluoresce in thin brain sections.[14] Using this method, Dahlström and Fuxe (who studied with Hillarp) were the first to show that the greater part of the R-complex (olfactostriatum and corpus

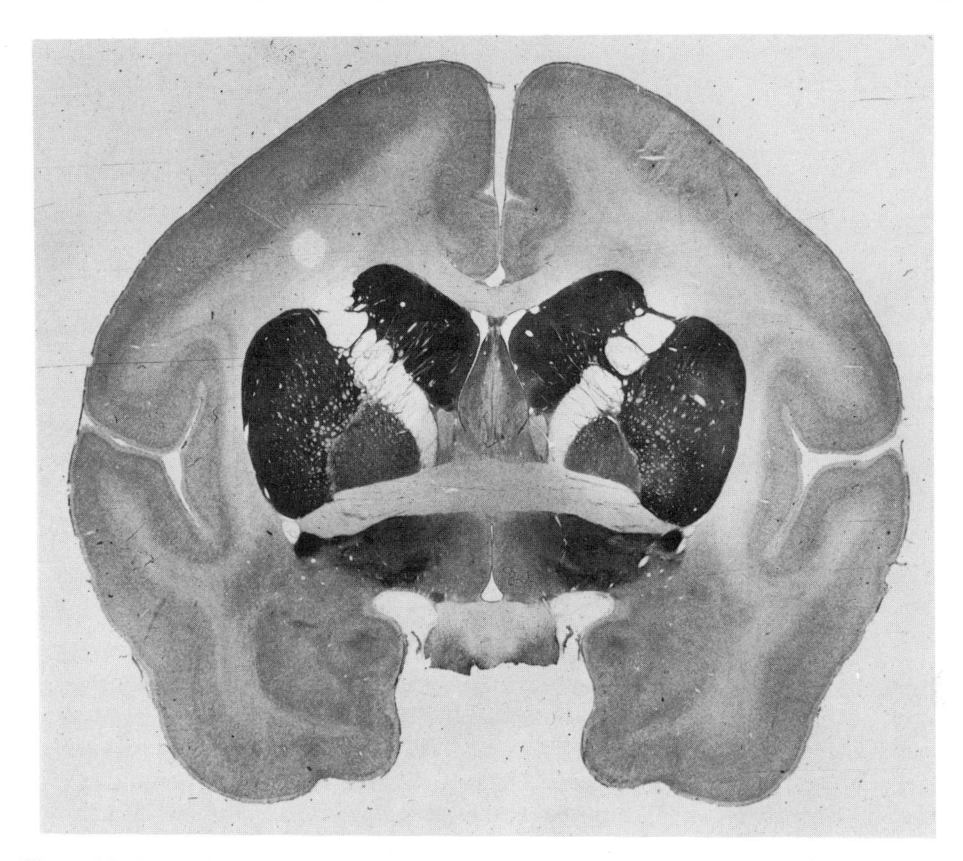

Figure 4-5. Section from brain of squirrel monkey, showing the dark, sharply demarcated coloration of parts of the striatal complex produced by a stain for cholinesterase. Clockwise on the right-hand side, one identifies the caudate nucleus between twelve and one-thirty; the putamen between two and three-thirty; and caudal part of the olfactostriatum between four and five. The lateral olfactory tract marks the lateral border of the olfactostriatum. Note absence of staining of the claustrum. This finding, together with the negative dopamine histofluorescence of the claustrum, provides evidence that it is not part of the corpus striatum. From MacLean (1972a).

striatum) lights up with a bright green fluorescence because of the presence of dopamine.[15] The dopamine in these structures is derived from cells of the ventral tegmental area and substantia nigra (pars compacta) in the ventral midbrain, the axons of which ascend and arborize among cells of the olfactostriatum and corpus striatum.[16] The same structures singled out by the cholinesterase stain in Figures 4-5 and 4-6 would also glow a bright green because of the presence of dopamine.[17] The earlier work on histochemical mapping of monoaminergic systems had been done on nonprimates. On the basis of our findings in monkeys, Jacobowitz and I inferred that the prototypical pattern of organization of these systems has been faithfully retained in the evolution of primates.[18] The distribution of dopamine nerve endings in the R-complex of a monkey is shown in Figure 4-7A, B, and C. Figure 4-7D indicates the location of the cells in the ventral midbrain that send their processes to the striatal complex. The labeling of the cell groups conforms to that of Dahlström and Fuxe for the rat.[15] Ungerstedt and others have shown in the rat that the axons from the A9 group in the substantia nigra innervate the corpus striatum (caudate and putamen), while those from the A10 group in the ventral tegmental area arborize in

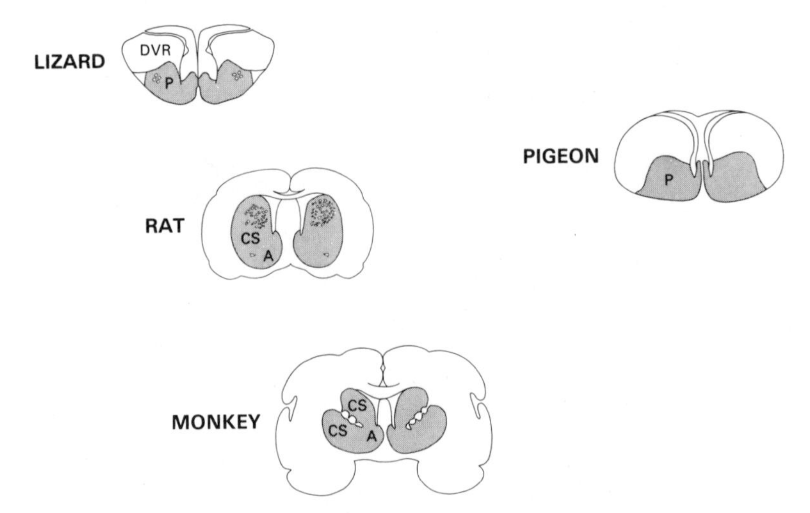

Figure 4-6. Effect obtained upon applying a stain for cholinesterase to brain sections of a lizard, rat, monkey, and pigeon. Combined with other evidence, the stain reveals that the striatal complex is a common denominator of the brains of terrestrial vertebrates. Abbreviations: A, N. accumbens; CS, corpus striatum; DVR, dorsal ventricular ridge; P, paleostriatum. Redrawn from Parent and Olivier (1970) with substitution of lizard for turtle.

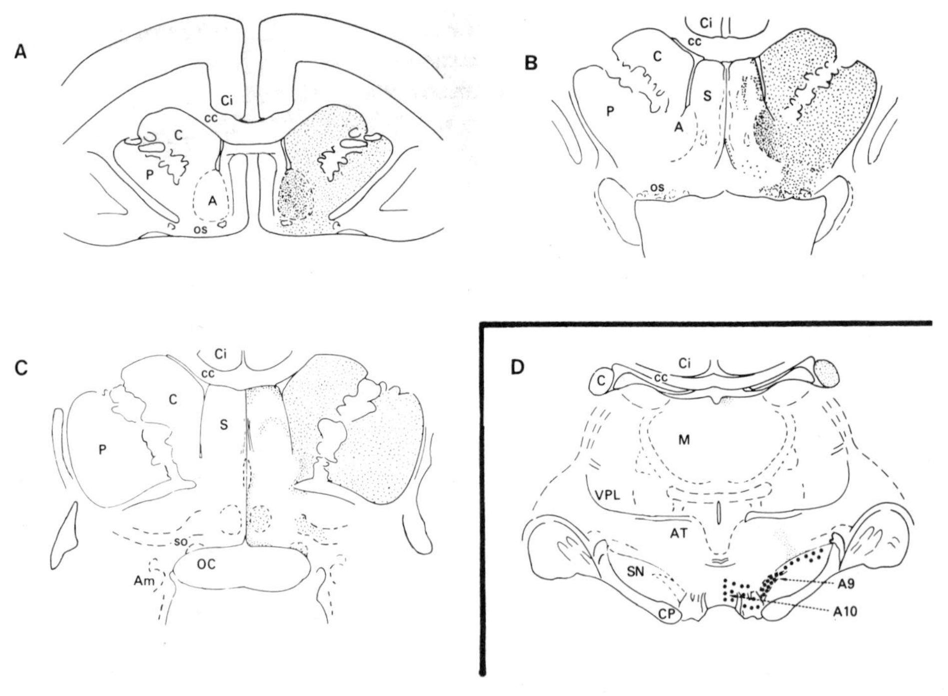

Figure 4-7. Location of dopamine-containing cells and their terminations in the striatal complex of a pygmy primate (*Cebuella pygmaea*). The cells are denoted by the large dots in D, while the terminals are represented by the fine stipple in A, B, C. The so-called A10 group of the medial part of the ventral tegmental area of the midbrain (D) innervates the olfactostriatum (A, B, C). The olfactostriatum (OS) also includes the nucleus accumbens (A). The A9 group of cells in the area compacta of the substantia nigra projects to the corpus striatum composed of the caudate and putamen (C and P). Adapted from Jacobowitz and MacLean (1978).

the olfactostriatum (nucleus accumbens and olfactory tubercle)[19] (see also section on anatomical connections).

Serotonin

In 1953, Twarog and Page[20] reported that serotonin could be found in brain extracts. A year later, Gaddum and co-workers confirmed this finding, stating that most of the brain's serotonin existed in the hypothalamus.[21] Other workers subsequently found that serotonin also occurred in high concentration in the corpus striatum.[22] With the histo-fluorescence technique, the presence of serotonin in the corpus striatum is *masked* by the great amount of dopamine. In contrast to dopamine, serotonin gives a fleeting, yellow fluorescence.

Shortly after the report in 1954 that serotonin was present in high concentration in the hypothalamus,[21] Gaddum and others discovered that the hallucinogenic (since called psychodelic) agent LSD (lysergic acid diethylamide) competitively blocks the action of serotonin.[23] These findings generated considerable excitement because many people then, as now, regarded the hypothalamus as the ''center'' of the emotions. Gaddum and others suggested that the symptoms induced by LSD may be due to its interference with the normal action of serotonin.[24] When it turned out that the bromo derivative of LSD, which also blocks the action of serotonin, failed to produce psychotic symptoms, it seemed to wreck the hypothesis. Nevertheless, since breakdown products of serotonin and other amines can, in very small amounts, produce psychotomimetic symptoms, the possibility that some disorder of amine metabolism may result in psychoses continues to be regarded as an attractive hypothesis.

Opiate Receptors

In Chapter 24 we shall consider neural mechanisms underlying ''eureka'' feelings. The wave of excitement created by the disclosures about serotonin seems, in retrospect, like a ripple compared with the one generated by the discovery of opiate receptors and opiatelike substances in the brain. These findings are most relevant to the present account because a number of these substances are found in particularly high concentration in the striatal complex. In 1973 there were three independent reports of the discovery of opiate receptors in the brain.[25] In the forebrain Pert and Snyder[26] found them in particularly high concentration in the corpus striatum. The opiate receptors presumably serve to influence brain function when they match up with an opiatelike substance.

Endorphins

Why are opiate receptors present in the brain? The possibility suggested itself that they interact with some naturally occurring opiate-like substance involved in neural mechanisms underlying pain, mood, or some other state. In 1975, Hughes[27] reported the isolation of an opiatelike substance in the brain. Shortly afterwards he and his colleagues provided the additional information that the active ingredients were pentapeptides—one called methionine enkephalin and the other leucine enkephalin.[28] Curiously enough, *the concentration of enkephalins in the globus pallidus part of the R-complex is several times greater than in any other cerebral structure.*[29] Enkephalins have been found localized to the external segment,[30] a fact easy to remember by associating the ''e'' of *e*xternal and *e*nkephalin. Soon after the discovery of the enkephalins, peptides of greater molecular weight, and also having an opiatelike action, were isolated from the pituitary gland.[31] The name *endorphins* was originally given to these peptides, but now the term is also used to include the enkephalins.

In the words from one monograph, "it would be hard to overestimate the excitement generated among pharmacologists" by the discovery of naturally occurring opiatelike substances in the brain.[32] The enkephalins were so simple in structure that they "were immediately synthesized and studies commenced in an amazing number of laboratories."[32] Already, the picture has become greatly complicated by the isolation of many other neuroactive peptides, some of which are also found in the intestinal tract and elsewhere.

Substance P

In the same study in which Gaddum and his associates investigated the distribution of serotonin in the brain, they used a method by which a compound called substance P could also be assayed.[21] They found a high concentration of substance P in the caudate nucleus and in the hypothalamus. Substance P refers to a vasodilating agent discovered by Euler and Gaddum (1931) and found to be present in brain and intestine.[33] The active principle was in the precipitate of their standard preparation. "This standard preparation," the authors explain, "we call P."[34] It appears now that the "P" was well chosen, because the active principle has turned out to be a peptide[35] occurring in several parts of the nervous system and postulated, among other functions, to be involved in pain mechanisms.[36] As opposed to enkephalins identified with the external segment of the globus pallidus, substance P is found in the medial segment.[30]

Iron

In 1922, Spatz[37] drew attention to the high concentration of iron in the globus pallidus and substantia nigra. It is associated with the surrounding glia and exists partly in the form of ferritin.[38] Iron can also be detected within nerve cells. When a stain for iron is used, the globus pallidus stands out in cross section like a bluish green ball. In a comparative study while working in the Laboratory of Brain Evolution and Behavior, Switzer and Hill emphasized that the pallidal staining extends into parts of the olfactostriatum that may be regarded as "paleopallidum."[39] There is no apparent explanation of the high concentration of ferric iron in the pallidum. It is known to accumulate with age.[40] Hill has shown in rats that the pallidal and nigral content of iron is about 1.5 times greater in females than in males.[41] As illustrated in Figure 4-8, the difference is so marked that with a new modification of Perls's method it is readily apparent on gross inspection. In addition, Hill has demonstrated that the amount of pallidal and nigral iron almost doubles at proestrus[42] and also shows a great increase on day 8 of pregnancy.[43] In discussing possible metabolic mechanisms, she emphasizes that both iron and enkephalins are found in highest concentration in the pallidum. Switzer suggests that there may be some metabolic interconnection between iron and γ-aminobutyric acid (GABA).[44] Found in especially high concentration in the globus pallidus,[45] GABA belongs to the class of neurotransmitters having a suppressive effect on the activity of nerve cells.

In summary, histochemical studies have shown that the greater part of the R-complex can be quite sharply delineated in reptiles, birds, and mammals on the basis of the

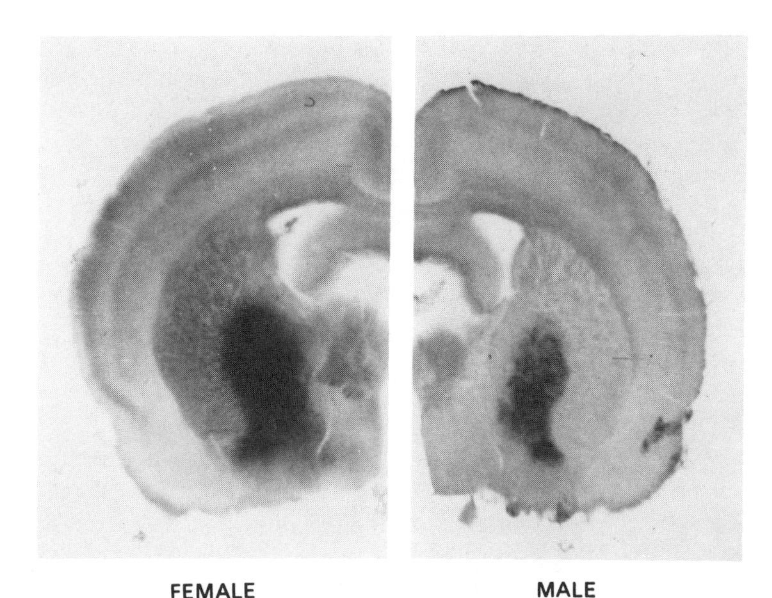

Figure 4-8. A histochemical demonstration that the iron content of the globus pallidus and olfactostriatum is greater in female rats than in males. The hemisphere section of a female is on the left; the one on the right is from a male of the same litter. Sections prepared by a diaminobenzidene intensification of Perls's histochemical method for iron. From Joanna Hill (unpublished).

distribution of large amounts of acetylcholinesterase and dopamine. In addition, different parts of the R-complex are of special interest because of high concentrations of serotonin, opiate receptors, enkephalins, GABA, and iron. The relevance of glutamic acid to corticostriate excitation will be noted under hodological aspects.

The Fork in Avian and Mammalian Evolution

Before considering embryological evidence as to what constitutes the R-complex in reptiles, birds, and mammals, it is helpful for orientation to point out major differences in the evolution of the forebrain in these animals. From an evolutionary standpoint the reptilian brain is of particular interest because it suggests how developments in a transitional region described by Elliot Smith[46] (1871–1937) may have tipped the scales so that "some animals evolved in the direction of birds, while others went the mammalian way."[47] The shaded area in Figure 4-9 shows the location of the paleostriatum in the brain of the tuatara (*Sphenodon punctatum*). As noted earlier, in both reptiles and birds it lies below the *dorsal medullary lamina*. This lamina of white matter covers it like a capsule. Above the lamina is a large mass of gray matter that in 1916 J. B. Johnston (1868–1947) termed the dorsal ventricular ridge,[48] now commonly referred to in writings as the DVR (see also Chapter 17). The arrow in Figure 4-9 points to the transitional region that Smith believed responsible for the inward growth of the superficial pallium and the formation of what he accordingly termed the hypopallium enveloping the DVR of Johnston. If the DVR were imagined as a mountain ridge, then the transitional region in

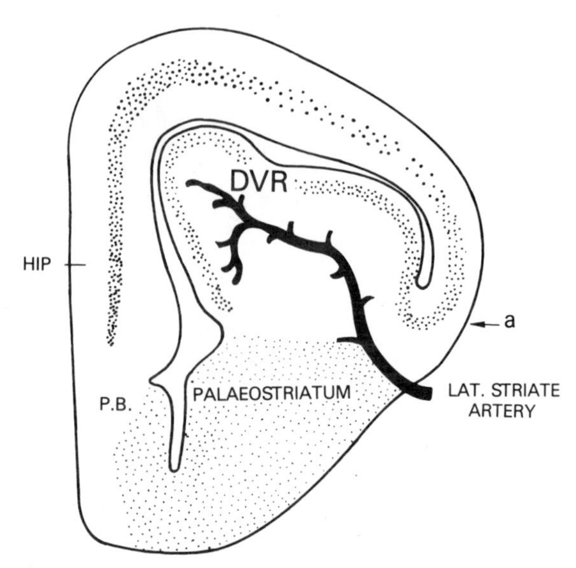

Figure 4-9. Hemispheric section of a reptilian brain (after Elliot Smith) used in the present context to discuss the divergence in the evolution of the forebrain in birds and mammals. The dotted area labeled palaeostriatum identifies the striatal complex, which constitutes the basic part of the forebrain in reptiles, birds, and mammals. The arrow (a) points to a transitional zone between the outer cortex and the ventricular bulge denoted as DVR. Smith regarded this site as a pivotal zone of growth. If such an interpretation were correct, one might imagine that in birds its proliferation led to a piling up of ganglia upon ganglia in the DVR, whereas in mammals it accounted for the ballooning outwards of the cerebral cortex. DVR has become a conventional abbreviation for "dorsal ventricular ridge," the name given to this part of the brain by Johnston in 1916. HIP identifies the rudimentary archicortex of the hippocampal formation. P.B., paraterminal body. Smith pointed out that the lateral striate artery is always found in this position in reptiles, birds, and mammals. Slightly modified representation of Smith's (1918–19) drawing of the brain of the tuatara.

question might be compared to a turbulent volcanic zone. In reptiles, and to a much greater extent in birds, its internal eruption resulted in a piling up of ganglia upon ganglia in the DVR, whereas its outward mushrooming in mammals was responsible for the cortex forming the lateral convexity of the cerebral hemispheres. In birds the ganglia above the R-complex pile up to a much greater extent than in reptiles. Despite these divergent cerebral developments in birds and reptiles, on the one hand, and in mammals, on the other, the R-complex in all three is "as much the bedrock of the forebrain as the Laurentian shield is to the North American continent."[47]

Ariëns Kappers (1877–1946) observed that the structure that he originally named the *paleostriatum*[49] shows a greater development in Squamata and Crocodilia than in Chelonia and a still greater development in Aves. Accordingly, he referred to the expanded peripheral part as the *paleostriatum augmentatum*[50] and the central portion as the *paleostriatum primitivum.* He presumed that these subdivisions were comparable to the medial and lateral segments of the globus pallidus. He regarded the ganglionic mass overlying the paleostriatum as a new part of the striatal complex, terming it the *neostriatum*[49] and suggesting that it corresponds to the mammalian corpus striatum (caudate and putamen). Subsequently, many workers tended to look upon the entire DVR of birds as an overgrown corpus striatum. Since the findings with the histochemical methods described above, however, neither the neostriatum nor any of the other subdivisions of the DVR (see Chapter 12) can be considered as enlargements of the corpus striatum. Rather, it is the paleostriatum augmentatum that appears to correspond to the corpus striatum.[51] In view of the hard-won information about corresponding parts of the R-complex in terrestrial vertebrates, the present tendency to adopt the outworn term *neostriatum* for the mammalian corpus striatum[52] seems like a backward step.

In contrast to the great development of the DVR, the cortex in birds has but a small representation and is poorly developed. The situation is about the same in reptiles. Anatomical and histochemical findings suggest that the rudimentary cortex in reptiles and birds corresponds to the archi- and "paleocortex" of mammals (see Chapter 17). In recent years several authors seem to have been persuaded that the DVR is the equivalent of the mammalian neocortex, and they may even refer to it in quotation marks as "neocortex." Such a view stems from Nauta and Karten's argument that structures may be "homologous" if they are similarly connected in different species.[53] In other words, evidence of homology rests on like connections between pools of neurons, regardless of the cytoarchitecture of the pools themselves. Thus, despite their great dissimilarity in structure, the DVR and neocortex would be considered "homologous" because each contains structures that relate to certain parts of the visual system. As one proponent of DVR/neocortex equivalence has commented, the hypothesis has "proved sufficiently procrustean to make possible an explanation for each and every set of observations."[54] It will be recalled that the word *procrustean*—referring to something "designed to secure conformity at any cost"—derives from the name for the mythical robber of Attica, who either stretched his victims or cut off their legs in order to make them conform to the length of his iron bed.

The question of whether or not some other mammalian structure might be the counterpart of the DVR is better deferred until embryological findings have been reviewed. Since the DVR projects to the paleostriatum in reptiles and in birds (see Chapter 11), questions of this kind are by no means outside the present focus of interest on the R-complex. There has as yet been no explanation for such a marked difference in the evolution of the forebrain in birds and mammals. It is possible that the piling up of ganglia upon ganglia in the DVR of birds reflects their winged way of life, facilitating a three-dimensional orientation to their environment. Apart from bats, and most recently humans, there are no terrestrial vertebrates that are called upon to navigate freely in an airy medium. It is self-evident that the need to recognize objects such as trees and the like from every angle—both from above and at ground level—greatly adds to the complexity of orientation and navigation. Exceeding the task of homing to nests in such indiscriminate locations as hay fields and dense thickets, numerous species of birds must accomplish migrations to specific locations thousands of miles away.

Detailed information is lacking about the architecture of ganglionic structures such as the DVR. James W. Papez (1883–1958) suggested that cells of such ganglia may be arranged in long cords,[55] with the nerve processes arborizing along them somewhat like fingers on the stops of a clarinet. Highly organized in their arrangement, but folded upon themselves somewhat like a nest of caterpillars, their true nature would not be suspected after they were cut into thin brain sections and viewed through a microscope. Perhaps such long, folded cords of cells would provide a picture of three-dimensional space that could not be achieved to the same extent by the short columns of cells making up the layers of cortex.

Embryological Evidence

Ernst Haeckel (1834–1919) popularized the view that ontogeny recapitulates phylogeny.[56] Although there are well-known exceptions to his biogenetic law,[57] it articulates certain compelling parallels. Are there embryological findings that help to identify an evolutionary linkage of the striatal complex in reptiles, birds, and mammals? Over the years comparative neuroanatomists have reached general agreement that certain parts of the diencephalic and telencephalic walls of the mammalian embryo develop into structures corresponding to the paleostriatum of reptilian and avian forms.

The once superficial ectoderm of the neural plate becomes the inside of the neural tube after the neural ridges grow upwards and meet at the midline (see Chapter 3). Hence, it continues to be referred to as epithelium—specifically, neuroepithelium—in spite of its new inside position. The neuroepithelium is the origin of the cellular elements that migrate and become the neural and supporting structures of the brain. The neuroblasts (literally, nerve germs) multiply locally before beginning their migration and differentiation into mature cells with dendrites and axonal processes. Using the rapid Golgi method, Morest has presented evidence that the neuroblasts establish contact with the internal and external limiting layers by means of primitive processes that serve as tubular elements for the nucleus and perikaryon to migrate to the final location of the differentiated cell.[58] Nuclei can be identified because they do not stain, while nucleoli (commonly two of them) appear as small yellow bodies within the nucleus. In the case of cortex the layers develop from within and become added to toward the surface (the so-called "inside-out" sequence), suggesting in some respects the building upwards of a brick wall. In the ganglionic masses of the striatal complex, however, the cells may assume their final position in an "outside-in" sequence somewhat as if one were to fill a room with balloons that floated first to the ceiling and then gradually filled in the space below.

In the following account, reference will be made to incipient elevations (also known as swellings or eminences) at certain places in the ventricular walls. These initial elevations reflect the beginning of active proliferation of neuroblasts.

Reptilian Counterpart

Kallen is one of the few anatomists to have included lizards in his embryological studies of the reptilian forebrain.[59] As illustrated in Figure 4-10A, he depicts two elevations in the developing lateral wall of the telencephalon. Using Holmgren's terminology for the turtle,[60] he refers to the lower elevation as the "floor area" and the upper elevation as the "roof area." Several years earlier (1923) Marion Hines had described similar elevations in the tuatara (*Sphenodon punctatum*).[61]

It should be emphasized that the "roof" and "floor" areas are *not* to be confused with the alar and basal plates that lie dorsal and ventral, respectively, to the sulcus limitans of His that runs from the caudal end of the neural tube to the diencephalo-mesencephalic boundary "near the mammillary recess."[5] These latter two plates are identified, respectively, with sensory and motor functions. *As many authors have pointed out, the fact that the telencephalic part of the striatal complex is entirely rostrodorsal to the basal plate speaks against the traditional categorization of it as part of the motor apparatus.*

According to Kallen, the floor area can be subdivided into three "bands," corre-

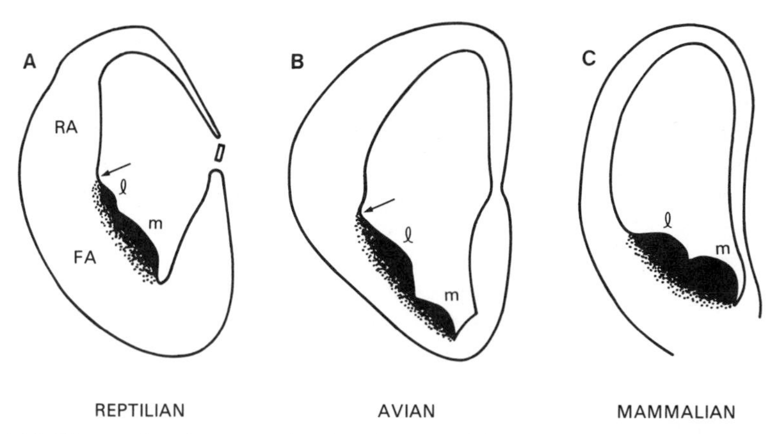

Figure 4-10. Diagrams focusing on two elevations (l and m) of the developing striatum in reptilian, avian, and mammalian embryos. The reptilian embryo shows ventral and dorsal parts of the lateral telencephalic wall respectively identified as "floor area" (FA) and "roof area" (RA) according to the terminology of Holmgren. These terms have no relevance to the basal and alar plates of the neural tube. Middle and right-hand drawings correspond to avian (sparrow) and human embryos. Drawings adapted from illustrations in Kallen (1951), Durward (1934), and Sidman and Rakic (1982).

sponding, as was indicated in his illustrations, to the paleostriatum, olfactostriatum, and septum. As will be considered later on, the roof area appears to be the site of development of the pallial region and the DVR. At the present time there is the need to apply modern techniques to the study of neurogenesis in various reptiles (see below). Improved methods of incubation promise to ease the difficulties of obtaining sequences of lizard eggs of known age.

Avian Counterpart

Durward's findings in the common sparrow serve to illustrate the development of elevations of the ventricular wall in avian forms that resemble those of the reptile.[62] As diagrammed in Figure 4-10B, a groove (arrow) appears in the lateral wall that compares to the one in reptiles forming the boundary between the roof area and the floor area (Durward refers to it as the "neo-paleostriatal groove"). This boundary region is an important one because it marks the site at which the dorsal medullary lamina will form, separating the paleostriatum below from the DVR above. On the external surface this boundary zone also marks the place of entry of arteries (Figure 4-9) found as a constant feature in the brains of reptiles, birds, and mammals[46] (see below). Later on in development (stage D) the two components of the paleostriatum—primitivum and augmentatum (see above)—become evident.

Formerly, knowledge of the time of cell division in neurogenesis depended on the identification of dividing cells (mitotic figures) in thin brain sections. Since 1960 the combined use of tritiated thymidine and autoradiography has made it possible to identify not only the approximate "birthdate"[63] of a cell, but also its place of final destination in the developing brain. Thymine "occurs uniquely in DNA, and thymidine is incorporated efficiently and exclusively into DNA."[64] Cells in the premitotic process selectively assimilate thymidine. Accordingly, [^{3}H]thymidine can be used to label cells in the process of division. Brain sections that are dipped in a photographic emulsion and exposed for several days show silver grains superimposed on dividing cells. Cells undergoing a single division after a single injection ("pulse") of [^{3}H]thymidine are heavily labeled by silver grains, whereas those proceeding to divide two or three times will appear lightly labeled.

After injections of [^{3}H]thymidine into chicken eggs on successive days of incubation, Tsai *et al.* found that the birthdates of the cells forming the paleostriatum primitivum occurred on day 4, whereas those for the augmentatum ranged from the 5th to the 7th day.[65] In 1938 Kühlenbeck had suggested that some of the cells of the primitivum are of diencephalic origin that migrate "along the lateral forebrain bundle" into the telencephalon.[66] The findings of Tsai *et al.* were compatible with this possibility, but did not provide conclusive information.[65] As noted earlier, the reptilian and avian primitivum is generally regarded as corresponding to the globus pallidus of mammals (see also below), whereas the augmentatum is the counterpart of the corpus striatum.

The olfactostriatum, including the olfactory tubercle and nucleus accumbens, is an unquestioned part of the telencephalon. As mentioned earlier, in mammals the cells in the deeper part of the tubercle may be characterized as "paleopallidal." Cells here become labeled at the same time as those in the primitivum. Because of the outside-in positioning of cells noted above, it appears that cells in the region of the nucleus accumbens are among the last of the striatal complex to be labeled.

Tsai *et al.* draw particular attention to the "compartmentalization" of the avian telencephalon that takes place during neurogenesis.[67] The respective compartments extend radially from the medial to the lateral surface of the brain and are separated by laminae. After proliferation in the neuroepithelium the cells migrate to the place of their final destination, filling in the compartments progressively in an outside-in sequence. Significantly, in regard to the question of derivation and correspondence of the striatal structures in the various classes of land vertebrates, the earliest and most distinctive compartmentalization appears on each side of what becomes the dorsal medullary lamina. As Tsai *et al.* state, "A dramatic example of this compartmentalization is found in the independent proliferative behavior of the neuroepithelium sectors above and below the forerunner of the lamina medullaris dorsalis. . . ."[68]

Mammalian Counterpart

In a study of age-graded pouch young of the opossum, Morest found that the "same features observed in corticogenesis" characterize those seen in the corpus striatum.[58] Of special significance, however, was his observation that in the striatum the migration of a cell ceases when its axon begins to sprout. Where the cell is at that time becomes its "final location." Morest is noncommittal in regard to the question of the "outside-in" sequence mentioned in connection with the bird. He notes that the earliest differentiation of striatal neurons occurs in "the ganglionic eminence" at the level of the anterior commissure and that rostral and caudal to this level there is a "developmental gradient."

The observations of Smart and Sturrock on the mouse serve to illustrate the development of ventricular elevations in placental mammals.[69] As in reptiles and birds, two elevations characterize the formation of the striatal complex (see Figure 4-10C). Gestation in the mouse covers a period of 19 days. An initial swelling in the lateral wall of the telencephalon appears on the 11th day of gestation, *straddling the telo-diencephalic boundary* (see below). As confirmed by histological examination, this swelling signals the beginning of cellular proliferation in the striatum. On the 12th day a second swelling appears just above it (Figure 4-10C). Hence, the two swellings are referred to as the ventral and dorsal elevations. Unlike other parts of the brain, the proliferation of neuroblasts in the striatum takes place mainly in the subependymal, rather than ependymal (neuroepithelial), layer. By the 15th day of gestation, the sulcus between the ventral and dorsal elevations is barely visible, and by the time of birth it has disappeared.

In extending the original [^{3}H]thymidine observations of Sidman and Angevine,[70] Smart and Sturrock found that differentiation within the corpus striatum itself is completed between the 14th day of gestation and the 5th postnatal day.[69] They confirmed, in general, the findings of Angevine and McConnell[71] that neurogenesis in the striatum first occurs predominantly in the caudolateral portion and then shifts toward the rostromedial parts. The findings were suggestive of the "outside-in" sequence described above in regard to the avian brain. Smart and Sturrock provided the additional valuable information that myelination of the corpus striatum begins on the 10th postnatal day and is largely completed by the 180th day.[69]

In regard to the globus pallidus, Smart and Sturrock point out that its primordium is coextensive with the ventral part of the medial elevation that incorporates part of the diencephalon in the wall of the third ventricle.[69] They make no specific mention of the rostroventral extension of the pallidum, which, as noted earlier, may be regarded as the paleopallidal part of the olfactostriatum. As in the case of the corpus striatum, Angevine and McConnell observed a prolonged period of origin of the cells of the nucleus accumbens in the mouse, extending into postnatal development.[71]

In the rhesus monkey the period of gestation is about 165 days. Brand and Rakic observed that the large cells of the corpus striatum develop between the 36th and 43rd embryonic day and the small cells between the 36th and 80th day.[72] They found no "gradient" in the location of the cells. It should be noted that whereas the corpus striatum develops during the latter third of pregnancy in the mouse and other rodents, its formation is virtually completed in the monkey by midpregnancy.[72]

As illustrated in Figure 4-10C, the picture of the developing elevations on the ventrolateral wall of the human embryo is similar to that already described for a reptile, bird, and small mammal. A medial elevation appears in the 7.5-mm embryo, followed by a second lateral elevation in the 15-mm embryo. In the 45-mm embryo, Hewitt has observed a third intermediate elevation.[73] All three elevations appear to enter into the formation of the caudate and putamen. In the 135-mm fetus, the globus pallidus is "clearly defined."[73] Part of the medial elevation has a tonguelike extension into the wall of the diencephalon. In line with evidence presented by Spatz,[74] many authorities agree that at least part of the globus pallidus is derived from the diencephalon.[75] Information appears to be lacking about the ontogeny of the olfactostriatum in the human brain.

Chemoarchitectonic Aspects

Based on a brief description by Ramón y Cajal,[76] Papez noted in his textbook *Comparative Neurology* (1929) that the histological structure of the corpus striatum is "very characteristic and primitive" and that the "caudate and putamen are composed of tubular clusters of large quantities of small radiating cells with short axis cylinders."[77] In the light of recent findings with autoradiography and histochemistry, his choice of the word *clusters* has proved to be most fitting. In 1974, Jay Angevine (one of Papez's former students) and J. A. McConnell reported that in the mouse embryonic striatal cells labeled with [^{3}H]thymidine show up as "clusters."[71] Subsequently (1978), Brand and Rakic described a similar picture in the monkey.[72] Graybiel and Hickey found that clusters of nerve cells labeled in the cat during embryonic development were coextensive with acetylcholinesterase-*poor* patches (or so-called "striosomes"[78]) of the adult animal. A "remarkably precise match" was also demonstrated between the thymidine-labeled clusters and enkephalin-rich zones.[79] As yet it is not known what relationship thymidine-labeled cell clusters have to acetylcholinesterase-rich patches seen in fetal brains. Such patches have been observed in the striatum of human fetuses as young as 16 weeks.[80] Postnatal findings on the rat have revealed that initially "islands of acetylcholinesterase activity" occur predominantly in the lateral part of the striatum and that within a period of 15 days there is a lateral-to-medial spread of such activity to the rest of the striatum.[81]

Studies on the ontogeny of monoamine systems have also shown a patchlike distribution of dopamine varicosities. In three reports by different investigators in 1972,[82] such patches were referred to as "islands." Tennyson *et al.* found that in the rabbit such islands appear by the 28th day of gestation.[83] In fetal cat brains obtained between 49 and 57 days of gestation, Graybiel and co-workers observed that dopamine "islands" were coextensive with patches rich in acetylcholinesterase.[84] By means of the histofluorescence technique the entire dopaminergic pathway has been demonstrated in the mouse on the 13th embryonic day[85] and in the rat on the 19th day.[86]

The accumulating findings of anatomical and neurochemical "compartmentalization" of the mammalian striatum (see also below) recall the provocative statement by Hughlings Jackson in 1873 "that the corpus striatum is a mass of small corpora striata."[87] Apropos of the comparative questions under consideration, it remains to be seen whether or not a similar compartmentalization will be found in reptilian and avian forms.

A Mammalian Counterpart of the DVR?

Gregory has emphasized that every one of the 28 bones of the human skull "has been inherited in an unbroken succession from the air-breathing fishes of pre-Devonian times."[88] Currently, as expounded by Ebbesson,[89] there seems to be some inclination to attribute the evolutionary development of various structures in the brain to an elaboration of anlage existing in earliest vertebrate forms. Regardless of such speculation, one might ask if there is any structure in the mammalian brain that may share similar origins with the DVR of reptiles and birds. The evolution of the brain's vasculature offers a possible clue. Elliot Smith called attention to vessels found at the same locus in the brains of reptiles, birds, and mammals that penetrate the brain at a point just medial to the lateral olfactory tract where it adjoins the piriform cortex.[46] At this site in reptiles and birds there arises a main vessel that courses through the dorsal medullary lamina that separates the paleostriatum from the overlying DVR. Smith called it the *lateral striate artery* and depicted it (see Figure 4-9) as supplying the DVR. Based on the study of arterial injections, Shellshear (a student of Smith) concluded that the corresponding vessel in mammals supplies the claustrum[90] and that Smith had thus wrongly interpreted it as being homologous in all three classes of animals. At the same time, he pointed out that Smith regarded the claustrum as "homologous" to the DVR! If the claustral interpretation were correct, then the mammalian external capsule would appear to correspond to the dorsal medullary lamina. In the past the claustrum has been variously interpreted as (1) one of the basal ganglia closely associated with the corpus striatum,[91] (2) the deep layers of the insular cortex,[92] (3) a continuation of the lateral amygdala,[93] and (4) a structure independent of the cortex and the striatum.[94] In any event the interpretation that it represents an addition to the corpus striatum seems no longer tenable; as I have pointed out elsewhere, and noted in the legend of Figure 4-5,[95] it does not show a positive reaction with respect to cholinesterase or dopamine (see Chapter 18 for additional data on the claustrum).

Anatomical Structure and Connections

As yet there is insufficient knowledge of the cytoarchitecture and ultrafine structure of the R-complex to draw comparisons among reptiles, birds, and mammals. Although the same does not apply to the *connections* of the R-complex, the question of corresponding anatomical pathways within the three classes of animals can be more meaningfully dealt with when discussing the outcome of the neurobehavioral studies. Consequently, the present anatomical summary, based on findings in mammals, is intended to serve both as general background and as an introduction to the final topic of this chapter—namely, the question of function of the striatal complex.

Cellular Constituents

As already noted, the corpus striatum (caudate nucleus and putamen) and globus pallidus constitute by far the greater mass of the striatal complex. When viewed in histological sections, the caudate and putamen have a similar homogeneous appearance, being comprised primarily of small cells, along with a scattering of larger cells about twice the size of the small ones. Estimates of the ratio of small to large cells in different parts of the striatum have ranged from 20 to 1[96] to greater than 50 to 1.[97] But, as we have seen, more recent studies have shown that the homogeneous appearance of the striatum is deceiving and that instead, it appears to contain subunits that can be identified both anatomically and chemically.

In 1873, Golgi reported his chromate of silver stain for revealing the body and processes of nerve cells (see Chapter 18). Upon applying this stain to the corpus striatum, Ramón y Cajal identified four different types of cells.[76] In a recent Golgi study of the striatum in the monkey, Fox and co-workers found that small neurons with numerous spines on symmetrically radiating dendrites were the predominant type of cell,[98] comparing to Kemp's observation in the cat that such neurons comprise about 95% of the population.[99] They concluded that the large aspiny neurons, presumably corresponding to the large neurons seen in Nissl-stained preparations, give rise to small myelinated axons that account for the projections from the striatum to the globus pallidus and substantia nigra. In a reexamination of this matter, Pasik *et al.* arrived at an opposite conclusion. In the striatum of the monkey they identified six types of neurons—two of the spiny type, three aspiny varieties, and the neurogliaform cells of Ramón y Cajal and other authors.[100] According to their findings, all of the aspiny neurons have *short* axons. Their aspiny type II cell would correspond to the large neurons seen in Nissl-stained sections. In their view it is the spiny type neurons that are the source of the long axons issuing from the striatum.

If so, the long-standing question as to how the relatively few large cells of the corpus striatum could provide the myriad of fine fibers projecting from the striatum would no longer be a concern.

The cells of the globus pallidus are predominantly large in size, fusiform in shape, and have a prominent Nissl substance.[101] The olfactostriatum, consisting of the nucleus accumbens and the olfactory tubercle, is regarded as a phylogenetically older part of the striatal complex. The cells of the nucleus accumbens resemble those of the corpus striatum, while those in the deeper part of the tubercle are similar to those of the globus pallidus. The superficial portion of the tubercle and rostrobasal part of the septum contain clusters of deeply staining neurons (islands of Calleja), some comprised of medium-sized cells, others packed with tiny neurons suggestive of glia.[102]

Finally, the so-called substantia innominata (''unnamed substance'') encompassing the basal nucleus of Meynert forms an extensive mass of gray matter belonging to the striatal complex. This substance contains widely scattered, deeply staining large cells that are vividly colored by a stain for acetylcholinesterase. In the primate brain, cells of this type are found within the internal and external medullary laminae of the globus pallidus, as well as in the satellite collections of cells underneath the pallidum and within the basal nucleus of Meynert.[103]

Inputs to Striatal Complex

In attempting to identify the anatomical connections of a structure, it is helpful if one can first trace the course of its inputs, particularly the inputs, if any, from specific receptive (''sensory'') systems. Except for direct olfactory innervation of parts of the olfactostriatum, there is as yet no precise information about inputs to the striatal complex from other sensory systems. A summary of the findings on olfactory connections will be found in Chapter 18. As to be discussed below, the results of electrophysiological studies have helped to buttress anatomic inferences about afferent routes for other sensory modalities affecting striatal function.

The very fine caliber of most of the afferent fibers to the striatal complex has been a major obstacle in determining their origin, course, and termination. Some headway began to be made in the 1940s when Paul Glees introduced a modified Bielschowsky (1869–1940) silver stain for tracing degenerating nerve fibers. Then followed the well-known Nauta–Gygax silver method and its later modifications (see Chapter 18 for brief historical review). The findings with the silver methods have been extended and supplemented by the application of techniques involving physiological transport of substances that afford a detection of nerve cells and fibers. One of the main advantages of these methods is that they entail little destruction of nervous tissue and thereby greatly reduce uncertainty of interpretation because of complicating damage to fibers of passage. One of these methods involves injections of micro-amounts of radioactive nutrients that nerve cells assimilate and transport to their terminals. An autoradiogram reveals the presence of the transported substance within the nerve fibers and their terminals. The other principal method involves the injection of an enzyme, horseradish peroxidase (HRP), which, after being preferentially assimilated at nerve terminals and transported back to the cell bodies, gives a color reaction when treated with a chromogen (see Chapter 18 for further details).

The following account provides a summary regarding inputs to the striatal complex from five main sources—namely: (1) the central tegmental tract, (2) monoaminergic systems, (3) the ancient intralaminar structures of the thalamus, (4) widespread areas of the neocortex, and (5) limbic structures.

Central Tegmental Tract

The central tegmental tract, one of the most "ancient" conducting systems in the brainstem, comprises a large, poorly defined bundle of fibers of diverse origin that contain both ascending and descending fibers that ramify along its course from the level of the thalamic tegmentum to the inferior olive. In the midbrain the bulk of its constituents are located ventrolateral to the central gray. Nauta and Kuypers have provided evidence that ascending components include fibers from reticular cell groups in the medulla, pons, and midbrain and that at least some of midbrain origin terminate in the corpus striatum.[104]

Ascending Monoaminergic Systems

Mention has already been made of studies in the rat showing that ascending dopaminergic fibers from the A9 group of cells in the substantia nigra and the A10 group in the ventral tegmental area arborize, respectively, within the corpus striatum[19] and the olfactostriatum.[19] A dopaminergic nigrostriatal pathway was also demonstrated early on in the cat and monkey.[105] Some of the ascending nigral fibers sweep over the subthalamic nucleus and then, as illustrated in the monkey in Figure 4-11, radiate out through the globus pallidus to reach the putamen. Other nigral fibers ascend through the thalamic tegmentum and then follow the internal capsule (Figure 4-11) to reach the caudate nucleus. Silver,[106] HRP,[107] and autoradiographic[108] methods have provided additional details about nigrostriatal projections, not all of which arise from dopamine cells. Ascending fibers from the ventral tegmental area apparently follow the medial forebrain bundle to reach the olfactostriatum. The fibers delivering serotonin to the striatum arise from dorsal raphe cells in the midbrain (see also Chapter 18).[109]

Intralaminar Afferents

The intralaminar nuclei comprise phylogenetically ancient groups of cells associated with the internal medullary lamina of the dorsal thalamus (see Chapter 18 for further details). Parent has provided additional evidence that some of these cell groups correspond to those in the reptilian brain.[110] The central lateral nucleus and the paralaminar part of the medial dorsal nucleus receive a somatosensory input from the spinothalamic pathway.[111] Long ago Walker called attention to probable inputs from the tectum to medial thalamic nuclei,[112] providing a presumed source of information from the visual and auditory systems. Findings with HRP are indicative of such inputs, as well as inputs from the vestibular nuclei of both sides.[113] It is most significant in regard to vagal inputs and somatovisceral functions that the autoradiographic method has revealed a projection from the parabrachial nuclei of the isthmus to several intralaminar nuclei (see electrophysiology below and also Chapter 18).[114] Experiments involving retrograde degeneration originally provided evidence of laminar-striate projections in the rat,[115] rabbit,[116] and monkey.[117] Such findings were confirmed by the use of silver methods.[118] In the primate, Mehler showed that the centromedian nucleus, representing an expanded part of the intralaminar group, projects mainly to the putamen.[119] The retrograde labeling method afforded by HRP has been particularly useful in helping to dispose of long-standing

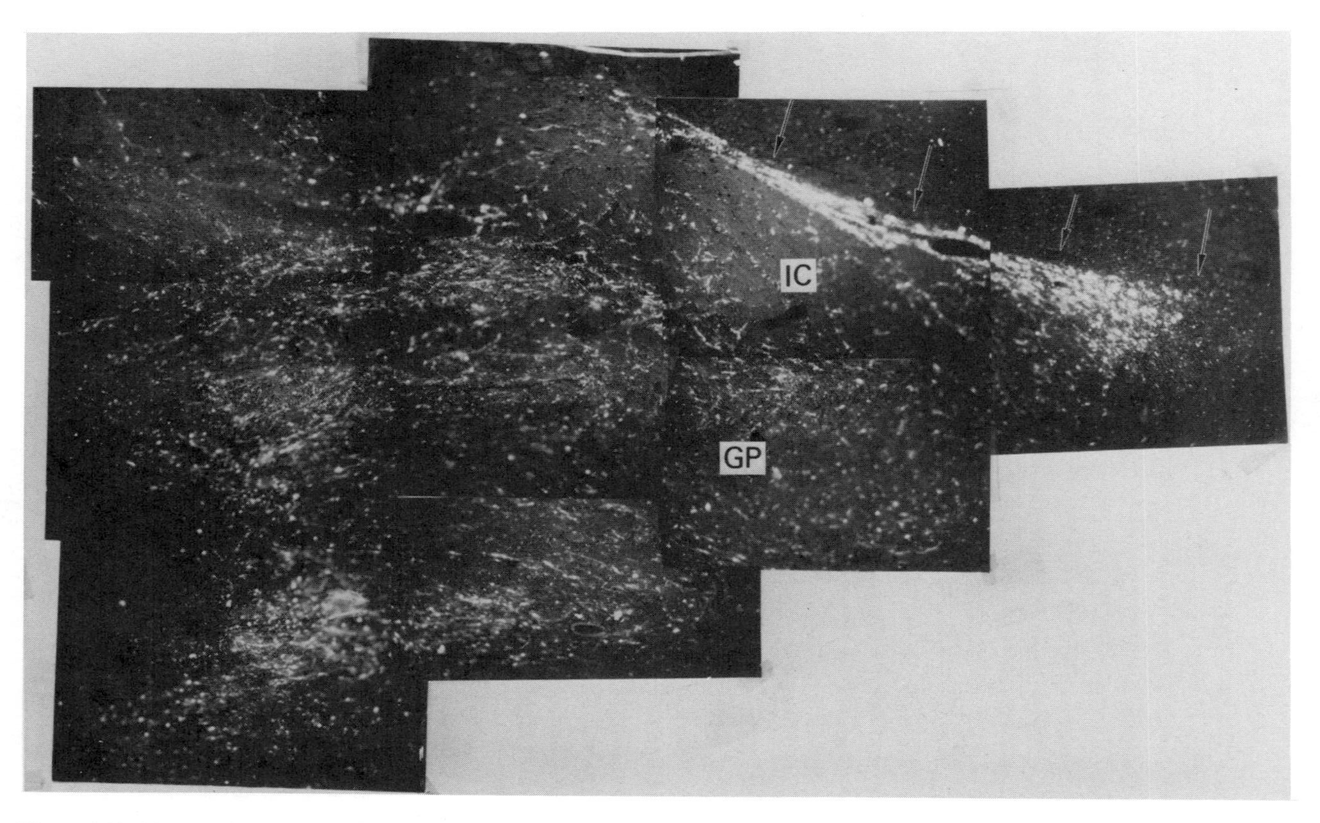

Figure 4-11. Montage showing histofluorescent picture of swollen, dopamine-containing fibers of the nigrostriatal pathway subsequent to lesions produced in caudate nucleus and putamen of a squirrel monkey (*Saimiri sciureus*) by 6-hydroxydopamine. Note collection of fibers (arrows) in internal capsule (IC) destined for the caudate nucleus and the radiating fibers in the globus pallidus (GP) headed for the putamen. From MacLean (1978b).

uncertainties as to the location of intralaminar cells projecting to the striatum[120]; labeled cells are found in the paracentral, central medial, central lateral, centromedian, and parafascicular nuclei. With autoradiographic methods it has been shown that projections from the region of the centromedian nucleus have a patchlike termination in the caudate and putamen, respectively, of the cat[121] and monkey.[122]

Neocortex

In 1914, Wilson noted that his failure to demonstrate corticostriate fibers[123] was in agreement with the observations of Ramón y Cajal and of Dejerine[124] among others. His study seemed to give the coup de grace to any lingering belief in such connections. When Glees reexamined this question with his new silver stain (1944), he found evidence of corticostriate connections (from the so-called suppressor areas) via collaterals from corticofugal fibers descending in the internal capsule.[125] In 1961, Webster reported that every part of the rat neocortex projects in an orderly manner onto the striatum.[126] Comparable results were later obtained in the opossum,[127] rabbit,[128] cat,[129] and monkey.[130] In 1975, Kunzle reported his autoradiographic findings that projections from the motor cortex of area 4 terminate in the striatum in patchlike distributions reflecting a somatotopic organization.[131] Follow-up studies indicated a convergence of projections from the somatosensory cortex.[132] It now appears from autoradiographic investigations by different workers that virtually all areas of the neocortex project in a patchlike manner to the striatum.[133] The findings of Yeterian and Van Hoesen suggest that reciprocally related areas of "association" cortex innervate overlapping parts of the striatum.[134] Van Hoesen *et al.* emphasize that "*major parts of the corticostriate system arise from areas of cortex which cannot be conceptualized as motor*"[135] (italics added).

Beginning with the observations of Carman *et al.*,[128] evidence of cortical projections to the contralateral striatum has been found in the opossum,[127] rat,[136] rabbit,[128,136] cat,[129,136] and monkey.[131,132] Findings with recent techniques[131] favor the original interpretation that the "sensorimotor" areas,[136] and not other cortical areas, are the primary source of contralateral connections.

The question remains unsettled as to whether the corticostriate projections represent collaterals from corticofugal fibers within the internal capsule. Both Glees[125] and Webster[129] presumed they were collaterals because the striate fibers branch at right angles from the main descending axons. On the basis of an HRP study and other evidence, Royce has suggested that most, if not all, of the corticostriate projections derive from collaterals.[137] In addition to the innervation via the internal capsule, the caudate and putamen, respectively, receive cortical afferents coursing through the subcallosal bundle of Muratoff[138] and the external capsule.[139]

Limbic Structures

Comparative neurologists have long recognized the close anatomical relationship of parts of the limbic cortex with the olfactostriatum[13] (see Chapter 18). Heimer and Wilson have emphasized that the piriform cortex projects primarily to the olfactory tubercle, whereas the hippocampus projects to the nucleus accumbens.[140] Nauta was the first to show with a silver stain that the cingulate cortex has robust connections with the caudate nucleus.[141] The same has been found for the amygdala (see also Chapter 18).[142]

Outputs of Striatal Complex

In considering outputs of the striatal complex we will deal successively with those from the (1) olfactostriatum, (2) corpus striatum, (3) globus pallidus, and (4) substantia innominata.

From Olfactostriatum

As Heimer and co-workers have demonstrated, the superficial part of the olfactory tubercle and the nucleus accumbens innervate cells in the deep part of the tubercle that represents the so-called paleopallidum.[140] The latter in turn projects via the olfactopeduncular tract to the hypothalamus and midbrain (see Chapters 14 and 18 for further details).

From Corpus Striatum

The corpus striatum (caudate and putamen) has two main outputs—one to the globus pallidus and the other to the substantia nigra. At present the relative number of fibers passing to each of these structures is not known. Making lesions in the putamen, Wilson showed that fibers from striatal cells in the monkey collect in bundles leading to the globus pallidus.[123] He referred to the bundles as "pencils,"[143] and hence the expression "Wilson's pencils." He followed them to the external medullary lamina of the pallidum and from there traced them as "narrow pencils" into the medial pallidal segment. The converging pallidal pencils are more commonly referred to as radial fibers[144] because of their resemblance to the spokes of a wheel. Adinolfi and Pappas[145] confirmed the original findings of Verhaart[146] that the majority of the fine fibers from the corpus striatum are 1 μm or less (mean 0.630 μm). Employing electron microscopy, Fox and co-workers observed that the radial fibers terminate on every aspect of radiating dendrites of the large cells of the globus pallidus.[147] They suggested that some fibers may be branches of axons that continue as finely myelinated or unmyelinated projections to the substantia nigra.

In his silver studies on the monkey, Szabo found that the caudate and putamen project upon the globus pallidus and substantia nigra in an orderly topographical manner.[148] In the case of the globus pallidus, the caudate projections end predominantly in its rostral and dorsal portions,[148,149] while those from the putamen terminate in the more caudal and ventral parts.[150] In the case of the substantia nigra, the caudate projections distribute mainly to the part of the nucleus located rostral to the level of the exit of the third nerve,[148,149] while those from the putamen terminate in the portion caudal to that level.[148,150] There is also a medial-to-lateral organization of the projections from the caudate and putamen, such that they are laid down in the pallidum and substantia nigra in a medial-to-lateral sequence.[148–150] The projections to the substantia nigra end predominantly in the pars reticulata, which, as Spatz pointed out,[151] resembles the globus pallidus in its cells and structure. Apropos of "sparing of function" owing to the dual projections from the corpus striatum to the globus pallidus and reticulate part of the nigra, it should be noted that large lesions of the globus pallidus would not interrupt a sizable component of caudatonigral fibers running in the so-called lamina limitans forming the ventrolateral margin of the internal capsule and the dorsal margin of the pallidum.[152] The same would be true in regard to numerous fibers from the putamen and tail of the caudate that pass through or skirt the caudal part of the pallidum.

Pallidal Projections

The corpus striatum is the main source of inputs to the globus pallidus. The pallidum is subdivided by an internal medullary lamina into internal and external segments (Figure 4-3). In primates, an accessory medullary lamina may partially divide the rostral part of the medial segment.[153] In subprimates, the entopeduncular nucleus cupped within the cerebral peduncle corresponds to the internal segment of the pallidum.[154] As illustrated in Figure 4-3, there are separate outputs from the two main pallidal segments. The lateral segment projects to the subthalamic nucleus, which in turn projects mainly back to the medial segment.[155] The neural outflow from the medial segment divides into two main streams, one to the thalamus and the other to the subthalamic and midbrain tegmentum. Kuo and Carpenter have shown that the rostrolateral part of the medial segment projects largely to the ventral anterior nucleus (pars principalis) of the thalamus, while the caudal part of the same segment splays out into the ventral lateral nucleus,[153] including its medial part (see Chapter 14 for further details). The centromedian nucleus is the third main thalamic destination.[156] Part of the tegmental outflow from the medial segment innervates the nuclei in the field of Forel in the subthalamic tegmentum, while another important contingent of fibers fans out above the substantia nigra and descends to the nucleus tegmenti pedunculopontinus.[156]

Finally, it has been demonstrated by silver,[156] autoradiographic,[157] and HRP methods[158] that the medial pallidal segment projects to the epithalamus—specifically, the lateral nucleus of the habenula. This projection, which mainly follows the course of the stria medullaris, ranks phylogenetically as one of the oldest pathways in the forebrain.

From Substantia Innominata

Nauta and Mehler have provided evidence that the sublenticular part of the substantia innominata projects to the posterolateral hypothalamus.[159] Reference has already been made to the large acetylcholinesterase-positive neurons in the substantia innominata, including the neurons in the basal nucleus of Meynert. These cells project to several limbic structures and to widespread areas of the neocortex.[160]

Nigral Outputs

Although the substantia nigra is not defined as part of the striatal complex, its outputs are summarized here because of the above-mentioned similarities of the pars reticulata to the globus pallidus. The substantia nigra borders upon so many conducting pathways that there was little confidence about the destination of its projections until improved tracing techniques were applied. Three main projections have been identified. The nigrostriatal projections have already been described. Another projection that may also prove physiologically to be of major importance is the nigrotectal pathway.[161] It projects to the deeper layers of the superior colliculus,[161] and, according to some reports, to the reticular formation.[162] Finally, physiological transport techniques have given better definition to nigral projections to the thalamus—specifically, the magnocellular part of the ventral anterior nucleus, the medial part of the ventral lateral nucleus, and the paralaminar portion of the medial dorsal nucleus.[163]

Neurochemical Aspects of Hodology

Hodology refers to the subspecialty of neurology that is concerned with the knowledge of pathways of the nervous system (*hodos,* Greek for "path"). As mentioned earlier, neurochemical research has revealed a growing number of substances found in unusual amounts in the striatal complex that appear to be associated with particular pathways (for brief review, see Ref. 99). Of particular interest are acetylcholine; the three mentioned monoamines; GABA; substance P; and the enkephalins. One should also mention glutamine, which some authors claim to have identified with corticostriate fibers from the frontal region and which is believed to have an excitatory effect on striatal cells.[164] The question of neurochemical transmission can be more meaningfully discussed following the presentation of the neurobehavioral findings (see Chapter 15).

Summarizing Comment

The striatal complex of mammals consists largely of two contiguous, telencephalic structures (the olfactostriatum and corpus striatum) together with adjoining, interdigitating structures at the junction of the telencephalon and diencephalon. Although having several similarities in structure and chemistry, the olfactostriatum and corpus striatum are distinguishable on the basis of their inputs and outputs.

In addition to descending connections from the olfactory apparatus and the limbic cortex of the piriform area and hippocampus, the olfactostriatum (olfactory tubercle and nucleus accumbens) receives ascending inputs from the brainstem. It projects mainly to the hypothalamus and midbrain. The corpus striatum receives inputs from (1) the reticular system and elsewhere via the central tegmental tract and medial forebrain bundle; (2) the substantia nigra; (3) intralaminar nuclei; (4) the limbic cingulate cortex and amygdala; and (5) widespread areas of neocortex. The cascading neural outflow from the corpus striatum originates with two major projections—one to the globus pallidus and the other to the substantia nigra. From the globus pallidus one stream of fibers arising in the internal segment feeds into certain parts of the ventral and medial thalamus, while another stream distributes to the subthalamic and midbrain tegmentum (Figure 4-3). Still a third stream leads to the habenula. Large numbers of fibers from the lateral segment converge on the subthalamic nucleus, which in turn projects back to the internal segment. In addition to its supply of afferents to the corpus striatum, the substantia nigra projects to the superior colliculus and specific parts of the ventral and medial thalamus. Parts of the substantia innominata have probable connections with the posterolateral hypothalamus, as well as projections to limbic structures and widespread areas of the neocortex.

Except for the connections of the olfactory apparatus with the olfactostriatum, there is no precise anatomical information about inputs from other sensory systems to the striatal complex. Some electrophysiological data relevant to this matter will be summarized at the end of the next section dealing with the question of striatal function.

THE QUESTION OF FUNCTION

As noted in Chapter 2, more than 150 years of investigation has failed to reveal specific functions of the striatal complex. The prevailing clinical view that it is solely involved in motor functions is difficult to reconcile with numerous clinical and experimental observations that large cavernous lesions of different parts of the striatal complex result in no obvious motor disability. Moreover, as will also be discussed, electrical stimulation of extensive parts of the striatal complex elicits no specific motor effects. In his well-known *Text-Book of Physiology,* published in 1900, E. A. Schäfer commented: "*The corpus striatum is generally believed to act as a centre for the higher reflex* movements . . . but the experimental grounds of this belief are still lacking"[165] (italics added). Sixty years later, in one of the most extensive reviews available, Crosby, Humphrey, and Lauer arrived at a similar conclusion: "*At present . . . in spite of numerous experimental studies and a wealth of clinical observations, there is no clear understanding of the precise functions of the caudate–lenticular complex per se* as distinct from the other brain areas with which it is in functional connection . . ."[166] (italics added). In 1979, the editors of a conference on the corpus striatum expressed the view that "in the face of now available evidence it seems impossible to maintain that the motor functions of the NS [corpus striatum] are either elementary or direct."[167]

Despite the negative evidence, one finds it stated in textbook after textbook that the striatal complex is part of the motor system dominated by the motor cortex of the neocortex. Tilney and Riley's authoritative textbook of neurology illustrates that this view was well established in neurological circles by 1930. "The corpus striatum," they wrote, "is now regarded as a fundamentally motor organ."[168] In the following account, it will be evident how such ideas developed, but there will be no attempt to explain why they have continued. Rather, the purpose is to provide the background and rationale of the present investigation. The emphasis to be placed on negative evidence should not be mistaken as promoting the argument that the striatal complex has nothing to do with motor functions. To propose this opposite extreme would be as unreasonable as to claim that the central processor of a computer has nothing to do with the printed output.

Methods Used for Investigating Cerebral Functions

In using animals to identify the role of various brain structures in different forms of behavior, the experimenter relies largely on three methods: (1) eradication of brain tissue, (2) focal electrical or chemical stimulation, and (3) obtaining measures of the spontaneous and elicited activities of cerebral nerve cells by a variety of techniques.

Eradication of Brain Tissue

In the attempt to discover the function of a specific structure, it is customary to destroy all or part of it and then attempt to assess what functions are retained or lost. This approach is somewhat like that of a mechanic who tests the operation of a machine by observing its performance both before and after removing a certain part. It is commonly held that ablation experiments give information only about the performance of the remaining brain, not the function(s) of the part destroyed. This argument is in line with Hughlings Jackson's classical concept that the nervous system represents a hierarchy of levels and that a loss of structures at higher levels gives release to the activity of those at lower levels.[169] At the same time Jackson recognized that disordered function of a structure at a higher level, such as in the case of a cortical epileptogenic focus, may induce abnormal activity not only at lower levels, but also at the same level.[170]

As studies on the visual system illustrate, if adequate testing is performed, specific functional deficits can be pinned to the loss of certain nerve structures. For example, destruction of nerve cells or fibers along the course of the visual pathway from the retina to the "visual" cortex may, upon examination of the visual fields, be manifest as blind spots (scotomata). At a far different level of complexity, the clinical study of aphasia provides another illustration of how destruction of brain tissue results in symptoms indicative of the function of the missing parts—specifically, the role of certain cortical areas in linguistic functions. In view of such considerations, the investigator conducting ablation experiments seeks to administer appropriate tests for revealing the presence and nature of a deficit. (See also a subsequent comment on "negative symptoms.")

Brain Stimulation

The second major experimental method involves electrical or chemical stimulation of the brain. For example, the experimenter may systematically explore parts of the brain in an attempt to identify the role of various structures in motor functions, observing the effects of different parameters of electrical stimuli on the elicitation of skeletomuscular responses. In other kinds of experiments it may be the purpose to learn whether or not brain stimulation of certain structures affects an animal's performance of a specific test. It should be noted that with a few exceptions, electrical stimulation of the brain fails to elicit coordinated patterns of behavior or, in the case of diagnostic stimulation in human beings, reproducible complex subjective states. Just as surges of line current during an electrical storm disrupt the function of an electronic device such as a computer or television set, so are meaningless electrical stimuli disruptive of normal brain function.

Silent Areas

Various parts of the striatal complex have been regarded as among the silent areas of the brain because of the failure of electrical stimulation to elicit behavioral changes. In clinical cases in which brain stimulation may be performed for diagnostic purposes, it is found that silent areas may be silent with respect to both behavioral changes and subjective symptoms.

What accounts for silent areas? A brief consideration of this question requires a word about different cellular types. Cortex and other nervous tissue generally contain two main types of cells. The so-called type I cells of Golgi[171] are fairly large and have long axons that join the white matter and connect with cells in neighboring or distant structures. The great majority of neurons are small cells with short axons (type II of Golgi) that do not leave the local area. Ramón y Cajal (1852–1934), the noted Spanish neuroanatomist, regarded the unusually large number of small cells in the human cortex as "the anatomical expression of the delicacy of function of the brain of man."[172] It is believed that such cells usually have a selective, suppressive ("braking") action on the larger type I cells.[173]

It may happen that longer and more intense electrical stimulation overrides inhibiting mechanisms and results in a self-sustained discharge of the larger output cells. When these output cells connect quite directly with motor cells, there may be sustained convulsive movements, as is the case when stimuli of too great intensity are applied to the motor cortex. If, however, the impulses from the discharging output cells propagate to structures without motor function, there may be no visible or subjective sign of the discharge.

Penfield and Jasper, for example, describe the case of a patient in whom diagnostic stimulation was applied to the cortex just posterior to the sensorimotor cortex.[174] This stimulation resulted in a self-sustained, slowly spreading "epileptic" discharge. There was, however, no sign of movement, and the patient experienced no symptoms. If it had not been demonstrated that the patient was unable to distinguish between two pressure points on the hand of the opposite side, it might have been concluded that the epileptiform discharge was without effect. The same authors describe another case in which a spontaneous epileptic discharge spread slowly backward from the frontal pole.[175] During the 3 min while it involved the so-called silent areas, the patient experienced no symptoms, was able to engage in conversation, and could count backward without error.

Recording Neural Activity

Because of bioelectrical changes of the kind just described, it is important when employing brain stimulation to obtain additional information by recording changes in the electrical activity of the brain. Inferences regarding function of various structures may be derived by observing correlations between certain stimuli (or actions) and neural activity.

In the following account, I will deal successively with attempts to demonstrate striatal functions by means of brain stimulation and by the production of cerebral lesions. Under each heading I will review first the experimental observations on animals and then cite relevant clinical findings. In conclusion, I will summarize the results of electrophysiological studies relevant to "sensory" and "motor" functions of the striatal complex and then mention findings with a noninvasive technique (positron-emission tomography) that bear on the question of striatal motor function in human beings.

Results of Stimulation

Findings in Mammals

Although it was known before the publication of Galvani's classical memoir of 1791[176] that nerve and muscle could be excited by electric shocks, and although Du Bois-Reymond had demonstrated "negative variation"[177] ("action current" of Hermann, 1868[178]) of nerve in response to an electrical stimulus, it was not until the 1870s that electrical stimulation was effectively used in experiments on the brain. In the meantime, it had often been the practice to apply mechanical stimulation, using a needle to prick different parts of the brain. In 1824, the French physiologist Pierre Flourens (1794–1867) described his observations on mechanical stimulation of the brain, noting the absence of any movement upon probing the corpus striatum.[179]

Electrical stimulation of the brain did not become a standard procedure until after 1870, when two German investigators, Fritsch and Hitzig, reported their findings that application of interrupted galvanic current to the frontal cortex of dogs resulted in movements on the opposite side of the body.[180] The strength of the current was just sufficient to produce a feeling when the electrode was applied to one's tongue. Using an improved stimulator providing faradic (alternating) current, Sir David Ferrier soon confirmed their findings.[181] In addition, Ferrier described the effects of electrically exciting the corpus striatum. "The results of stimulation of the corpora striata in monkeys, cats, dogs, jackals and rabbits,'' he wrote, ''are so uniform as to admit of being generalised together.''[182] Specifically, he observed that stimulation ''of the corpus striatum causes general muscular contraction on the opposite side of the body. The head and body are strongly flexed to the opposite side, so that the head and tail become approximated, the facial muscles being in a state of tonic contraction, and the limbs maintained in a flexed condition.''[182] Ferrier concluded by stating, ''In the corpus striatum there would thus appear to be an integration of the various centres which are differentiated in the cortex.''[182] Ferrier's authority was such that it is probably on the strength of his conclusions, more than those of any other investigator, that the corpus striatum came to be regarded as part of the motor system.

Subsequently, several physiologists conducted experiments on mammals that led them to conclude that the motor effects resulting from electrical stimulation of the corpus striatum were owing to a spread of current to the internal capsule, the great bundle of fibers leading to and from the cerebral cortex. Kinnier Wilson reviewed the literature on this subject in his much-quoted paper on the anatomy and physiology of the corpus striatum published in 1914.[183] He cited particularly the experiments of Minor (1889), Ziehen (1896), Stieda (1903), and Bechterew (1909), who, after removing the cerebral cortex in various animals and waiting for degeneration of the internal capsule to occur, found that stimulation of the caudate nucleus was no longer effective in inducing motor responses. In Bechterew's experiments, however, stimulation of the lenticular nucleus (putamen and globus pallidus) was said to elicit contralateral movements.

Based on his own observations, Wilson was in agreement with those who contended that the corpus striatum was electrically inexcitable. For his experiments he took advantage of the newly available Horsley–Clarke stereotaxic instrument for systematically exploring the lenticular nucleus with stimulating electrodes. He did not ''exclude pyramidal or corticospinal paths by previous section and subsequent degeneration.''[184] Rather, he said, ''[It] was . . . an advantage to have their integrity unimpaired, as the contrast between stimulation of the nucleus and of the capsular fibres was rendered so striking thereby as by itself to constitute satisfactory proof of the difference in electrical excitability between the two.''[184] He concluded that ''from the strictly motor point of view the corpus striatum seems to have been progressively shorn of its possessions. . . . that whatever function the corpus striatum once possessed there is no experimental evidence in apes to show that it exercises any motor function comparable to that of the motor cortex. . . . It is electrically inexcitable, and comparatively large unilateral lesions do not give rise to any unmistakable motor phenomena.''[185] Finally, Wilson insisted that the human corpus striatum is incapable of acting vicariously for the ''motor'' cortex: ''Its true motor function has gone, never to return.''[186]

On the basis of experimental work to be considered shortly, it would appear that Wilson's completely negative findings were probably owing to anesthesia and that some degree of the contralateral turning of the body described by Ferrier may indeed have resulted from stimulation of the striatum. Chronologically, the next stimulation experiments deserving consideration are those of Mettler and his co-workers reported in 1939.[187] As the authors pointed out, their findings were of unusual interest ''because stimulation of the corpus striatum by previous investigators had consistently failed to give definitive results.''[188] Veering from the usual procedures, they showed in anesthetized cats that stimulation of the corpus striatum inhibited movements induced by excitation of the motor cortex.[187] Subsequently, other workers, including Hodes *et al.*[189] and Liles and Davis,[190] essentially confirmed their findings.

Physiologists were well aware that the effects of anesthesia may account for the failure to elicit motor or other responses by stimulation. Around the middle of the present century more and more workers began to follow the lead of Swiss physiologist Walter Rudolf Hess[191] in attempting to devise humane methods of stimulating the brains of awake, unrestrained animals. Animals could be prepared under anesthesia for chronic experiments by implanting electrodes or guides for inserting electrodes under waking conditions into the substance of the brain which itself is without sensation. In 1957, Forman and Ward reported the results of stimulating the caudate nucleus in cats with chronically implanted electrodes.[192] They observed that stimulation of the caudate with 60 per sec alternating current resulted in turning and circling toward the opposite side, as well as fore- and hindleg responses. However, in preparations in which the frontal cortex had been removed 9 or more days before, there were no complicating movements of the opposite extremities. Hence, the authors concluded that caudate stimulation was effective in eliciting turning of the body to the opposite side and circling, but that evoked movements of the extremities were owing to the spread of current to the internal capsule. As opposed to earlier reports, they emphasized that inhibition of movements was never seen under any conditions. Six years later Laursen reported comparable findings.[193] *In regard to the absence of inhibition, however, it should be noted that slow-frequency stimulation was not systematically tested in either of the two mentioned studies* (see below).

The above authors attributed the inhibition of movements seen with fast repetitive stimuli to a spread of current to the internal capsule. Such a spread perhaps accounted for the inhibition of approach and consummatory responses that Rubenstein and Delgado observed in awake, freely moving monkeys upon stimulation of the head of the caudate nucleus.[194] The brain diagrams of their paper indicate that the effective loci were close to the internal capsule.

Delgado and his co-workers observed the effects of radiostimulation of the globus pallidus in freely interacting monkeys moving in a small enclosure.[195] Repeated stimulation was described as having a quieting effect on the dominant monkey insofar as it would not interfere with a submissive animal's obtaining and eating banana pellets.

Psychological Testing

In attempting to assess the role of the caudate nucleus in functions other than movement, Buchwald and co-workers tested the effect of bilateral stimulation of the caudate nucleus in cats on the performance of a visual discrimination.[196] They observed that such stimulation reduced the accuracy and prolonged the reaction time in performing a partially learned task. There was no effect on an ''overtrained'' task.

Electrical and Chemical Stimulation

At this point, with respect to both psychological and motor effects, it is of interest to call attention to parallel effects induced by either electrical or chemical stimulation of the caudate nucleus in cats. In the 1950s, while investigating the functions of the hippocampus,[197] I used a method that made it possible to compare the effects of electrical or chemical stimulation of the same locus of the brain and to observe the behavioral and electroencephalographic changes in freely moving animals. I focused particularly on the effects of stimulation with acetylcholine and its muscarinic and nicotinic congeners. Just as in the case of the hippocampus, there were persuasive reasons for testing the effects of cholinergic stimulation of the caudate nucleus: (1) the high cholineacetylase and cholinesterase content of the caudate nucleus suggested that acetylcholine might play some special role in its function; (2) since acetylcholine was known to excite nerve cells, but not fibers, its use might be expected to avoid a major complication of electrical stimulation—namely, activation of fibers of passage (e.g., internal capsule); (3) by the same token, it would circumvent the possible suppressive effects induced by antidromic impulses; and (4) it might elicit patterned responses. In order to avoid the inevitable reflux along the outside of the probe that occurs when chemicals are injected in solution or in a viscous substance, the agents were deposited in solid form.

Using cats as experimental subjects, Stevens, Kim, and I compared the effects of electrical and chemical stimulation within the head of the caudate nucleus.[198] In addition to looking for specific and general changes in behavior, we also tested the effects of such stimulation on conditioned avoidance behavior. For the latter purpose nine cats were trained in a shuttle box in which a buzzer served as the conditioned stimulus and a shock from the floor grid was the unconditioned stimulus. The animals could avoid a shock to their feet if, at the sound of the buzzer, they climbed over a barrier to the other compartment of the shuttle box.

The results with the electrical stimulation will be described first. When the caudate was stimulated with balanced, biphasic square-wave, 1-msec pulses at 60 per second and at intensities ranging from 0.7 to 1.4 mA, the subject engaged in contralateral circling. Under such conditions it might be expected that the animal would be stimulus-bound and unable to respond appropriately to the sound of the buzzer. Nevertheless, the cat was able to interrupt the circling and jump quickly into the other compartment. With stimulation at slower frequencies (4–12 per sec) cats appeared to develop a state of quietude. Under these conditions there was no circling; rather it was typical for the cat to assume a sphinxlike posture. These quieting effects were entirely comparable to those reported by Akert and Andersson in 1951.[199] Two cinematographically documented cases in our study serve to illustrate how stimulation at a slow repetition rate interfered with the performance of the conditioned avoidance test: One cat attended to the sound of the buzzer and would raise a forepaw as though about to cross the barrier, but would then stop, look about, and fail to cross. In the other example, the cat would hiss when the buzzer sounded and would hiss again and jump about aimlessly upon being shocked. Immediately upon cessation of brain stimulation it would cross the barrier, behaving as though it had been aware of the correct response, but unable to execute it.[198]

The major effects of cholinergic stimulation with carbachol were remarkably like those resulting from electrical excitation.[198] Carbachol has both the muscarinic and nicotinic action of acetylcholine and exerts prolonged changes because it is not destroyed by acetylcholinesterase. Before describing its behavioral effects, it should be noted that the insertion of an empty chemode into the caudate nucleus is sufficient in itself to cause contralateral circling. This "injury" effect occurs after a latency ranging from a few seconds to 3 min; the circling usually does not persist for longer than 1 min. Carbachol exerted a two-phase effect. During the first phase, for a period of 10 to 20 min, the cat would continue to circle in a direction opposite to the side of stimulation. For example, if the drug was deposited in the left caudate nucleus, the cat circled to the right. Just as in the case of circling during electrical stimulation, the cat would respond to the sound of the buzzer by crossing the barrier to the opposite side of the shuttle box. After the circling phase, cats became silent and assumed a sphinxlike posture and would continue in that state for a period of 4 to 24 hr. In terms of a malfunctioning electrical device, it was as though a large and important part of the animal's neural apparatus had been "shorted out." As with electrical stimulation at a slow frequency, there was interference with the performance of the conditioned avoidance test. In the majority of instances the animals either appeared to be unaware of the sound of the buzzer or, if seemingly attentive, "failed to respond appropriately."[198] In one case there was a loss of both the conditioned and unconditioned response for 6 hr.

Before concluding the summary of these experiments it should be emphasized that the deposit of crystalline cholinergic agents in the caudate nucleus fails to induce seizure potentials typically seen after the application of the same chemicals to cortex (see Chapter 19).[198] In keeping with such observations is the notorious failure of electrical stimulation to precipitate seizure afterdischarges in the caudate nucleus. Here I might add that, as in the cat, I have never been able to elicit afterdischarges by electrical stimulation of numerous points in the caudate nucleus of squirrel monkeys.

Summarizing Comment

In a monograph entitled "Corpus Striatum" (1963), Laursen prefaced his own investigations with a review of the experimental work on mammals, including the studies that have just been summarized.[193] Apropos of electrical stimulation, he concluded that circling toward the opposite side can be attributed to excitation of the corpus striatum, but that the other motor effects (e.g., flexion of the opposite forelimb) are due to a spread of current to the internal capsule. It is significant to add that during the circling, the animal is not "stimulus-bound," but as the conditioning experiments show, can interrupt its turning movements. Finally, it is evident from the experiments employing cholinergic stimulation or electrical stimulation at a slow, pulse-repetition rate, there may be an alteration of striatal function that interferes with the initiation and execution of learned kinds of performance.

Reptiles and Birds

Electrical stimulation of the R-complex in reptiles and birds has proved to be almost totally ineffective in eliciting specific behavioral responses. In a review of the findings in reptiles, Goldby and Gamble concluded that it is probable that no motor effects can be obtained by the stimulation of striatal structures.[200] In 1978 Distel reported the effects of brain stimulation in the freely moving green iguana lizard (*Iguana iguana*). In stimulating the R-complex, Distel obtained only one positive response: near the outermost border, stimulation resulted in head nodding like that of the signature display.[201] Most investigators would agree that in those studies in which motor effects were elicited, sufficient attention had not been given to controlling for spread of current to other structures (including the dura mater) in small-brained animals. In pigeons, Åkerman observed bowing with stimulation at a number of points in the paleostriatum, as well as in the hyperstriatal structures.[202] Under natural conditions, bowing occurs among males either when defending their territory against an intruder or when initiating courtship behavior. Anschel's study on the turkey may be cited as an example of a contemporaneous study in which there was a failure to elicit responses by electrical stimulation of the avian paleostriatum.[203]

Clinical Findings

There was a period beginning more than 30 years ago when neurosurgeons destroyed the medial globus pallidus or related structures for the treatment of Parkinson's disease. Brain stimulation and electroencephalographic recording were used to obtain information about the location of the probe used for destroying nerve tissue. In selecting a site for the "therapeutic" lesion, the neurosurgeon favored a locus at which stimulation either enhanced or reduced a parkinsonian tremor. On the basis of extensive experience, Jung and Hassler concluded that "electrical stimulation of the human pallidum does not result in any motor effects but causes a partial blocking of voluntary movements and arousal or adversive [turning] effects."[204] They described the overall picture as follows:

> Conscious patients, operated upon under local anesthesia, lose contact with their environment during pallidum stimulation and are unable to perform complex movements or to speak accurately . . . . During stimulation most of the patients consistently showed a tendency . . . to look to the contralateral side which could be overcome however by visual fixation. Some of the patients displayed anxiety and restlessness during stimulation of the internal pallidum at higher frequency or at voltages above threshold, described a constricting or hot feeling in the chest and occasionally a feeling of vital anxiety in the left chest; some of the patients even screamed anxiously as the stimulation was repeated.[205]

In an earlier report Hassler had described an orderly representation of the body in the globus pallidus,[206] but Orthner and Roeder, on the basis of their own numerous observations, concluded that all motor effects were attributable to the spread of current to the knee of the internal capsule.[207]

Van Buren *et al.* described the effects of stimulation of the striatum and neighboring white matter in 30 patients. The most common response was an arrest of voluntary movements such as counting with the hand.[208] Because of the high intensity of current used to obtain such responses, they could not exclude the possibility of spread of current to the neighboring white matter. Subsequent to such arrest reactions, the patients might show poststimulation confusion, inappropriate or garbled speech, and overt mood changes. Nine patients were observed to smile during the time of speech arrest. Two patients reported that the testing situation was amusing. One patient laughed aloud at the time of stimulation and later said, "[T]his is fun!" In eight patients there was an arrest or decrease of respiration that occurred during the expiratory phase. Such changes persisted throughout the period of stimulation. A patient could, however, breathe upon command. There was a single case in which stimulation resulted in crying, accompanied by a feeling of loneliness and confusion. The relevance of striatal function to laughing and crying will be dealt with when describing investigations of the functions of the midline frontal limbic and neocortex (Chapters 21 and 28).

Rosner *et al.* found that with current intensities below 2 mA, stimulation of the caudate nucleus never resulted in reliable changes of speech or motor function.[209] With current intensities above 12 mA, patients complained of nausea and showed a reduction or an arrest of motor function.

Summarizing Comment

Although there is no exact anatomical confirmation of the location of electrodes in the clinical studies, there are indications that stimulation of either the corpus striatum or the globus pallidus does not result in specific movements. Stimulation in the region of the pallidum, however, appears to result in turning to the opposite side, as well as interference with speech and the performance of complex movements. There is some evidence that stimulation of the corpus striatum also interferes with phonation and speech. Finally, there are indications that striatopallidal stimulation may induce disturbances in affect or mood and in awareness, being accompanied in some instances by respiratory changes and subjective somatovisceral symptoms.

Effects of Lesions

Findings in Mammals

Although numerous experiments have been conducted on the striatal complex since the early part of the last century, there are but relatively few in which there was a satisfactory histological confirmation of the location and extent of lesions. Except for work of historical interest, the following account will focus on studies in which an effort was made to obtain a verification of the locus and extent of lesions.

Corpus Striatum

In 1823, Magendie reported his observations on the effects of gross extirpation of the corpus striatum and overlying cortex in rabbits.[210] Upon completion of the removal on the second side, a rabbit would run forward as if ''emporte'' by an irresistible impulse. A hundred years later Mettler described comparable behavior in cats in which both the frontal cortex and caudate nuclei were excised.[211] Upon removal of the second caudate in his chronic preparations, the cats made leaping movements and engaged in ''obstinate progression,'' an expression used by Bailey and Davis to characterize the persistent pushing of ''thalamic'' animals into an obstruction.[212]

A study by Villablanca and Marcus appears to be the only one in which bilateral excision of the caudate nucleus was accomplished without significant injury to the internal capsule or without profound prostration of the animal.[213] Their approach for aspirating the entire head and body of the caudate was made medially just above the corpus callosum. Ten cats with bilateral removal of the caudate survived for a period of 6 or more months. They were able to stand or walk within 1 to 3 days, but required tube feeding for an average of 8 days. After that time the most striking alteration of behavior was characterized as ''compulsory approaching'' in which the subject would approach and follow a moving person, cat, or object as if drawn by a magnet. Other notable changes were a marked passivity; exaggerated kneading and treading with the forepaws; increase in purring; and hyperreactivity. Otherwise the investigators regarded their subjects as being remarkably free of neurological and behavioral deficits. These findings, together with psychological testing, led them to conclude that the caudate nuclei do not appear to have an essential role in the control of elementary sensorimotor or cognitive functions.

Upon becoming professor of physiology at the Yale University School of Medicine in 1930, John F. Fulton (1899–1960) launched an extensive program in which particular attention was focused on the physiology of motor performance in primates. For the brain surgery performed in hospitallike operating rooms, he used neurosurgical techniques that he had learned on the service of Dr. Harvey Cushing. Ten years later in a review of the outcome of investigating the functional interrelation of the cerebral cortex and basal ganglia in 34 monkeys and 3 chimpanzees, he and Kennard stated, "Neither unilateral nor bilateral lesions of the caudate nucleus or putamen were found to have any visible effect on motor performance in monkeys or chimpanzees with otherwise intact nervous systems."[214] Although there was meticulous examination of the motor performance of these animals, the anatomical findings were illustrated only by gross specimen. Years later in that same laboratory George Davis found that monkeys with lesions of the head of the caudate nucleus developed hyperactivity that was significantly greater than in the control animals.[215] There were no other changes in motor performance.

Globus Pallidus

Kinnier Wilson appears to have been the first to produce isolated lesions in the globus pallidus or putamen by the use of a stereotaxic instrument. None of his animals—either with a small or large lesion—"showed any sign of motor impairment that careful and repeated examination could detect."[216] In all of Wilson's experiments, the lesions involved only one side. Thirty years later Ranson and Berry reported experiments on monkeys in which they used a stereotaxic apparatus to produce large bilateral lesions in the globus pallidus.[217] Elsewhere, in a summary of their experiments on the basal ganglia, Ranson said:

> The monkeys with lesions in the caudate nucleus or in the putamen showed no symptoms. Those with bilateral lesions in the globus pallidus showed only slight impairment of motor functions. . . . In spite of the large size and bilaterality of these lesions there was only such impairment of motility as might be attributed to involvement of the internal capsule. There was no rigidity, or tremor. Athetosis or choreiform movements were never seen.[218]

Effects on Performance of Psychological Tests

Conditioning procedures and other forms of psychological testing have also been used in an attempt to identify the functions of the striatal complex. Here I will mention only a few representative studies. Thompson (1964) tested the effects of various cortical and subcortical lesions on the behavior of rats trained to avoid the arm of a T-maze in which they had just previously been shocked.[219] He considered the test comparable to one used for monkeys in performing a delayed response. Four animals with "extensive caudate lesions" attained performance levels "indistinguishable from the control." Rosvold and co-workers, however, found that caudate lesions altered the performance of the delayed alternation test in monkeys,[220] and that the nature of the deficit differed according to which of two interrelated parts of the caudate and frontal cortex were affected: (1) The orbitofrontal cortex projects to the ventrolateral sector of the caudate nucleus: Monkeys with lesions of either of these structures showed an inability to suppress perseverative response tendencies. (2) The lateral frontal cortex projects to the anterodorsal sector of the head of the caudate nucleus: Monkeys with lesions in either of these structures developed a deficit in dealing with spatial aspects of a problem, as evident by the inability both to perform a delayed alternation test and to associate a sound cue with a baited cup. On the basis of these results, the authors concluded that reciprocally related parts of the caudate nucleus and frontal cortex share similar functions.

Kirby and Kimble found that lesions of the caudate in rats interfered with both passive avoidance and active avoidance behavior.[221] In cats, however, Knott *et al.* observed that bilateral caudate lesions resulted only in a transient loss of conditioned approach responses.[222] In similar experiments, Gomez *et al.* observed a significant decrease in resistance to extinction of a previously learned avoidance response.[223] If the ablation involved 80% or more of the caudate nuclei, the cats deteriorated into a state of ''stupidity'' and developed ''obstinate progression.''

Reptiles and Birds

The relatively few neurobehavioral studies on reptiles and birds with direct relevance to the R-complex will be referred to in Chapters 11 and 12.

Summarizing Comment

The findings on the effects of striatal and pallidal ablations in mammals indicate that neither unilateral nor bilateral lesions affect specific movements. There are indications that bilateral destruction of the caudate nucleus results in hyperactivity and a persistence in ''following'' or progressive movements. The results of ablation experiments, like those involving stimulation, provide evidence that interference with striatal function affects learned kinds of performance.

Clinical Findings

Clinical observations have probably been more responsible than the experimental work on animals for promoting the view that the striatal complex is primarily motor in function. At this juncture it is historically significant to recall the distinction that has been made between pyramidal and extrapyramidal systems. Thomas Willis referred to the rounded masses of white matter along the midline of the ventral medulla as pyramids.[224] The pyramids were identified with the so-called ''voluntary'' nervous system because they contain the direct corticospinal tracts. In 1898, Prus introduced the term *extrapyramidal* to refer to cortical pathways to the spinal cord that traveled outside of the pyramids.[225] Subsequently, that term became applied to nonpyramidal systems believed to be involved in motor function. Wilson, who described the disease of the lenticular nucleus (putamen and globus pallidus) named for him, referred to the condition as an extrapyramidal disease.[226] He himself deplored the use of the expressions ''voluntary'' and ''involuntary'' nervous systems, because he attributed both voluntary and involuntary movements to both systems.[227] [Years later (1964) Bucy and co-workers were to show that the pyramidal tract was not essential for voluntary human dexterity.[228]] Thanks to his 1923 monograph on extrapyramidal diseases,[229] the German neurologist Alfons Jakob (1884–1931) was influential in crystallizing the increasingly popular view that the pyramidal and extrapyramidal systems are involved *only* in the motor aspects of voluntary and involuntary functions. As he wrote, ''our extrapyramidal system'' is the ''efferent organ'' of the thalamus, translating the stimuli from that part of the brain ''into highly developed motor phenomena.''[229] Elaborating upon ideas expressed by the Vogts in their well-known work on the basal ganglia,[230] he characterized the extrapyramidal system as ''a center for the movements of expression,'' reactive flight and defense patterns, automatic postural adjustments, compulsive associated movements, and the like. The Vogts themselves had given further articulation to ideas about the functions of the corpus striatum that circulated during the latter part of the 19th century (see introduction of Ref. 183). In 1876, for example, Ferrier had expressed the view that the corpora striata are "centres" in which "automatic" and "habitual" movements are organized.[181] To sum up, it became the established view that the striatopallidal system is primarily a *motor* apparatus involved in the execution of "involuntary," associated movements such as walking, yawning, stretching, mimicking, laughing, crying, and emotional expression in general.

There are several diseases that affect predominantly one or more parts of the system of structures shown in Figure 4-3, and that are associated with marked disorders of movement. In some of these conditions there is also disease elsewhere in the brainstem, as well as in the cerebral cortex. As F. H. Lewy summed up the matter in a landmark symposium on the basal ganglia:

> There are no systematic diseases of the basal ganglia proper. There are rather widespread pathological processes in the central nervous system attacking with predilection various elements in the basal ganglia. The actual clinical manifestations of these diseases depend upon the distribution and combination of lesions in these parts of the brainstem with those in other parts of the alterative and central vegetative nervous system.[231]

In view of these considerations, I will restrict consideration here to those clinical conditions in which well-demarcated lesions have been found in one or more structures of the striatal complex. In Chapter 15 there will be the opportunity to comment on symptoms in such diseases as Huntington's chorea and Wilson's disease in which there is extensive disease of the corpus striatum, as well as in Parkinson's disease in which a drastic effect on striatal function is attributed to a loss of dopamine-containing cells in the substantia nigra.

Corpus Striatum

Gowers, a noted English neurologist (1845–1915), appears to have been one of the first to call attention to the absence of symptoms in conjunction with discrete damage to the corpus striatum. "If the lesion is confined to the gray substance of either nucleus [caudate or putamen]," he wrote in his famous textbook, "there are usually no persistent symptoms motor or sensory."[232] He elaborated by saying, "I have seen a narrow vertical band of central softening, extending from the anterior to the posterior extremity of the lenticular nucleus, when no trace of hemiplegia could be detected before death, and a careful history had elicited no account of any previous paralysis."[232] Based on his extensive clinical experience, Denny-Brown has observed, "Not uncommonly a softening of the greater part of the putamen or whole striatum is found in patients without any involuntary movements having been noted during life."[233] As Biemond has commented, "It remains a mystery . . . why in many cases lesions, and even sometimes large lesions of the corpus striatum, do not give rise to any extrapyramidal symptom. It seems probable . . . that certain conditions have to be fulfilled before a lesion of the corpus striatum can result in chorea (or athetosis)."[234]

In the 1940s, the growing physiological knowledge of the "motor systems" induced some neurosurgeons to attempt to alleviate disabling motor disturbances by interrupting the abnormal flow of nerve impulses within the circuitry of the pyramidal or extrapyramidal systems. Meyers reported a case of a 39-year-old trolley car motorman whose caudate nucleus on both sides was extirpated for the treatment of tremor.[235] There was no resulting motor disability, and the tremor remained unchanged.

Globus Pallidus

As Denny-Brown points out, "One of the most striking lesions of the basal ganglia is the symmetrical necrosis of the globus pallidus found in patients who have suffered an episode of coma as a result of inhalation of coal gas or carbon monoxide. . . ." Curiously enough, as he also notes: "The patient commonly recovers from the initial coma, and may be discharged from hospital apparently completely relieved, or with only slight mental confusion, for a period of one to three weeks. . . . Often the delayed relapse is exhibited as the onset of an *akinetic mute state with rigidity of all four limbs* . . ."[236] (italics added). He gives an illustration of one such case, a 58-year-old man who was asphyxiated and lay in a coma for one day. Eight days later he was able to speak, read a newspaper, and travel from Florida to Boston. On the ninth day he lapsed into an unresponsive state, and his arms and legs became flexed and somewhat rigid.[237]

Alexander describes a similar case that, he points out,

> is particularly instructive, because of the well observed free interval . . . in which physical and neurological examinations were entirely negative in spite of the fact that the patient must have had a severe bilateral pallidal lesion at that time, i.e., a lesion which had completely destroyed the major part of either globus pallidus. The only subjective complaints of the patient at that time were general malaise, weakness and a burning sensation in the stomach.[238]

According to Alexander, the pathological examination in such cases may reveal lesions outlined by old scars, with no sign of recent or older progression. Denny-Brown, however, suggests that the reappearance of grave symptoms may be the result of an extension of the initial lesion, stemming, perhaps, from incrustations in the small pallidal vessels.[237]

In 1952, I. S. Cooper was performing a left cerebral pedunculotomy for the treatment of tremor and rigidity when there was inadvertent tearing of a vessel that proved in retrospect to be the anterior choroidal artery.[239] In order to control the bleeding it was necessary to occlude the blood supply in this vessel with a silver clip. Surprisingly, there was an improvement in the patient's symptoms. The anterior choroidal artery supplies the medial part of the globus pallidus, anterior hippocampus, lateral geniculate body, and other structures. The globus pallidus is particularly vulnerable to an occlusion of this vessel, because it has no collateral blood supply. The salutary outcome in the above case led Cooper, and subsequently others, to perform pallidectomy for the treatment of tremor and rigidity in Parkinson's disease. In summarizing the results of this operation, Jung and Hassler wrote:

> Following almost complete unilateral destruction by coagulation of the pallidum . . . yawning, increasing drowsiness, closing of the eyes, impairment of contact with the environment, arrest of spontaneous speaking, sleep or even an acute brief state of disorientation or amentia are observed, but later disappear. Transitory euphoria may appear in the postoperative period. . . . All these changes are much more pronounced after bilateral almost complete coagulation of the pallidum. Following this procedure performed in two stages, some patients first go into a confusional state with loss of orientation in time and space, with loss of capacity to identify persons and the environment and occasionally with severe hallucinations. . . . *Here again, it is interesting to note that bilateral destruction of the pallidum does not produce any motor symptoms* [italics added]. In a few cases a slight akinesis relative to speech, respiration and swallowing movements appears.[240]

The authors then conclude: "Experience with neurosurgical therapy of parkinsonism involving production of symmetrical almost complete bilateral lesions in the pallidum indicates that it may lead to such unfavorable psychological changes that most neurosurgeons think it advisable to avoid this type of operation.''[241] In an attempt to avoid the so-called psychoorganic syndrome some surgeons placed a lesion in the globus pallidus on one side and in the ventral lateral nucleus on the other. Krayenbühl performed such bilateral operations in 28 of his 263 cases, and bilateral thalamic procedures in 23. He and his associates observed that ''impairment of speech was the most striking complication after the second operation. . . .''[242] In over half of the cases, they commented, ''[S]peech was slightly or considerably worse than before the second operation; it was more slurred, dysarthric, or low and aphonic.''[242] A ''psycho-organic syndrome'' occurred in 30% of the cases with lesions involving the pallidum on one side and the thalamus on the other, one of them being characterized by compulsive crying.

Before surgery was largely abandoned for the treatment of Parkinson's disease, it was generally the opinion that lesions in the ventral lateral nucleus were more effective than pallidal lesions for the parkinsonian symptoms of tremor and rigidity, and, moreover, that this type of operation should be reserved for patients with symptoms confined to one side of the body. Significantly, Fager observed that few patients were aware that such operations affected the spontaneity and ''drive'' of the hand on the opposite side. ''For example,'' he noted, ''a hand may become lazy or quiet or may require constant coaxing to activity although it may not be weak.''[243] Based on the overall results of surgery in his series of 1500 cases, Cooper remarked: ''The role of the thalamus in motor activity likewise appears difficult to define at this time. One may interrupt pathways from the globus pallidus, red nucleus, and cerebellum to the thalamus, as well as the thalamo-cortical and cortico-thalamic circuits, without causing either motor weakness or faulty co-ordination upon the patient.''[244]

Summarizing Comment

Exclusive of cases involving disease of multiple structures, the clinical literature indicates that large unilateral or bilateral lesions of the corpus striatum, such as for example cystic lesions, may exist without motor or other symptoms. Unlike the lacunae associated with lesions of the visual system and alluded to earlier, a discrete lesion of the corpus striatum would appear to be unaccompanied by any identifiable void (see a subsequent comment on ''negative symptoms''). It also appears that up to a certain limit, the same assessment applies to discrete unilateral or bilateral lesions of the globus pallidus. There will be occasion to comment further upon this question when the findings in my experimental studies on the pallidum are described. In the two concluding chapters on the R-complex it will be pertinent to discuss certain symptoms observed clinically in diseases involving extensive parts of the striatal complex and/or related structures.

Evidence Based on Recording Neuronal Activity

Under the present heading I will summarize electrophysiological evidence relative to ''sensory'' and ''motor'' functions of the striatal complex.

Question of Inputs from Sensory Systems

As noted above, with the exception of the olfactory apparatus, anatomical studies have not yet clearly identified inputs from other sensory systems to the striatal complex.

Although some authors assume that the striatal complex receives information from sensory systems mainly via the cerebral cortex, it would seem likely that in the course of evolution other pathways were laid down prior to an extensive development of the limbic and neocortex. As will be summarized, electrophysiological findings support inferences based on anatomical findings that caudal intralaminar nuclei convey somatosensory information to the corpus striatum. It is also likely that these same nuclei serve to relay visual and auditory information from the tectum, as well as information from the vestibular apparatus.

Corpus Striatum

In 1952, Bonvallet *et al.* reported that olfactory, gustatory, vagal, somatic, visual, and auditory stimulation evoked slow potentials in the caudate nucleus.[245] Their findings with respect to the exteroceptive systems were essentially confirmed by Albe-Fessard *et al.*,[246] who obtained long-latency responses in animals anesthetized by chloralose, as well as in awake animals immobilized by Flaxedil. In an accompanying study they showed that the same forms of stimulation also evoked discharges of single cells in the caudate nucleus.[247] A convergence of afferent systems was evident by the finding that somatic, visual, and auditory stimulation evoked discharges of the same unit. In awake animals units could not be "driven" if stimuli were administered more often than once every 20 sec. Somatic stimulation included electrical excitation of muscle nerves. The authors pointed out that in view of the alleged motor functions of the corpus striatum, it was surprising that group 1 fibers of the proprioceptive system were ineffective in activating units.[247] On the basis of experiments involving application of electrical stimulation and ablation, they concluded that sensory information reached the striatum via the reticular formation and the posterior intralaminar nuclei.[246]

In experiments on immobilized, unanesthetized cats, Segundo and Machne found that single shock excitation of the sciatic, median, and peroneal nerves activated two-thirds of the units in the lentiform nucleus (putamen and globus pallidus).[248] Olfactory, auditory, or visual stimulation was "far less effective." Vestibular stimulation, however, affected the firing rate of a third of the units responding to somatic stimuli. Vagal excitation "modified the responses" of 25% of the units activated by somatic stimuli.[248] In contrast to the putamen, the globus pallidus was rarely affected by vagal stimulation.

In recording unit responses to vagal volleys (triple shocks) in awake, sitting squirrel monkeys, Radna and MacLean found that 28 (29%) of 97 units recorded in the putamen responded to vagal excitation, but not to the control somatic stimulus.[249] Seven units responded to both vagal and somatic stimulation. It was of particular interest to find that in association with the entrainment of respiration by vagal volleys, 6% of the tested units in the putamen gave a periodic discharge that appeared to correlate with the respiratory rhythm. Most of the small population of tested pallidal units were responsive to vagal volleys. Response latencies of units in the putamen and pallidum ranged from 6 to 200 msec, "values indicative of both rapidly and slowly conducting afferent pathways."[250] One potential pathway for impulses reaching the corpus striatum ascends via the solitary nucleus, parabrachial nucleus, and posterior intralaminar nuclei (see Chapter 18).

In a review of sensory functions of the corpus striatum, Krauthamer has emphasized various indications of a close relationship of the corpus striatum to polysensory systems.[251] He points out that "polysensory neurons" are "characterized by partial or complete convergence of different sensory modalities onto a single unit, wide and often bilateral receptive fields, and an inability to follow repetitive sensory stimulation at rates exceeding 2–6 stimuli per sec.''[252] He notes also that in contrast to ''specific sensory units,'' the activity of polysensory cells is greatly depressed by barbiturates, but may be unaffected or enhanced by chloralose. He presents evidence that the central lateral and centromedian nuclei are a main locus of polysensory neurons conveying information to the caudate, citing, inter alia, recent anatomical studies with HRP showing that the deep layers of the superior colliculi, as well as the vestibular nuclei, project to the centromedian nucleus.[251]

The Question of Correlations with Movement

It is well known that when recording the action potentials of single cells in so-called ''motor'' areas of the cerebral cortex or related thalamic nuclei, the great majority of cells can be shown to discharge in relation to somatoskeletal movements. For example, in a study of units in thalamic areas projecting to the ''motor'' cortex, the discharge of 81% was related to movement.[253]

In 1973 DeLong reported the results of a study in which he tested the hypothesis that the striatum is particularly involved in the execution of slow movements.[254] He identified 187 units that were movement-related, 45% of which responded during slow (ramp) movements and 10% during quick (ballistic) movements. In their summing up of a conference on striatal functions, Albe-Fessard and Buser commented that, with the exception of DeLong's data, the results of other investigators have been ''fairly disappointing'' in showing that striatal cells participate ''massively in the performance of a simple motor task.''[255] In testing 307 units in the caudate nucleus and putamen of monkeys, Rolls *et al.* found that only 7% showed responses related to particular movements.[256] Five percent were affected by ''sensory events,'' while the largest percentage (40%) seemed to be affected by some cue related to the task or some ongoing activity in the environment. Niki *et al.* reported that ''of the thousands of units encountered and tested,'' only 14 in the caudate head and body showed activity correlated with movement in the performance of a delayed alternation test.[257] These units discharged about 200 msec before the onset of the response seen in the myographic record. On the basis of such findings and their own data, Rolls *et al.* were inclined to the view that the striatum ''performs some considerably more sophisticated functions than might be suggested by simply viewing it as a 'motor' structure.''[258]

In squirrel monkeys Travis and co-workers tested the effects of a positive and negative stimulus with respect to a food reward. They found that 31 of 97 pallidal units showed a reduction in rate of discharge during (1) food searching, (2) food grasping, or (3) consummatory behavior.[259] In one of their studies only 4% of the population of pallidal units showed a discharge in connection with movement.[260]

In macaque monkeys DeLong[261] obtained quite different results from those of Travis *et al.* in the squirrel monkey.[259] In one set of experiments involving two monkeys, 19% of a population of 551 units showed a variation in the rate of discharge during alternating push-pull and side-to-side movements.[261] The percentage of responding units in each pallidal segment was the same. An unspecified number of units in the substantia innominata and so-called ''border units'' near the laminae showed a striking increase in the discharge frequency when the juice reward was delivered. DeLong suggested that the units discharging in relation to feeding may have been of the type described by Travis *et al.*, especially as they had indicated that half of their units were located ''near the inside boundary'' of the globus pallidus.[259] In each study, however, the methods used for histological verification provided only an approximate location of the units.

Summarizing Comment

It is evident from this brief review of electrophysiological findings that the great majority of units in the corpus striatum fail to respond to any recognizable stimulus and that only a small percentage appear to discharge in relation to specific movements. The types of responses for the remainder suggest that the striatum plays a role in as yet undefined psychological functions. If, as inferred, the globus pallidus is more clearly implicated in motor performance than the striatum, the relatively small percentage of its responding units indicates that it would have a lesser capacity to affect specific movements than the so-called ''motor'' nuclei of the thalamus.

Findings with Positron-Emission Tomography

Noninvasive techniques involving magnetic resonance or positron-emission tomography (PET scan) promise to be of increasing value in obtaining a picture of cerebral structures and ascertaining their metabolic functioning under various conditions in human beings. For example, Roland *et al.* obtained PET scans in ten young volunteers, so as to compare the blood flow in various cerebral structures during rest and during the performance of a simple finger exercise.[262] The authors found that local increases in blood flow were particularly associated with an increase in metabolism in synaptic regions, rather than with discharge in nerve cells. ''That is why,'' they explained, ''the globus pallidus, which is said to have a high rate of action potentials at rest, did not show particularly high flow during rest.''[263] During the performance of the finger exercise there was a 28% increase in the cerebral blood flow in the motor hand area, a bilateral increase of 30% in the supplementary motor areas, and a 10% increase in the premotor areas. In the thalamic region which projects to the cortical motor areas, there was a 10% increase in cerebral regional blood flow, while in the striatal complex the flow increased about 15% in the caudate–putamen and as much as 30% in the contralateral globus pallidus. The authors commented upon the need for obtaining a better resolution between the gray and white matter.

Concluding Comment

In the first part of this chapter, the striatal complex, which constitutes a fundamental part of the forebrain in reptiles, birds, and mammals, has been defined in terms of its phylogeny, ontogeny, anatomy, and chemistry. The greater part of the striatal complex is located in the telencephalon, comprising the olfactostriatum (olfactory tubercle and nucleus accumbens) and the corpus striatum (caudate nucleus and putamen). The part of the globus pallidus contiguous with the striatum may also be telencephalic in origin, but the medial segment, together with the substantia innominata, appears to arise from the diencephalon.

Not Only Motor in Functions

The remainder of the chapter has dealt historically with experimental and clinical work pertaining to the question of the functions of the striatal complex. Observations involving stimulation, ablation, and electrophysiology all attest to the continued uncertainty as to the nature of striatal function. Contrary to traditional views, the evidence indicates that the striatal complex is not *solely* a part of the motor apparatus under the control of the motor cortex. In a handbook article on motor mechanisms, Denny-Brown arrived at the following appraisal of striatal function:

> The only conclusion in terms of positive function of these structures appears to be that they normally interrelate at their various levels the several conflicting types of cortical and subthalamic mechanisms of movement so as to give priority to one or another type of behavior and thus integrate the total reaction of the organism to the environment. . . . The poverty of active response to electrical stimulation of the basal ganglia is consistent with such a mechanism. If our view is correct, the mechanism of natural activation and function of these structures will require a complex background of activity for its demonstration.[264]

In summing up the outcome of a 1979 workshop on the corpus striatum, Albe-Fessard and Buser comment: ''Everyone would agree that the caudate is certainly not a 'motor' centre in the common sense.''[265] Then they ask: ''Not being a 'motor' structure, is NS [neostriatum] a sensory structure?''[266] They conclude their discussion by suggesting that it plays some role in ''integrative operations which remain to be characterized.''[267]

An Experimental Approach with Respect to ''Nonevident'' Symptoms

As discussed earlier, neurologists have been inclined to attribute many of the symptoms associated with brain lesions to a ''release'' of nervous structures at lower levels. Symptoms of an active nature, such as a persisting writhing movement, are sometimes called positive symptoms. Jackson referred to manifestations of an opposite sort such as paralysis of an extremity or loss of consciousness, as symptoms of ''negative lesions.''[268] Such passive symptoms have also been known as negative symptoms. Purdon Martin has given particular emphasis to the presence of negative symptoms in ''extrapyramidal disease,'' singling out in particular inabilities due to a failure in the fixation of posture.[269] The headnodding of a parkinsonian patient provides an example. Martin distinguishes between such ''primary'' negative symptoms and ''secondary'' negative symptoms such as the loss of associated movements in walking seen in Parkinson's disease. He attributes (like other neurologists such as Wilson[227]) symptoms of the latter type to muscular rigidity. Because of the foregoing different meanings of negative symptoms, there is the need for still another expression to refer to an absence of a function that is not evident unless specifically sought. Both here and later on I will refer to such negative symptoms as ''nonevident'' symptoms, recognizing that, ideally, the word *symptom* should be reserved for a reported subjective manifestation, and the term *sign* for what the examiner observes.

In the light of the present review it is natural to ask whether or not in previous research on the striatal complex there has been a failure to examine animals under appropriate conditions for revealing its functions. Among other unknowns, have the conditions been adequate for disclosing nonevident symptoms? In studies on animals experimental neurologists have been largely concerned with whether or not stimulation or destruction of various parts of the striatal complex will affect bodily movements or other motor effects. Experimental psychologists have paid relatively little attention to the striatal complex. By tradition, their interests have tended to focus on cortical functions apropos of perception, memory, and learning, as well as emotion and motivation. In the usual psychological tests, animals are required to either manipulate inanimate objects, run mazes, bar press for nourishment, avoid noxious stimuli, or perform some unique task.

In our investigations, we have tried to learn whether or not experimentation on animals engaged in natural forms of behavior and interacting with other animals might reveal functions of the R-complex that would otherwise not be apparent. Moreover, it was expected that helpful insights might be obtained from comparative neurobehavioral studies on reptiles, birds, and mammals. For the experimental work on reptiles, it was important to learn as much as possible about the evolution of reptilian behavior and its various manifestations. The ancestral mammal-like reptiles (therapsids) are the subject of the next chapter. The subsequent five chapters deal with an analysis of reptilian behavior as typified by existing lizards.

References

1. MacLean, 1978d, p. 1
2. Colbert, 1966; Romer, 1966
3. Blinkov and Glezer, 1968, pp. 349, 371–372
4. See Schiller, 1967
5. MacLean, 1973b
6. See Christ, 1969
7. Wilson, 1925a, p. 1
8. See Smith, 1901
9. Koelle, 1954
10. Parent and Olivier, 1970; Karten, 1969
11. Cairny, 1926
12. Ariëns Kappers, 1908
13. Ariëns Kappers *et al.*, 1936
14. Falck, 1962; Falck *et al.*, 1962; Carlsson *et al.*, 1962
15. Dahlström and Fuxe, 1964
16. Anden *et al.*, 1964; Ungerstedt, 1971
17. Juorio and Vogt, 1967
18. Jacobowitz and MacLean, 1978
19. Ungerstedt, 1971; Bjorklund and Lindvall, 1978
20. Twarog and Page, 1953
21. Amin *et al.*, 1954
22. Paasonen and Vogt, 1956; Paasonen *et al.*, 1957
23. Gaddum, 1953
24. Gaddum, 1954; Woolley and Shaw, 1954
25. Pert and Snyder, 1973a,b; Simon *et al.*, 1973; Terenius, 1973
26. Pert and Snyder, 1973a; Kuhar *et al.*, 1973
27. Hughes, 1975
28. Hughes *et al.*, 1975
29. Hong *et al.*, 1977
30. Haber and Elde, 1981
31. Teschemacher *et al.*, 1975; Li and Chung, 1976; Bradbury *et al.*, 1976; Guillemin *et al.*, 1976
32. McGeer *et al.*, 1978, p. 329
33. Euler and Gaddum, 1931
34. Euler and Gaddum, 1931, p. 80
35. Chang and Leeman, 1970
36. See McGeer *et al.*, 1978, pp. 322–328, for summary
37. Spatz, 1922b
38. Diezel, 1955
39. Switzer and Hill, 1979; Switzer *et al.*, 1982
40. Spatz, 1922b; Hallgren and Sourander, 1958
41. Hill, 1980
42. Hill, 1981
43. Hill, 1982
44. Switzer, personal communication
45. Fahn and Cote, 1968; Perry *et al.*, 1971
46. Smith, 1918–19
47. MacLean, 1978d
48. Johnston, 1916
49. Ariëns Kappers, 1908, pp. 323–324
50. Ariëns Kappers, 1922
51. Karten, 1969
52. Divac and Oberg, 1979
53. Nauta and Karten, 1970
54. Webster, 1979, p. 117
55. Papez, personal communication
56. Haeckel, 1876
57. Thomson, J. A., 1893; Thomson, K. S., 1988
58. Morest, 1970
59. Kallen, 1951
60. Holmgren, 1925
61. Hines, 1923
62. Durward, 1934
63. Angevine, 1965
64. Hughes *et al.*, 1958, p. 476
65. Tsai *et al.*, 1981a
66. Kühlenbeck, 1938
67. Tsai *et al.*, 1981b
68. Tsai *et al.*, 1981b, p. 304
69. Smart and Sturrock, 1979
70. Sidman and Angevine, 1962
71. Angevine and McConnell, 1974
72. Brand and Rakic, 1978
73. Hewitt, 1958
74. Spatz, 1922a
75. Schneider, 1949; Richter, 1966; Sidman and Rakic, 1982
76. Ramón y Cajal, 1909–11
77. Papez, 1929, p. 319
78. Graybiel and Hickey, 1982
79. Graybiel and Ragsdale, 1978
80. Graybiel and Ragsdale, 1980
81. Butcher and Hodge, 1976
82. Loizou, 1972; Olson *et al.*, 1972; Tennyson *et al.*, 1972
83. Tennyson *et al.*, 1972
84. Graybiel *et al.*, 1979
85. Golden, 1972
86. Olson *et al.*, 1972
87. Jackson, 1873/1958, Vol. 1, p. 46
88. Gregory, 1929/1967
89. Ebbesson, 1980

90. Shellshear, 1920–21
91. Landau, 1919
92. Holmgren, 1925
93. Smith, 1930
94. de Vries, 1910
95. MacLean, 1972a
96. Foix and Nicolesco, 1925
97. Namba, 1957
98. Fox *et al.*, 1971
99. Kemp, 1970, quoted by Fox *et al.*, 1971
100. Pasik *et al.*, 1979
101. Fox *et al.*, 1966, 1974
102. Ramón y Cajal, 1901–02/1955
103. Mesulam and Van Hoesen, 1976; Olivier *et al.*, 1970
104. Nauta and Kuypers, 1958; see also Bowden *et al.*, 1978
105. Bedard *et al.*, 1969
106. Moore *et al.*, 1971; Carpenter and Peter, 1972
107. Nauta *et al.*, 1974; Kuypers *et al.*, 1974
108. Carpenter *et al.*, 1976
109. Conrad *et al.*, 1974; Nauta *et al.*, 1974; Bobillier *et al.*, 1976
110. Parent, personal communication; 1986
111. Nauta and Mehler, 1966
112. Walker, 1938
113. McGuinness *et al.*, 1976; Krauthamer, 1979
114. Saper and Loewy, 1980
115. Powell and Cowan, 1954
116. Droogleever-Fortuyn and Stefens, 1951; Cowan and Powell, 1955
117. Powell and Cowan, 1956
118. Nauta and Whitlock, 1954; Mehler, 1966
119. Mehler, 1966
120. Jones and Leavitt, 1974; Kuypers *et al.*, 1974; Nauta *et al.*, 1974
121. Royce, 1978
122. Kalil, 1978
123. Wilson, 1914
124. Ramón y Cajal, 1909–11; Dejerine, 1895–1901
125. Glees, 1944
126. Webster, 1961
127. Martin and Hamel, 1967
128. Carman *et al.*, 1963
129. Webster, 1965
130. Kemp and Powell, 1971
131. Kunzle, 1975
132. Jones *et al.*, 1977; Kunzle, 1975
133. Goldman and Nauta, 1977; Yeterian and Van Hoesen, 1978; Van Hoesen *et al.*, 1981
134. Yeterian and Van Hoesen, 1978
135. Van Hoesen *et al.*, 1981, p. 218
136. Carman *et al.*, 1965
137. Royce, 1982
138. Mettler, 1942; Webster, 1965; Carman *et al.*, 1965
139. Nauta, 1964; Webster, 1965; Van Hoesen *et al.*, 1981
140. Heimer and Wilson, 1975
141. Nauta, 1953
142. Haber *et al.*, 1985
143. Wilson, 1914, p. 445
144. Papez, 1942, p. 32
145. Adinolfi and Pappas, 1968
146. Verhaart, 1950
147. Fox and Rafols, 1975; Fox *et al.*, 1975
148. Szabo, 1962
149. Szabo, 1970
150. Szabo, 1967
151. Spatz, 1924
152. Rundles and Papez, 1937; Ranson *et al.*, 1941; Szabo, 1970
153. Kuo and Carpenter, 1973
154. Smith, 1930; Ariëns Kappers *et al.*, 1936
155. Carpenter *et al.*, 1968; Carpenter and Strominger, 1967
156. Nauta and Mehler, 1966; Kuo and Carpenter, 1973
157. Nauta *et al.*, 1974, Carter and Fibiger, 1978
158. Herkenham and Nauta, 1977
159. Nauta and Mehler, 1966
160. Mesulam and Van Hoesen, 1976; Parent *et al.*, 1981
161. Faull and Mehler, 1976; Graybiel and Sciascia, 1975; Rinvik *et al.*, 1976; Jayaraman *et al.*, 1977
162. Rinvik *et al.*, 1976
163. Carpenter *et al.*, 1976
164. Johnson and Aprison, 1971; Spencer, 1976; Kim *et al.*, 1977
165. Schäfer, 1898/1900
166. Crosby *et al.*, 1962
167. Divac and Oberg, 1979
168. Tilney and Riley, 1930, p. 728
169. Jackson, 1884/1958, Vol. 2
170. Jackson, 1882/1958, Vol. 2, p. 43
171. Ramón y Cajal, 1891
172. Ramón y Cajal, quoted by Lorente de Nó, 1949, p. 308
173. e.g., Buchwald *et al.*, 1973
174. Penfield and Jasper, 1954, pp. 723–724
175. Penfield and Jasper, 1954, p. 725
176. Galvani, 1791/1953
177. Du Bois-Reymond, 1849
178. Hermann, 1868
179. Flourens, 1824
180. Fritsch and Hitzig, 1870
181. Ferrier, 1876/1966
182. Ferrier, 1876/1966, p. 161
183. Wilson, 1914
184. Wilson, 1914, p. 442
185. Wilson, 1914, pp. 482–483
186. Wilson, 1914, p. 489
187. Mettler *et al.*, 1939
188. Mettler *et al.*, 1939, p. 995
189. Hodes *et al.*, 1951
190. Liles and Davis, 1969
191. Hess, 1932
192. Forman and Ward, 1957
193. Laursen, 1963
194. Rubinstein and Delgado, 1963
195. Delgado *et al.*, 1975
196. Buchwald *et al.*, 1961
197. MacLean, 1957a,b
198. Stevens *et al.*, 1961
199. Akert and Andersson, 1951
200. Goldby and Gamble, 1957
201. Distel, 1978
202. Åkerman, 1966
203. Anschel, 1977
204. Jung and Hassler, 1960, p. 920
205. Jung and Hassler, 1960, p. 877
206. Hassler, 1961
207. Orthner and Roeder, 1962
208. Van Buren *et al.*, 1966
209. Rosner *et al.*, 1966
210. Magendie, 1823, 1841
211. Mettler, 1942
212. Bailey and Davis, 1942
213. Villablanca and Marcus, 1975
214. Kennard and Fulton, 1942, pp. 229–230
215. Davis, 1958
216. Wilson, 1914, p. 444
217. Ranson and Berry, 1941
218. Ranson and Ranson, 1942, pp. 70, 72
219. Thompson, 1964
220. Rosvold, 1968
221. Kirby and Kimble, 1968
222. Knott *et al.*, 1960
223. Gomez *et al.*, 1958
224. Willis, 1664
225. Prus, 1898
226. Wilson, 1912
227. Wilson, 1925a–d
228. Bucy *et al.*, 1964
229. Jakob, 1923, 1925
230. Vogt and Vogt, 1919a, 1920
231. Lewy, 1942/1966, p. 17
232. Gowers, 1888, p. 286
233. Denny-Brown, 1946, p. 273

234. Biemond, 1970, p. 195
235. Meyers, 1942
236. Denny-Brown, 1962, p. 57
237. Denny-Brown, 1962
238. Alexander, 1942, p. 476
239. Cooper, 1956
240. Jung and Hassler, 1960, pp. 877–878
241. Jung and Hassler, 1960, p. 878
242. Krayenbühl *et al.*, 1961, p. 441
243. Fager, 1968, p. 148
244. Cooper, 1961, p. 227
245. Bonvallet *et al.*, 1952; Dell, 1952
246. Albe-Fessard *et al.*, 1960a
247. Albe-Fessard *et al.*, 1960b
248. Segundo and Machne, 1956
249. Radna and MacLean, 1981a
250. Radna and MacLean, 1981a, see pp. 29, 43
251. Krauthamer, 1979
252. Krauthamer, 1979, p. 264
253. MacPherson *et al.*, 1980
254. DeLong, 1973
255. Albe-Fessard and Buser, 1979
256. Rolls *et al.*, 1979
257. Niki *et al.*, 1972, p. 345
258. Rolls *et al.*, 1979, p. 178
259. Travis *et al.*, 1968
260. Travis and Sparks, 1968
261. DeLong, 1971
262. Roland *et al.*, 1982
263. Roland *et al.*, 1982, p. 476
264. Denny-Brown, 1960, p. 793
265. Albe-Fessard and Buser, 1979, pp. 315–316
266. Albe-Fessard and Buser, 1979, p. 316
267. Albe-Fessard and Buser, 1979, p. 319
268. Jackson, 1890/1958, Vol. 1, p. 429; Jackson, 1894/1958, Vol. 2, p. 411
269. Martin, 1967

5

The Mammal-like Reptiles (Therapsids)

Advances in science depend in part on the recognition of similarities and differences of things. In this and the next five chapters, it is the purpose to derive an inventory of behavioral characteristics that will provide a basis for recognizing similarities and differences between reptilian and mammalian behavior. For clues as to the evolutionary transition, we must look first at the mammal-like reptiles.

The mammal-like reptiles (Synapsida) are of great human interest because they are so close to the roots of our family tree. Despite their genealogical significance, these long-extinct animals, particularly the forms known as the therapsids, have received little attention in books on evolution, and compared with dinosaurs and some other reptilian species, are relatively unknown. Commenting upon this matter, the late Alfred Romer noted that the relatively recent time "at which mammals took over from the reptilian dynasties" might lead one to think that the stock from which they sprang developed late in reptilian history.[1] "This," he said, "is exactly the reverse of the true situation."[1] In Permian times, long before the dinosaurs (Figure 4-1), the mammal-like reptiles populated the world in great numbers. Their remains have turned up on every continent, recalling Alfred Wegener's (1915)[2] original contention that 250 million years ago the earth formed a single land mass, which he called Pangaea (Figure 5-1). Between 1969 and 1971 fossils of mammal-like reptiles were found in Antarctica that were the same kind as those in the Karroo beds of South Africa.[3] Antarctica had been connected to South Africa, forming part of the great southern continent that Eduard Suess (1904)[4] called Gondwanaland (Figure 5-2) because the strata of the untilted table mountains of Africa and South America resembled those of the Gondwana region of India.

Renowned paleontologist Robert Broom has commented, "If we except the Pliocene age, . . . there is no period in the world's history so important as that from the Middle Permian to the Upper Trias, as it was during this time that a group of reptiles slowly evolved into more and more mammal-like forms, and ultimately gave rise to primitive but true mammals. And in the South African Karroo shales we have a nearly continuous history of the land animals of this important period, and the study of the various beds is like examining the pages of a book of history."[5]

One may wonder how it is possible to say anything about the behavior of animals that have been extinct for millions of years. The acquiring of such information is essentially the business of paleobiology. One finds, for example, that on the basis of evolutionary changes in their skeletons and teeth, it is possible not only to infer a good deal about the behavior of the mammal-like reptiles, but also to derive some clues about their metabo-

Parts of this chapter provided the substance of an article entitled "The Neurobehavioral Significance of the Mammal-like Reptiles" (MacLean, 1986c).

Figure 5-1. Pangaea. At the time of the earliest mammal-like reptiles the land masses forming most of our present continents appear to have been gathered together into one megacontinent called Pangaea. Note that what are now the Mediterranean Sea and Indian Ocean formed one great body of water called the Sea of Tethys, so named for the wife of Oceanus. The Cordilleran Sea lies west of what are now North and South America, with their coastline mountain ridges extending from north to south. Cordillera ("a little rope") refers to a mountain ridge. Note where India and Madagascar (M) were located. India has since moved through the Sea of Tethys and come to rest against the Himalaya Mountains. Drawing based on a reconstruction by Irving (1977).

lism and thermoregulation. By looking at the shape of the bones and the places for insertions of muscles and tendons, one obtains a picture of how a particular part of the anatomy was used. The pattern of the blood supply in the bones of extinct animals provides some clues as to whether the animal was "cold-" or "warm-blooded."

Since only a handful of paleontologists have given extensive study to the mammal-like reptiles, much remains to be learned about these intriguing animals. We still do not know, for example, the form and size of the brain in the most advanced mammal-like reptiles. In considering now the evolution and physical characteristics of the mammal-like reptiles, we shall concentrate on those aspects that provide essential background for identifying different types and comparing them with mammals.

Evolution of the Mammal-like Reptiles

In reviewing the reptilian ancestry of mammals, it is to be recalled that the Greek word *phylon* is the name for a tribe or race, and that *phylogeny* refers to the evolutionary history of an organism or group of organisms. The following zoological address for one of the most advanced therapsids, a cynodont regarded to be close to the phylogenetic bound-

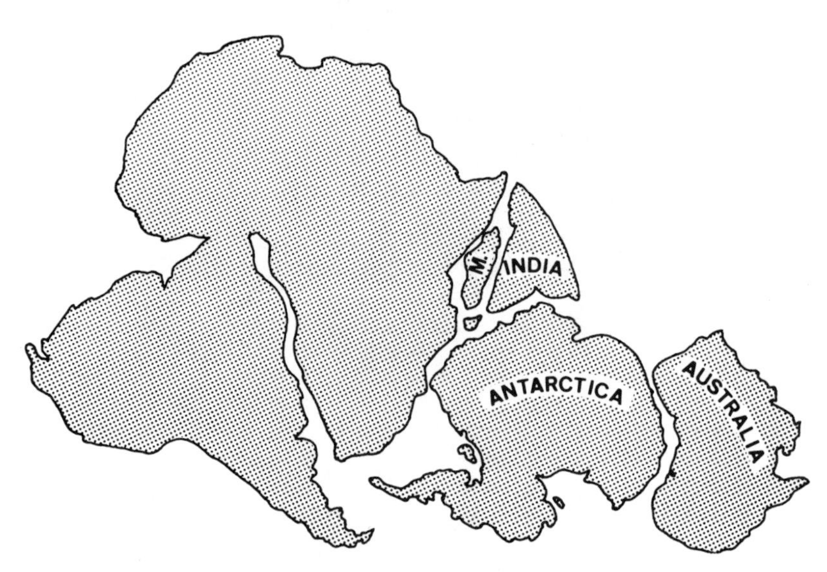

Figure 5-2. Gondwanaland. Based on geological similarities to Gondwana in India, Eduard Suess inferred the existence of a great southern continent, which he called Gondwanaland. In 1969 Colbert and others uncovered fossils of mammal-like reptiles on the slopes of Coalsack Bluff, Antarctica, that resembled those of the Lower Triassic beds of South Africa. A later expedition (see Figure 5-12A) uncovered further evidence of a link between Antarctica and Africa. With these new findings, it can now be said that mammal-like reptiles have been found on every continent, including those once contained in the great northern continent called Laurasia by duToit. Laurasia and Gondwanaland are believed to have been formed by the breakup of Pangaea. Redrawn after Palmer (1974).

ary between reptiles and mammals (Chapter 17), serves to illustrate the classification of a mammal-like reptile:

Class	Reptilia
(Subclass)	(Synapsida)
Order	Therapsida
(Suborder)	(Theriodontia)
(Infraorder)	(Ictidosauria)
Family	Diarthrognathidae
Genus species	*Diarthrognathus broomi*

Before tracing the evolution of the mammal-like reptiles from stem reptiles (cotylosaurs), it is necessary to point out a feature of the skull that is used to distinguish different reptiles.

Classificatory Features of Reptilian Skull

According to Osborn (1903), the "tendency to classify the Reptilia by the *structure of the temporal region of the skull*"[6] began in 1867, when Günther[7] published a descrip-

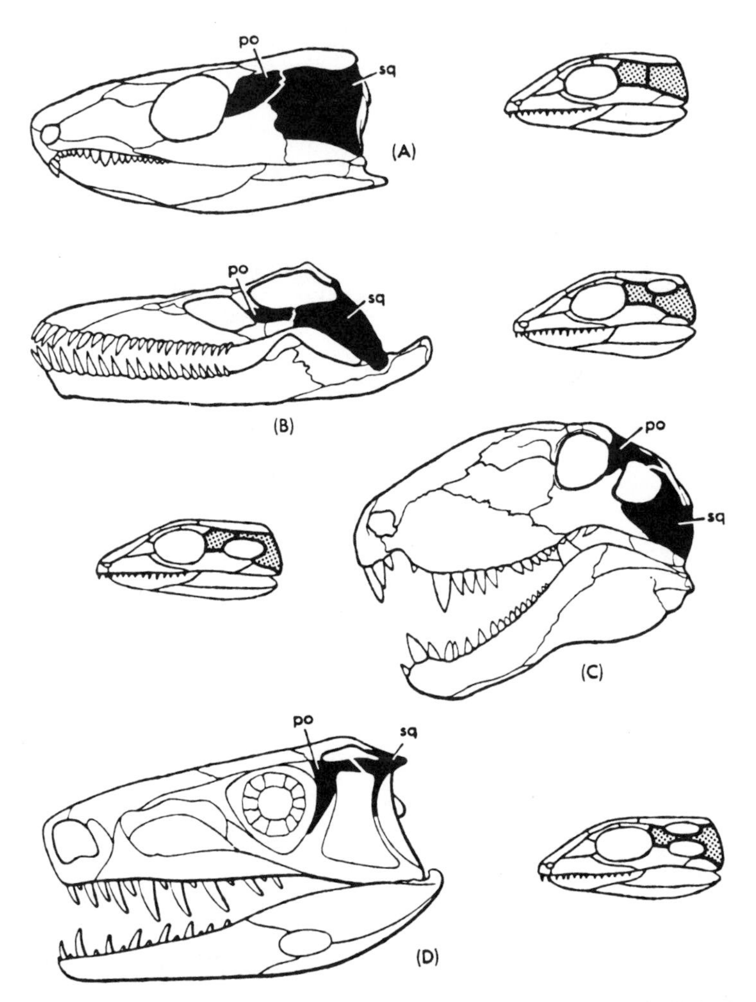

Figure 5-3. Distinctive skull features used in the classification of reptiles. The two main distinctions pertain to (1) the number of temporal openings (fossae, fenestrae) and (2) the relationship of the openings to the postorbital (po) and squamous bones (sq). In the cranial diagrams the two key bones are shown in stipple surrounded by heavy lines. Osborn (1903) used the term *apse* for the bony arch forming the roof of a fossa. He referred to the mammal-like reptiles (C) with a single arch above one opening as *synapsids*.

(A) The anapsid condition (absence of arched fossa) characterizes the stem reptiles and turtles.

(B) Colbert (1945) referred to an extinct group of reptiles (typified by marine forms) as *euryapsids* because of a wide cheek plate formed by the squamous bone and adjoining postorbital *below* the opening.

(D) The diapsid condition is exemplified by dinosaurs and lizards.

Slightly modified after Colbert (1969).

tion of differences in the "temporal arches" of Rhynchocephalia and Squamata. The classificatory distinctions arrived at over the next 75 years applied to the presence or absence of temporal openings and their bony arches. The present terminology can be traced largely to Cope (1892),[8] Osborn (1903),[9] Williston (1925),[10] and Colbert (1945).[11] In the primitive condition the dermal bony wall of the temporal region forms a solid roof. Cope[8] appears to have been the first to develop a theory of fenestration to explain how this roof became perforated and provided the site of origin of the temporal muscles participating in the adduction of the jaw. Like Günther,[7] he referred to the bony arches associated with these fenestrations (openings, fossae). Later, Osborn,[9] who had studied with Cope, introduced the Latin word *apsis* to refer to the arches over the openings, possibly having in mind the architectural meaning of the term as it applies to an arched-over niche or alcove.

As illustrated in Figure 5-3, there are two connecting bones—the posterior orbital and squamous—that, according to their arched relationship to the openings, provide the key to the taxonomic grouping of reptiles. The stem reptiles that have no temporal fossa and, hence, no temporal arch are classified as *anapsids* (Figure 5-3A). The same applies to the turtles and tortoises. Osborn referred to the mammal-like reptiles ''with single or undivided temporal arches'' above a single temporal fossa as *synapsids* (Figure 5-3C).[6] As opposed to them, he referred to reptiles with two openings as *diapsids* (Figure 5-3D). Included in this category are dinosaurs, crocodilians, rhynchocephalians, snakes, and lizards. Finally, an extinct group of reptiles, including aquatic forms known as placodonts, nothosaurs, and plesiosaurs (Figure 4-1), are distinguished by a single temporal opening bounded medially by the parietal bone and laterally by a broad cheek plate (Figure 5-3B). Colbert (1945) characterized this condition as *euryapsid* because of the underlying broad arch formed by the *wide* squamous bone and adjoining postorbital.

In advanced synapsids the temporal fossa undergoes great enlargement, with the result that the arch is so reduced in size as to resemble the zygoma characteristic of mammals. Hence, they are referred to as *therapsids* (''beast arch''), after the Greek word *therion,* signifying mammal in zoological nomenclature.

Forms Leading to Therapsids

In 1880 and 1882 Cope described an order of reptiles that he called Cotylosauria because he regarded them as ''stem'' reptiles. The oldest known reptile, *Hylonomus* (belonging to the suborder Captorhinomorph), was found in the Coal Measures of Nova Scotia (Figure 5-4A). As already noted, the stem reptiles are characterized as anapsid because of the absence of a temporal fossa.

The synapsids, as Romer has explained, can be broadly subdivided into primitive forms belonging to the order Pelycosauria and advanced types comprising the order Therapsida.[1] The pelycosaur *Varanosaurus* had a small temporal fossa, and as Romer notes, its proportions ''were not unlike those of many lizards'' (Figure 5-4B).[1] A more advanced form of pelycosaur is typified by the sphenacodonts best known by the genus *Dimetrodon* (Figure 5-5) with a long-spined ''sail'' rising from the back. There were other sphenacodonts without the sail, and it is probably from this stock that the therapsids evolved.

Romer characterizes the phthinosuchids, mainly found in the Permian beds of Russia, as the ''most primitive of therapsids''[1] (see Figure 5-6). They had a larger temporal opening than the pelycosaurs and possessed a lower canine tooth matched by an upper canine. In these respects, the features of the skull merit their classification as therapsids.

The Therapsids

The therapsid suborders Theriodontia and the Anomodontia (lawless-teeth) were the predominant terrestrial vertebrates during the late Permian and earliest Triassic times

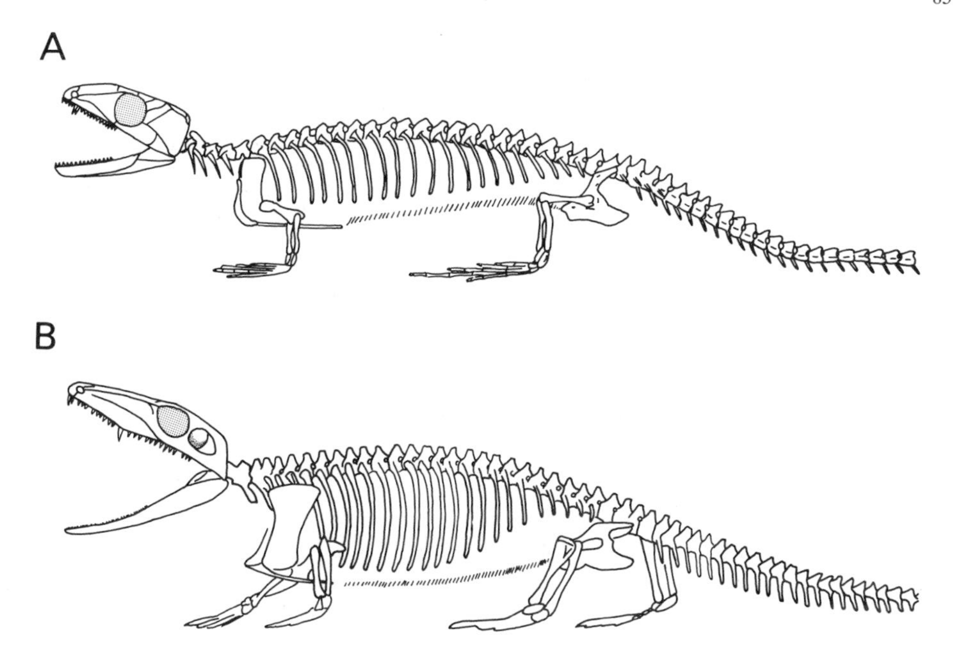

Figure 5-4. Skeletons of two early forms of reptiles.

(A) The oldest known reptile, *Hylonomous,* serves as an example of the so-called stem reptiles (cotylosaurs). Note the anapsid condition (see Figure 5-3A).

(B) A "primitive" mammal-like reptile, somewhat lizardlike in appearance, and called *Varanosaurus.* Belonging to the so-called pelycosaurs, it shows the beginning of the synapsid condition characteristic of the later mammal-like reptiles.

Redrawn, respectively, after Carroll (1964) and Romer (1966) but with the jaws agape.

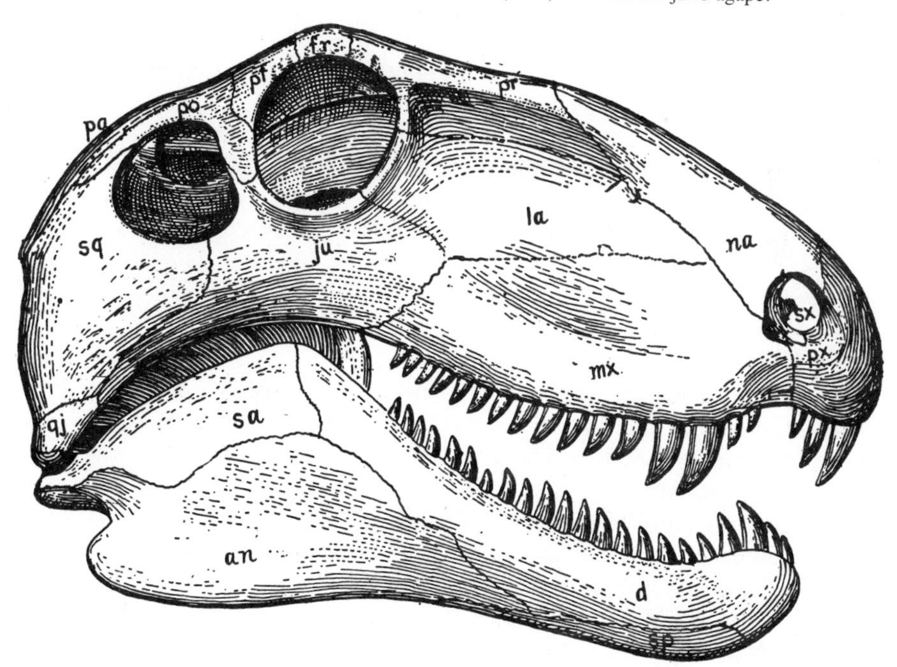

Figure 5-5. The skull of one of the advanced pelycosaur forms. This skull of *Dimetrodon* is shown because it illustrates an increase in size of the temporal recess, which along with other features marks these animals as being transitional between the stem reptiles and the therapsids. From Williston (1925).

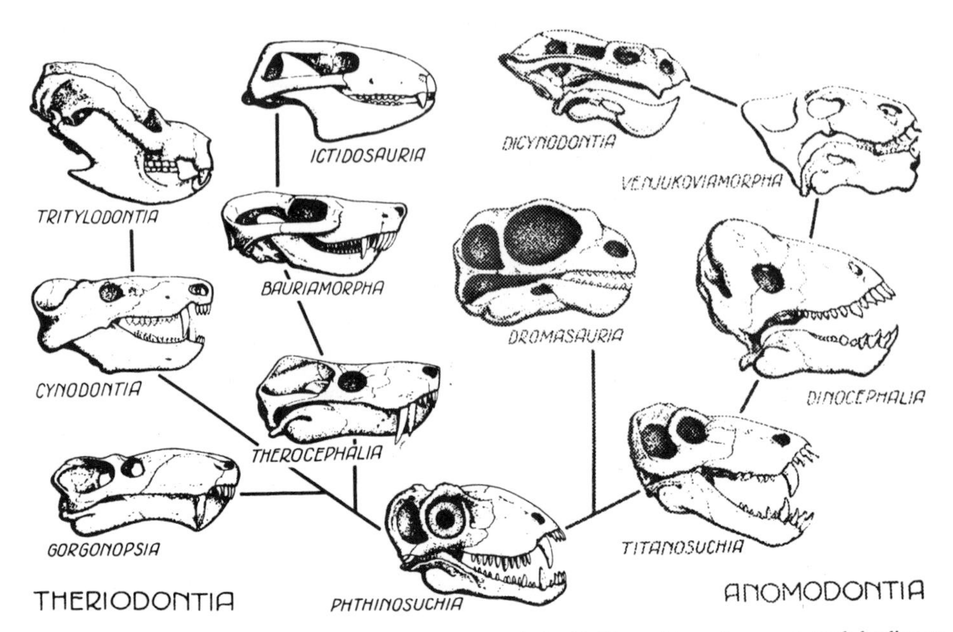

Figure 5-6. The "family tree" of therapsids. Romer regards the Phthinosuchia as the parent stock leading to "two great groups" of therapsids, characterized, respectively, as being predominantly carnivorous and herbivorous. Here the herbivorous line has been partly shaded out so as to focus attention on the carnivores from which mammals are derived. The progressive changes of the skull, teeth, and jaw toward the mammalian condition are described in the text. From Romer (1966).

(Figure 5-6). It gives one a feeling for their numbers to recall Robert Broom's calculation that there still lie hidden in the Karroo beds of South Africa the remains of more than 800 *billion* of these and other mammal-like reptiles.

The Theriodontia were carnivores, while the Anomodontia were predominantly herbivores. Since the mammals are believed to have been derived from the carnivorous variety, particularly the group known as cynodonts, the following discussion will be focused on the types of Theriodontia shown on the left-hand side of Figure 5-6.

Key Changes toward the Mammalian Condition

As a generalization, paleontologists point out that successive therapsids become progressively more mammal-like with respect to their locomotor skeleton, skull, teeth, jaw joint, and middle ear.[1,12,13] This is not to say that the changes were necessarily gradual and coordinated.

The successive types of therapsids and their development toward the mammalian condition are seen in marvelous parade in the successive zones of the Karroo beds of South Africa. Table 5-1 gives the names of the various types found in successive formations laid down in the Middle and Upper Permian and in the Triassic. The successive carnivorous therapsids illustrated on the left-hand side of Figure 5-6 are the therocephalians, gorgonopsians, cynodonts, bauriamorphs, and ictidosaurs together with the tritylodonts.[12]

Table 5-1. Theriodonts in Successive Formations of the Karroo Beds[a]

Period	Beds or *Zones*	Theriodonts	Series
Upper Triassic	Cave Sandstone Red Beds Molteno Beds	Ictidosaurs and tritylodonts	Stromberg
Middle Triassic			
Lower Triassic	*Cynognathus* *(Procolophon)* *Lystrosaurus*	Bauriamorphs Cynodonts (numerous) Small Cynodonts	Upper Beaufort
Upper Permian	*Cistecephalus*	Gorgonopsians (many) Therocephalians (few)	Lower Beaufort
	Endothiodon	Therocephalians (few)	
Middle Permian	*Tapinocephalus*	Gorgonopsians Therocephalians	Lower Beaufort
	Ecca Beds		

[a]Based on Broom (1932) and Colbert (1966). Except for *Cynognathus*, all zones are named for an anomodont. The *Tapinocephalus* zone is about 1000 feet in thickness. The spacing for the other zones and beds is roughly proportional.

The Body Skeleton

In contrast to the wide stance of earlier reptiles, the therapsids assumed a more upright posture, with the limbs supporting from underneath; the elbows were directed backwards and the knees forward.[1,13] These features, together with the less heavy structure of their bones indicate that some, at least, were capable of rapid and, perhaps sustained, locomotion. The relative slenderness of the long bones in some of the later therocephalians and of the gorgonopsians is suggestive of a canid alacrity (see Figure 5-7A). Most therapsids, however, had a heavyset bony structure that would have given them a robust body build reminiscent of bears or badgers (Figure 5-7B). The therapsid assumption of a more upright posture was accompanied by a change in the phalangeal formula. The therocephalians appear to be the first to have acquired the 2-3-3-3-3 mammalian phalangeal formula familiarly illustrated by our fingers and toes.[1]

Changes in Skull

The therocephalians also demonstrate another particular advancement toward the mammalian condition—namely, an increase in the size of the temporal fossa along with a reduction of the postorbitosquamosal arch (Figure 5-6). The mammalian likeness is further enhanced in the bauriamorphs in which the loss of the postorbital bar results in a continuation of the temporal fossa with the space for the eye (Figure 5-6). This is essentially the mammalian condition. In the bauriamorphs and ictidosaurs other cranial changes included the complete loss of the parietal foramen[12] that houses the parietal, or so-called third eye (see below).

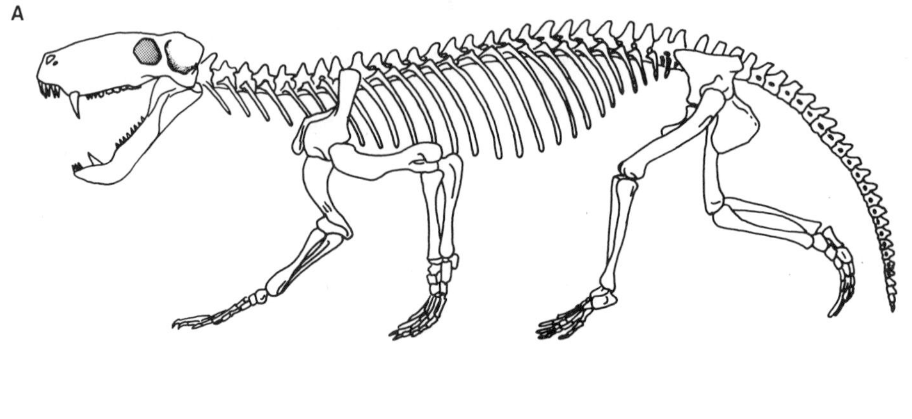

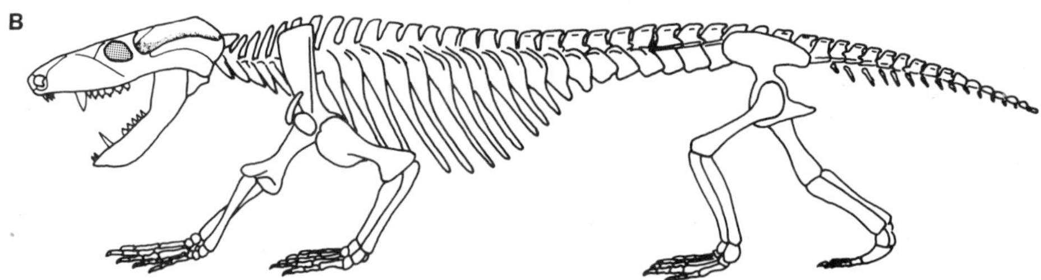

Figure 5-7. Examples of two carnivorous types of mammal-like reptiles. Note that the teeth and carriage of the body have a resemblance to the mammalian condition. See text for further details.

(A) A gorgonopsian called *Lycaenops* because of its wolflike appearance.

(B) A cynodont named *Thrinaxodon*. The name Cynodontia for mammal-like reptiles of this type refers to their doglike teeth.

Figures redrawn, respectively, after Colbert (1969) and Brink (1956) but with the jaws agape.

Dentition

Most early reptiles, like many existing forms, had teeth of conical shape that throughout life were alternately shed and replaced (polyphyodonty). In therapsids there was a change in this pattern in that various teeth became specialized. In therocephalians and gorgonopsians the front teeth became enlarged whereas the back teeth were either reduced in size or lost. In the cynodonts and bauriamorphs the teeth are differentiated into incisors for nipping or seizing; canines for tearing; and postcanine, molarlike teeth for shearing and crushing. In tritylodonts and ictidosaurs (Figure 5-6) the crowns of the molar teeth were comparable to those of some mammals. It appears, however, that no therapsids developed the diphyodont condition of mammals (deciduous and permanent dentitions).

Secondary Palate

Both the early therocephalians and gorgonopsians demonstrate a notable primitive feature in that they do not have a secondary palate. The later therocephalians may have had a membranous palate. The acquisition by the cynodonts and bauriamorphs of a bony secondary palate represents an important innovation because the partition allowed the animal to chew food into small pieces and to breathe at the same time without danger of aspiration. Reptiles without such a palate typically cease to breathe while they swallow their food in large chunks. Food chewed into small pieces has the physiological advantage that it can be more easily broken down by digestion and, hence, become a more ready source of energy.

The Jaw Joint

In 1932, Broom commented that the "ictidosaurians are so near to mammals that the only distinguishing character appears to be that they still retain the articular–quadrate hinge of the jaw."[14] In all reptiles the lower component of the jaw joint consists of an articular bone that hinges with the quadrate in the upper jaw. In mammals, on the contrary, the large dentary bone of the lower jaw articulates directly with the squamosal bone of the skull. In cynodonts the quadrate and articular bone become smaller and smaller until, with the ictidosaurs, they practically disappear. An ictidosaur that approaches the dividing line between reptiles and mammals was called *Diarthrognathus*, because of the double articulation of the jaw—namely, a close contact between the dentary and squamosal bone, as well as the reptilian hinging of the articulate and quadrate bones.[15,16]

Some Key Questions

In regard to changes in physiology and behavior during the evolutionary transition from therapsids to mammals, I will limit consideration to five key questions: (1) Were the therapsids warm- or cold-blooded? (2) Were they capable of communicative displays? (3) Was there evidence of egg laying or parental care? (4) Were the therapsids able to hear and to vocalize? (5) What was the size and form of the brain?

Cold-Blooded or Warm-Blooded?

Extant reptiles are said to be "cold-blooded" because if they stay in one place without moving, their body temperature approaches that of the external environment. Because of their poikilothermic condition, they must engage in behavioral thermoregulation, moving between warm and cool places in order to gain or lose heat. The question of whether or not the therapsids were ectothermic or endothermic is of signal importance because of the primary role of thermoregulation in metabolic processes underlying all aspects of an animal's biological functions and activities.

Since fossils of some advanced mammal-like reptiles have been discovered at high latitudes where they were potentially subject to cold conditions, some workers have proposed that they were protected by a coat of fur.[17] Based on skull markings indicative of foramina around the nasal orifice and mouth, Brink has suggested that therapsids may have had sweat glands and vibrissae.[18] The presence of such features would be indicative of a hairy integument. The fossil evidence of the existence of nasal turbinate bones has been regarded as another indication of endothermy, since such structures would have served to warm or cool inspired air.

The type of vascularization of bone affords inferences regarding the nature of thermoregulation in extinct forms. In a symposium entitled "A Cold Look at the Warm-blooded Dinosaurs,"[19] some authors advanced the argument that the dinosaurs were endothermic because the pattern of the haversian canals seen in cross sections of their bones is similar to that of birds and mammals. Bouvier, however, has pointed out that there is great variation in the degree of vascularization of bone in reptiles, birds, and mammals and that many species of birds and mammals do not have the haversian replacement.[20] In extensive comparative studies, Enlow and Brown found two types of vascularized bone in therapsid reptiles, one being comparable to haversian bone seen in many mammals.[21]

There is one other bit of evidence suggesting that the advanced mammal-like reptiles may have become endothermic. The argument rests on findings in lizards and involves the question of function of the midline third eye, often referred to as the pineal eye, but more appropriately termed the parietal eye. The eye owes its name to its midline location between the two parietal bones. A foramen in the skull leads from the eye to the pineal gland (Epiphysis cerebri), which in turn is connected to the habenula in the diencephalon. Among vertebrates three different conditions apply to the epiphysis that will be identified as follows[22]:

	Epiphysis	Parietal eye
Epi-0	*Absent*	*Absent*
Epi-1	Present	*Absent*
Epi-2	Present	Present

Roth and Roth have provided suggestive evidence that the Epi-0 condition existed in dinosaurs.[22] The crocodilians, sea cows (dugongs), and manatees (sirenians) are also characterized by the Epi-0 condition. The Epi-1 condition is found in most bony fish, birds, mammals, and in lizards living near the equator (primarily geckos and teiids).[23] The Epi-2 condition applies to the early sarcopterygians (fleshy fins) [which include the crossopterygians and the closely related dipnoans (lung fish)], amphibians, and to 11 of 18 families of lizards.[23] The crossopterygians are believed to be the ancestors of labyrinthodonts (amphibians with "labyrinthine" structure of individual teeth), which in turn gave rise to the stem reptiles. As was noted earlier, in the more advanced mammal-like reptiles, the parietal foramen disappeared.

It has been found in lizards that destruction of the parietal eye resets the thermostat for temperature regulation about two degrees higher.[24] Since these findings suggest that the epiphyseal complex is implicated in the thermoregulation of ectotherms, it has been suggested that the disappearance of the parietal eye in mammal-like reptiles may have reflected changes toward an endothermic condition.[22] This argument, however, is somewhat tempered by the recognition that there has been a loss of the parietal eye in many ectotherms that have survived into present times, including sarcopterygians.

The Question of Displays

It is one of the intriguing aspects of neurobehavioral evolution that a number of postures and autonomic changes seen in thermoregulation acquire symbolic significance in animal communication. For example, the piloerection and ruffling of feathers that serve, respectively, in mammals and birds to insulate against the cold may also serve to enhance the animals' size in aggressive or defensive encounters. Greenberg has cited references to four species of lizards that "use a similar kind of posture in thermoregulation as in a show of aggression."[25] The question arises as to whether or not a detailed study of muscular insertions in therapsids would indicate a capability to engage in communicative displays involving, for example, extension of a gular fold, sagittal expansion, and/or tiptoe extension of the limbs seen in displays of lizards.

The Question of Egg-Laying and Parental Care

Romer has referred to the amniote egg as "the most marvelous 'invention' in vertebrate history,"[26] regarding it as such because it could be deposited on land and mature to the adult form without passing through the risky aquatic larval stage. The egg of amphibious forms such as the frog or salamander is not of sufficient size to sustain the embryo until the adult form is reached. Figure 5-8B shows what has been described as the oldest known reptilian egg found in the lower Permian beds of Texas. Figure 5-8A identifies the main parts of the amniote egg. The word *amnion* refers to the innermost membrane of the sac enclosing the embryos of reptiles, birds, and mammals, being the diminutive of the Greek word *amnos* for lamb. The use of this term may have been

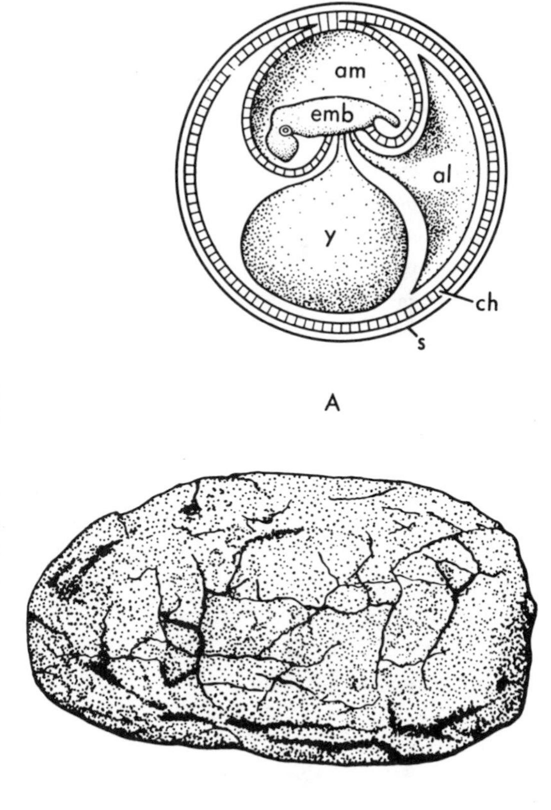

Figure 5-8. Diagram of the amniote egg alongside what has been alleged to be the oldest known reptilian egg.

(A) Cross section of reptilian egg, showing embryo (emb) floating within the amniotic sac (am), with the yolk sac (y) providing nourishment and the allantoic sac (al) serving as a receptacle for the excreta. These structures are, in turn, surrounded by the chorion (ch) and soft shell (s).

(B) Representation of a presumed fossilized egg ($2\frac{1}{2}$ inches long) in the lower Permian beds of Texas. Described by Romer and Price in 1939, it was thought to be the oldest reptilian egg. A recent reexamination by Hirsch (1979), however, has raised the question as to whether it is a fossilized egg.

From Colbert (1969).

suggested by the sight of a newborn lamb inside the glistening membranous sac. The amniote egg contains three sacs: (1) the amniotic sac; (2) the yolk sac, which provides nourishment for the embryo floating in the amniotic sac; and (3) the allantoic (sausage) sac, which balloons out from the cloaca (combined receptacle for the urinary and intestinal tracts). All three sacs are enclosed by the chorionic membrane, which in turn is surrounded by the shell of the egg. The shell of a reptilian egg is soft, having somewhat the feel of moist blotting paper.

Did the advanced mammal-like reptiles lay eggs? As yet, there is no good evidence to allow an answer to this question. One speculation requires comment about the egg-laying monotremes, which are the most primitive of existing mammals. They owe their name to a ''single hole'' forming a reptilianlike cloaca. They are represented by the duck-billed platypus [*Ornithorhynchus* (birdlike bill)] and the spiny anteaters [*Echidna* (viper)]. The platypus with its sensitive bill spends its life grubbing in streams, while the anteater holes up in the deep forest. They rank as mammals because they suckle their young on milk secreted by modified sweat glands corresponding to the mammae of higher mammals.[13] Judged by its less developed neocortex (which has a very large representation of the bill), the platypus is a more primitive mammal than the anteater.[27] Fossils of the monotremes cannot be traced back beyond the Pleistocene (Figure 4-1). Nevertheless, Romer believes that ''it is highly probable that they represent a line of descent from the mammal-like reptiles entirely separate from that of any other living forms.''[28] Since the monotremes lay eggs, it is usually assumed that the mammal-like reptiles did also.

Did the mammal-like reptiles care for their young or, like the contemporary Komodo lizard (see Chapter 8), did the young have to escape to the trees to avoid being cannibalized? In regard to the question of parental care, there is one bit of the positive evidence. In 1955 Brink reported a remarkable finding while he was examining a block of fossils discovered by J. W. Kitching in the ''old brickfield Donga,'' east of Harrismith, Orange Free State, South Africa.[29] As illustrated in Figure 5-9, he observed the skull of a tiny, immature cynodont (*Thrinaxodon liorhinus*) next to that of an adult. Explaining why the adult was probably a female, he conjectured that it ''could possibly be the mother of the tiny specimen.''[30] Since then, other paleontologists have agreed that the finding suggests the possibility that mammal-like reptiles may have shown parental care. At least, Colbert has commented, a discussion of this possibility can no longer be considered as idle speculation.

The Question of Audiovocal Communication

Could the therapsids vocalize, or were they mute like most extant lizards? How well could they hear? Ernest G. Wever, who has conducted extensive comparative studies on mechanisms of hearing, has commented: ''The lizard ear is of particular interest in relation to the problem of hearing in vertebrates, for this group of reptiles probably is more like the stem reptiles than any others now in existence.''[31] He points out that despite electrophysiological evidence of hearing in lizards, ''attempts to obtain behavioral indications of hearing have so far been unproductive.''[32] Some authorities have contended that the primitive mammal-like reptiles (pelycosaurs) were unable to hear, but Hotton (1959) has demonstrated by the use of models that these animals must have had auditory perception in the low range of frequencies.[33] Allin arrived at similar conclusions.[34] Since the articular and quadrate bones, which become the malleus and incus, respectively, of the

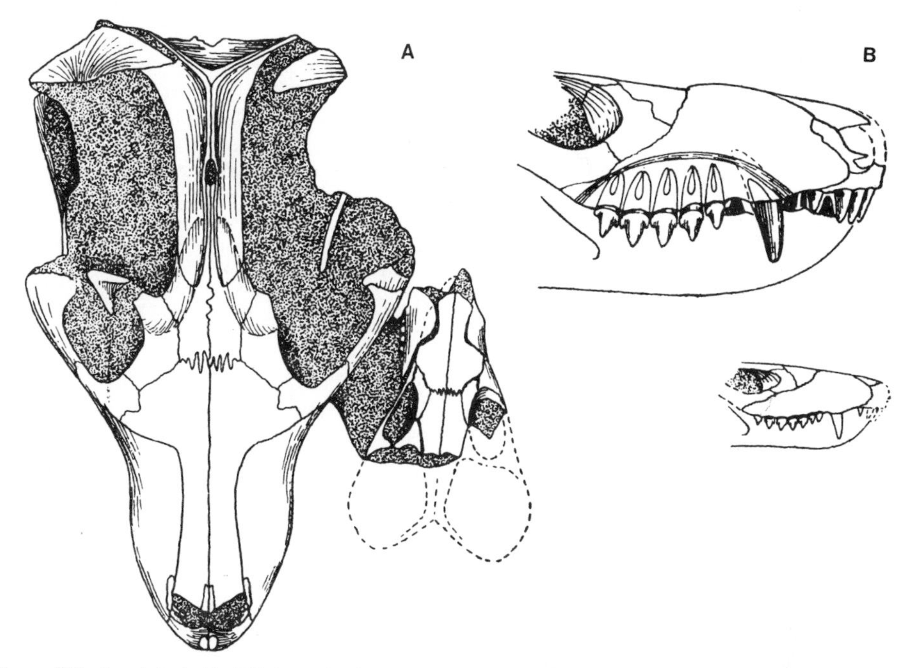

Figure 5-9. An adult skull of *Thrinaxodon liorhinus* found in the same nodule with a tiny specimen of the same species. On the basis of the bony structure, Brink (1955) inferred that the adult was a female, and he was also "inclined to believe that the elder is actually the mother of the young specimen. . . ." This finding has led to the speculation that these mammal-like reptiles engaged in parental care. The drawings in B show the profile of the tooth structure of the two specimens. From Brink (1955).

mammalian middle ear, were still part of the jaw joint in the advanced therapsids (Figure 5-10), it is probable that their perception of sound was confined to the low range of frequencies. If so, it is possible that, as in the case of lizards, vocalization did not play a significant role in their communication. This comment, however, must be weighed in the light of Westoll's suggestion that the recessus mandibularis of cynodonts functioned not only as a detector of airborne sound, but also as a vocal resonating amplifier.[35]

The Question of Brain Form and Size

In view of the pivotal position of the mammal-like reptiles in our ancestral past, it is disappointing that there exist so few cranial endocasts and reconstructions to give information about the size and form of the brain in these animals. Olson has described reconstructions of several fossil specimens,[36] but satisfactory endocasts are so rare that reference has been made principally to one pictured 75 years ago by D. M. S. Watson.[37] G. G. Simpson[38] used Watson's specimen from an intermediate cynodont for comparison with an endocast from an early mammal. The two are depicted in Figure 5-11. The cynodont endocast on the left reveals that, relative to their length, the cerebral hemispheres were quite narrow, being reminiscent of the shape of the amphibian brain. In a review of available material, Jerison[39] concluded that the brains of mammal-like reptiles were more reptilian than mammalian in appearance (see Quiroga[40] for contrary view). An

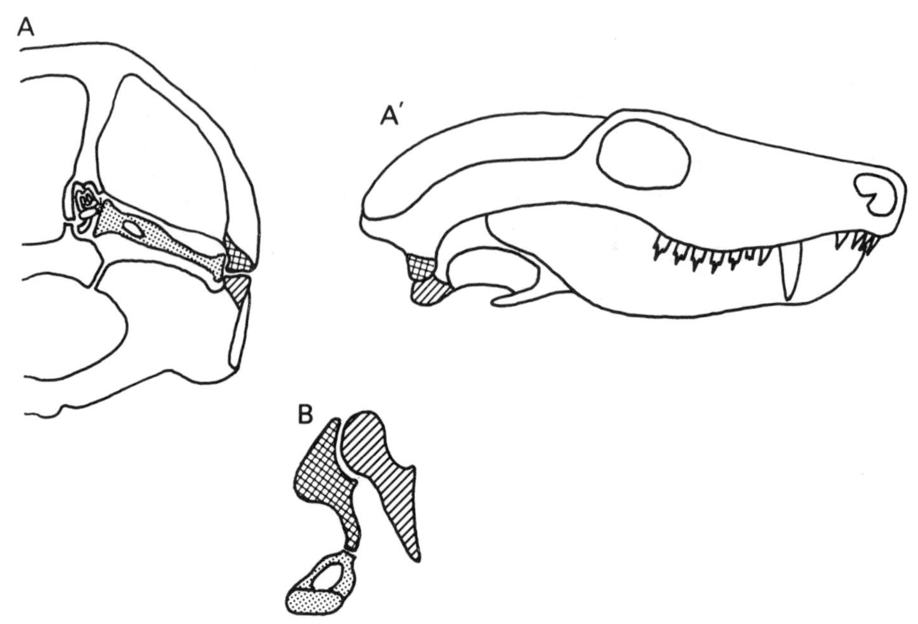

Figure 5-10. Evolutionary shifting of bones incorporated into the middle ear. As diagrammed in the cross section (A) and lateral view (A′) of the skull of a mammal-like reptile, two small bones of the jaw-joint, the articular and quadrate, become the malleus (hammer) and incus (anvil) of the middle ear of mammals (B). The two bones involved in this remarkable transformation are identified, respectively, by the slanted shading and cross-hatching. The stapes (stirrup) is shown in stipple. Redrawn after Romer (1966).

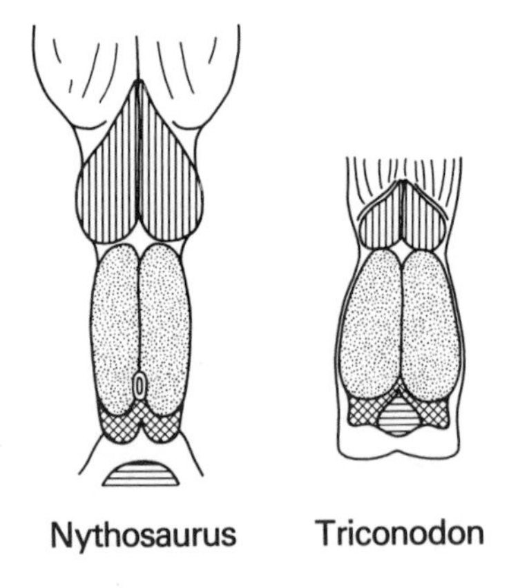

Figure 5-11. Drawings of cranial endocasts, depicting differences in size and form of the brain of a cynodont mammal-like reptile (*Nythosaurus*) and a primitive mammal (*Triconodon*). The forebrain (stipple) in each case is about 1.3 cm long, but note the relative widening of the hemispheres in the primitive mammal. Note also the disparity of the olfactory bulbs (longitudinal shading). The midbrain is shown in cross-hatching. The endocast of *Nythosaurus* indicated the presence of a pineal body. Redrawn after Simpson (1927).

explanation of the incipient widening of the mammalian forebrain will be considered in Chapter 17.

Apropos of the question of incipient maternal care, it should also be noted that the cynodont endocast reveals that the olfactory bulbs were large and broad (Figure 5-11). Fitch has pointed out that all brooding types of lizards are "secretive," and that olfaction plays an important role in maintaining the female's interest in her eggs.[41]

Why Their Extinction?

Late in Triassic times (about 190 million years ago) and continuing into early Jurassic times (see Figure 4-1), the mammal-like reptiles were on the road to extinction. Ferocious, and possibly cannibalistic, it is probable that they did not bring about their own extinction. Rather, it is generally believed that a swifter, more ferocious kind of reptile, the thecodonts (so named because of their encased teeth), began to outnumber them and to steal their niches.[1] The thecodonts were forerunners of the dinosaurs. Some of them had rather short arms and long legs and presumably could run swiftly. If they were as rapacious as pictured, they might easily have decimated less competitive animals.

Needless to say, the extinction of the mammal-like reptiles could have come about because of multiple events, with changes in climatic conditions being foremost among considerations. Until recently, continental drift was seldom suggested as a cause of climatic changes. Pangaea (Figure 5-1) is usually pictured with Antarctica at the South Pole.[42] Such presentations are based on "the assumption that Antarctica has moved very little from its original location when it was part of Pangaea."[43] In 1968, it was reported that mammal-like reptiles were found in Antarctica.[3] Figure 5-12 shows a photograph of one of these finds—an imprint of *Thrinaxodon,* a cynodont. This and other evidence of a mild climate suggests that we are far from understanding continental drift in Paleozoic times. A 30° change in latitude (1800 nautical miles) would account for marked alterations in temperature and climate. At a continental drift rate of 10 cm/year (about 4 inches), a 30° shift in latitude could occur in 33 million years. Irving has proposed that "exposure of the continental shelves, which house a high proportion of biological activity, could have been the major environmental change responsible for the widespread extinctions at the end of the Paleozoic."[44]

Summarizing Comment

Directional Evolution

Some students dislike the expression *directional evolution* because they read into it teleological connotations. But if it is used descriptively, its signification adheres more accurately to what is observed than some such term as *parallel*. Hence, this terminology will be used again in Chapter 17 and elsewhere. The successive zones of the Karroo beds of South Africa provide incontrovertible evidence of directional changes of the therapsids toward the mammalian condition, of which the following are particularly notable: With respect to the cranium, there is evidence of (1) progressive widening of the temporal fossa, with eventual loss of the postorbital bar; (2) disappearance of the pineal foramen; (3) acquisition of a secondary palate; (4) changes in dentition, including possession of

A

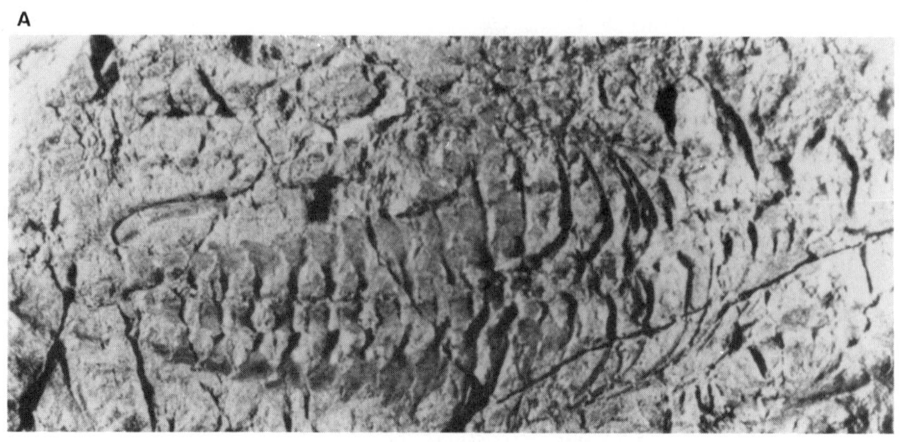

B

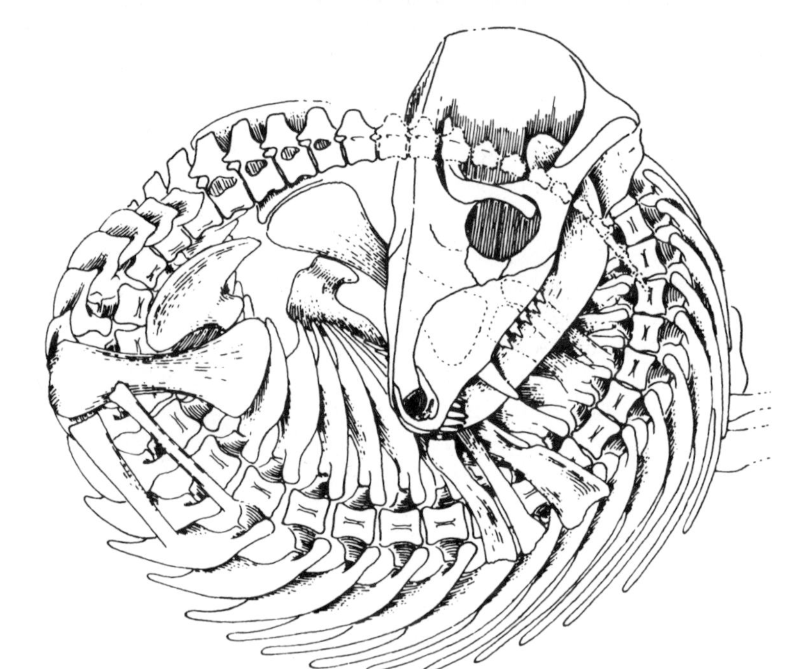

Figure 5-12. Remains of *Thrinaxodon* in Antarctica and Africa. See Figure 5-2 and text for geological significance of these finds.

(A) Imprint of a *Thrinaxodon* uncovered near McGregor Glacier, Antarctica, in 1970 by a member of an expedition including James Kitching, an authority on the mammal-like reptiles of the Karroo beds in South Africa. From Colbert (1972).

(B) A South African *Thrinaxodon*. It appears to have been captured by death in this dog-like, curled-up, sleeping posture. Hatchlings of some species of lizards (e.g., *Cnemidophorus velox*) may sleep in a less tightly curled posture (personal observations). From Brink (1958).

incisors, canines, and molariform teeth (without, however, the development of diphyodonty); (5) formation of a dentary squamosal articulation of the jaw but retention of a residual, typically reptilian quadrate and articular joint. Some skull markings are suggestive of the existence of vibrissae and sweat glands, the presence of which would be indicative of a hairy integument. In regard to the locomotor system, the most striking changes were (1) the assumption of a mammal-like posture, with the legs supporting the body from underneath, and (2) the acquisition of the mammalian phalangeal formula. Modifications of the masticatory apparatus would have afforded a more complete comminution of food required for rapid digestion and a quick source of energy, while the changes in the locomotor skeleton would have allowed greater suppleness and alacrity. Altogether, there are indications that the therapsids were approaching the warm-blooded condition of mammals.

Aside from the basic matter of endothermy, there are questions pertaining to the social and procreative behavior of therapsids that further study might help to answer. In regard to social communication, for example, a more detailed examination of fossil remains might yield clues as to whether or not the skeletomuscular system would have lent itself to the kinds of displays seen in extant reptiles. With respect to the question of parental behavior, it would be particularly relevant to search for more examples of associations between adult and immature therapsids.

Finally, there is the need for a far greater assortment of cranial endocasts to permit an assessment of encephalization in therapsids and early mammals.

Robert Broom (1866–1951) is perhaps better known for his helping Raymond Dart to establish *Australopithecus* as a legitimate human ancestor, than for his monumental work on the mammal-like reptiles. He originally went to Africa, however, for the sole purpose of obtaining more information about the evolution of reptiles. He encountered such an inexhaustible, movable feast that he never returned home.[45] According to his calculations, there still lie hidden in the Karroo beds of South Africa the remains of more than 800 billion mammal-like reptiles.[12] In a moment of optimism, Broom wrote: "If any intensive collecting is done in the next 20 or 50 years . . . we may then not only be able to trace the lines of evolution, but perhaps be able to see what has been the guiding or compelling force behind it all. America has given the world much of the evolutionary history of the horses . . . ; but South Africa will yet give a far fuller account of the evolution of the mammal-like reptiles, and clearly show every step that has led to the warm-blooded mammals."[46]

"Hundreds of thousands of pounds," he went on to say, "are spent investigating the mysteries of the milky way and the far distant nebulae, but a small fraction of this wealth spent in the further study of the Karroo Book of the Permian and Triassic History would probably yield results quite as important."[46]

Broom's projected 50 years have now gone by, but the challenge is more alive and promising than ever, waiting for the young at heart.

References

1. Romer, 1966, p. 173
2. Wegener, 1915
3. Colbert, 1972
4. Suess, 1904
5. Broom, 1932, p. 1
6. Osborn, 1903, p. 452
7. Günther, 1867
8. Cope, 1892
9. Osborn, 1903
10. Williston, 1925
11. Colbert, 1945
12. Broom, 1932

13. Colbert, 1969
14. Broom, 1932, p. 325
15. Crompton, 1963
16. Crompton and Hylander, 1986
17. Bakker, 1975
18. Brink, 1956
19. Thomas and Olson, 1980
20. Bouvier, 1977
21. Enlow and Brown, 1957
22. Roth and Roth, 1980
23. Gundy *et al.*, 1975
24. Roth and Ralph, 1976
25. Greenberg, 1978, p. 220
26. Romer, 1967, p. 1634
27. Bohringer and Rowe, 1977
28. Romer, 1966, p. 198
29. Brink, 1955
30. Brink, 1955, p. 75
31. Wever, 1965, p. 332
32. Wever, 1974, p. 163
33. Hotton, 1959
34. Allin, 1975, 1986
35. Westoll, 1943, 1945
36. Olson, 1944
37. Watson, 1913
38. Simpson, 1927
39. Jerison, 1973
40. Quiroga, 1979, 1980
41. Fitch, 1970
42. Dietz and Holden, 1970
43. Dietz and Holden, 1970, p. 39
44. Irving, 1977, p. 309
45. Keith, 1932
46. Broom, 1932, p. 309

6

Reptilian Behavior as Typified by Lizards

In Chapter 4 we considered the disappointing outcome of efforts over the past 150 years to clarify the functions of the R-complex. Contrary to the traditional view, there exists no *direct* evidence that the R-complex is involved primarily in motor performance. In attempting to elucidate its functions, was it possible that there had been a failure to ask the appropriate questions? Twenty-five years ago in making plans for the Laboratory of Brain Evolution and Behavior, it was the main purpose of my own research to test the hypothesis that the R-complex plays a basic role in organizing complex, species-typical behavior required for self-preservation and procreation. In experimental psychology, it has been customary to test animals in the *manipulation of inanimate objects*. As Ernst Mayr commented in the first Daniel S. Lehrman Memorial Symposium: "The classical experimental psychologists . . . almost invariably work only with a single species, and their primary interest has been in learning, conditioning, and other modifications of behavior. The phenomena they studied were to a large extent aspects of noncommunicative behavior, like maze running or food selection."[1]

In designing the proposed laboratory, there was a basis for expecting that experiments on animals living under seminatural conditions, and interacting with other animals, might reveal functions of the R-complex that would otherwise not be apparent. The neurobehavioral work to be described has dealt particularly with the role of the R-complex in species-typical, prosematic communication. As already explained, *prosematic* refers to rudimentary signaling in animals and human beings. Applying to any nonverbal signal—vocal, bodily, chemical—it circumvents the inappropriate use of the term *nonverbal communication* when referring to animals.

Identification of Basic Forms of Behavior

Our experimental approach has required that we not only develop improved methods and instrumentation (including computer-assisted event recorders) for quantifying behavior, but also obtain as complete a description as possible of the behavior of the different species used in our studies.

The Behavioral Profile

Different workers use various expressions for referring to a behavioral inventory, calling it a "behavioral profile," an "ethogram,"[2] or sometimes a "biogram." For purposes of analysis, animal behavior may be considered as being comprised of components, constructs, and sequences of constructs (Chapters 2 and 9). For the time being, we do not need to be concerned with focusing on any of these divisions or their subdivisions. It will turn out that simply by giving an overall description of reptilian behavior as typified by lizards, certain easily recognized patterns will precipitate out. One might think of a behavioral profile (i.e., ethogram) as comparable to a mountain range seen at a distance. Each recognized pattern of behavior would then correspond to one of the main peaks and its subpeaks. It is evident that if an experimental manipulation altered any particular form of behavior, it would be comparable to changing the shape of one of the main peaks or perhaps removing it altogether. In the following analysis of reptilian behavior (Chapters 6–10) the behavioral profile will, in terms of the metaphor, prove to consist of two main ranges. In one range are kinds of behavior that are linked together in the animal's *daily master routine* and *subroutines*. In the other range are behavioral patterns used in four main forms of social communication and referred to in lizards as signature, challenge, courtship, and submissive displays.

One of the surprising results of studying a variety of terrestrial animals is to discover how relatively few kinds of behavior one can identify in each and how most of these are common to all.[3] In Table 6-1 I have listed *special* forms of basic behavior observed in

Table 6-1. Special Forms of Basic Behavior

Behavior	
Selection and preparation of homesite	Domain
Establishment of territory	Domain
Use of home range	Domain
Showing place preferences	
Trail making	
Marking of territory	
Patrolling territory	
Ritualistic display in defense of territory, commonly involving the use of coloration and adornments	
Formalized intraspecific fighting in defense of territory	
Triumphal display in successful defense	
Assumption of distinctive postures and coloration in signaling surrender	
Use of defecation posts	
Foraging	
Hunting	
Homing	
Hoarding	
Formation of social groups	
Establishment of social hierarchy by ritualistic display and other means	
Greeting	
Grooming	
Courtship, with displays using coloration and adornments	
Mating	
Breeding and, in isolated instances, attending offspring	
Flocking	
Migration	

lizards and other reptiles that involve self-preservation and the preservation of the species. In Chapter 10 we shall, in addition, consider six general "interoperative" forms of behaviors that serve to bring into relation a number of the activities listed in Table 6-1.

Choice of Lizards as Experimental Subjects

From an evolutionary standpoint, it is curious that ethologists have paid little attention to reptiles, focusing instead on fishes and birds. Progress in science depends to a large extent on the recognition of similarities and differences of things. For our comparative neurobehavioral studies it has been essential to identify kinds of behavior that reptiles have in common with other land vertebrates. In this connection it must be reiterated that it is disappointing that none of the existing reptiles is in direct line with the mammal-like reptiles. As mentioned in the last chapter, one of the primitive mammal-like reptiles was so lizardlike in its skeletal anatomy as to be called *Varanosaurus,* after the monitor lizard (see Figure 6-1). Based on this and other paleontological considerations it was decided to

Figure 6-1. Williston's (1925) conception of the lizardlike appearance of *Varanosaurus,* a primitive mammal-like reptile (A), based on his reconstruction of the skeleton shown underneath (B). Describing the earlier stem reptiles, Carroll has repeatedly commented upon the similarity of various bony structures to those of present-day lizards (1964). In view of these and other considerations, one might say that of existing reptiles, lizards would come closest to resembling the progenitors of the mammal-like reptiles. Accordingly, they have been chosen for the neurobehavioral work to be described. From Williston (1925).

use lizards in our neurobehavioral studies. It also deserves noting that many present-day lizards occupy niches similar to those of the carnivorous mammal-like reptiles.

Domain

First and foremost of the activities listed in Table 6-1 are those that involve the selection and preparation of a homesite, the establishment of territory, and familiarization with a home range. In their totality, the homesite, territory, and home range comprise what will be referred to here as an animal's *domain*. Since these areas of activity are so fundamental to the behavior of reptiles and other animals, they require special consideration before presenting an overall picture of the behavior of lizards. The popularization of the territorial concept has had the effect of diverting attention from other important aspects of an animal's domain.

Homesite

For most lizards a homesite would apply to a nook, cranny, burrow, roost, or some other kind of shelter that the animal favors as a place to rest and to "sleep." For a territorial male such an abode is usually, but not invariably, within its own territory. Females and subdominant males belonging to the tenant's social group have their own favored shelters within the boundaries of the protected territory.

Territory

The question of the definition and role of territory has been subject to so much discussion and argument that it needs to be viewed in historical perspective. Although families, ethnic groups, and nations have contested over the possession of land and other belongings for countless generations, one might gain the impression from reading the behavioral literature that the important role of territory in human and animal affairs had not been recognized until the present century.

As Margaret Morse Nice noted in a review of "The Theory of Territorialism and Its Development" (1933), "The ornithologist with whom the concept of territory will always be linked is the Englishman, H. Eliot Howard."[4] Nice points out that although others such as Naumann in 1820 and Altum in 1868 had recognized the basic principles of territory, it was Howard's extensive observations, capped by the synthesis and explication in his book *Territory in Bird Life* (1920),[5] that led to the scientific recognition of the basic importance of territory. Howard believed that it is the function of territory to space pairs of birds of the same species so as to ensure an adequate supply of food both for themselves and for their young. It is the business of the male bird to stake out such a territory. Unless he can demonstrate himself fully in command of a particular area, he will not attract a mate. In a word, it was Howard's conclusion that the establishment of territory is an essential preliminary to mating and breeding.

Llewellyn Evans, who was perhaps the first to conduct a field study on territoriality of lizards, said of Howard's book that it had "probably done more than any other publication to stimulate interest in the subject of territoriality in higher vertebrates."[6]

What does territoriality mean? Students of behavior have had their own intellectual territorial struggle over the preferred meaning of the term. For example, Harris, whose observations on the territoriality of the rainbow lizard will be described in the next chapter, finds an early definition of *territory* by Tinbergen (1936) too narrow—namely, that territory is "an area which is defended by a fighting bird shortly before and during the formation of the sexual bond."[7] This definition would not apply to animals such as many lizards that have no particular sexual bond. Harris expresses a preference for Lack's definition (1939) of territory as "an isolated area defended by one individual of a species, or by a breeding pair against intruders of the same species, and in which the owner of the territory makes itself conspicuous."[8] In subsequent illustrations, we shall find that a lizard's territory, in addition to a preferred homesite, includes several places to which it can escape from predators, as well as a number of display posts from which it can declare its territorial ownership. Fundamentally, also, the territory provides a "core area" for obtaining food.

Home Range

The home range is an ill-defined area extending beyond an animal's territory. As will be evident in subsequent illustrations, the home range provides an extended area for foraging, hunting, obtaining water, and other needs. It differs distinctly from territory insofar as the animal usually does not defend any part of it. It is an area that provides less protection either because it is more open or because an animal foraging there is less familiar with places to which it might escape from predators.

The Territorial Potential

Since some animals live in areas where not even their home ranges overlap, it is sometimes concluded that they are not territorial. By the same token it may be assumed that such animals are not communal. If observed living in large untilled or desert areas where they are widely dispersed, the large Mexican black iguana (*Ctenosaura pectinata*) would give the impression of being nonterritorial and noncommunal. But as Llewellyn Evans has observed in his remarkable field study of these animals, the fortuitous combination of (1) the security of rock walls for dwelling places and (2) the bountiful supply of vegetables in a nearby field often brings these animals together in a communal situation. Under these conditions they become overtly territorial. This kind of "overnight" switch from a nonterritorial to a territorial animal is relevant to so many species including *Homo sapiens* that the present general remarks about territoriality should be illustrated by Evans's observations.[9]

Evans studied a colony of 22 lizards that was located on the wall of a cemetery near the village of Acapancingo, Mexico. Figure 6-2 shows his sketch of the wall surrounding the cemetery. As evident from Figure 6-3, the large egg-shaped boulders provided choice nooks and crannies for shelters and places to escape from predators. Evans observed that a social hierarchy developed among eight territorial males that established themselves on the north corner of the cemetery wall (upper right-hand corner in Figure 6-2) overlooking a bean field and an irrigation ditch. Evans identified them by their broken or regenerated tails and distinctive markings on the body. Each adult lizard tended to stay in the vicinity

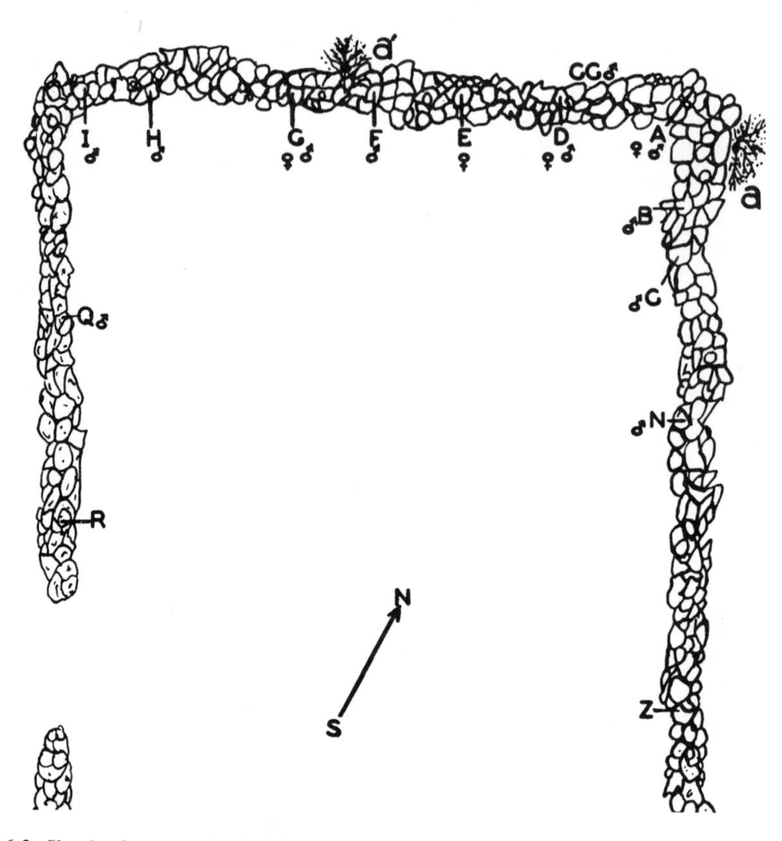

Figure 6-2. Sketch of stone wall of a Mexican cemetery where Llewellyn Evans observed the territorial and other social interactions of a group of black lizards (*Ctenosaura pectinata*). A social hierarchy existed among eight males at the north corner of the cemetery. The lines leading from the letters A to G identify their preferred basking rocks and lookout sites. The "tyrant" occupied the topmost rock at the corner (A), which was nearest the bean field and water supply (an irrigation ditch). The male identified as CC appeared on the scene after C was killed by a dog. See Figure 6-3 and text for further details. Redrawn from part of a figure from Evans (1951).

of a particular stone that served as a place for sunning and as a lookout site. The alpha male (identified by the letter A) held the corner nearest to the bean field. In the following account Evans refers to him as the tyrant. "The highest ranking male," Evans explained,

> preempted the right to trespass upon the footage of any other male that dwelled upon the wall. If he encountered any sign of opposition as he crossed a territorial boundary, he merely opened his jaws threateningly and passed on while his lesser rivals crawled down into the crevices until he had passed. The "tyrant's" nearest neighbors on the wall possessed the same right of trespass but to a very much more limited degree. They never passed over the "tyrant's" holding, which was located at the highest point of the wall, at the north corner, even though it was closest to the food supply.
>
> All members of the colony fed, unmolested by the "tyrant," upon bean seedlings in the nearby field and drank from the nearby stream. Each individual actually possessed no more "territory" than the narrow strip of wall that extended halfway between his lookout rock and that of his neighbors on either side of him. This small footage was defended against encroachment by all except the "tyrant" himself, and on rare occasions by a male who held a footage on the wall next to that of the "tyrant." All lesser males in the hierarchy respected one another's territorial rights and were never observed to trespass.[10]

Figure 6-3. A close-up of the north corner of the cemetery wall diagrammed in Figure 6-2. Numerals 1 to 5 identify the lizards referred to in Figure 6-2 as male A, male CC, male D, female A, and male B. As the "tyrant" (No. 1) withdraws into a crevice, the newcomer CC (No. 2) asserts himself by temporarily taking over the former's rock located topmost at the corner. See text regarding the territorial rule of the wall. The tyrant made a "daily 'tour of inspection.'" Drawing by Robert Gelhard of a photograph from Evans (1951).

Throughout the period of observations during the late winter and spring Evans observed that only the tyrant regularly threatened the others and preempted the right to temporary trespass on their resting areas. Evans points out that in *patrolling* the wall the tyrant was responding "in the ancestral way," doing what would have been his habit had he lived under isolated conditions in a desert area. By yielding to the tyrant, Evans explained, the lesser members gained the security of the rock wall as dwelling places and enjoyed the bountiful food in the neighboring bean field.

From the standpoint of the defense of territory and its relevance to "nativism," it is of interest that if the "tyrant" was seriously challenged by a powerful stranger, other high-ranking males also entered the fray. Evans had made similar observations on the anolis lizard. "It often happened," he observed, "that if a strange male was placed in a cage containing several males, the dominant male of the group would, of course, quickly begin to strut and to challenge the newcomer. However, one or two other large males would also strut and challenge the stranger. . . ."[11]

An Introductory Outline of the Daily Life of the Lizard

Before proceeding in the next two chapters to describe field studies on two representative types of lizards, it will be helpful for seeing relationships to be alerted to features of the two main aspects of lacertilian behavioral profile that, as said earlier, might be compared to two ranges of mountain peaks seen from a distance. First to be considered is the chain of activities in an animal's daily master routine and subroutines. With respect to the daily master routine, one can recognize over and over again an orderly succession of activities recognized as (1) a slow, cautious emergence in the morning; (2) a preliminary period of basking followed by (3) defecation; (4) local foraging; (5) an inactive period followed by (6) foraging farther afield; (7) return to the shelter area; and (8) retirement to a shelter or roost.

It has turned out that for different species of lizards the daily routine within a structured quasi-natural environment in the laboratory is similar to what takes place in the field. For the purpose of conducting systematic observations, focusing on particular details, and obtaining quantifiable data, a laboratory setting provides many advantages over conditions in the field. Here I shall summarize the findings in one of the few laboratory studies in which the investigator followed the activity of a group of animals around the clock, including what took place in their shelters. For his dissertation on the behavior of lizards, Neil Greenberg compared the behavior of blue spiny lizards (*Sceloporus cyanogenys*) in their natural habitat and in a confined quasi-natural habitat.[12] The genus name of these animals refers to their large femoral pores, and the species name to their blue color. The body of one of these animals would fit snugly into one's hand. The tail is about the same length. The females have a creamy white ventral surface and a dull green back, while the males have a blue-green back and blue patches on either side of their chest and belly.

The Daily Routine

Behavior in Shelter

Greenberg devised a ''column'' for watching lizards in their shelters without their being aware of him.[13] He was also able to monitor the body temperature of a number of his lizards.[14] As typical of many lizards, the blue spiny lizard has a concealed, favorite place to ''sleep.'' Greenberg found that there was often a relatively slow transition between the waking and ''sleeping'' state, both upon retiring at night and upon emerging in the morning. In the halfway state, a lizard assumed a posture suggestive of basking. During what appeared to be a state of sleep, it is found in the narrowest part of its shelter with its head wedged into a crevice. The question as to what arouses a lizard to morning activity is still unsettled. Primary emphasis was formerly placed on photic and thermal factors. Heath, however, observed that horned lizards (*Phrynosoma*) would emerge before artificial sunrise, and postulated that their arousal and preemergence activity depended upon an internal clock.[15] In Greenberg's study the time of emergence from the shelter appeared to be cued by light.

Emergence, Basking, and Defecation

In the morning when emerging from its shelter, a blue spiny lizard inches its way slowly and cautiously as though expecting at any moment to be seized by a predator. It proceeds to a preferred basking site where it adopts postures that maximize absorption of heat from the substrate or the rays from the artificial sun. Once its body has warmed to a near-optimum temperature, its next act is to empty its cloaca in an accustomed place near the basking site. In other words, like many kinds of mammals, it has a defecation post.

Drinking and Feeding

After defecation a blue spiny starts on its way to a favored perching place, pausing perhaps to take a drink of water. Like some other iguanids it is attracted only to moving water. Having attained its perch, it performs a brief signature display and then assumes a sit-and-wait posture, scanning the area for any moving prey. Its appearance is not so unlike that of a fisherman waiting for a fish to strike. Among both males and females, the peak of feeding activity is around the middle of the day. After an "anchored" sit-and-wait period of feeding, there is a period of afternoon inactivity. Then the lizards commonly drift farther afield and actively forage until their appetites are satisfied.

When going after elusive prey such as crickets, blue spiny lizards will approach stealthily with intervening pauses, recalling the stalking behavior of a giant Komodo dragon to be described in Chapter 8. Small prey are swallowed whole. Other prey too large for one gulp are shaken into pieces. Neighboring feeders may facilitate the process by pulling on the prey and attempting to steal it. Successful foraging by one lizard may attract others to feed in the same area. Some psychologists refer to the process that sets into motion such "isopraxic" activity (doing the same kind of thing) as "social facilitation."

Return to Shelter

As the day wears away, the females begin to return to their favored places in the shelter. Then with eyes closed they gradually settle down for the night with their heads pressed into a crevice. The males eventually do the same, but first there often seems to be a need for further basking, with the dominant lizard soaking up the warmth longer than all the rest. As with the blue spiny lizard, I have seen it happen over and over again in quasi-natural habitats that the dominant male in respective groups of dipsosaurus or rainbow lizards is always the last to be seen basking.

Night falls, morning breaks, and yesterday's cycle repeats itself—slow emergence, basking, defecation near the same spot, perching, watch-and-wait feeding, foraging, returning to the shelter. It should be recalled at this point that lizards are ectothermic animals, and must depend on both an external source of heat and exercise in order to maintain a desired temperature for engaging in their various activities. Somewhat analogous to a thermostat in a house, they appear to have a high and a low setting for maintaining the body at an optimum temperature for satisfying their various needs. The "selected" range of temperature is referred to as the *eccritic* temperature. Figure 6-4

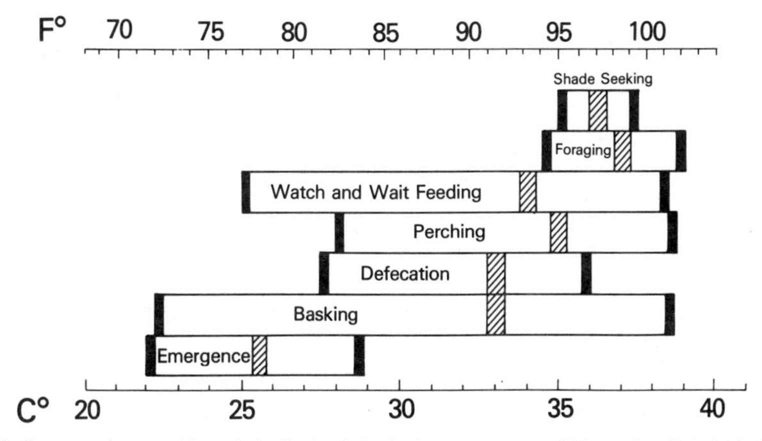

Figure 6-4. Range and mean (slanted shading) of the body temperature of blue spiny lizards during seven different kinds of activity. The figure is based on studies of Greenberg (1973, 1976) in which he used thermal radiotransmitters to record the body temperature of free-ranging lizards.

gives a composite of Greenberg's findings, showing the range and mean temperatures of his blue spiny lizards during their daily activities.

Communication by Displays

Except for feeding, we have mentioned no social interactions of the blue spiny lizard. In considering next their manner of species-typical, prosematic communication by different displays, we shall be dealing with a subject of central interest in connection with the experimental work to be described in Chapters 11 through 14. One might otherwise say that displays are part of a "body language" by which lizards communicate with one another. As already mentioned, there are four basic types of displays, and of these there are many variations among different species. The four main displays are easiest to identify in the visually oriented lizards. The blue spiny lizard under consideration provides an excellent example for illustrating the four basic types because its display patterns are so simple. It is now generally accepted that differences in the ways displays are performed provide lizards the means of recognizing members of their own species (conspecific recognition) and of differentiating between sexes. Carpenter appears to have been the first to attempt a detailed analysis of various lacertilian displays. He referred to them as display action patterns (DAPs).[16] For the blue spiny lizard we shall use Greenberg's analysis,[13] dealing successively with the (1) signature ("assertive"); (2) challenge ("territorial"); (3) courtship; and (4) submissive ("appeasement," "assentive") displays. Some of the displays are characterized by both dynamic and static modifiers, while others are manifest chiefly by dynamic components.[17]

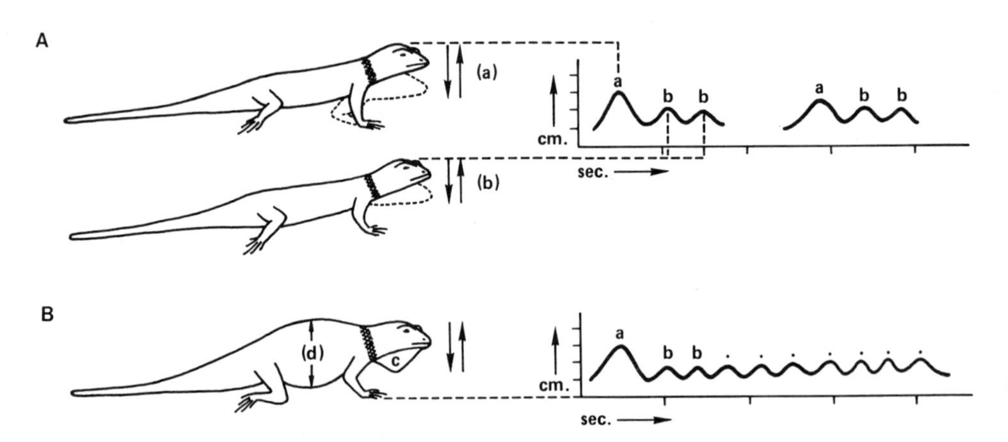

Figure 6-5. Features of the signature display (A) and challenge display (B) of the blue spiny lizard.

(A) The signature display consists only of dynamic components—a single pushup produced by flexion of the forelimbs (a), followed by two head bobs (b) (also called nods). The signature display may also be referred to as an assertive display.

(B) The challenge display includes both dynamic and static modifiers. In addition to an initial pushup (a), followed by several head bobs (b), there are two conspicuous static modifiers—extension of the gular fold (c) and sagittal expansion (d), produced by a side-to-side narrowing of the body.

Redrawn after Greenberg (1973, 1977).

Signature Display

Stamps and Barlow used the expression *signature bob* to refer to the particular head bobbing display of *Anolis aeneus.*[18] The term *signature* seems quite appropriate because it may be used to apply not only to a distinctive species-typical display, but also to individual intraspecific variations of the display. The display in question is also commonly referred to as the *assertion* display, a term introduced by Carpenter.[19] It consists mainly of dynamic modifiers. The signature display of the blue spiny lizard is perfect in its simplicity. As diagrammed in Figure 6-5A, it consists of a single pushup with the upper extremities (a), followed by two head bobs (b). It occurs in a variety of social and nonsocial contexts. It is commonly performed after a lizard moves to a new perch and at times when there may be no other lizards around. Jenssen and Rothblum have compared it to the song of a bird after flying to a new perch.[20] Sometimes its mute performance suggests a similarity to the subdued triumphant vocalization of geese, seeming to signify, "I did it."

In social situations the signature display is typically performed under three conditions. When it occurs upon the meeting of two or more lizards, it appears to serve as a form of greeting not too unlike the gestures used by human beings. It is usually the initial expression of a territorial male when encountering an intruder. And it is commonly the first overture of a male in the act of courtship.

Territorial Displays

If an adult blue spiny male trespasses on the territory of another, the tenant may at first respond with a simple "take-notice" signature display. If the intruder does not retire, the tenant will give him an emphatic warning by performing a challenge display—a term introduced by Carpenter.[19] The challenge display (see Figure 6-5B) includes both dynamic and static components. It begins with a pushup and is followed by as many as 12 head bobs. The static modifiers contribute to an expansion of the body profile effected by an extension of the gular (throat) fold ("c" in Figure 6-5B) and a narrowing of the body ("sagittal expansion" indicated by "d" in Figure 6-5B). The latter posture exposes the blue coloration on its chest and belly. If the intruder still fails to retreat, the territorial male will head for him in a loping gait, and as he draws near will turn the body sidewards somewhat in the manner of a football player attempting to block. Because of the sagittal expansion the sidewards approach serves to give full accent to an apparent increase in size. If the intruder stands its ground, there is a side-to-side "face-off" (Figure 6-6A), followed by nudging, pushing, and tail lashing. As the struggle continues with rapid circling, each contender attempts to pin down the other by a jawclamp on the neck or tail. If the fight is particularly vigorous, one animal may lose its tail. The struggle will go on until one member retreats or assumes a submissive bow characterized by a head down position.

To summarize, territorial displays in blue spiny lizards, as in other animals to be described, are of two main types that hereafter will be identified as "distant" and "close-in" challenge displays. In one species of anolis lizards (*Anolis limifrons*) Hover and Jenssen have identified as many as five gradations of challenge displays[21] that changed from one to another in a systematic way as the distance between the contestants decreased.

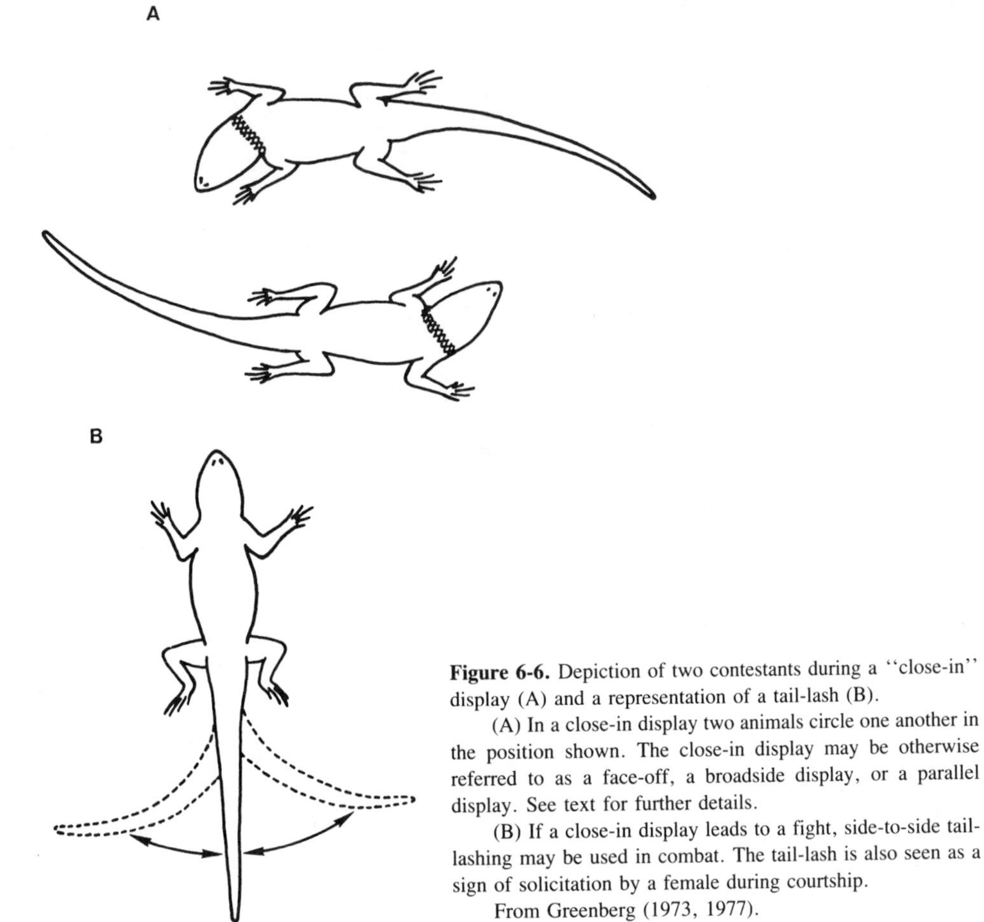

Figure 6-6. Depiction of two contestants during a "close-in" display (A) and a representation of a tail-lash (B).

(A) In a close-in display two animals circle one another in the position shown. The close-in display may be otherwise referred to as a face-off, a broadside display, or a parallel display. See text for further details.

(B) If a close-in display leads to a fight, side-to-side tail-lashing may be used in combat. The tail-lash is also seen as a sign of solicitation by a female during courtship.

From Greenberg (1973, 1977).

Courtship Displays

As applies to a wide range of animals, the courtship display of a blue spiny male lizard has some similarities to the challenge display.[12,13] One sees it performed by a sexually active male when a reproductive female comes within range and signals her interest in him by a swish of her arched tail. In response to her solicitation display (swinging tail in a wide arc, Figure 6-6B), the blue spiny male may perform a signature display and then launch into a full courtship display. The latter is characterized by a pushup and a series of head bobs performed while he lopes toward the female. The loping gait recalls the same kind of approach preceding a close-in challenge display. The courtship situation may also give the impression of a pugnacious encounter, because following the display, the male will aggressively give the female a number of side nudges and attempt to get a grabbing bite on her neck. There may also be a *partial* show of the static modifiers seen in the challenge display. The mating attempt becomes apparent when he wraps one leg around her tail and presses his vent against hers (see Figure 6-7). He then inserts one of his hemipenes. Whether the left or right hemipene is inserted depends upon

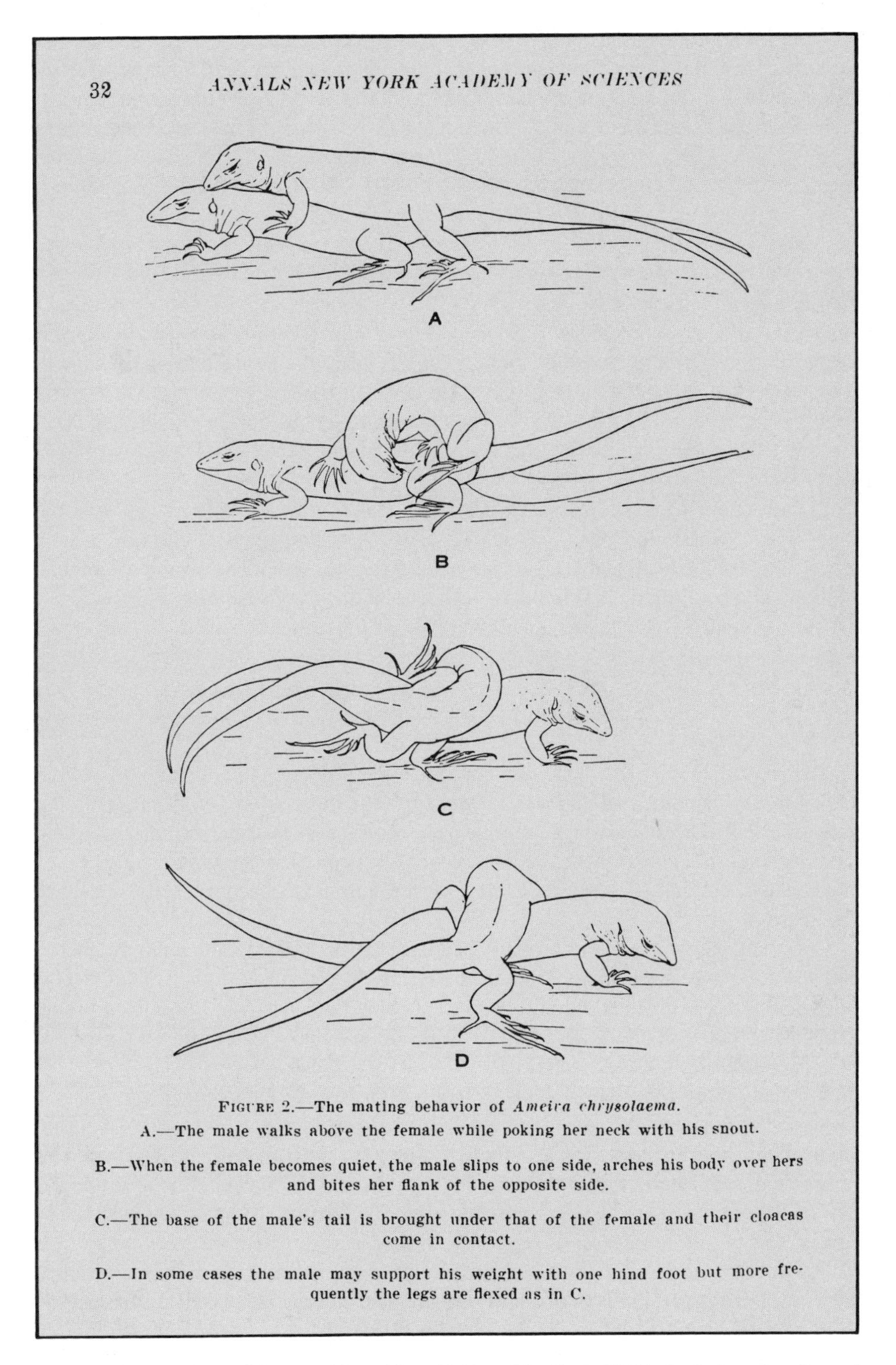
32 ANNALS NEW YORK ACADEMY OF SCIENCES

A

B

C

D

FIGURE 2.—The mating behavior of *Ameiva chrysolaema*.

A.—The male walks above the female while poking her neck with his snout.

B.—When the female becomes quiet, the male slips to one side, arches his body over hers and bites her flank of the opposite side.

C.—The base of the male's tail is brought under that of the female and their cloacas come in contact.

D.—In some cases the male may support his weight with one hind foot but more frequently the legs are flexed as in C.

Figure 6-7. Reproduction of figure and legend from Noble and Bradley's (1933) classical paper on the mating behavior of lizards.

the side of his approach. The copulatory embrace, sometimes accompanied by suggestive thrusts, is completed within a few seconds. Just as the male may drag its vent across the substrate upon approaching the female, it will rub the same region after separating from her, actions that might be variously interpreted as "marking" or "grooming" behavior.[12,13]

Submissive Displays

Because of their protective and survival value, submissive displays are recognized as an indispensable part of an animal's behavioral repertoire. Their value is that in signaling compliance, they serve to forestall, reduce, or terminate the punishing, and potentially deadly, actions of a dominating animal. One finds in the study of lizards that certain aspects of submissive displays are so subtle and inconspicuous that an observer doubtless often overlooks them. The variety, nature, and incidence of submissive displays are in need of systematic study.

In blue spiny lizards, Greenberg describes a *body down–alert* posture, which he equates with "submission" or "subordination."[13] With the legs held close to its sides, the lizard lowers the head and body to rest against the substrate. The posture of a female submitting to a courting male is the same, except that the head is held up. Descriptions of other forms of submissive displays will be given in subsequent chapters.

Summarizing Comment

The present chapter has dealt with introductory considerations regarding the domain and day-to-day activities of lizards, as well as four basic kinds of displays used in communication. The domain encompasses an animal's usual field of operations and includes a homesite, territory, and activity range. It becomes evident from the illustrations that circumstances may determine whether or not an animal becomes territorial in the usual sense.

Observations by Greenberg provide assurance that in a quasi-natural environment, a lizard will follow the same kind of master routine and engage in the same type of displays as are seen in the field. In addition, observations in a quasi-natural environment provide additional information about an animal's behavior within a shelter and about its thermoregulation during different activities.

The daily master routine is characterized by (1) cautious emergence from a shelter; (2) a period of basking, followed by (3) defecation in an accustomed place; (4) morning feeding within the territory; (5) a period of afternoon quietude; (6) afternoon feeding farther afield; (7) late afternoon basking; and (8) return to the shelter. The four principal types of displays are (1) the signature display; (2) territorial displays; (3) courtship displays; and (4) submissive displays.

Given this information about the daily routine and prosematic communication by means of four principal kinds of displays, we will next summarize two field studies—one on the rainbow lizard (*Agama agama*) and the other on the Komodo dragon (*Varanus komodoensis*). The rainbow lizard is representative of lizards that depend to a large extent on visual communication, while the Komodo dragon is illustrative of lizards with keen olfactory and chemical senses. We shall find the same kinds of behavior showing up in these animals as have been briefly described in the present chapter.

References

1. Mayr, 1977, pp. 15–16
2. Tinbergen, 1951, p. 7
3. Smith, 1969
4. Nice, 1933, p. 90
5. Howard, 1920; *see also* Howard, 1929
6. Evans, 1938a, p. 97
7. Quoted by Harris, 1964, p. 109
8. Harris, 1964, p. 109
9. Evans, 1951
10. Evans, 1951, pp. 23–24
11. Evans, 1951, p. 20
12. Greenberg, 1973
13. Greenberg, 1977a
14. Greenberg, 1976
15. J. E. Heath, 1962
16. Carpenter, 1961; 1967
17. Jenssen and Hover, 1976; Jenssen, 1978
18. Stamps and Barlow, 1972
19. Carpenter, 1962
20. Jenssen and Rothblum, 1977
21. Hover and Jenssen, 1976

7

A Day in the Life of a Rainbow Lizard

The scientific examination of behavior tends naturally to identify all the subtle aspects of a particular form of behavior—a "teasing out," as ethologists like to say, of all the various components. Accordingly, in writings on behavior, one finds special sections or chapters with such headings as "maternal behavior," "affiliative behavior," "communicative behavior," and the like. This means that the reader comes away with very little feeling of the day-to-day, around-the-clock activities of the subjects under study. It would compare to a physician's impressions upon visiting a patient in the hospital, as opposed to those of the nurse taking care of the same patient in long stretches. This is not to say, for example, that the recorded observations of many field studies would not allow a piecing together of the daily routine of the animals under investigation.

The tendency to compartmentalize behavior may help to explain why relatively little attention has been given to the special kind of intelligence required to organize and perform the activities in an animal's *daily routine* and *subroutines*—a performance that perhaps, more than any other, is vital to its health and survival. In this respect there comes to mind no neurophysiological study that has specifically addressed this question. As experimental subjects lizards would be ideal animals for study of the daily routine itself because their behavioral sequences are so explicit and predictable. The special problem would be to devise methods of cerebral intervention (see Chapter 11) that would circumvent disturbances in thermoregulation and in feeding that would of themselves interfere with the performance of the daily routine.

In parts of this and the following chapter I adopt a quasi-narrative style because it permits the piecing together of various published observations so as to give a picture of sequences of behavior of the daily routine and subroutines of certain lizards. Otherwise the descriptions adhere to the original observations. Since few people have the opportunity to observe lizards under other than torpid conditions, this manner of presentation may help to imagine them as they exist in nature. For the present account of the behavior of the rainbow lizard, I draw largely from the field studies of Vernon A. Harris as described in his monograph, *The Life of the Rainbow Lizard* (1964),[1] doing so against the background of our own observations on groups of animals of this same species living in habitats partly simulating natural conditions. While at the University of Ibadan in Nigeria, located in West Africa about 60 miles inland from the Bay of the Atlantic and 8° north of the equator, Harris conducted extensive field and laboratory studies on the behavior and physiology of the species of rainbow lizard taxonomically identified as *Agama agama agama*. Our own observations on these animals indicate that the same kind of behaviors develop in a quasi-natural environment. We observed small groups of rainbow lizards in a vivarium with an area of 1 × 3 m and adapted for simulating humidity and the daily trajectory of the sun from sunrise to sunset. It provided areas for basking and places for shelter and display posts. There was an overhead arrangement for supplying food without the intrusion of a person.

The Animal

The rainbow lizard, both physically and behaviorally, is one of the most colorful and beautiful of lizards. To observe the manners, rituals, and combats of these majestic animals, recalls one's readings about the days when knighthood was in flower. While in command of its own territory, the adult male (referred to as the cock lizard) is almost continuously in its reproductive colors of red and blue. The head is coral red, and there is a band of like color around the middle of the tail. The body and the greater part of the tail show gradations from blue to bluish white that from a distance give the appearance of a satiny sheen. The colors are owing to "red," "orange," and "white" cells in the epidermal layers underneath the scales.[1] Below these cells are the "black" cells (melanophores) that under the microscope suggest the configuration of a spider. When the cock is in his bright colors, all of the black pigment is packed within the cell bodies but with nighttime changes in temperature or under conditions of "emotional stress," the pigment moves out into the cells' processes that extend into the color layers, turning the animal's head and body "to a dark chocolate-brown."

The body of an adult male would almost fit into the palm, being about 15 cm (6") in length. The tail is once again as long. The female is smaller and less robust than the male. Her body is fawn with darker brown markings on the back. As Harris states, "Her beauty lies in a conspicuous pattern of pale sea-green spots that adorn her head and neck, and a vivid patch of yellow at the side of her body. . . ."[2] Juvenile males and females have essentially these same markings and color. They have a yellowish patch just behind the "arm-pit."

From Hatching to Maturation

It is said that reptiles come into the world equipped to do everything they have to do except to procreate.[3] The thrust of this statement is illustrated by the behavior of a common anolis lizard right after hatching in a holding tank with many other lizards of the same species. Upon emerging from the shell and seeing the other lizards gazing at him, it gives its characteristic signature display (personal observation). In a moment we'll encounter a like example in Harris's description of the rainbow lizard.

At the time of laying, the egg of a rainbow lizard is about as long and wide as a human fingernail. From the time that it is one of an underground clutch of five or six until hatching, it undergoes a four- to fivefold increase in weight.[1] The increase in size during incubation is possible because the shell is rubbery in consistency and can expand along with the intake of moisture from the soil and the growth of the embryo. Consequently, at the time the hatchling breaks out of the egg, it is only a little less than half the length of the adult animal. "[W]ithin a few moments of hatching," Harris observes, "the young lizards are as active and agile as their parents. They nod their heads in the same characteristic way and will occasionally be seen tilting the head down to pick up something with the short tongue."[4]

Despite their fairly large size, the agama hatchlings, like the young varanid lizards to be mentioned later (Chapter 8), must stay out of sight of the adult lizards (particularly the cock lizard) to avoid being cannibalized. The immature agama lizards spend the first 2 months of their lives living among dense herbage. By marking them, Harris discovered that they usually leave the territory of their parents and lead a wandering existence, seldom spending more than two nights in the same place. Because of their small size, they are not so dependent as adults upon basking. By the age of 4 months, they begin to associate with other juveniles while feeding and may join a territorial group when basking and roosting. Females reach sexual maturity in about 15 months and males in about 2 years.

It is evident from Harris's description that young males have a harder time in making a go of it than the females. This is because they encounter stern competition with territorial males, and must go out on their own to establish a territory if they are to become successful in procreation. Just as the territorial male rainbow calls to mind a knight in a bright tunic and dressed in mail, so one might think of the juvenile male as a young page training for knighthood while attached to the household of some lord, noble, or bishop. Harris's male protagonist lizard about to be described may be pictured at first as belonging to the retinue of a territorial male. In that situation he is with members of the retinue on their daily rounds of basking, foraging, and roosts with them at night. He becomes accustomed to giving the assentive nod whenever the cock lizard appears at one of his display posts and performs the assertive, signature display for all to see. At squire age he acquires a number of secondary sexual characteristics. There is an increase in the width of the base of his tail because of the development of hemipenes. The tail, which will be used as the main weapon in territorial encounters, becomes generally larger because of muscle growth. Orange pigment begins to collect in the superficial layers of the skin of his head and neck, heralding the time when the melanophores will withdraw the dark pigment from their processes and reveal the bright red and blue colors of his full maturation. In the meantime, if he or his companions of a like age were to don their colors prematurely, they would be immediately attacked by the cock lizard and continually harassed until they were driven out. Consequently, it often happens that there are two or three near-mature lizards within the territory of a dominant male living in the disguise of their drab coloration. Unlike what we observed in a small quasi-natural habitat,[5] Harris never saw a subdominant male persistently challenge and displace a cock lizard within its own territory. Rather, it seems to be the rule that a mature male living in disguise will finally set out alone and establish a territory elsewhere.

Harris's protagonist stemmed from a family of lizards that had long since left the country and became established in a lacertilian settlement near the University of Ibadan. In this settlement a cock lizard would have had the choice of trying to find a territory within the most densely crowded area where the population was about 40 lizards per acre, or he could have moved out to the outskirts where there were about 20 lizards per acre. It was usually an impossibility for a knight aspirant to establish himself in the oldest settlement, although by continually harassing one of the older inhabitants he might be able to acquire "half" a territory. The main requirements for a rainbow's territory are a roost for shelter and protection; a number of places for escape and for display posts; a good food supply; and, of prime importance, an area that is inviting to a reproductive female. Harris found that the size of a rainbow lizard's territory "is usually in the order of 600 to 800 square feet."[6]

Harris observed that once a near-mature male leaves the territory of his protector, his movements from place to place may be so frequent that one cannot keep track of him. This appears to have been so in the case of Harris's protagonist. By the time he makes his appearance in Harris's account he is disporting the resplendent colors of his full regalia, and he has already become established on an excellent piece of territory within an ill-defined domain (see Chapter 6) at a farm near the outskirts of town. His territory provides rafters for roosts, a high protective wall for basking, a large woodpile with a number of display posts, and many places for escape. To see him now with his retinue of 11 (a consort, two females-in-waiting, and eight juveniles) makes it difficult to imagine that the time is not far off when he will be fighting desperately to hold on to his territory and consort.

The Daily Routine, Subroutines, and Use of Displays

Let us imagine ourselves accompanying an ethologist, two students, and a photographer as they spend a day in the field recording observations on Harris's protagonist lizard and his retinue. Already on the basis of their hearsay acquaintance with the "red knight," the two students, Earl and Gwen, have dubbed him Sir Red, which they later shorten to Sir R. What we are about to see is pieced together from, and adheres in details to, descriptions in Harris's book.[1] In order to have a complete record of the day's events, we arrive at Sir R's domain before sunrise and wait for him and his retinue to emerge. So as to avoid the "observer effect" we hide ourselves behind a makeshift blind. Following the lead of the ethologist, the two students make a rough sketch of the surroundings and note the time in the column of their protocols reserved for clocking the day's events.

Morning Activities

After what seems a long wait, one lizard (a juvenile) makes its appearance at 6:50 a.m. By half past seven, when the sun has broken through the mist, another seven lizards have appeared and as though keeping an eye out for hawks, cautiously make their way to bask on the east wall of the territory. Subsequently, two more come down from the roost. Curiously, there is no sign of Sir R. With his reputation as a go-getter, one might expect that he would be an early riser. Minutes go by. Presently there is a scuffle and a drab-looking lizard appears. Gwen whispers that perhaps the scuffle signals that Sir R is about to appear. Our eyes remain glued on the site of emergence. Twelve minutes creep by, and our eyes become restless. At that moment Earl calls attention to the basking site where a remarkable change is taking place. The drab-looking lizard of 12 minutes ago is undergoing a transformation; it is Sir R blossoming into his resplendent reproductive colors of red, white, and blue.

The mood of the observers behind the blind is now almost like that of people at court, but not for long. The members of Sir R's retinue go their separate ways for attending to their emunctories. Sir R retires to his usual place on a narrow ledge and defecates over the edge. (Harris found that a number of his lizards were so constant in using the same defecation posts that he could take advantage of this situation for obtaining serial samples of their excreta and learning the composition of their diet.)

Now that the lizards have warmed themselves and attended to their emunctories, they go off to forage nearby in the long grass, feeding mostly on insects. Periodically they stop to bask on a concrete walk. It is almost ten o'clock, and the student Gwen reminds her preceptor that Sir R has not been seen for some time. The cameraman is annoyed because he is all set up in the foraging area for taking close-ups. The ethologist sends Earl to look for Sir R. He reports back that the "red knight" is patrolling his woodpile. Half apologetically the ethologist asks the photographer if he would mind folding up his equipment and following us to the woodpile. Proceeding cautiously, we find Sir R dashing about on his woodpile like a frenzied manager who cannot be in enough places at once. It turns out that he is trying to pursue two of the females while at the same time keeping an eye on another cock lizard that has appeared and posted himself about 30 feet from the woodpile. He alternates between chasing the females through the labyrinthine spaces of the woodpile and going to his topmost display post to transmit warning signals to the intruding male.

Diagnosing the Displays

In order to assure a correct understanding of the signaling in the various social interactions, we take a minute to review Harris's diagrams of the different displays of rainbow lizards. As noted in the preceding chapter, lizards engage in four main types of display: (1) signature; (2) challenge; (3) courtship; and (4) submissive displays. The displays of the rainbow lizard are harder to differentiate than those of the blue spiny lizard that were used as examples. There are two types of signature display. The first type used

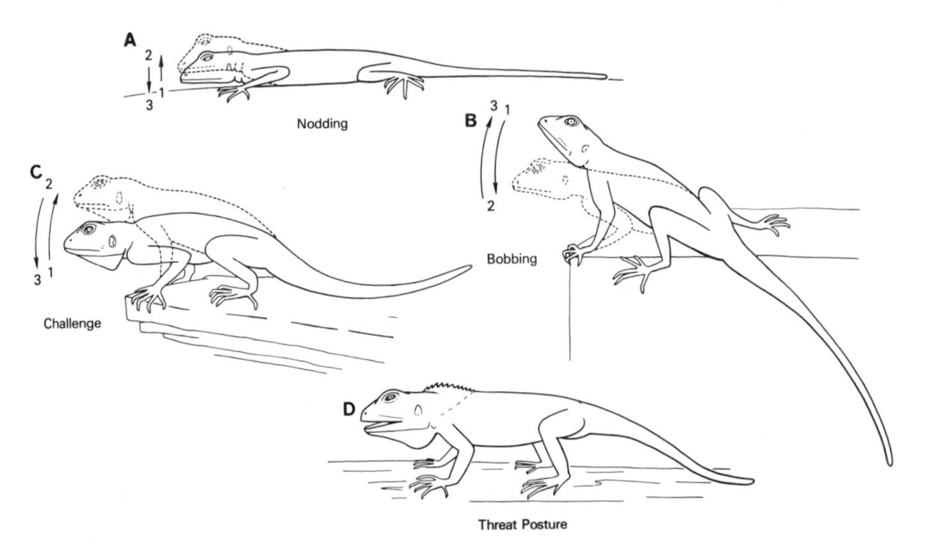

Figure 7-1. Harris's drawings and terminology for some of the displays of the rainbow lizard (*Agama agama agama*).

(A) The nodding display consists of a single up-and-down movement of the head. It is performed by newborn, juvenile, and adult rainbow lizards of both sexes. Depending on the situation, it appears to function as either a "signature" or an "assertive" display.

(B) A territorial rainbow lizard has his own "signature" display beginning with the head held high, followed by a flexion of the forelimbs and a return to the original stance. Harris refers to it as the bobbing display.

(C) As shown by the arrows and numeration, the dynamics of the "distant" challenge display are just the reverse of the bobbing display. In addition, an extension of the gular fold and an erection of the dorsal crest appear as static modifiers.

(D) In the "threat" of the close-in display, all four legs are extended. The close-in display also includes the dynamic and static modifiers of the "distant" challenge display.

Redrawn with minor modifications after Harris (1964).

by all rainbow lizards (newborns, juveniles, females, and males) is performed while the body is prone against the substrate. As illustrated in Figure 7-1A, it is characterized by an initial elevation and then a lowering of the head. Harris refers to it as the nodding display. In the nonsocial context this display conveys somewhat the impression of a punctuation mark, signaling the accomplishment and/or termination of an act, such as having moved to a new perch or having swallowed an insect.

The second type of signature display appears to be reserved for the territorial cock lizard, serving as a kind of privy seal. The pushup features that distinguish it from a challenge display are indicated in Figure 7-1B. In contrast to the challenge display (Figure 7-1C), it begins with the forelegs extended and the head held high. Then there is flexion of the forelimbs and a lowering of the head. The display terminates with a pushup that brings the head again into a high position. The entire sequence is performed somewhat slowly and deliberately. Harris identifies this special type of signature display as the "bobbing" display. The cock lizard performs this display as it patrols its territory and announces its presence from various display posts. The display seems to convey the message, "Take notice, I am here." The other lizards of his retinue respond with the simple nodding display, as though they were standing up to be counted. As will be illustrated later, it is also "usual for the male to give one or two bobs at the beginning of courtship."[7]

Without the differentiating features pointed out by Harris, one might conclude that the pushups used by a cock lizard in its signature and courtship displays were the same as those at the beginning of its "distant" and "close-in" challenge displays. But as diagrammed in Figure 7-1C, the challenge pushups are just the reverse, beginning with the head down and the forelimbs flexed. The rest of the dynamic sequence involves a pushup and a return to the original position. As also depicted in the figure, there are two important static modifiers that distinguish the challenge displays from the signature display—namely, an extension of the gular fold and an erection of the nuchal crest.

Defending Territory

Having reviewed the distinctive elements of the displays, we turn our attention again to the woodpile. The intruding lizard has now approached to within 15 feet of the woodpile, and tension is rising. At this very moment the cameraman with tripod over his shoulder is about to catch up with us. Fearing that the "observer effect" might disrupt the developing conflict, the ethologist signals him to stay back.

The continued approach of the intruder requires that Sir R desist all other activity and elevate himself to the topmost display post for performing a series of distant challenge displays (Figure 7-1C). He presents himself sideways to the intruder, an orientation that allows the display to be seen at the best advantage. But the young intruding cock lizard (later dubbed Sir Nemesis) keeps moving in and replying with his own corresponding display. Finally he crosses a critical threshold that requires the defending knight to take drastic action. Adding fuel to the fire, Sir R's consort, no longer being courted, comes out of the woodpile with one of the females-in-waiting, and both assume a prominent place to observe the combat. Sir R jumps down from the top display post and in a loping gait heads for the intruder to engage in a close-in challenge display. Just as he seems about to run into the intruder, he rises up on all fours (Figure 7-2) and squares off sideways somewhat like a football player in making a block. The whole performance, however, is somewhat awkward, recalling someone walking on stilts. Like all territorial males, he gives further

Figure 7-2. The orientation of two cock rainbow lizards in the close-in display. In this position they circle one another while walking with awkward, stilted steps interlaced with the pushups of the challenge display. Note the extended gular fold, the nuchal crest, and sagittal expansion. If a fight ensues, their bright colors disappear. The tails alone are used as weapons. Redrawn after Harris (1964).

accent to his staccato steps by alternately flexing and extending all four extremities. With his nuchal crest elevated, his gular fold extended, and his body flattened, he looks particularly big and menacing when seen from the side (Figure 7-1D).

The intruder fails to stand up to him and a chase ensues. Having seemingly dispensed of him, Sir R returns to his top display post and proclaims his victory by three head bobs. But peace is not for long. The intruder soon returns and again Sir R repeats his warning challenge. The situation once more calls for a face-off (Figure 7-2), in which the contestants perform repeated challenge displays with the whole body going up and down in the manner described. This time the intruder stands firm, and a fight results. The two go in circles while sidestepping, displaying full-body pushups, and jousting and lashing with their tails. There is no attempt at biting; only the tails are used as bludgeons. If too intent on watching the fight, one might fail to note that the combatants' bright red and blue

Figure 7-3. Mating posture of a female rainbow lizard. She raises her tail vertically and presents her rear quarters. See text for further details. Redrawn after Harris (1964).

colors had dissolved away and were replaced by a muddy brown that characterized the protagonist's appearance when first seen this morning. Suddenly the circling stops, the intruder dashes off, and Sir R chases after him. There will be no more fighting this morning. On his way back to the woodpile Sir R "casually picks up an ant or two" and his bright colors return.[1] He climbs to his topmost display post and again gives three bobs of triumph. His consort and attendant have in the meantime disappeared, and he begins to search for them in the woodpile.

Courtship

During the seeming pause in activity, the ethologist beckons the scowling cameraman to come forward. But suddenly again he must give him the signal, "Wait, stay where you are." The consort has appeared on the apron in front of the woodpile, and Sir R soon creeps out from a space between two logs. The consort arches her back with her four legs extended and her tail held in a vertical position (see Figure 7-3). This posture also reveals the yellow patch on her belly. Then she further accents her receptiveness by directing her head downward and expanding her body by inflation of the lungs. "This posturing is accompanied by turning movements, first to the right and then the left, performed with stiff staccato steps by the forefeet."[8] Attracted by this enticement display, Sir R responds with a simple down and up bob of the head and a pushup in the manner already described (Figure 7-1B). He then moves toward her with the same kind of loping gait that he used in approaching his rival. At this point, with her tail still erect, she turns her rear quarters toward him (Figure 7-3). Sir R grips the side of her neck with his teeth and wraps one leg around her tail, bringing his vent up underneath and pressing it against hers. Since on this occasion he has gripped her tail with his right leg, it will be the right hemipene that is inserted into her cloaca.

Harris has observed that the entire act of copulation may be completed within 30 to 90 sec.[1] The sequences just mentioned—the neck biting, leg embrace, approximation of the vents, and insertion of the hemipene closest to the female—are essentially the same in all lizards. Throughout this time the red knight *retains his bright reproductive colors.*

The Rest of the Day

It was ten o'clock when we left the rest of the retinue foraging in the outlying territory. It is now noon and as we return to continue our observations, we find no sign of life or activity. The lizards have retreated to favored hiding places out of the sun and are having a midday siesta. Toward 3 p.m., as the day beings to cool, they come out of their seclusion and roam farther afield than during their morning feeding. Like human shoppers who could not find everything they wanted at the local stores, they may travel some distance from the home territory to look for special delicacies. Then as the shadows begin to lengthen, they commence to return toward home. We check the time, and find it is half past five. Some linger to bask on the concrete walk. As the shadows creep across the walk, the members of Sir R's retinue climb to the top of the wall not far from the roost to sop up the last rays of the sun, filling up their heat tanks for the cool night ahead.

Now, to everyone's surprise, instead of proceeding directly to their roost underneath the roof, one by one the lizards descend to the ground. Since it is getting dark, it would be unlikely that they were looking for more food. Rather, there is some other singleness of purpose: one by one they walk across the open ground and start to make a difficult climb up a vertical supporting beam. Some jutting wooden ledges make the ascent as difficult for them as overhangs do for a mountain climber. Over and over again a number of them lose footing and fall to the ground. "Why," the students ask, "do they go to their roost by this roundabout and difficult route when all they would need to do is crawl directly to it from their basking site on the top of the wall?" The ethologist offers Harris's explanation. Presumably at some earlier time one older member of the group took this roundabout, but evidently safe route, and the others learned to follow.[1] This observation illustrates at a primitive level how an odd precedent may result in the adoption of a custom. It also serves as an example of a subroutine occurring within the daily master routine.

Just as this morning Gwen had called attention to Sir R's disappearance, she now points out again that there has been no sign of him. Tiptoeing away she finds him at another basking site. We learn from Harris's account that he is always the last to retire. It is relevant to note here that in our laboratory we have observed over and over again that the cock rainbow lizard living with a group in a quasi-natural environment is always the last to retire. As the artificial sun was setting, the cock lizard would be seen alone at his favorite basking site before crawling into a preferred nook within one of the cinder-blocks.[9]

Fighting among Females

After observing Sir R's skirmishes with the intruding lizard, the ethologist and his students began a discussion of how the conflict might be resolved. Before dealing with that question, it should be noted that fighting is not confined to males. Females also have altercations among themselves, but, according to Harris,[1] they do not fight over territory. Rather, it is usual to see them suddenly turn on one another in a feeding patch or when competing for the same basking site. There is head bobbing and extension of a small gular fold, together with a "squaring off" with their four legs extended. Unlike the males, they do not use their tails as bludgeons, but rather try to bite one another. Such sparring may conclude with one chasing the other.

The Denouement

It was the early part of February when we visited Sir R's domain. The domain, it will be recalled, comprises a homesite, a territory, and an indefinite extension of ground sometimes referred to as the home range or activity range (see Chapters 6 and 8). We learn from Harris that toward the end of February, fights were becoming common between the protagonist and his territorial neighbor, which the students had aptly named Sir Nemesis. The latter, however, "was always chased back to his own territory before he reached the woodpile."[10] By early March Sir Nemesis was becoming more persistent, with chases now occurring in the woodpile and fights in the open. Harris's account of the happenings on March 7 may be paraphrased as follows: On two occasions between 10 and 11 a.m. Sir Nemesis climbed to the *top* of the woodpile and displayed his orange and blue colorations—thus demonstrating that he was now master of the pile. Fighting between him and Sir R continued, but at a more frenzied pitch. Sir R would close in for combat without any of the normal, preliminary challenge displays. Curiously enough, Sir R would break off from combat in order to chase and court a female.

The usurpation of the top display post by Sir Nemesis was a sign to the entire retinue that Sir R was disestablished once and for all. The females and juveniles thereafter allied themselves to the new master. Less than a week later Sir R's consort was seen mating with Sir Nemesis. The latter had left his previous mate on his old territory. In Harris's experience, reproductive agama lizards rarely maintain a bond for longer than 1 year.[1]

On two occasions in our laboratory we have seen two territorial rainbow lizards goaded into fighting by younger males which made their initial challenges before betraying their reproductive colors.[5] The fatal outcome of these encounters will be described in Chapter 16.

Summarizing Comment

In day-long observations of Old World agamid lizards one finds the same master routine as in New World iguanid lizards—(1) cautious emergence in the morning, (2) basking, (3) defecation, (4) hunting within or close to the territory in the morning, (5) a period of inactivity, (6) hunting farther afield in the afternoon, (7) late afternoon basking near the shelter, (8) retirement to the roost. In addition, it has become evident that structured within the master routine there may be subroutines such as was illustrated by the ritualized, roundabout route that was taken from the basking site to the roost by most of Sir R's retinue. Finally, one finds that the agamid lizard engages in the same four kinds of displays that are typical of other lizards—namely, signature, challenge, courtship, and submissive displays. Apropos of the present observations it is of special interest to keep in mind the stilted, broadside parallel territorial display, which will be encountered again in the next chapter on the Komodo dragon and finally in the discussion of comparative aspects of displays in species as diverse as rodents and anthropoid apes (Chapter 16).

References

1. Harris, 1964
2. Harris, 1964, p. 34
3. Bellairs, 1970
4. Harris, 1964, p. 86
5. MacLean, 1978c
6. Harris, 1964, p. 154
7. Harris, 1964, p. 92
8. Harris, 1964, p. 105
9. Greenberg and MacLean, unpublished
10. Harris, 1964, p. 131

8

A Week in the Life of a Giant Komodo Dragon

Now that we have obtained an overall picture of the behavior of lizards that communicate largely by visual signals, it is of comparative interest to paint in some highlights of the everyday activities of a form of lizard that depends a great deal on olfactory and chemical communication. For this purpose, we choose the giant Komodo lizard, which, one suspects, might come closer than all the existing reptiles to resembling one of the large carnivorous mammal-like reptiles. The Komodo lizard is a timely choice because in the last few years this species has been extensively studied both in the field and in the laboratory by Walter Auffenberg, of the Florida State Museum, University of Florida.[1] As he has noted, most of what we know about the behavior of lizards has been derived from observations on members of the agamid and iguanid families. The Komodo dragon belongs to the family of varanid lizards that are found in southern Asia, Africa, and Australia, a distribution reminiscent of Gondwanaland.

On the basis of fossils found in Wyoming, it is evident that varanid lizards inhabited North America during the Eocene period, about 60 million years ago.[2] The varanids are familiarly referred to as monitor lizards. The generic name stems from the Latin word *varanus* meaning warning. Unusually wary and cautious, the Nile monitor will escape to a hiding place at the slightest provocation and thereby serves to warn anyone nearby of possible harm or trespass.

The giant Komodo lizard (*Varanus komodoensis* Ouwens) derives its name from the island of Komodo in the Lesser Sunda group belonging to the Republic of Indonesia. *Ora* is the local name for it. The ora was first brought to the attention of the scientific world in 1912 by P. A. Ouwens, a Dutchman, who obtained five specimens.[2] The Komodo dragon is also found on several neighboring smaller islands (principally Padar and Rintja) and along the coast of the large adjacent island of Flores.[3] The islands are almost 1000 km (625 miles) south of the equator, being about equidistant from Borneo to the northwest and Australia to the southeast. The islands are volcanic in origin and have a monsoon forest and savanna. During the dry season in August, surface temperatures may reach 75°C (167°F).[4]

The Animal

The giant lizard may grow up to 3 m in length and weigh as much as 90 kg (10 feet and 200 pounds).[5] Except for a bright yellow eye and bright yellow tongue in a bright red mouth, their color is brown and somber. They have shark-shaped teeth slanting backwards with serrated edges on the posterior aspect. In feats of strength and endurance, the Komodo dragon comes closest of all the lizards to resembling mammals. They are good climbers, effective swimmers, and can run at speeds of 30 km/hr (19 miles/hr).[5] They can run continuously for almost 1 km (0.6 mile), and they have been seen swimming as far as 4 km from shore. In keeping with such feats, the dragon heart has a more complete ventricular septum than any other lizard, approaching in appearance the four-chambered heart of mammals.

Because we are not in the habit of making this comparison, it cannot be overemphasized that a major distinction between most reptiles and mammals is the total lack of parental care of most reptilian species. Hatchling Komodo lizards quickly take to the trees and during their first year of life largely spend an arboreal existence that protects them from being cannibalized by adult monitors.[6] But a threat to their existence by their parents does not begin with hatching. Monitors are also known to eat their own eggs.[7] After young oras resume a terrestrial life, they may roam for several years before they establish themselves in what Auffenberg refers to as a "core area" with a surrounding "activity range."

Up until now we have dealt with lizards as though their lives could be illustrated according to the principles of dramatic composition, with the action taking place in the same location on a single day. In reading Auffenberg's account of the activities of a giant dragon identified as "Number 19W," the weekly cycle of the book of Genesis might be a more appropriate time base for describing this animal. In one of his hunting forays Number 19W was tracked for a period of 6 days before returning to his core area. Since more information exists about this particular dragon than any other, he will occupy center stage during the rest of the description of the behavior of giant Komodo lizards. As an aid to remembering his code number (No. 19W), he will be referred to as Tennine.

Field Observations on "Tennine"

When first caught for tagging, Tennine was a full-grown adult about 2.6 m (8½ feet) in length and weighing 39 kg (85 lb). He was the dominant male on the east side of the Leong Valley of Komodo and was "probably the sire of hundreds of skittering young."[4] Examined again after 8 months he had grown to 2.7 m (9 feet) and weighed over 45 kg (100 lb). "He could pull a 300-pound deer to the ground and swallow the entire head of a dead wild boar."[4] Figure 8-1 shows his dragonlike appearance as he charges forward like a steamroller.

Homesite

In his core area Tennine had scooped out a homesite under bushes at the top of the hill. The "foxhole" was about 1.8 m long and 0.6 m wide. In addition he had several other similar resting spots that afforded exposure to a good breeze when temperatures might reach as high as 50°C (122°F). When not off on a hunting expedition, one might expect to find Tennine spending the night at his homesite lying prone against the ground with his eyes closed.

Auffenberg believes that Komodo dragons establish a pair bond. When this occurs, however, they seem to keep their distance, preferring, so to speak, to sleep in separate

Figure 8-1. A giant Komodo dragon extensively studied by Auffenberg on the island of Komodo. Identified as No. 19W, he is referred to in the present text by the mnemonic name Tennine. Note the huge forked tongue with which he samples the substrate as he lumbers forward, kicking up the dust. From Auffenberg (1970).

rooms. Yet there is considerable overlap of their core areas. Tennine's core area overlapped approximately 80% of that of his preferred female No. 23W, hereafter identified simply as Twothree.

Territoriality

After an indefinite number of years, adult oras become residents of core areas. These core areas appear to be equivalent to the "territories" of other lizards described in the preceding chapters, and are encompassed by an activity range. Together, the homesite, core area, and activity range comprise the domain of a Komodo dragon.

Question of Marking

In previous illustrations of the behavior of blue spiny lizards, it was mentioned that these animals wipe their muzzles after feeding and drag the vent during courtship, but whether or not these same acts are used for marking a territory is not known. Before discussing this question in regard to monitors it should be noted that these animals, like the ground-dwelling teiids of the western hemisphere, have a well-developed olfactory apparatus, including the accessory olfactory bulbs associated with the vomeronasal organ (see Chapter 18 regarding neuroanatomy). The vomeronasal organ found in lizards and snakes is located forward in the roof of the mouth and is accessible to the tip of the tongue used for probing the substrate. The monitor has a huge forked tongue that it uses in exploring an area (see Figure 8-1). As the monitor moves along, it samples the ground alternately with the left and right fork of the tongue in a fashion recalling a blind person using a cane. The tongue is also constantly in motion during feeding and mating. Possessing such an antenna, in conjunction with a keen olfactory sense, the Komodo dragon is well equipped for detecting chemical and odorous markings of its environment.

Based on extensive observations, Auffenberg has concluded that an ora deposits fecal pellets as a means of marking its core area.[8] The whitish accumulations of these pellets near the lizard's outlook post serve as visual as well as olfactory markings. Pellets are also deposited along paths, and transient lizards show a great deal of interest in them, sampling them repeatedly with the tongue. Young monitors may perform an appeasement display after having sampled one. The appeasement display of a young lizard has many similarities to the aggressive display to be described later. Such a display is understandable in the light of observations that adult males have been seen to attack and devour young monitors without provocation. Auffenberg recorded two such instances in the case of Tennine.[9]

Auffenberg believes that fecal pellets also serve other communicative functions. The animals, he states, express "so much interest in the pellets that I am led to the conclusion that they constitute an important source of information regarding the sex, age, and breeding condition of another lizard."[8]

Question of Territorial Defense

It has not been established whether or not resident monitors will defend their core areas, but there is much evidence that adult males, at least, display aggressively to one another and if provoked to fighting, may inflict severe wounds with their sickle-bar teeth. We learn from Auffenberg's descriptions what the situation would probably be if a resident ora such as Tennine were to cope with an intruder. There are gradations in the intensity of the threat postures and displays. In some cases it would appear as though the show of a gaping mouth (Figure 8-2A) was enough to make an intruder back off. Unless it was pointed out, one might fail to note that there was also a simultaneous bowing of the tip of the tail (Figure 8-2B). If the intruder stands his ground, the defender will adopt a posture known as *Schrägstellung,* a term referring to a sideways posturing of the body so that the maximum surface shows in profile.[10] Except for the absence of head bobbing and pushups, this sideward display is very similar to that of the iguanids and agamids (Figure 8-2C). There is side-to-side narrowing of the body (sagittal expansion), erection of a roach (crest), an expansion of the gular (throat) region, and a stiffening of the tail. The elevated roach extends from the back of the neck to half way down the trunk. The inflated appearance of the throat is owing to an extension of the hyoid apparatus. Upon gular expansion, the monitor begins the now familiar stilted stepping of the close-in display that will be referred to again when we consider the broadside displays of mammals, including the gorilla. As Auffenberg describes it, the monitor adopts a "slow stiff-legged, stereotyped walk" (Figure 8-2C).[10]

The combination of an S-shaped curve of the neck and bowing of the tail, together with *hissing,* may signal impending attack. There is a lowering of the head, somewhat like a bull ready to charge, associated with an elevation and bowing of the tail. (The head-

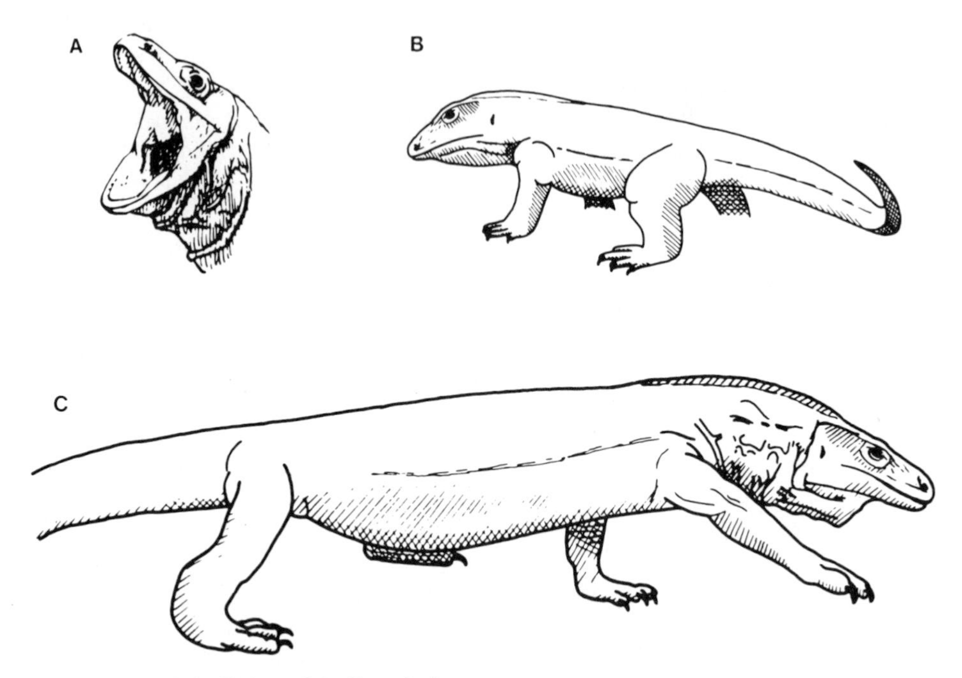

Figure 8-2. Agonistic displays of the Komodo dragon.

(A) Adult dragon threatening with wide gape.

(B) Bowing of tail seen as part of a threat or when about to strike.

(C) The close-in display of an adult is similar to the appeasement display of the juvenile shown here. The animal walks slowly in a stiff-legged, stilted manner. The angle of the right forelimb in this picture is reminiscent of the goose step. Note three static modifiers that have been described for other lizards—namely, elevated roach (nuchal and dorsal crests), extension of gular fold, and sagittal expansion.

Redrawn from selection from Auffenberg (1978).

down position, also seen in other ground-dwelling lizards, is said to be the same as the food searching position. One sees a similar head-down position in fighting, charging ducks.) If fighting ensues, there is striking with the tail, but the damage done with this weapon is not so severe as what occurs when rainbow lizards strike one another. Rather it is the monitor's sharp serrated teeth that may inflict near-mortal wounds. Tennine's unscarred hide is evidence that his imposing size has thus far saved him from any violent confrontations.

Without squaring off, oras of all ages seem to have a remarkable capacity to judge the size of another animal. The larger animal usually receives right-of-way.

The Daily Routine

We have already noted how Tennine spends the nights in his burrow on top of the hill lying with his body prone against the ground. On the basis of the previous descriptions of the blue spiny and rainbow lizards, one is prepared to predict his initial routine after the sun rises. Emerging slowly from his shelter, he proceeds to his basking site and warms up for his daily round of activities, the first of which is defecating at an accustomed site.

Most oras spend about half their time hunting within their core areas. Tennine is no exception. Except for prolonged hunting expeditions of the kind next to be described, he sticks to his core area in looking for food. Despite his great appetite he will not pass up even the smallest of grasshoppers. In making the rounds and patrolling his core area, he will check in, very much like a watchman, at every one of his lookout posts. If feeling overheated, he will stop at one of his resting sites where he is assured of shade and a good breeze. With the coming of darkness, he settles down in his accustomed burrow at the top of the hill.

An Unsuccessful Hunting Foray

The Komodo dragon appears to be unique among lizards in having an activity range in which it will go on extensive hunting expeditions and be away from its homesite for as long as 6 days. Thanks to Auffenberg's persistent efforts we are able to follow Tennine for a period of five days when he failed to bring down any of his favorite prey (see below). Figure 8-3 shows a map of his day-to-day progress. On no less than three nights he beds down in places well known to him from previous hunting expeditions. The first two days of hunting were apparently unsuccessful because on the third morning (position F on the map) he appears on the beach looking for dead seafood. On the fourth day he shows up next to a major game trail (point J) where Auffenberg had observed him on several previous expeditions. This time no deer passed by.

On the fifth and last day Auffenberg temporarily lost track of him at the point marked O on the map. When next seen he is at the bower of his consort Twothree in her core area. He gives her a halfhearted greeting display, touches her a few times with his tongue, and makes one abortive mount. Then, he lumbers off to his own burrow.

During Tennine's five-day absence, two young adult suitors had visited Twothree's bower and made amorous advances.[11] She dealt severely with both of them, and one received such a deep and festering wound from her serrated teeth that he was gravely incapacitated.

A Successful Deer Hunt

As familiar to those with a knowledge of *Varanus komodoensis,* the giant Komodo dragon is capable of remarkable feats as a hunter. The rusa deer (*Rusa timorensis*) appear to be its favorite prey. The stags are much larger than the does, "weighing up to 440 pounds."[12] In stalking and lying in ambush for these animals, a monitor demonstrates that it has a detailed knowledge of its own activity range and an excellent sense of time. Although Tennine captured no big game on the five-day foray just described, Auffenberg's research makes it possible to reconstruct what would have been his tactics and manner of killing during a successful deer hunt.[13] The information is invaluable because of the light it sheds on the ethology of deceptive behavior in terrestrial vertebrates, a much neglected subject (see Chapter 10). Auffenberg presents the following scenario for the hunted and hunters:

> . . . deer feed mainly at night on a variety of grasses, leaves and fruits. With the approach of dawn, they choose a resting place, usually in high grass, which serves as cover on a flat-topped ridge above the forested valleys. The deer lie there, awake and ruminating, until late morning

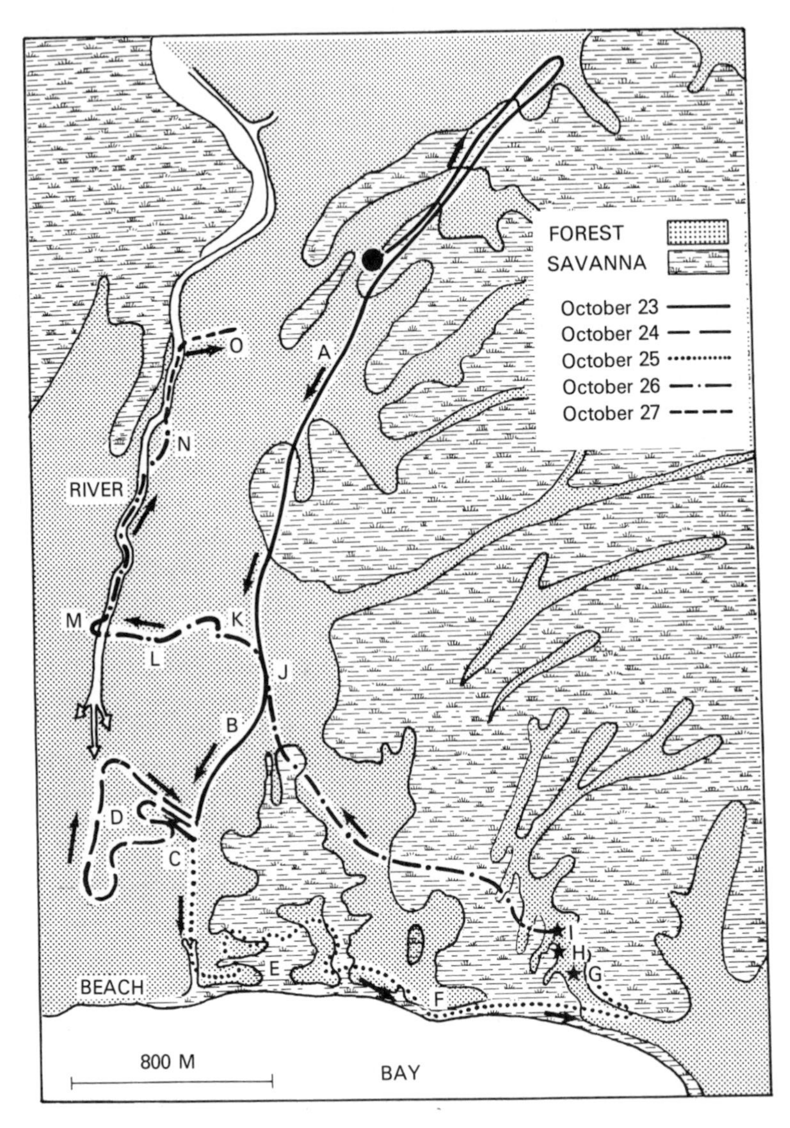

Figure 8-3. Ground covered by Tennine during five-day hunting foray.

October 23: Searching monsoon forest for sleeping deer. Spends night at C.

October 24: Hunting for deer and wild boar. Spends night again at C.

October 25: Searching beach for marine carrion. Spends night at G.

October 26: Looks for jungle fowl at H and I. Later lies in ambush at a familiar spot (J) along a game trail, waiting for deer. Spends night at an accustomed burrow (N).

October 27: Moves into an area known for its high density of deer. Observers lose track of him

Redrawn from Auffenberg (1978).

> when . . . they begin moving down into the thicker vegetation on the valley floors. This is when the monitors do most of their serious hunting, lying in wait along the game trails leading to the valley. Choosing one trail from the several on any slope, a monitor will lie in wait during the late morning hours, expecting a small troop of deer to move past.[12]

We may imagine that Tennine has again taken to the trail, but once more has positioned himself at the wrong spot. With the coming of afternoon and with no deer in sight, he turns off into the thickets and steals to a place where he knows from previous experience they may be resting. He has learned that at such times it may be possible to surprise and capture a sleeping deer. But on this occasion, too, he meets with no success. Being far from his core area, he beds down in an accustomed place for the night. He will try again in the morning when the deer descend to the valley floor.

Although out on the trail, Tennine begins the next day with his usual routine—basking and the emunctories. He proceeds to a favored ambush site at point J on the map in Figure 8-3. Well camouflaged by his own colors and the surrounding herbage, he hides himself about 1 m from the trail where he can lunge forward and grip the deer by one of its hind legs. Afterwards he remains so motionless that one might think that he was asleep or dead. Grasshoppers are around but he ignores these bits of food. At last, there is a rustle some distance up the trail, but Tennine does not hear it because like all Komodo dragons, he is hard of hearing. He may be sensitive, however, to some low-frequency vibrations transmitted through the substrate. Finally, the moment arrives! One deer passes by, and then another, and still another. Suddenly, Tennine lunges forward, grabbing a large stag by the hind leg. In the struggle to extricate itself, the stag pulls against Tennine's serrated teeth and severs several tendons, including the Achilles. Lamed by these lacerations, the stag falls sideways like a chair with a broken leg. With his victim momentarily paralyzed by fright and pain, Tennine lunges at its soft underbelly. Grabbing hold of the erupted intestines, he quickly eviscerates the deer with side-to-side lashing of his powerful jaws. He flails the intestines from side to side to free the contents and quickly gulps them down. As the deer goes into shock, the dragon's jaws plunge like a scooping digger into the thoracic cavity for the heart and lungs. The hunt is over. After consuming the viscera, Tennine settles down to slice into and devour the meat.

Although a giant monitor may show a one-track-mind persistence in pursuing and stalking a rusa deer, it may bend its purpose and go after larger or smaller game. A monitor has been known to attack and kill a water buffalo "weighing as much as 1,300 pounds."[14] It also preys on wild hogs, goats, rats, and jungle fowl.

A Scavenger Hunt

The Komodo dragons are scavengers as well as predators. Thanks to Auffenberg's observations, we are also provided a detailed account of how oras converge on carrion and devour it. On the day of such an occasion, Tennine is to be pictured as making the usual rounds of his core area. Upon reaching the brow of the hill, he suddenly comes to a halt. It is evident that he has picked up a scent. With the help of field glasses, and cued by the direction of his snout, we would size up the situation diagrammed in Figure 8-4. A number of resident dragons in the neighboring hills, as well as some transients, are converging toward a large carrion. Auffenberg describes the precipitating train of events: "During the first day after death, the volatile oils produced by the decomposition of a carcass are not [present] in sufficient amounts to attract any but the very closest lizards. In

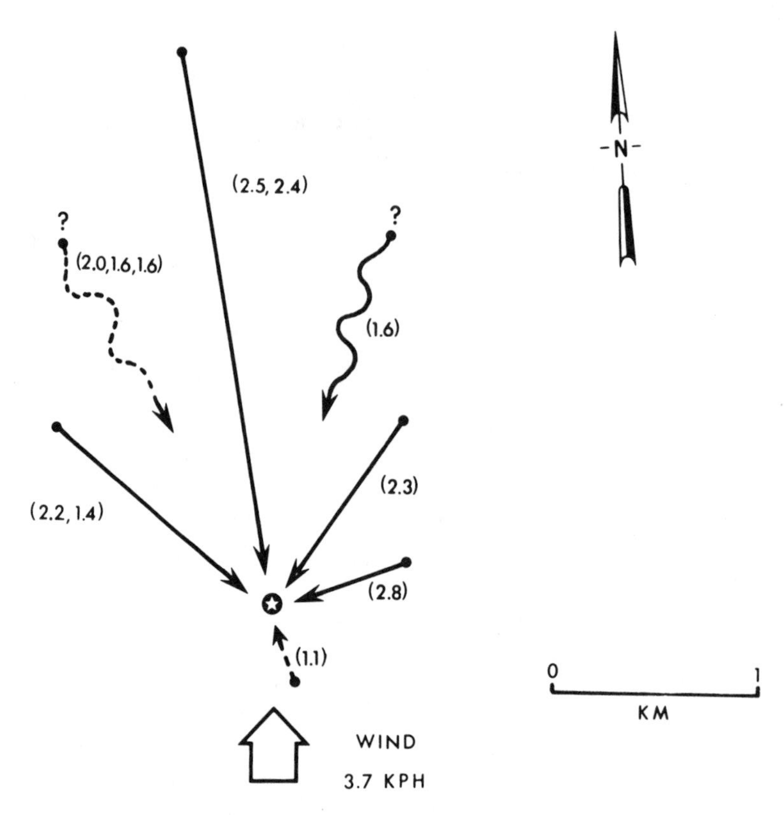

Figure 8-4. Diagram showing the convergence of Komodo dragons on carrion (star), November 21–23, 1969. Numerals in parentheses give the length of the animals in meters. Resident and transient monitors are indicated, respectively, by solid and dashed straight lines. Wavy lines apply to animals coming from unknown distances. Direction and speed of wind shown by large arrowhead. From Auffenberg (1978).

48 hours, however, the decay is sufficiently advanced that considerable scent is produced, and many monitors from over a large area begin to move toward the carcass."[15] The olfactory sense of the Komodo dragon is so sensitive that carrion may be detected as far as 8 km (5 miles) away.

For Tennine, as well as the others, carrion has an irresistible attraction. Without alerting his consort, he sets off immediately in the direction of the carcass. Twothree, picking up the scent later, will follow after. Once at the carcass, the dragons group themselves in a hierarchical arrangement in which size is the determining factor. Tennine and two other large residents of the area are at the center and feed on the choicest parts of the carcass. Next to them are some medium-sized monitors and peripheral to them a number of juveniles. During the rest of the day as many as 17 dragons are seen trying to get near the carcass. The younger dragons alternate between giving appeasement displays and dashing to the carcass at times when the mouths of the giant dragons are full. Auffenberg points out that if they attempt to feed at other times they may be eaten themselves. He emphasizes that at these scavenger gatherings the greater the number of dragons, the greater is the tension. His records show that as the number of lizards increases, there is a disproportionate increase in the number of appeasement displays and tongue flicks.[16]

Courtship at Carrion

In view of the reciprocal innervation of structures involved in oral and sexual functions (see Chapter 19) it is not surprising that the increased level of excitement among a group of monitors around carrion seems to result in a mixture of feeding and sexual activity. Except at times of meeting at carrion, there is little opportunity for close interaction among several monitors. Mating and copulation frequently take place in the vicinity of the carrion, involving either strangers or pairbonded individuals. Shortly after Tennine's consort, Twothree, arrived at the carrion, he made a rapid attempt to mount her. A little later while they were feeding side by side, he momentarily adopted a posture suggestive of a threat display, but under the circumstances it might have been an abortive manifestation of courtship. Between late morning and midafternoon, there were at least eight occasions when Tennine and Twothree alternated between feeding and mating behavior. Mounting sometimes occurred while both were at carrion and at other times at some distant, sequestered spot. As in the case of many long-established couples, the long acquaintance between Tennine and Twothree apparently accounted for the almost total absence of preliminary courtship displays. Rather, what took place were a number of chases between the carrion and the bush. In the words of Auffenberg's protocol: ''He returns to carrion, eats, and suddenly spins around and chases 23 W for 8 m.''[17] In 15 min there is another chase and he catches her. ''As she rests with belly and tail on the ground and head in an attentive pose, he rubs her entire back with the sides of his face.''[16] Then there is a short, complete mounting, after which he ''scratches her back with his claws,'' followed by another complete mounting ''with his front legs clasped around her body.'' Dismounting he nudges her and flicks his tongue in the region of her groin after which he rubs the sides of her face and her nether hip region. The protocol concludes at 3:02 p.m.: ''He scratches her on her hip with his snout, tries to stick his head underneath her belly again, scratches her on the side of her face with his right foot, licks the side of her face a number of times, and then both eat together. . . .''[18]

Successful mating between Tennine and Twothree occurred in late June and again in early October. On one of these occasions after frequent mountings, Twothree was observed to raise the base of her tail, after which Tennine made a rapid tail twist, placed his vent in contact with hers, and ''began a short period of feeble thrusting. They remained in this posture approximately 12 minutes.''[18]

Summarizing Comment

In summary, we find that the routine of a lizard with a well-developed olfactory and chemical sense is in many ways similar to that of lizards that depend largely on visual signals for communication. The Komodo dragon has a domain that includes a homesite, a core area corresponding to territory with several lookout and resting sites, and an activity range for hunting. The homesite is a favored place to spend the night. The giant varanid lizard emerges, basks, defecates, and seeks food in the same order as illustrated for the agamid and iguanid lizards. It differs insofar as that if on a hunting expedition, it may not return home to its core area for several days. It also differs in that its visual displays are less pronounced and in that it appears to establish a stronger pair-bond. Of striking interest is the similarity of the male's broadside display to that of other species of lizards and, as will be described, to various mammalian species, including the gorilla.

References

1. Auffenberg, 1970, 1972, 1978, 1981
2. Bogert, 1954, p. 1315
3. Auffenberg, 1981, Chapter 3
4. Auffenberg, 1970, p. 19
5. Auffenberg, 1972, p. 54; 1978
6. Auffenberg, 1972, p. 54
7. Auffenberg, 1981, Chapter 8
8. Auffenberg, 1978, p. 306
9. Auffenberg, 1978, p. 316
10. Auffenberg, 1978, p. 309; 1981, Chapter 12
11. Auffenberg, 1978, p. 325; personal communication
12. Auffenberg, 1972, p. 55
13. Auffenberg, 1972, p. 55; 1970, p. 20
14. Auffenberg, 1972, p. 56
15. Auffenberg, 1978, p. 316
16. Auffenberg, 1978, p. 310
17. Auffenberg, 1978, p. 326
18. Auffenberg, 1978, p. 328

9

Other *Special* Forms of Basic Behavior

In the last three chapters, the descriptions of the behavior of the black lizard, the blue spiny lizard, the rainbow lizard, and the Komodo dragon have illustrated that their daily lives are characterized by a master routine, along with subroutines, and that four principal kinds of displays are used for prosematic communication.

In using Table 6-1 as a checklist, one finds that up until now we have seen reptilian examples of 20 of the 25 special forms of basic behavior singled out for consideration. At the beginning, particular attention was given to an animal's base of operations referred to as its *domain*. In the process of becoming identified with a particular domain, an animal establishes within it (1) a homesite and (2) territory and, peripheral to these, (3) an ill-defined activity range. Continuing down the list, we note that in the various accounts we have also seen examples of (4) place preference behavior; (5) trail making; (6) marking of territory; (7) patrolling territory; (8) ritualistic displays in defense of territory; (9) formalized intraspecific fighting; (10) triumphful display; (11) behavior signaling submission or surrender; (12) use of defecation posts; (13) foraging; (14) hunting; (15) homing; (16) formation of social groups; (17) formation of social hierarchies; (18) greeting; (19) courtship; and (20) mating.

In the present chapter there are two main purposes in rounding out the picture of special forms of basic behavior. The first is to deal with the remaining five items of Table 6-1 that have been illustrated only in part or not at all—namely, those listed as grooming, breeding, hoarding, flocking, and migration. The second is to point out and discuss some additional aspects of territorial behavior, including the element of *strangeness*. In the next and last of the chapters devoted solely to reptilian behavior, I shall (1) describe six *general* forms of basic behavior and then (2) discuss the question of learning and memory in the light of protomentation.

It is worth noting at this point that in using the expression *basic behavior*, there is no intention to imply that all the forms of basic behavior in question originated with reptiles. Quite to the contrary, one finds many of them represented in invertebrates and "lower" vertebrates, as for example the territorial behavior of the stinging wasp[1]; the broadside display of fishes that heralds a similar display seen in terrestrial vertebrates[2]; and the homing of a tree frog, which may return each night to its homesite on a specific leaf.[3]

Five Other Special Forms of Basic Behavior

"Grooming"

In regard to grooming, nature perhaps devised the best of all methods when it made it possible for lizards and snakes to shed an old skin for a new one. During the period of shedding, a lizard will promote the process by rubbing, scratching, and pulling away the old skin with its mouth. An adult gecko devours its skin as soon as it is shed.[4] In observing green anolis lizards or blue spiny lizards living in experimental habitats, one often sees them help a shedding animal rid itself of the dead skin. As Greenberg has observed, this participation might be regarded as a form of "social grooming."[5]

An unambiguous form of self-grooming is seen when a lizard rubs its nuzzle upon the ground or on some object after feeding[6] or drags its vent over the substrate after defecation.[7] In Chapter 6 it was noted that in courtship the male blue spiny lizard may drag its vent on the substrate during its approach to a female and after copulation. Some observers might interpret this behavior as a kind of pre- and postcopulatory grooming, while others would perhaps regard it as a form of "marking."

Hoarding

Hoarding plays an important role in the lives of many mammals. Familiar examples are provided by hamsters, squirrels, and human beings. I have been unable to uncover any examples of hoarding in lizards or other reptiles with the possible exception of crocodiles. Mr. DePrato of the National Zoological Park has noted that crocodiles may deposit food in the mud and return at some time later to eat it (personal communication). He points out that lizards show a "built-in" type of hoarding insofar as they store fat in their tails.

Breeding

The word *breeding* applies not only to the production of young but also to the rearing of young. Hence, its definition includes parental or foster care. One of the great innovations with the evolution of mammals is the progressive attention and care that they give to their young. In describing the mammal-like reptiles (Chapter 5), it was possible to cite only one piece of evidence suggestive of parental care. As for existing reptiles, the ones that have been dealt with in the preceding chapters show no care of their young. Quite to the contrary, agama and monitor hatchlings must avoid adults (either by keeping in the deep underbrush or taking to the trees) in order to avoid being cannibalized. Females of some species of skink lizards, however, show some maternal interest insofar as they brood their eggs and pay attention to the hatchlings. Llewellyn Evans observed that females of the species *Eumeces obsoletus* regularly turn and clean their eggs with their tongues and, on some occasions, may move them to a more suitable site.[8] As the young are hatching, the mother will help to express them from their shells. Evans obtained motion pictures of a female and her young 10 days after the date of hatching. If the young disappeared, she would go searching for them, and upon finding them, would momentarily lick their vents.

It is of comparative interest that the crocodilians, which have an evolutionary affinity to birds, are the only existing reptiles that show a *prolonged* interest in their young. During the incubation period of 9 to 10 weeks, the female American alligator remains in the vicinity of the nest and will moisten it occasionally with her cloacal excretions.[9] While hatching, the young give a high-pitched grunt. Responding to this sound, the female helps to release the young from the eggs. Afterwards, she will make a small pool for them and guard them from predators, including other alligators. The young may remain with the mother for a year or more. There are accounts of female crocodiles engaging in a similar kind of maternal behavior. The mother may escort the young to the water and, swimming along with them like a duck, drive predators away.[10]

Collective Behavior (e.g., Flocking Behavior)

The expression *collective behavior* is used here to refer to the gathering together of animals (most of them strangers) in greater numbers than one finds in a territorial retinue. Because of the inference that lizards bear a closer resemblance to the mammal-like reptiles than do other existing reptiles, I have, whenever possible, used these animals to illustrate different kinds of reptilian behavior. In dealing next with collective and migratory behavior, however, we find that lizards fall way short in exemplifying the far extreme of these conditions.

As Evans points out, fishes, amphibians, turtles, and snakes are known to congregate during the nonbreeding season.[11] A fall in temperature is one of the factors conducive to such aggregations. Among lizards, those of the family Xantusiidae have separate home ranges during the breeding season, but are seen in aggregations in the colder months when they come into the open on warm winter days. Some of the hibernating snakes form the most cohesive aggregations of all, coming out of their dens in spring in so-called "snake-balls" while copulating.[12]

The largest aggregations of reptiles occur among sea turtles at the time of migrations leading to the deposit of their eggs on beaches. The Ridley turtles (*Lepidochelys olivacea*) of the Pacific Ocean periodically show up in waves of 10 to 30 thousand along beaches stretching from California in the north to Chile in the south.[13] At one beach it was estimated that over 120 thousand turtles nested over a four-day period.

Among lizards, one finds no aggregations that would begin to approach such numbers. Night geckos congregate by the hundreds.[14] Except for their vocalization, little is known about their social behavior. Of all lizards the marine iguana lizards (*Amblyrhynchus cristatus*) of the Galapagos Islands are best known for collecting in large groups. They apparently aggregate in flocks because of the year-round availability of seaweed as a source of food. In his book *Galapagos,* Eibl-Eibesfeldt describes his impressions upon first seeing these "dragons" of the sea:

> [O]n a rocky ridge that pushed out into the sea, there lay hundreds of lizards, about three feet long, basking on the black stones. The beasts were jumbled together, closely crowded in the sun's glow. . . . The tide was running out and as soon as the first seaweed-decked rocks were exposed, the marine iguanas, one after another, forsook their solarium. They slipped into the water [and swam] to the sea-kelp clinging to the rocks. Through my glasses I could see clearly how they browsed; they bit alternately with the right and left sides of their jaws—just like dogs gnawing a bone.[15]

Later we shall see that despite their swarming together on the rocks and beach, there is a territorial order to their society.

It was mentioned above that aggregations of some lizards occur in the nonbreeding season. The nesting season is another time when many members of some species gather together. Evans notes that both gecko and skink females often tend to aggregate at what appear to be favorable nesting sites where egg clusters or nesting cavities may be no more than a few inches apart.[11] Iguanids usually nest singly, but when conditions are favorable, there may be communal nesting (see below). Most of the nests are within an area of 50 m^2.[16]

Migratory Behavior

In collective behavior animals gather together in numbers ranging from small groups to great hordes. The numbers involved in migratory behavior may range from a single individual to very large aggregations. The hatchling rainbow lizards described in Chapter 7 might be characterized as migratory insofar as they are obliged to leave their original surroundings and to forage in the deep underbrush so as to avoid being cannibalized by their elders. In having to fend for themselves, they are hardly better off than the larval amphibians mentioned in Chapter 5. The young adult agama males living in the disguise of their brown colors in the territory of the cock rainbow lizard would qualify as migrants once they leave their group and travel from place to place looking for a site to establish their own territories. All of the transients in the population of Komodo lizards described by Auffenberg (Chapter 8) could likewise be characterized as migratory, since unlike the reproductive oras, they have no fixed core area and are continually on the move.

Among lizards there is nothing comparable to the massive migrations of the Ridley turtles already described or to the green turtles of South America that, with pinpoint navigation, migrate to Ascension Island.[17] In the Panama Canal Zone, Rand and Rand have observed yearly migrations of 150 to 200 female iguanid lizards to the same nesting site on the tiny island of Slothia.[16] Their radio-tracking experiments indicate that the migrations do not extend beyond a few miles.[18]

Other Aspects of Territorial Behavior

In Chapter 6, I used Llewellyn Evans's field study of the Mexican black lizards living on a cemetery wall as an illustration of territoriality among lizards. This particular study had the added value of illustrating that animals that are usually nonterritorial and noncommunal may under favorable conditions live closely together in small territories.

Retention of Territoriality under Conditions of High Density

In considering other dimensions of territoriality, we find that some lizards living under conditions of high density may continue to be territorial, whereas others cease to be territorial. After describing how the rocks swarmed with marine lizards (see above), Eibl-Eibesfeldt commented: "From my observations I noticed that strict order reigned among them. In every case a male inhabited a certain piece of rock which he shared with some rather smaller females, and he jealously watched over this [rock] as his own territory."[19] These territories might be separated by no more than 1.5 m. The iguanas appeared to be identified with a particular piece of ground, because if they were carried 400 yards away, they would return to the same place. If two neighbors found themselves too close, both would show a threatening attitude, raising themselves on stiff legs and strutting up and down with their dorsal combs elevated. But if the intruder failed to back off, fighting would occur. Eibl-Eibesfeldt's choice of words in describing the first encounter he witnessed is reminiscent of the "face-off" displays of the rainbow and Komodo lizards: "With stiff, stilted gait, the rivals stalked round each other, each one making himself appear as big as he could and endeavoring to show off by keeping himself broadside against the other."[20] Then with lowered heads, each kept butting the other until the intruder was pushed over sideways. The defender nodded his head up and down in triumph. The intruder made another comeback, but once more was defeated. This time he collapsed in an abject posture before the victor, his dorsal comb flattened and his legs stretched out.

Loss of Territorial Expression

If lizards with known territorial proclivities are crowded within a small space, they may show no antagonism to one another. In our laboratory we saw a striking example of such passivity when for a period of time a large number of desert lizards (*Dipsosaurus dorsalis*) were housed together. These sleek, sand-colored animals lay straddling one another in a languid state, somewhat as one would imagine Milton's stunned and fallen angels after their crash from paradise. Subsequently, when six of these animals were placed in a large, structured, quasi-natural habitat, they soon became very active, with the males chasing and fighting one another until one became dominant and tyrannized over all the rest. In their morphology, alacrity, and general behavior they are reminiscent of the rainbow lizards described in Chapter 7.

If two territorial male lizards are housed in a small space, they may seem to get along amicably together. Mayhew found that two Australian amphibolurus lizards lived peaceably in a 30 × 51 cm habitat. But when transferred to a 0.6 × 1.8 m cage, the smaller lizard continually attacked the larger one.[21]

Harassing Behavior

If lizards are kept under conditions in which subdominant adult males cannot escape from a tyrant, they may be harassed to death. This was the fate of the larger of the two amphibolurus lizards just mentioned; it was chased so continuously by "the tormenter" that it died two weeks later. In this case the tyrannical behavior of the smaller animal might be interpreted as an expression of its unrelenting urge to drive the bigger animal from its territory.

One finds that harassing behavior may also apply to just part of a lizard's territory. Under such circumstances it may call to mind vindictive behavior. In our laboratory, Greenberg devised an experimental situation that fortuitously led to such behavior (unpublished observations). The habitat was a 2.7 × 2.7 × 2.7 m cubicle with a viewing window above. A 30-cm-wide shelf at a height of 1.8 m extended the entire length of two opposing walls. A lizard had access to the shelves by two ramps at one end of the cubicle. The ramps came together at the floor level, giving the appearance of a "V." Midway on each shelf was a lamp for basking, one a red light and the other a white tungsten light. Greenberg introduced a mature iguana lizard (*Iguana iguana*) and allowed it to become adapted to the room for a period of several days. At first, it would spend time basking at either the red light or the white light. Eventually, however, it showed a preference for the red light and spent almost all of its basking time under it. Greenberg then introduced another male iguana lizard of about the same size. Since the red basking light was occupied by the original resident, the newcomer began to bask under the white light on the opposite shelf. It was not long, however, before the resident drove him from the white light and began to use it himself. Since the resident lizard now seemed to prefer the white light, the stranger switched to using the red light. But the resident lizard would not allow it to use either light. The newcomer was no sooner settled down to bask at one light than the resident would take the circuitous route down and up the ramps to drive it away. This constant seesawing resulted in numerous trips up and down the ramps by both lizards. Finally, the second lizard became so exhausted by the continual harassment that it lay sprawled at the junction of the two ramps on the floor of the cubicle. As in like cases in other animal societies, it would not have survived had it not been removed from an impossible situation.

The Element of Strangeness

Strangeness and familiarity might be regarded as opposite sides of the same coin. The recognition of what is strange or novel depends on a contrast with what is familiar. Familiarity, in turn, derives either from learned experience or from "innate" experience (see tropistic behavior, Chapter 10). When, as is usually the case, an animal recognizes upon the very first exposure what is strange, it might be said to be capable of first-trial learning. Lizards, living communally in cages or in territorial groups, have been shown, over and over again, to be capable of immediate recognition of a stranger.[22] The unanswered question regarding neural mechanisms underlying an animal's response to what is alien or different has so much human interest in regard to intolerance of aliens, blemishes, and the like, that it deserves further illustration here under the topic of territory.

It is unnecessary to elaborate upon previous illustrations of a territorial male's handling of intruders, but in the present context it is of particular interest to recall the two examples described in Chapter 6 of the "ganging up behavior" of communal groups of lizards against a newcomer. In each case, the subdominant males, which were lorded over by a "tyrant," joined him in driving away an adult male intruder. Evans has also observed that several spayed female lizards (*Anolis carolinensis*) living in the same cage and dominated by one or two others "all showed pugnacity toward an intruding female."[23] His observations suggest that an animal's disposition toward strangers is at least partially under hormonal influence. (It should be noted, however, that the females of some species of lizards such as *Anolis sagrei* may be naturally strongly territorial and combative.[24] The females of this species extend a small dewlap in challenge to an intruder. As Evans further explains, "If the intruder is an adult male, he replies to her challenge with his own dewlap display. She then shifts to her courtship pattern, withdraws her dewlap, and nods her head in reply. Her head-nod then causes the fighting pattern of the male to shift over to that of courtship."[25])

We have seen that juvenile lizards run the risk of being treated as "strangers" by adults of their own species, since they may be cannibalized by them. It is curious that juvenile lizards of such diverse types as *Dipsosaurus dorsalis* and *Varanus komodoensis* may, in the presence of adults of their own (or even another) species, adopt a posture that resembles part of the face-off display of mature animals. Carpenter, for example, describes how a juvenile desert iguana under such circumstances will stand "high on all four legs" and, with body compressed and dewlap extended, walk by a resting adult lizard in a stiff-legged manner.[26] Carpenter believes that these actions (unaccompanied by pushups) serve to protect the juvenile from attack. Auffenberg's description of the "appeasement display" of juvenile Komodo dragons could almost be substituted for that of dipsosaurus

(Chapter 8). Auffenberg attributes such displays to "stress."[27] But perhaps there exists here a situation in which "form" partly makes up for what is lacking in "size."

Among birds and mammals, immature animals with various kinds of blemishes (strangeness) may be killed or driven off from the home territory by continual harassment (see Chapter 29). I have neither observed nor heard of comparable behavior in lizards. According to Burden, if one of a number of Komodo dragons is wounded, "it is subject to attack."[28] The question of territorial behavior in connection with inanimate objects will be considered under tropistic behavior (Chapter 10).

Summarizing Comment

The first part of the present chapter has been devoted to illustrating five special forms of basic behavior in reptiles that were either not mentioned or barely touched upon in the preceding chapters. These behaviors are identified as grooming, hoarding, breeding, collective behavior, and migratory behavior (see Table 6-1). In considering grooming, hoarding, and breeding behavior, one recognizes marked differences between reptiles and mammals. Parental care of any kind, for example, is nonexistent in the case of most reptiles, and this may have also been true of the mammal-like reptiles.

The second part of the chapter deals with less commonly recognized aspects of territorial behavior, including the element of "strangeness." In the next chapter we find that the question of differences between familiarity and strangeness is particularly relevant to the consideration of protomentation and its associated learning and memory processes.

References

1. Lin, 1963
2. Chiszar, 1978
3. Goin and Goin, 1962
4. Bogert, 1954
5. Greenberg, 1978
6. Burden, 1928; Greenberg, 1977a; Auffenberg, 1978
7. Greenberg, 1977a
8. Evans, 1959
9. Goin and Goin, 1962
10. Bellairs, 1970
11. Evans, 1961
12. Bellairs, 1970, p. 405
13. Zahl, 1973
14. Marcellini, 1970
15. Eibl-Eibesfeldt, 1961, p. 67
16. Rand, 1968; Rand and Rand, 1978
17. Carr, 1965
18. Rand and Rand, 1978
19. Eibl-Eibesfeldt, 1961, p. 67
20. Eibl-Eibesfeldt, 1961, p. 68
21. Mayhew, 1963
22. Greenberg and Noble, 1944
23. Evans, 1938a, p. 98
24. Evans, 1938b
25. Evans, 1938b, p. 483
26. Carpenter, 1961, p. 403
27. Auffenberg, 1978
28. Burden, 1928

10

Six *General* Forms of Basic Behavior

We have yet to consider six general forms of basic behavior that I refer to as interoperative kinds of behavior[1] because they come into play in several different contexts, rather than being characteristic of a special form of behavior such as fighting or courtship behavior. As listed in Table 10-1, they are referred to as (1) routinizing, (2) isopraxic, (3) tropistic, (4) repetitious, (5) reenactment, and (6) deceptive behavior. The special forms of basic behavior that have been dealt with up until now immediately bring to mind a distinctive kind of activity. The same is not true of the interoperative forms of behavior. Deceptive behavior, for example, may not only show up in hunting, fighting, mating, and other special forms of behavior, but also appears in a number of different guises. Likewise, isopraxic and the other behaviors listed in Table 10-1 may be operative at one time or another in connection with the special forms of basic behavior listed in Table 6-1.

It should be pointed out that *interoperative* is used here in the intransitive sense. The behavior itself does not actively interoperate with other forms of behavior; rather it is the brain itself that is operational in this respect.

Six Interoperative Forms of Behavior

The six interoperative forms of behavior will be considered in the same order as given in Table 10-1.

Routinization

Of the six principal interoperative types of behavior, the one pertaining to routinization should be considered first because, as we have seen, a lizard's day is not only dominated by a master routine, but may also have quite rigidly structured within it a number of subroutines. The master routine consisting generally of cautious emergence, basking, defecation, etc., has been illustrated by the behavior of the blue spiny lizard, the rainbow lizard, and the giant Komodo dragon. Subroutines apply to habitual and individually typical acts that are carried out during the course of the master routine. The difficult and roundabout route that the group of rainbow lizards took upon retiring to their roost (Chapter 7) is an example of a subroutine. An illustration of a subroutine of a single individual is provided by Harris's account of a rainbow cock lizard that after taking over a new territory, continued to sleep at the same site in its former territory.[2]

Eibl-Eibesfeldt described marine iguana lizards in the Galapagos Islands that adopted subroutines totally unlike those of the main flock that grazed together at regular times in

Table 10-1. General ("Interoperative") Forms of Basic Behavior

Routinizing
Isopraxic
Tropistic
Repetitious
Reenactment
Deceptive

the sea. These particular animals became habituated to living near the house of Mr. Carl Angermeyer. They came to his house at a regular time each day for "breakfast." Every evening they would clamber up the walls of the house, to roost under the roof and let their tails dangle down. One female showed up regularly at half past five and climbed up the chimney.[3]

Whatever may be the neural mechanisms underlying both the master and subroutines, it is evident that they not only regulate the order in which the actions occur, but also the time of their occurrence. It is the regular *round-the-clock,* temporal sequencing of behavior that characterizes routinization and requires that it be considered apart from repetitious and reenactment behavior to be dealt with later. When subroutines become rigidly structured in their patterns and time of occurrence, they become known as rituals. Subroutines may creep into the master routine quite accidentally. It was as though the animal's brain had sized up the accidental circumstance as having survival value (see Chapter 16). It is the kind of circumstance that provides the basis for "precedent." Through some fortuitous circumstance, one of Harris's rainbow lizards took the difficult roundabout course in retiring to its roost. On the first occasion, the long route may have been taken to avoid some threat. Having proved to be safe, it was taken again, then again, and again. Younger lizards coming along learned to follow the older, more experienced animal. Precedent was established.

Isopraxic Behavior

The word *isopraxic* may seem a little forbidding the first time that it is encountered. It means simply "performing or acting in a like manner," "doing something the same way," "behaving in the same way." I found myself resorting to the use of the term when I needed a word that was purely descriptive and had no causal overtones.[1,4] There are no dictionary words that do the work of *isopraxis*. The word *imitation,* for example, not only fails to convey the desired meaning, but also has acquired so many "explanatory" connotations that it is of limited usefulness. For instance, one psychological book entitled *Social Learning and Imitation* begins by stating, "All human behavior is learned."[5] According to that doctrine, all human imitation would be ascribed to learning. The authors of this same book emphasize the well-known difficulty of teaching animals to imitate. Left to themselves, however, animals not only engage in species-typical pair or group activities of the same kind, but also may adopt some novel practice of one of their group (see Chapter 16). When animals engage simultaneously in the same kind of species-typical activity, it may be necessary to experiment with subjects brought up in isolation in order to find out whether or not such isopraxic behavior is learned. In the case of unanswered questions of this kind, it is evident why a descriptive word such as *isopraxis* is to be preferred to such an expression as *social facilitation,* which has overtones of a functional or other explanation. Such words as *mimic, ape,* and the like suffer from the same shortcomings as their synonym *imitate. Isopraxic* serves as an alternative to such a hard-to-pronounce word as *allelomimetic,*[6] referring to a group of animals behaving in a similar way.

Regardless of mechanisms, it is recognized that isopraxic behavior is invariably implicated in conspecific recognition and in most forms of communication involved in self-preservation and in the procreation of the species. In circular language one might define a species as a group of animals that had acquired the perfect ability to imitate themselves.[4] Just as isopraxic behavior is involved in conspecific recognition, so does it serve *in the opposite sense* to promote species isolation. In other words, isopraxis has its counterparts in *heteropraxis,* which may signal either a partial or a complete behavioral difference of one individual from another. When heteropraxis signals only a partial difference, it might serve, say, to communicate sex or age differences among animals of the same species. When two territorial male lizards of the same species are responding to one another with challenge displays, they are essentially engaged in isopraxic behavior, i.e., behaving in a like manner. Such behavior permits both the species and sexual identification that would not be possible if one of the animals had none of the identifying physical or behavioral characteristics. An adult male *Anolis carolinensis* will perform its signature display in the presence of a male *Anolis sagrei* (a lizard of about the same size and configuration), but never follows through with a challenge display.[7] Presumably, it does not receive any behavioral signal in return that would call for the aggressive territorial response.

In preceding chapters we have seen how motor components of displays, as well as the color and configuration of an animal's body, provide individuals of a social group of lizards a means of differentiating dominant and subdominant animals, and of making distinctions in regard to age and sex. When we consider tropistic behavior (see below), we shall encounter factors other than isopraxis and heteropraxis that may in turn elicit isopraxis or heteropraxis. Before that, we need to mention briefly group isopraxis and mass isopraxis.

Group Isopraxis

A clarification of neurological mechanisms underlying group and mass isopraxis promises to be as difficult as any problem in neurophysiology. Group isopraxis among lizards is illustrated by the simultaneous head nodding response of several juvenile and females to the display of a territorial male lizard. It occurs with as much unison as the group gobbling response of tom turkeys or the hand clapping response of human beings. An exaggerated form of group isopraxis among lizards can be demonstrated under the following artificial conditions: For the receipt of a new shipment of the common green anolis lizard, we have in our laboratory a large holding tank with a branch of a tree extending from opposite ends of the tank toward the middle of the habitat. It invariably happens that with each new shipment a number of males rise up from the ranks and vie for the control of one of the two branches. Because of the separation of the two branches, a lizard can gain control of no more than one branch. Eventually, there is a territorial male in command of each branch. Many other lizards are allowed to perch on these branches provided they do not challenge the authority of the alpha male. The latter is active much of the time patrolling the branch or the underneath substrate. Instead of barking orders like a top sergeant to keep discipline, each of the successful lizards gives its mute signature display characterized by head bobs and pushups and an extension of its colorful throat fan (see Chapter 11). On such occasions one may see a dozen or more subdominant lizards perform in unison a series of rapid assentive head nods.

The following examples of group isopraxis are among those already mentioned. Iguana lizards, which usually nest singly, behave in an isopraxic manner when, under certain conditions, 150 to 200 females meet in the same spot and simultaneously dig their nests not far from each other. The so-called social facilitation that occurs when one or two successfully foraging lizards precipitate feeding by a whole group is another example of isopraxic behavior. The convergence of a number of Komodo lizards at carrion is a third example.

Mass Isopraxis

Because of the untold number of individuals involved, it would be particularly important to understand the prosematic and neural factors at work in mass isopraxis of human beings, e.g., mass migrations that have occurred from time to time in the past; mass rallies; adoption by millions of people of fads and fashions that may not only sweep a country overnight, but also jump oceans (see Chapter 16).

The experiments in which Noble showed that schooling among fishes requires the presence of the forebrain, indicate that the basic neural mechanisms required for group and mass isopraxis are located in the forebrain.[8] Among lizards, we find nothing that compares to the mass isopraxis of turtles illustrated by the hordes of migrating Ridley turtles (Chapter 9). Isopraxic behavior of large aggregations of lizards is best illustrated by the marine lizards of the Galapagos Islands, which, as if on cue, enter the ocean to graze and then later come out of the ocean at about the same time to resume basking on the shore (Chapter 9).

Tropistic Behavior

In biology the word *tropism* (derived from the Greek word *tropos* meaning "a turning") is used to refer to an unexplained positive or negative response to a stimulus. The turning of a plant toward sunlight is an example of a positive tropism. The word is also used to signify an inborn (innate) inclination. *Tropistic behavior* is used here to refer to positive or negative responses of an animal to partial or complete representations, whether alive or inanimate. It includes behavior that ethologists attribute to "innate releasing mechanisms."[9] Lorenz referred to the unlearned responses of animals to certain sign stimuli as "innate motion patterns,"[9] now more commonly known as "fixed action patterns." In ethological writings, the stickleback fish is a favorite subject for illustrating "fixed action patterns." The red color on the belly of a dummy is sufficient for eliciting the fighting response of an adult male stickleback. A more intense response can be elicited by placing the dummy in a near-vertical position, a posture ordinarily assumed when a territorial male encounters another male near its territory. In addition to the identification of these two "primary sign stimuli"—color and direction of body—the ethological analysis has revealed that evocation of the fighting response depends on the "internal state" of the animal: sign stimuli are ineffective unless the animal is in a reproductive condition.

Comparable forms of tropistic behavior are commonly seen in lizards. Noble and

Bradley showed that if the blue belly of a male *Sceloporus undulatus* lizard was painted gray, the animal would be courted by other males.[10] If, on the contrary, a female's belly was painted blue, she would be attacked by males. Similar treatment of the blue throat of the European lizard, *Lacerta verdis,* had the same kind of effect.[10]

Under other conditions, form and movement, rather than color, are the critical factors in eliciting tropistic forms of behavior. Crews conducted experiments in which he tested the influence of masculine behavior on the egg production of female green anolis lizards (*Anolis carolinensis*). Under conditions in which females could observe the extended throat fan of a male lizard, there was a stimulating effect on ovogenesis.[11] If the salmon color of the throat fan was painted black, the movement of the extended fan still had a positive effect. If, however, the throat fan was immobilized, the displaying males were as ineffective as the castrated controls in stimulating ovogenesis.

In contrast to fishes and birds, there has been relatively little experimentation on the responses of reptiles to models. One of the few studies is that of Vernon Harris on the rainbow lizard.[2] In using a crude model that mimicked the head color and bobbing display of a cock lizard, he found that the orange-red color was more effective than body movement in eliciting the challenge display of a reproductive male. When using this model, however, for mapping the outer limits of the territory of a cock lizard, Harris found that both the color and the head bobbing were essential components. As the model was moved closer and closer to the lizard's topmost display post, it became increasingly effective in eliciting the challenge display (see Chapter 7). When moved in the opposite direction there appeared to be a critical distance beyond which the model had no effect.

In the course of his experimentation, Harris discovered that orange objects generally (orange pieces of paper, orange carpet, amber glass beads) tended to attract the attention of a cock lizard and were often eaten by him. One lizard investigated a bottle with an orange label so vigorously that it was overturned and broken. Harris concluded that only color (and only an orange-red color) made it possible for a cock lizard to recognize another territorial male.

Imprinting

Lorenz (1935) used the term *imprinting* to apply to the situation in which a young bird during a critical period attaches itself to the first creature it meets.[9] Imprinting is sometimes interpreted as a special form of learning that occurs only during a critical period in the development of the organism. It is probably more generally regarded as an *innate* response or, in terms of the present account, a tropistic response. Examples of imprinting in lizards are lacking. Enough examples have been given already to make it apparent that the kind of imprinting just mentioned would be fatal to such hatchlings as those of the rainbow lizard or the Komodo dragon, the probability being that they would end up in the parent's belly.

Repetitious (Perseverative) Behavior

In Chapter 16, we shall consider some of the human implications of the work on reptiles. For the moment, we will turn things around and use human illustrations for clarifying the meaning of two kinds of behavior next to be described—namely, repetitious behavior and reenactment behavior. Repetitious behavior will refer to the repetitive performance of a specific act. Lady Macbeth's compulsive hand washing would be an example of such a repetitious performance. Unless one were familiar with the play, her repetitious hand washing by itself would have little meaning. Reenactment behavior, on the contrary, applies to a repeated performance in which a number of actions are meaningfully related. The theatrical performance of the entire scene in which Lady Macbeth washes her hands would correspond to a reenactment. As used here, *reenactment behavior* will apply to a repetition of a ritualized sequence of acts on a single day or to periodic reenactment at intervals ranging from days to years. In human society, periodic reenactments are illustrated by weekly church services, yearly graduation exercises, and the like.

Among lizards, repetitious behavior is perhaps best illustrated by the performance over and over again of the signature and challenge displays used by a male in trying to establish or maintain dominance over other males. In such situations, it is not always body size that is the deciding factor.[12] Rather, it would almost seem that it was the rule of the game that the animal that performs the greatest number of displays eventually wins out. There is a need for systematic studies relevant to this point. The presence of a reproductive female appears to have the effect of greatly increasing the number of displays by contesting males. The displays might be likened to a series of exclamation marks. Courtship presents another situation in which displays by the male and female may become repetitious and seem to carry the message of a series of exclamation marks. Speaking to this point in another context, Morris has remarked that the telephone often succeeds in attracting attention because of the number, rather than the loudness, of its rings.[13]

Displacement Behavior

In 1940, Kortlandt[14] and Tinbergen[15] each dealt with forms of behavior since identified in translation as "displacement activities." In ethology, displacement behavior refers to repetitious acts by an animal that seem to be inappropriate for the occasion. A common example is the preening engaged in by birds when they are in a threatening or otherwise disturbing situation. In our laboratory, we have seen comparable behavior in territorial male anolis lizards. When an opaque divider between the habitats of two such males is removed, there may be "indecisive" periods in which the animals will repeatedly scratch themselves or rub their muzzles against their perch or on the substrate. In this connection, one is reminded of the disproportionate increase of tongue flicks observed when increasing numbers of Komodo monitors meet at carrion (Chapter 8). Under such conditions the tongue flicks might be regarded as a form of repetitious displacement behavior.

Reenactment Behavior

As explained, reenactment applies, just as in a play, to a series of actions that are repeated essentially the same way in each performance. The use of the expression *reenactment behavior* would be superfluous if it did not refer to something quite different from the reenactment of an animal's daily master routine such as the now-familiar routine of the lizard in emerging, basking, defecating, and so forth. Ritualized subroutines, however, occurring within the daily master routine would answer the description of a reenactment. The retirement-to-shelter behavior of the group of rainbow lizards described in Chapter 7, and referred to again in connection with routinization, would provide an illustration. Each evening these animals rigidly adhered to the custom of taking the same roundabout, difficult route to their roosts.

Among reptiles, cyclical reenactments are most dramatically illustrated by the nesting behavior of sea turtles. There are five kinds of sea turtles that, except for times of nesting, spend their lives grazing upon abundantly growing turtle grass in the warm parts of the oceans. Carr has identified female green turtles (*Chelonia mydas*) along the coast of Brazil that migrate to lay their eggs on Ascension Island—"a target 5 miles wide and 1400 miles away in the South Atlantic."[16] These reenactments occur in 2- or 3-year cycles. Four tagged turtles were recaptured on the same short section of beach.

In the preceding chapter, there was mention of the waves of 10 to 30 thousand Ridley turtles that returned to old nesting places along the beaches of the Pacific Ocean. By such standards, the migratory, egg-laying reenactments of lizards seem quite insignificant. In the study by Rand and Rand (see preceding chapter) the number of iguanid lizards nesting on the island of Slothia ranged only from 150 to 200.

Deceptive Behavior

Of the six interoperative behaviors under consideration, deceptive behavior would rank as one of the most essential for survival. This realization makes it all the more curious that so little attention has been given to an analysis of its behavioral and neural fabric. It deserves emphasis that deceptive behavior comes into play not only in obtaining the necessities of life (e.g., a home base, food, mate) but also in avoiding virtual elimination or death through the actions of others. Some students of behavior might contend that it requires more guise and ruse to avoid failure than to achieve success.

The hunting tactics of the giant Komodo dragon provide one of the best illustrations of deceptive behavior of which lizards or other reptiles are capable. As described in Chapter 8, the successful stalking of a deer requires that the monitor have not only a detailed knowledge of its own home range, but also an excellent sense of time. In addition, there must be a capacity for great patience, persistence, and endurance.

The predatory behavior of small lizards is difficult to analyze because their movements are so fast and the prey so small that the actions are as hard to detect as those in a speeded-up motion picture film. To see a blue spiny lizard "stalking" a cricket is to be reminded, in miniature, of a lurking Komodo dragon. The banded gecko (*Coleonyx variegatus*) of the American Southwest is said to stalk its prey with all the caution of a cat in catching a mouse, "pouncing on it with ludicrous ferocity."[17]

Some female lizards may perform enticement displays as though inviting courtship and mating by a male, but will then run away or fight the male off if he takes hold of her. What is the purpose of these deceptive solicitations? Do they serve to raise the level of the male's excitement, or are they calculated to attract other reproductive males to the scene? Noble and Bradley have noted that many female lizards "raise the tail and turn it to one side in non-sexual situations."[10]

A review of observations on reptiles other than lizards would increase the number of fragmentary illustrations of deceptive behavior. A discussion of "mimicry" would also provide further details about the use of deception by animals. Zoologists use the term *mimicry* to refer to superficial resemblances between animals or between animals and their environment, or indeed to parts of the same animal, a condition known as automimicry. Deceptive mimicry is illustrated by the tongue of a snapping turtle that a fish may mistake for a worm.[18] One form of snake has a marking on the tip of its tail that mimics the eye in the head.[19] Automimicry in the lizard is illustrated by the postocular dark spot that appears in the green anolis lizard while performing a series of challenge displays (see next chapter). It has the effect of making the eye appear larger in size.

Protomentation

It may be surprising that except in the case of imprinting there has been no mention up until now of the question of learning. Learning could hardly be listed on one of the tables as a form of behavior because it refers more strictly to mental processes attributable to the behaving entities of the brain. One might say that problem-solving behavior could have been listed, but except in the case of experimental intervention, it may not be apparent just what problem an observed animal is trying to solve. The predatory activity of a Komodo dragon would involve situations of the kind that we associate with problem-solving behavior. But since animals living under natural conditions are not usually presented with problems contrived by human beings, it would seem inappropriate to list problem-solving behavior along with naturally occurring, interoperative forms of behavior.

Although lizards come into the world programmed, so to speak, to perform most requisite behavior, they are thereafter continually in situations that bring learning and memory into play. Constrained as they are by a given daily master routine, the particular manner in which they go through this routine depends on learning and memory. They must, for example, learn inside-out their entire territory. Otherwise, they could not quickly retreat to places of escape when threatened by a predator. Without the learning and memory process, they could not find their way home. Without learning and memory the territorial lizard could not make known to himself and others his topmost display post. Learning and memory are requisite for the recognition of differences between members of a social group and between them and a stranger. As trenchantly illustrated by the Komodo dragon's keen sense of time and knowledge of its activity range, experiential learning and memory become an indissociable addition to an animal's natural capacity for the practice of deceptive behavior. One could hardly point to a behavior listed in Table 6-1 or 10-1 that did not require some degree of experiential learning and memory.

Reptiles have the reputation of being very poor learners. Lizards, for example, are said to be incapable of learning to crawl over a low, transparent barrier in order to get a mealworm.[20] Brattstrom has repeatedly pointed out that much of the formal testing on lizards has been done under environmental conditions in which the animals are so cold as to be almost anesthetized.[21] He has found that if given the opportunity to work for the reward of heat from a lamp, they may show themselves to be remarkably apt learners. These observations provide yet one more illustration of how important it is to take into account an animal's special attributes when devising tests of learning. Rats compare unfavorably with monkeys when administered tests involving visual or auditory cues, but, as Slotnick and Katz have shown, if they are allowed to make similar discriminations through the use of their olfactory sense, they may be capable of one-trial learning.[22] In this connection, and recalling the discussion of "strangeness" in the preceding chapter, one might even say that lizards are capable of "first-time" learning. For example, it almost invariably happens that when there is a confrontation of two territorial male lizards of the same size, it is usually the animal that intrudes on the other lizard's territory that either retreats or loses out in a fight. Students of behavior attribute the lizard's defeat to its recognition that it is on foreign territory. Since such an outcome is usually observed upon the first encounter, should not this be regarded as first-time recognition and retention, i.e., first-time learning? Similarly, one might say that when one lizard or members of a retinue immediately recognize another of their species as a stranger, and persist in this recognition, it is the equivalent of immediate learning.

In Chapter 2, I used the term *protomentation* to apply to rudimentary mental processes that underlie a meaningful sequential expression of prototypical patterns of behavior. Hereafter, it will be implicit that protomentation encompasses the capacity for experiential learning and memory associated with such behavior.

Concluding Comment

In addition to the *special* forms of basic behavior illustrated in the preceding chapters one is able to recognize six *general* kinds of basic behavior. They are designated as (1) routinizing, (2) isopraxic, (3) repetitious, (4) reenactment, (5) tropistic, and (6) deceptive behavior. Since they find expression through different interrelationships of *special* kinds of behavior, they are otherwise referred to as "interoperative" forms of behavior. Routinizing behavior is one of the most basic, because it manifests itself in a sequential linking together of activities comprising the daily master routine and a number of subroutines upon which the health and well-being of an animal depend. Isopraxic behavior (and its opposite heteropraxic) is essential for prosematic communication required for conspecific recognition and sexual identification, as well as all activities involving two or more individuals. Repetitious behavior has great value in social communication because it helps to assure that a message will be received and is attended to. Reenactment behavior is significant for sustaining customs having survival value, as best illustrated by periodic gatherings leading to reproductive activity. Tropistic behavior applies to positive or negative responses to partial representations, whether alive or inanimate, and would include "fixed action patterns" of ethology attributed to innate releasing mechanisms. Its manifestations are especially significant in aggressive, defensive, and reproductive behavior. Deceptive behavior may find expression in almost every effort to survive.

Finally, protomentation is again discussed with respect to cerebration that accounts for the ordered expression of special and interoperative forms of basic behavior. Deserving special emphasis, protomentation entails learning and memory connected with all such behavior.

References

1. MacLean, 1975a
2. Harris, 1964
3. Eibl-Eibesfeldt, 1961
4. MacLean, 1975b
5. Miller and Dollard, 1941
6. Scott, 1962
7. Greenberg, unpublished observations
8. Noble, 1936
9. Lorenz, 1935; 1937; *see also* Tinbergen, 1951
10. Noble and Bradley, 1933
11. Crews, 1978
12. Evans, 1936; Greenberg and Noble, 1944
13. Morris, 1957
14. Kortlandt, 1940
15. Tinbergen, 1940
16. Carr, 1965, p. 79
17. Bogert, 1954, p. 1321
18. Goin and Goin, 1962
19. Bellairs, 1970
20. Cookson, 1962
21. Brattstrom, 1978
22. Slotnick and Katz, 1974

11

Neurobehavioral Findings on a Lacertian Display

As stated in Chapter 4, it is unfortunate that for comparative neurobehavioral studies no extant reptiles are directly in line with the therapsids that are believed to be the antecedants of mammals. In the last five chapters the analysis of reptilian behavior has focused on lizards because they have been described as having a number of close resemblances to the early mammal-like reptiles.

The analysis of reptilian behavior reveals that there are at least 25 special forms of basic behavior that are also found in mammals. Those most notably lacking are (1) a combination of nursing and maternal care, (2) audiovocal communication for maintaining maternal–offspring contact, and (3) playful behavior.

Before proceeding in the next four chapters to summarize some comparative neurobehavioral studies, it is worthwhile to restate briefly the problem concerning the R-complex—namely, that despite more than 150 years of investigation, experiments involving stimulation or destruction of respective parts of the R-complex have failed to reveal its functions. As summarized in the historical review of Chapter 4, the findings by many workers that destruction of large parts of the R-complex on one or both sides may result in no apparent paralysis or other motor disability, are not in harmony with the traditional view that this basic part of the forebrain is primarily involved in motor functions.

In past experimentation, had there been a failure to ask the right questions? As pointed out, it has been customary to make neurobehavioral observations on animals living isolated in cages and subjected to psychological tests in which they perform in some artificial apparatus or manipulate inanimate objects. For our experiments it was decided to test the possibility that experiments on animals living under seminatural conditions, and interacting with other animals, might reveal functions of the R-complex that would otherwise not be apparent.

The promise of such investigation was heralded by some experiments performed by Fred T. Rogers, a student of C. Judson Herrick. In the early 1920s, long before *ethology* was a familiar word, he observed the effects of destruction of parts of the forebrain on naturally occurring behavior in pigeons, paying particular attention to the performance of mating. The outcome of his studies will be described in the next chapter.[1]

Influenced by Roger's experimental approach, I began my neurobehavioral studies by observing the effects of forebrain lesions on a highly predictable form of display behavior in male squirrel monkeys. As described in Chapters 13 and 14, I found that lesions of certain parts of the R-complex, but not of other forebrain structures, resulted in an elimination or fragmentation of the display.

For comparative purposes, the extension of this kind of study to lizards presented a problem beyond those complicating ablation studies in mammals and one not considered in discussing the pros and cons of various experimental methods in Chapter 4. In mammals, experience has shown that bilateral destruction of a particular brain structure is usually necessary in order to demonstrate a behavioral deficit. Since reptiles depend on behavioral thermoregulation (see Chapter 5) to achieve the requisite body temperature for all their functions, it was possible that a failure to perform a test following some bilateral forebrain lesions might be attributable to a torporous state like that associated with nighttime cooling. The resolution of this and other problems in conducting neurobehavioral experiments on display behavior in lizards is the subject of the rest of this chapter.

Effects of Brain Lesions on Challenge Display of Green Anolis Lizards

Curiously enough, almost no experimentation had previously been conducted on the effects of selective lesions of the forebrain on the behavior of lizards. All that was known was that the neural guiding systems for directed behavior were located forward of the midbrain. The effects of decerebration on lacertilian behavior were summarized in Chapter 3. As in the case of birds and "lower" mammals, decerebrate lizards retain the basic mechanisms for posture and locomotion, feeding, and so forth. Like all preparations in which there is an elimination of the three main evolutionary formations of the forebrain, the striking finding is the absence of spontaneous, directed behavior.

Selection of Experimental Subject

Several factors determine one's choice of a particular species of lizards for experimental purposes. The animals should be readily available in adequate numbers. They should be relatively easy to maintain in good health under seminatural conditions. The behavior under study should be performed with sufficient reliability that experimentally induced alterations can be shown to be statistically significant. The animals should be resistant to the stressful effects of handling and surgical procedures.

On the basis of our experience with several species of lizards (of four different families), we selected the common green anolis lizard (*Anolis carolinensis*) for testing the effects of lesions of different brain structures on display behavior. Often referred to as the American chameleon, the green anole is not to be confused with the Old World true chameleon (Chamaeleonidae), a much larger animal with independently moving eyes and a long protruding tongue that snaps out to catch insects. The average body length of an anolis lizard is about that of one's forefinger. The maximum snout–vent length is 75 mm. The brain is about the size of a small pea (average length 2.7 mm).

Behavioral Aspects of Study

Adult, green anolis male lizards engage in the four main types of display described in Chapter 6—namely, (1) the signature (assertive) display; (2) the "distant" and "close-in" challenge (territorial) displays; (3) the courtship displays; and (4) the submissive (assentive, appeasement) displays. Our own analysis of these displays is in essential

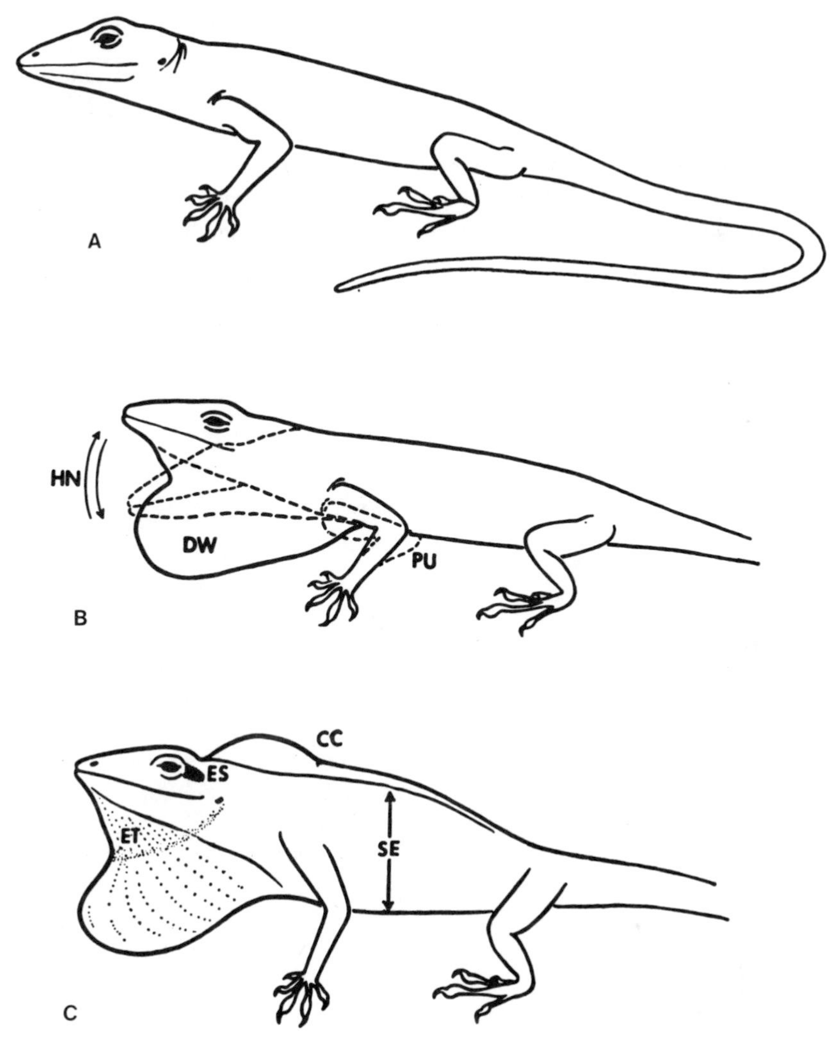

Figure 11-1. Features of the signature and challenge displays of the common green anolis lizard (*Anolis carolinensis*).

(A) The usual attentive posture.

(B) Diagrammatics of the signature (assertion) display. The signature display consists of three to five head nods (HN) and pushups (PU) along with an extension of the dewlap (DW). The broken lines indicate the excursion of the head and flexion of forelimbs during pushups. Note absence of static modifiers.

(C) Diagnostic features of the challenge display of adult male lizards. In addition to the dynamic components of the signature display, the challenge display has several static modifiers. The first to appear are the extended throat (ET) and sagittal expansion (SE), followed by an elevation of the nuchal and dorsal crests (CC). A darkly pigmented eyespot (ES) may appear after 2 to 3 min. See text for further details.

From Greenberg *et al.* (1979).

agreement with the detailed descriptions given by Greenberg and Noble[2] in 1944, although they did not, as in the present account, specifically refer to a "distant" and a "close-in" challenge display. As in the case of other lizards, there are "dynamic" and "static" modifiers of the displays. Figure 11-1 shows the diagnostic features of the displays.

Signature Display

The signature display consists of three to five combined head nods and pushups, together with a brief extension of a "crimson-colored"[3] throat fan (dewlap) occurring after the second head nod (Figure 11-1B). It has been noted that dominant males are frequently bright green when performing the display. Otherwise, one finds no specific mention of static modifiers in connection with the signature display. More descriptive data are needed, particularly in regard to pupillary changes and alterations in respiration and heart rate.

As emphasized in Chapter 6, the signature display occurs in both social and nonsocial contexts; Greenberg and Noble suggest that the performance of this display by "voiceless" lizards "takes the place of the vocalization so pronounced in birds and mammals."[4]

Challenge Displays

The "distant" and "close-in" challenge displays are the most specific of all of the displays because they take place only when two adult, territorial males are exposed to one another.

The "Distant" Display

The "distant" challenge display is characterized by the performance of the signature display in conjunction with two static modifiers—namely, sagittal expansion and extension of the gular (throat) fold (Figure 11-1C). Planimetric measurements of photographs reveal that sagittal expansion accounts for about a 30% increase in the apparent size of the body profile, while extension of the gular fold contributes an additional 20%.[5] About 2 to 3 min after the initial challenge, nuchal and dorsal crests appear (Figure 11-1C). The static modifiers are said to depend on the secretion of the male hormone.[2] About 4½ min after the initial challenge, a dark eyespot develops just posterior to the eye, secondary to an "expansion of melanophores."[3] The eyespot has the effect of making the eye itself seem larger.

The "Close-In" Display

The "close-in" challenge display usually takes place when the distant display fails to frighten away an intruding adult male. It will be recalled that in ethological language the "close-in" display is referred to as the "face-off" or "broadside" display. Dian Fossey uses the expression "parallel display" for a similar performance by gorillas[6] (see Chapter 16).

Upon engaging in the close-in display, two male anolians approach and assume positions parallel to one another at about 3 inches apart. All the static modifiers of the distant challenge display are present, and there is, in addition, a marked arching of the neck. The dynamic modifiers consist of several repetitions of the signature component of the display (i.e., combined head nodding and pushups) interspersed with whole-body pushups involving flexion and extension of all *four* legs. As two males circle one another in this parallel face-off, they move in a stiff, stilted manner that has been referred to several times in the preceding chapters. During these maneuvers they seem to have poor control of the rear legs.

Greenberg and Noble also observed the expression of territoriality and combativeness among female anolians.[2] Both the "distant" and "close-in" displays are similar to those of the male, except that the female lacks the erectile tissue for developing a nuchal and a dorsal crest.[2] The dynamic modifiers of the display are like those of the male, consisting of combined head bobs and pushups together with an extension of the diminutive throat fan. The throat fan is about one-third the size of that of the male anolian (see Figures 9 and 10 of Evans[7]). The static modifiers are the extended throat and flattening of the sides of the body (sagittal expansion). After repeated challenges, a dark eyespot may develop. If the intruding female does not retreat, there occurs a "close-in" challenge display resembling that of the male.

Courtship Display

Here we need consider only the courtship behavior that is precipitated by the encounter of a reproductive male with an adult, receptive female. Upon seeing the female, the male performs several signature displays, after which he approaches her while bobbing his head with "increasing rapidity."[8] The courtship approach gives the appearance of a "prancing strut."[8] *None of the static modifiers of the challenge display is present under these conditions.* The display of the female in signaling a state of receptiveness is of a subtle nature and consists simply of head nodding and a slight inclination (arching) of the head.

Submissive Display

The anolian submissive display is characterized by four slight up-and-down motions of the head.[2] It is performed by members of either sex and of all ages. It is the usual response to the presence of a dominant male. In contradistinction to the submissive nod, Greenberg and Noble describe a "wavering" nod ("a very gentle vertical oscillation of the head"[4]) that occurs when an anolian catches an insect.

Experimental Focus on the Challenge Display

In undertaking the studies to be described here,[9] we decided (1) to use adult male lizards and (2) to focus on the "distant" challenge display because of the following considerations:

1. In all of the displays having both dynamic and static modifiers, the number and distinctiveness of the modifiers are greater in the male than in the female.
2. The challenge displays are interpretively the least ambiguous of all the displays, occurring only when two adult males are exposed to one another. (Consistent responses cannot be obtained by exposure of an adult male to a mirror or to a dummy.) Courtship behavior, on the contrary, is not strictly sex- or age-related, while the signature and submissive displays may occur under a variety of social and nonsocial contexts.
3. For the initial studies the challenge displays were preferred to the courtship display because they involve fewer variables (both with respect to external cues and internal states) and are more predictable in occurrence.

4. The primary use of the "distant" display, rather than the "close-in" display, was more desirable because it did not risk the complications of physical contact.
5. Finally, as will be dealt with next, the concern mentioned in the introduction in regard to thermoregulation called for an experimental design requiring the use of the "distant" challenge display.

Experimental Design[9]

Reference has been made to reasons for producing bilaterally symmetrical forebrain lesions in neurobehavioral experiments on mammals. If, however, some such lesions in a lizard interfered with its behavioral thermoregulation, it might fail to display because of failure to achieve its eccritic temperature. A way out of this quandary was suggested by the visual connections of *Anolis carolinensis*. In this species the optic nerves leading to the hemispheres are almost entirely crossed.[10] This condition provides the possibility of conducting experiments by destroying structures on only one side of the brain. As illustrated in Figure 11-2, under such conditions a lizard might be expected to display to another animal when looking with the eye innervating the normal hemisphere, but fail to display when looking with the eye projecting to the damaged hemisphere. In a pilot experiment in which part of the R-complex was destroyed on one side, the outcome was quite dramatic: the operated animal paid no attention to the rival lizard when "looking" at it with the eye leading to the damaged hemisphere, but suddenly gave the full challenge display upon seeing it with the other eye.

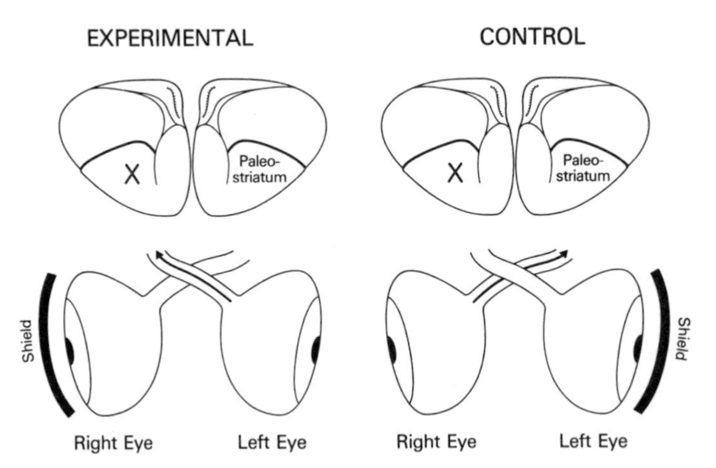

Figure 11-2. Design for obtaining control and experimental observations in the same brain-damaged lizard. In neurobehavioral research on lizards, it is important to minimize the possibility of interfering with the regulation of the body temperature. Since the optic nerves are almost completely crossed in the anolis lizard, the conditions are provided for destroying nerve tissue in one hemisphere and then testing the animal with either eye covered by a shield. Under control conditions (right side of figure) the lizard gives its usual distant challenge display upon seeing a rival territorial lizard with the eye projecting to the normal hemisphere. Although capable of seeing, there is a negative response when looking with the eye projecting to the hemisphere with part of the paleostriatum destroyed (left side of figure). Lesions elsewhere in the hemisphere are without effect (see Figure 11-4).

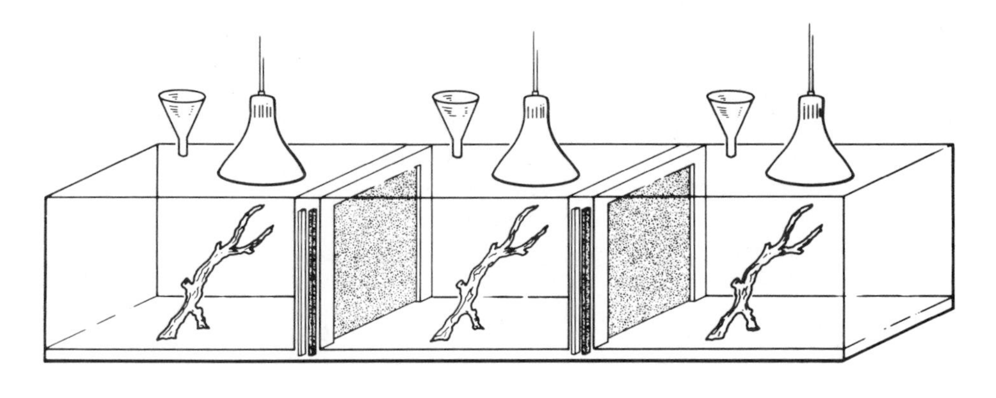

Figure 11-3. Habitat for testing the challenge display of an experimental animal in the presence of one or the other of two rivals. The experimental animal lives in the center cage, the rivals in the two end cages. Two removable panels, one opaque and one transparent, separate the modules. A test for the distant challenge display begins with the removal of one of the opaque panels. See text for other conditions. In order to keep the animals in a "primed" endocrine state, the overhead lights are automatically turned on for 14 hr a day. They also provide heat for basking. The overhead funnels serve as an unobtrusive means of delivering crickets and mealworms.

Behavioral Testing[9]

For the experiments involving the "distant" challenge display, we selected vigorous displayers from a large holding enclosure. As illustrated in Figure 11-3, the lizard chosen for surgery was provided a home base in the middle compartment of three adjacent compartments of equal dimensions. An antagonist male of the same size was placed in each of the end compartments. The three compartments were separated by removable opaque and transparent panels. There was a center perch in each compartment. An overhead 60-watt daylight-blue incandescent bulb provided light and heat for *14 hr* daily, a condition requisite for maintaining reproductive capability and a territorial disposition. The temperature in the compartments ranged from 22 to 37°C (72–99°F), while the humidity varied inversely with temperature from 72 to 96%. The animals were supplied with fresh water and were required to seek nourishment by preying on crickets and mealworms.

After a period of acclimation, the lizards were tested twice a day—once in the morning and again in the afternoon. The tests were 10 min in duration and began with the removal of the opaque screen between the experimental animal and one of the antagonists. The animals could see, but not touch, each other because of the transparent panel. The observers sat in a darkened part of the room and used a 20-channel inkwriter for recording the incidence and duration of the diagnostic components of the various display activities. Changes in an animal's posture and location in the habitat were also recorded.

In the formal testing procedure finally adopted, the experimental animal was tested with one or the other of its eyes covered with a soft dental plastic. In order to reduce the amount of handling, a patch was left on for 2 to 3 days and the lizard tested under these conditions.

If the experimental animal failed to perform a "distant" challenge after the first 5 min of a test, we attempted to provide the conditions of a "close-in" display by removing the transparent divider and prodding the antagonist to approach within striking range. We recognized that under these circumstances the interpretation of the outcome might be ambiguous because of the possibility of triggering a display through activation of other than the visual senses. For instance, there was the possibility of accidental bodily contact or stimulation of the chemical senses through tongue-touching. The second phase of the test was terminated after 5 min.

Surgical Procedure[9]

Because of the small size of our experimental subject and its tiny brain, it was necessary to devise special microsurgical techniques and to miniaturize some of the operating equipment. A brain atlas with measured locations of cerebral structures also had to be prepared.

An important part of any surgical procedure is the method for inducing anesthesia. Lizards, like newborn rodents, can be easily and quickly anesthetized by refrigeration. During the surgical procedure anesthesia is maintained by packing the lizard in ice.

As in comparable operations on larger animals, we use a stereotaxic instrument for holding the lizard's head in a standard plane so that an electrode can be pinpointed on a particular target in the brain.[9] The lizard's upper jaw is gently clamped against a rubber-coated bite plate and the head is centered with respect to the openings of the ears. The three-dimensional brain atlas provides measurements for lowering the electrode to a particular point in the brain. The plane of the bite plate determines the horizontal plane of the atlas, while the center of the surface of the parietal eye (third eye) is the zero reference for the transverse and midline planes. Before introducing the microelectrode, a small hole is drilled in the skull and a tiny slit is made in the dural wrapping of the brain. After the microelectrode is lowered to the desired point, the tissue surrounding the uninsulated tip is coagulated by passing a measured amount of electric current at a high frequency (500 kHz). Following the operation, the lizard begins to wake up a few minutes after it is removed from the ice.

Postoperative Procedures[9]

Testing of display behavior was usually resumed on the third day after the operation and on the average was conducted over a period of 42 days (range 31 to 82 days).

A crucial part of the experiment is the preparation of stained, serial sections of the brain for confirming the location and extent of the destroyed tissue. After completion of testing, the subject is deeply anesthetized by sodium pentobarbital and perfused through the beating heart with 0.9% saline, followed by 10% formalin in normal saline. After cranial decalcification by 5% formic acid the entire head is embedded in gelatin and serial frozen sections of the brain are cut at 50 μm. Alternate sections are stained for cells and fibers. Sectioning of the anolian hemispheres in the frontal plane yields about 55 sections. Microscopic study involved an evaluation of the extent of destructions, gliosis, and degeneration of myelinated pathways. A reconstruction of the lesions was obtained by projecting and drawing the brain sections at a magnification of 40$\times$. The volume of destroyed tissue was calculated on the basis of planimetric measurements of the extent of the lesions in the drawings of the serial sections.

Results

In this study the target of destruction in the R-complex was the paleostriatum. It forms the largest mass of the R-complex and has been regarded by many authors as the counterpart of the striatum and globus pallidus (see Chapter 4). The dorsal ventricular ridge of Johnston (DVR) (see Chapter 4) overlying the paleostriatum is composed of an anterior major division (ADVR) and a posterior minor division (PDVR). The minor division is believed to correspond to the amygdala of mammals[11] (see Chapters 4 and 18).

General Behavior

Altogether, 18 animals survived lesions in different parts of the paleostriatum and of the overlying DVR. None showed abnormalities in posture or locomotion. Regardless of which eye was covered, all were adept in jumping up to, or down from their perches. All sought water, and in addition to feeding on mealworms, all were able to catch, kill, and devour crickets. Like unoperated animals, they changed color from brown to green and vice versa.

Effects on the Challenge Display

It turned out that lesions of the paleostriatum, but not of the overlying ADVR or PDVR, affected the performance of the ''distant'' challenge display. Figure 11-4 shows the reconstruction of the lesions and outcome of the performance testing for representative cases. The left and right bar graphs indicate, respectively, the percentage of tests in which a particular behavior occurred when the animal was looking with the eye projecting to the intact and damaged hemispheres. From above downwards the four bars represent, respectively, challenge displays (CD); assertive displays (AD, alias signature displays); site changes (SC); and postural changes (PC). The Fisher exact chi-square test was used to assess significant differences in performance, depending on which eye was covered.

There were six animals in which the lesion was largely confined to the paleostriatum, with damage to the overlying ADVR not exceeding 2.5% of its total mass. In all cases there was some damage of the lateral forebrain bundle (LFB), varying from slight to severe. The LFB contains the brainstem fibers that ascend to and descend from the paleostriatum. In two cases with small coagulations in the paleostriatum (4–8%) there were no significant behavioral changes. In the remaining four cases there was a complete or almost total failure to perform the challenge display when looking with the eye projecting to the damaged hemisphere. In the case of lizard Ac-6, illustrated in Figure 11-4, the lesion was almost entirely beneath the dorsal medullary lamina, destroying 16% of the paleostriatum. This lizard performed the challenge display in 12 of 14 tests under control conditions, but in only 1 of 18 tests when looking through the eye projecting to the hemisphere with the damaged paleostriatum.

Lizards with destruction of tissue largely confined to either the ADVR or PDVR continued to perform challenge displays regardless of which eye was covered. The records for lizards Ac-37 and Ac-84 in Figure 11-4 show the results in two representative cases.

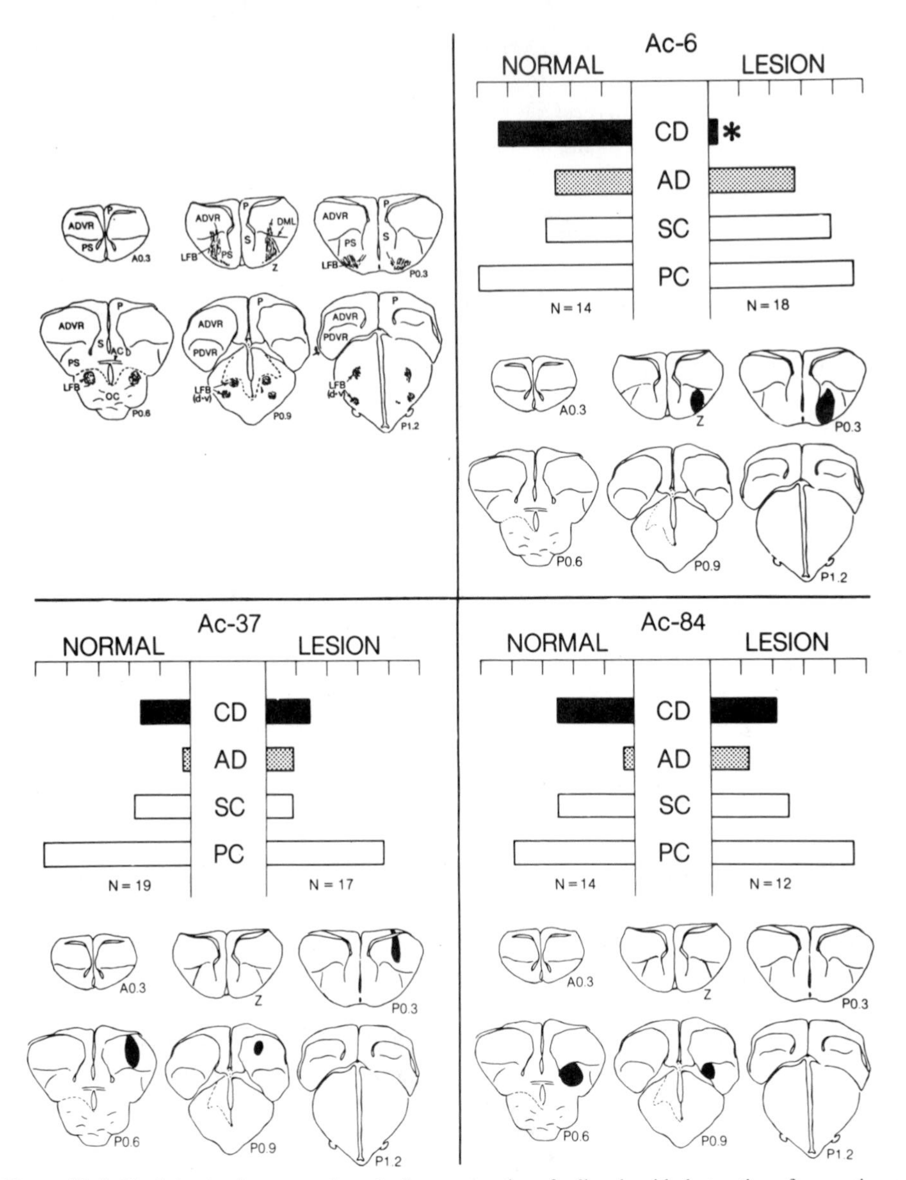

Figure 11-4. The behavioral scores and cerebral reconstructions for lizards with destruction of respective parts of one cerebral hemisphere at representative levels of brain atlas.

Bars on left and right show percentage of tests in which a particular behavior occurred when lizard was looking with eye projecting to the intact (NORMAL) or damaged (LESION) hemisphere. Black bars represent the challenge display. Shaded bars show percentage of tests in which signature (assertive) displays (AD) were observed. Open bars give percentage of test in which site changes (SC) and postural changes (PC) occurred.

Statistically significant deficits in the performance of the challenge display occurred only in lizards with damage involving primarily the paleostriatum (e.g., Ac-6). Lizards with damage predominantly in the anterior dorsal ventricular ridge (e.g., Ac-37) or posterior dorsal ventricular ridge (e.g., Ac-84) were unaffected. Asterisk refers to $P < 0.05$ with Fisher's exact test, two-tailed. *N* refers to number of trials.

The left upper diagram identifies key structures in the brain drawings. Numerals and letters at lower right of each drawing give distance in mm from anterior (A) and posterior (P) to zero (Z) frontal plane of brain atlas.

Abbreviations: AC, anterior commissure; ADVR, anterior dorsal ventricular ridge; DML, dorsal medullary lamina; LFB, lateral forebrain bundle (d + v, dorsal and ventral peduncles); OC, optic chiasm; P, preoptic area; PDVR, posterior dorsal ventricular ridge (''amygdala''); PS, paleostriatum; S, septum.

Excerpts of figures from Greenberg *et al.* (1979).

As already noted, PDVR is believed to correspond to the amygdala in mammals—a structure that will receive much attention when considering the anatomy and functions of the limbic system (Chapters 18–20).

As has been repeatedly emphasized, the signature display occurs in a variety of social and *non*social situations. Unlike the "distant" challenge display it does not depend on visual cues. In all of the animals the incidence of signature displays was about the same regardless of which eye was covered.

Discussion of the Results

The results of this pilot study indicate that the paleostriatum plays a pivotal role in the performance of the challenge display by anolian lizards. The word *pivotal* requires clarification in anatomical terms. As will be further dealt with in Chapter 14, the LFB is one of the major nerve trunks of the forebrain. It both leads to, and stems from, the paleostriatum. All of the descending fibers are believed to originate in the paleostriatum,[12] whereas the ascending fibers innervate the paleostriatum and parts of the overlying ADVR.[13] Recent neuroanatomical studies have shown that all of the fibers from the ADVR that enter the paleostriatum also terminate there.[12] These findings indicate that any influence that the ADVR might exert via the LFB on the challenge display would have to be mediated by the paleostriatum.

There is also neuroanatomical evidence that visual information reaches the ADVR via ascending fibers in the LFB.[13] Fibers from the optic tectum project to the nucleus rotundus of the thalamus, that in turn projects via the LFB to a part of the ADVR above the paleostriatum—namely, the ectostriatum. That the failure in performance of the challenge display did not result primarily from a visual deficit is evidenced by the ability of all of the animals with partial to extensive damage of the LFB to jump up and down from their perches, to obtain water, and to catch crickets. The same was true of lizards with lesions in part of the ADVR above the paleostriatum, as well as of animals with lesions in the PDVR.

Tarr reported that in the lizard *Sceloporus occidentalis,* large bilateral lesions of the PDVR resulted in decreased activity and the loss of all display behavior.[14] With our unilateral procedure, a moderately large lesion within the PDVR that interrupted most of its connections had no apparent effect on the challenge display. Greenberg *et al.* have since added further confirmation to our findings.[15]

Summarizing Comment

For a neurobehavioral study on display behavior in lizards we have reviewed the reasons for the selection of the common green anolis lizard as an experimental animal and the choice of the distant challenge display as the behavioral focus of interest. Of the four main types of displays, the challenge display is the most explicit because it occurs only when there is a confrontation of two adult, territorial males. Moreover, its static and dynamic modifiers make it the most distinctive of all the displays.

In order to avoid possible complications owing to disturbances of thermoregulation, an experimental method was adopted that made it possible to place a lesion in only one hemisphere and then test the lizard's performance with either eye covered. This procedure is possible because the optic nerves are almost entirely crossed in the green anolis lizard. Although demonstrably capable of seeing, territorial anolians almost invariably failed to display when looking at a rival lizard with the eye projecting to the hemisphere with partial destruction of the R-complex. The full challenge display was reinstated under control conditions. Lesions elsewhere in the cerebral hemisphere were ineffective.

The experimental method devised for the present study has the disadvantage of limiting testing to the visual sphere. For further study of other displays, and particularly for the important matter of experiments concerned with brain mechanisms organizing the daily master routine in lizards, it will be essential to find means to circumvent the limitations of the present experiments.

References

1. Rogers, 1922
2. Greenberg and Noble, 1944
3. Greenberg and Noble, 1944, p. 394
4. Greenberg and Noble, 1944, p. 407
5. Greenberg, 1977b, p. 192
6. Fossey, 1976
7. Evans, 1938b
8. Greenberg and Noble, 1944, p. 397
9. Greenberg *et al.*, 1979
10. Butler and Northcutt, 1971
11. Johnston, 1923
12. Hoogland, 1977; Ulinski, 1978
13. Butler, 1978
14. Tarr, 1977
15. Greenberg *et al.*, 1984

12

Avian Display

In birds I know of no studies that are exactly parallel to those on display behavior that have just been described in regard to lizards and the ones on squirrel monkeys to be dealt with in the next chapter. By this, I mean that there are no studies specifically directed at clarifying the functions of the paleostriatum. There are, however, in the older literature, two experimental studies on the courtship behavior of pigeons that deserve a brief chapter here because of what they contribute to the comparative knowledge of the functions of the paleostriatum.

As in the case of lizards, there have been relatively few neurobehavioral studies on birds. One reason for this may be that birds are often considered to be an anomalous evolutionary offshoot with respect to both their behavior and their cerebral anatomy. Many seasoned neuroanatomists admit to feeling quite lost when they are faced with identifying some of the nerve cell aggregations and pathways in the brains of birds. When caught in the situation in which they are asked their views about this or that structure in the bird's brain, they may look like a person who had stepped out of an automobile into the cockpit of an airplane and was asked to fly it. As we have seen in Chapter 4, developments in histochemistry have been of invaluable help in identifying what corresponds to the R-complex in the brains of reptiles, birds, and mammals (Figure 4-6). The stain for acetylcholinesterase and the histofluorescence technique for demonstrating dopamine have been particularly useful in delineating corresponding structures of the R-complex. In lizards and birds the R-complex is encapsulated above by the *dorsal medullary lamina* (compare Figure 4-6 with Figure 4-9). Before the availability of histochemical techniques, the dorsal ventricular ridge (DVR) lying above the dorsal medullary lamina was commonly regarded as corresponding to part of the corpus striatum of mammals. It is important to make this point here, because that was the view of the two workers whose experiments we are about to consider.

Although experimental psychologists in America may have been slow in recognizing the important contributions of ethologists in Europe, it might be said that the New World saw the beginnings of a tradition in neuroethology before ethology became a discipline in its own right. Fred Rogers, whose study on *pigeons*[1] we are about to consider, might be credited as one of the first to conduct neuroethological experiments. He was a student of C. Judson Herrick (1868–1960), who was widely known for his work in comparative neurology and his semipopular books on brain function and behavior. Notable among others who followed in this tradition were G. Kingsley Noble (1894–1940), Lester Aronson, and Frank Beach at the American Museum of Natural History.

Rogers's Findings on Mating Behavior of Pigeons

In explaining the reasons for his interest in studying the functions of the avian brain, Rogers cited previous experimental work that had failed to demonstrate either definite motor or sensory functions of the "striatum." "This," he stated, "has left this large mass of brain tissue a physiologic enigma. If neither simple motor nor simple sensory functions are to be attributed to it, it seemed that possibly a study of it as a correlating or associating mechanism might be of value. In examining such a view, it was thought that the instinctive reactions of the bird to environmental stimuli (food, other birds, etc.) would serve as suitable test objects."[2]

Although Rogers set out to observe the effects of lesions of different parts of the forebrain on (1) normal feeding, (2) self-protection, (3) mating, and (4) nesting reactions of domestic pigeons, his observations of the mating behavior came nearest to being carried out in a systematic fashion. In the present chapter, I will focus mainly on this aspect of his study.

As a guide for his behavioral observations, Rogers drew upon the work of Whitman[3] and his students Carr, Craig, and Riddle. For the neurobehavioral work on mating of male pigeons he used the following checklist: (1) pursuit of the newcomer, whether a female or another male; (2) pecking and fighting; (3) charging, cooing, and bowing; (4) wing-preening; (5) billing (insertion of female's beak in male's mouth followed by crop regurgitation by the male); (6) alternate wing-preening and crop regurgitation; and (7) mounting and copulating with the female after she stoops.[4] For the female he noted: (1) avoidance of the male; (2) preening of neck feathers of male; (3) wing-preening; (4) billing; (5) stooping; and (6) submitting to copulation.

Altogether Rogers presents protocols on seven animals in which he had made preoperative observations followed by periodic observations for as long as 2 years. There were four males and three females. He shows drawings of representative sections through the brains of six of these animals that had had varying amounts of the forebrain removed by suction.

As in the lizard, we will refer to the ganglionic mass below the dorsal medullary lamina as the paleostriatum. The bird most relevant to our interest in the functions of the paleostriatum was a male pigeon identified as Number 71. As indicated in the third column of brain diagrams in Figure 12-1, the posterior three-fifths of the cerebral hemispheres overlying the paleostriatum and adjacent ectostriatum had been removed. The structure corresponding to the mammalian amygdala was removed on one side and damaged on the other. Five and one-half months after operation, this male went through the entire courtship and copulatory sequence listed above. Apropos of neural mechanisms underlying parental behavior that will be considered in Chapter 21, it is of interest that this pigeon failed to take his turn in sitting on the eggs of his mate.

For purposes of control, Rogers had operated on a number of pigeons in which he had removed the entire forebrain. He found, like previous workers (see Chapter 3), that such preparations were characterized by the lack of spontaneous behavior, inability to initiate feeding or other kinds of behavior. He showed the reconstruction of the brain of one animal (Number 130) in which he had failed to remove the entire forebrain. This pigeon recovered the ability to feed itself and was observed to coo and strut in the presence of another pigeon. It would also "rush and jump" on a newcomer. As indicated in Figure 12-1 (column 2) it turned out that in this animal the paleostriatum and part of the adjoining ectostriatum had been preserved on one side. Rostrally, there was a remnant of DVR.

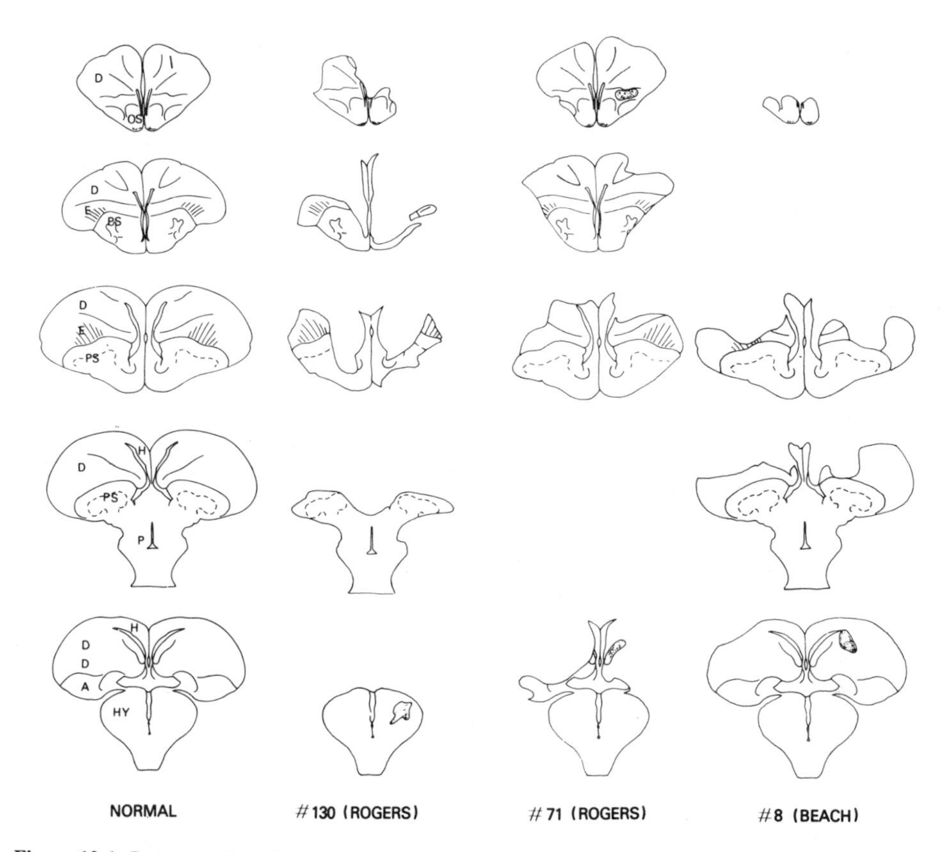

Figure 12-1. Reconstruction of pigeon brains from studies of Rogers (1922) and of Beach (1951), showing amount of forebrain destruction compatible with part or all of the courtship and mating activities of male pigeons. Reconstructions in columns 2, 3, and 4 are to be compared with column 1, showing corresponding levels of the normal pigeon's forebrain.

With only forebrain remnants including paleostriatum and ectostriatum on one side, Rogers's pigeon No. 130 (second column) recovered the ability to feed itself. It would also coo and strut in the presence of another pigeon and would "rush and jump" on a newcomer. Pigeon No. 71 (third column), with destruction of the *posterior* three-fifths of the cerebral hemispheres overlying the paleostriatum and ectostriatum, recovered the ability to go through the entire courtship and mating sequence. The findings on pigeon No. 8 (fourth column) from Beach's study supplements those of pigeon No. 71. After removal of the *forward* part of the brain overlying the paleostriatum, this bird also recovered its usual courtship and mating behavior.

Abbreviations: "D" is used as a blanket symbol for all the ganglia in the anterior dorsal ventricular ridge (DVR) except the ectostriatum (E). "A" identifies the part of the posterior DVR corresponding to the amygdala. H, hippocampal formation; HY, hypothalamus; OS, olfactostriatum; P, preoptic area; PS, paleostriatum.

Drawings in column 1 based on those in the Karton and Hodos atlas (1967); reconstructions in the remaining columns redrawn after illustrations of Rogers (1922) and of Beach (1951).

Comment on Rogers's Findings

Granted that Rogers's study includes only a small number of animals in which the brains were histologically examined, his experiments provide certain information that makes them of interest. They demonstrate, first of all, that adult pigeons fail to feed spontaneously or to show *elements* of courtship behavior following cerebral ablations unless the paleostriatum is preserved at least on one side. Second, they reveal that pigeons completely devoid of the rudimentary cerebral cortex and large amounts of the DVR overlying the paleostriatum are capable of performing the entire sequence of acts seen in courtship and copulatory behavior. Left unanswered was to what extent the remnant of the forward two-fifths of the DVR contributed to the organized expression of such behavior. Rogers himself was inclined to the view that the hyperstriatum (forming a large part of the DVR) is necessary for carrying out "the sequence of events in the mating cycle."[5]

Supplementary Findings

In 1951, Frank Beach published some findings on pigeons that partially extend and confirm those of Rogers.[6] Beach was primarily interested in learning whether or not treatment with male hormone (testosterone) would reinstate copulatory behavior of male pigeons in which lesions of the forebrain interfered with mating behavior. He systematically scored the presence or absence of different aspects of mating and copulatory behavior before and after surgical operations in which he sucked out varying amounts of the forebrain. The eight forms of behavior that he listed were: strutting; cooing; bowing; nibbling; wing signaling; billing; mounting; and copulating. It will be noted that there are three behavioral items on Beach's list that were not included by Rogers—namely, "strutting," "nibbling," and "wing signaling." "Strutting" refers to the male's characteristic rapid walking back and forth in front of, or around, the female. "Nibbling" is a useful descriptive addition because it distinguishes the gentle pecking seen in courtship from aggressive pecking. In Beach's words, it "refers to the act of pecking gently at the head of the partner, usually in the region around the eyes or on top of the head."[7] "Wing signaling" applies to a vertical up-and-down movement of the wings while they are held close to the body. It recalls the motion of a hatchling's wings as it attempts to keep its balance while being fed by the parent bird.

An analysis of Beach's study reveals three findings that either confirm or add to Rogers's observations. Beach presents the reconstruction of the brains of 16 animals. There were 4 in which the cerebrum was removed on one side, while in the remainder there was an attempt to destroy varying amounts of both hemispheres. The findings relevant to those of Rogers may be summarized as follows:

1. None of the four animals with the hemisphere removed on one side mated without treatment with testosterone. It may be inferred from the drawings that the left hemisphere was removed in two pigeons, while the right was excised in the other two. Pigeon Number 4, with presumably the left hemisphere removed, copulated 24 hr after the third injection of 1 mg testosterone and was thereafter observed to mate and copulate without benefit of treatment. Pigeon Number 5 copulated 24 hr after the fifth treatment. Since the left hemisphere was presumably ablated in the first case and the right hemisphere in the second, these findings would extend those of Rogers by showing that either the left or right hemisphere may sustain mating and copulation.

2. As in Rogers's cases, bilateral superficial lesions of the cerebrum (including the rudimentary cortex) did not interfere with mating and copulation.

3. There were four animals with extensive invasion of the DVR that failed to mate spontaneously. One of these pigeons (Number 8) regained the ability to mate and copulate after treatment with testosterone. Significantly, it also mated after a long period without treatment—once on the 75th day after the last injection and again on the 79th day. As

illustrated in Figure 12-1, this animal complements the findings on pigeon Number 71 in the experiments of Rogers, because most of the brain overlying the paleostriatum was excised except for the most posterior part. In Rogers's case, all of the gray matter overlying the paleostriatum had been removed except for the rostral two-fifths.

Summarizing Comment

To sum up, we may infer from Beach's experiments that the right and left hemisphere are equipotential with respect to their capacity to sustain all aspects of mating and copulatory behavior of pigeons. Most significantly with respect to the present question of localization of function, the combined evidence from Rogers's and Beach's studies indicates that the paleostriatum together with remnants of either the ADVR or PDVR is sufficient for the expression of the various manifestations of mating and copulatory behavior. The question of whether or not the paleostriatum by itself would be sufficient for the organized expression of these forms of behavior remains unanswered.

Because of complicating interference with the brain's blood supply and other sequelae, it is doubtful that an answer to the question just raised can be obtained by performing cerebral ablations in adult birds. Brain operations, however, on a developing fetus or newborn animal may be well tolerated. As to be described in Chapter 21, hamsters deprived of the entire neocortex at the time of birth, develop quite normally and show all of the naturally occurring behaviors typical of hamsters.

Collias[8] has reported that 1- to 2-day-old chicks with ablation of the entire cerebrum above the dorsal medullary lamina "were virtually normal" in regard to imprinting and in moving toward clucking sounds. The further application of this approach (including, if possible, intervention in the embryonic stage) in the study of other basic forms of behavior will hopefully extend the comparative knowledge of striatal function.

References

1. Rogers, 1922
2. Rogers, 1922, p. 25
3. Whitman, 1919
4. Rogers, 1922, pp. 31–32, 35
5. Rogers, 1922, p. 45
6. Beach, 1951
7. Beach, 1951, p. 37
8. Collias, 1980

13

Role of R-complex in Display of Squirrel Monkeys

It will help to give an understanding of how our work on display behavior developed, if I take a moment to explain my choice of squirrel monkeys for electrophysiological and neurobehavioral research. In 1957 when I joined the Laboratory of Neurophysiology at the National Institute of Mental Health, my prospective experiments concerning visual influences on limbic functions called for the use of a monkey. I did not wish to use the well-known rhesus monkey because, in addition to certain other disadvantages, its brain is too large. The reason for this statement is readily appreciated when one recalls that a cube with dimensions twice those of a smaller cube has a volume eight times as great. This means that much more time is required for studying and processing a large brain than one, say, with dimensions one-third or one-half as great. On the basis of my initial experience with squirrel monkeys (*Saimiri sciureus*) in 1951,[1] I decided to use this species for the work in question. These New World monkeys are indigenous of Central and South America. Not much bigger than a tree squirrel, the squirrel monkey has an oval-shaped brain that could almost fit into the shell of a chicken's egg. A brain of this size is of desirable dimensions for recording electrical activity and for anatomical study. Table 13-1 gives measurements that we obtained of the body and brain of squirrel monkeys used in our work.

Display Behavior of Squirrel Monkeys

In addition to preparing a brain atlas,[2] it was necessary for the neurobehavioral work to obtain a detailed behavioral profile of the squirrel monkey. For that purpose it was desirable to have the collaborative help of an ethologist. At that time, however, there were only a few workers in our country with an interest in ethology. In Europe, I learned of a young neurologist, Detlev W. Ploog, who had had an introduction to ethology at the Max-Planck Institute at Seewiesen when both Lorenz and von Holst were there in residence. In 1958 he came to work with us as a visiting scientist for a period of 2 years.

Our first colony of squirrel monkeys contained three adult males, one juvenile male, and two adult females.[3] It soon became apparent that in a communal situation, males of this species display the erect phallus under a variety of conditions. We identified four main kinds of displays that we referred to as (1) aggressive, (2) courtship, (3) greeting, and (4) submissive displays. What we called "aggressive" displays are comparable to the challenge displays of lizards, while at least one form of the "greeting" display would probably be the counterpart of a lacertian signature display.

Table 13-1. Sample Bodily and Brain Measurements of Squirrel Monkeys[a]

	Number of animals	Mean	Standard deviation
Weight	76	717 g	± 170.4
Body length	49	26 cm	± 2.47
Tail length	50	36 cm	± 2.14
Total length	48	63 cm	± 3.57
Head length	45	5.4 cm	± 0.22
Head width	45	4.1 cm	± 0.36
Brain weight	21	26 g	± 1.72
Brain length	29	4.9 cm	± 0.21
Brain width	30	3.5 cm	± 0.17

[a]From Carmichael and MacLean (1961).

I will first describe the aggressive, courtship, and submissive displays, and then consider "greeting" displays. I adopt this order because the neurobehavioral work focuses on the "mirror display," which I consider to be a variation of a greeting display. It also contains elements of the aggressive and courtship displays, making it an excellent paradigm for an investigation of neural mechanisms underlying prosematic communication.

Aggressive and Courtship Displays

We have seen that territorial adult male lizards engage in aggressive displays of two types referred to as the "distant" and "close-in" challenge displays. At the time we made our observations on the display behavior of squirrel monkeys, I had not had the opportunity to study lizards. What we called the "distant" display in the original paper on the displays of squirrel monkeys,[4] may be comparable to the distant challenge display of lizards. As illustrated in Figure 13-1, male squirrel monkeys use a similar display in both aggression and courtship.[3] In each case, the male approaches the other animal head-on, vocalizes, places one or both hands on its back, spreads one thigh, and thrusts the erect phallus toward the head or side of the other monkey. The recipient of the display may sit quietly with its head bowed, or make ducking movements with its head, as though dodging a blow.

During the course of systematic observations, it became apparent that the close-in display performed by one male to another is used as a means of establishing and maintaining dominance.[3] Caspar, the monkey that proved to be the alpha male in the first colony, displayed to all of the other males, but none displayed to him. He and two other males displayed to a fourth male called Edgar that proved to be of lowest rank. Edgar never displayed to them. A quantification of the different measures used in display and feeding studies revealed that the incidence of aggressive displays by various members of a group is a better indicator of social, hierarchic structure than is the outcome of rivalry for food.[4]

In observations on another colony of six members in which the sex ratio was reversed (i.e., two males and four females), we found that the animal judged to be the dominant female sometimes responded to a distant display of the dominant male by assuming a

Figure 13-1. Depiction of aggressive (challenge) display of a dominant male squirrel monkey to a lower-ranking male. The aggressive male approaches the other monkey, vocalizes, spreads one thigh, and directs erect phallus toward the submissive animal. A similar display is used in "courtship." From Ploog and MacLean (1963).

similar posture.[4] Rarely, a female would engage in a close-in display to another female or male. The female squirrel monkey has a large clitoris, and during a display there is prominent genital tumescence.

In the wild, squirrel monkeys (which run in troops of about 50) are said to be nonterritorial because they do not defend a particular area. Groups living together in a laboratory, however, will behave as if they regarded their habitat as their territory. One sees the aggressive display in its most dramatic form when an adult male is introduced into an established colony of squirrel monkeys. Within 15 sec the males will follow suit in approaching and aggressively displaying toward the stranger,[5] at the same time grinding their teeth. If the stranger does not remain quiet with its head bowed, both males and females attack the newcomer from behind, and it may be necessary to remove it immediately lest it be wounded. In one established colony there was a female that would foment a struggle in the following way: If a male monkey was introduced into the colony and followed the protective protocol of remaining quiet with its head bowed, she would rush up to him and pull him so that he would be thrown off balance. The slight movement required to regain his balance was sufficient to trigger an attack by the resident males.

Submissive Display

The bent-over posture and the bowing of the head that has just been described represents the submissive display of squirrel monkeys. On the basis of observations on several colonies, it became evident that the adoption of a submissive posture by adult male monkeys usually wards off any further aggression by the displaying male. Parenthetically, we found that covering a subdominant male with a loincloth had the effect of reducing the number of displays made to him by the dominant male, but the same garb worn by the dominant male did not significantly decrease the number of his displays.[6]

The animal Caspar already referred to became the dominant male in three different colonies. Because of his prowess and diminutive size, he earned for himself the additional name of a well-known politician. Whenever he was introduced into the company of a new group of animals, he would isolate himself and remain very still for a period of 24 to 48 hr. Upon making our observations on the second or third day, it was evident that overnight he had become the dominant male.

Adult male squirrel monkeys do not attack females or immature males. The nature of the protective signaling system is deserving of study.

Greeting Displays

In our original observations, we identified two distinct forms of distant display that we interpreted as greeting displays.[4] It is possible that some variations of these greeting displays may actually represent distant aggressive (i.e., challenge) displays. Here, I will describe only the distant display that proved relevant to the choice of a mirror display in testing the effects of brain lesions. It invariably happened that when a new monkey was introduced into the animal room, some of the males that were housed separately or in pairs would climb up the sides of their cages and perform a characteristic display. The same type of display also occurred among old inhabitants living in separate cages. The display could be precipitated by merely transferring an old inhabitant from one nearby cage to another. Hence, we began to regard it as a kind of greeting display.

During the display, the monkey climbs the side of its cage, flexes the head to one side, retracts the corners of the mouth while making high-pitched, peeping vocalizations, spreads one or both thighs, and makes thrusting movements with the fully erect phallus.[7]

Under some conditions a monkey will perform abortive (partial) displays in *non*-social contexts, recalling the multiple functions of the signature display in lizards.

Mirror Display

I have described one variety of squirrel monkeys that will regularly perform a greeting display upon seeing their reflections in a mirror.[7] Early in the course of our work with them, we had been aware that some males displayed to their reflections in a mirror, but it was puzzling why some had this proclivity while others did not. An explanation was most desirable because a test employing the mirror display would have special advantages for studies on brain mechanisms underlying species-typical, prosematic behavior.

With continued observation, it became apparent that we were dealing with two varieties of monkeys. As illustrated in Figure 13-2, the mirror displaying animal shown on the left is characterized by a grayish white, circumocular patch that forms a peak above the eye somewhat like a gothic arch. In the other variety of monkeys, which show little or no interest in their reflections, the patch is round like a roman arch (right-hand picture). As a means of ready identification, the two varieties were referred to as gothic-type and roman-type monkeys.[7] This convention has since been adopted by other authors.[8]

Figure 13-2. Two varieties of squirrel monkeys referred to as "gothic" (left) and "roman" (right) because of the pointed and rounded shape of the circumocular patch above the eye. While both varieties engage in the same type of aggressive and courtship display in the communal situation, only the gothic type will consistently display to its reflection in a mirror. See text for experimental significance and Figures 13-3 and 13-4 regarding vocal and chromosomal differences. From MacLean (1964a).

Hershkovitz has recognized the validity of the distinction in his extensive investigation of the taxonomy of squirrel monkeys.[9]

There are other behavioral features that distinguish these two varieties of squirrel monkeys. The gothic and roman types have a different way of scratching themselves during a display[7] and their vocalizations are also different[10] (Figure 13-3). Ma *et al.* have shown differences in the karyotypes of the gothic and roman varieties (Figure 13-4), as well as of the hybrid forms.[11] Here, parenthetically, is an illustration of how detailed behavioral observations can bring to light taxonomic distinctions and lead to genetic studies providing decisive chromosomal information.

Figure 13-3. Sound spectrograms of the isolation calls of gothic (left) and roman (right) squirrel monkeys. Such calls occur when monkeys become separated from one another. Consisting of a sustained tone, the dominant frequency is about the same for both types, but note the downward termination of the spectrogram of the gothic-type monkey, as opposed to the terminal rise for the roman-type. Horizontal time line corresponds to 500 msec. From MacLean (1985a); spectrograms provided by J. D. Newman.

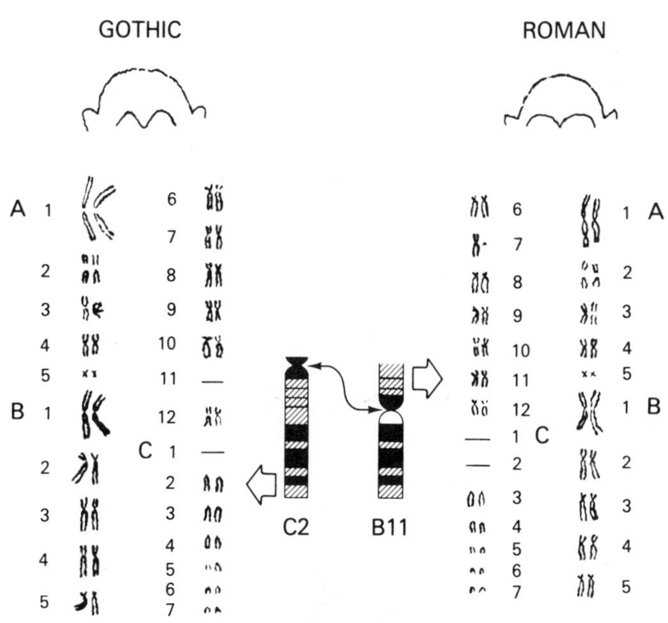

Figure 13-4. Differences in karyotypes of gothic- and roman-type squirrel monkeys. In addition to differences in appearance and behavior, these two varieties of squirrel monkeys have minor chromosomal differences. Their 44 chromosomes can be classified into three groups on the basis of the length and position of the centromeres. The differences become readily apparent by first looking at the acrocentric chromosomes in group C, which look like inverted v's. As illustrated by the enlargement at the center of the figure, Ma and co-workers postulate that the differences between the gothic-type monkey from Colombia and the roman-type from Peru is owing to a pericentric inversion of the B11–C2 chromosomes. In addition to the method of grouping, the demonstration of karyotypic differences depends on the staining of the G bands. Regarding geographic factors, the authors speculate that "differences in the number of acrocentric chromosomes in *Saimiri* may be due to individual migration followed by recombination and selection." Entirely redrawn after Ma *et al.* (1974).

It is basic to the neurobehavioral studies to be described that Ploog and his co-workers have observed that without exposure to any other animal than its own mother, an infant squirrel monkey will display to another monkey as early as the second day of life.[12] *This finding clearly indicates that the display is an innate form of behavior.*

Experiments on the R-complex

Having identified the variety of squirrel monkey that displays to its reflection in a mirror, I decided to use a mirror display test in an attempt to identify brain structures involved in prosematic communication. Since the mirror display incorporates behavioral manifestations of the aggressive and courtship displays, it also had the potential of revealing information about neural mechanisms underlying these forms of behavior. Moreover, as opposed to requirements for testing aggressive and courtship displays, the mirror display test has the special advantage that it does not involve the presence of another monkey and makes it unnecessary to control for olfactory, sexual, and other related variables.

Since an autonomic manifestation is one of the main components of the squirrel

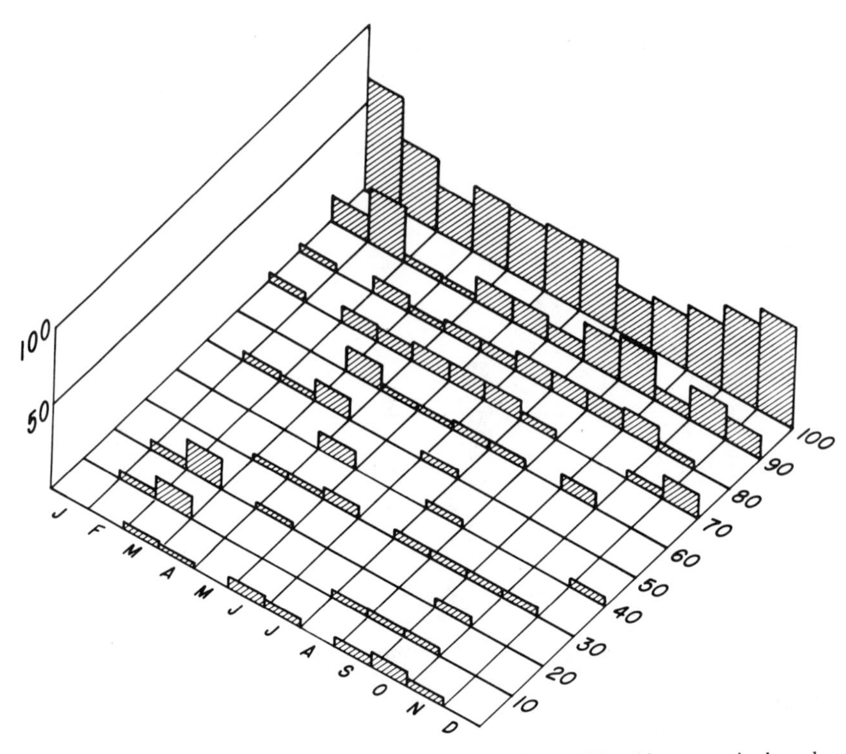

Figure 13-5. Array of histograms, showing results of mirror testing on 105 gothic-type squirrel monkeys during different months of the year. Criterion is the performance of the display in 80% of a set of 30 trials. About 80% of monkeys achieve this level of performance. The fewer number of monkeys achieving criterion in March is statistically significant, while the same is true of the greater number exceeding criterion in January. Consequently, seasonal factors must be taken into account in testing mirror display. Right-hand scale gives percentage values for scores, while the ordinate scale applies to percentage of animals achieving a particular score for respective months indicated by initials. From MacLean (1978b).

monkey's display, the research to be undertaken also afforded the opportunity to identify brain mechanisms involved in the integration of autonomic and somatic functions.

Systematic observations on 105 monkeys (Figure 13-5) showed that 81% gave a full display to a mirror in 80% of a series of 30 trials.[13] Analysis of the same data revealed that a statistically significant number of randomly chosen animals are less likely to display in March[13]. Hence, seasonal factors must be taken into consideration in using the mirror display test. It is quite possible that the March decline is owing to neuroendocrine factors. During the breeding season squirrel monkeys undergo a spermatogenic phase along with a "fatted state."[14] Our monkeys achieve peak weight during January and February.

Testing of the Mirror Display

We test animals twice a day in their home cages where they are screened from seeing other monkeys. (If a monkey is removed from its home cage for the purpose of testing, its performance is unpredictable.) A mechanical contrivance exposes a full-length mirror for 30 sec.[13] Typically, the monkey climbs the side of the cage toward the mirror, tilts the

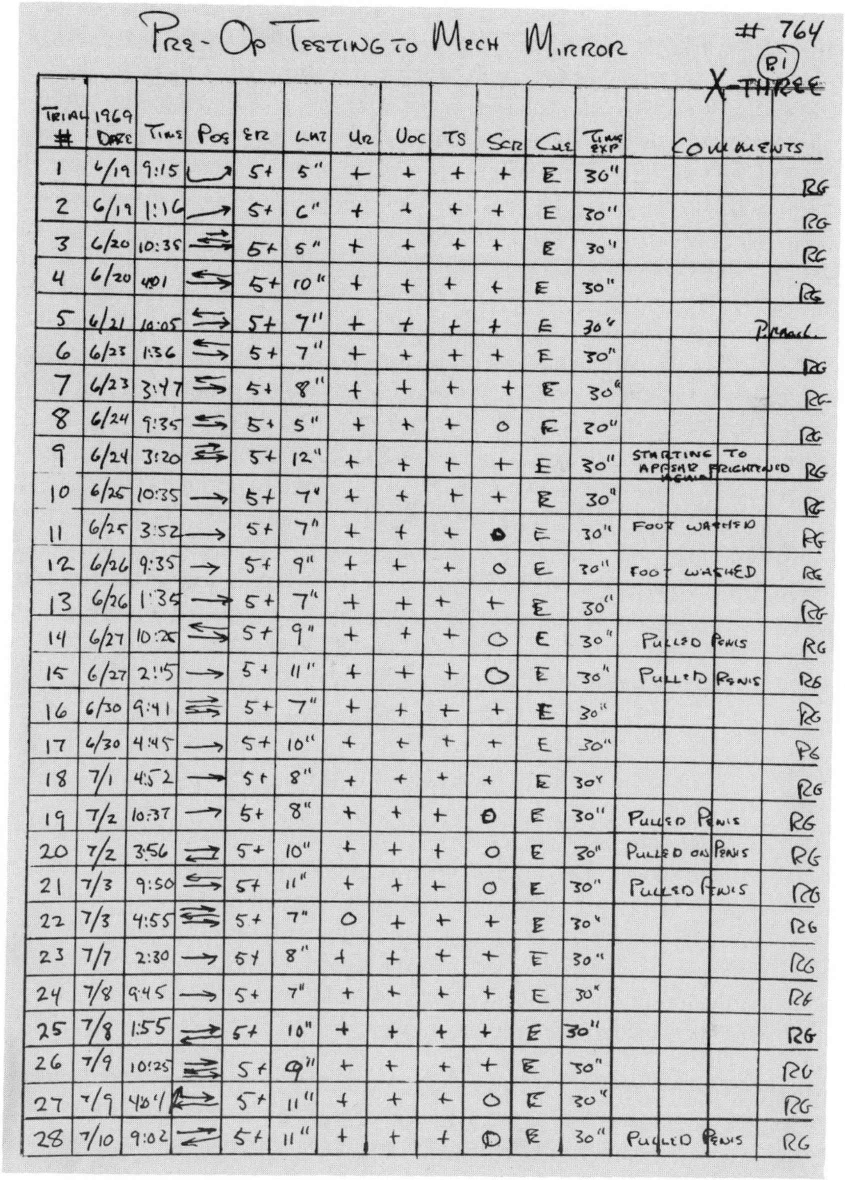

PRE-OP TESTING TO MECH MIRROR # 764 (R1) ~~X-THREE~~

TRIAL #	1969 DATE	TIME	POS	ER	LAT	UR	VOC	TS	SCR	CUE	TIME EXP	COMMENTS	
1	6/19	9:15	↪	5+	5"	+	+	+	+	E	30"		RG
2	6/19	1:16	→	5+	6"	+	+	+	+	E	30"		RG
3	6/20	10:35	⇉	5+	5"	+	+	+	+	E	30"		RG
4	6/20	4:01	⇄	5+	10"	+	+	+	+	E	30"		RG
5	6/21	10:05	⇄	5+	7"	+	+	+	+	E	30"		P. [illegible]
6	6/23	1:36	⇄	5+	7"	+	+	+	+	E	30"		RG
7	6/23	3:47	⇄	5+	8"	+	+	+	+	E	30"		RG
8	6/24	9:35	⇄	5+	5"	+	+	+	O	E	30"		RG
9	6/24	3:20	⇄	5+	12"	+	+	+	+	E	30"	STARTING TO APPEAR FRIGHTENED AGAIN	RG
10	6/25	10:35	→	5+	7"	+	+	+	+	E	30"		RG
11	6/25	3:52	→	5+	7"	+	+	+	●	E	30"	FOOT WASHED	RG
12	6/26	9:35	→	5+	9"	+	+	+	O	E	30"	FOOT WASHED	RG
13	6/26	1:35	→	5+	7"	+	+	+	+	E	30"		RG
14	6/27	10:25	⇄	5+	9"	+	+	+	O	E	30"	PULLED PENIS	RG
15	6/27	2:15	→	5+	11"	+	+	+	O	E	30"	PULLED PENIS	RG
16	6/30	9:41	⇉	5+	7"	+	+	+	+	E	30"		RG
17	6/30	4:45	→	5+	10"	+	+	+	+	E	30"		RG
18	7/1	4:52	→	5+	8"	+	+	+	+	E	30"		RG
19	7/2	10:37	→	5+	8"	+	+	+	O	E	30"	PULLED PENIS	RG
20	7/2	3:56	⇄	5+	10"	+	+	+	O	E	30"	PULLED ON PENIS	RG
21	7/3	9:50	⇄	5+	11"	+	+	+	O	E	30"	PULLED PENIS	RG
22	7/3	4:55	⇄	5+	7"	O	+	+	+	E	30"		RG
23	7/7	2:30	→	5+	8"	+	+	+	+	E	30"		RG
24	7/8	9:45	→	5+	7"	+	+	+	+	E	30"		RG
25	7/8	1:55	⇄	5+	10"	+	+	+	+	E	30"		RG
26	7/9	10:25	⇉	5+	9"	+	+	+	+	E	30"		RG
27	7/9	4:01	⇄	5+	11"	+	+	+	O	E	30"		RG
28	7/10	9:02	⇄	5+	11"	+	+	+	O	E	30"	PULLED PENIS	RG

Figure 13-6. Page from a protocol, illustrating manner of scoring display in squirrel monkeys. Testing twice a day, observer notes magnitude and latency of penile erection, as well as incidence of other components of the display—namely, vocalization (Voc), thigh-spread (TS), scratching (Scr), and urination (Ur). Arrows in column labeled "Pos" indicate position and direction of monkey's movements in cage during 30-sec exposure to a full-length mirror.

head to one side, vocalizes softly with the corners of the mouth retracted, spreads widely one or both thighs, and makes thrusting movements with the fully erect phallus. In more than half of the displays, a monkey will also eject a few spurts of urine and scratch different parts of the body with one hand. Formal testing of 115 monkeys has shown that vocalization and thigh-spreading occur almost as consistently as penile erection.[13] A display combining these three major manifestations is referred to as the *trump display*.[13] Destruction of some parts of the brain may *not* significantly alter the incidence of the display, but will result in a fragmentation of the trump display.

During a test we use the check-sheet shown in Figure 13-6 for recording in different columns (1) the monkey's horizontal and vertical movements in the cage; (2) the latency and (3) size of the erection; and the occurrence of (4) vocalization, (5) thigh-spreading, (6) urination, and (7) scratching. The last column is used for special comments, as well as the notation of less common features of the display such as "foot washing" (urinating on hands and rubbing soles of feet), rolling on back, and defecation.

After a monkey scores 80 to 100% in a series of 30 or more trials, it is scheduled for surgery in an operating room equipped for carrying out procedures with sterile surgical drapes and instruments. A monkey is deeply anesthetized during the operation. The procedure for destroying deep structures by electrocoagulation is essentially the same as that described for lizards in Chapter 11. Surgical aspiration with a small glass or metal sucker is used for removing cortical structures and certain other parts of the brain.

Postoperative testing is resumed as soon as the monkey becomes ambulatory and may be conducted for as long as one year. Each animal has its own "hospital" chart for entering progress notes and recording the results of tests. The use of the same kind of metal chart binders employed on hospital wards serves as a reminder to be ever vigilant and concerned about the care and comfort of the animals. Both before and after operation, animals are weighed and closely inspected once a week. In addition to a high-protein diet they receive supplementary doses of vitamins.[2]

Experimental Findings

Systematic observations on the effects of brain lesions on the mirror display have been made on more than 115 squirrel monkeys. In conducting these experiments, one discovers that the mirror display is remarkably resistant to elimination or modification. If certain parts of the R-complex and its projecting pathways had not been destroyed, the outcome of this investigation would have been largely negative. Lesions in different telencephalic structures of the paleomammalian and of the neomammalian brain had either no effect or only a transitory effect on the performance of the mirror display.

Negative Findings

One cannot have a full appreciation of the positive outcome of the experiments on the R-complex without a prior knowledge of the negative results of destroying many other parts of the brain.[13] For the purpose of condensing an extensive amount of negative data, I have listed in Table 13-2 many structures that can be partially or completely destroyed without altering or eliminating the display.

The superior colliculus, a phylogenetically old visual structure, is under the influence of the corpus striatum via the nigrotectal pathway arising in the zona reticulata of the substantia nigra. Figure 13-7B shows the section from an animal (R-2) in which all but a small and questionably functioning remnant of the superior colliculus had been aspirated. On the 16th postoperative day, this subject began to perform partial displays and by the 29th day gave the full display.

Table 13-2. Brain Destruction Compatible with Full Expression of Mirror Display of Adult Male Squirrel Monkeys[a]

Midbrain and thalamic structures
1. Virtually entire superior colliculus (down to level of central gray)
2. Brachium of superior colliculus
3. Central gray matter of midbrain (rostral to nucleus of 4th nerve)
4. Pretectal area
5. Stria medullaris[b]
6. Habenular nuclei
7. Greater part of centromedian nucleus (nucleus centri mediani)
8. Habenulopeduncular tract, including the part within the interpeduncular nucleus
9. Greater part of interpeduncular nucleus
10. Intermediate substantia nigra

Limbic structures
11. The entire rostral portion of septum
12. Subcallosal cingulate cortex
13. Pregenual cingulate cortex
14. Supracallosal anterior cingulate gyrus
15. Posterior cingulate gyrus
16. Retrosplenial cortex
17. Greater part of the hippocampus
18. Greater part of the mammillary bodies
19. Greater part of the anterior thalamic nuclei
20. Entire amygdala
21. Stria terminalis[b]
22. Fornix[b]
23. Mammillothalamic tract[b]
24. Cingulum[b]

Neocortex and related structures
25. Prefrontal cortex
26. All frontal neocortex rostral to level of genu of corpus callosum
27. Medial parietal cortex
28. Temporal neocortex rostral to vein of Labbé[c]
29. One-sided destruction of dorsal frontal sensorimotor cortex does not interfere with use of contralateral paretic leg in thigh-spread component of display
30. One-sided lesion of internal capsule has same outcome as noted in No. 29

[a]Bilateral destruction unless otherwise specified.
[b]Regardless of location of lesion along its course.
[c]In conjuction with bitemporal lobectomy; see text regarding inappropriate expression of the display.

The amygdala (once regarded as belonging to the corpus striatum) comprises a nuclear group within the frontotemporal division of the limbic system (Chapter 18). Monkeys with large lesions or virtual destruction of the amygdala on both sides continued to give the full display following surgery. Figure 13-8 shows the performance curve of an animal (Q-1) with a large lesion of the amygdala. On the day after operation and through-

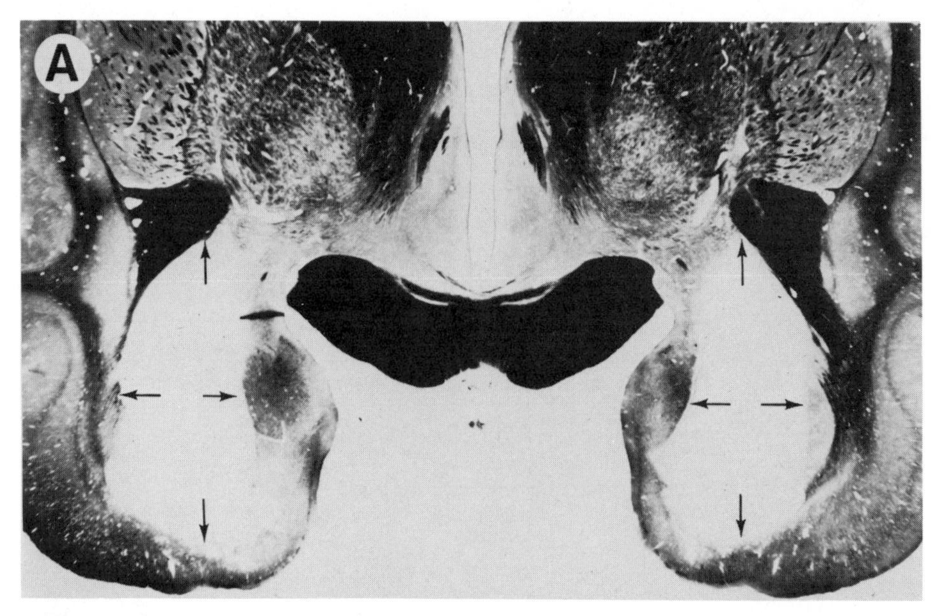

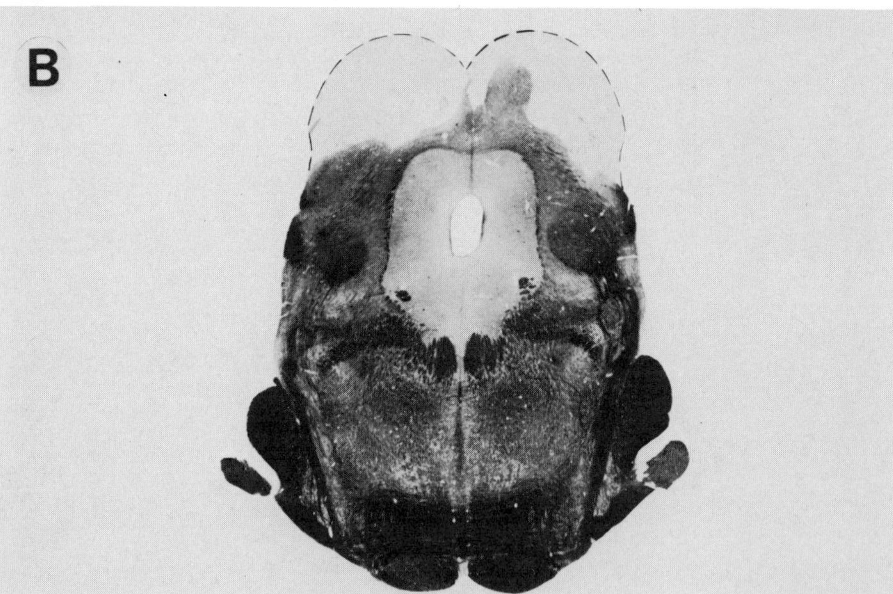

Figure 13-7. Brain sections from two different monkeys, showing in (A) destruction by electrocoagulation of most of the amygdala (monkey Z-4) and in (B) surgical aspiration of virtually the entire superior colliculus (monkey R-2). The white areas enclosed by the dashed lines in B indicate the part of the superior colliculus that was excised.

out the entire follow-up this monkey achieved a score of 100%. Figure 13-7A shows a section from the brain of another case in which virtually the entire amygdala had been shelled out.

In summarizing the work on lizards, it was mentioned that the PDVR corresponds to the amygdala of mammals. It will be recalled that a lesion of this structure failed to affect the challenge display, a result that tallies with the present findings.

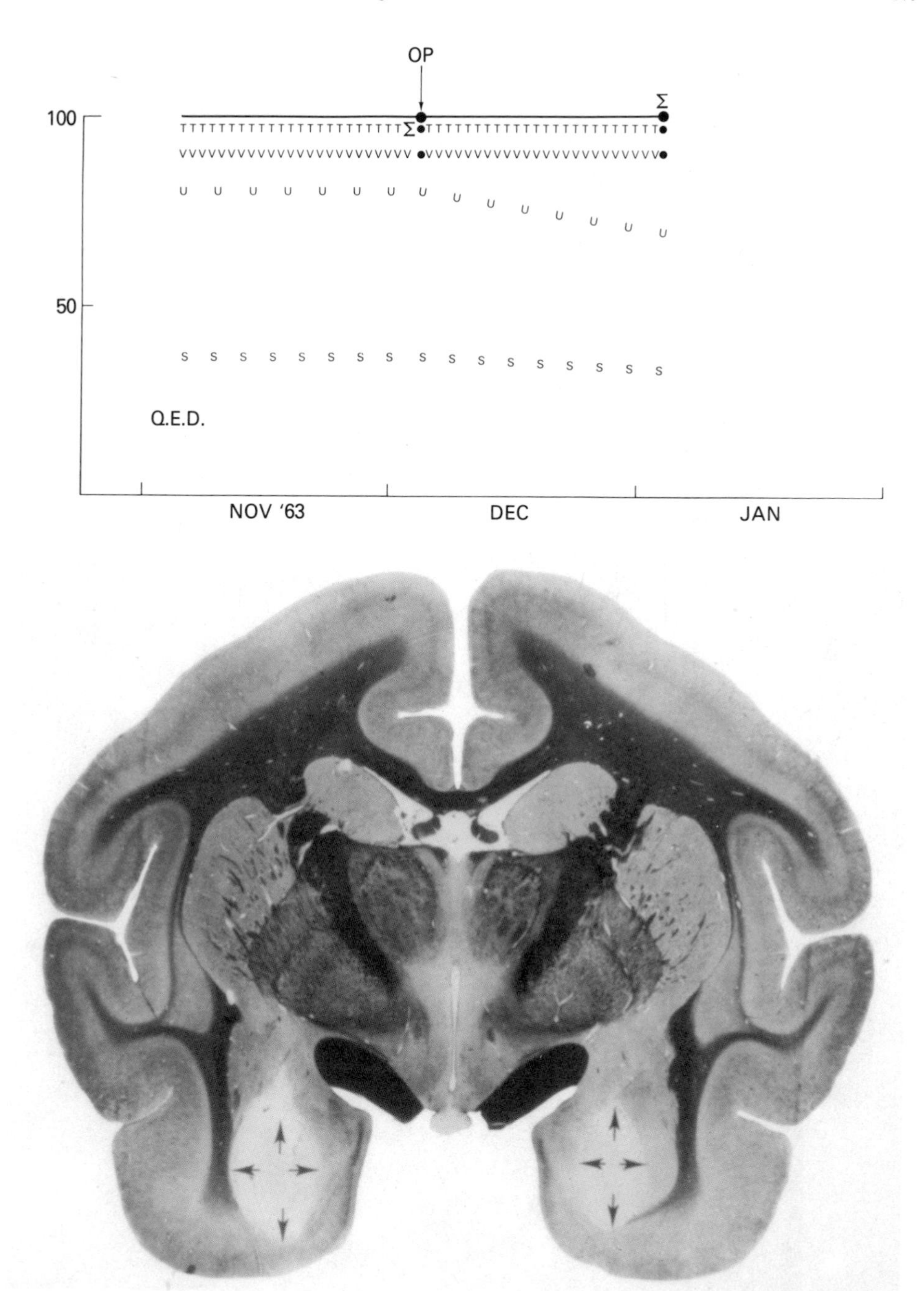

Figure 13-8. Performance curve and brain section from a monkey (Q-1) with a large bilateral electrocoagulation of the amygdala. On the day after surgery and thereafter, this monkey continued to perform a trump display in every test.

Labeling: Solid line represents penile erection, the central manifestation of the display. Curves for other two manifestations of the trump display (see text) are accented by continuous letters (V, vocalization; T, thigh-spread). Curves for minor manifestations are shown by faint lettering (U, urination; S, scratching). Each dot represents an accumulation of 30 trials. The larger dots call attention to the genital index (Σ), the average magnitude of the erection (see text). Ordinate and abscissa scales, respectively, represent percentage of tests and time in months.

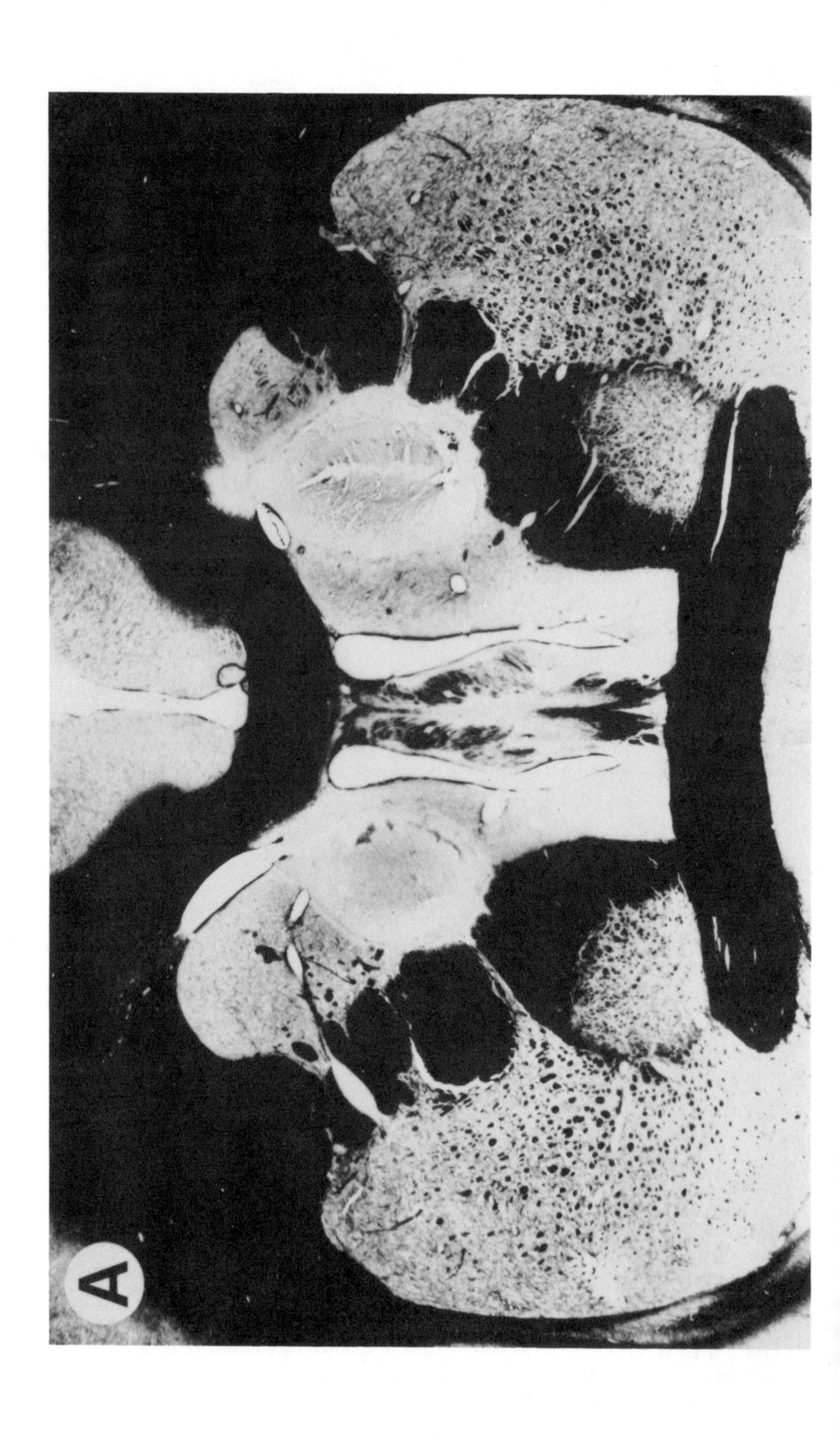
A

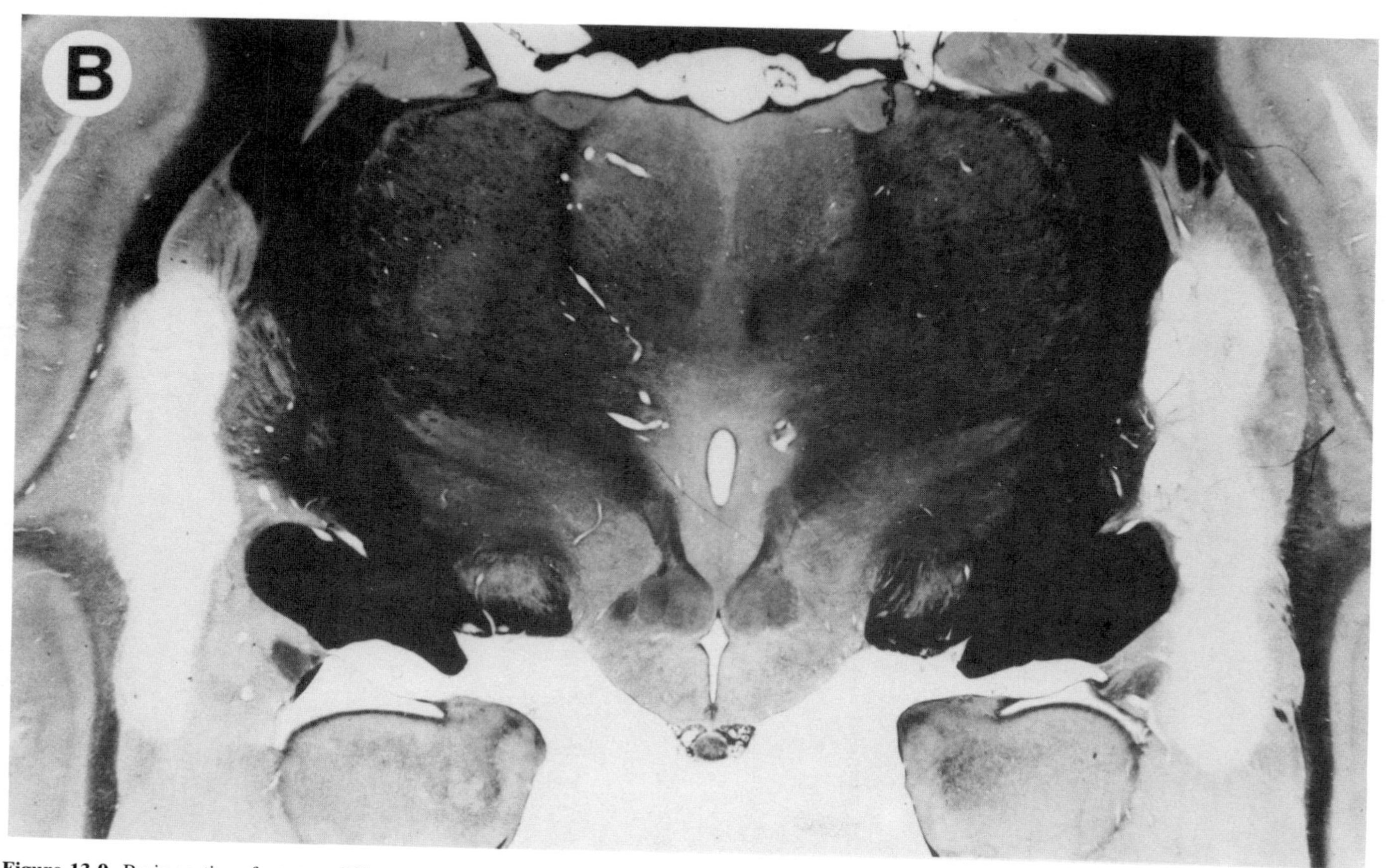

Figure 13-9. Brain sections from two different monkeys (E-1, W-1) showing large oval-shaped coagulations in the caudate nucleus (A) and coagulations in the posterior putamen (B). Such isolated lesions in the corpus striatum were without effect on the display. See Figure 13-10.

Although performed inappropriately, full display was observed following a bitemporal lobectomy that extended caudally to the vein of Labbé. In addition to the amygdala and the greater part of the hippocampal formation, the lobectomy included the neocortex of the temporal convolutions.

In regard to neocortical areas (Table 13-2) that are not essential for the performance of the display, mention should be made particularly of one animal in which one-sided destruction of the leg area of the sensorimotor cortex resulted in a partial paralysis of the contralateral leg. This monkey continued to display and to use the affected hindlimb in performing the thigh-spread component of the display.[13]

Lesions of the R-complex

The anatomy of the R-complex was reviewed in Chapter 4. The telencephalic part of the R-complex is largely represented by the corpus striatum (caudate and putamen) and olfactostriatum (nucleus accumbens and olfactory tubercle). The corpus striatum constitutes a much larger mass than the olfactostriatum. It projects primarily to both segments of the globus pallidus and to the reticulate part of the substantia nigra that histologically resembles the pallidum. The striopallidum is the focus of the experiments next to be described.

Lesions of Corpus Striatum

The corpus striatum is so situated that attempts to destroy large parts of it invariably result in damage to surrounding structures, including the massive internal capsule (see Figures 4-2 and 14-3) that partly divides the caudate nucleus and the putamen. Large bilateral isolated lesions of the head of the caudate or to the putamen (Figure 13-9A and B) have no apparent effect on the mirror display and result in no motor deficits. The same was true of an animal (W-1) in which during a two-stage procedure, 22 small coagulations were placed bilaterally in these structures between frontal levels AP 17 and AP 11 of the brain atlas.[2] In one experiment, however, in which a small electrocoagulation partially cut off the blood supply to the forward part of the corpus striatum and globus pallidus on one side, there was a transient catalepsy and enduring deficit in the performance of the mirror display.[15] The vascular lesion shown grossly in Figure 13-10 had the effect of destroying scattered small groups of cells throughout the rostral part of the right corpus striatum, and to a lesser extent, in the pallidum. Under the microscope the tissue had a freckled appearance. Despite the preservation of intervening tissue, the tiny lesions would have had the effect of interrupting communication in an extensive neural network. This finding would suggest that if a means could be found to produce widespread interruption of the striatal neural network on both sides, it might be possible to demonstrate marked changes in display and other forms of behavior.

The foregoing contrast in the effects of large focal lesions and widespread damage, calls to mind writings on this subject by the late Fred Mettler. In discussing past failures to demonstrate behavioral changes following destruction of various parts of the R-complex, he emphasized the factor of ''physiological safety'' as enunciated by Meltzer (1851–1920).[16] For example, up to a limit, the greater part of the liver or other vital organs may be removed without manifest impaired bodily function. In like vein Mettler proposed that a large part of a brain structure may be destroyed without an evident impairment of function. The situation would be somewhat analogous to the use of a broken mirror: the

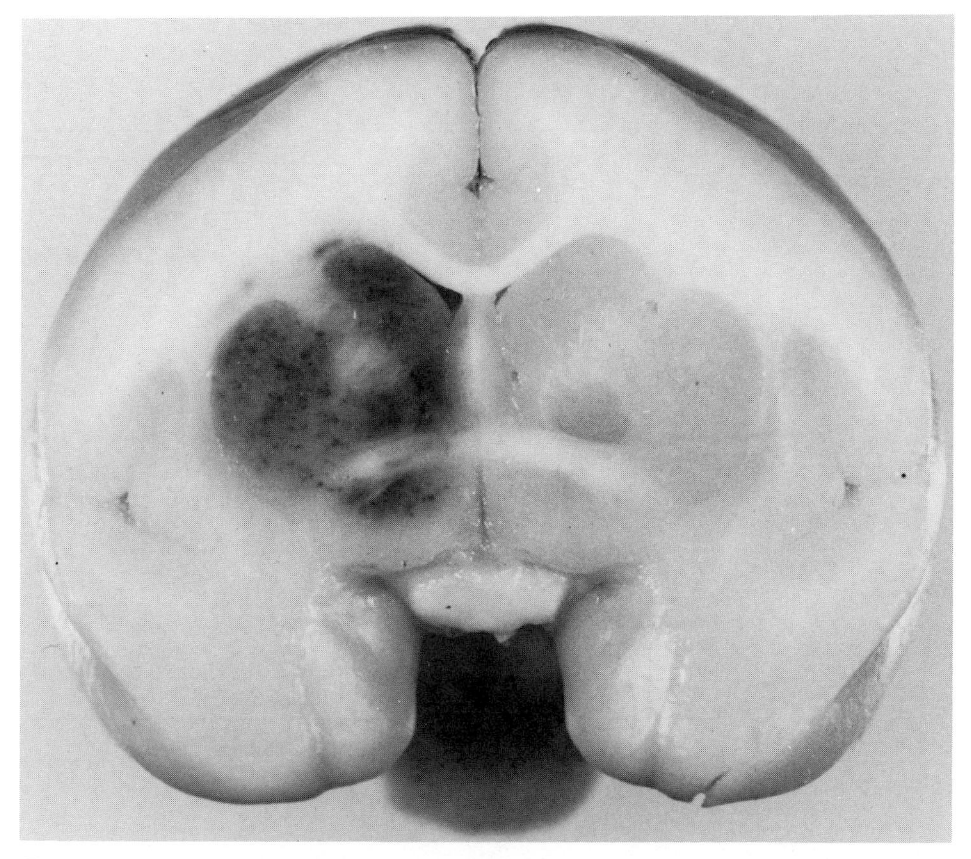

Figure 13-10. Freshly cut brain, in which the forward part of the striatal complex in the right hemisphere was riddled with tiny hemorrhages (left side of figure). The widespread lesions apparently resulted from interference with the blood supply during the insertion of an electrode into the ventromedial nucleus of the hypothalamus. After apparent recovery in all other respects, this monkey (J-2) performed only erratic and partial displays. It is of interest to compare the selective distribution of the blood supply to the striatal complex with the selective cholinesterase staining in Figure 4-5. From MacLean (1972a).

greater part of a mirror could be destroyed before it would no longer be useful for obtaining a reflection of things. A technological analogy would be a hologram, which some workers believe has attributes of a kind possessed by the brain.[17] From a small part of a hologram one can reconstruct the entire picture.

Olfactostriatum, Meynert's Nucleus, Substantia Innominata

In cases in which there were bilateral partial coagulations of the olfactostriatum,[18] substantia innominata, or basal nucleus of Meynert,[19] there were no deficits in the performance of the mirror display or any other noticeable behavioral effects.

Lesions of Globus Pallidus

Since the globus pallidus is the major destination of projections from the corpus striatum and since it receives relatively few afferents from other structures, it would be

Table 13-3. Volume of Left and Right Globus Pallidus (GP) Destroyed in Respective Animals[a]

Animal	Globus pallidus total volume (mm³)	Amount of destruction (% mm³)			
		Total[b]	Left GP	Right GP	Parts in common
P-3*	144	5 (7)	4 (3)	6 (4)	3 (4)
I-4	141	7 (10)	7 (5)	7 (5)	5 (7)
W-3	120	8 (10)	9 (5)	8 (5)	7 (9)
H-4	156	8 (13)	9 (6)	8 (6)	8 (13)
X-3***	135	16 (21)	17 (12)	14 (9)	10 (13)
F-4***	138	16 (22)	14 (11)	17 (11)	10 (13)
T-3	127	17 (22)	21 (15)	12 (7)	10 (12)
U-3***	114	19 (22)	20 (12)	17 (10)	2 (2)
K-4	111	21 (24)	26 (14)	17 (10)	13 (14)
A-4***	132	23 (31)	13 (9)	34 (22)	12 (16)
J-4***	113	24 (27)	25 (14)	23 (13)	15 (17)
L-4***	143	24 (35)	20 (14)	29 (21)	18 (26)
G-4***	150	25 (37)	9 (7)	39 (30)	5 (8)
V-3**	132	32 (43)	36 (21)	30 (22)	19 (25)
Mean	132 ± 14	18 (23)	16 (10)	19 (13)	10 (13)
		Nonlesioned animals			
H.C. 1	147				
H.C. 2	151				

[a]From MacLean (1978b).
[b]Listing is in rank order according to total amount of destruction. Calculations based on volumetric measurements for each side. Figures rounded.
*$p < 0.01$ at termination of experiment.
**$p < 0.0025$ at termination of experiment.
***$p < 0.0001$ at termination of experiment.

reasonable to expect that sufficiently large pallidal lesions on both sides might provide inferences about the functions of the striatum itself. I therefore conducted a series of experiments in which different parts of the globus pallidus were electrocoagulated in 14 animals. Following surgery, the performance of the display was affected in all but one animal.[13] There was a virtual elimination of the display in five monkeys (X-3, A-4, F-4, G-4, J-4); a persisting, statistically significant, decreased incidence in four (P-3, U-3, V-3, L-4); and a transient, significant decline in four (T-3, W-3, H-4, K-4). As noted earlier, vocalization, thigh-spreading, and thrusting of the erect phallus are three major manifestations of the display, constituting the trump display. Short of an elimination of the display, a fragmentation of the trump display was a diagnostic sign of pallidal damage, with a decrease in the incidence of thigh-spreading being the most sensitive indicator.[13]

The anatomical reconstructions revealed that for all animals combined, the lesions involved the entire rostrocaudal extent of the globus pallidus. Planimetry was used for estimating the amount of damage in each case. The analysis showed that both the location and the size of the lesions were determining factors in the outcome of an experiment. In all animals in which there was an enduring elimination or a significantly decreased incidence of the display, the lesions involved the rostral two-thirds of the medial segment. With destruction of about 16% or more of this part of the medial segment (Table 13-3), one may expect an enduring elimination of, or significant decline in, the performance of mirror

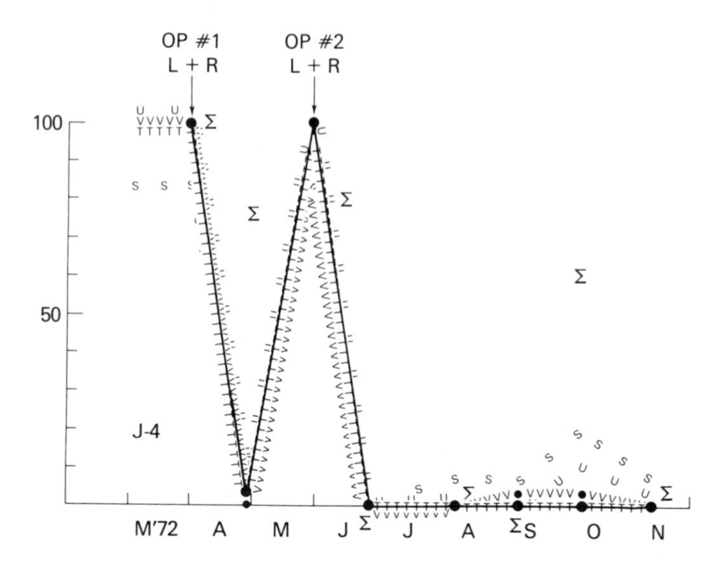

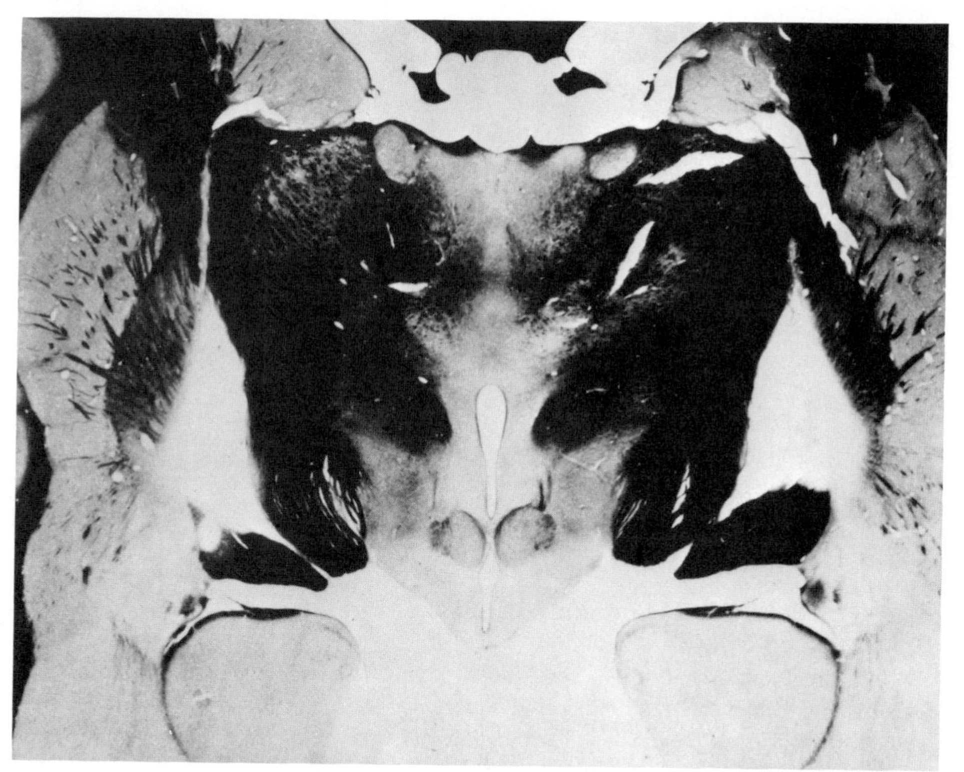

Figure 13-11. Performance curve of monkey J-4, together with a brain section showing bilateral electrocoagulations in the globus pallidus. The operations were performed in two stages (OP #1 and OP #2), with coagulation first of the posterior part of the globus pallidus, and then a part of the internal segment farther forward. After the second procedure there was a virtual elimination of the display continuing to the time of sacrifice nearly 6 months later. See Figure 13-8 regarding abbreviations and labels. Excerpted from MacLean (1978b).

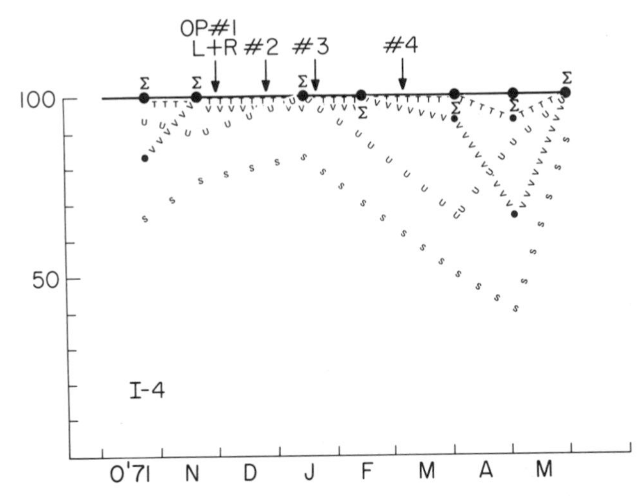

Figure 13-12. Performance curve of monkey I-4, illustrating close-to-perfect performance of the trump display throughout a period of 6 months despite four bilateral operative procedures. Such findings help to reinforce other evidence that deficits in the display following coagulations of the medial segment of the globus pallidus cannot be attributed to nonspecific effects of operative procedures and multiple penetrations of the brain. See Figure 13-8 regarding abbreviations and labels. From MacLean (1978b).

display. Figure 13-11 shows the performance curve and the location of the lesions in one animal (J-4) observed for $5\frac{1}{2}$ months following the second operation and in which there was a virtual elimination of the display.

In six of seven animals subjected to more than one operation, the initial bilateral lesions were placed in the posterior part of the globus pallidus. Such lesions had either no effect or only a transitory effect on the display.

Two animals (W-3, K-4) with lesions primarily involving the lateral segment eventually recovered the ability to display.

In regard to controls for nonspecific effects of operative procedures and multiple penetrations of the brain, it is significant that one animal (I-4) subjected to four operations in which lesions involved the caudalmost part of the globus pallidus, the olfactostriatum, and the tip of the temporal lobe continued to display in 100% of the trials throughout 6 months of testing (Figure 13-12).

Assessment of Complicating Factors

As reported in the original study there were a number of actual or questionable complicating factors that required consideration in assessing the effects of pallidal lesions on the performance of the display. The outcome of the analysis in regard to the main items is summarized under the headings of (1) transitory failure to eat and drink; (2) bradykinesia; (3) seasonal effects; and (4) motivation.

Transitory Failure to Eat and Drink

Eight animals with lesions involving the rostral two-thirds of the medial pallidal segment failed initially to eat or drink and required hand feeding for periods ranging from

3 to 9 days (mean: 6 days). Resumption of spontaneous eating and drinking usually occurred at about the same time. Thereafter all except three subjects maintained their weight or gained weight. Failure of ingestion was not a problem in those cases with lesions in the caudal or lateral parts of the pallidum. Based on the nutritional and weight data, as well as the appearance of the animals, it was unlikely that the display deficits could be attributed to a deterioration of health.

Bradykinesia

Six animals developed a persisting mild degree of bradykinesia. One of these (K-4) with lesions involving primarily the lateral segment recovered its ability to display. Hence, bradykinesia in itself is not sufficient to interfere with the performance of the display. (An in-depth discussion of this subject matter, including neural mechanisms, is presented in Chapter 14.)

Seasonal Effects

As noted earlier, significantly fewer gothic-type monkeys achieve criterion in the mirror display during March. There was only one animal (A-4) in which testing during March might have been a contributing factor in the loss of performance of the display.

Motivation

Reference must be made to the original paper for further details regarding the complicated question as to whether or not a loss of motivation might have accounted for the failure to display. Here it may simply be stated that all except one animal gave evidence of being ''motivated'' to approach the mirror once the test was started.

Summarizing Discussion

The present investigation has utilized the mirror display of one variety of squirrel monkey as a paradigm to test the hypothesis that the striatal complex is implicated in the organized expression of species-typical forms of communicative behavior. As background, findings in numerous experiments were summarized showing that many structures elsewhere in the forebrain, as well as in the optic tectum, are not essential for the performance of the mirror display.

The corpus striatum (caudate nucleus and putamen) constitute the largest and ''highest'' representation of the striatal complex. Initial experimentation revealed that large bilateral single or multiple lesions of the caudate and putamen have no demonstrable effect on the performance of the display. In one case, however, in which a disseminated lesion of the rostral putamen and caudate on one side was fortuitously produced by a vascular lesion, there was an enduring deficit in display behavior. Given the present inability to produce extensive destruction of the striatum without complicating damage to neighboring surrounding structures, is there an experimental alternative? Since the globus pallidus is the only known structure of the *forebrain* to receive striatal connections, it is a reasonable assumption that pallidal lesions might result in deficits providing clues as to the nature of striatal functions.

A systematic investigation has shown that destruction of certain parts of the globus pallidus results in an enduring elimination or fragmentation of the mirror display. The most effective lesions are those involving the rostral two-thirds of the medial segment. Lesions of the posterior globus pallidus are ineffective, while those involving primarily the lateral segment do not prevent eventual recovery of the display.

Anatomical Factors

Is there an anatomical explanation that would account for the different consequences of lesions in respective parts of the pallidum? The medial and lateral segments appear to receive similar connections from the corpus striatum (Chapter 4), but each segment projects to different regions.[20] As schematized in Figure 4-2, the *lateral* segment (which does not appear to be critical for the display) projects mainly to the subthalamic nucleus, which in turn innervates the medial segment.[21] The functions of the subthalamic nucleus are still unclear, but there is both experimental[22] and clinical evidence[23] that damage to this nucleus on one side results in hemiballismus, an expression for periodic attacks of violent movements of one side of the body suggestive of throwing a ball.

In regard to the posterior part of the globus pallidus, lesions in this area would destroy many of the connections of the lateral segment with the subthalamic nucleus, as well as bundles of fibers descending from the corpus striatum to the substantia nigra.

The medial segment of the globus pallidus, in contrast to the lateral segment, gives rise to a cascading fiber system that divides into two streams, one leading to nuclei of the thalamus and the other to structures of the thalamic and midbrain tegmentum. It is unnecessary to go into details at this time because in the next chapter we shall consider a simplified presentation of the anatomy prior to showing that destruction of pathways from the medial segment, results in the same kind of behavioral changes as damage to the segment itself. Such correlative findings add cogent evidence that the R-complex is basically involved in prosematic communication.

Complicating Factors

Chapter 14 will deal with a functional analysis of medial pallidal projections, and we shall find again that a number of animals develop a bradykinesia and experience a transient failure to eat and drink. These manifestations (somewhat more severe with brainstem lesions) will again require an evaluation in regard to their effect on display behavior. In addition, we shall encounter a new manifestation—hypothermia—that will likewise need to be assessed.

Relevant to the emphasis that was placed on the master routines and subroutines of lizards, it is of interest to comment on changes in the routine of the monkeys with pallidal lesions while they were being tested. The analysis of records of 15 subjects has shown that most monkeys adhere to a particular sequence of movements in the cage once a test of the mirror display has begun.[13] In the present experiments, an individual monkey usually engaged in two or three patterns of movement during the preoperative display tests. Postoperatively, there was a disorganization of the pretest performance in all but one animal. The number of different movement patterns increased on the average by almost a factor of two.[13]

Negative Aspects of Prior Experimentation

Reference was made in Chapter 4 to the negative findings of Ranson and his group in their experiments on the striatal complex. Lesions of the caudate nucleus and putamen in adult macaque monkeys resulted in no symptoms. Commenting on the outcome of experiments on the pallidum, Ranson and Ranson said, "[No] monkeys with large bilateral pallidal lesions showed the mask-face or the slowness and poverty of movement found in monkeys with bilateral hypothalamic lesions. It, therefore, seems probable," they concluded, "that damage to the hypothalamus itself, and not the associated interruption of the ansa and fasciculus lenticularis, was responsible for these symptoms."[24] Because of the negative behavioral findings and the earlier anatomical observations that lateral hypothalamic lesions resulted in a disappearance of neurons in the medial pallidal segment,[25] it is understandable that these investigators were disinclined to attribute any specific function to the globus pallidus. In Chapter 15 there will be occasion to comment on "negative symptoms" (or what I will otherwise call nonevident symptoms), recognized clinically in association with disease of the striopallidum. Manifestations of this nature may not be observed unless they are specifically looked for; this would have been the case in the display studies described in this chapter and in Chapters 11 and 12 had the display manifestations not been specifically looked for. We shall see more examples of nonevident symptoms in Chapter 14.

In conclusion, the results of the present study suggest that the medial globus pallidus is a site of convergence of neural systems involved in the species-typical mirror display of gothic-type squirrel monkeys. Together with the experiments described in Chapters 11 and 12, the findings indicate that in animals as diverse as lizards and monkeys, the R-complex is basically involved in the organized expression of species-typical, prosematic behavior of a ritualistic nature. Since the behavior in question involves conspecific recognition and isopraxis, the results also suggest that the R-complex is implicated in the psychological processes underlying these respective functions.

References

1. MacLean and Delgado, 1953
2. Gergen and MacLean, 1962
3. MacLean, 1962; Ploog and MacLean, 1963b
4. Ploog and MacLean, 1963b
5. MacLean, 1973a, pp. 48–49; 1978b, p. 190
6. Denniston and MacLean, 1961
7. MacLean, 1964a
8. Rosenblum and Cooper, 1968
9. Hershkovitz, 1984
10. Winter, 1969
11. Jones *et al.*, 1973; Ma *et al.*, 1974
12. Ploog *et al.*, 1967
13. MacLean, 1978b
14. DuMond and Hutchinson, 1967
15. MacLean, 1972a
16. Mettler, 1942, p. 201; Meltzer, 1906–07
17. Pribram, 1971
18. MacLean, 1978b, p. 179
19. MacLean, 1978b, p. 183
20. Nauta and Mehler, 1966; Carpenter, 1976
21. Carpenter and Strominger, 1967; Carpenter *et al.*, 1968
22. Carpenter and Strominger, 1967
23. Martin, 1927
24. Ranson and Ranson, 1942, pp. 70–72
25. Ranson and Ranson, 1939

14

Further Evidence Implicating Striatal Complex in Display Behavior

The globus pallidus receives most of its connections from the corpus striatum (caudate nucleus and putamen). Consequently, the finding that lesions of the medial pallidum result in an elimination or fragmentation of the mirror display provides inferential evidence that the corpus striatum itself is implicated in the expression of such prosematic behavior. Now if it could be shown, in addition, that lesions of the neural pathways leading out of the medial pallidal segment eliminate or alter the expression of the display, it would give further correlational support to the hypothesis that the striatal complex plays a basic role in prosematic communication. It is easy to propose such experiments, but to carry them out brings one face to face with some of the most tangled anatomy of the brain. This is not only because the main neural outflow from the medial segment divides into a number of nerve bundles, but also because these same bundles either lie next to, or interweave with, other neural pathways within the hypothalamus.

Basal Longitudinal and Transverse Forebrain Bundles

Basal Longitudinal Forebrain Bundles

In this chapter the main focus will be on the fiber system referred to in comparative terms as the lateral forebrain bundle. In the latter part of the 19th century neuroanatomists became aware of a basal forebrain bundle in the brains of all vertebrates. It was referred to as the basal longitudinal forebrain bundle[1] because it extended in a lengthwise direction from the forebrain to the midbrain.

Once there was recognition of the basal forebrain bundle, anatomists noted that it could be subdivided into a medial and a lateral bundle. In 1910 C. Judson Herrick, a noted American anatomist (1868–1960), used the latinization *fasciculus medialis telencephali* in referring to the medial component.[2] Later, Herrick identified three components—medial, intermediate, and lateral bundles.[3] The lateral forebrain bundle is characterized by relatively large and well-myelinated fibers, whereas those of the intermediate are generally smaller with less myelin. The fibers of the medial component are predominantly finely myelinated or unmyelinated. A section through the precommissural septum of a lizard provides an excellent illustration of the three prototypic bundles.

All three bundles contain both efferent and afferent fibers connecting the telencephalon with lower levels. Herrick's medial bundle is mainly comprised of fibers passing to and from the septal region.[3] His intermediate bundle corresponds to the olfactopeduncular tract that extends from the anterior olfactory nucleus to the midbrain and also contains fibers descending from and ascending to the olfactory tubercle and nucleus accumbens (see Chapter 18). The lateral bundle is associated with the striopallidum.

Herrick's medial and intermediate bundles constitute what is generally labeled in atlases as the *medial forebrain bundle*. About midway through the hypothalamus, however, this label is inadequate because in this region numerous fibers of the lateral forebrain bundle interweave in a transverse direction with the medial forebrain bundle.

Transverse Bundles

In addition to the three longitudinal bundles, there are six more or less vertically oriented transverse pathways associated with the hypothalamus that must be taken into account in assessing variously placed lesions. For the present account the anatomical picture may be simplified by visualizing the hypothalamus as a sailing vessel with six masts (Figure 14-1). The medial, intermediate, and lateral bundles would correspond to the lowermost (most medial), middle, and top decks. From fore to aft the six masts would represent the (1) fornix, (2) stria terminalis, (3) stria medullaris, (4) inferior thalamic

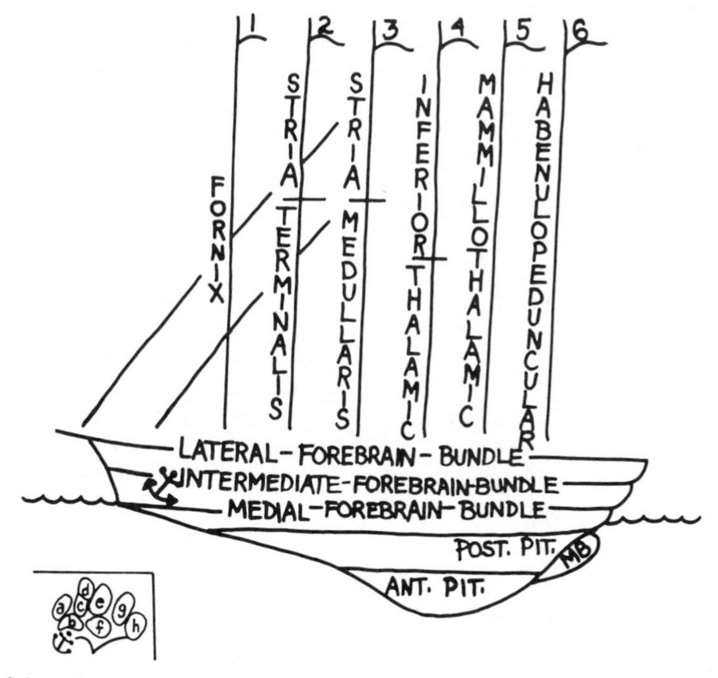

Figure 14-1. Scheme for representing nine main neural pathways of the hypothalamus and describing the effects of their destruction on the mirror display of gothic-type squirrel monkeys. When written out in the above manner, the three main forebrain bundles and the six main vertically oriented pathways give the impression of a sailing ship with three decks and six masts. For masts 5 and 6, add *peduncle* and *tract* to the pathway name. Anchor symbolizes optic chiasm and in the insert serves as reference for hypothalamic nuclei identified by letters: a, preoptic region; b, supraoptic nucleus; c, anterior hypothalamus; d, paraventricular nucleus; e, dorsomedial nucleus; f, ventromedial nucleus; g, posterior hypothalamus; h, mammillary body. Abbreviations: ANT. PIT., anterior pituitary; MB, mammillary bodies; POST. PIT., posterior pituitary.

peduncle, (5) mammillothalamic tract, and (6) habenulopeduncular tract. An innermost fiber system lining the hull in all directions is the periventricular fiber system.

For present purposes, it is unnecessary to give more than a summarizing statement about the anatomical connections of these pathways (for additional details see Chapter 18). Both the stria medullaris and the habenulopeduncular tract connect certain parts of the striatal complex and the limbic system with the evolutionarily ancient nuclei of the habenula. The stria terminalis connects the amygdaloid nucleus of the limbic system with the septum, preoptic region, and hypothalamus, while the fornix serves in a similar capacity for the limbic cortex contained in the hippocampal formation. The inferior thalamic peduncle contains fibers running between the medial thalamus and the hypothalamus, as well as the amygdala. The mammillothalamic tract is phylogenetically the most recent of the six vertically oriented pathways linking the hypothalamus with the third, and most recently evolved, subdivision of the limbic system (Chapters 18 and 21). It has its origin in the mammillary bodies.

Effects of Lesions *Exclusive* of Lateral Forebrain Bundle (LFB)

In the investigation of the role of the pallidal pathways in display behavior, observations were made on more than 50 animals in which lesions were placed at selected sites in the hypothalamus, ventral thalamus, subthalamus, and midbrain tegmentum.[4] In addition, lesions were placed in structures rostral to the hypothalamus that are known to project via the medial forebrain bundle—including the olfactory tubercle, substantia innominata, amygdala, and septum (see Chapter 13).

Lesions of Pathways Other Than LFB

Taking the fornix as the first target (Figure 14-1), one finds that its destruction at any level does not interfere with the full expression of display behavior. The same applies to the stria terminalis, the stria medullaris, the mammillothalamic tract, and the habenulopeduncular tract. Following lesions of the inferior thalamic peduncle, the monkey continues to show all the manifestations of the display, but is unable to achieve full erection. A similar deficit results from bilateral lesions along the course of the medial forebrain bundle. These effects on the genital component of the display might be expected in the light of the findings that electrical stimulation of either of these pathways elicits full penile erection[5] (see Chapter 19).

Destruction of Hypothalamic Gray Matter

In addition to the hypothalamic nerve bundles, it is necessary to take into account the effects of destroying different parts of the gray matter of the hypothalamus.

Anatomical Picture

Despite its great diversity of functions, the size of the hypothalamus is surprisingly small even in the largest of mammalian brains. In the human brain it would fit into the tip of one's little finger. The hypothalamus has been regarded as a "center" for the nervous control of body temperature, feeding, drinking, sleep, sexual behavior, and all forms of emotional expression.[6] In retrospect, it appears that some of the functions assigned to it are partially or wholly attributable to nerve bundles from other parts of the forebrain that pass through it or terminate within it. Through the infundibulum (literally funnel) pour all the cerebral nerve impulses and nerve-cell secretions affecting the functions of respective parts of the pituitary with respect to metabolism, growth, reactions to stress, and reproductive processes. With the development of improved techniques, endorphins and other neuroactive peptides have been added to the growing list of neurochemical agents identified in the hypothalamus.

For the present account the ship model is again useful for keeping in mind the hypothalamic nuclei, which, comparable to the various quarters of a boat, are named according to their position. First, let it be noted that in terms of the hull, the optic chiasm may be imagined as being located like an anchor, toward the bow (Figure 14-1). The wooden part of the keel would correspond to the infundibular region and neurohypophysis, and the lead part of the keel to the anterior pituitary; the mammillary bodies are in the position of the rudder. The "location" names for most of the hypothalamic nuclei are anterior, lateral, ventromedial, dorsomedial, and posterior[7] (see insert of Figure 14-1). The terms *medial* and *lateral preoptic nuclei* apply to collections of gray matter lying forward of the chiasm (anchor) and extending to the level of the anterior commissure. The lateral preoptic nucleus is one of the bed nuclei of the medial forebrain bundle. In the experiments to be summarized, the nuclei identified with the chiasm (supraoptic and supracommissural) and the neighboring paraventricular nucleus were included in various lesions, but not selectively destroyed.

Following bilateral lesions of the medial preoptic area and the adjoining part of the anterior hypothalamus, a monkey shows all manifestations of the display, but is unable to achieve full erection (e.g., monkey S-1). There is a similar outcome with lesions of the medial forebrain bundle that necessarily involve its bed nuclei at anterior, intermediate, and posterior hypothalamic levels. One case that will be discussed in another context serves as an illustration. In this monkey (W-4) the entire lateral part of the intermediate and posterior hypothalamus was destroyed. Figure 14-2 shows a section through the lesion at the level of the mammillary bodies, together with the performance curve. The monkey required hand feeding for a period of 17 days, after which it began to eat and behave normally. By the 58th day it began to display regularly during formal testing. All main components of the display were present except for the notable difference that throughout the ensuing 3 months of testing there was a statistically significant decrease in the genital index. It will be recalled that the genital index provides a measure of the animal's capacity to develop erection, being calculated by summing the values for the magnitude of each erection in a series of 30 trials and dividing by the number of occurrences. The relative impotence observed in this case was to be expected since electrical stimulation within this region in intact animals elicits full erection (see Chapter 19).

The dorsomedial hypothalamic area is another location where electrical stimulation elicits full erection. In one monkey (V-5) this area together with the dorsomedial nucleus was destroyed without damage to neighboring fiber bundles. Two days later and thereafter it would regularly vocalize and thigh-spread when tested, but never displayed erection over a period of $2\frac{1}{2}$ months. Then during the remaining 4 months of testing, it regularly achieved erection of 1+, and sometimes 2+, magnitude.

Following a bilateral lesion of the ventromedial nucleus of the hypothalamus that

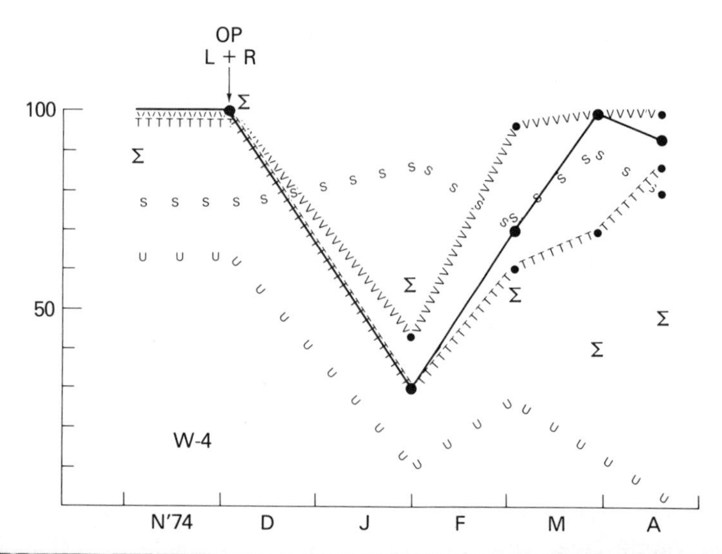

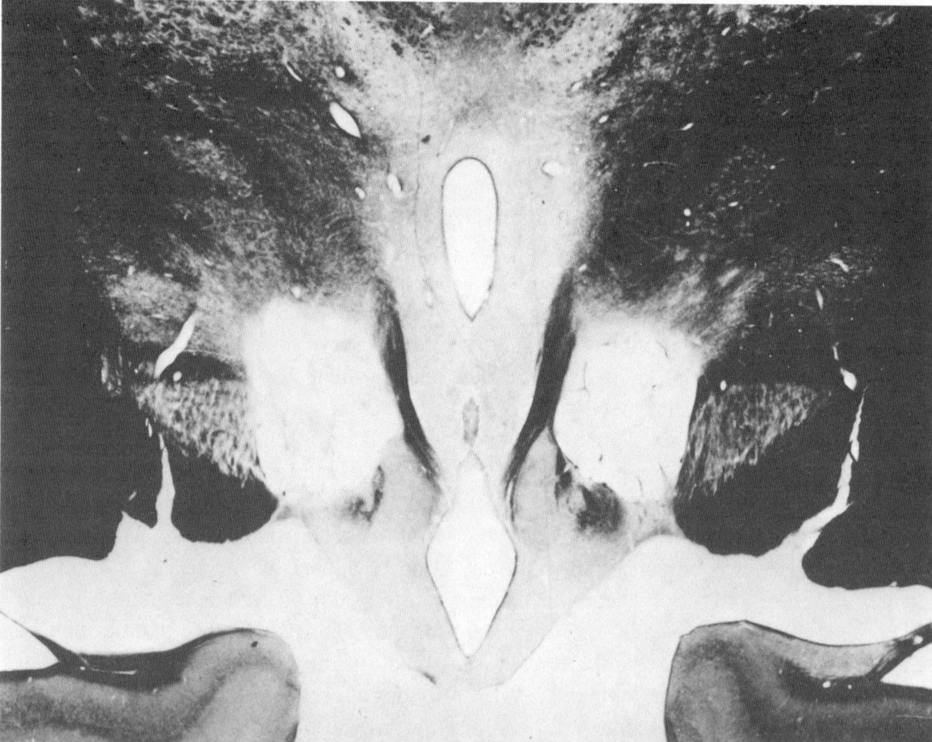

Figure 14-2. Performance curve and histological section, illustrating that large posterolateral coagulations of the hypothalamus are compatible with almost full recovery of the mirror display. In this case (monkey W-4), the medial forebrain bundle and intermediate forebrain bundle were almost totally destroyed at the level of the mammillary bodies. As noted in text, the decline in the genital index may be attributed to damage of the medial forebrain bundle.

Key to performance curves: solid line represents penile erection, the central manifestation of the display. Each dot indicates an accumulation of 30 trials. The curves shown by letters give percentage incidence of various components of the display: V, vocalization; T, thigh-spread; U, urination; and S, scratch. Σ refers to the genital index, a value of the average magnitude of erection for all trials in which it occurred. Ordinate and abscissa scales, respectively, represent percentage of tests and time in months.

Histological section from MacLean (1981).

extended into the tuberal region, one monkey (Y-5) resumed his typical trump display and continued performance at the 100% level throughout the subsequent 2 months of testing. Another monkey (F-3) with a larger symmetrical coagulation of the ventromedial nucleus that extended upwards to interrupt the ''pallidohypothalmic'' tract on each side recovered its usual propensity to display, but there was a significant reduction in the occurrence of the thigh-spread component. This monkey developed none of the angry manifestations described by Wheatley[8] in cats with a corresponding lesion. Neither of the monkeys showed changes in feeding habits or gained weight.

Monkeys with midline lesions of the mammillary bodies or of the interpeduncular nucleus revealed no changes in their display performance.

Lesions of Pallidal Projections and Associated Structures

We now consider functional aspects of a massive fiber system that has received relatively little consideration in studies on the hypothalamus. In the mammalian brain, the greater part of this fiber system represents subdivisions of the ansa lenticularis that are believed to correspond to components of the LFB in submammalian forms (see Chapter 4). First we deal with the effects of lesions of divisions of the fiber system that pass through the hypothalamus and then analyze the results of experiments in which an attempt was made to differentiate the effects of destroying branches of this fiber system and associated structures in the thalamus, the thalamic tegmentum, and the midbrain tegmentum.

Anatomy of Transhypothalamic Pathways

An account of the experiments now to be considered requires a further anatomical description than was given in Chapter 4. Elsewhere, I have referred to the main components issuing from the medial pallidal segment as *transhypothalamic pathways*.[4] This expression was used because of the lack of evidence that main or collateral fibers terminate in the hypothalamus. The term *transhypothalamic* leaves open the possibility of the discovery of fibers projecting to the hypothalamus, being no more exclusive of local connections than the name *transcontinental* for an airline.

In mammals the hypothalamic picture of these bundles becomes further complicated by the development of a fourth bundle associated with the evolving neocortex and identified as the internal capsule. The newly evolved bundle might be regarded as the ''far lateral bundle.'' It runs largely circumferential to the hypothalamus. As diagrammed in Figure 14-3, the internal capsule splits part of the lateral forebrain bundle so that large numbers of fibers issuing from the forward part of the globus pallidus must loop around the capsule before turning in a posterior direction to reach their destination. Because of this loop, this part of the LFB is called the *ansa lenticularis* [literally, loop or handle of the lens-shaped nucleus (globus pallidus + putamen)]. In transverse sections, the appearance of the loop fibers as they sweep through the lateral hypothalamus suggests the ''swish of a mare's tail'' (Figure 14-4). A large proportion of these fibers collect into a compact band, while the others remain loosely associated.

The drawing in Figure 14-3 illustrates the course of numerous other fibers originating from cells in the medial segment. They give an onion-layer appearance as they reach for

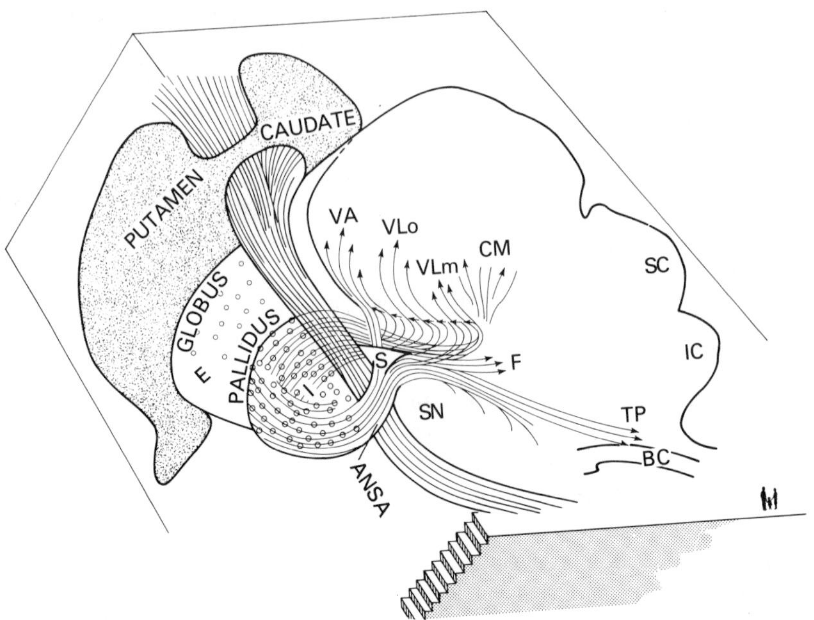

Figure 14-3. A three-dimensional diagram illustrating the roundabout course of two main projections from the internal segment (I) of the globus pallidus that at first diverge and then converge. One projection known as the *ansa* (loop) *lenticularis* hooks around the internal capsule and subsequently joins the other projection (*fasciculus lenticularis*) after the latter weaves through the capsule. Each of these two projections has two main destinations—one, to the thalamus, and the other, to the tegmentum. Some of the fibers reach the thalamus as though following the threads of a continuous sheet folded back on itself, while others travel downstream before making a reverse curve and ascending in the thalamic fasciculus (see arrows pointing to the left). The thalamic projections fan out in the centromedian nucleus (CM), the medial part of the ventral lateral nucleus (VLm), the oral part of the ventral lateral nucleus (VLo), and finally the principal part of the ventral anterior nucleus (VA). The tegmental projections descend mainly to the field of Forel (F) and after passing above the substantia nigra (SN) proceed to the tegmental pedunculopontine nucleus (TP) that hugs the brachium conjunctivum (BC).
Although not indicated, the caudate nucleus and putamen project to both the internal (I) and external (E) segments of the globus pallidus. As was illustrated in Figure 4-3, the external segment projects to the subthalamic nucleus (S), which in turn projects back to the medial segment. Other abbreviations: SC, superior colliculus; IC, inferior colliculus.

the internal capsule and then weave through it before taking a transhypothalamic route to the thalamus and tegmentum. As also schematized, many fibers from this contingent and from the ansa itself peel off in a rostrocaudal direction and travel medially toward the ventricle prior to reversing course to enter ventral parts of the thalamus. As Beck and Bignami[9] observed, it was as though they stemmed from a continuous sheet.

The contingent of fibers that weaves through the internal capsule is sometimes referred to as the *dorsal division* of the *ansa lenticularis.*[10] After penetrating the capsule numerous fibers of the dorsal division converge and form a rather thick band at the oral pole of the subthalamic nucleus.[11] This confluence is referred to as the *fasciculus lenticularis.*

In a rostrocaudal direction, toward the midlevel of the hypothalamus, numerous fibers from the ansa and fasciculus lenticularis join to form a compact band (Figure 14-5). Ranson and Ranson[12] pointed out that the ansa at this level is "more sharply defined in the monkey than in the human brain." Followed caudally, the compact band continues to

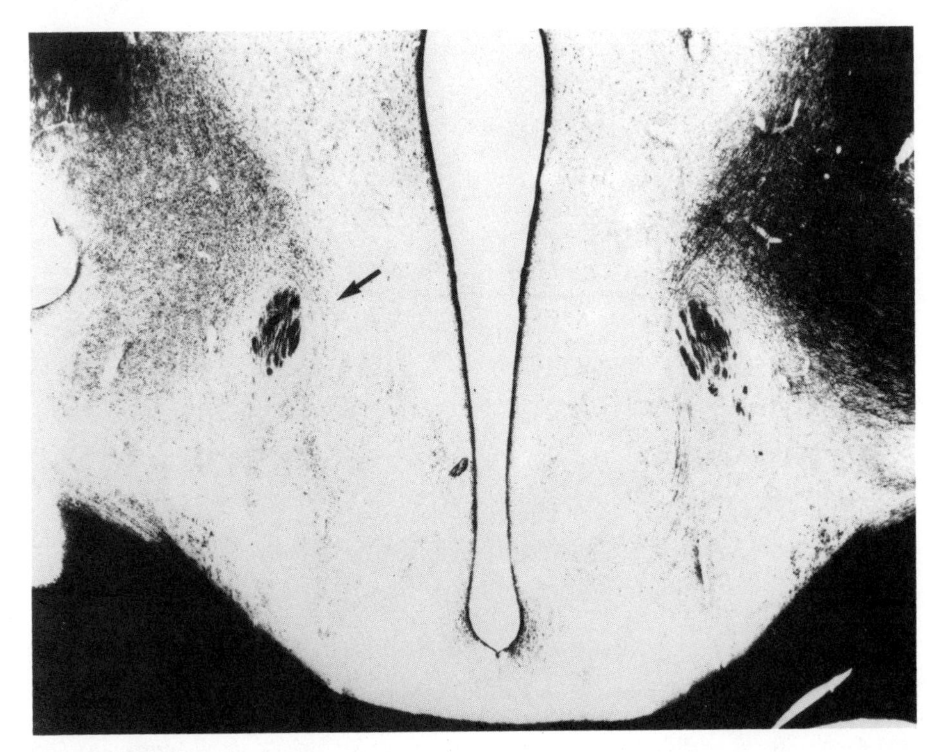

Figure 14-4. Hypothalamus of squirrel monkey, showing degeneration of the ansa lenticularis subsequent to a large electrocoagulation of the globus pallidus that also involved a large complement of cells projecting via the fasciculus lenticularis. The two dark, oval-shaped bundles of the fornix appear somewhat like two eyes, while the outline of the third ventricle in between resembles a nose. Compare the "pepper and salt" degeneration of the ansa lateral to the right fornix (left side of figure) with the dark black appearance of the ansa of the other side. The arrow points to degeneration in the so-called pallidohypothalamic tract. See text. From MacLean (1981).

the level of the mammillothalamic tract before ascending and joining the thalamic fasciculus. In fiber-stained sections, the ascending, darkly impregnated mass of fibers gives the appearance of the "menacing funnel of a tornado." Upon reaching the thalamic fasciculus, the fibers reverse course and run rostrally to peel off in the *medial* and *oral* parts of the ventral lateral nucleus and farther forward into the lateral, "principal" part of the ventral anterior nucleus. These particular structures belong to the so-called *ventral tier nuclei,* a term used by Carpenter in 1967.[13]

It is worth stating again that at the midlevel of the hypothalamus numerous fibers of the ansa system interweave with those of the medial forebrain bundle frequently labeled as such in brain atlases. In the squirrel monkey Emmers and Akert[14] place the label for olfactomesencephalic tract in this region (alternate name for Herrick's intermediate bundle). Near midlevel of the hypothalamus some fibers stemming from the fasciculus lenticularis loop over the medial border of the fornix, as well as pass between its fascicles. These fibers form the so-called pallidohypothalamic tract, which somewhat more rostrally appears to be coextensive with the crossed and uncrossed fibers of the dorsal supraoptic decussation of Ganser. As noted in Chapter 4, present evidence indicates that the pallidohypothalamic tract does not terminate in the ventromedial nucleus but rejoins other fibers of the ansa directed toward the tegmental area (field H of Forel) and midbrain.

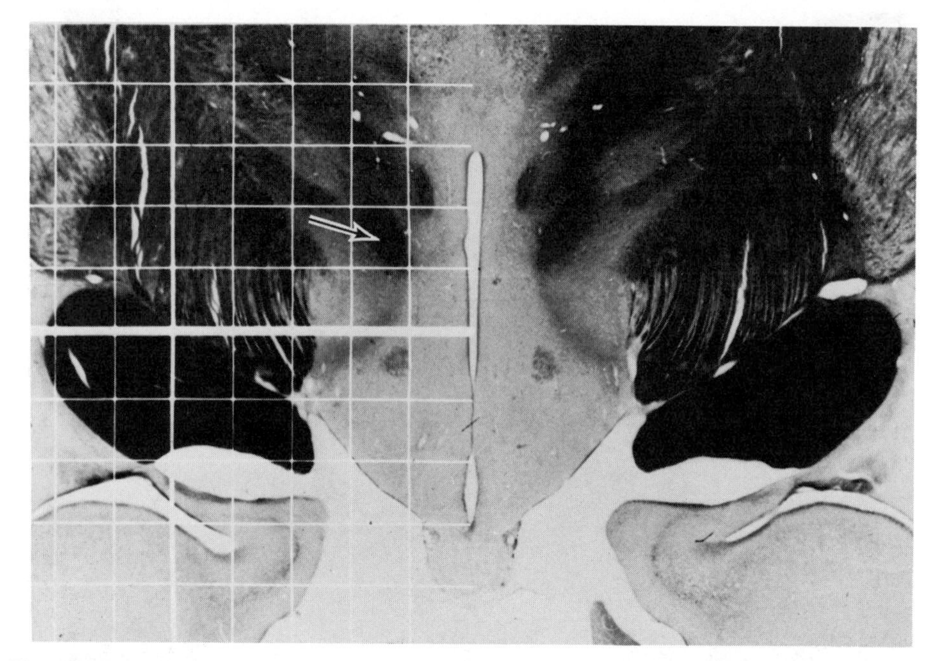

Figure 14-5. Picture of compact ansa as seen in brain atlas section at F8.5. Arrow on left points to compact part of the ansa. The fasciculus lenticularis (H2) appears just left of the arrow; H1 containing the thalamic fasciculus, is seen above it. The mammillothalamic tract is located above and medial to the arrow. Adapted from Gergen and MacLean (1962).

The study on the globus pallidus (Chapter 13) provided evidence of an orderly arrangement of fibers in the compact part of the ansal system: Lesions in the rostral pallidum resulted in gliosis of the ventrolateral part of the ansa; lesions of the intermediate pallidum were associated with a lateral wedge-shaped gliosis in the ansa; while in cases with damage in the caudal pallidum gliosis extended from the fasciculus lenticularis into the dorsomedial part of the compact ansa.[15] In all animals showing an enduring change in mirror display, gliosis was evident in the region of the dorsal medial nucleus of H1, and in some cases there appeared to be degenerative changes (transneuronal?) in the nerve cells of this nucleus.

Destruction of Transhypothalamic Pathways

Altogether there were seven animals (N-4, O-4, P-4, Q-4, C-5, D-5, and J-5) in which coagulations at rostral and intermediate levels of the hypothalamus destroyed a major part of the ansal system. In each case there was an enduring elimination of the mirror display. Figure 14-6 shows the performance curve and bilateral symmetrical coagulations in one case. The lesions destroyed the rostral compact part of the ansa lenticularis, extending from AP 10 to AP 8.5 and stopping short of the mammillothalamic tract. The monkey in this case (P-4) appeared so normal the day after surgery and during the acute postoperative period that there was some doubt as to whether or not coagulations had been produced. Unlike cases with coagulations of the ansal system extending farther caudally, this monkey fed and drank spontaneously during the acute postoperative period.

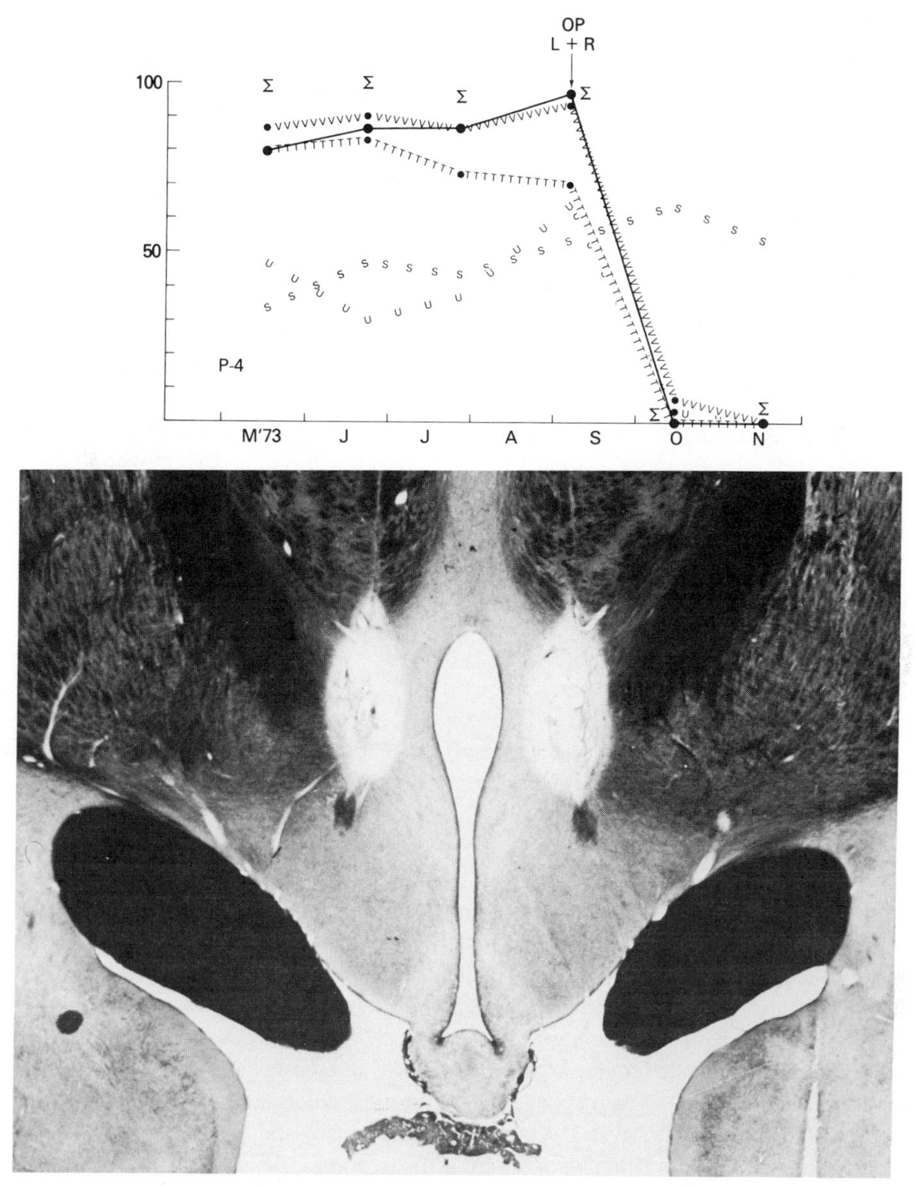

Figure 14-6. Performance curve and histological section, illustrating the overnight elimination of the mirror display following electrocoagulation of the ansa lenticularis. In this monkey (P-4) the tear-shaped coagulations destroyed the forward part of the compact bundle of fibers formed by the ansa lenticularis and the fasciculus lenticularis (see Figure 14-5). The monkey appeared so normal subsequent to the operation that there was doubt that an electrocoagulation had been made. As evidence to the contrary, this subject never displayed again. See Figure 14-2 for key to curves. Histological section from MacLean (1981).

The only evidence that lesions had been made was that the monkey ceased to display and, as shown in Figure 14-6, never displayed during 2 months of testing. It was also observed to lack its preoperative spontaneity of movement and facial expression. Squirrel monkeys commonly roll food pellets or grapes on the tip of their tails; this monkey would start to do so, but could not follow through.

Except for transient fragmented displays in one case (O-4), lesions somewhat more caudal and destroying both the ansa and mammillothalamic tract at AP 8 (O-4, Q-6) also eliminated the display. One monkey (T-4) in which similar lesions were intended, but in which the electrode on one side was medially too near the ventricle, continued its regular performance of the full display. Four other monkeys in which the thalamic fasciculus was also partially involved (N-4, C-5, D-5, J-5) failed thereafter to display.

In summary, the findings on these seven monkeys revealed that destruction of a major part of transhypothalamic pathways issuing from the medial segment of the globus pallidus eliminates the mirror display of gothic-type squirrel monkeys. These results substantially reinforced the earlier evidence that the medial pallidal segment is basically involved in the organized expression of the display.

Differential Effects of Pallidothalamic and Pallidotegmental Lesions

With the evidence that projections from the medial segment are essential for the performance of the display, it was important as a next step to learn whether the thalamic or the tegmental projections of the globus pallidus were differentially involved in the expression of the display.

Comparative Anatomical Picture

Because of the traditional clinical view that the striatal complex comprises part of the motor apparatus, it is usually assumed that its connections with the motor cortex via the thalamus are more fundamental than those leading the tegmental area. Both divisions of the thalamic and tegmental connections are phylogenetically ancient and can be clearly visualized as large fasciculi in reptilian and avian brains. Figure 14-7A shows in the lizard brain how the LFB divides into a dorsal and ventral branch known respectively as the dorsal and ventral peduncle. The dorsal peduncle contains efferent and afferent connections between the striopallidum and the thalamus, while the ventral peduncle provides a similar relationship with the thalamic and midbrain tegmentum. Figure 14-7B shows the parallel situation in the avian brain. This same prototypic arrangement exists in the mammalian brain, but the course of the pathways is altered because of the intrusion of the internal capsule as described above, and by the great enlargement of the neothalamus. Figure 14-3 is intended to give a schematic of the course and destination of the ansal projections beyond the hypothalamus in the primate brain.

The experiments involving the thalamic division will first be described, and then the anatomy and findings on the tegmental division will be summarized.

Lesions Involving the Thalamic Division

As diagrammed in Figure 14-3, in addition to the rather direct thalamic projections from the ''continuous sheet,'' those fibers from the ansa that join the thalamic fasciculus peel off in a caudorostral direction to innervate the centromedian nucleus, the medial and oral parts of the ventrolateral, and the principal part of the ventral anterior nucleus.

The centromedian nucleus is small in the brains of such animals as the cat and dog, but becomes progressively larger in the evolution of primates, reaching its greatest size in the human brain. It is a source of afferents to the corpus striatum, which in turn projects to the globus pallidus and then back to the centromedian, forming a complete neural circuit. The brain section in Figure 14-8A is from a case (Y-2) in which a large part of the centromedian nucleus was destroyed on each side. There was no postoperative change in the monkey's near-perfect performance of the display (see below for changes following lesion shown in Figure 14-8B).

It is virtually impossible to destroy selectively the sites of the pallidal projections to the ventral lateral (pars oralis et medialis) and ventral anterior (pars principalis) nuclei. Consequently, an attempt was made to denervate these nuclei by coagulating both the thalamic fasciculus and the spray of fibers entering the ventral parts of the nuclei. In one animal (X-4) partial lesions were produced in the ventral anterior and ventral lateral nuclei and the associated thalamic fasciculus. There was also destruction of the greater part of the zona incerta at these levels, as well as some of the dorsal projections of the ansal system. The monkey recovered its ability to display after the second week and thereafter never failed to display in the twice-daily tests conducted over a period of 3 months. In a second monkey (V-4) the coagulations involved the same structures, but extended farther ventrally into the dorsal projections of the ansal system. As shown by the performance curve in Figure 14-9, during the following 6 months there was at first an elimination of the display followed by a period of sporadic fragmented displays. Then during the final 2 months there was recovery to the extent that the monkey displayed in every test, but there was a fragmentation of the trump display, with the thigh-spread component occurring less than 60% of the time.

What would be the outcome if thalamic lesions were placed at the site where the compact ansa becomes continuous with the thalamic fasciculus? In one monkey (S-5) a symmetrical supramammillary electrocoagulation largely eradicated this target, but also destroyed the origin of the medial longitudinal fasciculus (MLF) and part of the posterior hypothalamus. Except for the loss of the vocal component attributed to the lesion of MLF, this subject continued to display at or near the 100% level of performance. Another monkey (U-5) with lesions of these same structures, but with a more ventral extension of neural damage, began to display regularly after it recovered from hypothermia (see below), but like monkey S-5 had a marked deficit in the vocal component.

In summary, the findings in three of the foregoing cases were indicative that the display does not depend on an intact thalamic fasciculus or intact ventral tier nuclei.

Lesions of Pallidotegmental Projections

Since the pallidotegmental division appears to compensate for bilateral lesions involving the pallidothalamic division, does the reverse condition apply? Negative findings in this regard will be described after a brief review of the projections of the pallidotegmental division.

Anatomical Picture

The word *tegmentum* (cover, roof) refers to the parts of the brain overlying the cerebral peduncles, the great mass of fibers stemming predominantly from the neocortex. Here we are concerned with both the thalamic and the midbrain tegmentum. The most

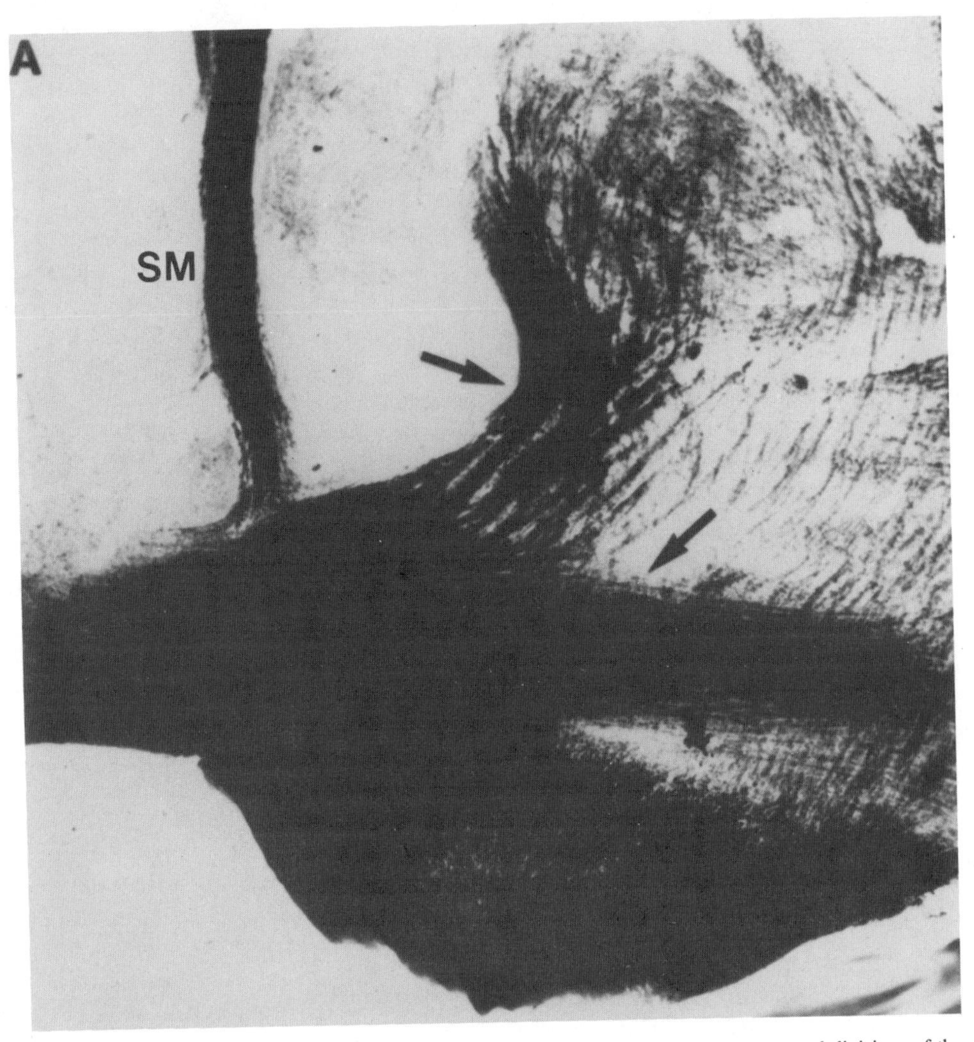

Figure 14-7. Sections from reptilian and avian brains, illustrating the thalamic and tegmental divisions of the lateral forebrain bundle that are also represented in mammals.

(A) Section from the brain of a green anolis lizard (*Anolis carolinensis*). Left arrow points to thalamic division (dorsal peduncle) of the lateral forebrain bundle connecting with the nucleus rotundus, and right arrow identifies division (ventral peduncle) headed for the tegmentum. SM, stria medullaris.

(B) From brain of turkey (*Meleagris gallopavo*). Left and right arrows point, respectively, to the thalamic and tegmental divisions of the lateral forebrain bundle.

From MacLean (1981).

central part of the thalamic tegmentum is known as the field of Forel (1841–1912), after the anatomist who first described this region and referred to it as *die Haube* (cap, hood),[16] which has since served as the initial for identifying the central (H), dorsal lateral (H1), and ventral lateral (H2) parts of the field. As yet there is no consensus about the distribution and number of pallidal projections to the central field (H) and to the dorsal medial nucleus in H1. According to Nauta and Mehler[17] the main pallidotegmental projection is to the

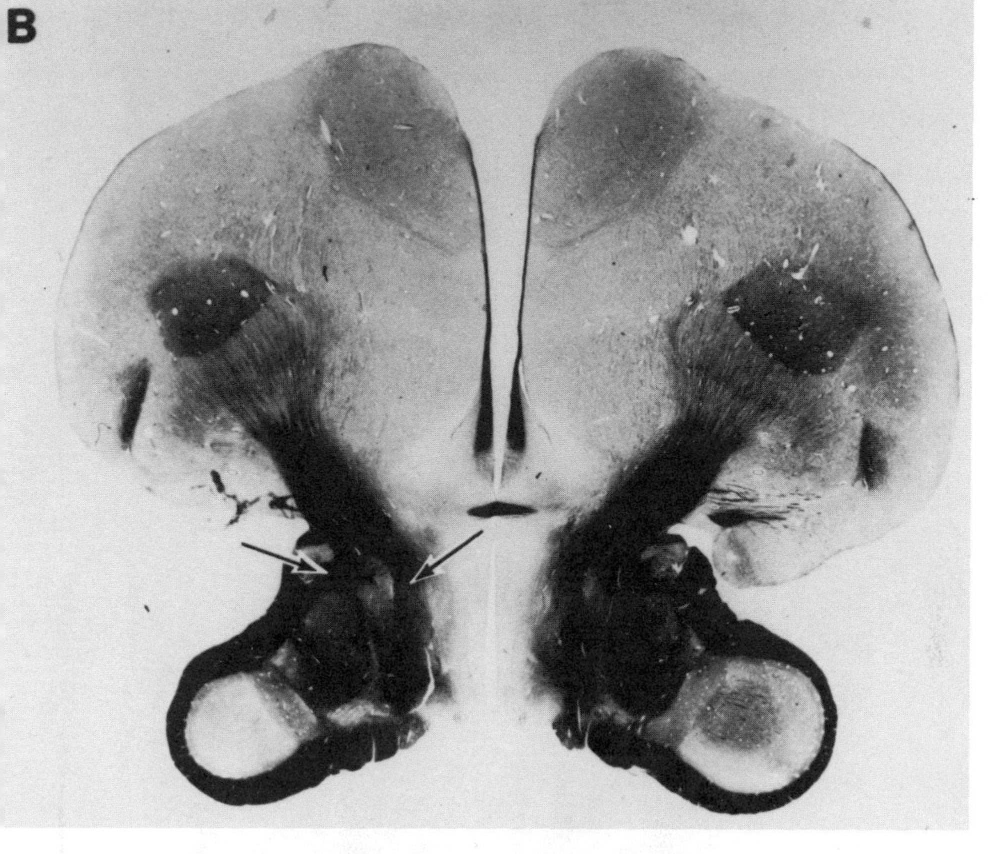

Figure 14-7 *(Continued)*

tegmental pedunculopontine nucleus of the midbrain that lies partially embedded in the brachium conjunctivum. They refer to it as the "ventrolateral pallidotegmental fiber system" and state that the fibers "first become identifiable where they separate from field H2 and collect dorsal and medial to the subthalamic nucleus."[18] As the fibers descend ventrolateral to the red nucleus, they partly mingle with the medial lemniscus. Some fibers turn upwards into the central tegmental region; others distribute to the compact part of the substantia nigra; the remainder terminate "almost exclusively" in the tegmental pedunculopontine nucleus.[17]

Lesions of Tegmental Structures

In regard to the question posed above, the findings in one animal indicated that the system of structures associated with the thalamic fasciculus might for a time compensate for damage to tegmental projections, but not permanently. In this animal (I-5) a large symmetrical lesion was placed in the supranigral tegmentum (Figure 14-10) rostral to the red nucleus. As shown by the section through the rostral part of the lesion (Figure

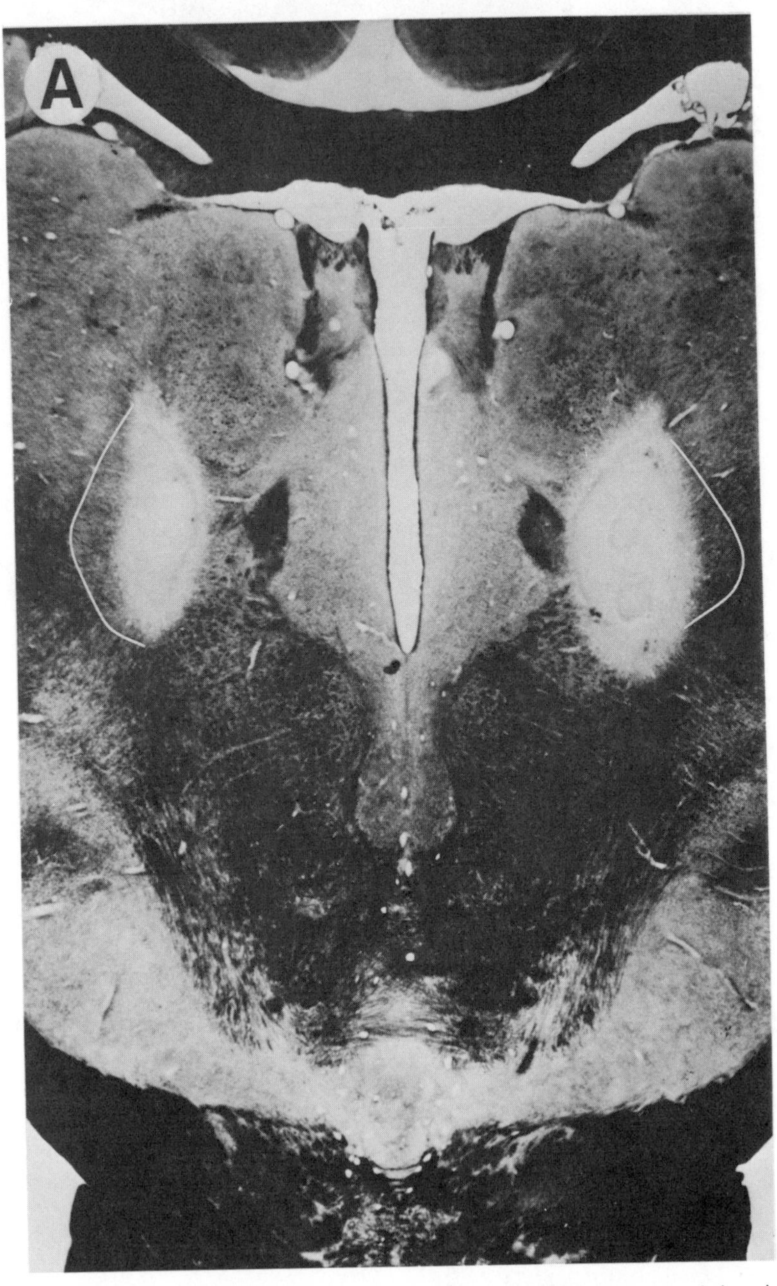

Figure 14-8. Sections from two different squirrel monkeys (Y-2, L-5) showing coagulations in the phylogenetically newer (A) and older (B) parts of the parafascicular–centromedian complex that is a source of fibers to the corpus striatum. Coagulations in the older, but not the newer, parts interfere with the expression of the mirror display.

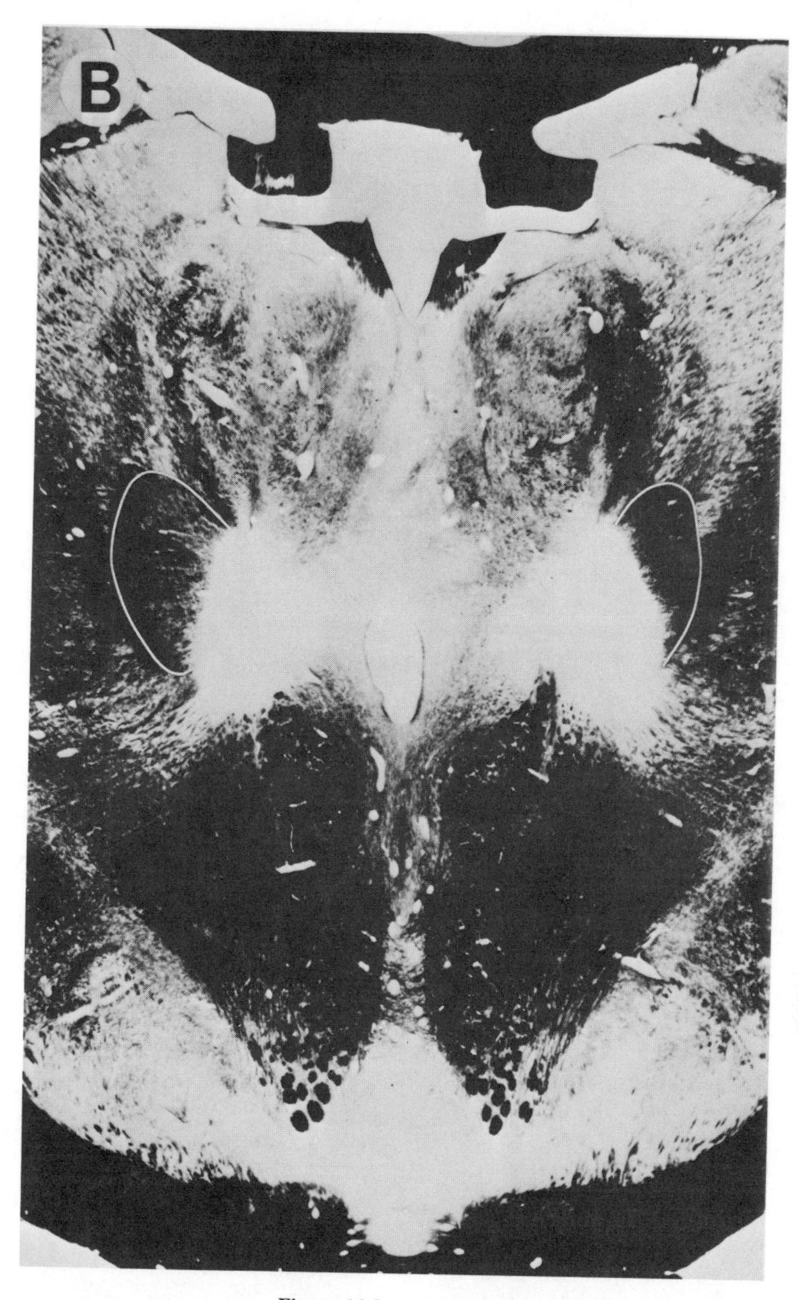

Figure 14-8 *(Continued)*

14-10A), the destruction did not extend sufficiently far forward to involve the continuation of the ansal system with the thalamic fasciculus. This monkey continued to display regularly for 5 months. Then, after a period of irregular performance, there was virtual elimination of the display during the last 6 months of testing. This finding suggests that

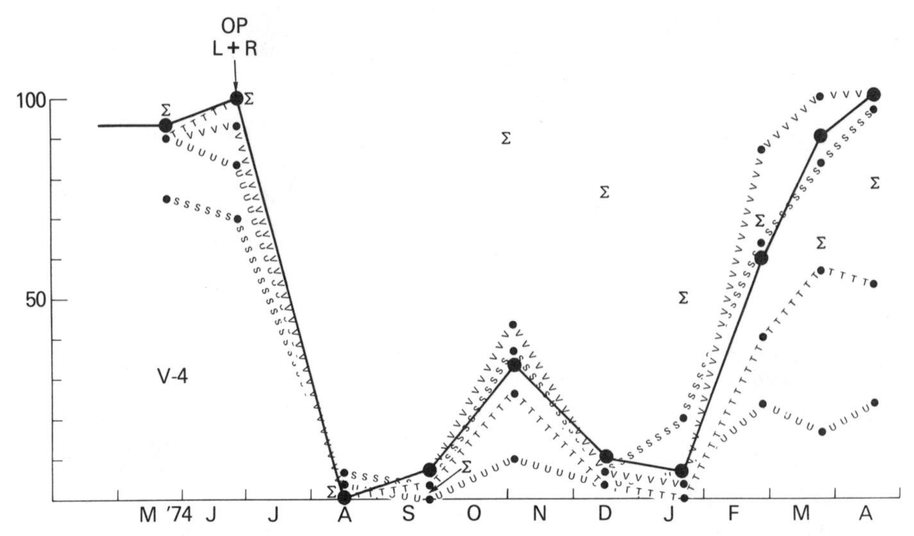

Figure 14-9. Performance curve of display of monkey V-4 with bilateral lesions of the thalamic fasciculus, ventral parts of nucleus ventralis anterior, nucleus ventralis lateralis, zona incerta, and dorsal projections of the ansal system. Note the appearance of sporadic displays 2 months after surgery. After 8 months, displays could again be elicited regularly, but there was a fragmentation of the trump display, with the thigh-spread component occurring less than 60% of the time. See legend of Figure 14-2 for key to curves. From MacLean (1981).

pallidothalamic mechanisms may have accounted for temporary retention of the display. The near-total absence of displays during the last 6 months of testing may have been owing to progressive gliosis around the site of the lesion and/or retrograde and transsynaptic degeneration of nerve cells associated with the effective neural circuit.

In another monkey (H-5) in which a supranigral lesion spanned the tegmentum somewhat more caudally (Figure 14-11), there was but one full display and 19 fragmented displays during 14 months of testing. In every test, however, the monkey climbed the side of the cage toward the mirror and showed interest in its reflection. The lesion in this case extended more ventrally than in I-5, possibly destroying pallidotegmental fibers fanning out over the substantia nigra, as well as largely obliterating the "intranigral fasciculus"[19]; it also infringed upon the rostral pole of the magnocellular part of the red nucleus. The lesion could also have involved nigrotectal projections (see below). Two other animals (U-4, A-5) with tegmental lesions that impinged on the dorsal part of the substantia nigra gave only sporadic and usually fragmented displays.

In one monkey (E-5) bilaterally symmetrical lesions were placed in the lateral tegmentum that, upon histological examination combined with volumetric measurements, proved to have destroyed two-thirds of the subthalamic nucleus on each side. Although this monkey suffered motor deficits characterized by alternating jerky approach and withdrawal movements, it regularly vocalized and displayed full erection upon seeing its reflection in a mirror, and by the end of 11 months of testing there was recovery of the thigh-spread component to nearly criterion.

As has been noted, lesions involving the interpeduncular nucleus and habenulopeduncular tract had no apparent effect on the display. Other evidence suggests that cerebellar projections, the red nucleus, and the medial lemniscus are not essential for the display. One monkey (W-5), for example, with complete destruction of the red nucleus and ventral part of the decussation of the brachium conjunctivum, together with the adjoining parts of the substantia nigra, had a bilateral intention tremor and Wilson-like athetosis, but continued to meet criterion in its display.

As was noted, the bulk of the ventrolateral pallidotegmental projections are believed to terminate in the tegmental pedunculopontine nucleus. Attempts to destroy the greater part of this nucleus were unsuccessful. One monkey with a coagulation of part of this nucleus adjacent to the superior cerebellar peduncle, as well as part of reticular formation, continued to perform the trump display at near the 100% level throughout 7 months of testing.

Lesions Involving Monoaminergic Systems

Anatomy

The course of the ascending dopaminergic nigrostriatal pathway was partly described and illustrated in Chapter 4. Fibers ascend to the striatum by three main routes: One group passes through the field of Forel and follows the ansa lenticularis, while a second group enters H2 above the subthalamic nucleus. Upon reaching the pallidum, fibers of both groups radiate out (see Figure 4-11) and gain access to different parts of the putamen and caudate nucleus. (There is some evidence that the pallidum itself receives ''very sparse'' collaterals.[20]) Fibers of a third contingent congregate along the medial margin of the internal capsule and follow it upwards to reach the caudate nucleus.

The noradrenergic system (described and illustrated in Chapter 18) appears to provide little innervation to the striatal complex. The fibers arise from catecholamine containing neurons in the medulla, pons, and midbrain and collect in ventral and dorsal bundles that become confluent at the thalamomidbrain junction and ascend via the medial forebrain bundle (see Figure 18-4). The serotonergic system arises from cells in the dorsal raphe nucleus, as well as those in ventral raphe and superior central nucleus of Bechterew (see Chapter 18). The largest complement of fibers ascends in the medial forebrain bundle.

Effects of Lesions

As noted in a brief report,[21] destruction of the dopaminergic nigrostriatal pathway in the squirrel monkey by the injection of 6-hydroxydopamine near the rostromedial pole of the substantia nigra results in tremor and bradykinesia resembling advanced Parkinson's disease. As described later, several animals with lesions of the ansal system, like those with lesions of the pallidum (Chapter 13), developed varying degrees of a Parkinson-like bradykinesia. When the lesions were placed near or in the rostromedial pole of the substantia nigra, hypothermia was an additional complication (see below).

The case of monkey W-4, described earlier, with large bilateral lesions of the posterolateral hypothalamus that destroyed virtually all of the medial forebrain bundle (Figure 14-2), would indicate that at least the somatic components of the display are not dependent on noradrenergic or serotonergic systems. Further support for this inference in regard to serotonin is provided by a subject (O-5) with virtual destruction of the superior central nucleus of Bechterew that continued its near-perfect performance of the display throughout 4 months of testing. In this case there would also have been an interruption of

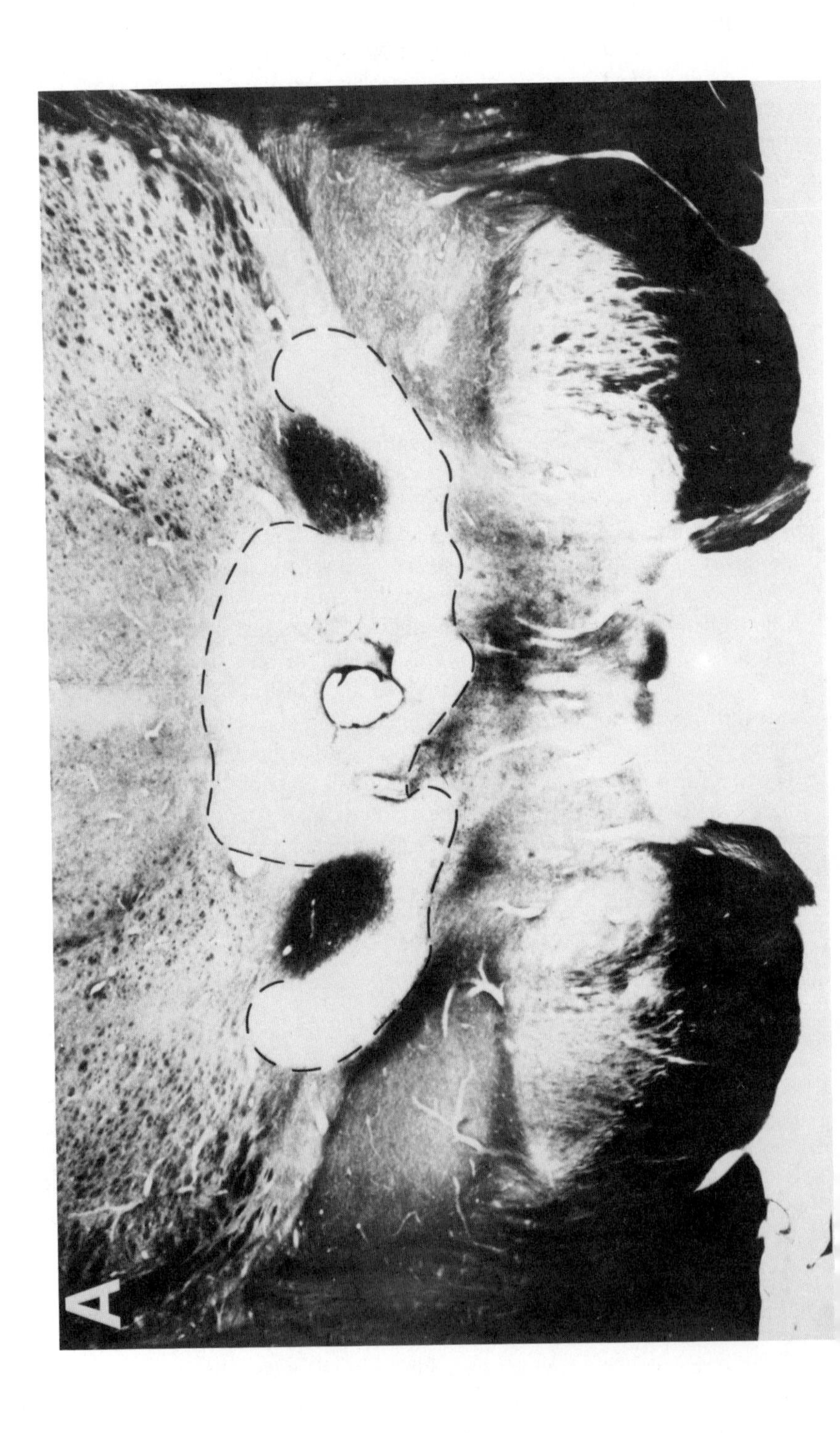
A

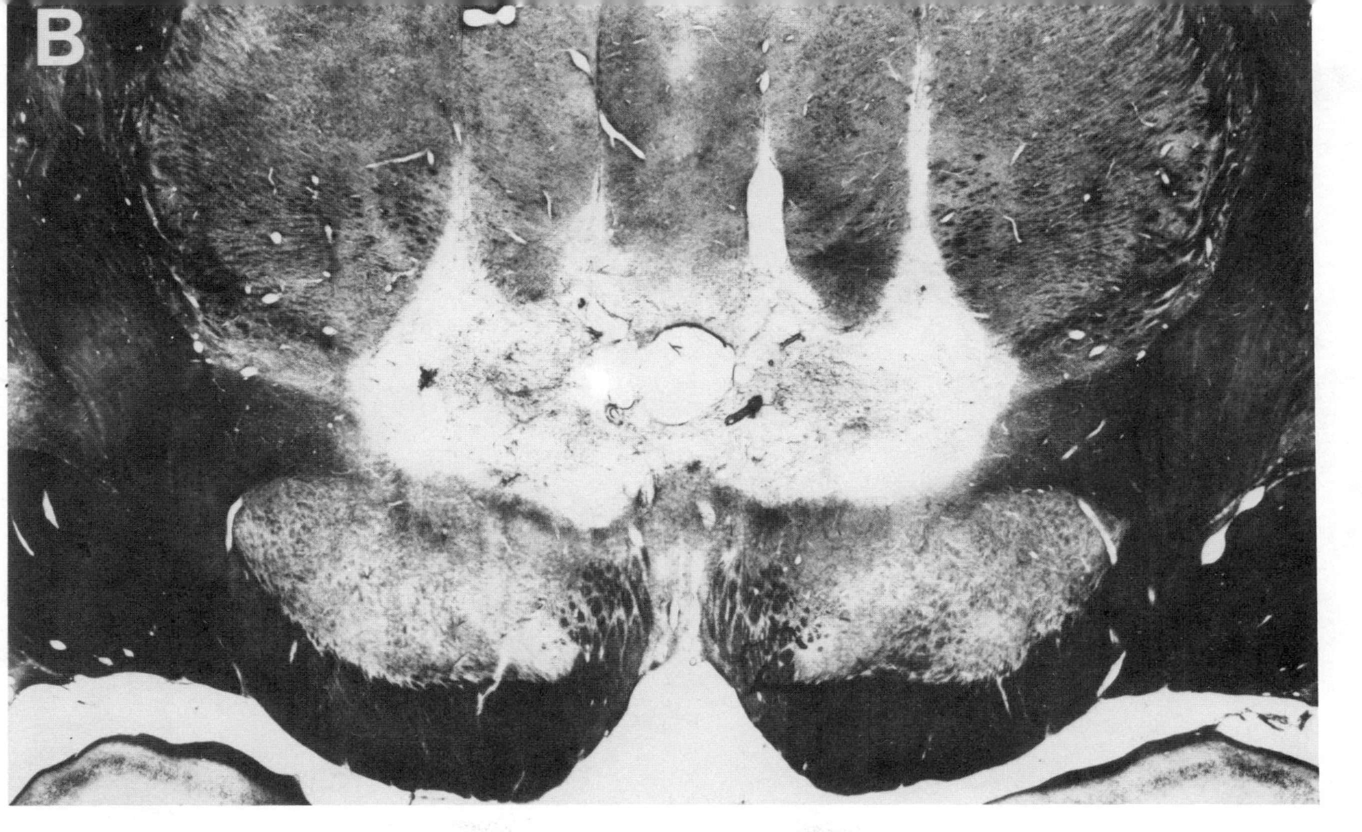

Figure 14-10. Two sections from brain of the same subject, showing extensive tegmental lesions that eventually resulted in the elimination of mirror display.

(A) The most forward part of the electrocoagulations spared the recurrent bundle of fibers forming the thalamic fasciculus and appearing here as two black islands of tissue. It is possible that sparing the thalamic division of the ansa system sustained the performance of the mirror display for several months.

(B) Appearance of the electrocoagulations at a more posterior level in the tegmentum. Fibers running just underneath the coagulations and just above the substantia nigra presumably preserved temperature regulation in this animal. See Chapter 21 regarding alterations of isolation cry.

From MacLean (1981).

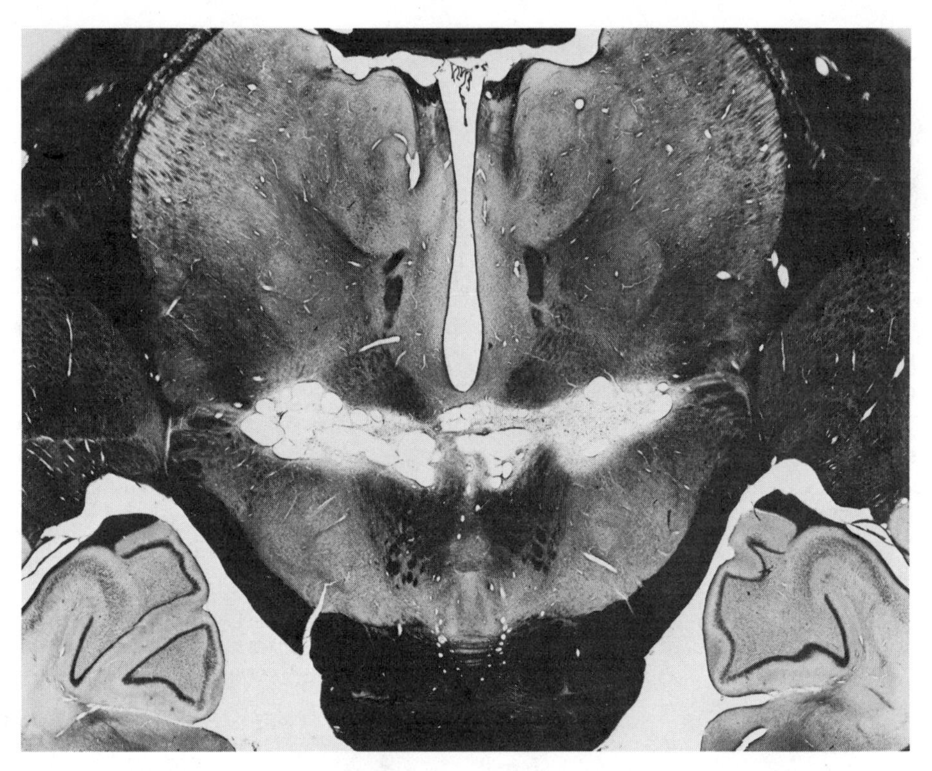

Figure 14-11. Section showing a wing-shaped tegmental coagulation that eliminated mirror display behavior and also resulted in abnormal vocalization. Note that the coagulation is more posterior than that in the animal illustrated in Figure 14-10. The monkey in this case (H-5) produced persistent abnormal isolation peeps, including harmonic components. The anatomical analysis of cases with tegmental lesions suggests that the vocal abnormalities were the result of injury to gray matter and fiber systems below the posterior floor of the third ventricle. From Newman and MacLean (1982).

some fibers from the dorsal raphe nucleus. It is of added interest in this case that there was retrograde degeneration of the deep tegmental nucleus of Gudden, a finding indicative that this structure is likewise not essential for the expression of the display.

Lesions of Central Tegmental Tract

On the basis of certain anatomical studies,[17] there are indications that some pallidotegmental fibers enter the medial longitudinal fasciculus, but there is no evidence of a contribution to the central tegmental tract. In the midbrain, the central tegmental tract courses ventrolateral to the central gray matter. It can be followed through the pons to the inferior olive. As described in Chapter 4, this phylogenetically ancient and poorly defined pathway of the brainstem contains massive numbers of ascending and descending fibers, including an unknown number of afferents destined for the thalamic tegmentum and corpus striatum.

Bilateral lesions of the central tegmental tract were produced in two animals. Monkey M-5 with such lesions at the level of the fourth nerve nucleus gave a genital display in 50 to 95% of the trials during 9 months of testing. The penile index ranged from 2.9 to 4.3. The incidence of vocalization averaged around 50%, while the thigh-spread component seldom exceeded 15%. In N-5, with lesions of the tract at the level of the third nerve, the score for genital display was usually below 50%, with the genital index declining to 2.7, while toward the termination of 10 months of testing the vocal and thigh-spread components occurred in only 6% or less of the trials. In this animal volumetric measurements revealed that about 75% of the magnocellular part of the red nucleus was destroyed (but see W-5 above). In both animals the lesion extended into the ventrolateral part of the central gray matter. In this respect it is to be noted that one monkey (P-5) with a lesion destroying the central gray matter ventral to the aqueduct at the level of the third nerve, after performing fragmented displays for 5 months, reachieved criterion in the sixth month.

Some ascending fibers in the central tegmental tract terminate in the parafascicular nucleus. In one monkey (L-5) with lesions largely involving this nucleus (Figure 14-8B) there was a statistically significant decline and fragmentation of the display.

Finally, mention should be made of one monkey (T-5) in which a large symmetrical lesion of the inferior colliculus also extended into the central tegmental tract. This animal become avocal. Followed for a period of nearly 8 months, it displayed full erection in nearly 100% of the trials and recovered from a significant decline in the occurrence of the thigh-spread. But vocalization occurred only two times in eight sets of 30 trials.

Until more is known about the origin and destination of fibers contained in the central tegmental tract, the question of whether or not the changes in display behavior of these monkeys are owing to a disruption of afferent or efferent fibers must remain unresolved. Based on present knowledge, it would appear that direct pallidotegmental fibers do not join the tract.

Assessment of Complicating Factors

Animals with lesions involving the transhypothalamic and tegmental pathways from the medial segment of the globus pallidus appeared somewhat slowed up and poorly groomed. A number of monkeys gradually took on the appearance of old animals. The behavior of others recalled that of the macaque monkeys described by Ranson in 1939 with large posterolateral destruction of the hypothalamus.[22] They seemed to become tamer and had an apparent inability to recognize harmful situations. A few seemed somewhat somnolent during the first few postoperative days. The three most common complications that will now be discussed were (1) failure of spontaneous eating and drinking, (2) hypothermia, and (3) bradykinesia.

Failure of Spontaneous Eating and Drinking

In the original study of the 22 animals with lesions involving primarily the trans-hypothalamic pathways and the tegmental region, all but two required hand feeding for an average period of 8 days.[4] The range extended from 2 to 17 days. The two exceptions were one animal with destruction of the most rostral part of the ansa and another with destruction of the most dorsal part of the ansa system. Recovery of the ability to *feed* and to *drink* usually occurred about the same time. Thereafter the monkeys seemed to eat and drink normally and maintained their preoperative weight. Two animals (S-5, U-5) operated on since the original report had lesions involving the junction of the ansa with the thalamic fasciculus. Each, respectively, was eating and drinking by postoperative day 3 and 6. These findings on ingestion correlated with those of the earlier study in which coagulations involved the medial segment of the globus pallidus (Chapter 13). As will be recalled, eight monkeys with such lesions required feeding for an average period of 6 days.

Both clinically and experimentally it has long been recognized that lesions of the hypothalamus may be associated with gastric ulcers. In the present cases one animal (D-5) with an extensive lesion of the thalamic tegmentum died on postoperative day 54 from a bleeding, gastric ulcer.[4]

Hypothermia

Ten of the twenty-two animals with lesions of the transhypothalamic pathways or tegmental pathways developed a transient or enduring inability to prevent a decline in body temperature from the normal level.[4] In six cases (N-4, O-4, Q-4, T-4, A-5, I-5) there was a recovery from the hypothermic condition within a period of 14 days. In the three cases (U-4, D-5, J-5) with an enduring hypothermia, there was either a complete destruction of the tegmental area corresponding to the field of Forel or an interruption of fibers passing over the surface of the substantia nigra.

Two monkeys (S-5, U-5) operated on since the original report have provided information that helps to narrow down the location of structures where damage results in hypothermia. The symmetrical retromammillary lesions in these two cases could almost be superimposed, except that in U-5 the ventral margin extended nearly to the dorsomedial border of the substantia nigra, whereas that for S-5 was about 0.75 mm distant. Subject S-5 had no problem in thermoregulation, whereas U-5 registered a morning temperature that did not exceed 35°C (95°F) for 29 days. This animal, like another subject (U-4) [sic] with hypothermia, developed a type of behavioral thermoregulation, and like a lizard would move in and out of a heat source. At first there might be as much as a 6°C difference (34°C to 40°C) between the a.m. and p.m. rectal temperature. Upon recovery of thermoregulation after 36 days, U-5 recovered its regular ability to display (see above). Comparison of its lesion with that in monkey U-4 revealed an overlap in an area just dorsomedial to the substantia nigra.

Clark *et al.*[23] concluded that a thin ribbon of fibers passing over the dorsal surface of the substantia nigra was requisite in the cat for maintaining normal temperature. As judged by a comparison of their Figure 1 and Figure 14-11 of the present study, my own findings in monkeys would be in agreement with their conclusion. The monkey (H-5) with the lesions shown in Figure 14-11 initially developed hypothermia (c. 34°C), but recovered thermoregulation by day 9. Although there may have been initial edema, the lesion itself did not extend into the area described above.

The findings in one experiment, however, would not support the belief of Clark *et al.*[23] that the critical fibers had their origin in the posterolateral hypothalamus. In the case (W-4) illustrated in Figure 14-2, virtual destruction of the posterolateral hypothalamus failed to produce even a temporary hypothermia. Consequently, the question arises as to the possibility that both nigrostriatal and pallidotegmental fibers may be implicated in neural mechanisms that prevent a decline in body temperature.

A fall in body temperature might be expected to result in a sluggishness that would interfere with the disposition to display. However, one animal (U-4) proved capable of giving a full display when its rectal temperature was only 34°C (93°F). It would therefore appear that hypothermia in itself is not sufficient to prevent the expression of the full display.

Bradykinesia

Sooner or later all of the animals with lesions of transhypothalamic pathways or tegmentum gave the impression of moving somewhat slowly and deliberately, lacking the usual alacrity of squirrel monkeys in climbing and exploring. There was the exception that they showed a momentary rapid jumping to catch a grape dropped from above and also showed defensive, lightning-speed mobility when given a brief trial of defending themselves when introduced to an established group of squirrel monkeys. In the study on the globus pallidus, five of the animals developed a mild degree of bradykinesia[15] (Chapter 13). One of them with lesions involving primarily the lateral segment, recovered the ability to display despite the presence of bradykinesia.

It seems probable that the slowness of movement is due to an interruption of dopamine-containing fibers ascending from the substantia nigra to the corpus striatum. In the early 1960s it had been shown in macaque monkeys that lesions interrupting the dopaminergic nigrostriatal pathway result in bradykinesia and tremor.[24] As noted above, the same symptoms develop in squirrel monkeys after injection of 6-hydroxydopamine in the region near the rostromedial pole of the substantia nigra.[21] In view of the multiple routes of ascending nigrostriatal fibers, lesions involving either the subthalamic tegmentum or transhypothalamic pathways might have interrupted sufficient axons from dopamine-containing cells to induce bradykinesia. Similarly, the interruption of such fibers radiating through the pallidum might have accounted for slowness of movement in some cases with lesions of that structure (see Chapter 13). Apropos of the possibility that bradykinesia itself might interfere with the expression of the display, it deserves emphasis that in the present series of experiments the monkey (U-4) judged to be the most bradykinetic was able sporadically to perform full displays.

Summarizing Discussion

Since the globus pallidus is one of the two major destinations of the projections from the corpus striatum, and since these projections are almost exclusively the only projections to the pallidum, the results of the experiments described in the preceding chapter justify the inference that the striopallidum itself is basically involved in the integrated performance of socially communicative displays. Based on fractional ablations, the critical efferent channels arise from the rostral two-thirds of the medial pallidal segment.

The experiments described in the present chapter provide further correlative evidence of the role of the striopallidum in displays. By the process of elimination it was found that the pallido-transhypothalamic pathways, rather than any of the other major pathways associated with the hypothalamus, are essential for the integrated expression of the display. Destruction of the transhypothalamic pathways from the medial pallidal segment resulted in an enduring interference with the display. Also, as in the case of medial pallidal lesions, there was a temporary elimination of spontaneous feeding and drinking.

In addition, it was found that the pallidal projections to the thalamic and midbrain tegmentum are more essential than those to the ventral thalamus for sustaining performance of the display. This finding deserves special emphasis because neurologists have traditionally emphasized the functional role of pallidothalamic projections almost to the exclusion of those to the tegmentum. This bias can be attributed to the recognition that the ventrolateral nuclei have direct connections with the supplementary motor areas of the neocortex.

Differential Effects of Lesions on Components of Trump Display

The trump display has been defined as consisting of the three main components of the display—namely, 4–5+ penile erection, vocalization, and thigh-spreading. In cases in which there was not an elimination, but rather a fragmentation of the display, the anatomical findings provided some clues as to structures contributing to the performance of respective components of the trump display.

Penile Erection

Genital erection would rank as the central component of the display because in criterion performance, it may occur somewhat more consistently than vocalization and thigh-spreading. As described, lesions involving the inferior thalamic peduncle, as well as those destroying the medial forebrain bundle at different loci along its course, had no enduring effect on the performance of the display, but resulted in a persisting, significant decline in the magnitude of penile erection. There was a similar outcome following lesions of the medial preoptic nuclei and adjoining anterior hypothalamus. Coagulation of the dorsomedial hypothalamic area had the most profound effect on the genital component: although monkey V-5 with such a lesion would regularly vocalize and thigh-spread when tested, erection failed to occur for a period of $2\frac{1}{2}$ months and thereafter never exceeded a magnitude of 1–2+ during the remaining 4 months of testing. The lesions in every case involved sites at which electrical stimulation was known to be highly effective in eliciting penile erection.

Vocalization

In the course of the present experiments it became apparent that some of the animals with tegmental lesions developed abnormal sounding vocalizations. Recording of sound spectrograms in these animals led to a special study in which it was observed that damage to structures at the thalamomidbrain junction affected particularly the production and structure of the separation cry. Such cries perhaps represent, evolutionarily, the earliest and most basic mammalian vocalization. The significance of these particular findings will be dealt with in Chapter 21. In the present study it was found that small, symmetrical lesions (Figure 14-12) just rostral to the critical zone for the separation cry, resulted in a virtual elimination or drastic reduction in incidence of the display vocalization. This vocal manifestation has some characteristics resembling the separation cry. The overall analysis suggested that the vocal deficits may have resulted from a coagulation of fibers originating in or near the origin of the medial longitudinal fasciculus. It will also be recalled that in cases with lesions involving the central tegmental tract there was a marked and significant decline in the incidence of vocalization.

Thigh-Spread Component

Short of complete elimination of the display, a fragmentation of the display marked by a significant decline in the thigh-spread component, is the surest indication of damage either to the internal segment of the pallidum[15] (Chapter 13) or to its projecting pathways. At the present time it would be safer to list structures that appear to be unnecessary for the expression of the thigh-spread than vice versa. At the midbrain level the list of unessential

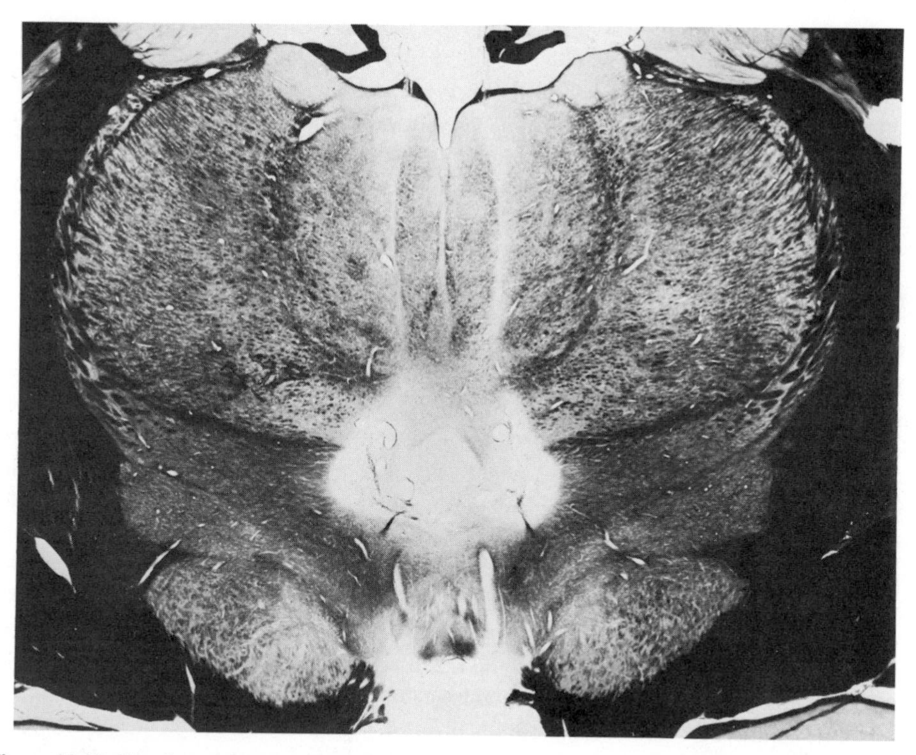

Figure 14-12. Site of symmetrically restricted lesions that resulted in virtual elimination of the display vocalization. In both this animal (S-5) and U-5 with similar lesion there was recovery of the other components of the trump display. The selective effect on vocalization is attributed to damage in the region of the origin of the medial longitudal fasciculus, which is just rostral to lesions affecting the isolation call.
From MacLean (unpublished).

structures would include the superior colliculus, central gray matter, red nucleus, brachium of the superior conjunctivum, medial and intermediate substantia nigra, superior nucleus of Bechterew, interpeduncular nucleus, habenulopeduncular tract, and (see Chapter 13) pyramidal fibers descending from the leg area. Given these data, it would appear likely that the integration of the main components of the display occurs in the midbrain tegmental reticulum. The capacity for full display following virtual elimination of the superior colliculus speaks against a primary role of nigrotectal projections.

Recapitulation

In concluding this final chapter on the neurobehavioral work dealing with displays, the following brief recapitulation will recall the thrust and highlights of the investigation:

For communication, terrestrial vertebrates engage in four main kinds of displays that may be characterized as (1) signature, (2) challenge, (3) courtship, and (4) submissive displays. In addition to dynamic modifiers, such displays may include static modifiers largely dependent on autonomic function. For the comparative work on primates, squirrel monkeys were chosen because the highly predictable mirror display of the gothic-type squirrel monkey is especially suitable for systematically testing the effects of cerebral lesions on the somatic and autonomic components of a species-typical display. The mirror display of these monkeys combines features of their challenge, courtship, and signature displays. The most constant features of the display—penile erection, vocalization, and thigh-spreading—constitute the "trump display." Based on systematic testing of more than 120 monkeys with lesions of various parts of the brain, it has become evident that the medial globus pallidus is a site of convergence for neural systems required for performance of the mirror display. Lesions within the rostral two-thirds of the medial segment or of its projecting pathways eliminate, or result in a statistically significant decline in, the performance of the display. Pallidotegmental projections are more vital than those to the thalamus for sustaining performance of the display. Electrocoagulations of the medial preoptic area, the dorsomedial hypothalamic area, or the medial forebrain bundle interfere with the development of genital tumescence, but do not otherwise affect the performance of the display. Small hypothalamic lesions involving fibers originating in or near the medial longitudinal fasciculus selectively interfere with the display vocalization. A decline in incidence of the thigh-spread component is the most sensitive indicator of lesions of the medial pallidum or of its respective pathways. Bilateral lesions of other pathways such as the fornix, stria medullaris, mammillothalamic tract, and habenulopeduncular tract have no apparent effect on the performance of the display.

To emphasize again what was said in the last chapter, the comparative neurobehavioral studies on animals as diverse as lizards and monkeys indicate that structures of the R-complex are essential for the performance of communicative displays. Moreover, since these displays involve conspecific recognition and isopraxis, a probable role of the R-complex in these psychological functions must also be considered.

In the next chapter dealing with clinical conditions relevant to functions of the striatal complex, there will be occasion to consider the role of the striopallidum in integrating the performance of the daily master routine and subroutines. The material covered in that chapter will also call for a further consideration of chemical mediators in striopallidonigral functions.

References

1. Ganser, 1882
2. Herrick, 1910
3. Herrick, 1948
4. MacLean, 1981
5. MacLean and Ploog, 1962
6. Haymaker *et al.*, 1969
7. Rioch *et al.*, 1940
8. Wheatley, 1944
9. Beck and Bignami, 1968
10. von Monakow, 1895; Mettler, 1948
11. Gergen and MacLean, 1962, p. 81
12. Ranson and Ranson, 1939
13. Carpenter, 1967
14. Emmers and Akert, 1963
15. MacLean, 1978b
16. Forel, 1872
17. Nauta and Mehler, 1966
18. Nauta and Mehler, 1966, p. 37
19. MacLean *et al.*, 1963a, p. 289
20. Lindvall and Bjorklund, 1979
21. MacLean, 1975c
22. See Ranson, 1939, for summary
23. Clark *et al.*, 1939
24. Poirier, 1960; Poirier and Sourkes, 1965

15

Some Relevant Clinical Findings

Before discussing some of the human implications of the neurobehavioral research (Chapter 16), attention should be called to certain symptoms of patients with diseases that affect the functions of the striatal complex. As defined in Chapter 4, the striatal complex consists of the olfactostriatum, corpus striatum (caudate and putamen), globus pallidus and satellite collections of gray matter. In clinical neurology, these structures are described as belonging to the *basal ganglia* (see Figures 4-2, 4-3, and 14-3). It has been emphasized in preceding chapters that large parts of the striatal complex may be destroyed on one or both sides with no apparent effect on bodily movements. As Denny-Brown has remarked, "Not uncommonly a softening of the greater part of the putamen or whole striatum is found in patients without any involuntary movements having been noted during life."[1] The clinical view that the basal ganglia have primarily motor functions is partly owing to the recognition that disease of related structures (subthalamic nucleus, substantia nigra) may result in severe disturbances in motility of parts or all of the body. In regard to our present interest in the analysis of functions, there is the added complication that diseases affecting the basal ganglia may also be associated with degenerative changes elsewhere in the brainstem, as well as in the cerebral cortex.

Three Relevant Clinical Conditions

Making allowance for the mentioned complicating factors, we will single out three well-known diseases that affect the functions of the striatal complex and identify certain manifestations that are relevant to the question of brain mechanisms involved in basic forms of behavior and in the regulation of the daily master routine and subroutines. Named after the physicians who described them, the nervous disorders are called (1) Parkinson's disease (paralysis agitans), (2) Huntington's chorea (hereditary chorea), and (3) Sydenham's chorea (rheumatic chorea; St. Vitus' dance). They will be considered in that order. Then after a brief consideration of childhood autism, the remainder of the chapter will deal with some neurochemical data relevant to current views about the involvement of the striatal complex in schizophrenia and manic–depressive illness.

Parkinson's Disease

In 1817 Parkinson (1755–1824) published a description of a disease called *paralysis agitans* (*shaking palsy*).[2] The disease may appear in the later decades of life as a result of

The author used the material on Huntington's and Sydenham's chorea presented in this chapter as the basis for comments on the role of the daily master routine and subroutines in his Adolf Meyer Lecture (MacLean, 1985a).

unknown causes, or occur as a consequence of a viral encephalitis, poisoning with certain heavy metals, or a prolonged lack of oxygen. It is characterized by shaking movements of one or more of the extremities (usually the upper); loss of emotional expression (masklike facies not described by Parkinson); generalized muscular stiffness and slowness of movement; stooped posture and shuffling walk.[1]

In 1895 Brissaud cited a case in which Parkinson's disease was associated with tuberculous destruction of the substantia nigra.[3] His impression that disease of the nigra "might well be" the cause of paralysis agitans[4] was supported by Tretiakoff's findings on 54 cases that were published in 1919.[5] Many authorities held contrary views, and in the late 1940s I remember the argument going back and forth as to whether Parkinson's disease was due to degenerative changes in the globus pallidus or substantia nigra. At the present time, there is a consensus that the most constant finding in this disorder is loss of nerve cells in the substantia nigra,[6] with the manifestations occurring after disappearance of 60 to 75% of the neurons.[7] But it is only since the discovery of the monoaminergic systems of the brain that the motor symptoms have been found to be attributable to the loss of dopamine-producing cells in the substantia nigra.[8] As mentioned in Chapter 4, the nerve fibers of these neurons ascend and ramify somewhat like a vine among the nerve cells of the corpus striatum.

"Automatic" Behavior

In the years encompassing the First World War, the European neurologists Oskar Vogt (1870–1959) and Cecile Mugnier Vogt (1875–1962) conducted extensive studies of the striatal complex. They concluded that this system of the brain is basically involved in the control of expressive movements, such as gesture and mimicry, as well as in the regulation of the so-called automatic associated movements that occur in walking, running, and other activities.[9] It was, however, the tradition then, as now, to regard the striatal complex as a motor system under the control of the cerebral cortex, having "no mind of its own."

The late Paul Yakovlev, in his seminal paper "Motility, Behavior, and the Brain," emphasized the role of the basal ganglia in the "outward expression of internal (visceral) states or emotions, such as facial mimicry, vocalizations, speech, gestures, body attitudes, muscle tone and posture. . . ."[10] Later he stressed that the central feature of Parkinson's disease is an inability to carry out these very functions, calling attention, in particular, to cases illustrating impaired performance in responding to *internally experienced needs.*[11] He mentions first a woman "in her early fifties" confined to a hospital for the chronically ill. She was so incapacitated that she could not feed herself, freezing when her fork would come about halfway to her mouth. Nevertheless, she was very helpful on the wards in feeding *other* patients. He mentions another patient (a man in his early thirties) who remembered as his earliest symptom, an inability after urination to make the "habitual 'shove' with the hip to adjust the penis on the proper side in his trousers."[12] This same patient also complained "of unwanted procrastination and inefficiency in adjusting his posture on the toilet seat." Another male patient "volunteered the complaint of having lost spontaneity in his postural adjustments to his mate during intercourse."[13] "My wife," said the patient, "has to take charge of the whole . . . business. . . ." "Yet in spite of this impediment of automatic postured adjustments," Yakovlev points out, "the libido, the potency, and the ultimate orgasmal movements were, apparently, normal."

In summary, these clinical observations indicate that the striatal complex is implicated in the organized expression of basic behavior that is *initiated by internally derived cues* and depends on an integration of viscerosomatic functions.

Huntington's Chorea

Chorea is the Greek word for dance. Clinically, it refers specifically to rapid, jerky movements of the extremities. In 1872 George Huntington (1850–1916), a native of Long Island, described a rare, familial form of chorea that develops in the third decade of life and invariably leads to insanity.[14] Inheritance of the disease has since been found to depend on autosomal, dominant genetic transmission.[15] Huntington explains how he first became aware of the disease: ". . . [R]iding with my father on his professional rounds, I saw my first case of 'that disorder', which was the way in which the natives always referred to the dreaded disease. . . . We suddenly came upon two women, mother and daughter, both tall, thin, almost cadaverous, both bowing, twisting, grimacing. I stared in wonderment, almost fear. What could it mean? My father paused to speak with them and we passed on. . . . From this point on, my interest in the disease has never wholly ceased."[16] In addition to the grimacing, writhing gesticulations, and progressively violent jerking movements, victims of this dread disease give the appearance of walking as though intoxicated. The disease invariably leads to insanity.

In Huntington's chorea there is characteristically a shrinkage of the corpus striatum (caudate nucleus and putamen) (Figure 4-3), with a marked loss of the small nerve cells and infiltration by glia. But as is so often the complication in the clinical interpretation of neurological disorders, the disease process involves other cerebral structures: there may be neuronal damage in the globus pallidus and thalamus, as well as degenerative changes in the cerebral cortex, particularly the frontal areas.

Disorders of "Routine" in Huntington's

The evolution of arithmetic is said to have been retarded by our ancestors' inability when counting on the fingers to see the empty space (the "sunya") existing between the fingers. In neurobehavioral work one encounters a somewhat parallel situation: there seems to be a difficulty in recognizing what is "absent." Moreover, as has been emphasized in the preceding chapters, the design of neurobehavioral experiments has commonly made it impossible for the experimenter to be aware of unexpressed behavior. In his mongraph on the basal ganglia the English clinician J. P. Martin has insightfully employed the term *negative symptom* (in the Jacksonian sense) in calling attention to certain postural deficits of patients (Chapter 4).[17] An experimental example of a "negative symptom" is provided by the fragmentation of the trump display of squirrel monkeys (Chapters 13 and 14) that is most commonly evident by the absence of the thigh-spread component.

In the present context, I consider "negative symptoms" of quite a different nature that might alternatively be referred to as *nonevident symptoms* (Chapter 4). To start with, emphasis will be given to the loss of routines by patients in the early stages of Huntington's disease. Then it will be described how in other conditions the daily master routine may be interfered with by the excessive expression of subroutines. In preceding chapters the observations on reptilian behavior have drawn attention again and again to the importance of daily master routines and subroutines in self-survival and procreation. Every day lizards engage in no less than seven sequences of behavior that are performed with almost clocklike regularity. But there are times when courtship activity or the defense of territory provoked by an intruding male will greatly interfere with the daily master routine. In clinical neurology, as in neurobehavioral investigations, relatively little attention, it seems, has been given to identifying brain mechanisms underlying the integrated performance of the daily master routine. Such an ordering of activities would require a special intelligence capable of regulating what has been described as interoperative behavior (Chapter 10; see also Chapter 16).

In a study of 18 patients in the early stages of Huntington's chorea, Caine *et al.* found that the inability to remember, plan, and organize their activities was more troubling to them than the unpredictable movements associated with the disease. "As we watched the patients on the ward," the authors remark, "it became clear that they rarely initiated independent activity."[18] "Left alone, most seemed content to sit and do nothing; many watched television for hours. . . . When presented with planned, specified tasks, however, they would participate eagerly, with sustained interest for whatever time the organizing staff member remained."[19] Although the authors do not mention routine, what they seem to be saying is that the patients lacked the steam for initiating their own routine, but both welcomed and followed the superimposed routine provided by the staff.

In their own environment at home it was also evident that these patients had difficulty in executing *subroutines*. One woman, aged 43, had difficulty with a number of household tasks that required organization. She could perform each step separately, but was unable to maintain an orderly sequence. Preparing meals became impossible. As she explained, "I haven't been able to prepare a Thanksgiving dinner for five years, even though I know how to do it."[20] A dentist reported that one of his initial symptoms was finding himself unable to remember a step in a dental procedure. "I would be standing there with a tool in my hand and with the patient's mouth open, and I wouldn't remember what to do next. I had done these things for years. Later on I would remember."[20]

In line with traditional views, the authors suggest that the various manifestations that have been described were the result of a loss of *cortical* "executive" functions. But is it not possible that the incapacity to pursue a spontaneous routine may have been the result of diffuse striatal damage and that the ability to carry out a prescribed routine reflects compensatory cortical function?

As indicated above, the inability to establish a master routine, as well as the deficits in carrying out subroutines, might be regarded as "negative symptoms." We will next consider symptoms of a positive kind—namely, an overdoing of subroutines—that interfere with adherence to a master routine.

Sydenham's Chorea and Subroutines

Thomas Sydenham (1624–1689), known as the English Hippocrates, described a form of chorea once commonly seen among young girls and associated with acute rheumatism of childhood and adolescence. The choreiform movements affect the muscles of the face, tongue, and extremities. The disease process in the brain is poorly understood.[21] There may be an inflammatory condition around the cerebral blood vessels, and these changes, together with the loss of nerve cells, may be most pronounced in the corpus striatum (caudate nucleus and putamen). But the substantia nigra, subthalamic nucleus, and cerebral cortex may also be affected.

The disease is better known to the layperson as *St. Vitus' dance,* a name attached to it by Sydenham, and suggested by Paracelsus's use of the expression in referring to epidemics of dancing mania that swept Germany and the Netherlands during the 14th, 15th, and 16th centuries.[21] These epidemics were believed to occur under the influence of religious excitement, and pilgrimages were made to various shrines in the search for relief of the symptoms. The most famous shrine was that of St. Vitus in Zabern. It has since been suggested that the epidemics of dancing may have been a complication resulting from eating rye grain contaminated by an ergot-producing fungus.

In William Osler's (1849–1919) little-known monograph on Sydenham's chorea, one finds case histories that illustrate how morbid subroutines may drastically interfere with adherence to the daily master routine.[22] Under these conditions the disruption in routine may be ascribed to "positive," rather than to "negative" symptoms. It is possible that in Sydenham's chorea the inflammatory condition has an irritative effect on the corpus striatum. As opposed to Huntington's chorea, the loss of nerve cells is minimal. It must be emphasized that in the illustrations to follow, as well as in the cases mentioned above, there is no postmortem information confirming the location of cerebral disease.

Two of Osler's cases are particularly relevant. The first is that of a girl, 13, with twitchings of the muscles of the face and neck, who developed several varieties of compulsive, ritualistic behavior. Before putting on clean underclothes she had to count so many numbers that there was great difficulty in getting her dressed. After brushing her teeth, she had to count to one hundred. She would not brush her hair except at the extreme tips. She would enter her house only by the back door, would knock three times on the edge of the window nearby, and three times on the door before unlocking it. Before getting into bed at night she would lift each foot and tap nine times on the edge of the bed.

The second case is that of a girl aged 12 whose symptoms recall a syndrome described by Gilles de la Tourette, a member of Charcot's group in Paris.[23] In addition to echoing what others said (echolalia) and an obsessive use of obscene words (coprolalia) she had a compulsion to keep touching objects. It is evident that the continual touching of objects would entail a constant interference with a person's daily routine. The condition calls to mind the disruption in the routine of animals that neglect the pursuit of all other activities except pressing a bar to receive electrical stimulation of the brain.[24]

In addition to echolalia, echopraxia, and coprolalia, the Gilles de la Tourette syndrome is characterized by various tics. The compulsion of the girl just mentioned to touch objects might be regarded as a "behavioral tic." Since drugs known to have an action on the corpus striatum also alleviate the symptoms in question, there is presently an inclination to believe that the Gilles de la Tourette syndrome is owing to disease or malfunction of the striatal complex.[25] There has been at least one case in which there was a marked decrease of nerve cells in the corpus striatum.[26]

Osler gives examples of other forms of perseverating behavior, which, unlike the repetitive, "displacement activities" of animals mentioned in Chapter 10, seldom desist for long. He mentions, for example, a young child recovering from chorea minor, who in picking up anything, first smelled and blew upon it.[22]

Childhood Autism

In his essay on chorea, Osler describes other forms of abnormal behavior of a rhythmical nature and involving nodding, rocking, spinning, and head banging.[22] Such activities immediately remind today's reader of childhood autism.

In 1943, Leo Kanner described what was regarded as a new clinical syndrome in a paper entitled "Autistic Disturbances of Affective Contact."[27] I briefly mention the condition here because of its relevance to the question of isopraxis. Before doing so, it should be emphasized that as yet there are neuropathological findings on only a few of the questionable cases. The manifestations of autism are usually present at a very early age. As Rimland (1964) has emphasized, parents first become aware that something is wrong when (1) the baby does not show the anticipatory movements of holding out its arms when about to be picked up or (2) fails to cradle in the parent's arms when being held.[28] Upon being cuddled, the infant twists and squirms so much that one is fearful of dropping it. At kindergarten age, the autistic child characteristically insists on being left alone and engages in highly ritualistic forms of play. At other times, the child lapses into a "subroutine" of endlessly rocking, or, more distressing to watch, repeatedly banging its head. There seems to be a total incapacity (interpreted by some as willful stubbornness) to learn everyday things, such as tying shoes through the so-called imitative process. In a parallel fashion there appears to be an inability to identify with the feelings of others. As Eisenberg and Kanner comment, "This amazing lack of awareness of the feelings of others . . . runs like a red thread through our case histories."[29] As an illustration they describe a 4-year-old boy playing on a beach who goes straight to his goal, walking over people and their belongings or whatever else lies along his path. The authors conclude that the primary disturbance in infantile autism might be described as a disturbance in social perception.

Anyone who has devoted much time to these children realizes that, just as they may have a large vocabulary conjoined with an inability to communicate meaning, they may also show indications of strong emotional feelings that cannot be appropriately expressed. It was as though nature had given them all the bits and pieces of humanity without the means of putting them together. Considering the diverse symptoms associated with chorea, one wonders if in cases of autism something may have gone wrong in the original wiring diagram of the corpus striatum. Or might there have been some intrauterine or natal embarrassment of the arterial or venous circulation that could have led to a diffuse, dropping out of nerve cells that would be difficult to detect in a routine histological examination of the brain? In line, however, with the traditional emphasis on the cerebral cortex in human behavior, past speculation about the nature of disturbed brain function has centered on "the failure of cortical integration of affective and cognitive components of behavior"[30] in childhood autism.

Apropos of the symptomatology of chorea, it should also be pointed out that obsessive compulsive behavior is another distinguishing feature of childhood autism. "Thus," Eisenberg and Kanner remark, "a walk had always to follow the same prescribed course; bedtime to consist of a particular ritual of words and actions; and repetitive activities like spinning, turning on and off lights, and spigots, or flushing toilets, could preoccupy the child for long periods. Any attempt to interfere with the pattern would produce bursts of rage or episodes of acute panic."[31] Reminiscent of the Gilles de la Tourette syndrome, there may be parroting of speech, repetition of stored phrases, and affirmation by repetition.

Neurochemical Considerations

In Chapter 4, reference was made to the growing list of substances that are believed to be implicated in the transmission of nerve impulses in respective parts of the R-complex. Those having received foremost attention are acetylcholine, dopamine, serotonin, GABA, substance P, glutamic acid, and enkephalins. Although existing in a concentration many times less than that of dopamine,[32] norepinephrine must enter the present discussion. Of all these agents that may influence the discharge of nerve cells, acetylcholine might be considered the main workhorse. I say this because it is the one compound that when applied to different parts of the brain can be most depended upon to elicit behavioral responses associated with electroencephalographic changes seen in recordings obtained with macroelectrodes[33] (see Chapters 4 and 19). Despite the many functions that have been ascribed to the monoamine systems of the brain, workers are quick to admit ignorance regarding the nature of their actions. Pre- and postsynaptic dopamine receptors in the striatum are respectively designated as D1 and D2.[34] Dopamine that attaches itself to D2 receptors is presumed to inhibit the action of neurons that release acetylcholine (see below). There has been no clarification as yet about the role of norepinephrine and serotonin in the R-complex. The same would apply to the peptides—substance P and enkephalins. GABA is recognized generally as having a suppressive action on the discharge of nerve cells.

In the following brief discussion, I focus attention on acetylcholine and dopamine because an assortment of evidence suggests that modifications of their action on the R-complex may be conducive to some manifestations of the diseases already discussed, as well as to symptoms of toxic and endogenous psychoses.

Actions of Dopamine and Acetylcholine

It used to be thought that dopamine served only as a precursor of norepinephrine and epinephrine. The year 1957 saw the first of a series of reports that radically altered this view and became of momentous significance in regard to the neurochemistry of the striatal complex. In that year Montagu published his finding of the presence of dopamine in the brain,[35] a discovery that was soon confirmed by Carlsson *et al.*[36] In 1959 Bertler and Rosengren reported that dopamine occurred in highest concentration in the corpus striatum,[32] a condition that was to be visually demonstrated when Dahlström and Fuxe applied the histofluorescence technique of Falck and Hillarp (see Chapter 4). The initial reports by the Swedish workers led two Austrian pharmacologists, Hornykiewicz and Ehringer,[37] to assay the postmortem concentration of dopamine in the human brain. Compared with "normal" brains, there was a significant decrease of dopamine in the corpus striatum of patients with a diagnosis of Parkinson's disease.

As noted in Chapter 4, the dopamine in the corpus striatum and olfactostriatum is derived from nerve fibers that ascend from cells in the ventral midbrain (Figure 4-7). When dopamine is released at the nerve terminals, it is pictured as attaching itself to dopamine receptors[38] of striatal neurons and thereby preventing their discharge. It is proposed that such striatal cells ordinarily excite other cells through the release of acetylcholine. In line with such a scheme, the following considerations would be compatible with the explanation that a deficiency of dopamine results in "runaway" activity of cholinergic cells that would, inter alia, result in the discharge of GABAergic striatal cells projecting to the pallidum and pars reticulata of the nigra.

1. Electrophysiological studies have shown that the iontophoretic application of dopamine to striatal cells usually depresses their activity,[39] whereas the application of acetylcholine appears to excite most of the tested cells.[40]
2. Symptoms of parkinsonism manifested by bodily rigidity, tremor, impaired

speech, and the loss of facial expression, allegedly occur with the loss of two-thirds or more of the dopamine cells in the substantia nigra.

3. A picture of advanced parkinsonism appears in monkeys if the majority of dopamine cells in the substantia nigra are destroyed by the local application of 6-hydroxydopamine (Chapter 14) or by the systemic administration of methylphenyltetrahydropyridine (MPTP).[41]

4. Neuroleptic drugs such as chlorpromazine that presumably attach to dopamine receptors and block the normal action of dopamine, induce parkinsonian symptoms.

5. As demonstrated by Cotzias *et al.*[42] (in following a lead of Hornykiewicz), the oral administration of L-dopa to patients with Parkinson's disease substantially alleviates their symptoms. L-Dopa (which passes the blood–brain barrier) provides the brain with a next-to-final molecule for making dopamine.

6. Finally, as long recognized, drugs such as atropine and scopolamine that interfere with the action of acetylcholine alleviate the symptoms of Parkinson's disease.

Significantly, in regard to parkinsonian symptoms induced by neuroleptic drugs, Snyder and co-workers have shown that chlorpromazine (a drug introduced in 1952 by Henri Laborit and co-workers[43]) has a molecular conformation resembling dopamine.[44] It has been suggested that some of the beneficial effects of this and other like drugs in the treatment of psychotic illnesses may be due to blocking of dopamine receptors and a consequent slowing down of a person's behavior and mental processes. In 1957 on a visit to the psychiatric service at the University of Vienna, I vividly remember being conducted through the hospital and mistaking the ward for psychotic patients for the one with neurotic patients. The tranquilized psychotic patients, sitting quietly in their chairs, seemed hardly different from the visitors in the anteroom. Although the use of ''antipsychotic'' drugs such as the phenothiazines has resulted in a remarkable reduction in the number of patients in mental hospitals, some patients do not share the generally expressed enthusiasm regarding the therapeutic benefits of these agents. The characterization of their subjective feelings gives an added dimension to conditions existing in asylums before Pinel began to inveigh against the use of straitjackets. It was as though they felt constrained by two straitjackets–one physical and one mental.

Questioned Role of Dopamine in Psychoses

In view of the present focus on dopamine, it should be noted that some clinicians contend that an overproduction or deficiency of catecholamines may be involved in ''schizophrenia'' and in manic–depressive illness. A brief comment on these possibilities requires a statement regarding the production and inactivation of catecholamines which are generally regarded as neurotransmitters (but see below).

Release and Inactivation of Catecholamines

Dopamine and norepinephrine are synthesized, respectively, in the cells of dopaminergic and adrenergic systems (Chapters 4 and 18). Present evidence indicates that in each case the monoamine is stored in vesicles located in synaptic terminals.[45] The substances are released by rupture of the vesicles (exocytosis) into the synaptic clefts—

the spaces between the terminals of the nerve cell and parts of the neighboring nerve cell to be acted upon. Axelrod and co-workers have demonstrated that most of the released substance is inactivated by a "reuptake" in the nerve terminals where they are "repackaged" in vesicles.[45] Some inactivation is also accomplished by the enzyme monoamine oxidase *inside* the cell and by the enzyme catechol-*O*-methyltransferase *outside* the cell. Hence, the catecholamines appear to be disposed of by at least three mechanisms. (When acetylcholine is released, it is destroyed within 0.5 msec by cholinesterase.) It is usually assumed that the release of catecholamines is triggered by nerve impulses traveling within the same nerve cells in which the substances are stored. Jacobowitz, however, has suggested that in the case of noradrenergic systems, at least, metabolic factors may be responsible for their release. He regards the adrenergic systems that innervate widespread areas of the brain (Chapter 18) as making available a supply of norepinephrine for any particular group of cells whenever it is needed in conjunction with other substances for affecting neural transmission.[46] If so, norepinephrine would more appropriately rank as an adjuvant transmitter or modulator, rather than as a transmitter in its own right.

It is believed that several of the drugs variously known as psychoactive agents, psychic energizers, antidepressants, and the like, owe their effects to alterations in the release, reuptake, and enzymatic inactivation of catecholamines. Some examples of current interpretations are as follows: Amphetamine, which has a molecular similarity to norepinephrine, releases norepinephrine from nerve cells, blocks its reuptake, and inhibits monoamine oxidase. As to be noted, it also releases dopamine. Monoamine oxidase inhibitors used as antidepressants interfere with the enzymatic inactivation of catecholamines, while the tricyclic antidepressant drugs such as imipramine retard the uptake of these substances.[45]

Connell, Kety, and others have suggested that the amphetamine-induced toxic psychosis might serve as a model for workers attempting to uncover neurochemical factors underlying paranoid schizophrenia.[47] The amphetamine psychosis has similarities to this particular form of schizophrenia, one notable exception being that visual hallucinations are more prominent than auditory hallucinations. In the light of such a model, Snyder has developed the argument that overactivation of the striatal dopaminergic system may partially account for the symptoms of paranoid schizophrenia.[48] In comparing the activating effects of dopamine systems in animals and human beings, he emphasizes that one of the manifestations of the amphetamine psychosis is a stereotyped "picking at the skin" that is equivalent to amphetamine-induced stereotypies in animals, as exemplified by the sniffing, licking, and gnawing in rats. The crux of his argument depends on evidence of differences between the effects of dextro- and levoamphetamine, respectively, on locomotor activity and stereotypies. He reviews evidence that increased locomotor activity results from activation of noradrenergic systems of the brain, whereas stereotypies are owing to activation of the striatal dopaminergic system. For example, after destruction of the nigrostriatal dopamine system, amphetamine elicits increased locomotor activity, but not stereotypies. In this regard, he points out that dextroamphetamine is several times more effective than the levo-isomer in inducing increased motor activity, whereas levoamphetamine is almost as effective as the dextro-isomer in producing stereotypies. Snyder then makes a major point regarding the dopamine hypothesis, citing findings by Angrist *et al.* that dextro- and levoamphetamine are about equally potent in inducing a psychosis.[48] As further clinical support of the dopamine hypothesis, he points out the effectiveness of such dopamine blockers as phenothiazines as antidotes to amphetamine psychosis.[48]

Manic–Depressive Illness

Apropos of the present focus on dopamine, it should be noted that some clinicians also contend that an overproduction or deficiency, respectively, of this substance may be involved in manic–depressive illness.[49] Animal experimentation provides some suggestive parallels. In Chapter 14 it was pointed out that monkeys with partial to extensive injury of the ascending dopamine system may appear slowed (''depressed'') and prematurely aged. On the contrary, it is striking to observe in some animals how a drug such as apomorphine, which is presumed to have a stimulating effect on dopamine receptors, results in some form of hyperactivity. Randrup and co-workers have reviewed the activating effects of apomorphine in reptilian (turtle[50]), avian, and mammalian forms.[51] In our laboratory we have found that the effects are more profound in some birds than in mammals. For example, turkeys treated with apomorphine show an incessant running in and out of a flock for periods up to 4 hr.[52] A caged parrot shows no such running activity, but it goes into a state in which it vocalizes continuously for periods up to 40 min.[52] A hyperactive state similar to that of turkeys can be induced in dogs if their dopamine receptors have been previously ''sensitized'' by chronic treatment with a ''dopamine blocker'' such as pimozide.[53]

Concluding Summary

In this chapter, Parkinson's disease and two forms of chorea provide illustrations of behavioral and psychological changes that may occur as the result of disease affecting the functions of the striatal complex and its related structures in the brainstem. A discussion of such case material must be tempered by uncertainties regarding the nature and extent of the disease process, not only in the systems under consideration, but also in ''unrelated'' structures.

In Parkinson's disease, in which striatal structures receive insufficient dopamine from cells in the midbrain, there may be an impairment of *internally cued* somatovisceral functions. Illustrations are given of deficits in coordinating the acts of feeding, defecation, and copulation. A consideration of how striopallidal diseases shed light on mechanisms underlying prosematic behavior will be deferred until limbic and neocortical functions are dealt with.

In the two discussed forms of chorea we encounter examples of the crippling consequences of the inability to perform the master routine and various subroutines, as well as the disruptive effect of an inability to discontinue aberrational subroutines. Some of the symptomatology of Sydenham's chorea suggests parallels with childhood autism.

The chapter concludes with a consideration of certain hypotheses regarding the involvement of the striatal complex and the nigrostriatal dopamine system in schizophrenia and manic–depressive illness. There is evidence that drugs that block dopamine receptors of striatal cells make possible a discharge of cholinergic neurons that activate mechanisms having a braking effect on motor functions. The efficacy of such drugs in alleviating psychotic symptoms may be partly owing to a similar mechanism of action that is subjectively and outwardly manifest by a ''braking'' on psychic and motor functions. Animal experimentation provides some support of the clinical inference that manic and depressive states may be attributable, respectively, to overactivity and underactivity of the nigrostriatal dopaminergic system.

References

1. Denny-Brown, 1946, p. 273
2. Parkinson, 1817
3. Brissaud, 1895
4. Haymaker and Schiller, 1970
5. Tretiakoff, 1919
6. e.g., Earle, 1968, p. 11
7. Pakkenberg and Brody, 1965; Lloyd, 1977
8. Hornykiewicz, 1963, 1966
9. Vogt and Vogt, 1920
10. Yakovlev, 1948, p. 327
11. Yakovlev, 1966
12. Yakovlev, 1966, p. 294
13. Yakovlev, 1966, p. 295
14. Huntington, 1872
15. Vessie, 1932
16. Huntington, 1909; quoted in Haymaker and Schiller, 1970, pp. 453–454
17. Martin, 1967
18. Caine *et al.*, 1978, p. 382
19. Caine *et al.*, 1978, p. 380
20. Caine *et al.*, 1978, p. 379
21. Haymaker, 1956, p. 341
22. Osler, 1894
23. Gilles de la Tourette, 1885
24. Olds, 1958
25. Shapiro, 1970; Van Woert *et al.*, 1976
26. Pakkenberg and Brody, 1965
27. Kanner, 1943
28. Rimland, 1964
29. Eisenberg and Kanner, 1958, p. 6
30. Eisenberg and Kanner, 1958, p. 23
31. Eisenberg and Kanner, 1958, p. 4
32. Bertler and Rosengren, 1959
33. MacLean, 1957a,b; 1957b
34. Kebabian and Calne, 1979
35. Montagu, 1957
36. Carlsson *et al.*, 1958
37. Ehringer and Hornykiewicz, 1960; Hornykiewicz, 1963
38. Carlsson and Lindqvist, 1963
39. McLennan and York, 1967; Aghajanian and Bunney, 1977
40. Bloom *et al.*, 1965
41. Langston *et al.*, 1983; Ballard *et al.*, 1985; Burns *et al.*, 1985
42. Cotzias *et al.*, 1967
43. Laborit *et al.*, 1952
44. Horn and Snyder, 1971
45. For reviews see: Axelrod, 1965, 1971, 1974
46. Jacobowitz, 1979
47. Connell, 1958; Kety, 1959
48. Snyder, 1972
49. Bunney and Davis, 1965; Schildkraut, 1965
50. Andersen *et al.*, 1975
51. Randrup and Munkvad, 1974
52. MacLean, 1974
53. Unpublished observations

16

Human-Related Questions

Since a primary goal of the present investigation is to obtain further insights into epistemics (knowledge of the subjective brain), this might seem to be an opportune time to launch into a full discussion of the human implications of the behavioral and neurobehavioral observations that have been described. However, we are not yet in a position to consider how functions of the striatal complex might be modified or elaborated upon through interaction with the limbic and neomammalian formations (Figure 2-2). It should be noted, in particular, that information regarding the role of the limbic system in generating subjective apperception is a prerequisite for arriving at a meaningful characterization of psychological functions and the three main forms of mentation referred to in Chapter 2 as protomentation, emotional mentation, and rational mentation (ratiocination). Consequently, it will be the sole purpose of the present chapter to call attention to the comparative aspects of the work that suggest behavioral or neurobehavioral parallels between animals and human beings.

Introductory Comments

The Problem of Comparison

When ethologists draw parallels between animal and human behavior, they may be criticized for equating animals and human beings. Comparative neurologists are subject to the same kind of criticism when they give emphasis to anatomical and biochemical similarities of different parts of the brain in animals and human beings. In neither case is it the intention to equate animals and humans. Rather it is regarded as a reasonable assumption that if particular brain tissue from a variety of species conforms generally in its constituents, construction, and connections, it may have corresponding functions. This will recall what was said in Chapter 4 when discussing the meaning of homology.

The Question of Human Proclivities

The term protomentation has been used to apply operationally to rudimentary mental processes underlying special and general forms of basic behavior (Tables 6-1 and 10-1), including four basic forms of prosematic communication. As explained in Chapter 10, the term also serves to cover the experiential learning and memory entailed in rudimentary behavior. When considering the categorization of psychological functions in Chapter 23, it will be suggested that cerebration underlying propensities and compulsive behavior may be regarded under the same rubric. The "innate" sound of that statement is obviously quite grating to a psychology subscribing to the Lockean and Pavlovian principles discussed in Chapter 2. Reference was made there to a psychological text in which the first sentence reads: "All human behavior is learned." An elaboration on that belief appears in the writings of a social anthropologist who says, "In fact, the whole notion of predetermined forms of behavior in man is outmoded, for man's uniqueness, among other things, lies in the fact that he is free from all of those predeterminants which condition the behavior of non-human organisms. . . ."[1] If, as claimed, all human behavior is learned, then it must be explained why human beings with all their intelligence and culturally determined behavior, continue to do all of the ordinary things that animals do and show the same kinds of proclivities.

Source Material

The special and general forms of basic behavior listed in Tables 6-1 and 10-1, respectively, will serve as an outline of the topics to be discussed. In going down the lists, one is quickly reminded of the dearth of systematic observations for a number of items. Take, for example, the fourth item of Table 6-1—"place-preference" behavior. Exclusive of "father's chair" and like examples from the popular media, one would in most instances have to resort to biographical material for illustrations. A case in point would be that of a well-known American neurologist who was accustomed throughout his long career to arrive at an exact time and to take a particular seat at the weekly staff conferences.[2] This is not to discount the value of such biographical details, but, as I will comment on toward the end of this chapter, they cannot take the place of systematically acquired observations on human ethology. Given these constraints, we will glean what we can from the behavioral literature.

Special Forms of Basic Behavior

In Table 6-1, the special forms of basic behavior are so listed that those primarily concerned with self-preservation appear first, while those usually associated with social and procreational activities are given below. The same order for grouping topics will be used in the following discussion.

The Question of Territoriality

To judge by some writings, there continues to be heated debate as to whether or not human beings are naturally territorial.[3] In the strict ethological sense, territoriality refers to an animal's demonstrated determination to protect a particular piece of ground. In Chapter 6 we found that the discussion of territory becomes more manageable if it is dealt with in terms of an animal's domain, which in addition to territory encompasses a home site and an ill-defined activity range. Take, for example, the question of home site and territory in the case of roving animals. Given the parameters of domain, we may conceive of the roving animal as having a transportable home site and territorial "surround." Gorillas roam in a domain in which a new nest each night represents a home site and the immediate space around the group is the counterpart of territory.[4] The space occupied by the group will be defended against unwanted intruders.

In an operational sense, the domain (with a potential home site, territory, and activity range) represents a place of refuge, a place to hunt for food, and a place to mate and to breed. It might otherwise be conceived of as a space required by an organism for both self-preservation and the preservation of the species. As noted in Chapter 6, Eliot Howard in his memorable study of bird life concluded that for many species of birds, the male must show his ability to establish and protect a certain area before he will be considered as a suitable mate. As exemplified by the rainbow lizard (Chapter 7), a similar situation may apply to reptiles. Ethologists have provided many illustrations showing that the same holds true for many species of mammals. In the usual sense, the Uganda kob cannot be regarded as a territorial animal, but at times of mating the bulls return to breeding grounds where they must compete for, and hold on to, leks in order to earn the attraction of a female and the opportunity to mate with her.[5] A lek (literally, play; a gathering) is for the kob a special area of green turf about 20 m in diameter.

Phenotypic Expression of Territoriality

Lindegren has given renewed emphasis to the argument that the availability (or the lack) of a certain food may be more responsible than any other factor for the development of speciation.[6] This matter deserves consideration in the present context because food and the conditions of its availability may also be a determining factor in regard to the phenotypic expression of territoriality or nonterritoriality of different species. The term *phenotype,* coined by Johannsen in 1911,[7] refers to the visible characteristics of a plant or animal, as opposed to the unseen makeup of the so-called genotype. As a simple example of phenotypic expression, Karl von Frisch cited the European dandelion that grows tall in the valleys, but is stunted in the mountains.[8]

In Chapter 6 it was pointed out that black lizards may exist as nonterritorial (territorial animals, depending on the conditions under which food is available. Like foraging cattle, monkeys and apes living in tropical forests must keep on the move in order to obtain a supply of food. But even under these conditions, monkeys such as the Red howler or apes such as chimpanzees and gorillas may confine themselves to a "moving territory" within a domain that they will try to keep other groups of their own species from entering. There exists a social space—or as one might say in Calhoun's words, "a conceptual space"[9]—that other animals are not allowed to enter. A troop of baboons roaming the savanna does not defend a particular territory, but there may be vicious assaults on intruding, unwanted strangers.[10]

As long as human beings were predominantly hunters, it may be presumed that they led a roaming existence. At the times of the glacial invasions the roaming existence probably amounted to migration. Under stable climatic conditions, it may be imagined that prehistoric human beings settled down in domains. Bernard Campbell has suggested that it was the women, remaining at a homesite with the children while the men were off hunting, who learned to plant seeds and harvest the crops.[11] This would mean that women, and not men, were the inventors of an agriculture that heralded the rise of civilization. Whenever it was that there were gardens to tend, it is probable that it became the fashion to establish boundaries and that walls or fences were a by-product of an agricultural existence.

Territorial Marking

Animals such as the cat and dog, with a well-developed sense of smell, mark their territories with urine. The message, so to speak, is "stay away." We have seen, on the other hand (Chapter 13), how the squirrel monkey thrusts the erect genital as part of an aggressive visual display. Wickler has described so-called sentinel monkeys in troops of baboons and green monkeys in Africa, which sit at lookout sites with their thighs spread and a display of partial erection while the rest of the animals feed or take a siesta.[12] He regards this display as an "optical marker of boundaries," warning other monkeys not to intrude.

It is a long leap from monkeys to human beings. Do comparative observations have any human relevance? In mythology the gods Pan, Priapus, Amon, Min, and others are all identified with fertility and often portrayed with an enlarged or erect phallus that is superstitiously endowed with the power of protection. In Asia Minor phallic images identified with Priapus were placed at vantage points for the protection of orchards.[13] In primitive cultures in different parts of the world, house guards (stone monuments showing an erect phallus) have long been used to mark territorial boundaries.[14] It was as though a visual, urogenital symbol is used as a substitute for the olfactory, urinary, territorial markings of macrosmatic animals. Vandalism and graffiti would seem to be a form of visual marking. A refined type of visual marking is typified by signing a guest book upon ceremonial occasions.

The human use of symbols affords unlimited boundaries to protected "conceptual space." Even the symbols themselves may reach large proportions, as, for example, the mile-long numeral "1" that an American division carved in a Vietnam forest. In addition to our personal space and domestic space about which we feel possessive and protective, we incorporate into the scheme of belongingness the boundaries of a town, city, county, state, country, offshore areas, and now in modern times the territory of outer space. To this list may be added the space that we assign to schools, churches, clubs, and the like. Administrators in the same business firm try to avoid overlap in areas assigned to salesmen. Friends in the same professions shun overlapping competition. Many teachers and scientists have the reputation of establishing intellectual and research territories and protecting them with all their might. If human beings are not born with some degree of territorial proclivity, it is remarkable that there is so much preoccupation with trespass and no-trespass and that in every advanced culture complex legal systems and a whole body of law have evolved for settling disputes regarding ownership of lands and possessions.

Esser has observed dominance hierarchy and territoriality among institutionalized boys with severe mental retardation and learning handicaps.[15] If all human behavior is learned, it is curious that these children with no apparent instruction establish miniterritories that will be recognized and readopted even after a year of separation.

Patrolling of Territory

A territorial animal usually has a routine for patrolling its territory. Chapter 8 included a description of how the Komodo dragon patrols its core area. Patrolling was also mentioned in connection with the black lizard (Chapter 6). It is now recognized that mammals may also patrol their territories or extended parts of their domains, as is well exemplified by wolves.[16] With the aid of radio-tracking, Peters and Mech have found that wolves may regularly cover and mark defined areas within a range of 125–130 km^2.[16] In an artificial habitat rodents such as hamsters[17] or rats[18] can be observed patrolling their "territories." Goodall and her colleagues have reported patrolling on the part of the chimpanzees living in the wild.[19] "Perhaps," they remark, "the most striking characteristic of patrolling chimpanzees is the silence which they maintain for well over 3 hours."[20] Even a charging display will be performed in silence. Among human beings, regular patrolling of territory is best illustrated by military groups.

Comparative Aspects of Challenge Displays

There appear to be some carryovers from animal to human displays that are so subtle that they apparently have escaped notice of expert ethologists, making it seem all the more remarkable that, if as claimed, "everything that human beings do as human beings, they have had to learn from other human beings" (Chapter 2). Note has been made earlier that certain features of the aggressive displays of mammals have a striking similarity to the "close-in" challenge display of territorial lizards. The lacertilians rise up on all fours and present themselves sideways while stepping in a stilted, staccato manner that makes them appear off balance. Some rodents perform a similar broadside display, but it happens so rapidly that observers may fail to notice it. Barnett has described the broadside display of rats as follows: "The back is maximally arched, all four limbs are extended and the flank is turned towards the opponent. While in this attitude the rat may move round his victim with short, mincing steps, still presenting his flank."[21] Stonorov has described the stereotyped, stiff-legged display of the brown or so-called "grizzly" bear (*Ursus arctos*): With canines showing and ears flat, the bear walks "with its head down and muscles tensed, and its front knees appear to be locked."[22]

I had been unaware that the "challenge" display of two adult, rival gorillas incorporated lacertilian features until Dian Fossey presented a seminar at our laboratory and acted out what she refers to as the "parallel display" of two silverbacks.[4] When she mimicked their sideways presentation and their walking with stilted, awkward steps, I was immediately reminded of the close-in display of certain lizards (Chapters 6–9). The so-called parallel display of gorillas had been earlier referred to by Schaller as the "strutting walk."[23] Phrased in his words, the gorilla displays the side of the body; the arms are bent outward at the elbow, giving them a curious curved appearance and making the hair on the forearm look impressive; the body is held very stiff and erect, the steps are short and abrupt, and except for brief glances, the head is turned slightly away from the opponent. Reminiscent of voiceless lizards, the silverbacks that strut within 10 feet of each other *utter no vocalizations during the display*. A similar strutting walk is seen in courtship.[23]

The chest beating displays of the great apes would also be appropriate to consider at this time, but because of their strong emotional overlay and frightening vocalizations, they will be discussed in Chapter 25, when the topic will present itself in connection with the emotional manifestations of limbic epilepsy.

In the case of chimpanzees, Lawick-Goodall has described a bipedal swagger that appears to correspond to the strutting display of the gorilla. The chimpanzee "stands upright and sways rhythmically from foot to foot with his shoulders slightly hunched and his arms held out and away from the body, usually to the side."[24] Her description calls to mind the posture and movements of a Japanese wrestler.

As in the case of lizards, the stilted, staccato steps of the displays of the great apes seem to carry the message of a series of exclamation marks. The *Schrägstellung* gait of the Komodo dragon shown in Figure 8-2C calls to mind the goose step of a military parade. The question naturally arises as to whether the striking similarity between the challenge displays of animals as diverse as lizards and gorillas represent "convergent" or "parallel" evolution. Among different species the sideways presentation and the stilted, staccato steps have such an uncanny resemblance that it would almost seem that the challenge display had been genetically packaged and handed up the phylogenetic tree of mammals.

Gajdusek, in an article on Stone Age Man, has called attention to the parallel between the display behavior of squirrel monkeys and certain rituals of Melanesian tribes. Referring to our observations on squirrel monkeys he says, "I have noted a quite similar presentation and display in both spontaneous and socially ritualized behavior in some New Guinea groups. It is similarly used to express both aggression and dominance. . . . When frightened, excited, elated, or surprised, groups of Asmat men and boys spontaneously meet the precipitating event by a penile display dance, which involves much the same sequence as the presentation display of the squirrel monkey."[25] Gajdusek also comments on the use of greatly elongated phallocrypts characteristic of the highland cultures of western and central New Guinea. Eibl-Eibesfeldt has described and presented cinematographic documentation of genital display and thigh-spread among Bushmen adolescents.[26] In his *Three Contributions to the Theory of Sex,* Freud commented that the child is above all shameless, and during its early years it evinces pleasure in displaying its body and especially its sexual organs.[27]

In analyzing the mirror display of the gothic-type squirrel monkey (Chapter 13), I found that the full trump display could be elicited by the reflection of a single eye.[28] It was as though the eye and the genital acquire an equivalent meaning through generalization. The aggressive association of eye and genital manifests itself in the fear of the evil eye. In Italy, less than 200 years ago, amulets showing an erect phallus were worn as a protection against the evil eye.[29] Some patients with a diagnosis of schizophrenia are said to be thrown into a panic if someone catches their eye, recalling that the word *panic* derives from the god Pan, who found amusement in terrifying travelers (i.e., strangers). These various considerations suggest the possibility that "primitive man may have learned that by covering himself he reduced the unpleasant social tensions arising from the archaic impulse to display and that this, rather than modesty, has led to the civilized influence of clothing."[30]

Static and Dynamic Modifiers

R. W. G. Hingston, in his book on the use of color and adornments in animals and human beings, has called attention to a number of static and dynamic features of human threat displays.[31] On the basis of the direction of the hair follicles in the neck, he speculates that the human male once sported a mane. Hair tufts are part of the display trappings of certain birds (e.g., the common turkey) and mammals. Hingston cites evidence that among primitive peoples the raising of the arm and showing the axillary hair is used as a threat. A hair tuft is part of the ensemble of a Scottish kilt.

Those who subscribe to the dictum that all human behavior is learned have as yet failed to provide a satisfactory explanation of why people seem to mimic the ways of animals by exaggerating their size and further calling attention to it by use of color and manner of carriage. Apart from everyday examples provided by the uniformed services, one can go to the very halls of learning for illustrations, citing academic processions that, in addition to showy caps and gowns, are characterized by the somewhat stiff and off-balance swagger of the participants, giving the impression that everyone is out of step. Ironically, some will explain that the human inclination to ''dress up'' is as *natural* as the desire to eat.

Symbolic Equivalents

Smith, Eibl-Eibesfeldt, and others have pointed out that in different cultures throughout the world, the tongue may be protruded as a form of threat.[32] A noted American boxer was photographed giving such a display upon being weighed in. It is of comparative interest that the white-lipped tamarin marmoset, instead of displaying erection as in the case of the squirrel monkey, protrudes the tongue to the level of the forehead, both under conditions of threat and courtship.[33] In a study in which we mapped the brain for genital responses, we found that protrusion of the tongue was elicited along a course close to sites at which excitation induced genital-related manifestations.[34] Morris has commented on the various human gestures of the hand used as expressions of aggression and hostility, some of them symbolically representing the genital.[35]

The Question of Power and Size

The striving for territory (or what Ardrey has called the ''territorial imperative''[36]) is but one manifestation of the struggle for dominance that is everywhere manifest in nature. What is the origin of this life force? Why does it find more forceful expression in some individuals than in others? What are the genotypic determinants? What are the phenotypic determinants?

As World War II was to remind us, the *will-to-power* became the driving force in Nietzsche's philosophy. The idea of the will-to-power and the superman (Übermensch) was an indissociable part of Nietzsche's revelation in August 1881 regarding the doctrine of the eternal recurrence.[37] According to his autobiographical reflections in *Ecce Homo,* ''this highest formula of affirmation'' first dawned on him near that ''holy spot,'' Sils Maria, at an altitude ''6000 feet beyond man and time.''[38] It was shortly afterwards that *Zarathustra* began to pour out of him. Nietzsche concluded that the will-to-power is the basic life force of the entire universe. ''Thus life . . . taught me,'' he wrote.[39] In Nietzsche's superman, we hear the echoes of Aristotle's ''great-souled man,'' who being so far superior to other human beings ''is justified in despising other people.''[40] Similarly, Nietzsche's superman has the draconian right of riding roughshod over other people. As one of his interpreters explained, ''All that proceeds from power is good, all that springs from weakness is bad.''[41]

In the world of animals, one will hardly find the struggle to overpower more dramatically expressed than in the behavior of some lizards. As noted in Chapter 7, the resplendent colors and aggressive encounters of the rainbow lizard resurrect images of Arthurian knights. In a contest, once the gauntlet is thrown down, the aggressive displays give way to violent combat and the struggle is unrelenting. In our laboratory we have twice seen territorial males humiliated in defeat. They lost their majestic colors, lapsed into a kind of depression, and in each case died 2 weeks later.

It is one thing to describe the will-to-power as it occurs in nature, but quite another to offer any explanation for it. In Chapter 10 I mentioned our holding areas for new shipments of the common green anolis lizard and explained how it invariably happens that some anolians rise up from the ranks and strive to achieve dominance over the animals perched on one or the other of the two main branches. Among lizards the spoils usually go to the animal of largest size (see Chapter 11). They seem to have an uncanny way of recognizing a larger animal. As noted in Chapter 8, a Komodo dragon immediately sizes up an approaching stranger and steps aside if it is larger. Evans[42] and others,[43] however, have shown that size is not the only factor. The territorial lizard on its home ground appears to hold advantage over an intruder. Recalling what takes place in a political arena (see "repetitious behavior" below), it may be the number of displays, rather than size *per se,* that decides a winner in a contest.

Lawick-Goodall describes a chimpanzee named Mike, which was catapulted overnight to the top-rank position in his group after he had discovered some empty gasoline cans and had terrified the others by kicking them against each other and making them bang.[44] As human beings have learned to recognize so well, the use of color, various trappings, and size of retinue may help to make up for what a particular individual lacks in size.

Submissive Behavior

The emphasis thus far on the challenge (territorial) display might give the impression that it is more important than the other displays. Ethologists have made it popularly known, however, that a passive response (a submissive display) to an aggressive display may make it possible under most circumstances to avoid unnecessary, and sometimes mortal, conflict.[45] Hence, it could be argued that the submissive display is the most important of all displays because without it numerous individuals might not survive.

Courtship Behavior

As ethologists repeatedly emphasize, the struggle for territory is, in the lives of numerous species, a necessary first step for courtship, mating, and breeding (Chapter 6). Although pertinent cross-cultural material doubtless exists in various archival collections of film, there do not appear to be any written accounts of systematic ethological studies of human courtship. It would seem that one would have to go to the popular literature or attend the theater in order to put together an ethogram on human courtship. Musical comedy provides exaggerated examples of different aspects of human courtship—the swagger and puffed-out chest of the male, the hip-swinging walk of the female.

The condition known as automimicry is common in nature, as illustrated by markings on the head that make the size of the horns of a stag or ram seem larger; colorations that enhance the size of the eye; hair tufts or white streaks on the side of the cheek that amplify the size of the canines.[46] Morris suggests that the breasts of women represent a form of automimicry, giving in face-on meetings the appearance of buttocks.[47]

It is of comparative interest that one can trace back to reptiles a rear-end display of the female (see Figure 7-3) that suggests a similar presentation seen in the pygmy marmoset,[48] Old World monkeys,[49] and the great apes.[50] The posterior display may be used by female primates as a defiant or "put-down" gesture.[49] Chaucer provides a human example in his Miller's Tale.

Eibl-Eibesfeldt has made a world-wide search for common denominators of human expression in various cultures. He has found one ubiquitous signal that he relates to flirtation among women and men.[51] It consists of an upward jerking movement of the eyebrow as the person glances sideways.

Formation of Social Groups

On the basis of extensive surveys and an analysis of spatial factors, Calhoun has concluded that for many species of mammals the optimum number of individuals in a social group is around 12.[52] Corresponding data on reptiles do not exist. As evident in some of the field studies on lizards (Chapters 6 and 7) the number of animals sharing a particular territory would not be out of line with Calhoun's figure of 12. Even among flocking marine lizards, one finds that they divide up into small groups of allied individuals when basking on the shore (Chapter 9).

Somewhat like fraternities of wild turkeys,[53] small numbers of male chimpanzees roam together in social groups.[54] Members of such groups play together, hunt together, and patrol their domain together. There always seems to be one high-ranking male. Except during times of mating, the females and the infant and adolescent chimps keep together in "families." Birute Galdikas has observed a parallel situation in orangutans.[55]

Although chimpanzees are regarded as biologically closer than gorillas to human beings, gorillas have a social structure that is more like that of human beings. Beginning in 1967 Dian Fossey studied seven groups of mountain gorillas, ranging in size from 6 to 20 individuals and containing on an average 14 members.[4] Such a group depends on a fully mature, adult male, called a silverback. He keeps with him a number of reproductive females and their offspring, including infants and juveniles. There is usually also one young adult male called a blackback. Daughters of silverbacks do not remain as part of the "family" but are taken (captured) by silverbacks of other groups. It happens rarely that a lone silverback will fight to take over the retinue of another silverback. Seventy-five percent of such encounters result in serious injury. Fossey found two skulls of silverbacks with an embedded canine of another silverback. When a silverback takes over a retinue, it will kill the infant of a female, usually biting first into the skull and then into the belly. There is, however, no cannibalism under these circumstances. Among gorillas Fossey knew of only one case in which cannibalism may have occurred—an infant apparently eaten by its mother and older brother.

As yet there have been no neurobehavioral investigations specifically dealing with the question of the role of the R-complex in the formation of social groups, the establishment of social hierarchy, selection of leaders, and the like.

General Forms of Basic Behavior

In extending the present discussion to six general forms of behavior, I will, as in the case of the special kinds of basic behavior, give some representative examples of the possible human relevance of the behavioral and neurobehavioral work. As explained in Chapter 10, *general* forms of basic behavior may be "interoperative" with respect to *special* kinds of basic behavior.

Routinizing Behavior

As discussed in the last chapter, certain diseases of the striatal complex suggest that the striopallidum is fundamentally involved in orchestrating the daily master routine and subroutines. It is now of interest to consider some everyday happenings that are routine-connected.

As has been emphasized, reptiles are slaves to routine, precedent, and ritual. Obeisance to precedent often has survival value (Chapter 10). If, for example, a particular crevice served as an escape from a predator on one occasion, it may do so again. Harris's rainbow lizards with their accustomed roundabout way to their roost provides another example (Chapter 7). As Lorenz has observed, "If one does not know which details of the whole performance are essential for its success as well as for its safety, it is best to cling to them all with slavish exactitude."[56]

In his book *Wild Animals in Captivity,* Hediger cites several examples of how mammals may observe a clocklike regularity in adhering to a certain path. He mentions a tree porcupine (*Erethizon*) that Shiras observed for 7 years: "Its habits . . . were of almost clocklike regularity. Evening after evening it would appear between seven and eight o'clock on its trail on a shore of a lake, where it was often photographed."[57] Hediger comments in passing, "Unfortunately, investigations about such vital matters as the complete daily cycle in the life of the animal . . . have been strangely neglected."[57]

We are aware of our inclinations to follow favored routes from one place or another. Hediger refers to a study by Hinsche in which 800 school children were asked to give details about their walking to school.[58] Most of them followed quite a definite path, keeping, for example, to the right or left of a pillar, walking under projecting eaves of a roof, or each time jumping over a manhole cover. "If they failed to observe these rites [sic] of the road scrupulously, they thought that it would bring back luck, e.g., low marks at school."[59]

Many people recognize that they have a tendency to engage in particular acts that were successful in getting them out of bad situations. In the course of time, many of these acts become established as rituals and are incorporated into the daily routine. An outside observer might regard such rituals as superstitious acts. Scientists have the reputation of being critical of ordinary people who do superstitious things. But who in a research setting has not seen scientists perform superfluous pet maneuvers in trying to replicate an experiment?

For someone not in the legal profession it may not seem clear why a particular case cannot be decided on its own merits. Instead, much time and money are spent in searching for precedents. Why should it be that one must uncover the case of some obscure individual living years ago and in some remote place to prove that one's own case has some merit? Lawyers will explain that "the law" likes to be evenhanded, to be as fair as possible to all parties. This, they claim, is best assured by trying to find a similar case that might have been decided by a particularly renowned judge or by one of the highest courts. The greater the authority, the greater the weight of the judgment. What they fail to emphasize is that whoever sits in judgment derives great reassurance if it can be shown that the ruling on a similar case *survived* an appeal.

Break in Routine

Lorenz relates an incident in the life of a pet goose Martina to illustrate disturbed behavior that may result from a break in a long-established routine. Since she was a week old, Martina had been accustomed to climb the main stairway of the Lorenz house to reach her place of sleep. Subsequent to an early fright, it had been habitual for her to make a right-angled turn before approaching the stairs. One evening, in a hurry, she forgot to make the turn. Upon reaching the "fifth step" she suddenly seemed to panic, let out a "warning cry," and fluttered back to the foot of the stairs to make her usual turn.[60] That part of the brain's intelligence that enforces routine has a powerful means, it seems, of making known when there has been a break in routine. Although limbic and neocortical functions have not yet been dealt with, this does not preclude stating that infractions of protoreptilian routines may commonly result in great emotional and intellectual perturbations, as witness the distress that many people experience when the regularity of their day-to-day work is interrupted by the change of pace during the weekend or the intrusion of a holiday. The stress generated by an actual or threatened change in routine is many times compounded when entire organizations are involved, as exemplified at the institutional level by the upheaval resulting from proposed drastic alterations in the curriculum at a school or university. For occasions of an irregular nature, precirculation of agenda, ceremonial rituals, and the like represent cultural measures for preserving a semblance of routine and dissipating anxiety. As supreme commander of the allied armies in Europe and later as president of the United States, Eisenhower had the reputation of becoming extremely upset and angry if he was not provided the agenda before a meeting.

Isopraxic Behavior

In Chapter 10 an explanation was given for using the neutral term *isopraxis* as a substitute for such expressions as *imitation, mimicry, social facilitation,* and the like. *Isopraxis* (meaning, simply, "behaving in the same way") allows one to add any further qualifications as to whether or not the behavior occurs naturally or is learned. It was emphasized that isopraxis and its opposite, heteropraxis, are fundamental forms of prosematic communication for species recognition and making a distinction between sexes. Curiously enough, except for echolalia and echopraxia, one can find hardly any reference to isopraxic conditions (e.g., imitation) in neurological textbooks. In most current texts the word *imitation* or one of its synonyms are not listed in the indexes. As noted in the preceding chapter, echolalia and echopraxia may be conspicuous symptoms of the Gilles de la Tourette syndrome, which some neurologists ascribe to malfunction of the corpus striatum. In 1912 Kinnier Wilson described a familial disease in which there is progressive degeneration of the putamen and globus pallidus.[61] Wilson gives brief case histories, including those of four patients he himself examined. In view of his terse accounts, it is of interest that he felt obliged to mention that one of his patients, a 17-year-old boy (Case #3), *made no acknowledgment of a good-bye,* implying that there was a deficit in social communication. In terms of the present investigation, there was a failure of isopraxis in prosematic communication.

As opposed to neurology, the psychological literature is replete with articles and books on imitation. In their book, *Social Learning and Imitation,* Miller and Dollard would contend that there is no natural disposition to imitate, that everything human beings do has to be learned by trial and error.[62] At the same time, they call attention to the well-recognized difficulty of teaching animals to imitate one another. It might otherwise be said that animals find it difficult to do things that are not natural to them. Left to themselves they show a remarkable capacity for imitating one another. They may even take to imitating animals of a species not too different from themselves. We had, for example, three squirrel monkeys born in our laboratory that did backward somersaults, something that one is not accustomed to see in animals coming from the wild. This behavior originated with two squirrel monkeys, Cain and Abel, who imitated the backward somersaults of a capuchin monkey living in the same quarters. A female squirrel monkey born later and named Naamah picked up the knack from Cain and Abel.[63] A colony of rhesus monkeys in Japan became famous for having adopted the custom of one of its young females of dunking sweet potatoes in seawater before eating them.[64] A group of chimpanzees in Louisiana imitated one of their members by urinating in their hands and drinking the urine.[65] Dian Fossey has pointed out that the gesture of scratching is apparently reassuring to gorillas. In her final and successful effort to establish contact with a young adult gorilla (a blackback) she used the social signal of scratching in her overtures.[66] Shortly afterwards, she was able to reach out and touch the hand of the gorilla.

If all human imitation is learned, it is remarkable that mentally retarded children with severe learning disabilities have a well-known propensity (at the very first exposure) to mimic the actions of others. In micrencephalic children there is a great reduction in the development of the neocortical formation. Nevertheless, the majority of micrencephalics reveal an "absence of any sensory defect," "a general vivacity," a gift for keen observation, and a very marked "power of mimicry."[67] Advantage is taken of their imitative ability to introduce order (routine) into their lives. On the contrary, autistic children provide an example of the devastating effects of an inability to duplicate the actions of others (Chapter 15).

Gajdusek describes an encounter with a Stone Age tribe that had never seen Western people before. He was interested to observe that whenever he scratched his head or put his hand on his hip, the whole tribe did the same.[68] If, as some claim, all human *imitation* is learned, it is curious that members of an unschooled tribe were immediately able to mimic his every gesture without resort to trial and error. It has been suggested that such imitation may have some protective value by signifying, "I am like you."[68]

Although there is only a distant prospect of understanding the anatomical, biochemical, and functional factors that account for the attractive and repulsive forces involved in isopraxis and heteropraxis, the neurobehavioral work that has been discussed may be regarded as one small step toward an identification of underlying neural mechanisms. We have seen that several of the displays that have been described involve isopraxis, as, for example, two territorial anolians each using the challenge display in a confrontation. The same would be true of the mirror-displaying, gothic-type squirrel monkeys used in the neurobehavioral studies (Chapters 13 and 14). Since destruction of parts of the R-complex in both lizards and monkeys interferes with display behavior, it might be inferred that the corresponding structures in each species are implicated in conspecific recognition and the expression of isopraxic behavior.

Tropistic Behavior

Tropistic behavior is characterized by positive or negative responses to partial or complete representations, whether alive or inanimate (Chapter 10). In Chapter 10, we saw how a cock rainbow lizard responds to a colored, bobbing dummy as though it were another territorial male. Fabricius described a rooster that regularly mated with a feather on the ground.[69] Such responses need not depend on visual representations. I have seen tom turkeys perform the entire copulatory act *in vacuo* upon walking onto an area of crushed stone or coarse, dry straw.[48] Presumably, the material under foot that triggered the response gave the impression of the sharp pinfeathers on the back of a hen turkey. The hide of a sow on a wooden dummy is sufficient to elicit the cooperation of a hog for the collection of semen for artificial insemination. For human beings a fetish may induce sexual activity. In Chapter 23 reference will be made to a patient with psychomotor epilepsy who would look at a safety pin in order to precipitate a limbic seizure and sexual gratification.

In Chapter 10 "fixed action patterns" and "imprinting" were cited as examples of tropistic behavior. Sensitive to the criticism of traditional psychologists, ethologists have tended to sidestep the question of "instinctive" responses in human beings. When pressed to give examples of innate human responses, they commonly recite a list of behavioral developments in infants, including references to critical times in the life of the child when it begins to smile, sit up, walk, speak, and so forth.[26] Much has been made of the infant's tropistic responses to the features of the face, with the infant at first smiling in response to two (or even three) round circles representing the human eyes, and later requiring more and more detail of the human face for elicitation of the smiling response.[26]

Since there is no opportunity to observe human beings growing up in isolation, it is not evident how it can ever be established that beyond the age of infancy there exist naturally occurring tropistic responses and other kinds of propensities. Consequently, when making a case for human tropistic propensities, students of human ethology may draw upon illustrations from the visual and performing arts, commercial advertising, and various other sources. Most people would agree that cubistic painting owes some of its appeal to the portrayal of archetypal patterns and partial representations, such as, for example, a Picasso painting showing the human figure in two dimensions, with the eyes and buttocks in the same plane.[70] Clinically, perhaps the best examples of tropistic responses are to be found in patients' reports of what they see in the inkblots of a Rorschach test.[70]

The present ignorance of brain mechanisms makes it premature to ask many obvious questions. For example, what are the neural mechanisms that conjoin tropistic and isopraxic behavior and so galvanize the collective human mind that overnight there is a round-the-globe spread of fashions in dress (e.g., hair styles, "Levis"), games (e.g., Hula-Hoops), and reading (e.g., Haley's *Roots*)? What neural mechanisms account for the interplay of verbal and prosematic rhetoric so that the collective human mind seems suddenly receptive to some particular "movement" and inclined to mass demonstrations? What are the cerebral structures especially attuned to the seasonal, ecological, economic, and other factors that account for periodic and nonperiodic migrations?

In the study of neural mechanisms of perception, it is usually implicit that the primary goal is to learn how animals achieve the recognition of well-defined holistic patterns, as though this aspect of perception was what mattered most to the organism. As yet, little consideration has been given to the fundamental question of the opposite sort—namely, what accounts for complex behavioral responses to partial representations of the kind mentioned in this and preceding chapters?[71] Some information about brain structures implicated in tropistic responses may come from a least suspected quarter—namely, electrical recordings of the activity of single nerve cells. This statement requires some qualification. Recordings of individual cells of the visual system have revealed an assortment of units that respond to specific aspects of stimulus objects such as edges, contrast, orientation, directional movement, color, and so forth. Such findings have tended to generate a jargon in which reference is made to "sophisticated neurons" that are "edge detectors," "motion detectors," and the like,[72] somewhat as though a single nerve cell possessed subjective properties of Leibnitz's monads and were especially constituted to recognize only one or two types of stimuli. *Rather, it is the neural network to which the cells belong that accounts for the selective responses.* In the squirrel monkey one finds some cells in the limbic cortex and neocortex that will respond only to species-typical vocalizations of another squirrel monkey.[73] In view of such findings it is possible that the presentation of partial representations with respect to other sensory systems might activate the cells within neural networks "genetically tuned" to particular partial representations. As noted, the reflection of a single eye may be sufficient to elicit the full "greeting display" in the squirrel monkey.[28] What might be the effect of such a visual stimulus on cells of the striatal complex and other cerebral structures?

Perseverative Behavior

In discussing the displays of lizards (Chapter 10) emphasis was given to the value of perseverative behavior for reinforcing and assuring communication of signals. This led to the mention of Morris's views regarding "typical intensity" and his comment that when trying to reach someone by telephone, it is not the loudness, but rather the number of rings that succeeds in bringing the party to the phone. For the election years 1960 and 1972, Masters analyzed 4536 photographs in selected print media of the two main contenders for the presidency of the United States. In each case the "winner had a larger share of the coverage than the loser."[74]

As defined in Chapter 10, displacement behavior (alias adjunctive behavior) applies to repetitious acts that seem inappropriate for a particular occasion, as exemplified by a bird's preening in a threatening situation. Since there are indications that displacement reactions are strongly conditioned by emotional factors, the present comment upon the possible underlying mechanisms would also be appropriate for the discussion of limbic functions. For example, I have observed displacementlike reactions such as enhanced grooming in cats following agonistic forms of behavior elicited by electrically induced afterdischarges of the hippocampus (limbic archicortex) (Chapter 19).[75] Such afterdischarges propagate to parts of the R-complex. It would appear that in the case of both (1) naturally induced stress and (2) artificially induced stress by brain stimulation, there is a residual spillover excitation, bringing restorative mechanisms into play. Such phenomenology suggests a reciprocal innervation of mechanisms of "stress" and of "repair" that compares to the reciprocal innervation of muscles.

Human "displacement" reactions during uneasy moments may become more manageable when they are recognized for what they are—e.g., grooming and cleaning reactions such as scratching the head, rubbing the face or hands, clearing the throat, picking the nose, biting nails, spitting, and so forth.[76] A well-known conductor remarked that on the day of a concert, "I insist on being scrupulously clean. Even if I have had two showers already, . . . I take another. It's a ritualist approach."[77] At the institutional level displacement propensities may take the form of such time-honored procedures as appointing an ad hoc committee. It seems to be understood in universities, as well as in government, that at any one time the number of existing committees is a measure of existing tension.

Reenactment Behavior

Reenactment behavior, by definition, involves a precedent. In human affairs, observances of various kinds and ceremonial reenactments occupy such a prominent place as to require no illustrations.

More often than not, an occurrence that becomes a precedent for the establishment of a subroutine happens only once, as, for example, the celebration of a person's birthday, Bastille Day, and the like. In other instances success in tracing the origins of a particular custom may be as unlikely as uncovering the precedent for the egg-laying reenactment of the iguana lizards on the islet of Slothia described in Chapter 10. Whereas the observance of Groundhog Day was probably based on superstition, the lifesaving customs of cooking corn in some connection with calcium carbonate or the ban against eating uncooked pork appear to have insinuated their way into the collective human consciousness through a cerebral learning process quite unlike that of Pavlovian conditioning, and perhaps more akin to what induces "bait shyness."

Deceptive Behavior

Ever since predation became a way of life, deceptive tactics have been indispensable to both hunter and hunted. Almost nothing is known about brain mechanisms underlying deceptive behavior, but it is quite probable that very basic circuitry will be found in the R-complex. Stimulation in the region of the transhypothalamic pallidal projections discussed in Chapter 14 has been found to elicit stalking behavior in the cat.[78] Metaphorically, the stalking behavior of some recent presidential assassins could be compared to that of a Komodo dragon.[79] As described in Chapter 8, giant Komodo lizards will relentlessly stalk a deer for days at a time or wait in ambush for hours, activities that require detailed knowledge of the terrain and a good sense of time.

Deceptive behavior is no respecter of animals or persons. A book reviewer made the following comment about one of the best known philosopher-mathematicians of our times: Although he protested that "truth was the divinity he had mainly served, the sad fact seems to be that in his dealings with women, he was almost compulsively deceitful."[80] Twice in a generation an extensive web of deceit has been exposed at the highest level of government. In the first instance, as was repeated over and over again in Count One of the indictment: ". . . the conspirators would by deceit, craft, trickery, and dishonest means defraud. . . . The conspirators would give false, misleading, evasive and deceptive statements and testimony."

If people have learned through culture that "honesty is the best policy," why is it that they are willing to take enormous risks to practice deception? Why do the games that we teach our young place such a premium on deceptive tactics and terminology of deception? How can pupils be expected to come off the playing fields and not use the same principles in competition and struggle for survival in the classroom?[76]

Concluding Comments

Since it is inherent in the comparative, evolutionary approach to look for similarities and differences, both this and the preceding chapter have used the neurobehavioral studies on animals as background for considering *special* and *general* ("interoperative") forms of basic behavior in human beings. Some of the illustrations, such as the phyletic preservation of certain elements in challenge gestures, call into question a widely expressed view that all human behavior is learned.

In the case of human beings, the present best hope of obtaining further knowledge of the functions of the striatal complex lies in the discovery of correlations of psychobehavioral changes with disease of its respective parts and connecting systems. Unlike other disciplines, the field of medicine is in the unfortunate position that advances in knowledge usually depend on an analysis of conditions resulting in human illness and suffering. All the more for that reason, there is the obligation to wrest from human misery information that will contribute to the relief of suffering and prevention of disease. In regard to possible disorders of the striatal complex, there are two obvious needs: The first is to obtain more complete case histories, particularly in the light of what is known and being learned about comparative neuroethology. Such ethological knowledge is of key importance in being alert to the presence of "negative symptoms" that have been otherwise referred to as nonevident symptoms (Chapter 15). Clinics of neurology and of psychiatry stand in a unique position to help acquire a much-needed, fine-grained analysis of human ethology. A second obvious need is to obtain a much more thorough examination of the brain than has been possible in the past in cases available for postmortem examination. With the aid of computer technology, together with the many improved histological and neurochemical techniques that can be applied, one can foresee the day when the entire brain can be viewed in any desired plane, as well as quantitatively scrutinized in regard to cell losses, cell changes, and the like. I vividly remember an occasion 2 years before his death when the German neurologist Oskar Vogt said to me, "The reason other people have not seen the changes in the corpus striatum [and elsewhere] that we have described is because they have not examined serial sections throughout the entire structures." Finally, it is not to be overlooked that current developments in noninvasive techniques (e.g., magnetic resonance and positron emission tomography) promise to contribute to the knowledge of striatal functions.

In the preceding chapters, it has been repeatedly emphasized that the traditional view that the striatal complex is primarily involved in motor functions represents an oversimplification. It has been a primary purpose of the present investigation to test the hypothesis that the striatal complex plays an essential role in regulating the basic forms of behavior under consideration, including the control of the master routine and subroutines, as well as the four main kinds of prosematic communication.

The initial neurobehavioral studies described in Chapters 11, 13, and 14 are indicative that in animals as diverse as lizards and monkeys, the R-complex plays a basic role in prosematic communication. Other experiments to be described later on in connection with the limbic system provide further inferential evidence of the part played by the R-complex in maintenance activities of animals, including the regulation of the daily master routine and subroutines (Chapter 21). This matter has already been touched upon in the discussion of clinical case material in the preceding chapter, as well as when mentioning the altered routines of squirrel monkeys in the mirror display test subsequent to partial destruction of the globus pallidus (Chapter 13).

It deserves emphasis that in the evolution of primates the increase in mass of the R-complex keeps pace with the enlargement of the neothalamus.[81] As discussed in the following chapter, primates appear to have stemmed from insectivore-like animals. On the basis of volumetric measurements, Stephan has concluded that the striatum has "undergone a distinct enlargement" during evolution.[82] His "progression indices" for apes are 6.5 times the value for the basal insectivores, whereas the human index is 14 times as large. It is therefore evident that contrary to what some have claimed, the striatum is in no sense a "vestigial" structure,[83] or as Herrick has said, "a relict . . . preserving important but very sharply circumscribed functions."[84] Nor can it be presumed that with progressive encephalization its functions have been altogether usurped by the neocortex.[83] Certainly, there is no existing evidence to support Kinnier Wilson's prediction in 1914 that "the superman of the future will have no corpus striatum at all."[85]

As explained in the first two chapters, a primary purpose of the present investigation of the triune brain is to obtain information that will contribute to the knowledge of the subjective self ("epistemics"). Thus far in considering the protoreptilian formation, the nature of the material has allowed almost nothing to be said about the question of subjectivity. As unfolds in later chapters, the human capacity to "tune in" on the subjective self appears to have depended in large measure on the evolutionary development of the limbic system.

References

1. Montagu, 1956, p. 42
2. Walsh, 1971
3. Montagu, 1976
4. Fossey, 1976
5. Buechner, 1961
6. Lindegren, 1966
7. Johannsen, 1911
8. Frisch, 1964
9. Calhoun, 1971
10. Washburn and DeVore, 1961
11. Campbell, 1979
12. Wickler, 1966
13. New Larousse Encyclopedia of Mythology, 1968; Knight, 1865
14. Knight, 1865; Wickler, 1966
15. Esser, 1968, 1973
16. Peters and Mech, 1975
17. Murphy *et al.*, 1981
18. Calhoun, personal communication; Hill, personal communication
19. Goodall *et al.*, 1979
20. Goodall *et al.*, 1979, p. 26
21. Barnett, 1963, p. 87
22. Stonorov, 1972, p. 92
23. Schaller, 1963, pp. 235–236
24. Lawick-Goodall, 1968, p. 276
25. Gajdusek, 1970, pp. 58–59
26. Eibl-Eibesfeldt, 1971
27. Freud, 1948
28. MacLean, 1964a
29. Knight, 1865
30. MacLean, 1962
31. Hingston, 1933
32. Smith *et al.*, 1974; Eibl-Eibesfeldt and Wickler, 1968
33. Smith *et al.*, 1974; MacLean, unpublished observations
34. MacLean *et al.*, 1963b
35. Morris, 1979
36. Ardrey, 1966
37. Kaufmann, 1968, p. 207
38. Nietzsche, 1908/1969
39. Nietzsche, 1888/1968
40. Aristotle, 1908–52
41. Forster-Nietzsche, 1954
42. Evans, 1936
43. Noble and Bradley, 1933
44. Lawick-Goodall, 1971, pp. 115–116
45. e.g., Lorenz, 1966
46. Guthrie and Petocz, 1970
47. Morris, 1967
48. MacLean, unpublished observations
49. Zuckerman, 1932
50. Lawick-Goodall, 1968; Galdikas, 1978; Nadler, 1975
51. Eibl-Eibesfeldt, 1970, pp. 416–420
52. Calhoun, 1964
53. Watts and Stokes, 1971
54. Lawick-Goodall, 1968, 1971
55. Galdikas, 1978
56. Lorenz, 1966, p. 72
57. Hediger, 1950, p. 16
58. Hediger, 1955
59. Hediger, 1955, pp. 20–21
60. Lorenz, 1966, pp. 68–70
61. Wilson, 1912, p. 354
62. Miller and Dollard, 1941, p. 1
63. MacLean, 1975e
64. Miyadi, 1964
65. Gajdusek, quoted by MacLean, 1975e
66. Fossey, 1971
67. Tredgold and Soddy, 1963, p. 273
68. Gajdusek, quoted by MacLean, 1973c, p. 118
69. Fabricius, 1971
70. See MacLean, 1973b, p. 119
71. MacLean, 1975a
72. Michael, 1969
73. Sudakov *et al.*, 1971; Newman, 1979
74. Masters, 1981
75. MacLean, 1957c; MacLean *et al.*, 1962
76. See MacLean, 1978c
77. Bernstein, 1976
78. Wasman and Flynn, 1962
79. MacLean, 1975b
80. Clark, 1976
81. Blinkov and Glezer, 1968
82. Stephan, 1979
83. Wilson, 1914, p. 482; Herrick, 1926, p. 123
84. Herrick, 1926, p. 123
85. Wilson, 1914, p. 482

The Limbic System with Respect to Thymogenic Functions

17

The Limbic System in Historical Perspective

In Part III of the present study, we shall be dealing with the anatomy and functions of a development within the forebrain that reflects an inheritance from early mammals. This development consists of the phylogenetically old cortex and the structures of the brainstem with which it has primary connections. As illustrated in Figure 17-1, most of the evolutionarily old cortex is contained in a large, annular convolution that Broca[1] called the great limbic lobe because it "forms a border around" the brainstem. It was Broca's great contribution to demonstrate that this convolution forms a common denominator in the brains of all mammals. In 1952 the term *limbic system* was introduced as a designation for the limbic cortex and directly related structures of the brainstem.[2] Some authors refer to the existence of a limbic system in the brains of birds and reptiles, but it is to be emphasized that the cortical areas in these two classes of animals are at best rudimentary and poorly developed. Moreover, as Clark and Meyer[3] have pointed out, structures comprising the evolutionarily newest part of the limbic system (identified in the present study as the thalamocingulate division) "have no representation in the reptilian brain."

The behavioral differences between reptiles and mammals might be expected to be reflected in differences in their cerebral development. As noted in Chapter 5, there are indications that the evolution of the mammal-like reptiles may have seen a transition from a cold-blooded to a warm-blooded condition. In any event, the development of endothermy represents a fundamental innovation that distinguishes mammals (and birds) from reptiles. Because of its key role in metabolic processes, thermoregulation may be said to affect every aspect of an animal's biological functions and activities.

In addition to endothermy, there were three cardinal behavioral developments that characterize the evolutionary transition from reptiles to mammals—namely, (1) nursing, in conjunction with maternal care; (2) audiovocal communication for maintaining maternal–offspring contact; and (3) play.[4] Because of this unique family-related triad, one might say that the history of the evolution of the limbic system is the history of the evolution of mammals, while the history of the evolution of mammals is the history of the evolution of the family.

Apart from these considerations, the limbic system has unrivaled interest for an investigation concerned with paleopsychic processes. As mentioned in Chapter 2, experimental and clinical findings of the past several decades have provided evidence that the limbic system derives information in terms of emotional feelings that guide behavior required for self-preservation and the preservation of the species. Clinical findings provide the best evidence that the limbic system is involved in emotional behavior and, indeed, the only evidence that it underlies the subjective experience of emotion. Scarring

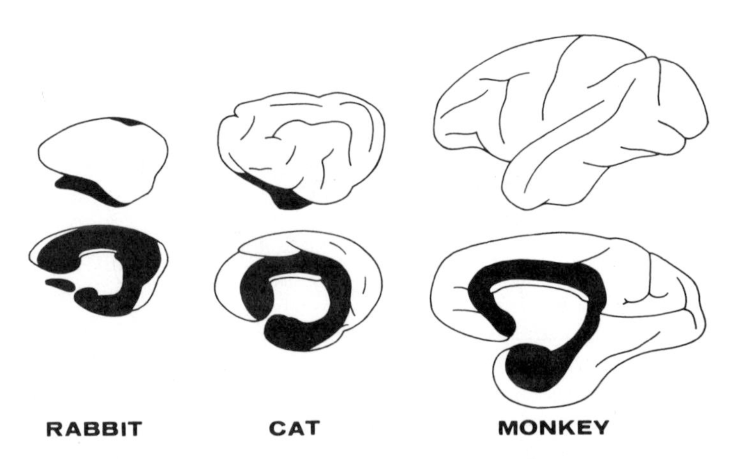

Figure 17-1. Brains of three representative animals, illustrating that the limbic lobe (shown in black) is found as a common denominator in the brains of all mammals. Most of the phylogenetically old cortex is located in the limbic lobe. The limbic cortex and structures of the brainstem with which it has primary connections constitute the so-called limbic system.[1] The neocortex (shown in white) undergoes progressive expansion during the relatively recent evolution of primates. After MacLean (1954).

of the limbic cortex from whatever cause may result in epileptic storms in which the mind lights up with vivid emotional feelings that, in one case or another, involves affects ranging from intense fear to ecstasy. As will be discussed, these affects not only play a basic role in guiding behavior, but are also essential for a sense of personal identity and reality that have far-reaching implications for ontology and epistemology. A consideration of the underlying neural mechanisms leads to the question raised in the introduction as to whether or not we as intellectual beings are forever faced with a delusional impasse.

In some writings, the limbic system seems to be regarded as though it functioned globally in generating diffuse emotional feelings. In the chapters to follow, however, anatomical and functional findings will be reviewed that suggest that the limbic system may be subdivided into three main corticosubcortical subdivisions that may account for the generation of emotions germane to their respective functions. The rest of the present chapter will be devoted to a consideration of (1) possible factors contributing to the mammalian condition; (2) the evolution and nature of the limbic cortex; and (3) developments leading to the limbic system concept.

Possible Factors Contributing to the Mammalian Condition

It will be recalled that Romer[5] regarded the amniote egg as "the most marvelous 'invention' in vertebrate history" because it could be deposited on land and permit the embryo to mature to the adult form without passing through the risky aquatic larval stage (Chapter 5). But this apparent advantage may be a potential hazard for lizards. Since the mother lays her eggs and goes off and leaves them, the hatchlings may later be looked upon as foreign and fair prey. Hatchlings of the giant Komodo lizard must take to the trees for the first year of life to avoid parental cannibalism (Chapter 8). For the same reason, young rainbow lizards must hide in the deep underbrush (Chapter 7). However, as will be recalled from Chapter 9, female skink lizards may brood their eggs and pay some attention to the hatchlings. As the young are hatching, the mother may help to express them from their shells. If the young wander away, she may search for them, and upon finding them, lick their vents.

Development of Maternal Interest and of Placentation

Although the fossil record provides no evidence, it is generally assumed that the mammal-like reptiles were oviparous. This judgment is partly owing to the inference that the therapsids were ancestral to the egg-laying monotremes. Although fossils of the monotremes cannot be traced back beyond the Pleistocene, Romer[6] argues that it is highly probable that these animals "represent a line of descent from the mammal-like reptiles."

Fitch[7] suggests that brooding may be an alternative for viviparity. He points out that all brooding lizards are secretive and that olfaction plays an important role in maintaining the female's interest in her eggs. Propinquity to eggs and to hatchlings may be a precondition of maternal interest and care. Under inclement conditions some oviparous lizards will retain their eggs until the time of hatching.[7] There is some evidence that there was a cooling trend toward the end of the Triassic.[8] In the case of therapsids, it might be imagined that under periodic cold or arid conditions, some of the smaller ones might have sought underground cover and developed viviparity. Retention of eggs in the oviducts may result in the formation of membranes for the transfer of nutrients and waste material,[7] a process that suggests a mechanism leading to placentation.

Development of Nursing

The large olfactory apparatus evident in the cranial endocast of the cynodont illustrated in Figure 5-11 suggests that at least some of the therapsids would have been capable of the olfactory-dependent kind of maternal attention that has been described. As illustrated by monotremes, the other common reproductive feature in mammals that was just mentioned—placentation—is not a necessary antecedent to the mammalian condition characterized by nursing. How might nursing itself have developed? In raising this question, it should be noted that a number of paleontologists take exception to the term *mammal-like reptiles* for the most advanced therapsids, contending that it would be more appropriate to refer to them as *reptilianlike mammals* (e.g., Van Valen[9]). If that were so, a shorter informal designation could be obtained by changing the "l" in *mammals* to an "r" and referring to them as *mammars*.

Since the last century it has been generally assumed that milk glands are derived from sweat or sebaceous glands.[10] As noted in Chapter 5, the advanced therapsids may have developed vibrissae and associated glands. The presence of such would be indicative of a hairy integument. Evidence adduced for vibrissae also suggests the existence of a muscular lip (not present in extant reptiles) making it possible for the young to suckle. It has been proposed that the young staying near the mother for warmth might have been drawn to nuzzle around the ventral side (possible brood patch) and thereby happened upon nourishment from glandular secretions.[11]

The use of the term *mammals* for the successors of the therapsids implies that nursing and care of the young originated with these long-extinct animals. In an anthropological sense, the invention of nursing is to be regarded as no less important than that of the amniote egg. As Romer has commented, the nursing habit "tended strongly to the establishment of a family group."[12] Moreover, he notes, "[T]he continued association of parent and young due to the nursing habit marks the beginning of education," contending that it may not be too much of an exaggeration "to say that our modern educational systems all stem back to the initiation of nursing by the ancestral mammals."[12] The remains of the earliest mammals (including isolated teeth, skull fragments, and limb bones) have been found in the Rhaetic beds—a name stemming from the Rhaetic Alps and referring geologically to the uppermost Triassic beds laid down about 180 million years ago. The two most primitive types of mammals known as *Morganucodon* and *Kuehneotherium* were found in Wales.[13] Desmond has commented that in these two miniature types "we have evidence of milk and permanent molars" and that "since milk molars imply parental care, we are, by definition, dealing with mammals."[14]

Jaw Articulation and Audiovocal Communication

Aside from the teeth, the most distinguishing mammalian feature is a direct articulation between the dentary and squamosal bones. But it should be noted that in the earliest known mammals just mentioned, there was a functional articulation with the small articular bone. Unlike therapsids, the lower jaw makes a triangular excursion in chewing, bringing the grinding molars together on one side. A difference in the occlusion of the teeth distinguishes the Morganucodontidae from the Kuehneotheriidae.[15] The latter are known only from isolated upper and lower teeth and jaws.[15] According to Crompton and Jenkins,[16] there appears to be "general agreement" that members of the Morganucodontidae were ancestral to the nontherian mammals (including triconodonts, multituberculates, and modern monotremes) while the Kuehneotheriidae were the original stock for all Theria (including the symmetrodonts, pantotheres, marsupials, and placentals). In mammals stemming from these two lines there was incorporation of the quadrate and articular bones into the middle ear,[16] providing an improved sense of hearing.

It is commonly assumed that the tiny mammals with affinity to cynodonts survived the onslaught of predatory dinosaurs by living in the dark floor of the forest. They may also have been nocturnal. Under these conditions it is evident that the development of audiovocal communication would have been an invaluable additional means of maintaining maternal–offspring contact, as well as contact of mature members of a group. As will be a special topic in Chapter 21, the so-called separation call may represent the most primitive and basic mammalian vocalization.

Directional Evolution

It is remarkable that "the incorporation of the quadrate and articular in the middle ear must have occurred independently in therian and nontherian lines."[17] Olson has commented: "The most striking feature of the history of the mammal-like reptiles is the independent acquisition of mammal-like characters by the various therapsid lines with the resultant continuing increase of mammalian habitus."[18] He lists many examples, including the development of a secondary palate, broadening of the basal part of the brain, the formation of a double occipital condyle, and reduction of phalangeal formula. Olson points out the difficulty of reconciling "directional evolution" with current genetic-selective concepts of evolution. Where do we stand, he asks, with respect to the "apparent dualism" in the therapsids? "Are we dealing with two kinds of systems, one for now and one for the future?"[19]

Shands has remarked that the invention of a written language has made it possible for a society to "invent the future."[20] In a like vein the evolutionary changes that have just been considered, as well as such mammalian developments as endothermy, placentation, and nursing, could be regarded as inventions. Given the invention, it need not be the "natural selection" of a particular trait or mutation that leads to survival. Rather, the new invention may provide many species the "natural selection" of several different avenues to survival that previously did not exist.

The Evolution and Nature of the Limbic Cortex

In considering the paleomammalian brain, we are immediately faced with the question concerning the progressive evolution of cerebral cortex, because it is the expansion and differentiation of the *cortex* that most clearly distinguishes the brains of mammals from those of reptiles and birds. As noted in Chapter 4, cortex is found in all existing reptiles and birds, but it is rudimentary and poorly differentiated. In the extinct transitional forms between therapsids and mammals it is presumed that the cortex began to balloon out and become further differentiated. On the basis of examination of the external features of the skull of a tritylodont, Kuhne has inferred that the brain was "not much larger than in undoubted reptiles."[21] Hopefully, future efforts to obtain endocasts of these and other relevant forms will settle this issue. As illustrated in Chapter 5 (Figure 5-11), the cranial endocast of a triconodont reveals that the brain in this primitive mammal was wider than that of a cynodont, but the question to be raised as to what structures accounted for this enlargement must remain speculative.

It is not possible to throw light on this question by reference to any living animal, because there are no *existing* mammals that can be identified with either triconodonts or their middle Mesozoic contemporaries (see Figure 4-1). The marsupials, which are osteologically distinctive because of an internal angular process at the posterior margin of the jaw,[22] are characterized by a marsupium (sac), in which the very immature young are carried. Lacking a true placenta they are called *metatheria,* an expression suggesting that they were transitional to the eutheria, i.e., true mammals. But as diagrammed in Figure 4-1, the fossil record indicates that the marsupials and placentals branched off from the pantotheres about the same time in the late Jurassic.

The skull of the common opossum (*Didelphis virginiana*) is almost a perfect match in size and shape for the dawn opossum, Eodelphis.[23] Colbert[24] refers to the opossum as a "living fossil" insisting that this is no figure of speech, because this animal "has changed very little during the long lapse of time from the Cretaceous period to the present."[24]

A nocturnal prowler (as its antecedents may have been), the opossum feeds on vegetation, insects, and small fleshy animals. As Evans[25] has noted, the threat posture of an opossum is like that of reptiles, entailing an uptilted snout and widely opened mouth (see Figure 8-2A). The same characterization applies to its manner of attack with a downward slashing of the teeth. Gregory has commented that the skull of a modern opossum "is at first strangely similar to that of one of the mammal-like reptiles of the far-off Triassic" period.[26]

Until 1924 the opossum was the only "living fossil" for giving a clue regarding the evolutionary development of the brain in primitive mammals. Lacking a corpus callosum the marsupial brain is significantly different from that of a true mammal. In 1924 Roy Chapman Andrews and his colleagues turned up fragments of tiny skulls in Cretaceous beds of Mongolia that led to identifying another "living fossil" in the line of true mammals. The skulls were reconstructed and described by William K. Gregory (1876–1970) and George Gaylord Simpson.[27] Gregory was a foremost authority on teeth, having been an assistant of H. Fairfield Osborn (1857–1935), who had elaborated upon the original scheme of Edward Drinker Cope (1840–1897), an American paleontologist whose name will forever be identified with the discovery of dinosaur fossils and the taxonomy of reptiles. Based on his reconstructions, Gregory[28] concluded: "All the evidence . . . indicates that the remote ancestors of the line leading to all the higher mammals, including man, were small, long-snouted mammals, of insectivorous habits and not unlike some of the smaller opossums and insectivores in the general appearance of the head." The tree shrew of Borneo (of the genus *Tupaia,* native word for squirrel), which has been variously regarded as an advanced insectivore and a primitive primate, has, like the opossum, earned for itself the characterization "living fossil."[29] Regardless of labels, paleontologists believe that it is from such stock that the primates arose, with the line leading to the lemurs, tarsiers, and so on.

But the examination of the brain of the tree shrew shows that the development of its cortex stands about halfway between that of the highest insectivores and the lower primates.[30] Of existing true mammals the English hedgehog or the Madagascar tenrec would be more deserving of the term "living fossil."[30] As illustrated in Figure 17-2, the "neocortex" in these forms, as well as in primitive marsupials, is of small dimensions compared with the evolutionarily old cortex (see below), giving the brains somewhat the appearance of a blunted pyramid. Tilly Edinger[31] has described endocasts of the earliest camels showing similar disproportions. Because of this and other evidence that the neocortex is a relatively late evolutionary development, it is possible that skulls of more primitive mammals would show little, if any, impression of neocortex. These remarks lead to a consideration of the different types of cortex that must be taken into account in defining the cortical counterpart of the paleomammalian brain.

Three Main Forms of Cortex

The anatomical consideration of cortex immediately invites a brief comment on the functional significance of the evolution of the gray matter forming the covering (i.e., "rind") of the cerebral hemispheres. In mammals it is characterized by a distinct layering of the cellular elements. It is popularly taught that the cerebral cortex accounts for learning and memory. At best this must be a partial truth, undoubtedly reinforced by illustrations of the "stupidity" of animals such as lizards and turkeys having only a rudimentary cortex. A half-page abstract by J. H. Cookson[32] is repeatedly cited in textbooks as evidence of the failure of lizards to learn a simple task. Over a 3-month period his lizards failed to learn that they could snatch a mealworm if they simply crawled over the half-inch glass barrier of a petri dish. But dependent as they are on "ancestral learning" and "ancestral memories," lizards with their R-complex and only a rudimentary cortex have a great capacity for learning. As noted in Chapter 10, they are able to learn the features of their territory inside and out, to recognize strangers at first sight, and so on. Apropos of the transparent barrier, such intelligent animals as dogs and horses find

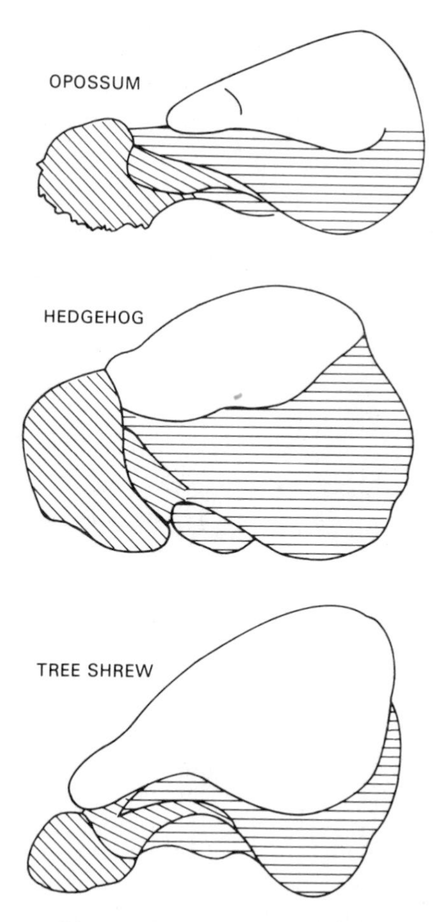

Figure 17-2. Relative development of the cortex above and below the rhinal fissure in a basal marsupial, a basal insectivore, and an animal transitional between insectivores and primates. Horizontal shading identifies the prepiriform and piriform areas in the opossum (*Didelphis virginiana*), hedgehog (*Erinaceus europaeus*), and tree shrew (*Tupaia glis*). The shading slanted at 45° identifies the olfactory bulb, while that at an intermediate angle delineates the lateral olfactory tract. The suprarhinal cortex, largely represented by neocortex, is shown in white.

it difficult to learn to deviate from a straight-on approach so as to obtain some reward directly in front of them. It would seem, therefore, that in addition to memory and learning identified with cortical function, emphasis should be given to the "unlearning" of what the species have learned to do "naturally."

Origin and Use of Cortical Terms

For explaining the evolution and nature of the paleomammalian cortex largely contained in the limbic lobe, it will be helpful to define the origin of the term *pallium* and the designations for three main forms of cortex—archicortex, mesocortex, and neocortex—that is not dealt with in standard texts.

The word *pallium* means cloak or mantle. According to Elliot Smith,[33] Reichert introduced this word in 1859 for the purpose of distinguishing the covering of the human embryonic hemispheres from underlying structures of the brainstem. *Pallium* is a particularly apt word for the covering of the hemispheres in reptiles and birds in which a layering of the cellular elements is either lacking or only rudimentary. The diagram in Figure 17-3 shows the traditional parcellation of three pallial areas in the reptilian brain. The labels "PIR" and "HIP" refer to the piriform area and hippocampal area. The names for these areas in reptiles are taken from nomenclature for convolutions in mammals.

Thanks to an influential paper by Smith[33] that appeared in 1901, there has been the tendency to regard the general pallium as the anlage of the neocortex. Proposing a "natural subdivision of the cerebral hemispheres," Smith referred to the "pyriform" (original spelling) and hippocampal areas as the "old pallium," whereas he suggested "neopallium" for the rest of the cerebral mantle. In his paper we find not only the origin of the term *neopallium* (alias *neocortex*), but also the precedent for including too much of the cerebral mantle under this designation (see below).

Sometime later Edinger, a German neurologist and anatomist, translated Smith's expression *old pallium* as *archipallium*,[34] implying that it represented the first form of cortex. Since some neurologists believed that the hippocampal area was the evolutionarily oldest cortex, they used the term to apply exclusively to that area. Subsequently, when Smith[34] found himself credited for the "invention" of a term that he despised, he was so incensed that he published a short note in 1910 entitled "The term 'archipallium,' a disclaimer." He not only railed at the practice of applying the term exclusively to the hippocampus, but also objected to it as a poor and redundant name for the hippocampus. In the meantime, to add to the confusion, Ariëns Kappers[35] introduced the word *paleopallium* to refer to the piriform cortex. In 1935, Raymond Dart, who had been a former student of Smith, described the identification of two kinds of neopallium in the reptilian brain, one of which he called "parapyriform" and the other "parahippocampal" (see below).[36] Here was the germ of the concept of "successive waves of circumferential differentiation" as developed by Abbie[37] and later elaborated upon by Sanides[38] in terms of a concentric growth ring hypothesis.

Thanks to the above sequence of naming, the term *neocortex* carried over for the characterization of cortical areas in mammals that, according to such authorities as Ariëns Kappers, Huber, and Crosby,[39] "are really transitional areas between neocortex and the archicortex." In particular, they had in mind limbic areas of the cingulate gyrus that Campbell,[40] von Economo and Koskinas,[41] M. Rose,[42] J. Rose and Woolsey,[43] Yakovlev,[44] and others have shown to be distinctive from neocortex.

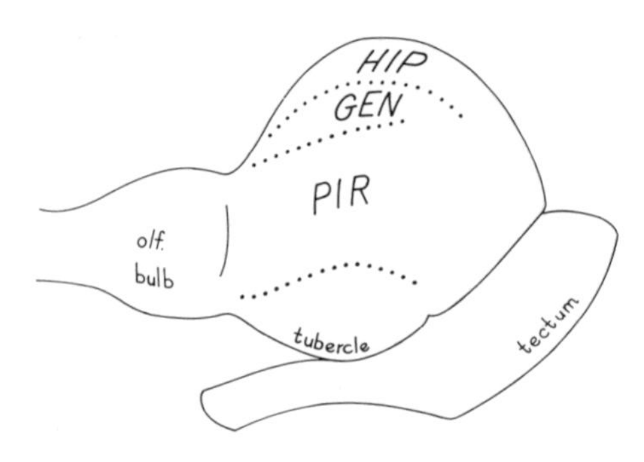

Figure 17-3. Diagram of the brain of a green turtle, illustrating the locations of the different pallial areas of the cerebrum in a reptile. The piriform (PIR) and hippocampal (HIP) areas are regarded as rudimentary forms of the like-named areas of mammals. See text for significance of the intermediate "general" (GEN) pallium. The olfactory tubercle makes up the rest of the mantle in reptiles. Other labels: olf. bulb, olfactory bulb; tectum refers to optic tectum. Redrawn after Papez (1929).

Mesocortex: The Growth Ring Hypothesis

As noted, Abbie, in line with the original observations of Dart, developed the concept that the cortex evolved by "successive waves of circumferential differentiation." Basing his argument on histological studies of the marsupial[45] and monotreme brain,[37] he proposed that the borders of the piriform and hippocampal areas provided the starting lines of successively evolving cortical areas. In further developing this thesis Sanides referred to it as the "growth ring hypothesis." Although recognizing the transitional nature of the cortex adjoining the piriform and hippocampal areas, Abbie followed Smith and Dart in referring to it as *neo*pallium. Sanides, however, in giving particular emphasis to its transitional features, characterized it by the Vogts' term *proisocortex,* i.e., rudimentary neocortex. His proisocortex includes the insular and cingulate cortex that arise, respectively, in the parapiriform and parahippocampal areas. Sanides[38] calls attention to three generalizations in regard to distinctive features of transitional cortex: (1) accentuation of layer II with respect to staining and cell density; (2) "bandlike layer V of dense medium-sized pyramidal cells"; and (3) presence in layer II of tufted (lophodendritic) cells with dendrites extending into layer I, but having only poorly developed basal processes (extraverted neurons of Morgane[46]; see Chapter 21).

The classification of cortex according to Maximilian Rose, of the Vogt school, would be compatible with the growth ring hypothesis. Based on his embryological studies, Rose[42] concluded that most of the cerebral cortex can be subdivided into three principal types consisting of two, five, or seven layers, which he referred to as "bi-, quinque-, und, septemstratificatus."

The two-layered cortex included the so-called archicortex. Rose characterized most cingulate and adjacent areas as *quinquestratificatus*. Since he considered such cortex to be transitional between the first and third types, he had otherwise referred to it as *mesocortex,*[47] a term that Yakovlev[44] and collaborators[48] subsequently applied to transitional cingulate cortex.

"Bauplan" of Cortical Afferents and Efferents

In a lecture on the development of the cerebral cortex and its different layers (1928), Ariëns Kappers[49] used the embryonic brain of the armadillo to illustrate three main evolutionary kinds of cortex. In essence, he was stating a general principle of the kind that German neurologists had in mind when using the expression *bauplan*—a building plan. He explained that the termination of the olfactory fibers in the superficial layers of the piriform and hippocampal areas predetermines the architecture of these most primitive forms of cortex and the adjacent areas. He pointed out that the most primitive condition is represented by the hippocampal formation where a receiving layer of granule cells only partially overlaps a layer of pyramidal cells with their outgoing fibers (Figure 17-4A). The situation would be analogous to a split-level house in which "telephone" lines for incoming calls entered at the ceiling and the answering lines went out underground. A more advanced condition would be represented by transitional cortex in which the receiving layer of cells is entirely superimposed above the effector layer. In this case the situation would be like that of a two-story house where, again, the incoming lines enter at the ceiling level of the receiving layer, while the outgoing lines are channeled under-

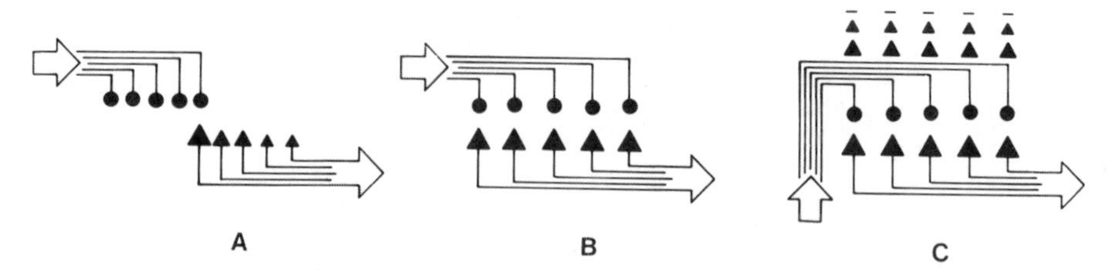

Figure 17-4. Schematic of the arrangement of the granular and pyramidal cell layers in (A) archicortex, (B) mesocortex, and (C) neocortex. Solid circles and triangles represent, respectively, granule and pyramidal cells. See text for other details.

ground (Figure 17-4B). In the neocortex, there is also a superimposition of the receiving layer on the effector layer. But like a building to which more floors can be added, there are, in addition, two layers of pyramidal cells above the receiving layer (Figure 17-4C). The two additional floors characterizing the neocortex are referred to as the supragranular layers. In terms of the building analogy, the neocortex has the distinctive difference that the lines carrying both incoming and outgoing messages enter and leave from underneath the building.

The evolutionarily old cortex is generally regarded as consisting of the archicortex and the transitional mesocortex. Lorente de Nó[50], who studied with Ramón y Cajal, singles out three features that distinguish the phylogenetically old cortex from the neocortex—namely, (1) termination of the main system of incoming fibers in the superficial layers; (2) the relatively poor development of the supragranular layers; and (3) the relative paucity of cells of short axon, i.e., the Golgi type II cells. In the next chapter on neuroanatomy it will be explained that the main afferents to the archicortex and mesocortex terminate in the outer half of layer I, while association fibers terminate in the lower half.

Cortical Evolution in the Light of Experimental Micrencephaly

Quiroga[51] has described an endocast from a relatively advanced South American therapsid, *Probainognathus jenseni,* and on the basis of its morphology has speculated about the location of the development of the "neocortex" in this animal. The shape of the cast reminded me of the cerebral hemispheres of rodents in cases of experimental micrencephaly in which neocortical development is massively reduced. In rats micrencephaly can be produced by injecting a pregnant female with a single dose (20 mg/kg) of the alkylating agent methylazoxymethanol (MAM) on the 15th day of gestation.[51a] MAM is the active compound of cycasin, the toxic ingredient of cycad seeds. MAM is converted *in vivo* to diazomethane, which kills dividing cells by alkylating the purine and pyrimidine bases in their nucleic acids.[51b] During the 2- to 24-hr effective period of the drug, rapidly dividing neuroblasts in the cortical matrix of the fetus are destroyed. There is a far greater diminution of the cortex and disruption of the cortical layers in the posterior half than in the frontal half of the brain (personal observations). Interestingly enough, the distribution of the cytoarchitecturally distinctive cortical areas corresponds roughly to the pattern shown in Figure 17-3. A large part of the hippocampal formation is preserved, with the dorsal portion showing, as in the diagram, on the superior part of the hemisphere and the ventral portion adjacent to the piriform area. The greatly diminished limbic cingulate mesocortex occupies a position next to the dorsal hippocampus comparable to Dart's "parahippocampal" area, while the location of the diminished limbic entorhinal area is approximately that of the "parapyriform" area. The entorhinal cortex (Brodmann's area 28), which occupies the posterior part of the hippocampal gyrus in the mammalian brain, is the major source of afferents to the hippocampus. Together, the cingulate and entorhinal areas would occupy a position corresponding to the posterior general pallium in Figure 17-3. The greatly reduced frontal neocortex would lie in the rostral part. Given this picture, the widening of the brain seen in the therapsid–mammalian transition (Figure 5-11) might be interpreted as a reflection of the development of limbic parahippocampal, cingulate, and entorhinal cortex, rather than of the neocortex.

The described cortical distribution in experimental micrencephaly does not conform to Quiroga's[52] inference regarding the location of the "neocortex" in *Probainognathus,* but it provides a certain parallel to what Sanides[38] has suggested with respect to the development of different cortical areas in therapsids.

The Great Limbic Lobe

In 1878, Paul Broca[1] (1824–1880), a French physician and pioneer anthropologist, reported a landmark finding based on an examination of the brains of a large variety of mammals. As noted earlier in the introduction to this chapter, it was his special contribution to provide evidence that a large convolution, which he called the *great limbic lobe,* is found as a common denominator in the mammalian brain (see Figure 17-1). As he explained, "The name . . . I have adopted indicates the constant relationship of this convolution with the border of the hemisphere; it does not imply any theory; . . . it is applicable to the brains of all mammals, to those that have a true corpus callosum, as well as to those in which the corpus callosum is absent or rudimentary. . . ."[53] Broca[1] mentioned that this lobe had been described some 40 years previously by Gerdy (1838), who called it the annular convolution, and by Foville (1844), who referred to it as "la circonvolution de l'ouret"—a pull-string purse.

Broca used the brain of the otter as his first illustration (Figure 17-5) to show the location and boundaries of the limbic lobe. He pointed out that the limbic lobe is walled off from the surrounding cortex by the rhinal fissure and by an extensive furrow that he called the *limbic fissure.* The limbic fissure may be nonexistent or only a slight indentation in the brains of lower mammals, while in higher forms it may be discontinuous. Elliot Smith[54] concluded that the constant parts of the limbic fissure are the genual (kneelike) sulcus located rostral to the corpus callosum and the splenial sulcus behind (Figure 17-6). For the inconstant sulcus in between he used the term *intercalary sulcus,* signifying its intercalated position. When the intercalary and genual sulci are connected, as in primates, they are referred to as the callosomarginal sulcus, or, as is now more usual, the cingulate sulcus. In some subprimate forms (e.g., cat and dog), the intercalary sulcus extends forwards and upwards to join the cruciate sulcus in the sensorimotor area of the neocortex. In carnivores and herbivores, the intercalary sulcus and its caudal continuation as the anterior calcarine sulcus of Smith corresponds to the splenial sulcus of Krueg.[55] In such animals, the retrosplenial stem of the splenial sulcus (corresponding to the forward part of the calcarine sulcus of primates) is an important landmark because it establishes the boundary between the primary visual cortex and the limbic cortex (Chapter 21). The

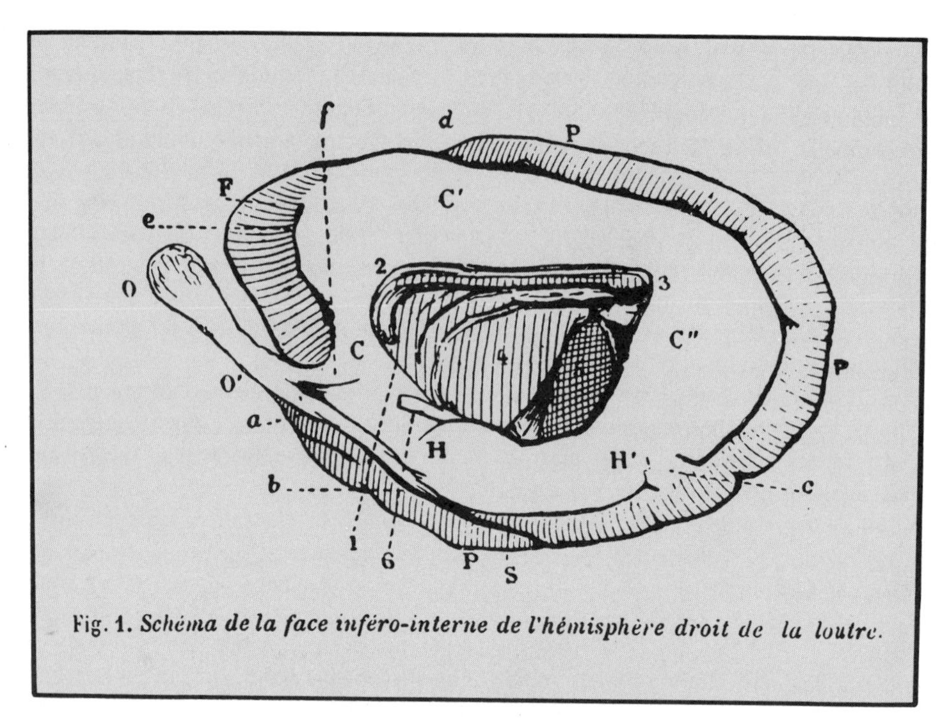

Figure 17-5. Representation of medial surface of the otter's brain used as the first figure of Broca's paper of 1878 for illustrating the location and configuration of the great limbic lobe (unshaded areas identified by H's and C's). ''O'' identifies the olfactory bulb.

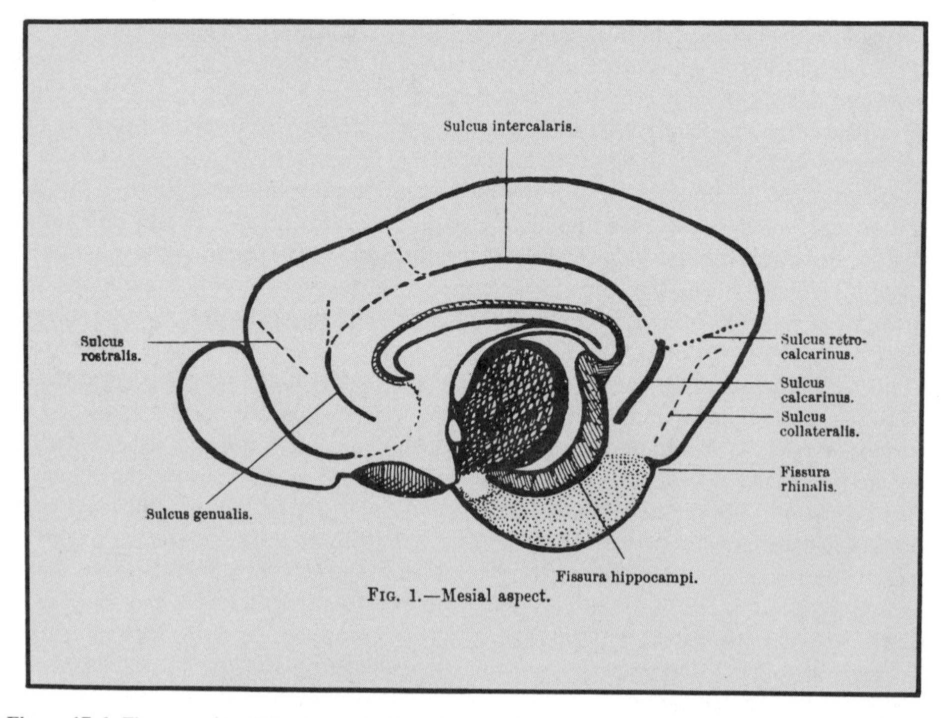

Figure 17-6. Figure used by Elliot Smith (1902) to describe the main fissures and sulci on the medial surface of the brain. See text for further details.

rhinal sulcus appears to be a constant feature in the brains of all mammals. These details about the sulci are important because of their usefulness in delimiting the boundaries of the limbic lobe in different mammals. The microscopic study of cortex was in its infancy at the time of Broca. Subsequent studies revealed that most of the evolutionarily old cortex is contained in the great limbic lobe.[56]

The great limbic lobe contains a number of subdivisions illustrated in Figure 17-7. The respective numerals overlying different parts of the lobe refer to cortical areas identified by Brodmann[57] of the Vogt school as having a distinctive cytoarchitecture. His numerical system for labeling them was simply a shorthand of convenience used by him and his colleagues. The pathways (tracts) from the olfactory bulb lead to the forward part of the limbic convolution. Calling this position 9 o'clock and going clockwise, one successively encounters area 25 in the preseptal part of the cingulate gyrus; areas 24 and 23 in the anterior and posterior cingulate; area 29 in the parasplenial cingulate; areas 35 and part of 36 in the posterior hippocampal gyrus; and area 28 in the anterior hippocampal gyrus. Area 16 of the insular cortex overlying the claustrum, and areas 14 of the medial postorbital and 13 of the postorbital cortex are not shown. *Cingulate,* meaning girdle, applies to the part of the limbic lobe engirdling the corpus callosum—hence the term *cingulate gyrus.* The hippocampal gyrus (alias parahippocampal gyrus) is so named because it forms a continuation of the underlying, infolded hippocampus. The latter convolution (which contains the greater part of the archicortex) was so named by the Italian anatomist Arantius (1587) because it reminded him of a sea horse. Almost 150 years later (1732), Winslow, in his famous textbook, compared the two hippocampi to "Ram's horns."[58] Ten years after that, Garengeot rerendered this comparison, saying, ". . . [F]or this reason, then, they are named the *horns of Ammon.*"[59] As will be

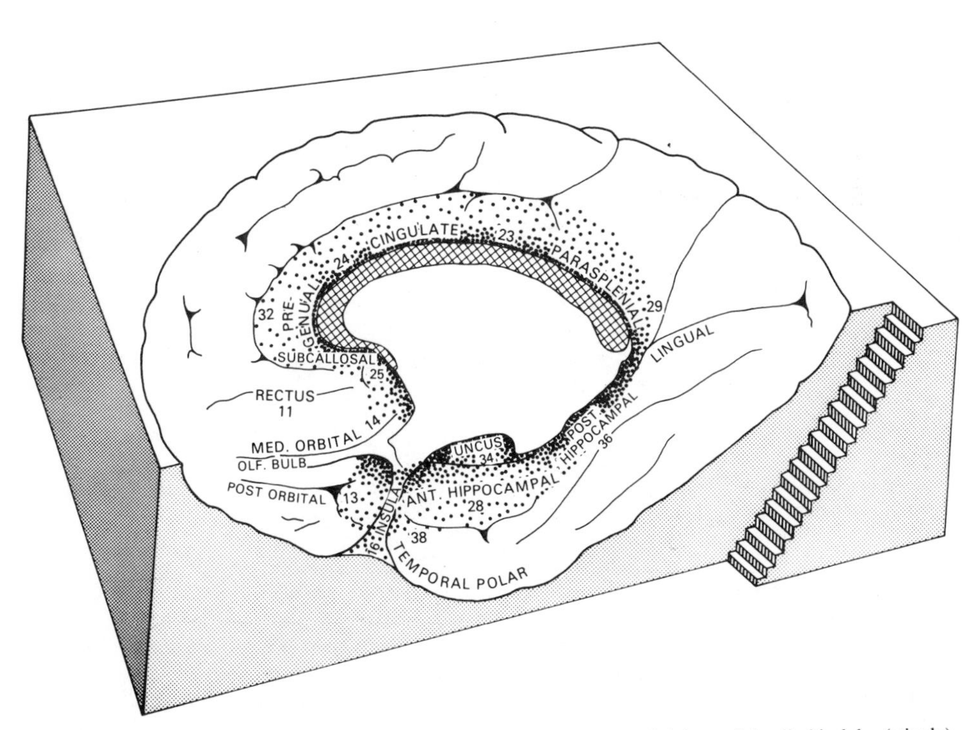

Figure 17-7. Three-dimensional view of the human brain showing subdivisions of the limbic lobe (stipple). Numbers identify cytoarchitecturally distinctive cortical areas.

mentioned in the next chapter, this additional naming after the Egyptian deity has contributed some confusion to the areal parcellation of the hippocampus. Together, the hippocampus and hippocampal gyrus give somewhat the appearance of a folded wide ribbon. As depicted in Figure 17-8, a cross section through the folds of the left hippocampal formation suggests a figure "2." The partially exteriorized dentate gyrus makes up the rest of the hippocampal formation. In cross section, it resembles somewhat a letter "C" cupping the upper part of the "2" corresponding to the hippocampus. The rostral half of the hippocampal gyrus is commonly referred to as the piriform lobe (former spelling "pyriform") because in animals with a well-developed olfactory apparatus it has a pear-shaped appearance (pirium = pear; the former stem "pyr" = fire; Smith said the philologic error was of such long standing that it would be "pedantic" to change it.[33])

In terms of the growth ring hypothesis, the limbic lobe may be imagined as comprising two major concentric rings of cortex, the inner ring representing the archicortex and

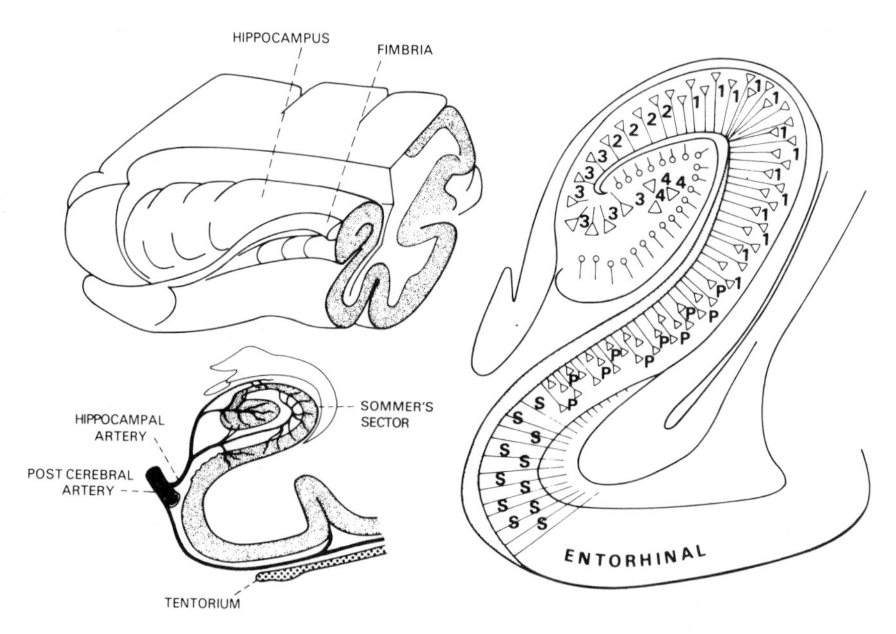

Figure 17-8. Drawings illustrating gross morphology, cytoarchitecture, and vascularization of the human hippocampus. The block of tissue in the upper left portion of the figure, together with the cytoarchitectural cross section at the right, illustrates that the hippocampus is folded upon itself in the ventricle somewhat like a wide, folded ribbon. In its externalized part, it becomes continuous with the hippocampal (parahippocampal) gyrus. In the right-hand diagram the circles identify the granule cells of the dentate gyrus, while the triangles represent pyramidal cells. The numerals refer to cytoarchitectural areas identified by Lorente de Nó as CA 1, CA 2, CA 3, and CA 4 (CA, cornu Ammonis). The prosubiculum (P) and subiculum (S), characterized by the presence of the perforant pathway, make up the rest of the hippocampus. The diagram of the blood supply to the hippocampus is based on one by Lindenberg (1955). As explained in a later chapter, the vessels leading to Sommer's sector (corresponding to area CA 1) are particularly significant in regard to neuronal disease resulting in psychomotor epilepsy.

the outer ring the transitional mesocortex. The body of the hippocampus contains most of the archicortex. In the brain of a marsupial such as the opossum without a corpus callosum (Figure 17-9A), or in the brain of a placental mammal with relatively little neocortex (Figure 17-9B), the body of the hippocampus hugs a considerable arc of the brainstem. But with the progressive expansion of the neocortex in higher mammals, the archicortex is displaced and stretched out by the enlarging corpus callosum that connects the neocortex of the two hemispheres. The attenuated part overlying the callosum comprises the indusium griseum and the associated medial and lateral striae of Lancisi (Chapter 18). As diagrammed in Figure 17-9C, in higher primates the bulk of the archicortex contained in the hippocampus becomes folded into the inferomedial part of the temporal lobe.

Neurochemical Distinctions

There are also neurochemical differences that distinguish not only the archicortex from the mesocortex, but also the combined limbic cortex from the neocortex.

In the first place, autoradiographic studies reveal a clear distinction between the limbic cortex generally and the neocortex.[60] As illustrated in Figure 17-10, we found that

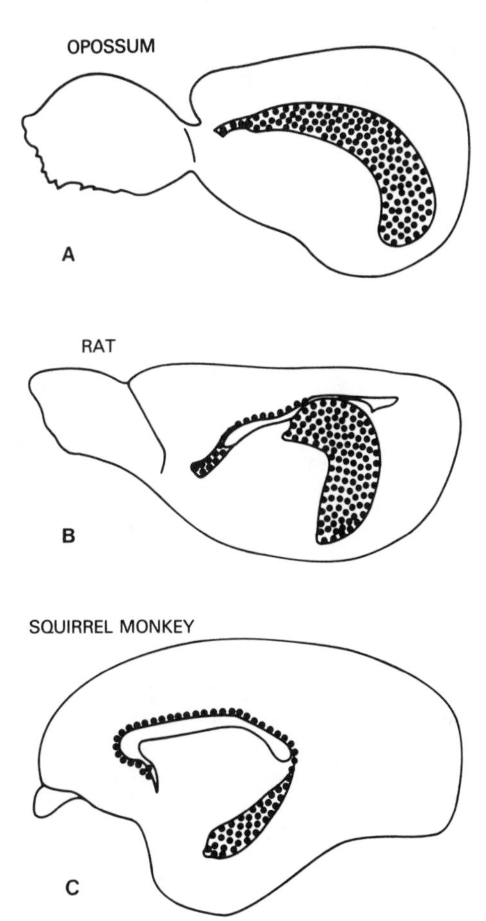

Figure 17-9. Drawings of sagittal sections through olfactory bulb and cerebrum, showing in representative animals how the progressive enlargement of the corpus callosum affects the configuration and location of the hippocampus (bold stipple). The top drawing shows the hippocampal configuration in an animal such as the opossum without a corpus callosum. In a rodent such as the rat the moderate development of the corpus callosum results in considerable attenuation of the forward part of the hippocampus, whereas in primates, as typified here by the squirrel monkey, the hippocampus is stretched thin over an extensive distance, so that the main part is displaced into the temporal lobe.

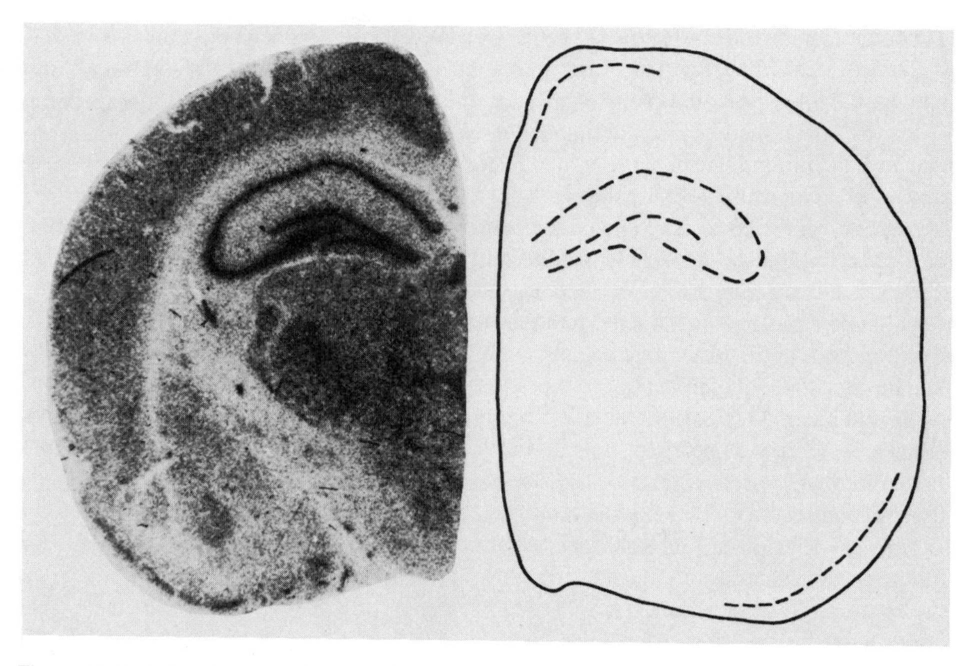

Figure 17-10. Autoradiogram of the brain of a normal rat, showing that limbic cortex has a greater uptake of ^{35}S-labeled L-methionine than the neocortex. Dashed lines in drawing on right identify from above downwards the limbic cortical areas of cingulate gyrus, hippocampus, and piriform lobe. Adapted from a study by Flanigan *et al.* (1957).

in autoradiograms from rats injected with ^{35}S-labeled L-methionine, the cellular layers of the limbic cortex appear darker than those of the neocortex. This observation, together with other findings, indicate that the limbic cortex has a higher turnover of protein than the neocortex.

Second, there is evidence that the chemistry of the archicortex is distinctive from that of the surrounding ring of transitional cortex. A little more than 100 years ago, Camillo Golgi[61] (1843–1926) described his epoch-making silver chromate staining method, which in later modifications made it possible to stain the entire nerve cell and its processes and which, in the hands of Ramón y Cajal,[62] led to the recognition that nerve cells exist as separate entities. Ramón y Cajal,[63] like Golgi, found that the silver chromate method was most effective in staining cells of the hippocampal formation. Here, perhaps, was the first evidence that the chemistry of the archicortex differs from that of other cortical structures. In illustrating their topistic theory, the Vogts[64] described a number of clinical conditions suggestive of a distinctive chemistry of different parts of the hippocampus. Experimental evidence supporting their supposition turned up in experiments in which Coggeshall and I[65] found that in the mouse the chemical 3-acetylpyridine (an antimetabolite of the antipellagra vitamin niacin) selectively destroys neurons of areas CA4 and CA3 of the hippocampus, as well as dentate cells innervating these areas. Maske[66] and others[67] subsequently showed that the chelating agent for zinc stains bright red the mossy fiber system leading from the dentate cells to CA4 and CA3. Other chemical aspects, including those concerning monoamines and endorphins, will be considered in subsequent chapters.

Development of the Limbic System Concept

The So-Called Rhinencephalon

In his writings, Broca[68] emphasized the connections of the olfactory apparatus with the forward part of the great limbic lobe, and on the basis of this anatomical relationship speculated that the entire convolution was involved in olfactory functions. As Professor E. A. Schäfer[69] of Edinburgh pointed out, other continental neurologists—notably Schwalbe (1881), Zuckerkandl (1887), and Retzius (1897)—"followed" Broca in this interpretation. Schäfer himself in the 1895 edition of Quain's *Anatomy* wrote: "The combined olfactory and limbic lobes may be spoken of collectively as the *rhinencephalon*."[70] But then he added: "[T]he limbic lobe unquestionably subserves others functions . . . since it is present even in those mammals (Delphinidae) which are devoid of an olfactory sense. . . ,"[71] the same point that Obersteiner[72] had made in 1890 in his well-known textbook of anatomy.

The persistence of the term *rhinencephalon* for referring to the great limbic lobe did so much to deter interest in this part of the brain that it is worth mentioning the origin of the word. According to Elliot Smith,[73] "the term 'rhinencephale' was originally applied by St. Hilaire [French naturalist] to a type of uniocular monsters without any direct reference to a region of the brain. . . . Richard Owen [English biologist, 1804–1902, and first curator of the British Museum] subsequently introduced the term . . . to distinguish those parts of the brain . . . known as the olfactory bulb and the olfactory peduncle." In 1890, Turner[74] (1832–1916), professor of anatomy at Edinburgh and a colleague of Schäfer, extended the meaning of this designation to include the lobus hippocampi (i.e., piriform lobe). As we have just seen, Schäfer and others included the rest of the limbic lobe. By the end of the 19th century the extended definition of *rhinencephalon* had become so entrenched in textbooks that the concept became a collective idée fixe passed on from generation to generation.

But it was mainly the functional implications of the term *rhinencephalon* that smothered interest in the great limbic lobe. Anatomists recognized that the olfactory bulbs are small in human beings. In 1890, Turner[74] had introduced the terms *microsmatic* and *macrosmatic* in reference, respectively, to animals with a small and large olfactory apparatus. According to some interpreters, the recognition of the human microsmatic condition, together with improved hygienic conditions in the 19th century, contributed to the belief that the sense of smell was unimportant in human beings. However that may be, the rhinencephalon was treated like an unwanted child in the teaching of neuroanatomy in medical schools. As one neuroanatomist commented in a book entitled *The Human Brain: From Primitive to Modern,* the rhinencephalon "probably has not contributed greatly to the evolution of the human brain and will, therefore, not be considered further."[75] In some authoritative texts one still finds the term *rhinencephalon* as a caption for a discussion of all of the structures in the limbic lobe.[76]

Question of Nonspecific Olfactory Functions

Despite the disparagement of the olfactory sense, some authors suggested that olfaction in some vague way may play a role in emotional experience, memory, and other functions. But as we shall see, it was not until 1937 that the entire limbic lobe was discussed in such terms. In his Croonian lectures published in 1919, Elliot Smith[77] commented that the olfactory structures, in a manner unlike that of other exteroceptive systems, may serve to provide an "affective tone" linking together anticipation and consummation into one experience and, in so doing, provide a germ for memory. As we have seen, he did not include the cingulate sector of the limbic lobe as part of the rhinencephalon. Failing to recognize the transitional nature of the cingulate cortex, he lumped it together with the neopallium, which he regarded as "the dominant organ of the body," "the organ of the mind."[78] Dart, as was also noted, regarded the cortex adjoining the piriform and hippocampal areas as neopallium. He speculated that the "para-pyriform neopallium controls the muscular display of skill," while the parahippocampal neopallium "governs . . . reactions associated with the expression of the emotions."[79] In an article on the functions of the olfactory parts of the brain, Herrick suggested in a concluding comment that they may serve as a "nonspecific activator of all cortical activities," as well as influence the "internal apparatus" accounting for general bodily attitude, disposition, and affective tone.[80] On the basis of clinical case material, two German neurologists, Kleist[81] and later Spatz,[82] emphasized that injury or disease of the frontotemporal basal cortex may account for changes in emotional behavior and otherwise affect the total personality.

Papez's Proposed Mechanism of Emotion

All of the views just mentioned applied either to nonolfactory functions of the rhinencephalon, in the restricted sense of Turner, or of the immediately adjacent structures. Turner's definition of the rhinencephalon excluded the hippocampus[74] (see above). Elliot Smith, however, argued that "the hippocampus has an unquestionable right to be grouped . . . in the 'smell-brain,' whether we call such a complex 'rhinencephalon' or not."[83] The argument over this question has continued ever since. In a detailed, provocative review appearing in 1947, Brodal claimed that the hippocampus had nothing to do with the sense of smell,[84] but this view is contradicted by recent anatomical and electrophysiological findings (see Chapter 26).

The question has also recently come up as to whether or not the hippocampus should be included in the limbic lobe itself. Stephan points out that Broca was not explicit about this point.[85] But its inclusion was implicit in Broca's definition since it forms the innermost cortex *surrounding* the brainstem. Moreover, it is also implied in Broca's characterization of the great limbic lobe in the original paper of 1878 and in a follow-up publication in 1879 in which he referred to the great limbic lobe as an animalistic ("brutale") brain and to the *rest of the cerebral mantle* as the "intellectual brain."[86]

James Wenceslas Papez (pronounced "papes"; see Chapter 4) appears to have been the first to single out the rhinencephalon in the broad sense of Broca and to ascribe to it important nonolfactory functions. Specifically, he developed the argument that certain limbic structures commonly included in the rhinencephalon are basically implicated in the experience and expression of emotion. In a paper entitled "A Proposed Mechanism of Emotion" published in 1937, Papez[87] began by calling attention to work that had shown that the hypothalamus (see Chapter 14) was essential for the *expression* of emotion. He argued that the *experience* of emotion depended on cortical function. He then drew attention to pathways leading to the diencephalon and from there to the structures of the telencephalon. He referred to the pathways destined for the corpus striatum (see Chapter 4 on the R-complex) as the "stream of movement." He characterized the pathways to the neocortex as the "stream of thought." Finally, he referred to those leading to the rhinencephalon (including the cingulate gyrus and hippocampal formation) as the "stream of feeling." This last inference was based on the strong connections of the limbic cortex with the hypothalamus, which, as mentioned, was believed to be essential for the expression of emotion.

Papez then cited a number of diseases involving the limbic lobe that were either manifest by emotional excitement or the opposite state in which there was a loss of emotional feeling and expression. He pointed out, for example, that in cases of rabies in which the virus shows a predilection for the hippocampus, patients may suffer from extreme terror. He noted, on the contrary, that in cases in which tumors press on the cingulate gyrus and depress its functions, patients are often apathetic and have a loss of memory. He summed up his argument in the following oft-quoted words: "It is proposed that the hypothalamus, the anterior thalamic nuclei, the gyrus cinguli, the hippocampus and their interconnections constitute a harmonious mechanism which may elaborate the functions of central emotion, as well as participate in emotional expression."[88]

In developing his argument, Papez took pains to point out connections by which information from sensory systems might be conveyed to the cingulate gyrus via the mammillary bodies. The cingulate gyrus, he noted, "may be looked on as the receptive region for the experiencing of emotion . . . in the same way as the [visual area] is considered the receptive cortex for photic excitations coming from the retina."[89] It is clear from this statement that with respect to emotional experience, Papez conceived of the "viewer" as residing in the cingulate gyrus. "Radiation of the emotive process," he continued, "from the gyrus cinguli to other regions in the cerebral cortex would add emotional coloring to psychic processes occurring elsewhere."[89]

The Limbic System

Sometimes my name is undeservedly conjoined with Papez's when reference is made to his theory of emotion. In a recent appreciation of Papez, I have explained how 10 years after the publication of his paper, I went to see him in order to learn some additional anatomical details.[90] My introduction to him happened in this way: In 1947 I had received a USPHS Fellowship for working with Dr. Stanley Cobb, a neurologist and neuropathologist who was instrumental in establishing the Department of Psychiatry at the Massachusetts General Hospital. A short experience in the practice of medicine had impressed on me the great importance of psychosomatic factors in contributing to illness. Through Dr. Cobb I arranged to conduct some electroencephalographic research in the Brain Wave Laboratory under the direction of Dr. Robert Schwab. There I developed an improved electrode for recording the electrical activity at the base of the brain.[91] This work led to a diagnostic electroencephalographic examination of patients with so-called psychomotor epilepsy. As will be dealt with in Chapters 22–25, patients with this form of epilepsy experience one or more of a wide range of vivid emotions at the beginning of their seizures, i.e., during the aura. In combined recordings of the basal and standard electroencephalograms, Arellano and I found that in some cases the maximal bioelectrical disturbance (manifest by "spike" potentials) occurred in the medial basal region of the temporal lobe, suggesting an origination in or near the hippocampal formation.[92]

About that same time I happened upon Papez's neglected paper on emotion and was struck by its obvious relevance to the emotional symptomatology of the patients in our study. In conjunction with their emotional feelings, patients with psychomotor epilepsy may experience symptoms identified with one or more of the sensory systems. Except for olfactory connections, there was almost no indication in Papez's paper how the hippocampal formation would receive information from other sensory systems. Papez had noted that an abstruse structure known as the subcallosal bundle (Muratoff's bundle) might provide an ''associational link between the general cortex and the hippocampal formation.''[93] The ''general'' cortical areas were located in the frontal, cingular, and parietal regions. With only these inputs, how was one to explain the various visceral, gustatory, somatic, auditory, and visual symptomatology that patients with the above kind of epilepsy experienced at the beginning of their attacks? It was this kind of question that took me to Ithaca in 1948 to discuss the problem with Dr. Papez.

On the strength of my visit to Papez and accumulated new information since his original article, I prepared a paper for presentation at a departmental seminar that in 1949 was published under the title ''Psychosomatic Disease and the 'Visceral Brain.' Recent Developments Bearing on the Papez Theory of Emotion.''[94] Figure 3 of that paper shows the sketch I used to summarize suggestive evidence of overlapping inputs to the hippocampal formation from all of the intero- and exteroceptive systems. In my discussion with Papez I was particularly interested in learning from him about the possible inputs from the visual, auditory, and somatic systems.[90] Since then our own neuronographic findings and anatomical studies of others have shown stepwise cortical connections of the sensory sytems with the hippocampal formation (Chapter 18). Moreover, in experiments involving a recording from single nerve cells in awake sitting squirrel monkeys, we have obtained evidence of inputs to *peri*hippocampal cortex from olfactory, gustatory, somatic, auditory, visual projections, and the hippocampus itself from olfactory and vagal pathways (Chapter 26).

The ''visceral brain'' paper was perhaps significant for introducing a few new ideas. First of all, it suggested how a phylogenetically old part of the brain, found as a common denominator in mammals, might receive information from all the sensory systems. In regard to the hippocampus itself, this would indicate that it was not an autonomous little factory of its own, manufacturing the raw materials of emotion out of thin air.

The multisensory inputs were also discussed as possibly relevant to somatovisceral symptoms of patients with psychosomatic illness who seem to have the tendency to experience happenings in the outside world as though they were inside. I suggested that in addition to the overlapping of incoming impressions from interoceptive and exteroceptive systems, the crudity of the analyzing mechanism of the visceral brain might account for the ''seemingly paradoxical overlapping'' of affective reactions, as well as the state of confusion between external and visceral awareness that allows outside situations to be experienced as though they were inside. In terms of Freudian psychology, I suggested that ''the visceral brain is not at all unconscious (possibly not even in certain stages of sleep), but rather eludes the grasp of the intellect because its animalistic and primitive structure makes it impossible to communicate in verbal terms.''[95]

In conclusion, it was suggested that although our intellectual functions are mediated in the newest and most highly developed part of the brain, ''our affective behavior continues to be dominated by relatively crude and primitive system.''[96] ''This situation,'' I noted, ''provides a clue to understanding the difference between what we 'feel' and what we 'know.' ''[96]

I used the expression *visceral brain* as a means of avoiding the narrow implications of the term *rhinencephalon*. In its original 16th century meaning, *visceral* applies to strong inward feelings and implicitly the accompanying visceral manifestations. Later when I joined the Department of Physiology at Yale (1949), I found that the term created misunderstanding because in physiological parlance the word *visceral* applies only to glands and hollow organs, including the blood vessels. Consequently, I resorted to Broca's descriptive term *limbic* and used the expression *limbic system* when referring to the cortex of the limbic lobe and the structures of the brainstem with which it has primary connections. This explains how the term *limbic system* was introduced into the literature in 1952.[97]

Summary

The present chapter has dealt successively with (1) certain distinctive behavioral and morphological changes marking the transition from mammal-like reptiles (therapsids) to mammals, (2) the evolution and nature of the cortex associated with the paleomammalian formation of the brain (limbic system), and (3) a commentary on developments leading up to the limbic system concept.

Before delving into the functions of the limbic system, we will need a more complete anatomical framework for reference. The next chapter provides a review of what has been learned to date about connections of the limbic lobe with structures of the brainstem. We shall find that for present purposes the limbic system can be provisionally regarded as comprising three main subdivisions.

References

1. Broca, 1878
2. MacLean, 1952
3. Clark and Meyer, 1950, p. 342
4. MacLean, 1985a
5. Romer, 1967, p. 1634
6. Romer, 1966, p. 198
7. Fitch, 1970
8. Frakes, 1979
9. Van Valen, 1960
10. Long, 1969
11. Guillette and Hotton, 1986; Duvall, 1986
12. Romer, 1958, p. 72
13. Parrington, 1941, 1971; Kermack *et al.*, 1968
14. Desmond, 1976, p. 94
15. Crompton and Jenkins, 1979
16. Crompton and Jenkins, 1973
17. Crompton and Jenkins, 1973, p. 139
18. Olson, 1959, p. 348
19. Olson, 1959, p. 349
20. Shands, 1977
21. Kuhne, quoted by Van Valen, 1960
22. Romer, 1966
23. Gregory, 1967, p. 48, Fig. 27
24. Colbert, 1969, p. 267
25. Evans, 1958
26. Gregory, 1967, p. 48
27. Gregory and Simpson, 1926
28. Gregory, 1967, p. 52
29. Gregory, 1967, p. 53, Fig. 30
30. Stephan and Andy, 1964
31. Edinger, 1966
32. Cookson, 1962
33. Smith, 1901
34. Edinger, quoted by Smith, 1910
35. Ariëns Kappers, 1909
36. Dart, 1935
37. Abbie, 1942, pp. 532–533
38. Sanides, 1969
39. Ariëns Kappers *et al.*, 1936, p. 1479
40. Campbell, 1905
41. von Economo and Koskinas, 1925; von Economo, 1929
42. Rose, 1927
43. Rose and Woolsey, 1948
44. Yakovlev, 1948 (p. 331, "mesopallium")
45. Abbie, 1940
46. Morgane *et al.*, 1986
47. Rose, 1926, p. 129; Rose, 1927, p. 67
48. Yakovlev *et al.*, 1960, p. 629
49. Ariëns Kappers, 1928
50. Lorente de Nó, 1949
51. Quiroga, 1979
51a. Spatz and Laqueur, 1968
51b. Johnston and Coyle, 1979
52. Quiroga, 1980
53. Broca, 1878, p. 391
54. Smith, 1902
55. Cited by Smith, 1902
56. e.g., Ariëns Kappers *et al.*, 1936, pp. 1578–1579, 1586–1592
57. Brodmann, 1909
58. Lewis, 1923–24
59. Lewis, 1923–24, p. 225
60. Flanigan *et al.*, 1957
61. Golgi, 1873
62. See Cannon, 1949, "Life of Cajal"
63. Ramón y Cajal, 1891
64. Vogt and Vogt, 1922, 1953
65. Coggeshall and MacLean, 1958
66. Maske, 1955
67. Fleischhauer, 1964
68. Broca, 1878, 1879
69. Schäfer, 1900, p. 765
70. Schäfer, 1895, p. 160
71. Schäfer, 1895, p. 161
72. Obersteiner, 1887/1890
73. Smith, 1901, p. 438
74. Turner, 1890
75. Lassek, 1957, p. 65

76. Crosby *et al.*, 1962
77. Smith, 1919
78. Smith, 1901, p. 453
79. Dart, 1935, p. 15
80. Herrick, 1933, p. 14
81. Kleist, 1931
82. Spatz, 1937
83. Smith, 1901, p. 446
84. Brodal, 1947
85. Stephan, 1964
86. Broca, 1879, pp. 420, 425
87. Papez, 1937
88. Papez, 1937, p. 743
89. Papez, 1937, p. 728
90. MacLean, 1978a
91. MacLean, 1949a
92. MacLean and Arellano, 1950
93. Papez, 1937, p. 732
94. MacLean, 1949b
95. MacLean, 1949b, p. 348
96. MacLean, 1949b, p. 351
97. MacLean, 1952

18

An Anatomical Framework for Considering Limbic Functions

On Sunday, June 13, 1886, Bernard von Gudden lost his life while trying to save King Ludwig II, the mad king of Bavaria, from drowning himself in Starnbergersee.[1] Gudden (1824–1886) was a neuroanatomist and professor of psychiatry at the University of Munich. In proceeding now to construct an anatomical framework for considering functions of the limbic system, we will very early turn up two monuments to Gudden—one concerning a method and the other a nuclear complex named for him.

Synopsis of Neuroanatomic Methods

In some respects the work of a neuroanatomist can be compared to that of a detective. Consequently, an acquaintance with the methods of detection adds interest to tracking down the connections of the limbic system that have been under investigation for more than 100 years. The purpose is to obtain a picture of subdivisions of the limbic system in analyzing the nature of limbic functions.

Classical Methods

Progress in the microscopic examination of cerebral structures and their connections depended on improved methods for fixing the brain, obtaining thin serial sections, and discovering satisfactory stains for nerve cells and fibers. By the early 1870s sliding microtomes for cutting serial brain sections of equal thickness were in use in the laboratories of Gudden, Meynert, and others. August Forel (Swiss neuroanatomist and psychiatrist, 1848–1931), working in Gudden's laboratory, helped to devise a microtome for obtaining whole sections of the human brain.[2] The use of carmine [introduced by Gerlach (1820–1896) in 1858[3]] had become popular for staining nerve cells.[4]

Until the middle of the present century anatomists relied principally on four methods for demonstrating connections of the brain. Identified, respectively, with the names of Gudden, Golgi, Marchi, and Nissl, the four methods will be outlined in the same sequence as their dates of publication.

The Gudden Method

In 1870 Gudden reported his remarkable finding that cutting a cranial nerve in a newborn animal such as the rabbit resulted in a clearly visible disappearance of structures connected with it.[5] As we shall see later on, the induced atrophy may affect serial linkages of neurons in a backward, as well as forward direction. In other words, the degenerative effect is able to express itself transsynaptically in either an anterograde or a retrograde direction, a result that was once regarded by some as a refutation of the so-called neuron doctrine that nerve cells (together with their processes) constitute separate entities.

The Golgi Method

In 1873 the Italian anatomist Golgi described a method that randomly stained a small population of cells in the brains of young animals.[6] Tissue previously immersed in an osmium dichromate solution is placed in a weak silver nitrate solution. The success of the stain depends upon the impregnation of a cell and its processes with silver chromate. As Ramón y Cajal commented in 1893, "Golgi's rapid method produces splendid results, but only in young animals."[7] He then explained that Cox's method employing sublimate of mercury must be used for visualizing the medullated parts of an axonal process. By tracing the course of an axon in the brain of small animals much can be learned about neural connections. In recent years the Golgi method has proved useful for learning where a degenerating pathway terminates on the dendrites of particular neurons.[8]

It should also be noted that the application of variations of the Golgi stain made it possible for Ramón y Cajal to provide the best available evidence that nerve cells (together with their processes) constitute separate entities.[9] It was largely on the basis of Ramón y Cajal's findings that Waldeyer (Berlin anatomist, 1836–1931) in 1891 expounded the "theory of isolated units" and crystallized this concept by coining the term *neurone* (νευρών).[10] It was to be another 15 years, however, before Ramón y Cajal conducted the additional studies on the degeneration and regeneration of nerve that won general acceptance of the "neuron doctrine."[9]

The Marchi Method

Somewhat reminiscent of directional evolution mentioned in the preceding chapter, several workers during the last quarter of the 19th century were independently beginning to regard nerve cells as separate entities rather than parts of a reticular syncytium. Already by 1850, Waller (English anatomist, 1816–1870) had described the degeneration occurring in nerves distal to a cut (Wallerian degeneration).[11] The situation was comparable to the death that occurs beyond a break in the branch of a tree. As Obersteiner typographically emphasized in his textbook (1887/1890): "THE NERVE FIBRE DIES WHEN CUT OFF FROM THE CELL OF WHICH IT IS A PROCESS."[12]

Until 1885 there had been no adequate means of staining degenerating nerve fibers. In that year Marchi and Algieri published their observation that such degeneration could be revealed in tissue mordanted with a chromic salt and then treated with osmic acid.[13] Marchi (1851–1900) had been a student of Golgi. For the next seven decades neuroanatomists relied primarily on the Marchi method for demonstrating degenerating nerve fibers in the central nervous system.

The Nissl Method

In 1892 the Heidelberg anatomist Nissl (1860–1919) published his discovery as a medical student (1884) regarding the efficacy of an aniline dye to stain certain components of nerve cells that have since become known as Nissl bodies.[14] In addition, he described the dissolution and fading of these bodies (Marinesco's chromatolysis) that occur following the section of a nerve. (The Nissl bodies constitute the so-called rough endoplasmic reticulum that provides a place of meeting and attachment for ribosomes and messenger RNA to "direct" transfer RNA in its role of lining up amino acids for the formation of cellular proteins.) Later by using his method to detect retrograde degeneration following cortical lesions, Nissl located thalamic cell groups that project to specific areas of the cerebral cortex.[15] As opposed to Gudden's method, the Nissl method lends itself to experiments on adult animals.

Methods of Recent Times

In relatively recent times further knowledge of the brain's connections resulted first from improved silver methods and then, quite lately, from techniques taking advantage of physiological processes within nerve cells.

Improved Silver Methods

In 1904 Max Bielschowsky (German neuropathologist, 1869–1940) reported an ammoniacal silver method for impregnating neurofibrils[16] that has been the crux of several recent techniques for revealing neuronal degeneration. In 1946 the Oxford anatomist Paul Glees advocated his own modifiction of the Bielschowsky stain as a supplement for the Marchi method.[17] It had the special value of revealing very fine pericellular terminal degeneration. Despite Glees's use of an ammonia–alcohol solution for dissolving the greater part of the myelin, stained normal fibers tended to obscure the presence of degeneration. Furthermore, in some areas such as the ventromedial nucleus of the hypothalamus, the picture under normal conditions resembled degeneration. A landmark advance occurred in 1954 when Nauta, in collaboration with Gygax, published a method that effectively suppressed the staining of normal fibers and thereby brought the degenerating fibers into relief.[18] The success of the Nauta method led to a widespread resurgence of investigations on cerebral connections. The method, however, was unsatisfactory for demonstrating terminal degeneration on cells, and it failed to impregnate some fiber systems. Later on, Fink and Heimer,[19] Voneida and Trevarthen,[20] and others introduced modifications that, when applied at critical survival times of the experimental preparation, solved the problem concerning terminal degeneration. Subsequently in 1969, de Olmos described what appears to be the most sensitive silver method of all—his so-called cupric-silver method.[21] The use of electron microscopy, which became progressively applied to the study of the nervous system during the 1950s, has made it possible to demonstrate the deposits of silver in degenerating terminals at the ultramicroscopic level.[22]

Autoradiography

A major disadvantage of methods described above is that they are dependent on the production of lesions. When a lesion is made within a certain structure of the brain, it commonly happens that fibers of passage are also destroyed—i.e., fibers unrelated to the targeted tissue. Since the early 1970s, two methods have found extensive application because, except for certain artifactual situations, they circumvent such complications.

One method involves autoradiography, and the other the use of the enzyme horseradish peroxidase (HRP). In each case the demonstration of pathways depends on physiological transport mechanisms rather than on degenerative processes.

In applying the autoradiographic method, radioactive nutrients consisting of amines are injected in micro-amounts into a particular collection of nerve cells where they are assimilated by the cells and transported to the nerve terminals.[23] A picture of the location of the radioactive substances is obtained by dipping serial brain sections into a photographic emulsion and then allowing an exposure time of several days.[24] In a counterstained section, silver grains precipitated by radioactive elements show up as black particles over the cells or fibers, whereas under darkfield examination they mirror the reflected light and appear as white particles.

The HRP Methods

The HRP method depends on the uptake of this enzyme by nerve terminals (pinocytosis) and its transportation back to the cell body.[25] The presence of HRP in the cell is demonstrated by employing a reaction involving hydrogen peroxide and a chromogenic aromatic amine.[26] The first description of the application of the HRP method for tracing *cerebral* connections appeared in 1972.[27] Originally diaminobenzidine was used as the chromogen.[27] This produced a brown color that was sometimes hard to detect, unless examined in conjunction with darkfield illumination. Since then Mesulam has introduced the chromogen tetramethylbenzidine, which has the triple advantages of (1) being a more sensitive marker, (2) producing a clearly visible blue product, and (3) not being a carcinogenic substance.[28] The sensitivity of the method has been further enhanced by the finding that conjugation of HRP to a nontoxic lectin (wheat germ) not only limits its spread at the site of application, but also greatly increases its uptake.[29] The HRP method has also proved useful for identifying the destination of fibers revealed by anterograde transport.

Histofluorescence and Other Techniques

In Chapters 4 and 14 reference was made to the Falck–Hillarp method for identifying the cells and axonal processes of monoaminergic systems. Currently a number of retrograde cell markers giving different fluorescent colors are being used to trace the connections of different cell populations of the same nucleus, as well as to identify their collateral projections.[30] Extensive use is also being made of the immunohistofluorescence techniques to identify neurons associated with particular hormones, peptides, and the like.[31] By combining these latter techniques with the HRP method it may be possible to learn the connections of such neurons.[32]

LIMBIC INPUTS

Just as people in a city can be identified with certain districts according to nodes in the network of communications, so can large populations of nerve cells be similarly grouped. Viewed in this light, the limbic cortex falls into three main sectors identified with nodal cell stations located in the (1) amygdala, (2) septum, and (3) thalamus. These three corticosubcortical subdivisions, in turn, are connected with cell stations nearer to the heart of the neuraxis. The three subdivisions will be referred to as the amygdalar, septal, and thalamocingulate divisions.

In arriving at this parcellation we will begin at the lower brainstem and trace the course of inputs leading to various parts of the limbic cortex.

Two Basic Kinds of Inputs

It has been traditional to separate the functions of organisms into those that maintain the internal environment and those that account for interaction with the external environment. Following the introduction of the Golgi method in 1873, neuroanatomists discovered that by making serial sections through the whole body of small vertebrates, they could reconstruct the entire peripheral and central nervous system in a way that could not be achieved by skillful manual dissection. In following up Strong's[33] "first successful application of this method," C. Judson Herrick found that he could identify neural systems associated with the two major functions of the organism—namely, "(1) those concerned with adjustment to environment, the *somatic* functions, and (2) those concerned with the maintenance and reproduction of the body itself, *visceral* functions."[34] We shall deal first with receptive systems signaling information from the internal and then with those from the external environment, using Sherrington's terms *interoceptive* and *exteroceptive*[35] to apply, respectively, to inputs from the so-called visceral and somatic systems.

Interoceptive Inputs

Except for findings in the rat[36] there is no other evidence that the three highest level subdivisions of the limbic system receive direct connections from the visceral nuclei of the medulla. Consequently, we will begin the present account at a more rostral locus anatomically identified as the isthmus and indicated by the shading in Figure 18-1. As reviewed in Chapter 3, early in brain development there become manifest three vesicles, which shortly thereafter become subdivided into five. The term *isthmus* refers to the unexpanded segment connecting the two vesicles that become the midbrain and pons. In deciding upon the name *isthmus* for the constricted segment, His[37] (Swiss anatomist, 1831–1904) possibly had in mind the settlements and thoroughfares of the Isthmus of Corinth connecting Peloponnesus with the mainland of Greece. The isthmus has since proved to be a pivotal site at which the forebrain establishes a linkage with neural mechanisms involved in the integrated control of somatovisceral functions.

Herrick was one of the first neuroanatomists to investigate ascending interoceptive systems within the brainstem. In 1914 he reported the finding of an "ascending secondary visceral tract" in the larval salamander,[38] which he regarded as the amphibian equivalent of a tract that he had described earlier in fish (1905). He noted that the tract consisted of "non-myelinated fibers which are derived from the neurones of the visceral sensory lobe of the same side and which terminate in a . . . secondary visceral nucleus in the isthmus. . . ."[39] He commented that the tract was clearly comparable to "the secondary vagus bundle of Mayser" (1881) found in fish.[40] The visceral sensory lobe is the equivalent of the solitary nucleus (nucleus tractus solitarii) of mammals. The gustatory components of the 7th and 9th cranial nerves project to the rostral part of the solitary nucleus, while the remaining part is innervated by the vagus nerve (so-called "wandering" or "great visceral nerve"). About three-quarters of the vagal axons are afferent fibers,

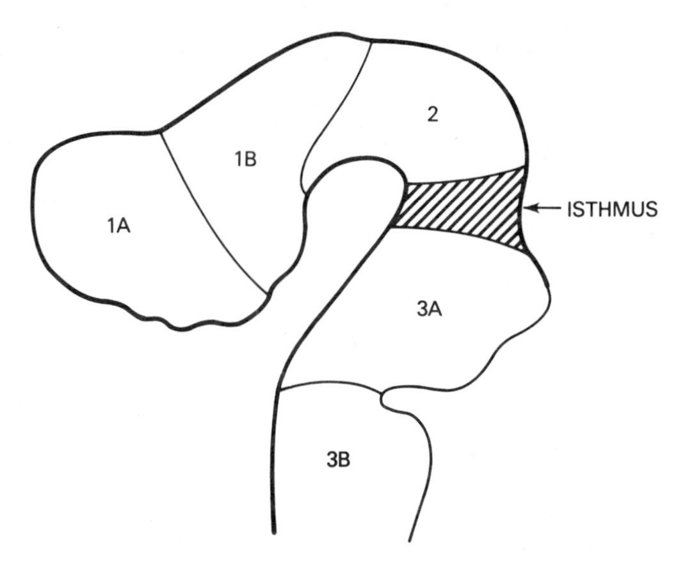

Figure 18-1. Five-vesicle stage of the brain with focus on the isthmus region (shaded). As reviewed in Chapter 3, the term *isthmus* applies to the constricted region between the midbrain and pons (mesencephalon and metencephalon). As illustrated in Figure 18-2, it is the site of five nuclear groups providing projections to the limbic system. Drawings adapted from ones appearing in His (1904) and in Villiger (1931).

conducting impulses from the pharynx, upper intestinal tract, heart, lungs, and other viscera.[41] Kuru has described what he refers to as the "pelvic sensory vagus"[42] that terminates in nucleus paraalar adjoining the lateral aspect of the solitary nucleus.[42]

Herrick emphasized that the superior visceral nucleus, which he also called the *visceral-gustatory nucleus,* lies in close proximity to the nuclei innervating the musculature of the head and mouth. This and other considerations led him to infer that "the isthmic apparatus is the chief regulator of the musculature concerned with feeding."[43] He further commented that "it is evident that control of feeding reactions is not the only function of the isthmic sector,"[43] noting that Aronson and Noble had shown that lesions in this region disturbed or completely abolished spawning responses in the frog.[44] As though foretold in Herrick's early studies on a larval amphibian, it has since been found that the isthmus plays an important role in somatovisceral, alimentary, respiratory, and genitourinary functions (see below and Chapter 19).

Although Herrick referred to the isthmus as a "transitional sector," he emphasized that it "occupies a strategic position between primitive bulbo-spinal mechanisms and the higher cerebral adjusters."[45] In fish the midbrain serves primarily as the goal-seeking, orienting brain (Chapter 3). In terrestrial vertebrates, the evolving forebrain assumed this role and in the process established connections with the strategic isthmic region. Much of the anatomical evidence for such connections was not forthcoming until very recently.

Four Isthmic Cell Groups

There are four main groups of cells associated with the isthmus that must be considered in regard to inputs to the limbic system. These are (1) the dorsal and ventral tegmental nuclei of Gudden; (2) the locus ceruleus; (3) the parabrachial nuclei; and (4) the raphe nuclei comprising the superior central nucleus of Bechterew and dorsal raphe nucleus.[46]

The last three cell groups have sparked great interest because of the discovery that they are partly characterized by the presence of monoamine-containing neurons. Figure 18-2A indicates the course of afferent pathways to be discussed. The locations of the nuclear groups themselves are shown in Figure 18-3. Starting with the Gudden group at 12 o'clock, we will describe them successively in a clockwise direction.

Gudden's Nuclei

The dorsal and ventral tegmental nuclei are named after Gudden. The dorsal nucleus is located in the ventral part of the central gray matter at the caudal end of the aqueduct of Sylvius, while the ventral nucleus lies ventral to it and underneath the medial longitudinal fasciculus (Figure 18-3). The latter is poorly developed in the monkey, and is not found in the human brain.[47] In 1884 Gudden reported that the ventral tegmental nucleus degenerated in young rabbits following the destruction of nerve tissue in the mammillary region.[48] Using Gudden's method, Ruth Bleier has confirmed and extended his findings.[49] With silver-degeneration techniques it has been shown that the ventral nucleus, as well as the dorsal tegmental nucleus of Gudden just above it, projects to the mammillary bodies by way of the mammillary peduncle.[50] Some fibers join the medial forebrain bundle (MFB) and reach the septum (Figure 18-2A).

The just-mentioned findings are of interest in the light of Morest's observation that a lesion of the vagal sensory nucleus results in degeneration that can be traced to the dorsal tegmental nucleus of Gudden.[51] In addition to its projections via the mammillary peduncle, the dorsal tegmental nucleus contributes fibers to the dorsal longitudinal bundle of Schütz that ascend within the central gray and terminate in the intralaminar nuclei of the thalamus, including the central, central lateral, and paracentral nuclei.[52] It will later become evident how nerve impulses traveling these different routes reach the limbic cortex.

Locus Ceruleus

The remaining three isthmic cell groups to be considered are the ones shown to have monoamine-containing cells.[53] The locus ceruleus and the parabrachial nuclei are distinguished by the presence of catecholamine-containing cells that in each case has been identified as norepinephrine.[54] Called the *blue place* because of its distinctive bluish color in freshly cut brain, the locus ceruleus lies caudal to the dorsal tegmental nucleus of Gudden and is partially roofed over by the ascending root of the fifth nerve (Figure 18-3). The catecholamine-containing cells of the locus were designated as the A6 group by Dahlström and Fuxe.[53] Ungerstedt (1971) was the first to show experimentally that the fibers from these cells collect into a dorsal bundle (Figure 18-4) and ascend "just laterodorsal to the *fasciculus longitudinalis*" up to the level of the mammillary bodies, after which its fibers turn ventrolaterally and join the medial forebrain bundle to reach the septum.[55] At this level some of the fibers continue into the cingulum (Figure 18-4). Both the fornix and the cingulum appear to be potential pathways by which the cerulean ascending fibers reach the hippocampus and cingulate gyrus. Ungerstedt's findings are supported by the work of Pickel *et al.*, who employed a combination of techniques, including autoradiography.[56] A study by Ricardo and Koh is significant regarding the long-suspected role of the locus ceruleus in visceral functions.[36] They injected radiolabeled proline and leucine into the solitary nuclei of rats and according to their Figure 2L traced fibers into the locus ceruleus.

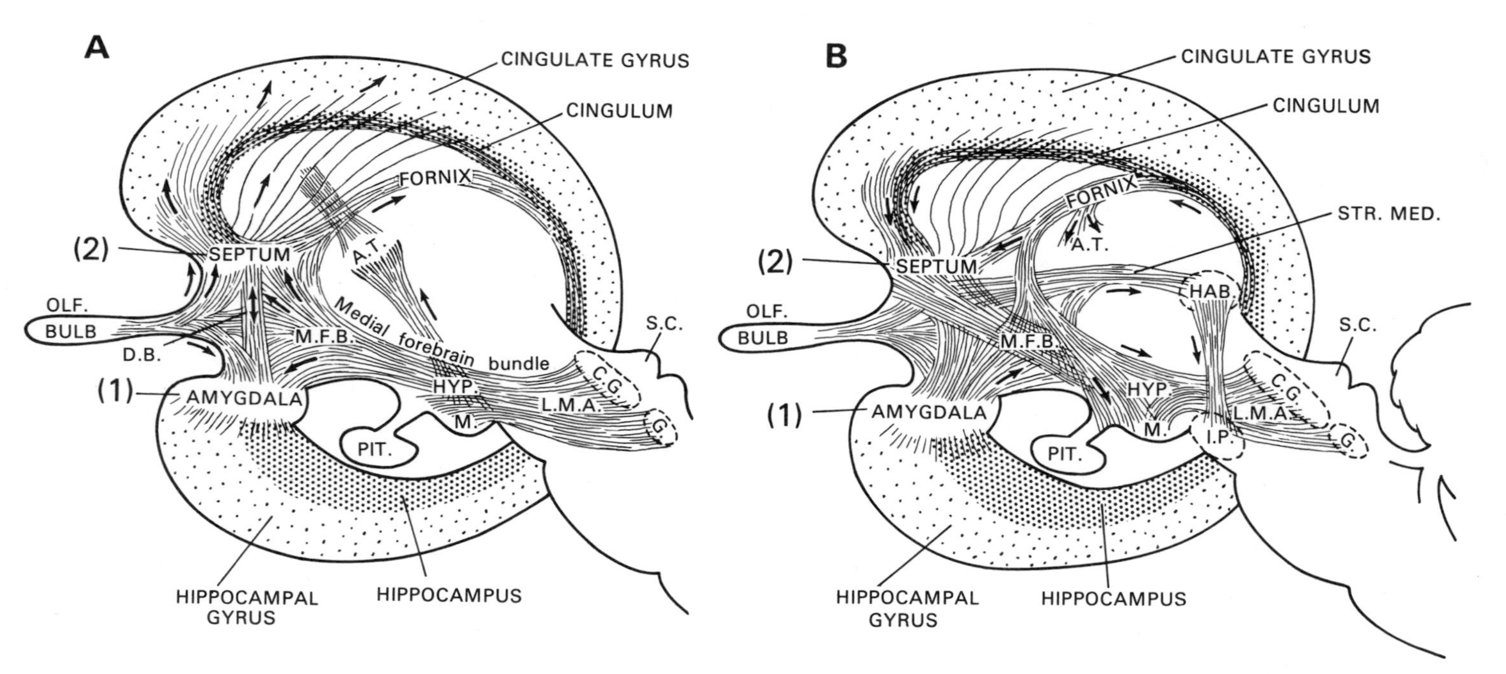

Figure 18-2. Diagrams of main afferent (A) and efferent (B) pathways of the limbic cortex. In recent years it has been shown by new anatomical techniques that, in addition to the classically recognized ascending projections from Gudden's nuclei (G.), four other nuclear groups of the isthmus supply a rich innervation to limbic structures. The medial forebrain bundle (M.F.B.) is a major conduit for these projections as well as those ascending from the ventral midbrain and hypothalamus. Following this major route, numerous fibers connect with nuclei in the amygdala and septum that project to the limbic cortex while others lead directly to the limbic cortex. Note relationship also to downstream connection from the olfactory apparatus. In contrast, observe that a large ascending pathway from the mammillary bodies (M.) to the anterior thalamic nuclei (A.T.) and to the cingulate cortex bypasses the olfactory apparatus. As illustrated in B, descending pathways from the limbic cortex in general run parallel to ascending connections. Other abbreviations: PIT., pituitary; S.C., superior colliculus. Redrawn from MacLean (1957).

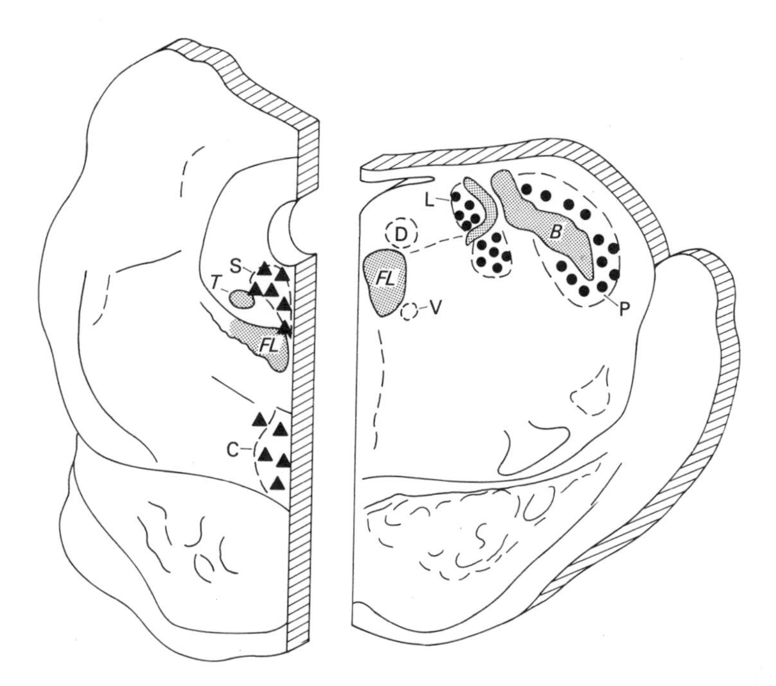

Figure 18-3. Five main nuclei of the isthmus region having extensive connections with the limbic system. In a clockwise direction, beginning at 12:00, they are (1) dorsal (D) and ventral (V) nuclei of Gudden; (2) locus ceruleus (L); (3) parabrachial nuclei (P); (4) superior central nucleus of Bechterew (C); and (5) supratrochlear nuclei (S) (dorsal raphe nucleus). Brain slice on left is at caudalmost level of the midbrain, being just rostral to the slice on right at the beginning of the pons. Drawings made from tracings of sections of squirrel monkey's brainstem. Other abbreviations: *B*, brachium conjunctivum; *FL*, medial longitudinal fasciculus; *T*, trochlear (4th) nerve. MacLean (unpublished).

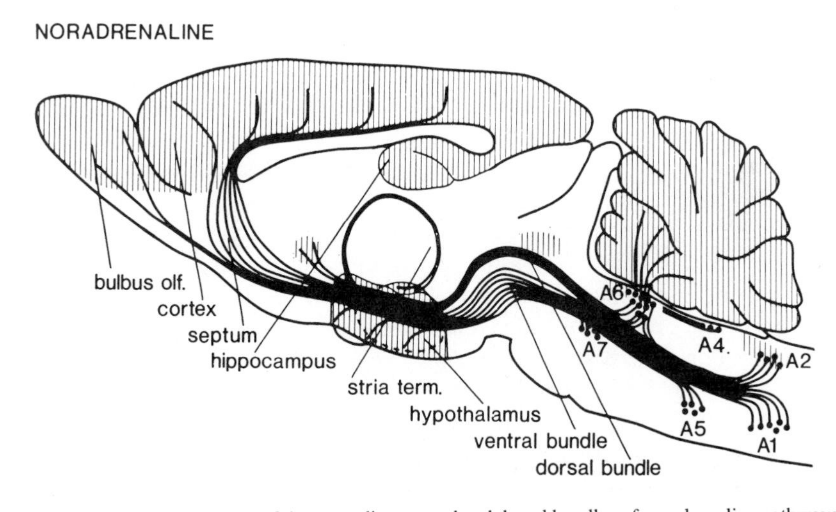

Figure 18-4. Ungerstedt's diagram of the ascending ventral and dorsal bundles of noradrenaline pathways to the cerebrum that he identified in the rat brain. The labeling of noradrenaline cell groups conforms to the original scheme of Dahlström and Fuxe. From Ungerstedt (1971).

The Parabrachial Group

The next isthmus group with norepinephrine-containing cells comprises the parabrachial nuclei. It is the lateral parabrachial nucleus that is of interest here. It forms a cuff of densely packed cells around the lateral aspect of the superior brachium of the cerebellum (Figure 18-5). It has fewer norepinephrine-containing cells than the locus ceruleus.[57] Herrick's identification of a visceral-gustatory nucleus in the isthmus region (see above) served as an impetus for Norgren and Leonard to explore the dorsal pontine region of the rat for units responding to stimulation of the gustatory component of the 7th nerve (chorda tympani).[58] They identified cells in the intermediate part of the cuff that responded to gustatory stimulation and found that a lesion in this area resulted in degeneration leading to the medial apex of the ventromedial posterior nucleus of the thalamus and to the ventral forebrain. Subsequently, Norgren found that injection of tritiated proline into the parabrachial nucleus resulted in the labeling of axons ascending to these same areas and that in addition to the "thalamic taste area" there was dense labeling of terminals in the central nucleus of the amygdala and bed nucleus of the stria terminalis.[59]

In the rat,[60] cat,[61] and monkey,[62] the use of autoradiography (and also the HRP method in the rat) has shown that the vagal component of the solitary nucleus projects heavily to the lateral parabrachial region, which in turn projects to the same thalamic and amygdala structures identified earlier by Norgren.[59] These findings are illustrated by our own material on the squirrel monkey. Figure 18-6B shows the autoradiographic labeling that occurs in the lateral parabrachial nucleus subsequent to the injection of tritiated leucine in the posterior part of the solitary nucleus (Figure 18-6A), while Figure 18-6C reveals the retrograde labeling of cells in this same nucleus following injection of HRP in the amygdaloid region (unpublished observations).

Ungerstedt has shown that axons of norepinephrine cells in the parabrachial region of the rat join the so-called ventral bundle (Figure 18-4) and then follow the medial forebrain bundle.[55] The terminals are found throughout the hypothalamus and, more rostrally, the subcommissural part of the nucleus of the stria terminalis.[55]

Raphe Group

The word *raphe* pertains to a seam that, in the present sense, refers to the midline "seam" between the two sides of the midbrain. In the isthmus region, the ventral raphe nucleus is also called the superior nucleus of Bechterew. The part of this nucleus below the decussation of the brachium conjunctivum has a ball-shaped appearance. Many cells in this nucleus contain serotonin, as evidenced by a transient yellow histofluorescence in Falck–Hillarp preparations.[63] Since the dorsal raphe nucleus appears somewhat like a fountain spraying over the trochlear (pulley) nucleus of the 4th cranial nerve (see Figure 18-3) it is sometimes referred to as the fountain nucleus. It is also called the supratrochlear nucleus. It has numerous serotonin-containing cells. Dahlström and Fuxe designated the serotonin cells of the dorsal raphe as the B7 group, and those of the ventral central group as the B8 group.[53] Using a combination of neurochemical and neuroanatomical techniques, Moore and Halaris showed that the superior nucleus, in particular, projects throughout the hippocampus.[64] Other workers found that the dorsal raphe nucleus projects to the amygdala region.[65]

Using the HRP method, Pasquier and Reinoso-Suarez have provided evidence that some of the monoaminergic nuclei of the isthmus project to different segments of the

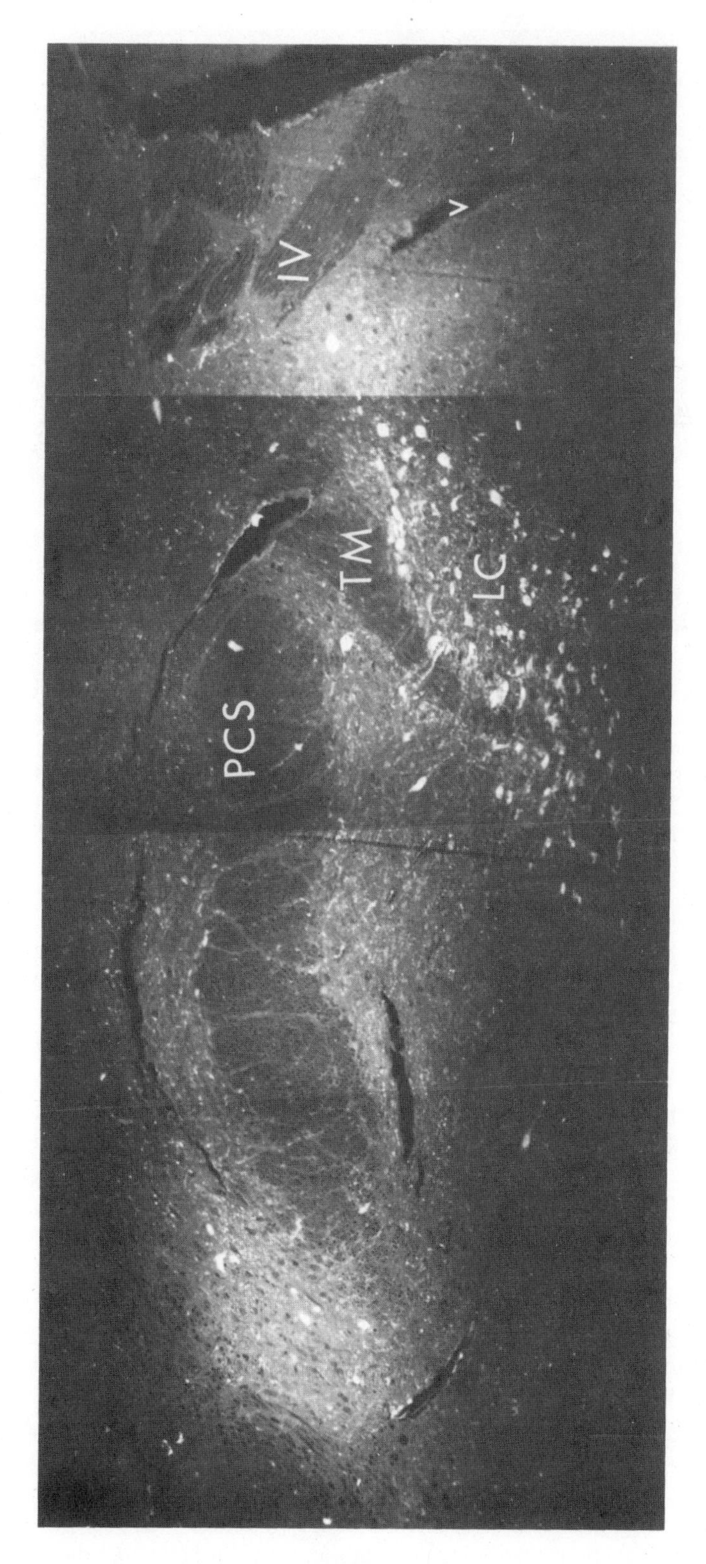

Figure 18-5. Montage showing catecholamine-containing cells in the locus ceruleus and parabrachial region of a pygmy marmoset. The neurons within and neighboring the locus ceruleus correspond to the cell group that Dahlström and Fuxe labeled A6 in the rat, while those of the parabrachial regions (more numerous in the monkey than in the rat) correspond to group A7. Note that some cells are located between the fascicles of the brachium. Note also catecholamine-containing varicosities. Labels: IV, trochlear nerve; LC, locus ceruleus; PCS, superior cerebellar peduncle; TM, ascending root of the 5th nerve. From Jacobowitz and MacLean (1978).

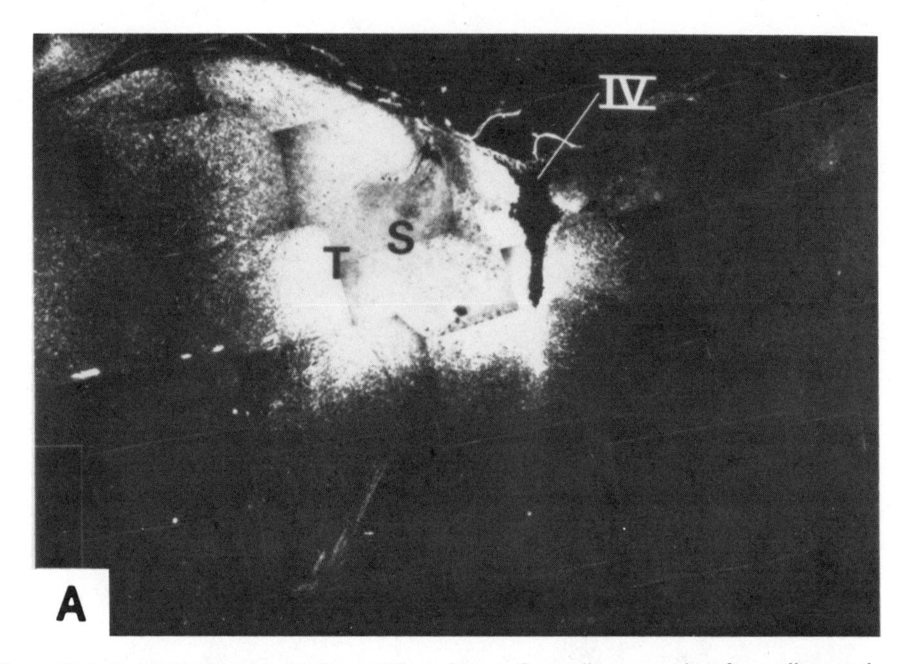

Figure 18-6. Darkfield microhistographs providing evidence of ascending connections from solitary nucleus to lateral parabrachial nucleus and from that site to the amygdala of the squirrel monkey. Autoradiograms in A and B, respectively, show site of injection of 0.1 μl of [^{3}H]leucine in solitary nucleus (S) and evidence of transport to terminals forming a cuff around the lateral border of the brachium (BC) where lateral parabrachial nucleus (P) is located. The HRP preparation in C shows retrograde labeling of numerous cells in the same parabrachial area following an injection of HRP in the amygdalar region. Other abbreviations: IV, fourth ventricle; L, locus ceruleus; T, tractus solitarius. From unpublished material of C. D. Conrad, A. Orr, and P. D. MacLean.

hippocampus[66] that, for descriptive purposes, I refer to here and later as the *proximoseptal* and *proximoamygdalar* segments. The caudal part of the dorsal raphe nuclei appears to project predominantly to the proximoseptal segment, while the rostral part distributes to the proximoamygdalar segment. The superior central nucleus, on the contrary, appears to terminate about equally in both segments. The inferred differential distribution of the dorsal raphe nucleus to the hippocampus is easy to remember if it is recalled that the rostral part toward the head of the animal projects to the proximoamygdalar segment, which is involved in oral functions, while the caudal raphe cells innervate the proximoseptal segment, which is identified with sexual and procreative functions (see Chapter 19). As will be explained, there are also segmental differences with respect to other hippocampal inputs, as well as outputs (see below, and electrophysiological findings in Chapter 26).

In their ascending course, fibers from both groups of the isthmic raphe nuclei appear to join the medial forebrain bundle (see Figure 18-2A). Various experiments suggest that the serotonin-containing cells of the raphe nuclei may be involved in procreational functions,[67] as well as thermoregulation[68] and sleep.[69]

Recapitulation

Briefly summarizing, the isthmus appears to be an important region for the integration of somatovisceral functions that in lower vertebrates is primarily under the influence

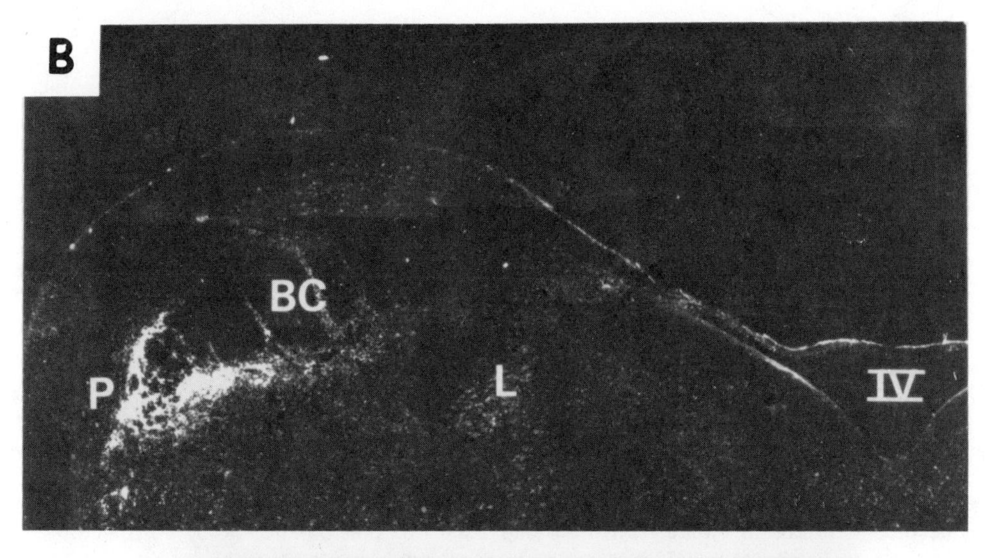

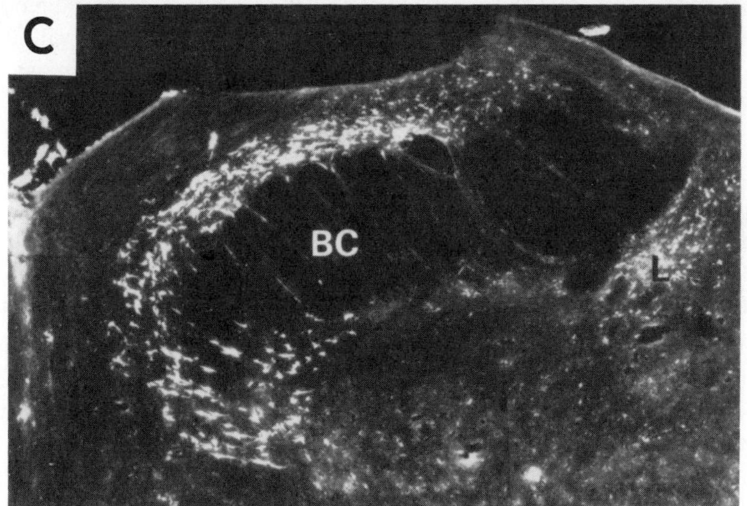

Figure 18-6. (*Continued*)

of the midbrain and that in the evolution of terrestrial vertebrates becomes extensively connected with the forebrain. On the basis of comparative studies, Parent has concluded that the "overall organization" of the isthmic and other monoaminergic systems "appears to be similar in the brains of reptiles, birds, and mammals."[70]

Hypothalamo-Mesencephalic Junction and Hypothalamus

The next cell groups believed to be involved in interoceptive transmission are located at the junction of the ventral midbrain and hypothalamus.

Junctional Cell Groups

As already noted in Chapter 4 dealing with the R-complex, there are dopamine-containing cells in the ventral tegmental area of the midbrain and the region bridging the interpeduncular fossa that innervate the olfactostriatum. Their axons run in the medial forebrain bundle. Other nerve cells in this same region can be grouped with those in the posterolateral hypothalamus and supramammillary region.[71] Their axons ascend in the medial forebrain bundle. Some terminate in the amygdala[72] and septum,[73] while others pass directly via the fornix to the hippocampus.[74] It has also been reported that cells in the peripeduncular[75] region and in the substantia nigra project to the amygdala.[76]

Hypothalamus

It was mentioned in Chapter 14 that groups of hypothalamic cells along the course of the medial forebrain bundle constitute bed nuclei of that bundle. Some cells of these nuclei innervate the septum[77] and amygdala,[78] as well as the hippocampus directly.[79] The dorsomedial nucleus of the hypothalamus is also a source of connections with the amygdala and the hippocampus.[80] The ventromedial hypothalamic nucleus projects to the medial part of the amygdala.[81]

Mammillothalamic Connections

Finally, for the present summary, great emphasis must be placed on the projections of the medial mammillary nucleus to the anterior thalamic group of nuclei. The mammillothalamic tract branches shortly after leaving the mammillary body. The main branch innervates the anterior nuclei of the thalamus. After producing lesions in different parts of the anterior nuclei, Powell and Cowan found retrograde degeneration in different parts of the medial mammillary nucleus.[82] Their conclusions about the orderly manner of mammillary projections are most simply explained by reference to Figure 18-7. Pars medialis and pars lateralis, respectively, of the medial mammillary nucleus project to the anterior medial and anterior dorsal nuclei, whereas pars posterior innervates the anterior ventral nucleus. The smaller branch of the mammillothalamic tract curves backwards and runs underneath the medial longitudinal fasciculus to reach the tegmental nuclei of Gudden.[83] Since these nuclei project via the mammillary peduncle to the mammillary bodies (see above), there exists a feedback loop between the tegmental isthmus and the hypothalamus.

Inputs from the Thalamic Nuclei

Until 1954, probably few anatomists would have suspected that the thalamus projected to any part of the limbic lobe except the cingulate convolution. Since then it has become evident that, with the possible exception of the piriform cortex, all parts of the lobe receive thalamic connections.

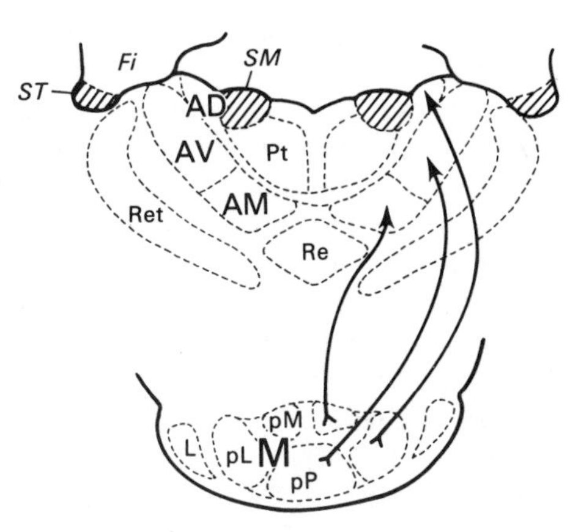

Figure 18-7. Projections from medial mammillary nucleus (M) to anterior thalamic nuclei in rat. Medial, posterior, and lateral parts (pM, pP, pL) project, respectively, to the anterior medial, ventral, and dorsal nuclei (AM, AV, AD). Guillery found connections from the mammillary peduncle to the medial and lateral parts of M, but not the posterior part; whereas the fornix innervates the lateral and posterior parts, but not the medial. Such findings indicate a convergence of ascending and descending systems on the lateral part projecting to the anterior dorsal nucleus. Note absence of projections from lateral mammillary nucleus (L). Pt, Re, and Ret identify the thalamic nuclei (parataenial, reuniens, reticular) surrounding the anterior group. *SM,* stria medullaris. Redrawn after figure from Powell and Cowan (1954).

Intralaminar and Midline Thalamic Nuclei

The so-called intralaminar and medial thalamic nuclei rank among the phylogenetically most ancient cell groups of the dorsal thalamus. In the mammalian brain, the intralaminar cell groups and connecting fiber systems assume somewhat the shape of a bird's nest with a partition down the middle. With the evolution of the limbic cortex, its associated thalamic nuclei develop as swellings within the matrix of the nest or along its lining. On the contrary, most of the thalamic nuclei connected with the neocortex evolve as progressively larger swellings external to the nest. The parvocellular part of the medial dorsal nucleus projecting to the prefrontal cortex is an exception: it takes on the appearance of two large eggs inside the nest and on either side of the partition. For the present account the term *nidal* will be used as an abbreviated expression to refer to the intralaminar and midline thalamic nuclei.

Nido-Hippocampal Connections

In Chapters 4, 13, and 14 we referred to intralaminar connections with the R-complex. In regard to limbic structures, Nauta and Whitlock in 1954 were the first to report connections with the hippocampal formation.[84] These connections arise in the nucleus reuniens, which is so named because it is located where the diencephalon becomes reunited above the third ventricle. (In terms of the nest analogy, it lies at the bottom of the nest.) Their findings were confirmed and extended by Herkenham, who as one of Nauta's students, injected tritiated leucine into reuniens and traced an extensive system of connections that run forward to join the cingulum and then follow it caudally to reach the entorhinal area in the parahippocampal gyrus, as well as all parts of the hippocampus.[85] In both subprimates and primates, the same nucleus has been found to project to the amygdala.[86] Other midline nuclei (nucleus paraventricularis *thalami* [sic] and parts of centralis) also innervate the amygdala.[87]

Nido-Insular Connections

The cortex of the anterior insula is by definition limbic because it "borders upon" the brainstem (see Figure 17-7). It consists of the cortex overlying the claustrum that

Brockhaus referred to as claustrocortex.[88] It is to be expected that the newer anatomical techniques will resolve conflicting views regarding thalamic projections to this part of the limbic cortex. Based on retrograde degeneration studies in monkeys, Bagshaw and Pribram inferred that the thalamic connections of the insula stem from nucleus reuniens (alias nucleus medialis ventralis),[89] whereas Roberts and Akert concluded that their origin is in the ventral posterior inferior nucleus and the small-celled part of the ventral posterior medial nucleus.[90]

Nido-Orbital Connections

On the basis of retrograde degeneration studies in the monkey, Pribram *et al.* concluded that the medial magnocellular part of the medial dorsal nucleus projects to the limbic part of the oribitofrontal cortex (see Figure 17-7).[91] Nauta confirmed their findings with the use of his silver method.[92] Figure 18-8 shows the degeneration that he traced forward from a midline lesion in the medial dorsal nucleus. The degeneration follows the inferior thalamic peduncle into the anterior commissure from which it appears to drop like rain into the orbitofrontal cortex, including area 14 of the posterior part of gyrus rectus and area 13 of the posterior orbital area.

Nido-Cingulate Connections

Within the internal medullary lamina at the rostral brim of the nest, there are three nuclear swellings that are primarily associated with the cingulate cortex. They are classified as the anterior group of thalamic nuclei and on the basis of their location in rodents are referred to as the anterior medial, anterior ventral, and anterior dorsal nuclei (see Figure 18-7). The association of the anterior group, as well as certain midline thalamic nuclei, with the mesocortical areas of cingulate gyrus is the primary reason for regarding this constellation of structures as a special subdivision of the limbic system. In regard to the evolution of mammalian behavior, it bears repeated emphasis that there appears to be no clear counterpart of this subdivision in the reptilian brain (Chapter 17).

Differences in terminology for the areas of cingulate cortex in primates and nonprimates will be dealt with in greater detail in Chapter 21. Rose and Woolsey refer to most of the rostral cingulate cortex of nonprimates as the anterior limbic area, while they subdivide the posterior cingulate cortex into cingular and retrosplenial areas (Figure 18-9; see also Figure 21-1).[93]

In 1950, Le Gros Clark and Meyer summed up as follows what was then known about the connections of the anterior nuclei with the cingulate cortex in the rat and cat: ". . . [T]he anteromedial element of the anterior nucleus is connected with the anterior limbic region (Area 24), the anteroventral element sends its fibers to the posterior limbic region, at least mainly to Area 23, while the smaller anterodorsal element projects still further back to the retrosplenial region of the cortex [Area 29]."[94]

Leonard has provided evidence that the medial frontal cortex in the rat, including the anterior limbic cortex, is co-innervated by the medial dorsal nucleus (MD) and the anterior medial nucleus (AM).[95] Her findings have been confirmed by other workers.[96]

In carnivores the most detailed information derives from HRP and autoradiographic studies on the cat.[97] In summarizing these connections it should be pointed out that the posterior part of Rose and Woolsey's retrosplenial area in the cat and other carnivores includes Brodmann's area 30. This dorsal "dysgranular" retrosplenial area (RSd in

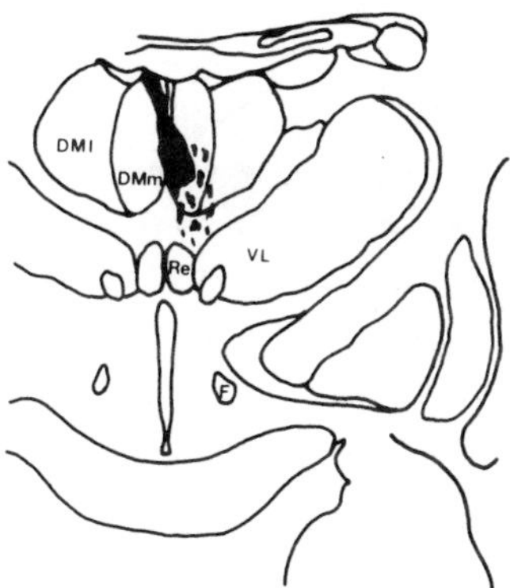

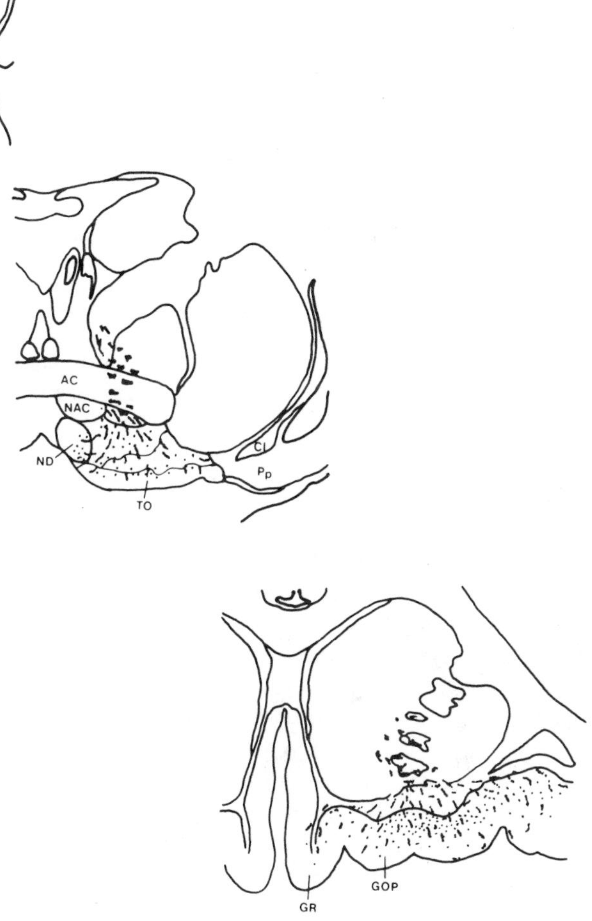

Figure 18-8. Drawings from a study by Nauta on the macaque, illustrating the degeneration traced from the medial dorsal nucleus to the gyrus rectus and posterior orbital gyrus (cf. Figure 17-6). From the site of the lesion (solid black) the degeneration follows the inferior thalamic peduncle to the level of the septum where it fans out in a forward direction and gives the appearance of dripping through the anterior commissure onto the cortex of the posterior gyrus rectus and posterior orbital area. Abbreviations: AC, commissura anterior; Cl, claustrum; DMl, nucleus medialis dorsalis thalami, pars lateralis; DMm, nucleus medialis dorsalis thalami, pars medialis; F, fornix; GR, gyrus rectus; GOP, gyrus orbitalis posterior; NAC, nucleus commissurae anterioris; ND, nucleus diagonalis Brocae; Pp, cortex prepiriformis; Re, nucleus reuniens; TO, tuberculum olfactorium; VL, nucleus ventralis lateralis thalami. Redrawn after Nauta (1962).

Figure 18-9) forms a quarter-moon-shaped border around the ventral retrosplenial granular area (RSv, corresponding to area 29). The anterior dorsal nucleus projects almost entirely to the ventral granular area.[97] The anterior medial nucleus has heavier projections to this area than to the anterior limbic area (LA), whereas the anterior ventral nucleus has a greater projection to the dorsal "dysgranular" area (area 30) than to the cingular area (C, corresponding to area 23). The lateral dorsal nucleus lying just caudolateral to the anterior group and peripheral to the internal medullary lamina, is a neothalamic nucleus and projects more extensively to the posterior cingulate than to the anterior cingulate cortex.[97] The limbic cortex of the so-called infralimbic area appears to receive fibers from nucleus reuniens.[93]

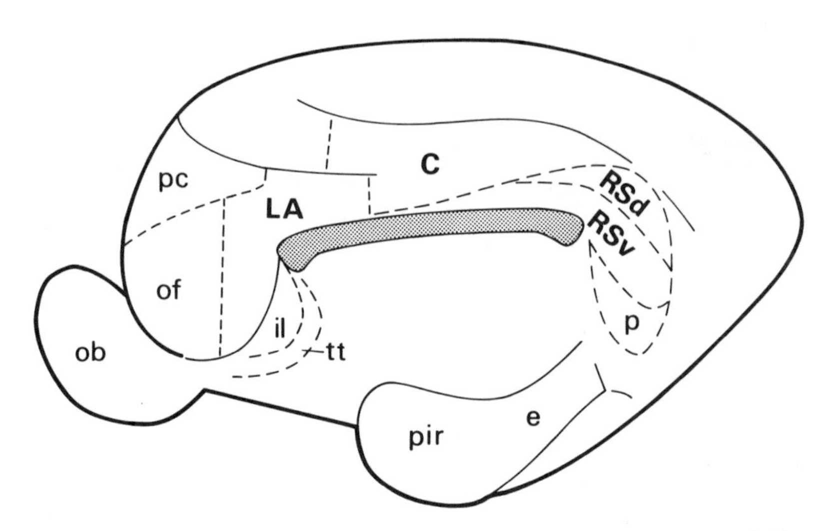

Figure 18-9. Cytoarchitectural areal subdivision of cingulate cortex in cat. The anterior limbic area (LA) corresponds to area 24 in the primate brain, while the ''cingular'' area (C) is comparable to area 23. The retrosplenial area of Rose and Woolsey is subdivided here into a dorsal dysgranular (RSd) and ventral granular (RSv) part according to representations by Robertson and Kaitz (1981). Other abbreviations: e, entorhinal area; il, infralimbic; ob, olfactory bulb; of, orbital frontal cortex; p, presubicululm; pc, precentral area; pir, piriform cortex; tt, taenia tecta. Redrawn after Rose and Woolsey (1948).

As noted in Chapter 17, it appears to be the rule that secondary olfactory projections end in the outer half of layer I, while tertiary afferents terminate in the lower half of layer I. A parallel situation would seem to apply to afferents innervating the cingulate meso-cortex. According to Oskar Vogt's scheme of 1910, the molecular layer may be subdivided into layers 1a, b, and c.[98] B. Vogt *et al.* state that layer 1a is the primary site of termination of fibers from the anterior thalamus.[99] In the anterior limbic area (areas 32 and 24; see Figure 21-1) the authors found that lesions of the medial dorsal nucleus (complicated by damage also to the anterior medial nucleus) terminated in layers 1a and 1b, whereas callosal fibers were distributed to layers 1b and 1c. In the cat, Robertson and Kaitz observed that the projections from the anterior thalamic group terminate in the outer half of layer I and in the superficial layers II and III.[97] The lateral dorsal nucleus, in keeping with its neothalamic origins, projects to granular layer IV; terminal labeling also appears in layer I and part of layer III contiguous with layer IV.[97]

In the monkey the level of the central sulcus marks a cytoarchitectural dividing line between the rostral and caudal parts of the cingulate gyrus, that is comparable to that for the neocortex. The gross morphological subdivisions of the primate cingulate gyrus, as well as the numbers of the cytoarchitectural areas, were described and illustrated in Chapter 17 (see Figure 17-7). Pribram and Fulton found that ablations limited to the anterior cingulate cortex (sparing the white matter) resulted in degeneration in adjoining parts of the anterior ventral and anterior medial nuclei.[100] Invasion of the underlying white matter, or extension of the ablation into the posterior cingulate gyrus, resulted in a greater amount of degeneration in the anterior ventral nucleus. Ablation of the pregenual cingulate cortex was associated with an increase in degeneration in the anterior medial nucleus, while the added excision of the subcallosal cortex resulted in degeneration in the most ventral part of this same nucleus and in nucleus parataenialis.

The use of autoradiographic and HRP methods has contributed further information about the thalamocingulate connections in the macaque monkey. In addition to projections from the anterior medial nucleus, the anterior cingulate cortex receives connections from other nidal nuclei, including the central, superolateral central, and central lateral.[101] Other sources are the ventral anterior and medial dorsal nuclei. Both the anterior medial and anterior ventral nuclei project to the posterior cingulate cortex.[101] Vogt *et al.* conclude that "all cingulate cortex is a common projection zone" for nucleus anterior medialis.[102] They do not include area 32 (see Figure 17-7) as part of the cingulate cortex because it does not receive anterior thalamic afferents. Further details reported in current studies will be described in connection with investigations on the functions of the midline frontal limbic and neocortex (Chapter 28).

Neothalamic Connections

Just lateral to the nidal brim and easily mistaken as a caudal extension of the anterior ventral nucleus is the lateral dorsal nucleus. Being extranidal, it is to be regarded as a neothalamic nucleus. As exemplified by the canid[103] and human brains,[104] this nucleus may achieve unusually large dimensions in some species. Yakovlev *et al.* appear to have been the first to demonstrate that the lateral dorsal nucleus projects to the cingulate gyrus.[105] Their findings have been confirmed by studies employing autoradiography and HRP.[106] The lateral dorsal nucleus projects to the greater part of the cingulate gyrus, as well as to the medial parietal cortex above it.

Termination in Cingulate Layers

In the preceding chapter mention was made of the difference in which main afferent systems terminate in limbic and neocortical areas. The main afferents to the limbic cortex terminate in the superficial layers (supragranular layers), whereas those to the neocortex end predominantly in the granular layer (layer IV) (see Figure 17-4). Recent observations on the laminar distribution of afferent projections to the cingulate cortex appear to reflect its transitional, "hybrid" structure. The afferents from the paleothalamic anterior group terminate largely in the supragranular layers, while the neothalamic lateral dorsal nucleus projects mainly to the granular layer IV.[107]

Inputs from Amygdala and Septum

With the embryonic subdivision of the prosencephalon into the telencephalon and diencephalon (see Figure 18-1), two important nuclear groups of the limbic system develop on the telencephalic side of the junction. One group constitutes the amygdala (almond), the other the septal (partition) nuclei. Each has a thalamuslike relationship with respective parts of the limbic cortex. In Figure 18-2 they are shown, respectively, as being the hub of two major subdivisions of the limbic system. As diagrammed, they owe their pivotal position to their close relationship with the olfactory apparatus, serving as centers of convergence for olfactory impulses from above with those conveyed by other systems from below.

In the brain of a fish or reptile, the amygdala and septum lie in close proximity. In the mammalian brain the two structures are drawn apart *pari passu* with the expansion of the temporal lobe. In the process one of the strong connecting pathways between the amygdala and septum—namely, the stria terminalis—is stretched out along a groove between the thalamus and caudate nucleus. Because of its arc-shaped path, the stria might be viewed somewhat like a bridge with its supports embedded on opposite shores of a river, forming a span from the amygdaloid nuclei on one side to the septum on the other side. At the junction of the telencephalon and diencephalon, the septum blends into the dorsal thalamus above, and into the preoptic region and anterior hypothalamus below. Once regarded as part of the telencephalon, the preoptic area is now generally described as part of the hypothalamic division of the diencephalon.[108]

Amygdalar Connections

Anyone who lived through the anatomically lean years of a few decades ago almost has a feeling of disbelief that direct connections have been demonstrated between such distant areas as the isthmus region and the telencephalon. As we have seen, cell groups in the isthmus project to the amygdala and septum, which, in turn, innervate parts of the limbic cortex.

On the basis of comparative studies, the amygdala can be subdivided into a medial and lateral group of nuclei.[109] The medial group, which is regarded as phylogenetically older than the lateral group, consists of the cortical, medial, and central nuclei, as well as an ill-defined anterior area. The lateral group comprises the accessory basal, basal, and lateral nuclei.

As early as 1923, J. B. Johnston had indicated the presence of connections between the amygdala and hippocampus.[110] His observations were based on the examination of normal material. In a 1965 monograph on the piriform lobe, Valverde cited the experimental findings by himself and others that failed to demonstrate such connections.[111] The application of autoradiographic and HRP methods has provided evidence in the rat and cat of projections from the basolateral amygdala to the ventral subicular areas of the hippocampal formation[112] (i.e., the proximoamygdalar segment). The lateral nucleus projects to the rostroventral part of the entorhinal area.[112] In addition, connections have been found with the cortex of the frontotemporal region associated with the uncinate fasciculus, including the posterior orbital,[113] anterior insular,[113] piriform,[114] and temporal polar[113] cortex (see below). There are also projections to the subcallosal cingulate cortex.[113] The combined application of autoradiographic and HRP methods in rhesus monkeys has indicated projections from the basolateral, basomedial, and basal accessory nuclei to the orbital prefrontal cortex, gyrus rectus, and anterior cingulate gyrus.[115]

Septal Connections

Although the septum can be variously subdivided into a number of cellular groups,[116] it will be sufficient both here and later on to focus principally on the medial and lateral nuclei, as well as the nucleus of the diagonal band and the different parts of the bed nucleus of the stria terminalis.

Daitz and Powell were apparently the first to provide convincing evidence of septal connections with the hippocampus. In 1954 they reported that section of the fornix in the rat, rabbit, and monkey resulted in retrograde degeneration in the medial nucleus of the septum.[117] Their findings have since been repeatedly confirmed by other methods.[118] It appears that numerous septohippocampal fibers are cholinergic in nature.[119] Here again we find evidence of difference in innervation of respective segments of the hippocampus. Using the method of retrograde transport of HRP, Segal and Landis observed that injections of HRP into the dorsal hippocampus (proximoseptal segment) of the rat resulted in labeling of the more medial cells of the medial septal nucleus, while injections into the ventral hippocampus (proximoamygdalar segment) were associated with labeling of the more lateral cells of the same nucleus.[120] Using cytochemical tracing techniques, Alonso and Kohler have demonstrated reciprocal connections between the septum and entorhinal cortex.[121]

In addition to the hippocampal formation, the nucleus of the diagonal band projects to the anterior limbic area.[122] The bed nucleus of the stria terminalis projects to the "amygdala-hippocampal area," as well as to the medial and central nuclei of the amygdala.[122]

Summarizing Comment

Up until now we have been concerned with inputs to the limbic cortex that appear to derive in part, at least, from interoceptive (visceral) systems. It has been noted that the solitary nucleus, which is innervated by the vagal and glossopharyngeal nerves, as well as by a gustatory component of the facial nerve, projects to nuclei of the isthmic region, which, in turn, project to nuclei at more rostral levels that connect with the limbic cortex. Some of the monoaminergic neurons in the isthmus region project directly to respective parts of the limbic cortex. At the highest nuclear level there appear to be three primary funnels for neural traffic into the limbic cortex—namely, one through the amygdala, one through the septum, and one primarily through the anterior and midline thalamic group. Specifically, (1) the amygdala innervates the cortex of the frontotemporal region, including the proximoamygdalar segment of the hippocampal formation that also derives afferents from the septum; (2) the septum, which supplies afferents to the hippocampal formation at all levels, is the main source of afferents to its proximoseptal segment. The anterior and certain other nidal thalamic nuclei provide the major input to the greater part of the cingulate cortex.

Exteroceptive Inputs

The exteroceptive systems are those that convey information primarily from the external environment. They are identified with the classically recognized five major "senses." The olfactory and gustatory senses are sometimes referred to as "mixed" senses because they are attuned to both the external and internal environment (in the latter case, the upper respiratory and alimentary passages). We have already considered gustatory inputs in connection with interoceptive inputs.

Olfactory Inputs

Although it was well known that the olfactory bulb projected to the limbic lobe, the anatomical techniques used during the first half of this century were not sufficiently refined to give clear-cut information about the sites of termination. Thanks to improved silver degeneration techniques, and especially to the introduction of autoradiographic and HRP methods, the projections of the olfactory apparatus have been worked out in remarkable detail during the past decade.

In reptilian and mammalian forms with a well-developed olfactory apparatus there are basically two independent olfactory systems. The main system conveys nerve impulses from the olfactory epithelium in the nasal passages, while the second system transmits impulses from a special structure called the vomeronasal organ encased as a paired structure in the forward part of the roof of the mouth [also known as Jacobson's organ after the Danish anatomist (1783–1843) who described it in 1809[123]] As diagrammed in Figure 18-10, each system projects to nonoverlapping areas in the limbic lobe. We will deal first with the main olfactory system.

Main Olfactory Input

It has long been known that, peripherally, the sense of smell depends on receptors in the olfactory epithelium that project to the main olfactory bulb. The area of olfactory epithelium in the nasal passages has a yellowish-brown coloration. Very fine unmedullated fibers arise from the receptors and collect somewhat like skeins of yarn before

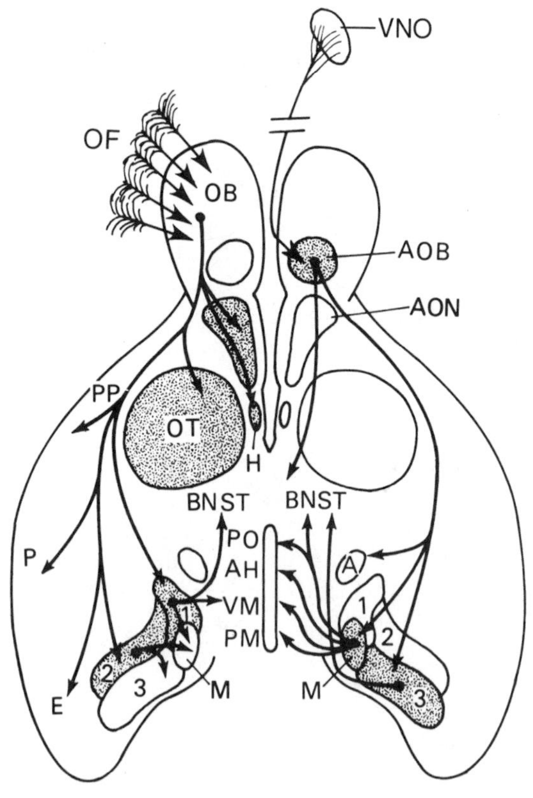

Figure 18-10. Diagram contrasting the projections of the main olfactory system (left-hand side) with those of the accessory olfactory system (right-hand side). The shaded areas serve to accent the different sites of termination for all structures except the prepiriform (PP), piriform (P), and entorhinal (E) cortex. Note particularly the differences in regard to the three subdivisions of the cortical nucleus of the amygdala. The main olfactory bulb projects to the anterior (1) and posterolateral (2) division, whereas the accessory olfactory bulb projects to the posteromedial part (3). Note also the difference in regard to the innervation of the medial nucleus (M) of the amygdala. It is suggested that this nucleus may serve to interrelate the two systems since it receives information from the olfactory bulb indirectly via the posterolateral cortical nucleus. Other abbreviations: A, nucleus of accessory olfactory tract; AH, anterior hypothalamic area; AOB, accessory olfactory bulb; AON, anterior olfactory nucleus; BNST, bed nucleus of stria terminalis; H, hippocampal rudiment; OB, main olfactory bulb; OF, olfactory fila; OT, olfactory tubercle; PM, premammillary nucleus, ventral part; PO, preoptic area; VM, ventromedial nucleus of hypothalamus; VNO, vomeronasal organ. Adapted from drawings by Kevetter and Winans (1981a, b) and Wysoki (1979).

passing through the small holes of the cribriform plate to reach the olfactory bulb. The fibers terminate in the glomeruli in the superficial part of the olfactory bulb, making contact with the dendrites of mitral cells, so called because of their resemblance to a bishop's miter. There are also contacts with the "tufted" cells.

The mitral and tufted cells represent second-order olfactory neurons whose myelinated axons project to the structures shown diagrammatically on the left-hand side of Figure 18-10—namely, (1) the anterior olfactory nucleus, (2) hippocampal rudiment, (3) olfactory tubercle, (4) prepiriform cortex, (5) piriform cortex, (6) rostral part of the entorhinal cortex, (7) anterior cortical, and (8) posterolateral cortical nuclei of the amygdala.[124]

The anterior olfactory nucleus, so named by Herrick,[125] has the appearance of rudimentary cortex wrapped around the peduncle of the olfactory bulb. As in the case of all the other olfactory structures just mentioned, it has a molecular layer in which the outer half is identified as layer IA and the inner half as layer IB.[126] It should be emphasized, as first recognized by White,[127] that second-order olfactory neurons characteristically terminate in layer IA. Third-order neurons, on the contrary, terminate in layer IB.[128] For example, third-order neurons of the olfactory nucleus innervate the same areas as the direct projections from the bulb, but with the important difference that the terminals are traced to the inner half of the molecular layer (IB).[129] The axons of the third-order neurons of this nucleus account for the large size of the anterior commissure in macrosmatic animals and provide the contralateral connections of the olfactory system, including those to the main olfactory bulb and prepiriform cortex.[129] This nucleus is also an important source of subcortical connections leading to the nucleus of the diagonal band and, via the medial forebrain bundle, to the lateral hypothalamus as far posteriorly as the nuclei gemini[129] of Lundberg.[130]

In the brains of some macrosmatic animals the hippocampal rudiment might be confused as a caudal continuation of the anterior olfactory nucleus. With improved techniques it has at last been possible to establish with certainty that the main olfactory bulb projects to the rudiment (layer IA) in several species (e.g., Figure 18-11).[131] In the monkey, however, there continues to be uncertainty as to whether the bulb does[132] or does not[133] innervate the rudiment. In any event, it has become clear that third-order neurons of the entorhinal area convey olfactory information to the hippocampus (see Chapter 26).

Vomeronasal System

A vomeronasal system is found in snakes and ground-dwelling lizards, as well as in most mammals except higher primates where it may be rudimentary or absent.[134] Despite this ubiquitous representation, the nature of its functions remained elusive until recent years. In a systematic series of experiments, Halpern and Kubie and their collaborators have shown that interference with vomeronasal function eliminates mating behavior of male snakes, as well as their ability to follow a scented trail.[135] It is probable that a similar outcome would be obtained in lizards such as the Komodo dragon, which, as mentioned in Chapter 8, has a well-developed vomeronasal system and which uses its giant forked tongue in mating and tracking. Winans and co-workers have convincingly demonstrated in hamsters that the vomeronasal organ is sensitive to odorous nonvolatile substances and that interference with its function eliminates the male's ability to discriminate vaginal secretions.[136]

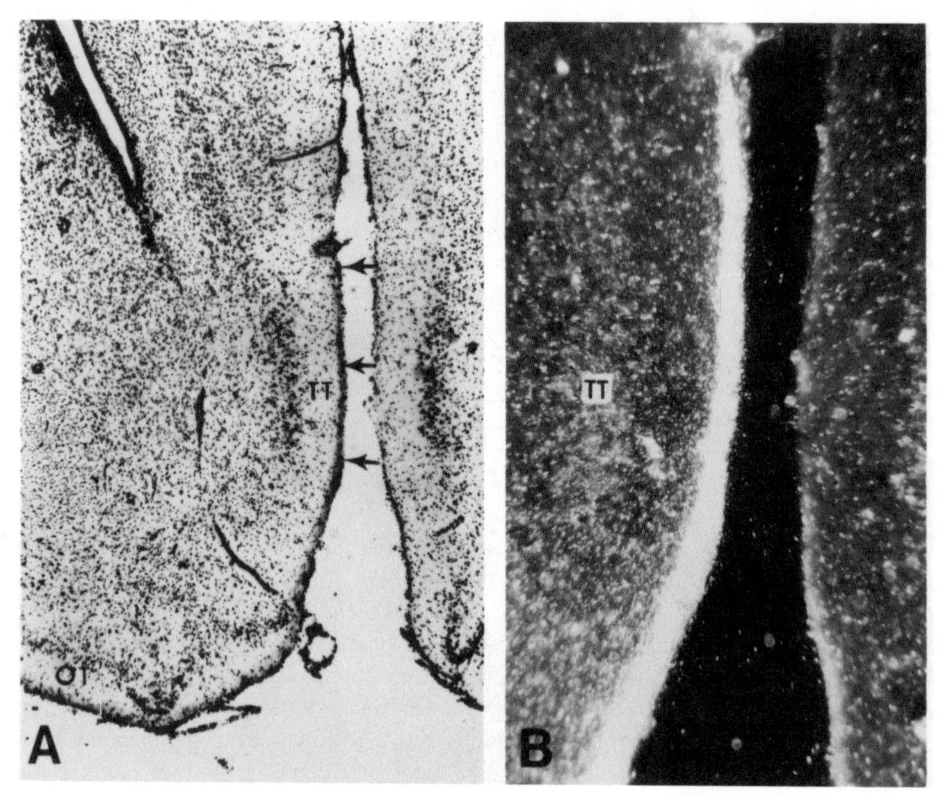

Figure 18-11. Autoradiographic picture of distribution of projections of olfactory bulb to outer half of molecular layer of hippocampal rudiment (taenia tecta, TT) in rabbit.

(A) Brightfield picture with arrows pointing to dark band of silver grains.

(B) Same as A as seen under darkfield. Elliot Smith (1895), examining Golgi material from the platypus brain, was the first to provide evidence of such connections. The debated issue, however, as to whether or not the olfactory bulb projected via a medial tract to the hippocampal rudiment of different species was not finally settled until the development of modern techniques. As noted in text, the termination of fibers in the outer half of the molecular layer is typical of all structures receiving direct connections from the main olfactory bulb.

From Broadwell (1975a).

On each side, the fibers from the receptors of the vomeronasal organ collect into a single bundle (Jacobson's nerve) that runs along the medial surface of the main olfactory bulb. In histological sections one can easily follow the nerve to the accessory bulb, which forms a cup in the dorsal, posterior aspect of the main olfactory bulb. The fibers terminate in glomeruli that are less structured than those of the main bulb.

The autoradiographic and HRP methods have been invaluable in distinguishing the origin and termination of the neural projections from the main and accessory bulbs. Broadwell has shown that the accessory olfactory tract runs as a separate bundle just underneath the lateral olfactory tract.[137] As illustrated in the right-hand diagram of Figure 18-10, he and others[138] have confirmed the earlier findings of Winans and Scalia[134] that the fibers terminate in particular parts of the amygdala—namely, layer IA of the posteromedial cortical nucleus and the medial nucleus—and that there is no overlap of this projection with that from the main olfactory bulb to the anterior and posterolateral cortical nuclei.

Just as the main and accessory bulbs terminate in different subdivisions of the amygdala, so do these nuclei have differential projections. In their autoradiographic study, Kevetter and Winans have found that the posteromedial cortical nucleus innervated by the accessory olfactory bulb projects to the postcommissural part of the bed nucleus of the stria terminalis (Figure 18-10), whereas the anterior cortical nucleus innervated by the main olfactory bulb projects to the precommissural component of the bed nucleus, as well as to the ventromedial nucleus of the hypothalamus.[139] The medial amygdaloid nucleus innervated by the accessory olfactory bulb projects to the medial part of the bed nucleus of the stria terminalis, as well as to the adjoining region of the medial preoptic and anterior hypothalamic areas; the ventromedial nucleus of the hypothalamus; and the ventral premammillary nucleus. The authors suggest that since the medial nucleus receives fibers from the posterolateral cortical nucleus innervated by the main olfactory bulb (Figure 18-10), it may serve as an interrelating nucleus for the two olfactory systems. Herkenham and Pert have demonstrated by histochemical methods that structures associated with the vomeronasal system are characterized by a high content of opiate receptors.[140]

Inputs from Other Exteroceptive Systems

As yet anatomical information about inputs from other sensory systems (somatic, auditory, visual) is fragmentary and can best be discussed in Chapter 26 in which electrophysiological findings help to flesh out the skeletal picture.

LIMBIC OUTPUTS

In proceeding to consider limbic outputs, we find additional evidence compatible with a parcellation of the limbic system into three main subdivisions—namely, the (1) amygdalar, (2) septal, and (3) thalamocingulate subdivisions. Two considerations stand out in particular: (1) the amygdalar and septal divisions are notably distinguishable from the thalamocingulate division because of their cascading projections to the hypothalamus, and (2) the hippocampal formation must be evaluated in regard to respective outputs from all three subdivisions. Moreover, there is an undetermined amount of overlap between the respective subdivisions. As discussed in the next chapter, such overlap of the amygdalar and septal divisions may help to account for overlapping of oral and sexual functions identified, respectively, with these subdivisions. Both here and in Chapter 26 we encounter instances of how electrophysiological findings help to distinguish these two subdivisions.

In the following account, we deal successively with (1) the amygdalar, (2) septal, and (3) thalamocingulate divisions.

Outputs from the Amygdalar Division

The cortical areas associated with the amygdala include the so-called frontotemporal cortex and the proximoamygdalar segment of the hippocampal formation. For the reason mentioned above, we will deal later under a separate heading with the question of hippocampal projections, confining attention here to outputs from the frontotemporal cortex and from the amygdala itself.

Preliminary comment and caveat. In order to avoid repetition in the exposition, it should be stated that there is only patchy information about the anatomical connections of most of the cortical areas of the frontotemporal region. It should also be stated as a caveat that the cortical proximity to underlying structures (for example, the piriform cortex and underlying amygdala) offers special difficulties in attempts to make selective lesions of the cortex or, as in the case of the newer techniques, to inject substances without complicating diffusion of the materials to neighboring areas.

Frontotemporal Cortex

The frontotemporal cortex includes the posterior orbital, anterior insular, prepiriform, piriform, rostral entorhinal cortex, as well as the temporal polar cortex adjacent to the rhinal sulcus. Here is an instance in which electrophysiological findings have been helpful in defining subgroups of anatomical structures. As illustrated in Figure 18-12,

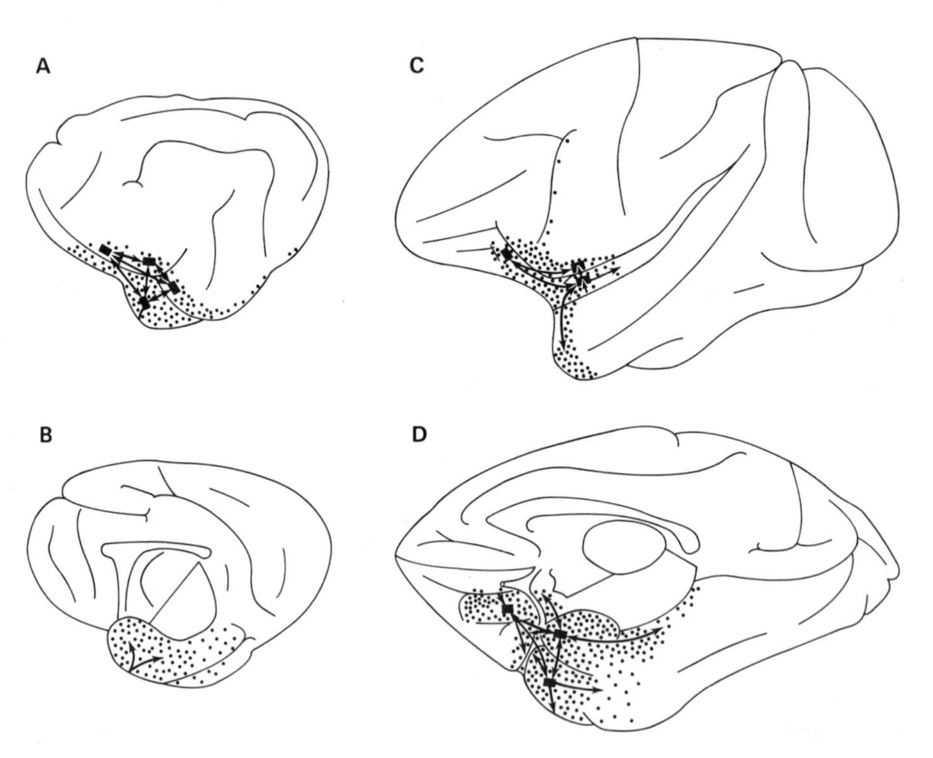

Figure 18-12. Interrelated cortex of frontotemporal region demonstrated in cat (A, B) and monkey (C, D) by strychnine neuronography. Drawings represent summary based on several experiments. Application of a small patty containing 1% strychnine at sites indicated by rectangles results in "firing" in the region indicated by stipple. Such neuronographic methods may be helpful in establishing boundaries of functionally integrated regions that would not be evident by neuroanatomical studies alone. The delimited region corresponds in the main to cortical areas fed by the amygdala and interrelated by the uncinate fasciculus. Adapted from studies by MacLean and Pribram (1953) and Pribram and MacLean (1953).

strychnine neuronography has revealed reciprocal connections among all the cortical areas just mentioned, a finding that serves to distinguish the frontotemporal sector from neighboring cortical regions.[141]

Posterior Orbital Cortex

The posterior orbital cortex in the macaque brain is bounded rostrally by a shallow sulcus. Not included in Brodmann's scheme,[142] it was identified by Walker in 1940 as *area 13*.[143] In mapping corticocortical connections by strychnine neuronography, it is remarkable that after application of strychnine to this area there is no propagation of potentials rostral to this sulcus (see Figure 18-12D).[144] In subprimate forms this area probably corresponds to the rostral part of the composite gyrus that is continuous with insular cortex.

Nauta[92] has confirmed the original findings of Wall *et al.*[145] that the posterior orbital cortex projects to the lateral hypothalamus. Of other brainstem connections he describes those with the corpus striatum; globus pallidus; lateral preoptic region; certain parts of the reticular, ventrolateral, paracentral, and medial dorsal thalamic nuclei; fields of Forel; red nucleus; and ventrolateral tegmentum of the midbrain.[92]

Anterior Insula

The cortex of the limen insulae (Figure 17-7, area 16) and anterior insula overlying the claustrum is by definition limbic because it comprises phylogenetically old cortex bordering the brainstem. Except for its cortical and amygdaloid connections little is known about the projections of the anterior insula. Mufson *et al.* have observed that it projects to almost all nuclei of the amygdala.[146] Its projections to cortical areas correspond to those of the frontotemporal cortex defined by strychnine neuronography (see above and Figure 18-12).

The *claustrum* underlying the anterior insula is an anomalous cellular structure. Among the possibilities as to its nature mentioned in Chapter 4, it was suggested that it may represent a mammalian counterpart of the dorsal ventricular ridge (DVR) of reptiles and birds. Several studies have indicated that the claustrum projects rather widely to the neocortex.[147] An HRP study has shown that cells in the ventral part are connected with the proximoamygdalar subiculum.[148] In *Saimiri* monkeys, injections of WGA-HRP into the subcallosal cortex (area 25) result in labeling of numerous cells of the claustrum lying along the lateral border of the lateral striate vessels.[149]

Piriform Cortex

Using silver methods, several authors have traced degeneration from the prepiriform cortex along the trajectory of the medial forebrain bundle.[150] In their experiments concerned with the projections of the piriform cortex itself, Powell *et al.* (1965) were satisfied that their superficial lesions did not invade the amygdala because there was no complicating degeneration of the stria terminalis that is seen when the lateral or basal amygdaloid nuclei are damaged.[151] In their silver preparations, they identified cortical connections with the subcallosal cingulate and lateral entorhinal areas. Subcortical connections were traced to the nucleus of the diagonal band, lateral preoptic area, and cellular elements along the entire course of the medial forebrain bundle, including nucleus gemini. They concluded that "the hypothalamus may be regarded as the principal subcortical projection site" of "the phylogenetically older cortical areas."[152] More emphatically, they state their agreement with Lundberg's findings[153] of "no direct projection to the hypothalamus from any area of neocortex."[152]

Heimer and Wilson arrived at quite different conclusions, claiming that "the major projections from the piriform cortex . . . terminate in extrahypothalamic rather than hypothalamic regions."[154] Their results indicated that the piriform cortex innervates predominantly the superficial and deep parts of the olfactory tubercle, whereas the hippocampus projects to the nucleus accumbens.

Anterior Entorhinal Cortex

This cortex will be considered along with the remainder of the entorhinal cortex discussed below.

Temporal Polar Cortex

The agranular temporal polar cortex (area 38) lying adjacent to the rhinal sulcus is regarded as transitional or mesocortex (alias proisocortex). In the brain of a monkey in which a shallow lesion was made in this area, Whitlock and Nauta traced degeneration to the putamen, pulvinar, zona incerta, fields of Forel, peripeduncular nucleus, and superior colliculus.[155]

Projections from Amygdala

The connections of the cortical and medial nuclei of the amygdala were described above in the section dealing with the main and accessory olfactory systems. In autoradiographic studies in the rat and cat, Krettek and Price have made the following observations regarding other amygdala connections[156]: (1) What they refer to as a special amygdalohippocampal area projects to the bed nucleus of the stria terminalis, the "shell" of the ventromedial nucleus of the hypothalamus, and the premammillary nuclei. (2) Both the basal lateral and basal medial nuclei project to the olfactostriatum and bed nucleus of the stria terminalis. The former also connects with the lateral hypothalamus, while fibers from the latter terminate in the ventromedial nucleus of the hypothalamus and the premammillary nuclei. (3) The central nucleus projects "via" the stria terminalis to the lateral hypothalamus and the lateral substantia nigra, as well as to the lateral parabrachial nucleus of the isthmus. (4) The lateral nucleus appears to have no connections with structures outside of the amygdala. (5) Finally, the endopiriform nucleus of Loo[157] projects to the ventral putamen and lateral hypothalamus. Comparable to one view of the relationship of the insula and claustrum, the endopiriform nucleus is sometimes regarded as representing deep layers of the prepiriform cortex.[113]

In regard to thalamic projections, Krettek and Price found evidence in the rat of amygdaloid projections to the medial dorsal nucleus, whereas in the cat, such connections originate in the endopiriform nucleus.[156] Using HRP, Mehler was unable to confirm reciprocal connections between the medial dorsal thalamic nucleus and the amygdala that Nauta had described in the monkey.[72]

Outputs from Septal Division

In addition to the cortex of what is referred to as the proximoseptal hippocampal segment (see also Chapter 19 and Figure 19-1), the cortical representation of the septal division will be tentatively regarded as including the indusium griseum overlying the corpus callosum, the taenia tecta, and adjacent part of the infralimbic area of Rose and Woolsey diagrammed in Figure 18-9. In the monkey Rosene and Van Hoesen reported that the subicular region of the segment referred to here as the proximoseptal segment, like other levels of the hippocampal subicular areas (Figure 18-13A), projected to the hilar limbic cortex shown in Figure 18-13B.[158]

Caudal Parahippocampal Gyrus

The parahippocampal gyrus represents an exteriorized fold of cortex that is continuous with the hippocampus (see Figure 17-8). In a rostrocaudal direction it includes the

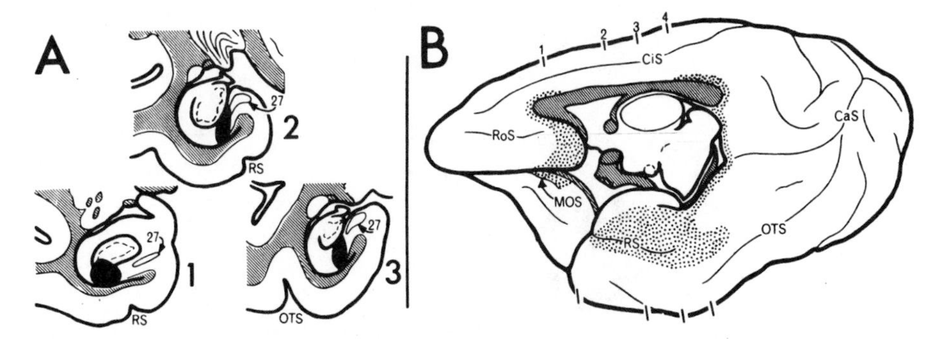

Figure 18-13. Illustration from study by Rosene and Van Hoesen (1977), showing limbic and perilimbic areas receiving projections from subicular areas (A). In A, numerals refer, respectively, to subicular fields in rostral, middle, and caudal levels of hippocampal formation. Cortical areas within the posterior gyrus rectus, subcallosal gyrus, suprastrial strip, the intermediate and posterior parts of the hippocampal gyrus, together with the proximoseptal segment would represent the cortical extent of the septal subdivision of the limbic system. See original paper for other illustrations identified by numerals. Abbreviations: CaS, calcarine sulcus; CiS, cingulate sulcus; MOS, medial orbital sulcus; OTS, occipital temporal sulcus; RS, rhinal sulcus; RoS, rostral sulcus; 27, area number for presubiculum.

piriform, entorhinal, and postrhinal parahippocampal cortex. As already noted, the most rostral part of the entorhinal cortex belongs to the amygdalar division because in addition to direct projections from the olfactory bulb, it receives afferents from the amygdala and other associated structures.

Caudal Entorhinal Cortex

Both the rostral and caudal entorhinal cortex are included in Brodmann's area 28 (see Figure 17-7). Because of its unique banded appearance in microscopic section (see Figure 18-15B), the entorhinal cortex is easily identified in most mammals. Brodmann gave it this name because it lies interior to the posterior stem of the rhinal sulcus.[159] Ramón y Cajal had previously referred to it as the spheno-occipital ganglion, specifying it as a cortical area between the spheno-olfactory and occipital cortex.[160] He also referred to it as the angular ganglion because of the angular shape of its borders in the brains of small macrosmatic animals (Figure 18-14). First Ramón y Cajal (1901)[160] and then Lorente de Nó (1934)[161] emphasized that the entorhinal cortex is a major source of afferents to the hippocampus. The areas of termination will be considered below when dealing with the hippocampus.

Postrhinal Parahippocampal Gyrus

As illustrated by the sagittal section of a monkey's brain in Figure 18-15 the caudal part of the entorhinal area terminates abruptly at the same level as the termination of the

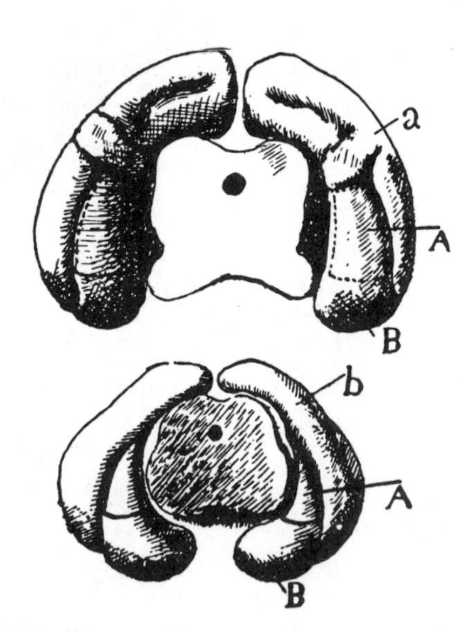

Figure 18-14. Illustration from Ramón y Cajal, showing the angular area of the hippocampal gyrus that he referred to as the angular ganglion, or otherwise as the spheno-occipital ganglion. The upper and lower drawings, respectively, show its appearance in the cat and rabbit. Lying medial to the caudal stem of the rhinal sulcus, it corresponds to what Brodmann identified as the entorhinal area (area 28). In the primate brain it occupies an intermediate position in the hippocampal gyrus; the cortex in the posterior, ''angular'' part of the gyrus represents an expansion of occipitotemporal limbic and neocortex in a region transitional with classically recognized visual areas of the neocortex (see Chapter 26). From Ramón y Cajal (1904; 1954, Fig. 100).

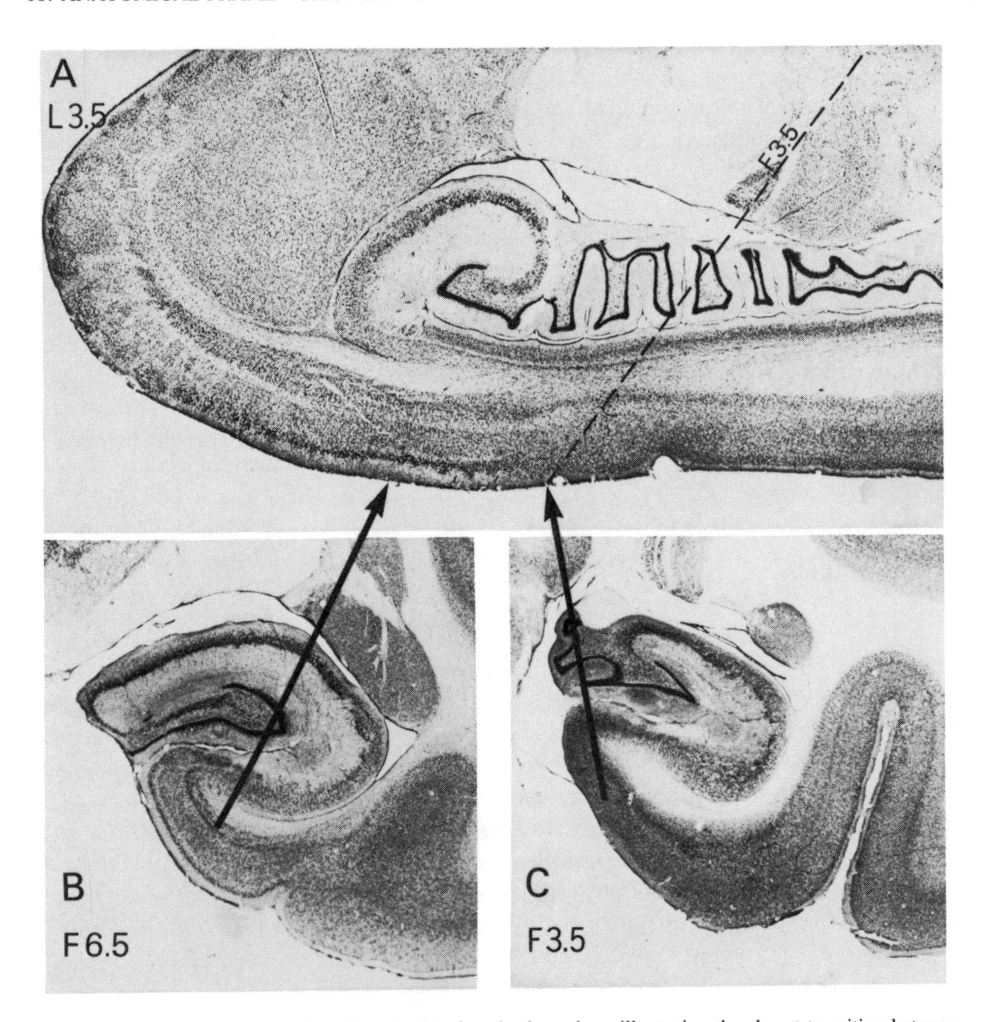

Figure 18-15. Nissl-stained sections from brain of squirrel monkey, illustrating the abrupt transition between the entorhinal cortex of the intermediate hippocampal gyrus and the cortex of the posterior hippocampal gyrus. (A) Sagittal section 6.5 mm from midline, with arrow pointing to the abrupt change in cellular appearance of the two areas. (B, C) Frontal sections at the same level as the arrow in A, showing the marked difference in appearance of the cellular layers. For comparison of the layers of the entorhinal area in the monkey with those in the mouse, see the drawing from Lorente de Nó in Figure 18-16B. From MacLean (unpublished).

posterior stem of the rhinal sulcus. It is replaced by the cortex of the postrhinal parahippocampal gyrus, which laterally becomes transitional with the temporal neocortex. The ill-defined area of transition falls within area 36 of Brodmann.[162] The caudal part of the hippocampal gyrus tapers into a microgyrus separated laterally from the parasubicular cortex of the lingual gyrus by a shallow groove formed by the rostral stem of the anterior calcarine sulcus. The groove marks the beginning of the lingual gyrus. In microscopic sections it appears as an invagination of molecular layer, while the deep white matter at this level assumes a bicuspid appearance (e.g., see Figure 18-18D). The cortex of the posterior hippocampal gyrus is also a source of afferents to the hippocampus. Further details will be given in Chapter 26 when discussing its role in visual functions.

Infralimbic Area

The cortex of the taenia tecta and adjacent infralimbic area corresponds in part to Elliot Smith's parolfactory area. On the basis of her observations on the rat, Domesick concluded that the limbic cortex including this area is the only part of the cingulate gyrus that projects to the hypothalamus, as well as to the substantia innominata and ventral tegmental area.[163] Other connections include those to the subthalamus, zona incerta, pretectal nuclei, superior colliculus, central gray, and pontine nuclei.[164]

Outputs from Archicortex

The projections from the archicortex will be considered below under the heading of "hippocampal outputs."

Septal Outputs

The medial and lateral septal nuclei, together with the nucleus of the diagonal band and the bed nucleus of the stria terminalis, account for most of the brainstem projections from the septum.[122] According to an audioradiographic study in the rat, virtually all of these nuclei project via the medial forebrain bundle to the medial and lateral preoptic areas, anterior hypothalamus, lateral hypothalamus, mammillary bodies, and paramammillary nuclei, as well as to the ventral tegmental area.[122] Both the medial septal nucleus and the bed nucleus of the stria terminalis project to the dorsal and medial raphe nuclei at the level of the isthmus in the midbrain. The bed nucleus also contributes fibers to the locus ceruleus. The question of connections with the habenula and with the medial dorsal nucleus will be dealt with below when this same query arises in regard to the fornix.

Outputs from Thalamocingulate Division

Based largely on Ramón y Cajal's findings with the Golgi method on normal material, it was believed for many years that the cingulum constituted primarily a pathway of projection from the cingulate cortex.[160] Using Nauta-stained material from rats, Domesick[165] confirmed Krieg's (1947)[166] impression that the cingulum is largely composed of afferent fibers from the thalamus. In reaching the cingulum, the thalamic fibers follow the anterior thalamic peduncle and then ascend along the outside of the fascicles of the internal capsule before piercing the corpus callosum. They then form the cingulum and follow it to reach their destination in the midline cortex. Above the corpus callosum, the cingulum forms the medial component of the thalamocortical stratum identified as the *external medullary lamina*.[165] The cingulate fibers of projection do not run with the cingulum, but, on the contrary, cross through it to reach the deeper stratum known as the *internal medullary lamina*. In this lamina they turn forward and laterally before penetrating the corpus callosum, after which they sweep through the caudate–putamen and around the stria terminalis to enter the thalamus via the lateral thalamic peduncle.[165]

Outputs from Cingulate Areas

In the rat, both the anterior and posterior cingulate cortex project to the anterior ventral nucleus.[165] Of other intralaminar and midline thalamic projections, particular mention should be made of anterior cingulate connections with the anterior medial and medial dorsal nuclei.[164] In regard to possible visual functions (Chapter 26), it should be noted that both the rostral and caudal cingulate cortex project to the pretectal region and superior colliculus.[165] Both cingulate areas project to the caudate nucleus, as well as to the subthalamus, central gray, and tegmentum of the midbrain.[165]

In the cat, Kaitz and Robertson found that in most instances cingulate cortical areas project back to the same thalamic nuclei from which they receive connections (see findings under Inputs).[167] The notable exception was that the dorsal dysgranular retrosplenial cortex is the only cingulate area that projects to the anterior dorsal nucleus. As mentioned, this nucleus innervates mainly the *ventral granular* retrosplenial area.

Autoradiographic tracing in the monkey has shown that there are reciprocal connections between the anterior and posterior cingulate gyrus and that both regions project to the anterior medial and anterior ventral nuclei.[101] The thalamic projections from each region differ insofar as the posterior cingulate establishes connections with the medial pulvinar, whereas the anterior cingulate connects with the caudal part of the medial dorsal nucleus, as well as with the paracentral and other nidal nuclei. Both cortical regions project to the corpus striatum and ventral part of the claustrum.[101] In two studies the findings regarding pontine projections were in disagreement as to sites of origin and termination.[168] Whereas Baleydier and Mauguiere found no labeled terminals in the pons following anterior cingulate (area 24) injections, Vilensky and Van Hoesen reported that not only area 24, but also areas 25 and 32, projected to the peripheral zone of the medial pontine gray matter.[168] Both studies were in agreement regarding posterior cingulate projections to the lateral pontine gray matter.

Hippocampal Outputs with Respect to the Three Limbic Subdivisions

For purposes of naming and classifying in biology, the oldest given name usually takes precedence over newer terms. In describing distinctive features of the hippocampus in Chapter 17, we mentioned how it originally received its name from Arantius in 1587 and then its later designation *Ammon's horn* (*cornu Ammonis*) from Garengeot in 1742. For cytoarchitectural purposes we will need to refer to the later designation *Ammon's horn* by which it was commonly known in the 19th century.

Older Terminology

By the end of the 19th century it had become customary to refer to the cortical layers in the infolded ventricular part of the hippocampus as the *upper blade* and the part toward the subiculum (a support such as a pedestal) as the *lower blade*. The clinical recognition that the lower blade (Sommer's sector, see Figure 17-8) is frequently the site of nerve cell loss in cases of epilepsy probably helped to perpetuate this terminology (see Chapter 22). Finding these terms confusing, Ramón y Cajal attempted to clarify the anatomical picture by introducing his own designations. Since his terminology continues to be used in the literature, it requires a brief explanation. In his monograph on the structure of Ammon's horn appearing in 1893, he said: "To avoid confusion, we shall . . . designate as the superior region of Ammon's horn that superior segment in which the fibers unite outwardly in the posterosuperior bundle of white matter; and we shall call the inferior region that inferior segment in which fibers accumulate in the fimbria."[169] (Figure 18-16A). The fimbria (meaning fringe, as the fringe of a scarf) represents the continuation of the fornix fibers where they both leave and enter the hippocampus. Since Ramón y Cajal's description applied to a particular plane of section in the rodent brain, his terms *superior* and *inferior* represented a reverse of the designations *upper* and *lower* for the corresponding

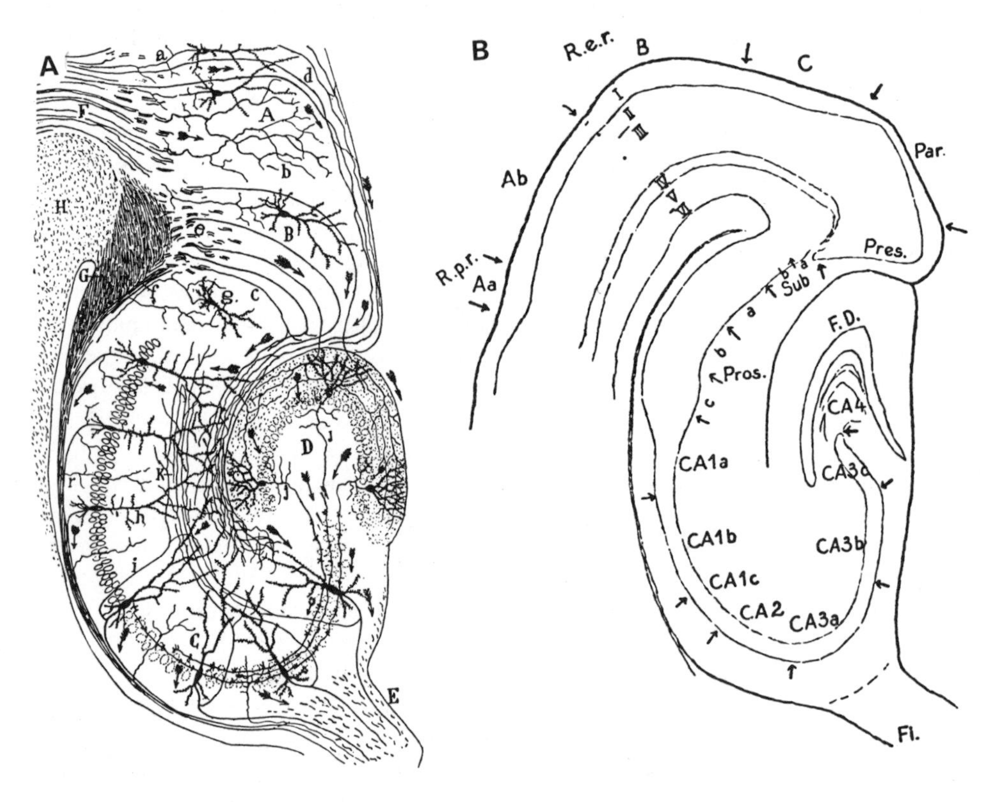

Figure 18-16. Drawings from Ramón y Cajal (A) and Lorente de Nó (B), showing respectively, the Golgi picture and the cytoarchitectural parcellation of the hippocampus. Both are drawings from sections of the mouse brain. Ramón y Cajal referred to the fimbrial part of the hippocampus (Lorente de Nó's areas CA4 and CA3) as the inferior segment (superior blade of the older terminology applied to human hippocampus). Note that the giant pyramids of these areas are characterized by the recurrent Schaffer's collaterals. Ramón y Cajal's superior segment (lower blade of human hippocampus) comprises areas CA2 and CA1, the prosubiculum and subiculum.

Abbreviations for panel A: A, angular ganglion; B, subiculum; C, Ammon's horn; D, dentate gyrus; E, fimbria; F, cingulum; G, commissural angular band; H, corpus callosum. Of the lowercase letters, note particularly the *h* identifying the apical dendrite of a medium pyramid in the superior segment and the *i* identifying a Schaffer collateral in the lower segment. J identifies an axon from a granule cell in the dentate gyrus projecting to the proximoapical dendrites of the giant cells in the inferior region.

Abbreviations for panel B: A, B, C, subfields of regio entorhinalis (R.e.r.); CA1–CA4, subfields of cornu Ammonis; F.D., fascia dentata; Fi., fimbria; Par., parasubiculum; Pres., presubiculum; Pros., prosubiculum; Sub, subiculum. Roman numerals I–VI identify layers of the entorhinal cortex. Ramón y Cajal (1901; 1955, Fig. 41, p. 78) and Lorente de Nó (1934, Fig. 2a, p. 116).

parts of the human brain. Hence, his terminology seems to have compounded the confusion.

Elsewhere Ramón y Cajal stated, ''Ammon's horn can be considered as a curvilinear palisade of pyramidal cells,''[170] explaining that those of medium size with thin axons reside in the superior region, whereas the giant pyramids with thick axons are located in the inferior region subjacent to the fimbria. He then commented, ''It is as if two layers of giant and medium pyramids of a motor convolution, instead of being superimposed on each other, were arranged in a single plane''[170] (see Figure 17-4A).

The terminology for the different layers of the hippocampus was well established by 1890. Most of the terms can be traced to Kupffer (1859)[171] and Meynert (1872).[172] The names of the main layers used in current descriptions are listed in Obersteiner's textbook of 1887 as follows: stratum moleculare, stratum lacunosum, stratum radiatum, stratum cellularum pyramidalium, and stratum oriens (of Meynert).[173]

Newer Terminology

In 1934, Lorente de Nó explained that an areal subdivision of ''the central part of Ammon's horn'' became significant subsequent to Ramón y Cajal's demonstration that the unmyelinated, mossy fiber system issuing from the granular cells of the dentate gyrus (see Figure 18-16) terminates on the apical portions of the giant pyramidal cells in the inferior region of the horn (superior blade).[174] He himself points out that Ammon's horn is characterized by the presence in its radiate layer of the long dendritic processes of the large and medium-sized pyramidal cells (cf. Figure 18-16A and B). The radiate arrangement of the dendrites of these cells suggests the wire spokes of a wheel. Lorente de Nó subdivided Ammon's horn into four main areas on the basis of the types of pyramidal cells, their location, and connections.[161] He allocated the giant cells cupped by the dentate layer and the ones subjacent to the fimbria, respectively, to areas CA4 and CA3 (Figure 18-16B). He subdivided the smaller pyramids of the superior region (lower blade) into two areas labeled CA2 and CA1. The large axons of the giant cells of CA4 and CA3 divide into a main process that enters the fimbria and a thick recurrent collateral (Figure 18-16A). As Ramón y Cajal had observed,[175] ''The recurrent collaterals discovered by Schaffer[176] and confirmed by us rise to the stratum radiatum and lacunosum, and, entering the superior region of the horn, make contact with the processes and terminal tufts of the medium (or superior) pyramids. This associative fiber is analogous to the associative branches of bifurcation which we found in many regions of the cortex of small mammals'' (Figure 18-16A). In addition to their large size, it is the possession of a Schaffer collateral that most clearly distinguishes the giant pyramids of CA4 and CA3 from the medium pyramids in areas CA2 and CA1.

Lorente de Nó applied the Vogts' term *prosubiculum*[177] to the area of cells beyond CA1 where the pyramids pile up into a thick layer (see Figure 17-8; see also below and Figure 18-18).[161] Hence, the apical dendrites of this area do not present the radiate appearance of areas CA1–CA4. Otherwise the pyramidal cells are similar to those of CA1. The term *prosubiculum* for this area signifies its location ''before'' the subiculum. The subiculum is characterized by bundles of perforating fibers that stem from the entorhinal area and terminate in the molecular layers of the hippocampus and dentate gyrus (see below). The perforating fibers serve to mark the transitional zone between the hippocampus and hippocampal gyrus (alias parahippocampal gyrus).

In his diagram of the ''Ammonic System,'' Lorente de Nó appears to have included all the above-mentioned areas under his label "C.A." for Ammon's horn.[178] But in the text he states: "The limit of the C.A. towards the prosubiculum is perfectly clear; the C.A. ceases when the stratum radiatum disappears. . . ."[179] *His arbitrary establishment of this limit has led some anatomists to include under the term hippocampus only that part containing the areas labeled CA1–CA4* (see below).

Further Background Regarding Hippocampal Inputs

Before considering outputs from the hippocampus, it is necessary to complete the picture of what is known about its afferent connections. We have already considered those derived from the brainstem. Both Ramón y Cajal and Lorente de Nó were quite confident in denying the existence of afferents from the brainstem. Rather they gave emphasis to three main inputs from cortical regions—namely, (1) the spheno-occipital ganglion (angular ganglion, entorhinal cortex) via the perforant pathway; (2) the cingulate cortex via the cingulum; and (3) the indusium griseum via the supracallosal striae. In studying the connectivity they focused on the spheno-occipital ganglion, which they regarded as the principal source of afferents to the horn. Most of their detailed observations on Golgi material from mouse brains have been confirmed by several workers in a variety of species.[180]

In addition, the experimental work has added significant details regarding an orderly laminar distribution of various afferents terminating on cells in the hippocampus and dentate gyrus. As partly illustrated in Figure 18-17, the accumulated findings may be summarized as follows:

1. Afferents from the septum enter the hippocampal formation via the fimbria and are distributed to both the hippocampus and dentate gyrus.[181] Their endings are most conspicuous in the stratum oriens of CA3 and CA4, but they are also present in the radiate layer, as well as in the molecular and subcellular layers of the dentate gyrus.[181] Those in the molecular layer of the dentate gyrus are in the deepest part of the layer, hugging the cells.
2. The commissural connections from the opposite hippocampus (CA3 and CA4)[182] should be mentioned next, because they also are most prominently distributed in the stratum oriens and middle portion of apical dendrites,[183] while those ending in the dentate gyrus are in the deep part of the molecular layer just above those from the septum.[183]
3. The entire entorhinal area projects via the perforating fibers of the subiculum to respective areas of the hippocampus. The medial entorhinal area projection terminates in the intermediate part of the molecular layer of the dentate gyrus and in the deeper half of the combined lacunosum–molecular layer of CA3.[184] The lateral entorhinal area, on the contrary, projects to the superficial part of the molecular layer of the dentate gyrus and to the outer part of the combined stratum lacunosum–molecular layer of CA3.[184] There are also projections from the entorhinal area to the superior segment (lower blade), including CA1, the prosubiculum, and subiculum: (1) Those from the medial entorhinal area terminate on the most distal part of the apical dendrites of CA1 located farthest from the subiculum, (2) whereas those from the lateral entorhinal area have a corresponding termination on the pyramidal cells adjacent to the subiculum.[185] Preliminary

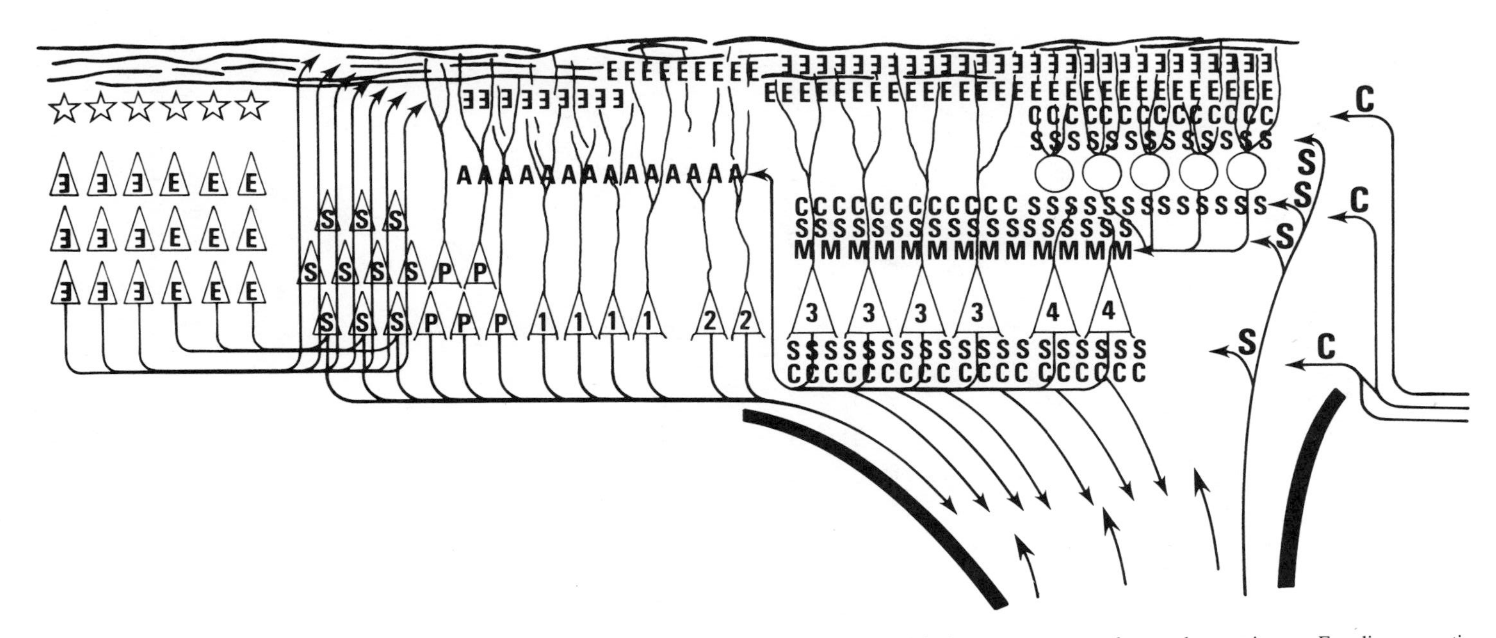

Figure 18-17. Schematic of the laminar distribution of inputs to the hippocampus form the entorhinal area, septum, and ventral commissure. For diagrammatic purposes, the gyri have been unfolded so that the cells lie on the same plane. At the far right the circles represent the granule cells of the dentate gyrus. Next to the left are the giant pyramids of CA4 and CA3, and then the medium pyramids of CA2 and CA1. Next in line are the medium pyramids of the prosubiculum and subiculum, while to the far left are the star cells of layer two and the pyramidal cells of layer three of the entorhinal area. The ribbons of different small letters identify the laminar distribution of different afferent systems. The E's and reverse E's, respectively, refer to the distribution of afferents of the medial and lateral entorhinal areas. The C's, M's, and S's identify, respectively, terminals from the commissure, septum, and mossy fiber system. The mossy fiber system accounts for the co-called stratum lucidum in Area CA3. See text for running account of the distribution of the various systems. From MacLean (unpublished).

findings indicate that layer II of the entorhinal area containing the star cells projects primarily to the dentate gyrus, whereas the pyramidal cells of the third layer innervate CA1.[185]

4. Apropos of the entorhinal projections to the terminal parts of the apical dendrites of CA1, it is of interest that thalamic afferents from nucleus reuniens have a similar distribution.[84]
5. Projections from the prefrontal cortex[186] and cingulate cortex[187] via the cingulum have been traced to the presubiculum, parasubiculum, and entorhinal cortex, but not to the hippocampus proper.
6. The sites of termination of the supracallosal striae from the indusium griseum have not yet been identified.[187]

Differential Hippocampal Projections

As now to be reviewed, there are not only differences with respect to the projections of proximoamygdalar and proximoseptal segments of the hippocampus to the brainstem, but also differences regarding the projections from the various cytoarchitectural areas.

Since the differences in the areal projections apply generally to those from the inferior and superior hippocampal regions (of Ramón y Cajal), respectively, it is pertinent at this point to call attention to the unusual degree of development of the latter region (inferior blade) in the human brain. As illustrated by Figure 18-18, the prosubicular and subicular areas of the caudal hippocampus give the appearance of having spread in a lateral direction with a resulting spillover of the subicular and presubicular areas onto the lateral aspect of the parahippocampal gyrus. Significantly (as will be noted in Chapter 26 when dealing with the functional neuroanatomy of the hippocampal formation), this overflow occurs near the place of meeting with the lingual gyrus and visual cortex within the rostral stem of the calcarine sulcus.

Although experimental work on the projections of the hippocampus may be regarded as having started with Gudden, it has been only since the middle of the present century that many of the finer details have been worked out. Prior to that time primary emphasis was given to hippocampal projections via the fornix to the mammillary bodies (see below), a pathway so large that it can be demonstrated in gross dissection. It is of historical interest to note, however, that Meynert, in his well-known chapter in Stricker's *Manual* of 1872, includes a figure showing the fornix looping around the mammillary bodies, and then ascending upwards to terminate in the anterior thalamic nuclei.[188]

Projections to Septum

Because of the traditional emphasis on fornix–mammillary connections (and perhaps also because of continued disinterest in the ''rhinencephalon'') the question of other hippocampal projections received little attention. In the early 1940s there appeared two Marchi studies noting hippocampal projections to the septum,[189] one of them by Fox showing well-illustrated degeneration in the lateral septal nucleus after damage to the proximoamygdalar part of the hippocampus.[190] Shortly after midcentury a series of papers revolutionized earlier views regarding the fornix. In 1952 Simpson reported that in the monkey ''each fornix column contains about 500,000 fibres before it enters the septum pellucidum, and is reduced to not more than 110,000 fibres in hypothalamus.''[191] The next year Daitz reported that whereas there were 2,700,000 fibers in the main body of

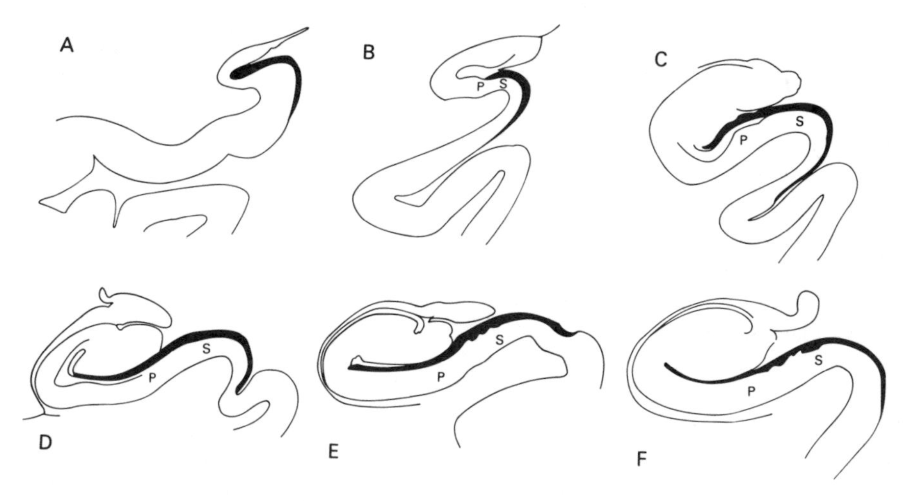

Figure 18-18. Drawings of sections through the human hippocampal formation, illustrating the lateral expansion of the prosubicular (P) and subicular (S) areas. Beginning at a caudal level below the splenium (section A) and proceeding rostrally, the drawings are of sections cut perpendicularly to the long axis and taken at regular intervals up to the level where the hippocampal fissure disappears. The superficial lamina in solid black represents the lamina medullaris superficialis, which is particularly well developed over the subiculum where the perforating fibers mark the transitional zone between the hippocampus and parahippocampal gyrus. Note that in proceeding *caudally* from F to E to D the *medial* margin of the lamina marks the beginning of the calcarine sulcus and the border with the forming lingual gyrus. Presumably because of the prosubicular expansion, the lamina spills over onto the lateral aspect of the parahippocampal gyrus to a much greater extent in the human than in the monkey. Followed caudally, the lamina continues upwards around the splenium of the corpus callosum and then forward in the lower bank of the cingulate gyrus where it gradually tapers off. From MacLean (unpublished).

the human fornix, the number was reduced to 1,200,000 in the postcommissural component.[192] Subsequently, Powell and co-workers confirmed these findings, noting that in primates, "the number of fibres in the pre-commissural fornix is approximately equal to the number in the post-commissural fornix."[193]

It has since been demonstrated in a variety of species that large numbers of fornix fibers terminate in the septum,[194] as well as in neighboring structures including the gyrus rectus, nucleus accumbens, and olfactory tubercle.[195]

Reasoning by analogy to the giant Betz cells of the motor cortex, anatomists were predisposed to believe that the large axons from areas CA4 and CA3 accounted for "long-distance" projections of the fornix. With the use of autoradiographic methods, however, it has been shown that the pyramids of CA4 and CA3 project primarily to the septum,[196] whereas it is the subicular areas (prosubiculum and subiculum) that innervate the mammillary bodies.[196] In regard to the important question of differential projections to the septum, it has been demonstrated by autoradiographic methods that the *proximoamygdalar segment projects to the ventral part of the lateral septum, whereas the proximoseptal segment innervates the dorsal part.*[196]

Projections to Hypothalamus

Fiber counts in the rat,[193] rabbit,[197] cat,[197] monkey, and human brains[193] have shown that about one-half to two-thirds of the postcommissural fibers of the fornix do not reach the mammillary bodies. The descending nonmammillary contingent includes fibers to three main destinations—namely, to the (1) rostral and intermediate hypothalamus, (2) thalamus, and (3) midbrain.

Ramón y Cajal was the first to describe a hippocampal projection to the tuber cinereum. He called it the "cornual band of the tuber cinereum"[198] (tractus corticohypothalamicus of Gurdjian[199] and of Krieg[200]). Significantly in regard to differential hippocampal projections, Nauta found in the rat that only the proximoamygdalar segment (his "caudal segment") projects to the tuberal region.[201] Autoradiographic studies have since provided evidence that these projections stem from the "subicular" areas of the proximoamygdalar segment.[202] Valenstein and Nauta were unable to find direct cornuotuberal connections in the cat and monkey, and concluded that hippocampal influence on the tuberal region in these animals probably depends on a multisynaptic pathway via the septum.[203] Using a newly modified silver method in the squirrel monkey, Poletti and Creswell, in our laboratory, identified a postcommissural "medial" and a precommissural "lateral" projection to the tuberal region.[204] With respect to the medial preoptic regulation of gonadotropic functions of the tuber, it is of great interest that they demonstrated for the first time in a primate that selective section of the fornix results in "dense fiber and terminal degeneration" in the medial preoptic area.[204] Most of the degeneration stemmed from the postcommissural fornix. In agreement with findings of other workers on a variety of species,[205] including primates, they traced degeneration into the perifornical nucleus, a structure that Hess and Brugger, on the basis of electrical stimulation, believed to be implicated in the expression of angry behavior.[206] The anterior and dorsal hypothalamus have also been found to receive hippocampal projections via the fornix.[207]

Since 1898 numerous studies employing the Marchi method in a variety of species have consistently shown degeneration in the postcommissural fornix leading to the medial mammillary nucleus.[208] These findings have been repeatedly confirmed with the use of silver[209] and autoradiographic[210] methods. In the rat Guillery found that degeneration could be traced primarily to pars posterior and pars lateralis of the medial nucleus.[211] As was illustrated in Figure 18-7, these two parts, respectively, project to the anterior ventral and anterior dorsal nuclei of the thalamus. Several studies have provided evidence of an orderly projection, either from different segments[212] or from different cytoarchitectural areas,[213] onto different parts of the mammillary bodies, but the findings have not been confirmed. The observations deserving major emphasis are the recent ones mentioned above demonstrating that it is the subicular areas (prosubiculum and subiculum), rather than any of the numbered areas (i.e., CA1–CA4), that project to the mammillary bodies. Swanson and Cowan report suggestive evidence that in the rat the postcommissural fornix from the subicular areas of the proximoamygdalar segment projects to the rostral and intermediate hypothalamus, whereas the subicular areas from the proximoseptal segment innervate the mammillary bodies.[214] Such a distinction would be of particular interest in the light of the evolutionary expansion (described above) of the subicular areas of the human hippocampus that is particularly evident in the proximoseptal segment neighboring the visual areas (Figure 18-18).

Projections to Thalamus

In 1881 Gudden described an "uncrossed fornix bundle" leading to the anterior thalamus.[215] This bundle curves upwards into the anterior thalamus immediately after the fornix enters the diencephalon. Except for a short note by Vogt in 1898, this bundle appears to have gone unnoted until after the middle of the present century. In a Marchi study, Vogt had observed degeneration in this bundle that he attributed to experimental damage of the septum.[216] In 1956, however, Nauta presented evidence that damage of any segment of the rat hippocampus resulted in heavy degeneration in Gudden's uncrossed bundle that could be traced not only to the anterior medial and anterior ventral nuclei, but also to the lateral dorsal nucleus.[201] In the squirrel monkey Poletti and Creswell found that selective section of the fornix resulted in "massive" degeneration in the anterior ventral nucleus and "dense" degeneration in the lateral dorsal nucleus.[217] Van Buren and Borke observed a similar picture in cases of human material in which vascular damage had destroyed parts of the fornix system.[218]

In terms of feedback loops and central processing devices, it is of special interest that in a comparative study, Powell *et al.* found that the number of fornix fibers entering the anterior thalamus about equaled those from the mammillothalamic tract.[193] Autoradiographic studies have indicated that it is the subicular rather than any of the numbered CA areas that project to the thalamus.[214] In one study there were indications that some of the hippocampal projections reached the anterior thalamus by a shorter route via the internal capsule.[219]

Ever since the beginning of experimental work on the fornix system there has been the recurring question of whether or not main or collateral fibers of the fornix enter the stria medullaris and terminate in the habenula. The present consensus appears to be that in studies showing degeneration leading from the fornix to the habenula, there were complicating lesions of the septum or stria medullaris. An autoradiographic study has demonstrated that the septofimbrial and triangular nuclei of the septum, as well as the medial septal nucleus, project to the habenula.[122] Septal projections terminate primarily in the medial nucleus of the habenula, whereas those arising from structures below the level of the anterior commissure (lateral hypothalamus and medial pallidum) are traced via the stria medullaris predominantly to the lateral nucleus of the habenula.[220] Cells in the triangular region also project via the stria medullaris and habenulopeduncular tract to the interpeduncular nucleus (see also Ref. 221).[122]

In a comparative study of the fornix system in the rat, guinea pig, cat, and monkey, Valenstein and Nauta observed that interruption of the fornix in the septal region of the monkey, but not in the other species resulted in "degenerating fibers of the stria medullaris which terminate in the medial, magnocellular division of the dorsomedial nucleus."[222] In retrospect, it appears probable that this degeneration resulted from damage to the septum or stria medullaris.[204] Septal projections to the medial dorsal nucleus have been described in various species, including the squirrel monkey.[223]

Projections to Midbrain

In 1958 Nauta gave emphasis to "the existence of various pathways by which the hippocampus . . . can transmit its influence to the mid-brain."[224] In addition to describing direct projections from the hippocampus to the ventral part of the central gray substance, he cited evidence of indirect connections with this and neighboring structures via the medial forebrain bundle, mammillotegmental tract, and habenulopeduncular tract. He singled out specifically the ventral tegmental area of Tsai, the interpeduncular nucleus, the superior central nucleus of Bechterew, and Gudden's dorsal and deep tegmental nuclei. He referred to the zone encompassing these structures and the central gray as the "limbic" midbrain region.[224] As witness Figure 18-2, this region soon became known as the limbic midbrain area of Nauta.

As Nauta pointed out, his finding with his silver method of direct projections of the fornix to the midbrain, was a confirmation of Edinger and Wallenberg's study of 1902 in which they employed the Marchi technique.[224] It is of unusual interest that in the elephant in which the mammillary bodies are greatly reduced in size, sagittal sections show a large fraction of fornix fibers leading directly into the midbrain.[225] What does it signify that, save for their massive bodies, animals as unlike as the elephant and whale have mammillary bodies that are greatly reduced in size? Does the reduction reflect a reduced number of nerve cells or reduced amount of myelin of the innervating fibers? The rock hyrax (*Hyrax syriaca*), which some authorities believe has a taxonomic relationship to the elephant, has large mammillary bodies and heavily myelinated limbic pathways (personal observations).

Summary of Hippocampal Projections

A summing up of the main findings on hippocampal projections will lead to a final comment regarding the question of limbic subdivisions.

1. The giant pyramids of CA4 and CA3 project primarily to the septum. Those from the proximoamygdalar segment innervate the more ventral parts of the septum, while those from the proximoseptal segment connect with the more dorsal parts.
2. The proximoamygdalar segment of the hippocampal formation is the origin of connections via the fornix with the tuberal region.
3. The subicular areas of the hippocampus (prosubiculum and subiculum) provide the main projection via the fornix to the mammillary bodies.
4. The subicular areas also appear to account for most of the hippocampal projections via the fornix to the anterior nuclear group of the paleothalamus, as well as to the lateral dorsal nucleus of the neothalamus.
5. The hippocampal projections to the mammillary bodies and anterior thalamic nuclei afford a dual influence on the cingulate gyrus, which in turn has feedback connections with the hippocampus via the cingulum and its terminations in respective areas of the hippocampal gyrus, as well as in the subiculum.
6. The hippocampus has indirect connections via the fornix with the medial dorsal nucleus, which in turn projects to orbitofrontal cortex. From the latter there are feedback connections with the hippocampus via fibers that travel in the cingulum and terminate in the cortex of the posterior part of the parahippocampal gyrus and in the entorhinal cortex of the rostral part.
7. In addition to the foregoing connections, the hippocampus projects through the fornix to the preseptal limbic cortex, the olfactostriatum, medial preoptic area, anterior hypothalamus, lateral hypothalamic area, ventral tegmental area, and central gray matter of the midbrain. In all of these connections there is the opportunity for overlap with the avalanche of direct projections from the amygdala and septum. This latter situation is to be contrasted with the absence of hypothalamic projections from the thalamocingulate subdivision.
8. Finally, as will be noted in Chapters 26 and 27, there are other outputs from the subicular areas of the hippocampus that follow nonfornical pathways.

Concluding Statement

As defined in the last chapter, the limbic cortex forms, essentially, two concentric rings of cortex around the hilus of the hemisphere, consisting of an inner ring of archicortex surrounded by a ring of transitional mesocortex. In gyrencephalic animals most of the limbic cortex is located within the boundaries of the rhinal and cingulate sulci. It would appear that the olfactory system, which approaches the telencephalon from the surface level, determines the prototypical pattern for the distribution of afferents to the archicortical and mesocortical areas of the cerebrum. Thus, the axons of the secondary neurons terminate in the outer half of the molecular layer on the extroverted dendrites of the superficial cell layers. Parts of the cingulate mesocortex represent a kind of hybrid insofar as paleothalamic afferents terminate in the outer layers, while fibers from the neothalamic nucleus LD distribute to layers III–IV, as well as to the molecular layer.

Three Main Limbic Subdivisions

Given this cytoarchitectural picture, one can identify three main nuclear groups located respectively in the amygdala, septum, and thalamus that are a source of afferents to respective parts of the limbic cortex. In this regard the amygdala and septum serve as telencephalic internodes for neural circuits relating the rostrally located olfactory apparatus with intero- and exteroceptive systems ascending from elsewhere in the nervous system.

The amygdalar division comprises the amygdala together with the frontotemporal limbic cortex that is brought into association by the uncinate fasciculus. In addition to reciprocal connections with the orbital, anterior insular, piriform, temporal polar, and rostral entorhinal cortex, the amygdala is interconnected with the proximoamygdalar segment of the hippocampus.

Of the three limbic subdivisions the septal division of the limbic system presents the greatest difficulty to define on the basis of anatomical grounds alone. Although the septum projects to the entire hippocampal formation, the so-called proximoseptal segment of the hippocampus appears to be primarily dependent on the septum for its supply of afferents, either as they arise directly from particular parts of the septal nuclei or from brainstem afferents that pass through the septum. The septal division also includes the cortex connected with the proximoseptal hippocampal segment—namely, the adjacent part of the entorhinal cortex, the postrhinal hippocampal gyrus, presubicular part of the lingual gyrus, and retrosplenial cortex. The division also encompasses the pericallosal hippocampal rudiment and a narrow strip of preseptal cortex continuous with it.

The thalamocingulate division is comprised of the mesocortical cingulate areas receiving afferents not only from the anterior thalamic nuclei but also from certain parts of the other nidal nuclei.

The experimental findings on function to be considered in subsequent chapters, together with electrophysiological data and further anatomical details, provide additional reasons for the parcellation of the limbic system into at least the three main subdivisions.

References

1. Forel, 1937, pp. 158–159
2. Forel, 1937, p. 93
3. Gerlach, 1858
4. Obersteiner, 1890
5. Gudden, 1870
6. Golgi, 1873
7. Ramón y Cajal, 1893/1968, p. 5
8. e.g., Scheibel and Scheibel, 1970
9. Ramón y Cajal, 1933/1954; see Cannon, 1949
10. Waldeyer, 1891
11. Waller, 1850
12. Obersteiner, 1887/1890, p. 20
13. Marchi and Algieri, 1885
14. Nissl, 1892a,b

15. Nissl, 1913
16. Bielschowsky, 1904
17. Glees, 1946
18. Nauta and Gygax, 1954
19. Fink and Heimer, 1967
20. Voneida and Trevarthen, 1969
21. de Olmos, 1969
22. e.g., De Robertis, 1956
23. Weiss and Holland, 1967; Droz and Leblond, 1963
24. Cowan *et al.*, 1972; Lasek *et al.*, 1968
25. Graham and Karnovsky, 1966
26. Straus, 1964
27. LaVail and LaVail, 1972
28. Mesulam, 1976, 1978
29. Gonatas *et al.*, 1979
30. Kuypers *et al.*, 1977; Van der Kooy and Kuypers, 1979; Bentivoglio *et al.*, 1979; de Olmos and Heimer, 1980
31. Sternberger, 1974
32. Bowker *et al.*, 1981
33. Strong, 1895
34. Herrick, 1948, p. 67
35. Sherrington, 1906
36. Ricardo and Koh, 1978; Pretorius *et al.*, 1979; Van der Kooy *et al.*, 1982; Terreberry and Neafsey, 1983
37. His, 1904
38. Herrick, 1914
39. Herrick, 1914, p. 377
40. Herrick, 1914, p. 373
41. Foley and DuBois, 1937; Hoffman and Kuntz, 1957
42. Kuru, 1956
43. Herrick, 1948, p. 190
44. Aronson and Noble, 1945
45. Herrick, 1948, p. 46
46. Riley, 1960, p. 657
47. Petrovicky, 1971
48. Gudden, 1884
49. Bleier, 1969
50. Guillery, 1956; Akert and Andy, 1955; Morest, 1961
51. Morest, 1967
52. Morest, 1961
53. Dahlström and Fuxe, 1964
54. Thoa *et al.*, 1977
55. Ungerstedt, 1971
56. Pickel *et al.*, 1974; Moore, 1978
57. Jacobowitz and MacLean, 1978
58. Norgren and Leonard, 1973
59. Norgren, 1976
60. Ricardo and Koh, 1978; Pretorius *et al.*, 1979
61. Loewy and Burton, 1978
62. Norgren, 1976; McBride and Sutin, 1976; Mehler, 1980
63. Dahlström and Fuxe, 1964; Jouvet, 1967; Jacobowitz and MacLean, 1978
64. Moore and Halaris, 1975
65. Ungerstedt, 1971; Bobillier *et al.*, 1975; Bowden *et al.*, 1978
66. Pasquier and Reinoso-Suarez, 1976
67. Meyerson, 1964; Tagliamonte *et al.*, 1969; Bachman and Katz, 1977
68. Myers and Beleshin, 1971; Quock and Horita, 1974
69. Jouvet, 1969
70. Parent, 1979, p. 282; *see also* Parent, 1986
71. Segal and Landis, 1974; Wyss *et al.*, 1979; Amaral and Cowan, 1980; Mehler, 1980
72. Mehler, 1980
73. Lindvall, 1975; Moore, 1978
74. Pasquier and Reinoso-Suarez, 1976; Amaral and Cowan, 1980
75. Jones *et al.*, 1976; Mehler, 1980
76. Ungerstedt, 1971; Fallon *et al.*, 1978; Nauta and Domesick, 1978; Mehler, 1980
77. Guillery, 1956; Swanson, 1976; Saper *et al.*, 1976, 1979
78. Swanson, 1976; Ottersen and Ben-Ari, 1979; Ottersen, 1980; Mehler, 1980
79. Pasquier and Reinoso-Suarez, 1976; Wyss *et al.*, 1979; Amaral and Cowan, 1980
80. Jones *et al.*, 1976; Mehler, 1980
81. Jones *et al.*, 1976; Saper *et al.*, 1976; Krieger *et al.*, 1979; Mehler, 1980
82. Powell and Cowan, 1954
83. Gudden, 1884; Van Valkenburg, 1912; Guillery, 1957; Nauta, 1958
84. Nauta and Whitlock, 1954
85. Herkenham, 1978
86. Herkenham, 1978; Mehler, 1980
87. Mehler, 1980
88. Brockhaus, 1940
89. Bagshaw and Pribram, 1953
90. Roberts and Akert, 1963
91. Pribram *et al.*, 1953
92. Nauta, 1962
93. Rose and Woolsey, 1948
94. Clark and Meyer, 1950
95. Leonard, 1969
96. Krettek and Price, 1977a; Vogt *et al.*, 1981
97. Robertson and Kaitz, 1981
98. Vogt, 1910
99. Vogt *et al.*, 1981
100. Pribram and Fulton, 1954
101. Baleydier and Mauguiere, 1980
102. Vogt *et al.*, 1979
103. Singer, 1962, plate 42
104. Riley, 1960, pp. 250–262
105. Yakovlev *et al.*, 1960
106. Baleydier and Mauguiere, 1980; Vogt *et al.*, 1979
107. Robertson and Kaitz, 1981
108. Christ, 1969
109. Johnson, 1923; Crosby and Humphrey, 1941
110. Johnston, 1923
111. Valverde, 1965
112. Krettek and Price, 1977c
113. Krettek and Price, 1977b; Amaral and Cowan, 1980
114. Cowan *et al.*, 1965; Lammers, 1971
115. Porrino and Goldman-Rakic, 1982
116. Stephan and Andy, 1964; Swanson and Cowan, 1979
117. Daitz and Powell, 1954
118. Mellgren and Srebro, 1973; Mosko *et al.*, 1973; Segal and Landis, 1974; Swanson and Cowan, 1979; Amaral and Cowan, 1980
119. Lewis and Shute, 1967; Lewis *et al.*, 1967; Mellgren and Srebro, 1973; Mosko *et al.*, 1973
120. Segal and Landis, 1974
121. Alonso and Kohler, 1984
122. Swanson and Cowan, 1979
123. Jacobson, 1811; see Cuvier, 1811
124. White, 1965; Scalia, 1966; Heimer, 1968; Price, 1973; Scalia and Winans, 1975; Broadwell, 1975a; Skeen and Hall, 1977; Kevetter and Winans, 1981a,b
125. Herrick, 1948
126. Vaz Ferreira, 1951
127. White, 1965
128. Heimer, 1968; Price, 1973
129. Broadwell, 1975b; Lundberg, 1960; Lohman, 1963; Lohman and Lammers, 1963
130. Lundberg, 1962
131. Adey, 1953; Clark and Meyer, 1947; Meyer and Allison, 1949; Scalia, 1966; Price, 1973; Broadwell, 1975a; Skeen and Hall, 1977; Kevetter and Winans, 1981a

132. Meyer and Allison, 1949
133. Turner *et al.*, 1979
134. Winans and Scalia, 1970
135. Kubie and Halpern, 1975, 1979; Kubie *et al.*, 1978
136. Powers *et al.*, 1979
137. Broadwell, 1975a
138. Broadwell, 1975a; Skeen and Hall, 1977; Kevetter and Winans, 1981a,b
139. Kevetter and Winans, 1981a,b
140. Herkenham and Pert, 1980
141. MacLean and Pribram, 1953; Pribram and MacLean, 1953
142. Brodmann, 1909
143. Walker, 1940
144. Pribram and MacLean, 1953
145. Wall *et al.*, 1951
146. Mufson *et al.*, 1979
147. Narkiewicz, 1964; Macchi *et al.*, 1981
148. Krettek and Price, 1977c
149. MacLean, 1988
150. Lundberg, 1960, 1962; Powell *et al.*, 1965; Leonard, 1969; Leonard and Scott, 1971; de Olmos, 1972
151. Powell *et al.*, 1965
152. Powell *et al.*, 1965, p. 806
153. Lundberg, 1960
154. Heimer and Wilson, 1975
155. Whitlock and Nauta, 1956
156. Krettek and Price, 1978
157. Loo, 1931
158. Rosene and Van Hoesen, 1977
159. Brodmann, 1909
160. Ramón y Cajal, 1901–02/1955
161. Lorente de Nó, 1934
162. Brodmann, 1909; Beck, 1934
163. Domesick, 1969
164. Nauta, 1953; Domesick, 1969
165. Domesick, 1970
166. Krieg, 1947
167. Kaitz and Robertson, 1981
168. Baleydier and Mauguiere, 1980; Vilensky and Van Hoesen, 1981
169. Ramón y Cajal, 1893/1968, p. 10
170. Ramón y Cajal, 1901–02/1955, p. 76
171. Kupffer, 1859
172. Meynert, 1872
173. Obersteiner, 1890, pp. 364–366
174. Lorente de Nó, 1934, p. 114
175. Ramón y Cajal, 1901–02/1955, p. 77
176. Schaffer, 1892
177. Vogt and Vogt, 1919b
178. Lorente de Nó, 1934, p. 114
179. Lorente de Nó, 1934, p. 118
180. Adey, 1952b; Raisman *et al.*, 1965; Hjorth-Simonsen and Jeune, 1972; Hjorth-Simonsen, 1972; Van Hoesen and Pandya, 1975; Steward and Scoville, 1976
181. Raisman *et al.*, 1965; DeVito and White, 1966; Mosko *et al.*, 1973
182. Gottlieb and Cowan, 1973
183. Blackstad, 1956; Raisman *et al.*, 1965; Gottlieb and Cowan, 1973
184. e.g., Hjorth-Simonsen, 1972
185. Steward and Scoville, 1976
186. Adey, 1952a; Nauta, 1964
187. Raisman *et al.*, 1965
188. Meynert, 1872
189. Gerebetzoff, 1942; Fox, 1943
190. Fox, 1943
191. Simpson, 1952
192. Daitz, 1953
193. Powell *et al.*, 1957, p. 435
194. Raisman *et al.*, 1966; Siegel and Tassoni, 1971a,b; Powell, 1973; Siegel *et al.*, 1974; Swanson and Cowan, 1975; Meibach and Siegel, 1977a; Poletti and Creswell, 1977
195. Swanson and Cowan, 1975; Poletti and Creswell, 1977
196. Swanson and Cowan, 1975; Meibach and Siegel, 1977a
197. Guillery, 1955; Powell *et al.*, 1957
198. Ramón y Cajal, 1901–02/1955, p. 138
199. Gurdjian, 1927
200. Krieg, 1932
201. Nauta, 1956
202. Swanson and Cowan, 1975; Siegel *et al.*, 1974; Meibach and Siegel, 1977a
203. Valenstein and Nauta, 1959
204. Poletti and Creswell, 1977
205. Allen, 1944; Sprague and Meyer, 1950; Nauta, 1956; Valenstein and Nauta, 1959; Votaw, 1960; Siegel and Tassoni, 1971a
206. Hess and Brugger, 1943
207. Powell and Cowan, 1955; Poletti and Creswell, 1977
208. Vogt, 1898; Edinger and Wallenberg, 1902; Allen, 1944; Gerebetzoff, 1942
209. Sprague and Meyer, 1950; Simpson, 1952; Powell and Cowan, 1955; Nauta, 1956, 1958; Raisman, 1966; Chronister and White, 1975
210. Swanson and Cowan, 1975; Meibach and Siegel, 1977a
211. Guillery, 1955
212. Simpson, 1952; Swanson and Cowan, 1975; Meibach and Siegel, 1977a
213. Raisman *et al.*, 1966
214. Swanson and Cowan, 1975
215. Gudden, 1881
216. Vogt, 1898
217. Poletti and Creswell, 1977; see also Powell, 1973
218. Van Buren and Borke, 1972
219. Meibach and Siegel, 1977b
220. Swanson and Cowan, 1975; Herkenham and Nauta, 1977
221. Valenstein and Nauta, 1959
222. Valenstein and Nauta, 1959, p. 352
223. Guillery, 1959; Cragg, 1961; Trembly and Sutin, 1961; Powell, 1966
224. Nauta, 1958
225. Diepen *et al.*, 1956

19

Functions of Amygdalar and Septal Divisions with Respect to Self-Preservation and Procreation

It is worthwhile at this point to recall that it is primarily the aim of the present investigation to obtain a better understanding of brain mechanisms underlying paleopsychic processes and prosematic behavior. The remaining chapters on the limbic system will therefore focus on animal experimentation and clinical observations that are relevant to these aspects of mentation and behavior. In this and the two succeeding chapters, I review experimental findings on the functions of the limbic system with reference to its three main subdivisions defined in Chapter 18. Since clinical case studies provide the best evidence that the limbic system plays a fundamental role in paleopsychic processes of an emotional nature, I will thereafter deal separately with that material.

Figure 19-1 shows schematically the three main subdivisions of the limbic system. The three main nuclear groups serving as hubs for neural communication between the limbic cortex and brainstem are identified by the large numerals 1, 2, and 3. As indicated, the two telencephalic nuclear groups identified with the amygdala and septum have a close relationship with the olfactory apparatus, whereas the diencephalic group is nodal for pathways that bypass the olfactory apparatus. As outlined in Chapter 2, experimental work indicates that the amygdalar and septal divisions are primarily involved, respectively, in oral- and sex-related functions required for self-preservation and the procreation of the species. The present chapter deals with ablation and stimulation studies that provide the basis of such conclusions. The following chapter rounds out the picture of cerebral representation of primal sexual functions, focusing first on the results of stimulating the third main subdivision of the limbic system, and then reviewing findings on the neurocircuitry of the brainstem that has thus far been discovered to be involved in sensorimotor aspects of genital function. This latter topic involves a consideration of neural mechanisms accounting for the integration of oral and sexual manifestations in agonistic behavior. The final chapter on the third subdivision (Chapter 21) deals with evidence leading up to the realization that this phylogenetically newest part of the limbic system includes a representation of a behavioral triad that distinguishes the evolutionary dividing line between reptiles and mammals—namely, (1) nursing and maternal care, (2) audiovocal communication for maintaining maternal–offspring contact, and (3) play.

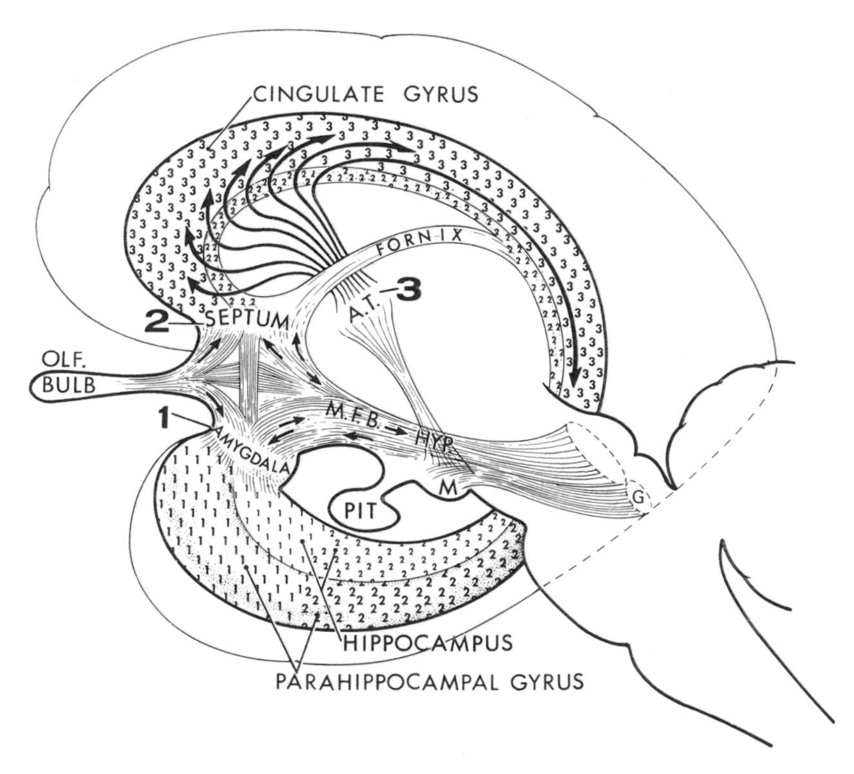

Figure 19-1. Three main subdivisions of the limbic system. The nuclear groups associated with the amygdalar, septal, and thalamocingulate divisions are respectively labeled with the large numerals 1, 2, and 3, while the cortical sectors primarily associated with them are overlain with the smaller corresponding numerals. The numerals overlying the archicortical areas are somewhat smaller than those identifying the rest of the limbic cortex. Abbreviations: A.T., anterior thalamic nuclei; G, dorsal and ventral tegmental nuclei of Gudden; HYP., hypothalamus; M, mammillary bodies; M.F.B., medial forebrain bundle; PIT, pituitary; OLF., olfactory. After MacLean (1958, 1973a).

THE AMYGDALAR DIVISION

The amygdala and hippocampus are sometimes dealt with as though they were two independent autonomous structures with their own particular functions. Conforming to the anatomical picture described in Chapter 18, Figure 19-2 helps to counteract this view, showing the hippocampus as an upright arch with one end structurally based in the amygdala and the other in the septum. As also noted in Chapter 18, the hippocampus appears to be organized along both axial and segmental lines. Lorente de Nó, for example, gave emphasis in one of his diagrams to three overlapping afferent systems innervating the hippocampus. Since neither the anatomical nor physiological findings provide a means of axially dividing the hippocampus, it will be the convention here to refer to the half toward the amygdala as the *proximoamygdalar segment* and the half toward the septum as the *proximoseptal segment*. Significantly, the neurobehavioral work to be described provides additional reasons for drawing a distinction between the two gross subdivisions. In describing the experimental work on the amygdalar division, I will deal first with the effects of ablation and then stimulation.

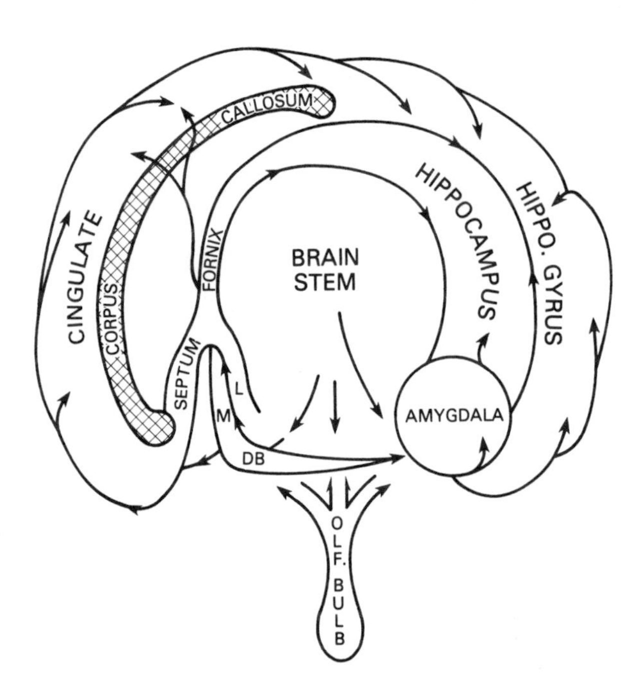

Figure 19-2. Upright view of the limbic lobe. This rotated presentation helps to emphasize that one end of the hippocampal arch is structurally supported by the amygdala and the other by the septum. Abbreviations: DB, diagonal band; L, lateral septal nuclei; M, medial septal nuclei; OLF., olfactory. After MacLean (1958).

Effects of Ablation

There are two reasons, in particular, for introducing the experimental work on animals with a summary of the results of a classical study by Klüver and Bucy.[1] In the first place, their study provides invaluable background for the work discussed in the present chapter because the operations performed on rhesus monkeys removed a large part of both the amygdalar and septal divisions. Second, the reports of their findings that appeared between 1937 and 1940[2] were to prove timely in giving support to the Papez theory and countering the established view that the great limbic lobe functioned primarily as a "smell brain."

Bitemporal Lobectomy

In the 1930s, the late Heinrich Klüver (1897–1979), professor of psychology at the University of Chicago, had conducted experiments testing the effects of the hallucinogenic agent mescaline on rhesus monkeys. He had found that the drug produced oral symptoms such as chewing and smacking of the lips seen in patients with "uncinate fits"[3] resulting from irritative lesions in the region of the uncinate gyrus of the limbic lobe (see Figure 17-7). He wanted to find out whether or not the same symptoms would occur after removal of the so-called olfactory cortex. Accordingly, he asked his friend, Paul Bucy, a neurosurgeon, to perform the operation.[4] Bucy decided that the operation could be most easily accomplished by a block removal of the temporal lobe on each side. As Klüver commented in retrospect, the resulting symptoms of such surgery "probably represent the most striking behavior changes ever produced by a brain operation in animals."[5] There were six main types of behavioral alterations that the authors identified as (1) emotional changes, (2) oral tendencies, (3) "psychic blindness," together with (4) a compulsive tendency to examine every visible object, (5) changes in dietary habits, and (6) changes in sexual behavior.[1]

Since wild, aggressive rhesus monkeys were used for this study, the most striking change observed initially was their apparent tameness and docility. In addition to submitting quietly to examination, they seemed to have lost their sense of fear and would as readily examine the "tongue of a hissing snake" as a piece of food. There was a loss of facial expression, and vocal signs of anger or fear were absent. Anyone who has observed such animals can vouch for these dramatic changes. On the morning following the surgery, the subject is characteristically sitting impassively at the front of the cage. The preoperative threat gestures—grimacing, ears moving backwards and forwards, and stiff, aggressive bounces—are no more in evidence. Klüver and Bucy gave emphasis to the apparent compulsion of these animals to attend to every visible stimulus and their repeated examination of any object—harmful or not—by smelling and mouthing it. Live animals, feces, broken glass, and metal objects are all treated in this manner, being submitted to touching with the lips, licking, and gentle biting. If permitted, such an animal would burn itself repeatedly by touching and mouthing an open flame. The authors noted that a piece of food and a nail may be picked up "hundreds of times" and "examined by the mouth." Ordinarily frugivorous, their monkeys postoperatively would eat raw meat or fish. After a period of 2 to 3 weeks, bizarre sexual changes developed and were characterized by continuous manipulation and oral stimulation of the genitalia, as well as by indiscriminate mounting of male or female monkeys. It should be added that inanimate objects may also be treated in a sexual manner (see Ref. 6). I have seen such monkeys "groom" a bottle brush and then attempt to mount it.

At the time of their experiments, Klüver and Bucy were unaware that Sanger Brown and E. A. Schäfer had 50 years earlier performed similar operations on monkeys and reported the same kind of findings.[7] [Sanger Monroe Brown (1852–1928) was a young Canadian who had gone to study at the University of London.] The description of one of their monkeys serves as a kind of retrospective confirmation of Klüver and Bucy's observations. Their account reads:

> Prior to the operations, [this monkey] was very wild and even fierce, assaulting any person who teased or tried to handle him. Now he voluntarily approaches all persons indifferently, allows himself to be handled, or even to be teased or slapped, without making any attempt at retaliation or endeavouring to escape. His memory and intelligence seem deficient. He gives evidence of hearing, seeing, and the possession of his senses generally, but it is clear that he no longer clearly understands the meaning of the sounds, sights, and other impressions that reach him. Every object with which he comes in contact, even those with which he was previously most familiar, appears strange and is investigated with curiosity. Everything he endeavours to feel, taste, and smell, and to carefully examine from every point of view. This is the case not only with inanimate objects, but also with persons and with his fellow Monkeys. And even after having examined an object in this way with the utmost care and deliberation, he will, on again coming across the same object accidentally even a few minutes afterwards, go through exactly the same process, as if he had entirely forgotten his previous experiments. His food is devoured greedily, the head being dipped into the dish, instead of the food being conveyed to the mouth by the hands in the way usual with Monkeys. He appears no longer to discriminate between the different kinds of food; e.g., he no longer picks out the currants from a dish of food, but devours everything just as it happens to come. He still, however, possesses the sense of taste, for when given a raisin which has been partly filled with quinine he shows evident signs of distaste, and refuses to eat the fruit. It is also clear that he still sees and hears. . . . He reacts to all kinds of noises . . . but shows no consequent evidence of alarm. . . .[8]

Klüver and Bucy's study included a total of 16 monkeys, 10 of which were still living at the time of the main report in 1939.[1] There were two females and six males that had undergone bitemporal lobectomy. The authors found that the behavioral changes did not appear unless the "rhinencephalic" structures (i.e., the amygdala and hippocampus) were included in the bilateral ablations. Bilateral excision of (1) the first temporal convolution, (2) the second and third temporal convolutions, or (3) the severance of connections of the temporal lobes with the frontal lobes, or (4) with the occipital lobes, failed to induce the typical symptoms.

In a brief report appearing in 1940, Bucy and Klüver described the anatomical findings in their first bitemporal lobectomized subject.[9] They stated that all but a fragment of the right amygdala and a remnant of the posterior hippocampal formation had been excised. Fifteen years later in an illustrated and detailed description of the findings in this same monkey—a female named Aurora—they remarked that the anatomical picture was not "essentially different" in other monkeys with bitemporal lobectomy.[10]

In summary, the experiments that have been described indicate that bilateral temporal lobectomy (including a large part of the amygdalar and septal subdivisions of the limbic system) produces a taming effect in wild, aggressive monkeys and a loss of emotional expression. There is an associated inability to distinguish between harmful and non-harmful situations, along with indiscriminate ingestion and bizarre sexual activity. In many respects, it was as though the animal suffered from a global agnosia. Klüver and Bucy referred to the subjects' compulsion to examine every object as *hypermetamorphosis* and the deficits in adaptive discrimination as *psychic blindness.*[1] In regard to visual perceptions and the question of psychic blindness, it was found that the bitemporal lobectomized animal was at least capable of visual discrimination, as, for example, distinguishing an angular from a curvilinear object. Hence, in line with the writings of von Monakow and others, the authors were led to suggest that under some conditions, agnosia may be a reflection of emotional indifference.[11]

Klüver–Bucy Syndrome in Human Beings

In 1955, Terzian and Ore of Padua described manifestations of the Klüver–Bucy syndrome in a 19-year-old epileptic Italian boy, who had undergone bitemporal lobectomy for the purpose of alleviating bouts of uncontrollable rage and violence.[12] Subsequent to a febrile illness at the age of 3, the patient had begun to suffer from epileptic attacks characterized by terrifying visual hallucinations and automatisms. In some attacks there were generalized convulsions. In the following years he developed personality changes along with "paroxysms of aggressive and violent behavior." During such outbursts he had attempted to strangle his mother and crush his younger brother underfoot. In the preoperative period in the hospital, he "enjoyed periods of lucidity lasting one or two weeks, in which he revealed a well-preserved memory, a serviceable character, and normal intelligence."[13] The electroencephalogram indicated an epileptogenic focus in the left temporal region and another focus in the contralateral lobe. Since there was no improvement in his condition after the removal of the left temporal lobe, the same operation was performed on the right. Subsequently, the patient developed behavioral changes that the authors regarded as similar to those of the Klüver–Bucy syndrome—including loss of emotional expression, "hypermetamorphosis," and changes in eating habits and sexual behavior. Formerly deeply attached to his mother, he afterwards referred to her formally as "Madam." Later he addressed his parents as mother and father, but never showed any affection for them. His voice became monotonous and lacking any emotional coloring. He would approach and stop any person he met and would take possession of all manner of objects. Although finding his way around the hospital, he was confused about its geographical location. He appeared to have a good sense of time, but was grievously disabled by a profound loss of memory, not only failing to recall what had recently happened, but also everything of his past. He developed an insatiable appetite, eating all kinds of foods, licking the dish incessantly, and repeatedly asking for more helpings. Fifteen days after surgery he became "exhibitionistic" and made some homosexual advances. In the mental hospital to which he was later committed, he was seen masturbating several times a day. Apropos of the functions of the striatal complex and release mechanisms discussed in Chapter 15, it is of interest that he became "very orderly and meticulous in his habits" and "showed a remarkable tendency to imitate the manners of the surrounding people (doctors, nurses) and would even reproduce professional gestures."[14]

This case is of additional historical interest because it provided one of the first clinical indications that destruction of limbic structures in the medial temporal lobe results in profound memory deficits (see Chapter 27).

Differential Effects of Lesions of Amygdalar Division

Since the dramatic findings of the Klüver–Bucy syndrome were published, several investigators have attempted to identify the specific structures accounting for the various manifestations. Figure 18-12 shows the interrelated cortex of the frontotemporal region that Pribram and I identified by strychnine neuronography and which, along with the proximoamygdalar segment of the hippocampus, represents the cortical sector associated with the amygdala. In neurobehavioral experiments on macaques and baboons, Pribram and Bagshaw attempted to aspirate the entire cortical area shown in Figure 18-12B, together with the underlying amygdala, so as to discover which manifestations of the Klüver–Bucy syndrome would appear.[6] This operation was undertaken in five immature macaques and a baboon. In three additional cases, they performed subtotal ablations, aspirating the posterior orbital and anterior insular cortex, respectively, in two macaques, and the temporal polar periamygdaloid cortex and amygdala in a baboon. On the basis of the description and diagrammatic reconstructions, it is not clear as to what extent the hippocampus contiguous with the amygdala was invaded. Since the temporal horn of the ventricle was used to demarcate the posterior extent of the lesion, it is possible that the sucker removed some of the rostral hippocampus or interfered with its blood supply. The five monkeys and one baboon with extensive frontotemporal ablations all showed most of the manifestations of the Klüver–Bucy syndrome, and the symptoms persisted throughout the 4 to 8 months that they were under observation. They appeared tame and docile; were attracted to every visible object; examined everything with their mouths; and would indiscriminately chew on cotton or bread soaked in quinine solution. Generally, they were submissive to monkeys, but if molested, would sometimes respond angrily and combatively. Klüver and Bucy[1] briefly noted that their bitemporal lobectomized monkeys did not vocalize for several weeks, or even months, and that when eventually they did, the sounds appeared different from those of normal monkeys. Pribram and Bagshaw make no reference to the presence or absence of vocalization in their animals and give no details regarding the animal's facial expressions and gestures. Two of the macaques (FT5 and FT6) showed some signs of hypersexuality.

Hyperactivity was also observed, and was significantly increased in the one subject in which it was quantitatively measured.

Based on tests for visual discrimination and acuity, Pribram and Bagshaw[6] concluded that the so-called psychic blindness and hypermetamorphosis of the Klüver–Bucy syndrome cannot be attributed to an animal's deficits in such perception. Moreover, adequate performance of the delayed response test indicated that the tendency to persist in examining every visible object does not result from a deficit in "immediate" memory (see Chapter 27).

On the basis of the subtotal resections of the frontotemporal region, the authors tentatively concluded that (1) destruction of the temporal polar cortex, the periamygdaloid cortex, and amygdala primarily accounted for emotional changes; (2) ablation of the posterior orbital cortex was responsible for hyperactivity (but see Davis's findings, Chapter 4); and (3) excision of the anterior insula and adjacent frontal operculum was the cause of taste deficits.

Ablation of Temporal Pole and Lateral Temporal Neocortex

As was noted, Klüver and Bucy found that bilateral removal of the second and third temporal convolutions did not produce any of the symptoms of bilateral lobectomy.[1] In 1961, Akert *et al.*[15] reported their findings in macaques that bilateral ablation of the temporal polar cortex, together with the lateral temporal neocortex (see Chapter 17 and Figure 17-7), resulted in certain manifestations of the Klüver–Bucy syndrome, of which tameness and oral tendencies were the most conspicuous symptoms. These manifestations largely disappeared after a period of 2 to 3 months. In all nine subjects, the entire area in question was consistently destroyed. In five, there was slight encroachment on the piriform cortex and amygdala. Since in all nine cases gliosis could be traced from the temporal polar cortex to the amygdala, the authors were inclined to attribute the Klüver–Bucy manifestations to destruction of the temporal polar region.

Partial Lesions in Infraprimate Species

Several investigators have observed that ablation of the amygdala and the surrounding cortex in species other than monkeys also results in docility and a failure to show aggressive and fearful reactions. Wild rats are noted for their resistance to taming and their persistent fearfulness and ferocity under caged conditions. They will viciously attack cagemates upon slight provocation. J. W. Woods found that electrolytic destruction or surgical ablation of the amygdalar complex in wild rats "uniformly" resulted in a permanent loss (up to 3 years) of aggressiveness.[16] Postoperatively, the animals would not escape even when their cages were left open for long periods; they would not fight with other rats; and they did not bite the experimenter even when provoked.

Schreiner and Kling conducted a comparative study in which they removed the amygdala and surrounding cortex in monkeys, domestic cats, bobcats (*Lynx rufus*), and agoutis (*Dasyprocta agouti*).[17] The latter two species are noted for their wildness and aggressive behavior. Subsequent to surgery, the wild animals failed under threatening situations to show their usual vicious demeanor, signs of fear, and escape activity. Indiscriminate hypersexuality developed in each species under study. In assessing such sexual changes, it should be pointed out that both domestic and wild animals may attempt to copulate with inanimate objects or members of another species. Such behavior, however, usually occurs in a familiar environment.[18] Schreiner and Kling found that under unfamiliar conditions, tomcats engaged in indiscriminate copulatory behavior. They presented an illustration of one tomcat mounting a chicken, a dog, and a monkey (see Figure 5 of Ref. 19).

Avian Forms

Phillips has described loss of ''wildness'' and prolonged escape responses in mallard ducks *(Anas platyrhynchos)* after lesions of the archistriatum (amygdala).[20] He has also observed that lesions of the medial archistriatum in lovebirds *(Agapornis roseicollis)* reduce their mobbing displays and vocalization provoked by unfamiliar objects.[21]

Partial Lesions in Human Beings

A few decades ago, during the high tide of psychosurgery, restricted operations on the temporal lobe were performed in an attempt to alleviate psychotic illnesses or to eliminate assaultive behavior in individuals subject to uncontrollable rage.[22] In 1954, Pool cited the case of a woman with a diagnosis of hebephrenic schizophrenia, who following a bitemporal anteromedial destruction of the temporal lobe was no longer assaultive.[23] She also showed some indiscriminate eating, such as chewing and swallowing a cigarette. Scoville *et al.* performed medial temporal ablations in 19 psychotic patients.[24] Except for one individual, who suffered from a profound anterograde amnesia, they concluded that there were no striking psychological or behavioral changes. H. E. Rosvold, who had the opportunity to examine some of these patients, observed that they showed a tendency to eat both food and nonfood objects.[25] One patient, for example, after eating a bag of potato chips, also ate the bag. Other individuals consumed such inedibles as toilet paper and orange peels.

In 1953, Scoville mentioned that in one of the cases in which aspiration of the medial temporal lobe extended 8 to 9 cm posteriorly from the temporal tip, there was a loss of memory for events occurring subsequent to the operation (anterograde loss of memory).[26] The patient has since been closely followed and subjected to extensive testing. The published findings will be discussed in Chapter 27 dealing with the question of the role of limbic structures in memory.

Role of Amygdalar Division in Social Behavior

Studies on the effects of cerebral lesions on the social behavior of animals are particularly plagued by the kinds of problems that characterized the croquet game in *Alice in Wonderland.* In that game flamingos served as mallets, hedgehogs as croquet balls, and soldiers as wickets. Under such conditions, one or two of these game props might change position and yet allow some semblance of the game to continue. But let there be movement of yet another and still another, the game becomes one of hopeless confusion. This analogy points up the difficulties that experimentalists have in designing experiments involving social groups: there are the difficult questions as to the ideal number of animals to include; the sex and age distribution; the preferred number of animals undergoing the experimental and control operations; and so on. Here I will summarize the findings of a number of studies on the effects of various temporal ablations on the social behavior of animals living under caged conditions and in natural habitats.

Caged Macaques

In 1954, Rosvold *et al.* reported what appears to be the first experiment on the effects of cerebral ablations on the behavior of a group of primates.[27] Their main purpose was to learn how the taming effect of an amygdalar ablation would influence the dominance hierarchy in a group of macaques. In the operation, it was intended to aspirate the medial temporal pole, periamygdaloid cortex, and amygdala. Since the sucker entered the ventricle, it is probable that the hippocampus was also damaged. Observations were made on a group of eight young male monkeys living in a relatively small cage of about 8 by 7 by 5 feet. The dominant animal, Dave, was the first to undergo surgery. Subsequently, this monkey fell from first to last position in the dominance hierarchy. A similar operation was then performed on a monkey, Zeke, after it had achieved the head position. Postoperatively, it became submissive to all animals except Dave and one other monkey. Finally, a third animal, Riva, underwent surgery after becoming the dominant animal. Until the end of the experiment, 2 months later, Riva maintained its dominance over all of the other animals. The authors point out that in this animal, in contrast to the other two, the basolateral nuclei were extensively spared on both sides. There was also no apparent injury to the hippocampus on one side. At this point some of the mentioned problems conducting studies on social behavior become plainly evident: in attempting to explain the maintained dominance of the third operated monkey, the authors were inclined to attribute the outcome to differences in the changed social situation rather than to differences in the extent of the brain lesions.

Dyadic Interactions

Kling observed the dyadic interactions of pairs of unoperated macaques with those of pairs in which the amygdala and periamygdaloid cortex were ablated.[28] In contrast to the normal animals, the operated pairs showed fewer aggressive interactions, and neither animal became "clearly dominant." In addition, they engaged in significantly more "rough-and-tumble play."

Maternal Behavior

In 1958, Klüver mentioned in a formal discussion that subsequent to bilateral temporal lobectomy, female macaques "may show a complete lack of maternal behavior."[29] He did not cite any particular cases. In a personal communication, Pribram told me that his collaborator, Muriel Bagshaw (in an unpublished study conducted over the course of 1 year), observed no changes in maternal behavior of macaques with bilateral ablation of the frontotemporal region. Since Bagshaw and Pribram attempted to avoid injury of the hippocampus in their monkeys, the question arises as to whether or not the discrepancies in their and Klüver's observations may be partly explained by this difference. (See Chapter 21 for further findings on the neural substrate of maternal behavior.)

As already described, bilateral excision of the temporal polar cortex and lateral temporal cortex results in transient manifestations of the Klüver–Bucy syndrome.[15] In 1973 Franzen and Myers reported some observations on the maternal behavior of rhesus monkeys in which these same temporal lobe areas were ablated.[30] They compared the maternal behavior of these animals with that of monkeys with prefrontal, anterior cingulate, and lateral temporo-occipital regions. As described in greater detail in an unpublished doctoral thesis by Kathryn Bucher,[31] the animals with temporal ablations were particularly remarkable because of the loss of protective retrieval behavior. They would, however, passively accept their infants and allow nursing.

Caged Squirrel Monkeys

In 1968 Plotnik[32] reported a study in which he used squirrel monkeys for the purpose of learning whether or not a species of New World monkeys would show the same kind of changes as the macaques in the study by Rosvold *et al.*[27] It was surgically intended to aspirate the inferior temporal gyrus, temporal pole, and amygdala while sparing the hippocampus itself. He made observations on four male squirrel monkeys (named Red, White, Blue, and Green) living in a small cage of about 3 by 3 by 2.5 feet. He tested them in group competitive situations involving positive reinforcement by food and negative reinforcement by electric shock. The most dominant animal was then subjected to surgery and the group retested. This procedure was repeated until all four monkeys had been subjected to the operation. Since each dominant subject showed a marked decrease in dominance and aggressiveness, the author concluded that his findings were essentially in agreement with those of Rosvold *et al.*[27] In addition, he noted that there was no hypersexuality and that, in regard to the question of taming, the monkeys continued to resist being caught and held. There was no mention of vocalization. Although not described, it would appear from the diagrammatic reconstructions that, in addition to the above-mentioned structures, the lesions intruded upon the rostral hippocampus.

As mentioned in Chapter 13, in our behavioral studies on squirrel monkeys we found that among a group of males a quantification of the number of displays by different members is a more reliable measure of dominance than the outcome of rivalry for food. Hence, it is disappointing that this measure was not included in Plotnik's experiments, particularly since, as noted in Chapter 13, extensive coagulations of the amygdala do not interfere with a monkey's capacity to perform the full trump display (see Figures 13-7A and 13-8). As was also mentioned in Chapter 13, the Klüver–Bucy monkey with all of the amygdala and greater part of the hippocampus excised can also perform the display, but may do so inappropriately.

Canids

Subsequent to their study on monkeys (see above), Rosvold and Pribram collaborated with John Fuller, an expert on canid behavior, in observing the effects of excision of the periamygdaloid cortex, amygdala, and rostral hippocampal formation on the individual and social behavior of dogs of different breeds.[33] Operations were performed on three cocker spaniels, two wirehaired terriers, and two beagles. Controls were littermates of the same sex. The three-sentence description of the anatomical findings states that the size of the lesion varied considerably; that except for some sparing of the periamygdaloid cortex in two animals, the amygdala and overlying cortex were completely destroyed; and that the hippocampal formation was destroyed except in its most posterior portion. The paper includes no anatomical reconstructions or histological illustrations. In both the "paired" and group dominance test, the operated dogs fell in dominance. One such terrier, however, would, if attacked, fight back and overpower the aggressor. Changes in sexual behavior were said to have been equivocal. A female cocker spaniel (No. 1192), with "one of the largest lesions," was sexually receptive, became pregnant, and successfully reared a litter of puppies. The authors concluded that the operation had the effect of making the animals less responsive to stimuli in general, but did not prevent appropriate responses if the stimulus was "persistent enough."

Rodents

Comparable to the study on canids just mentioned, Bunnell found that amygdalectomized rats initiated fewer social interactions than the controls and appeared to have a higher threshold to social stimuli.[34] In experiments on hamsters he and his co-workers observed the effects of amygdalectomy on (1) the aggressive member of a pair (11 pairs) and (2) on the submissive member of a pair (7 pairs).[35] The frequency of aggressive actions was reduced in the preoperatively aggressive animals, while the operated submissive ones showed fewer submissive responses. In a word, amygdalectomized hamsters "were both less aggressive and less submissive than normals." Contrary to findings in monkeys, there was no postoperative change in the social status of the dominant and submissive hamsters. The findings of decreased submissiveness in an amygdalectomized animal had not heretofore been reported.

Field Observations on Free-Ranging Monkeys

Because of the restricted cage conditions and the bare setting, the primate social studies that have been described cannot provide an answer as to how brain lesions might affect the behavior of animals living in the wild. As will become evident, however, attempts to observe brain-operated animals under free-ranging conditions may have unacceptable drawbacks of another kind. On the 40-acre island of Cayo Santiago off the coast of Puerto Rico, Dicks *et al.* studied the effects of medial temporal lobe ablations on five subjects from a well-established group of free-ranging rhesus monkeys.[36] The first two animals underwent surgery for ablation of both the amygdala and proximoamygdalar segment of the hippocampus. They were released to their natural surroundings 7 days after operation. The younger of the two monkeys was attacked by a less dominant member of his own group and died of severe wounds 4 days after his release. The other avoided both his own and other groups of monkeys. After once being driven into the sea, it was never seen again. The investigators continued their study with observations on four amygdalectomized monkeys and one control. Two operated immature animals returned to their mothers and were reassimilated into the group, and the control monkey was also accepted. The two other operated animals, however, suffered a fate similar to the original two, one dying 1 week after release and the other after 3 weeks. Since dead monkeys are rapidly devoured by land crabs, the brains of none of the ousted monkeys were recovered!

In an additional five animals from the Cayo Santiago group just mentioned, Myers and Swett ablated the temporal polar and anterior lateral temporal cortex as in the laboratory studies mentioned earlier.[37] Upon their release, the operated monkeys briefly "penetrated" their group, but quickly disappeared into the underbrush. Because of their solitary existence, it was difficult to keep track of them. All eventually succumbed after a period of 1 to 32 weeks, and again there was the unfortunate outcome that their brains were not recovered.

Kling *et al.* had a similar experience with amygdalectomized subjects in a group of 45 cercopithecine monkeys *(Cercopithecus aethiops)* at a reserve in Zambia.[38] All seven of the operated monkeys failed to rejoin their group, although the members of this group, contrary to the Cayo Santiago macaques, made friendly approaches. After 24 hr the operated animals disappeared and, with the exception of one that was spotted near Livingstone 6 months later, were never seen again.

In a review article Kling and Mass have emphasized that if any such studies are attempted again, it will be essential to use radio tracking devices for maintaining contact with the animals.[39] In summarizing the results, they comment that the monkeys sustaining bilateral ablation of the amygdala become social isolates regardless of whether they are attacked, as in the case of the rhesus monkeys of Cayo Santiago, or approached in a friendly manner, as in the case of the cercopithecine monkeys of Zambia. They note that in free-ranging monkeys with medial temporal ablations, one does not observe the Klüver–Bucy manifestations occurring under caged conditions. In conclusion, they point out that an important factor that may "confound" the results of laboratory studies on artificially composed groups of monkeys is that the subjects are not related by familial or group bonds and that the extent to which this affects their social interactions subsequent to cerebral lesions is unknown.

Summarizing Comment

Based on the two primate studies with the most complete anatomical work-up,[6,15] it would appear that different parts of the amygdalar division are involved in self-protective functions, food selection, and the regulation of spontaneous activity. Ablations including primarily the amygdala and the temporal polar cortex seem to account for the inability to discriminate between harmful and harmless situations. In the visual sphere, at least, the apparent tameness and loss of fear cannot be ascribed to an agnosia secondary to the loss of visual discrimination. Studies of the social behavior of primates under caged conditions indicate that extensive destruction of the amygdalar division results in an alteration of hierarchical relationships, with loss of dominance of the operated animal being the usual outcome. Although lacking anatomical confirmation, field studies provide evidence that large lesions of the amygdala and surrounding cortex are prejudicial to a monkey's survival in the wild.

Granted the insufficient description in many cases about the extent of lesions of the amygdala and surrounding cortex, experiments on infraprimate species show concurrence with those on primates in regard to "taming" effects and "loss of wildness," and in some cases, changes in dominance relationships. In some of the studies, hypersexual manifestations comparable to those of the Klüver–Bucy syndrome have been conspicuous. Finally, medial temporal operations in human beings have resulted in some manifestations of the Klüver–Bucy syndrome.

Effects of Stimulation

In considering next the effects of stimulation, we encounter many autonomic manifestations that accompany emotional expression. These effects result from limbic activation of the peripheral autonomic nervous system that innervates all the viscera, glands, blood vessels, and piloerector muscles of the body (Chapter 3). In the course of evolution several autonomic manifestations have been incorporated into an animal's behavioral repertoire so as to be symbolically employed in social communication. For example, piloerection that served originally as a thermoregulatory mechanism eventually acquired secondary value as a social signal. In social communication the very same autonomic manifestation may acquire a dual and opposite significance as, for example, the roughed feathers of a tom turkey that may be displayed during courtship and acquisition (assimilation) of a mate or when aggressively strutting and showing intent to drive away (eliminate) a rival male. In Chapter 23 dealing with an analysis of different forms of affects and their expression as emotion, I point out how the two divisions of the autonomic nervous system, as though mimicking their role in the assimilative and eliminative functions of an organism, serve in social communication to provide signals symbolic of an action intended to "assimilate" or "eliminate" another individual or individuals. In focusing now on some of the autonomic effects of limbic stimulation, it is to be kept in mind that emotional mentation represents the only form of psychological experience that, by itself, may induce pronounced autonomic activity.

Results of Acute Experiments

As early as the latter part of the 19th century, but most notably during the period from 1935 to 1950, several workers reported that stimulation of different parts of the limbic lobe in anesthetized animals variously resulted in respiratory, cardiovascular, gastrointestinal, and other autonomic-related responses. Since this work has been the subject of several reviews, I will draw upon an extensive comparative study by Birger Kaada published in 1951[40] to summarize the results of experiments of this kind. Kaada reviewed the literature up until that time, including a table citing some of the most significant papers. His own experiments were undertaken in 1948 while at the Laboratory of Physiology, Yale University School of Medicine, and included observations on cats, dogs, monkeys, and one chimpanzee. In his exploration of the limbic lobe he found two main areas of limbic cortex where stimulation elicited the kinds of responses in question. As indicated by the density of stipple in Figure 19-3A and B, the focus of one area is in the basal frontotemporal region and the other in the pregenual cingulate cortex. The stipple in the basal frontotemporal region coincides with cortex that is defined here as part of the amygdalar division of the limbic system. In Figure 19-3D, the areas enclosed by dashed lines indicate where stimulation elicited pyloric contractions, while the up- and down-directed arrows identify loci at which stimulation induced a rise or fall in blood pressure. The points marked with an asterisk and an S, respectively, refer to positive sites for piloerection and salivation. Kaada noted that pupillary dilatation, usually of slight degree, resulted from stimulation of the same regions involved in respiratory inhibition (Figure 19-3A and B). Illustrative recordings of blood pressure, respiratory, and gastrointestinal changes are shown in Figure 19-4. Finally, Figure 19-3C shows loci at which stimulation elicited mastication (encircled M) and vocalization (V). Kaada noted that vocalizations varied from coos to high-pitched sounds and cries, and were qualitatively the same for the active loci both in the anterior hippocampal gyrus and in the precingulate cortex. There were a few sites in the amygdala and elsewhere at which stimulation elicited defecation and urination. An important autonomic response that could not have been demonstrated effectively under anesthesia is that of genital tumescence, which will be considered later on.

In summary, the results of Kaada's stimulation studies and those of preceding workers demonstrated that the frontotemporal cortex of the amygdalar division now under consideration, as well as midline frontal limbic cortex to be dealt with later, is implicated in a wide range of vital visceral and viscerosomatic responses.

Observations on Chronically Prepared Animals

Findings in Cats

In connection with his studies, Kaada prepared eight cats with electrodes implanted in the amygdala and other parts of the brain.[40] On the following day, he tested the effects of electrical stimulation on the behavioral responses of the animals.[40] He found that stimulation at different sites within the amygdala resulted in sniffing, rhythmic chewing movements, and arrest of spontaneous movements, including respiration. In experiments begun in 1950, Delgado and I encountered additional kinds of responses, including forms of behavior seen in attack and defense.[41] We used a modified multielectrode needle that Delgado had devised for brain stimulation and electroencephalographic recording in chronically prepared, freely moving animals. A new kind of stimulator afforded independent control of the stimulus pulse width, stimulus frequency, and current. We made observations on seven cats and four squirrel monkeys in which one or two multielectrode probes were implanted in the frontotemporal region of each hemisphere. We usually applied bipolar stimulation through electrodes 2 mm apart. Figure 19-5, showing responses elicited at various loci in the cat, serves as a reference for summarizing the findings and comparing the results obtained by other investigators. As phrased in the original description, the responses in general were automatisms seemingly related to eating, feeding, or the animal's search for food and its struggle for survival. Responses in the feeding category included sniffing, chewing, chop licking, licking, swallowing, gagging, and retching. Such responses were elicited by stimulation at points throughout the rostrocaudal extent of the amygdala, but located predominantly in the region of the medial and basal nuclei and in the piriform area.

As in the case of alimentary manifestations, responses with components of defense and attack were elicited by stimulation at widespread points in the amygdalar region. Such responses included pupillary dilatation, piloerection, vocalization, ipsilateral retraction of the ear, ipsilateral snarling, pawing with extended claws, and protective movements of the head and trunk. Growling, and sometimes hissing, "could be obtained from a region extending from the olfactory tubercle through the medial part of the amygdala to the rostral part of the hippocampus."[42] It was shown that with repeated stimulation at a locus accounting for hissing and defensive pawing with extended claws, the response might change to one of attack. There were other instances in which the defensive reaction would be replaced by flight and attempts to climb out of the viewing stage (see locus for climbing in Figure 19-5C). At times when the cat was not moving, it was possible to obtain

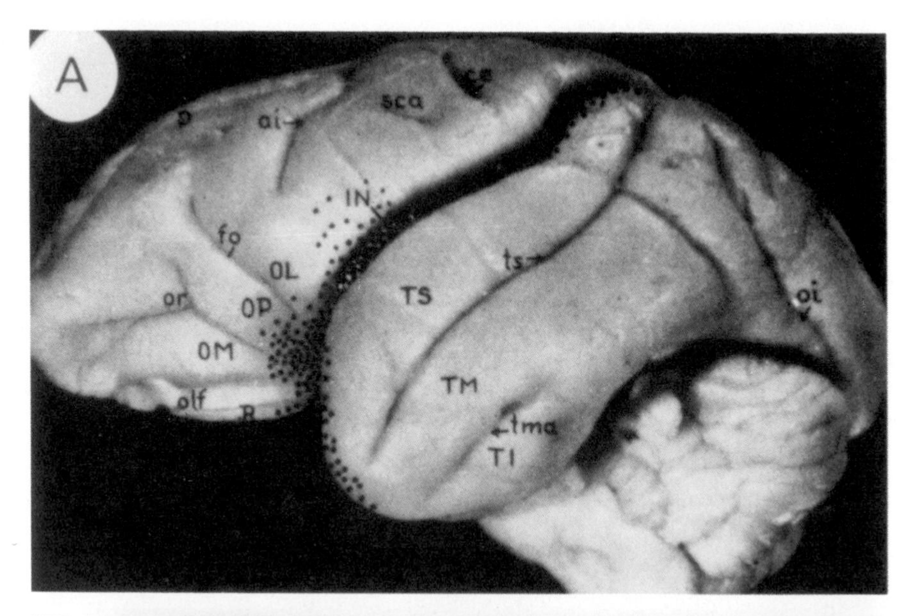

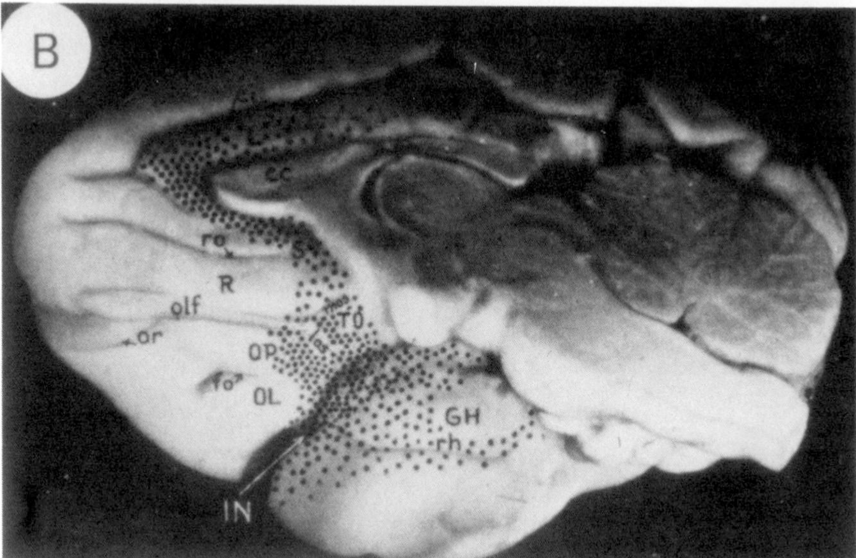

Figure 19-3. Kaada's illustration, showing loci in macaque's brain at which stimulation elicited autonomic and somatic responses.

(A, B) Views of lateral and ventromedial surfaces, with dots indicating cingulate and frontotemporal areas where stimulation resulted in inhibition of respiratory movements with arrest in expiration.

(C) Loci identified with vocalization (V) and mastication (encircled M).

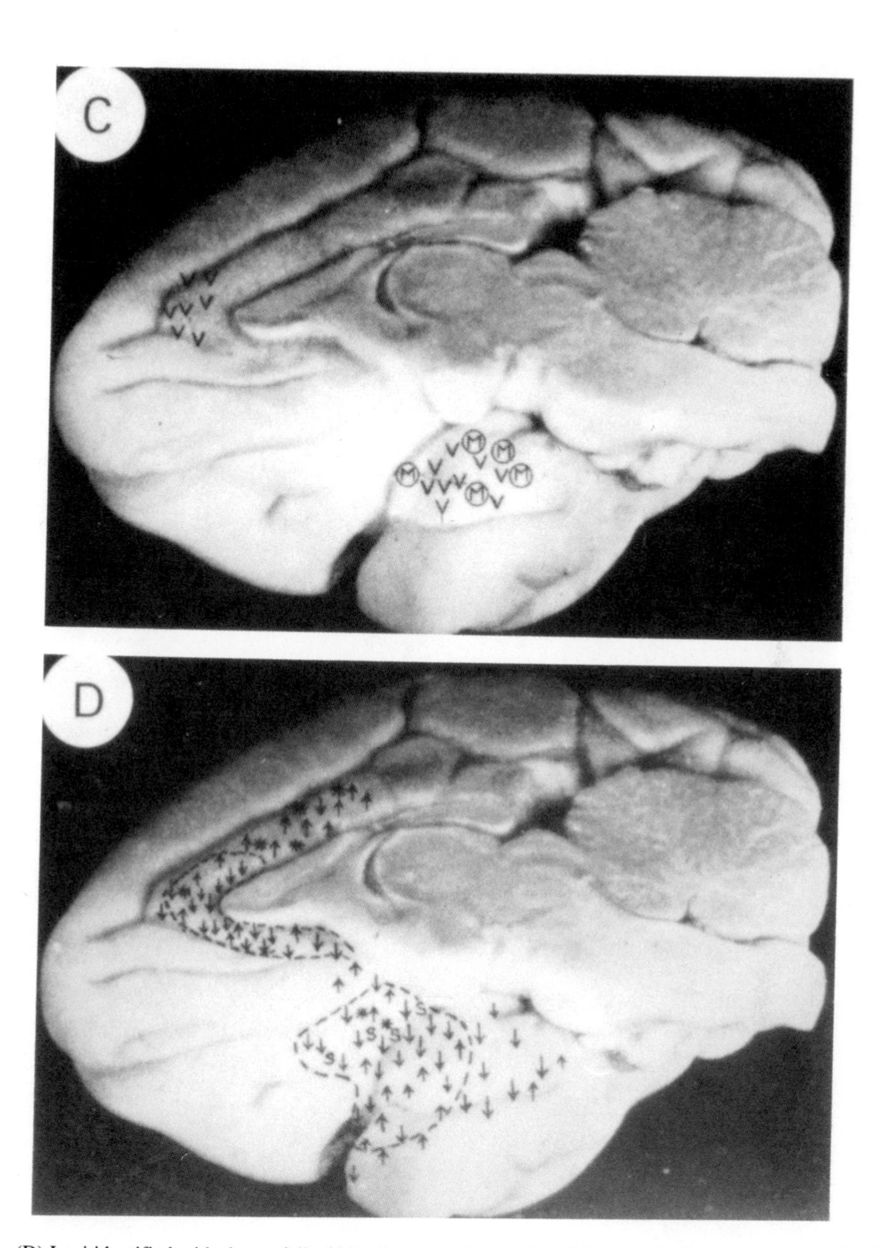

(D) Loci identified with rise or fall of blood pressure (upward- and downward-directed arrows); salivation (S); and piloerection (*).

Abbreviations: CC, corpus callosum; GH, hippocampal gyrus; IN, insula; L, anterior cingulate gyrus; OL, lateral orbital gyrus; OM, medial orbital gyrus; OP, posterior orbital gyrus; R, gyrus rectus; S, subcallosal gyrus; TI, inferior temporal gyrus; TM, middle temporal gyrus; TO, olfactory tubercle; TS, superior temporal gyrus.

From Kaada (1951).

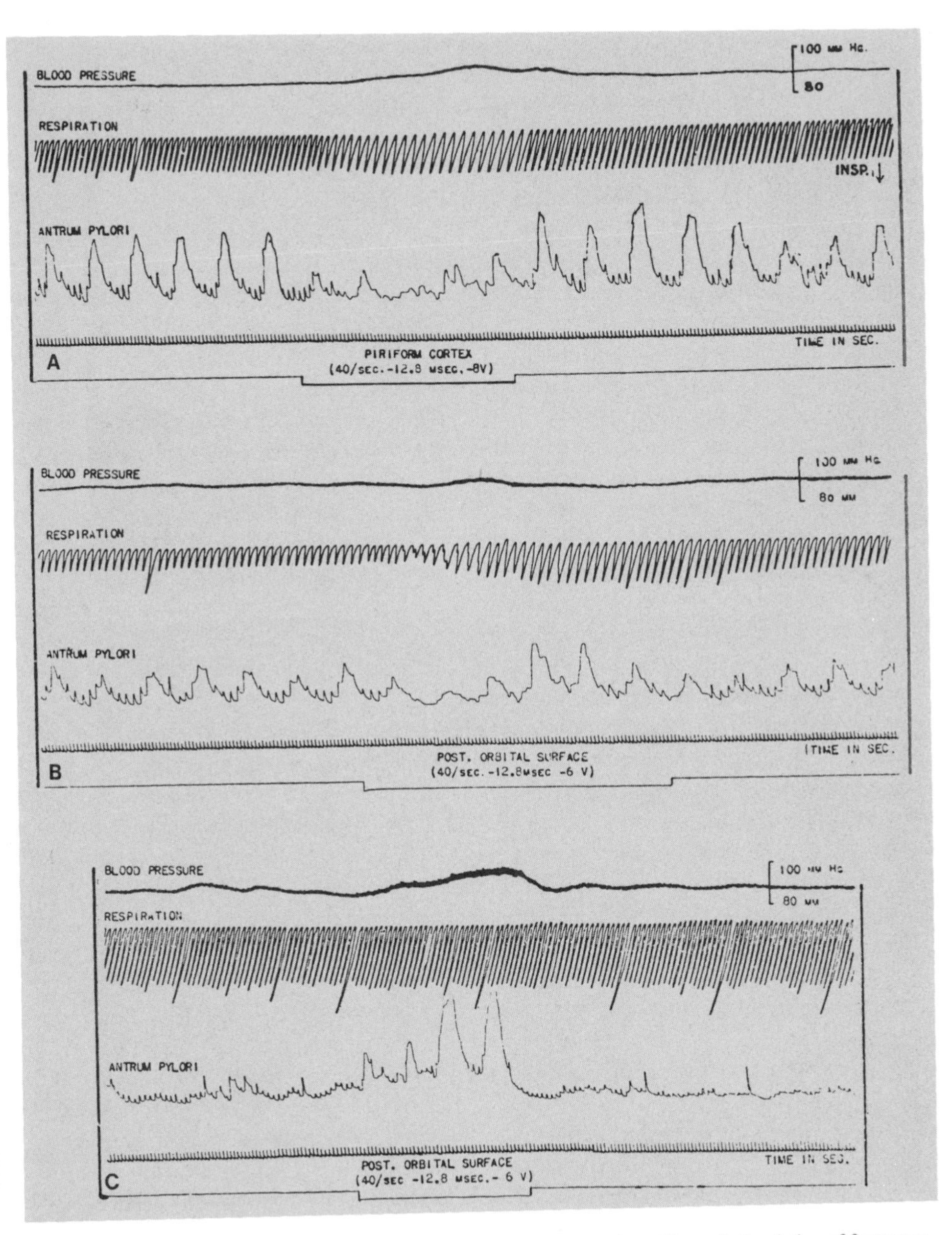

Figure 19-4. Recordings from Kaada's experiments on macaques, showing effects of stimulation of frontotemporal cortex on blood pressure, respiration, and pyloric activity. In each example the site stimulated is given above the last tracing showing duration of stimulation. Note that stimulation of posterior orbital cortex may elicit either inhibition (B) or augmentation (C) of pyloric motility. From Kaada (1951).

satisfactory recordings of cardiorespiratory changes. Figure 19-5 shows loci at which stimulation elicited slowing of the heart rate and acceleration or inhibition of respiration.

Stances and movements suggestive of pause, alerting, or searching were observed with stimulation involving the piriform area and lateral amygdala. Sometimes the arrest responses were dramatic, with the cat suddenly freezing in its tracks and maintaining that posture as long as stimulation was applied.

Gastaut and his colleagues in Marseilles had also stimulated the amygdala in awake, unrestrained cats and had observed "emotional" responses including those that they characterized as fear and anger, but did not specify the effective loci.[43] Magnus and Lammers, Fernandez de Molina and Hunsperger, Norris and Lanauze, and others have expressed general agreement with the localization of the effective loci for the defensive manifestations described here.[44]

Fernandez de Molina and Hunsperger dealt particularly with the question as to what amygdalar structures and pathways are involved in the defense reaction.[45] I was in Zurich during the year of their investigation (1956–1957) and witnessed a number of their experiments. Figure 19-6 shows the loci where they stimulated and elicited defense responses in the cat. The low-frequency stimulation used by them permitted observation of a step-by-step development of the response. Shortly after the onset of stimulation, the cat turned its head toward the observer, after which there was a widening of the eyelids; dilatation of the pupils; and increase in the rate and depth of respiration; a low growl; a laying back of the ears; a lowering of the head; and piloerection of the back and tail. At this point the cat might stand up with an arched back and continue to growl. Toward the end of stimulation the growls might be replaced by repeated hissing as the mouth was opened wide with a baring of the fangs. The cat would not attack, but might suddenly retreat in flight. The solid squares show the loci for responses combining growling, hissing, and flight. It is evident that the loci for the combined responses are found predominantly in the part of the amygdala traversed by the stria terminalis and in the bed nucleus of the stria in the septum (Figure 19-6B). The question as to what are the effective pathways for the response will be dealt with in the next chapter. In anticipation of a discussion of this matter, it is to be noted that in the amygdala the more lateral loci for the defense manifestations are near the fasciculus profundus.

Findings in Squirrel Monkeys

In the original study with Delgado we undertook for comparative purposes some additional experiments on squirrel monkeys.[41] The findings were generally similar to what had been observed in cats, but in a number of instances were more dramatically expressed. For example, during stimulation at a site in the tubercle that elicited chewing, the monkey was given a banana, which it proceeded to eat voraciously. The locus of stimulation was then changed to the piriform cortex; immediately the monkey stopped eating, made grimaces of distaste, spit out the banana into its hand, and threw it down. Defense reactions were also particularly vivid in nature. In one case, for example, bipolar stimulation in the ventricular area near the stria terminalis (see Figure 5A of Ref. 41) evoked piercing cries, a look of "alarm," ferocious biting of a leather glove, and well-directed escape.

The pilot studies just mentioned indicated that the squirrel monkey would be a

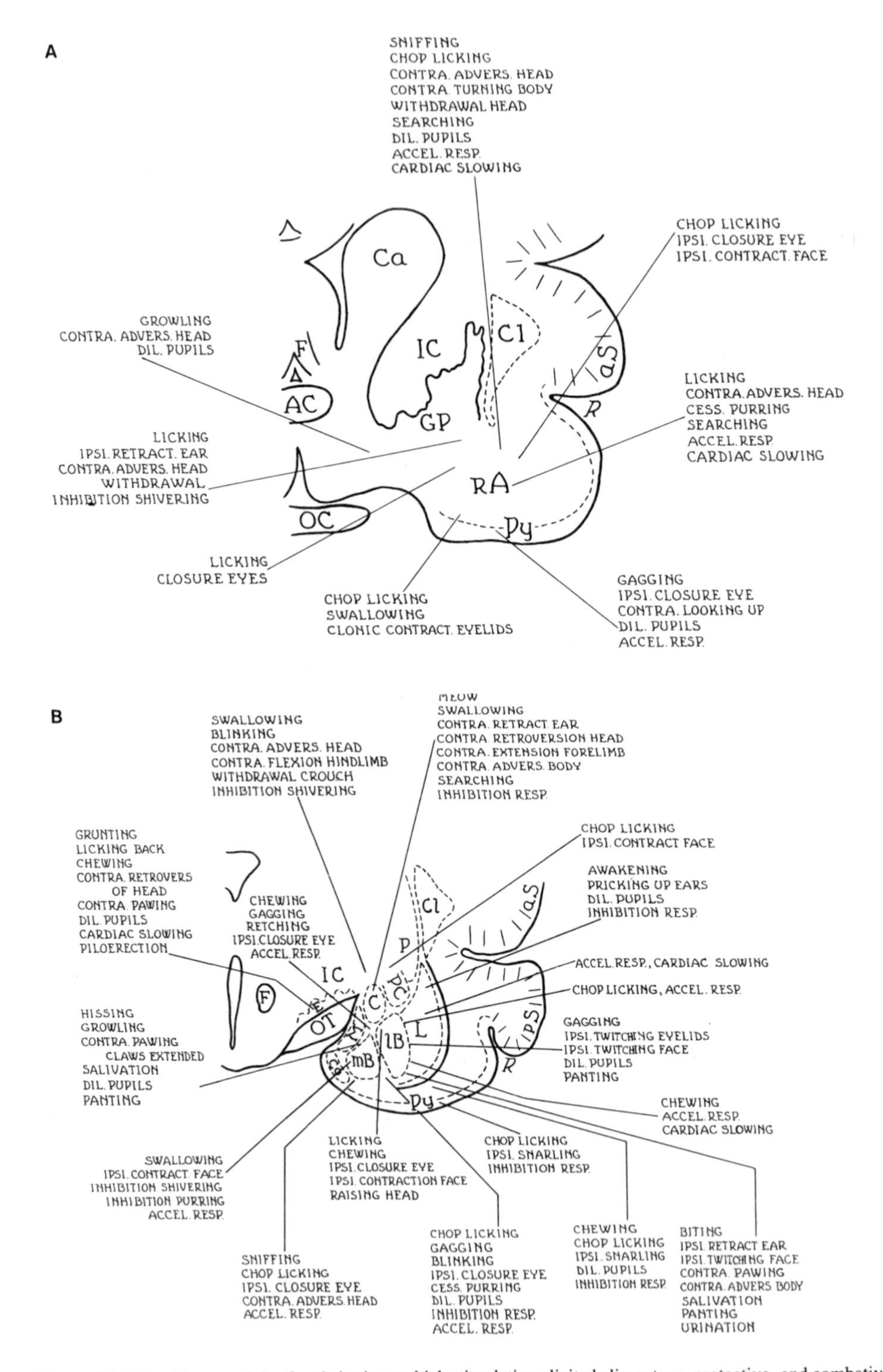

Figure 19-5. Loci in amygdala of cat's brain at which stimulation elicited alimentary, protective, and combative manifestations. A, B, C: Representative frontal sections through rostral, middle, and posterior amygdala.

Abbreviations for responses: accel., acceleration; advers., adversion; cess., cessation; constrict., constric-

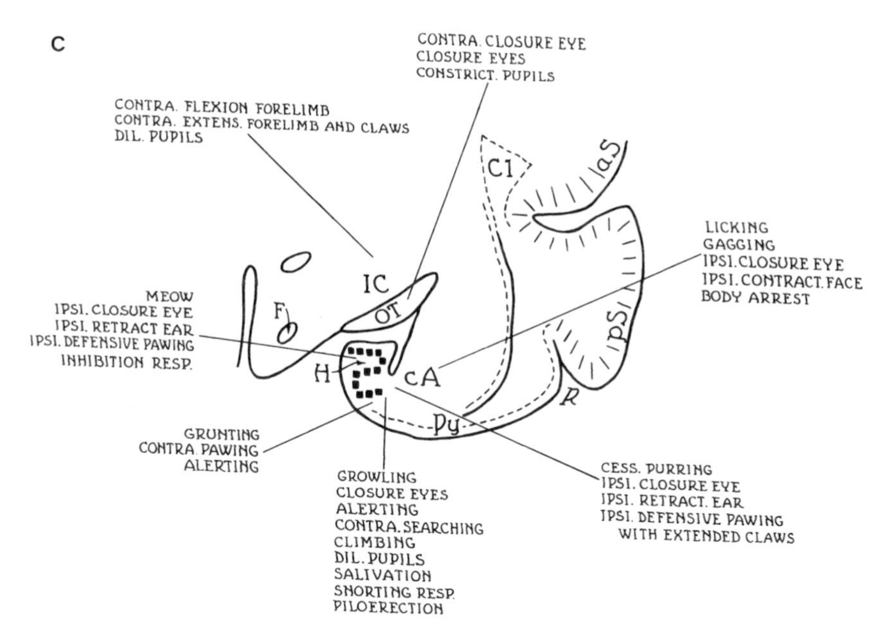

tion; contra., contralateral; contract., contraction; dil., dilatation; extens., extension; ipsi., ipsilateral; resp., respiration; retract., retraction.

Anatomical abbreviations: RA, rostral amygdala; cA, caudal amygdala; AC, anterior commissure; lB, basal nucleus of amygdala, lateral part; mB, basal nucleus of amygdala, medial part; C, central nucleus of amygdala; Ca, caudate nucleus; Cl, claustrum; Co, cortical nucleus of amygdala; E, entopeduncular nucleus; F, fornix; GP, globus pallidus; H, hippocampus; IC, internal capsule; L, lateral and M, medial nucleus of amygdala; OC, optic chiasm; OT, optic tract; P, putamen; PC, putamen–central amygdaloid complex; Py, piriform cortex; R, rhinal fissure; aS, anterior sylvian gyrus; pS, posterior sylvian gyrus.

From MacLean and Delgado (1953).

valuable experimental subject for neurobehavioral studies concerning anatomical and physiological questions of basic interest in regard to the evolution of primates (see Chapter 13). Consequently, I subsequently developed a method for a systematic step-by-step exploration of the brain with stimulating or recording electrodes in chronically prepared, awake, sitting monkeys.[46] Since this method was used in many of the experiments to be described here and in later chapters, I will give a brief account of its main features.

Method for Stimulation and Recording[46]

Figure 19-7 shows a view from above and from the side of the stereotaxic platform with guides for introducing electrodes into the brain in the same frontal plane as that for the stereotaxic brain atlas. As indicated the platform is placed above the scalp and is attached to four screws previously cemented in the skull. This device makes it possible to avoid open surgery at the time of an experiment and provides a closed system for lowering electrodes into the brain at any desired increment. In the amygdalar experiments, as well as in others to be described later, macroelectrodes are permanently implanted in the hippocampus for recording its bioelectrical activity. No more than eight exploring electrodes are introduced in one monkey, and no more than two experiments are conducted in

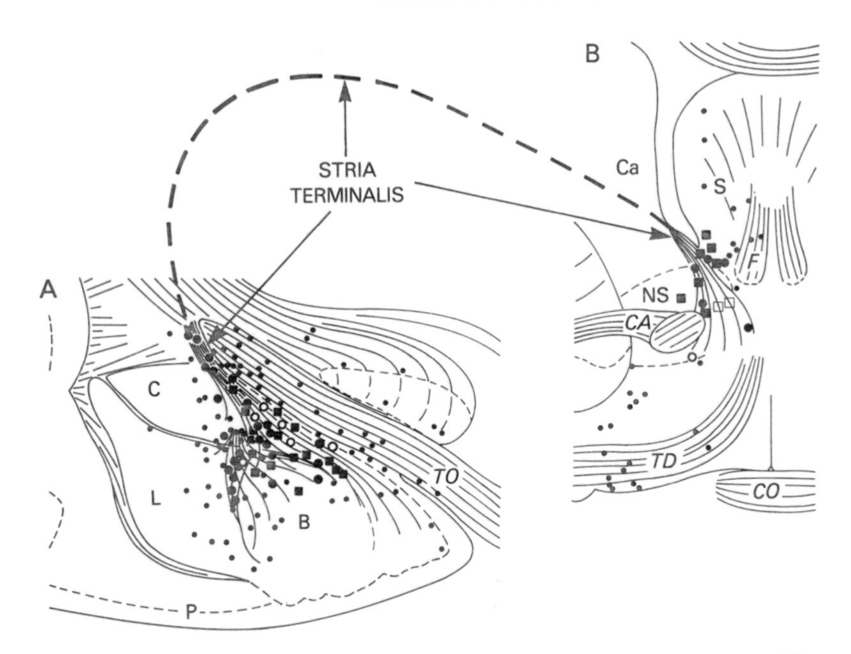

Figure 19-6. Amygdalar (A) and stria terminalis (B) circuit found by Fernandez de Molina and Hunsperger to be involved in defensive or combative reactions of the cat.

Key to responses elicited by electrical stimulation at loci denoted by symbols: large black dot—growling response; black square—growling followed by hissing, and sometimes flight; white square—primary hissing response; white circle—primary flight response; small dot—response other than agonistic manifestations.

Abbreviations: B, nucleus amygdalae basalis; C, nucleus amygdalae centralis; Ca, nucleus caudatus; *CA*, commissura anterior; *CO*, chiasma nervorum opticorum; *F*, fornix; L, nucleus amygdalae lateralis; NS, nucleus striae terminalis; P, piriform cortex; S, area septalis; *TD*, tractus diagonalis Broca; *TO*, tractus opticus.

After Fernandez de Molina and Hunsperger (1959).

1 week. During an experiment the monkey sits in a special chair and is given its favorite forms of liquid nourishment. Between experiments the monkeys live in their home cages and have complete freedom of movement. In a usual experiment involving electrical stimulation, two electrodes spaced 2 mm apart are lowered in 0.5- or 1-mm increments. Stimulation is performed at each point with the three different sets of parameters listed in Table 19-1. In the event of a local afterdischarge or a propagated afterdischarge to the hippocampus, there is a waiting period of 5 min before performing the next stimulation. This particular lapse of time is a compromise choice based on studies of the recovery cycle.[47] Once a track is explored, the electrode is fixed at a desired site so that it may be used again for stimulation and recording and also so that gliosis will outline the shank and tip for histological examination. Upon completion of the experiments, serial frozen brain sections alternately stained for cells and fiber are prepared for reconstructing the electrode tracks and plotting the points of stimulation. Accuracy is assured by blocking and sectioning the brain in the same frontal plane as the electrode tracks. Drawings of the sections with the electrode tracks are made on 1-mm graph paper at a 10 × magnification. The investigators themselves carry out all of these procedures except the mounting and staining of sections, which are performed by a histologist.

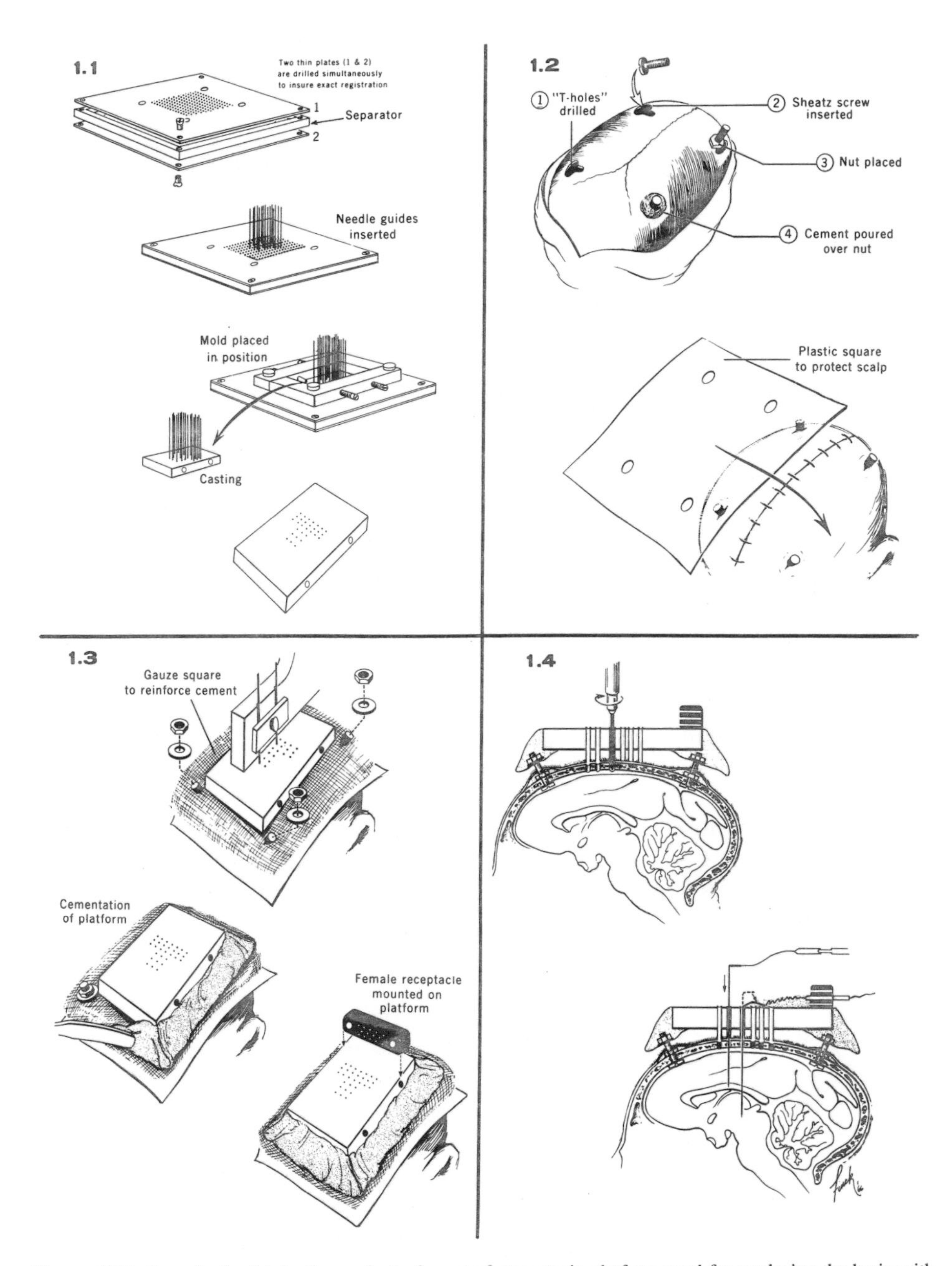

Figure 19-7. Steps in the fabrication and attachment of stereotaxic platform used for exploring the brain with stimulating and/or recording electrodes. In addition to providing an essentially closed system, the method makes it possible to obtain histological sections in the same plane as the electrode tracks. From MacLean (1967).

Table 19-1. Range of Stimulation Parameters Used in Exploration[a]

Pulse repetition rate	Pulse duration	Train duration	Peak current
4–10/sec	1–4 msec	30–60 sec	0–2 mA
20–30/sec	1.0–1.5 msec	10–20 sec	0–2 mA
60–100/sec	0.01–0.05 msec	10–20 sec	0–7 mA

[a]From MacLean and Ploog (1962).

Facial and Oral Responses

In experiments involving stimulation of the amygdala in nine squirrel monkeys, we systematically noted oral and facial responses, as well as specific movements of the head, extremities, trunk, and tail.[48] In Figure 19-8 I have used a circle with compass markings for denoting loci for facial and oral responses; the reference key shows the compass markings respectively identified with retching, contralateral contractions of the face, swallowing, chewing, licking, ipsilateral contraction of the face, and salivation. The use of the round symbol gives emphasis to the finding that stimulation within the amygdalar region elicits predominantly facial, oral, and alimentary responses. Ipsilateral contraction of the face was the most common response, occurring with stimulation at nearly 60% of the loci. Ipsilateral contractions were five times more frequent than contralateral contractions. In a study on macaques, Frost *et al.* had observed that upon stimulation of the amygdala, movements of the face always began on the ipsilateral side and that subsequent activation of the facial musculature on the contralateral side was abolished by interrupting the anterior commissure.[49] Of the oral responses noted in our study, chewing was the most frequent (34%) followed by salivation, licking, swallowing, and retching. Encircled V's refer to a chirping type of vocalization, while encircled *inverted* V's indicate a piercing type of vocalization or growls (see Chapter 21). The agonistic vocalization was accompanied by baring of the fangs and was usually elicited by stimulation near the origin of the stria terminalis. Loci for chirping types of vocalization were widely distributed, but as evident in Figure 19-8 (A9) were found predominantly along the course of fasciculus profundus.

Cardiovascular Responses

In our laboratory, Reis and Oliphant conducted a special investigation of cardiovascular changes elicited by stimulation in the amygdalar region.[50] They found that cardiac slowing, often complicated by extrasystoles, was the most common response. In some instances extrasystoles were recorded for as long as one-half hour after stimulation. It was demonstrated in anesthetized animals that such effects could no longer be elicited after vagotomy or atropinization. In awake sitting animals, a stimulus rate of 6 per sec was much more effective than a fast pulse repetition rate in eliciting bradycardia. Stimulation at an intermediate rate of 30 per sec ''indirectly favored'' the appearance of bradycardia because of the tendency of such stimulation to produce afterdischarges. Using data recorded in Reis and Oliphant's protocols, I have plotted in Figure 19-9 the anatomical loci identified with the greatest degree of cardiac slowing under the four conditions in question, i.e., with stimulation at 6, 30, and 100 per sec, or in association with afterdischarges propagated to the hippocampus. Afterdischarges induced at the site of the stimulating

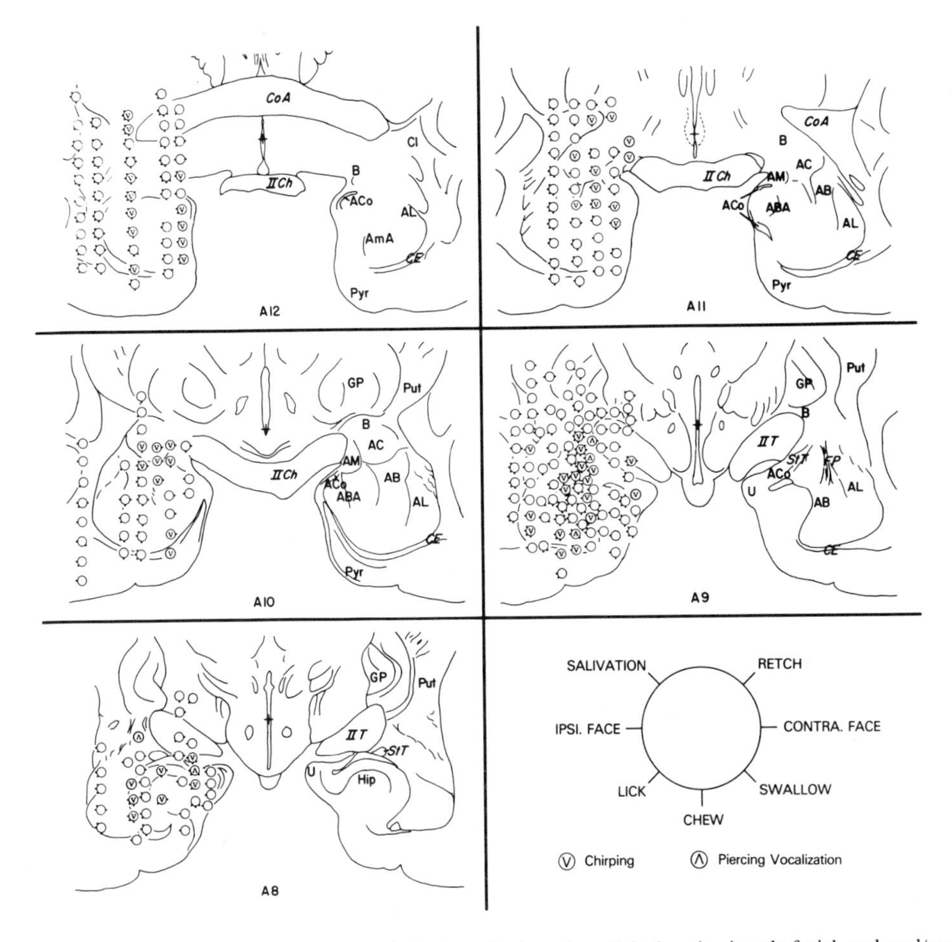

Figure 19-8. Loci at which amygdalar stimulation in squirrel monkey elicited predominantly facial, oral, and/or alimentary responses. Key to compass markings given in lower right-hand corner. Brain diagrams depict atlas drawings of the amygdala at millimeter intervals from A12 to A8.

Abbreviations: AB, nucleus amygdalae basalis; ABA, nucleus amygdalae basalis accessorius; AC, nucleus amygdalae centralis; ACo, nucleus amygdalae corticalis; AL, nucleus amygdalae lateralis; AM, nucleus amygdalae medialis; AmA, area amygdaliformis anterior; B, nucleus basalis; *CE*, capsula externa; *IICH*, chiasma nervorum opticorum; Cl, claustrum; *CoA*, commissura anterior; *FP*, fasciculus profundus; GP, globus pallidus; Hip, hippocampus; Put, putamen; Pyr, cortex piriformis; *StT*, stria terminalis; IIT, tractus opticus; U, uncus.

electrode were almost invariably associated with propagation of the discharge to the hippocampus. It is immediately apparent that the most marked cardiac effects occurred most often in association with afterdischarges.

In the same experiments it was demonstrated that bradycardia occurred independently of changes in blood pressure, often preceding a moderate rise in blood pressure by many seconds.[50] Hilton and Zbrozyna showed in cats that amygdalar stimulation resulted in an increase of blood flow to the striped musculature involved in the defense reaction and that the vascular response could be blocked by atropine.[51] Like the autonomic innervation of the integument, the vascular dilatation apparently represents a sympathetic response induced by cholinergic mechanisms.

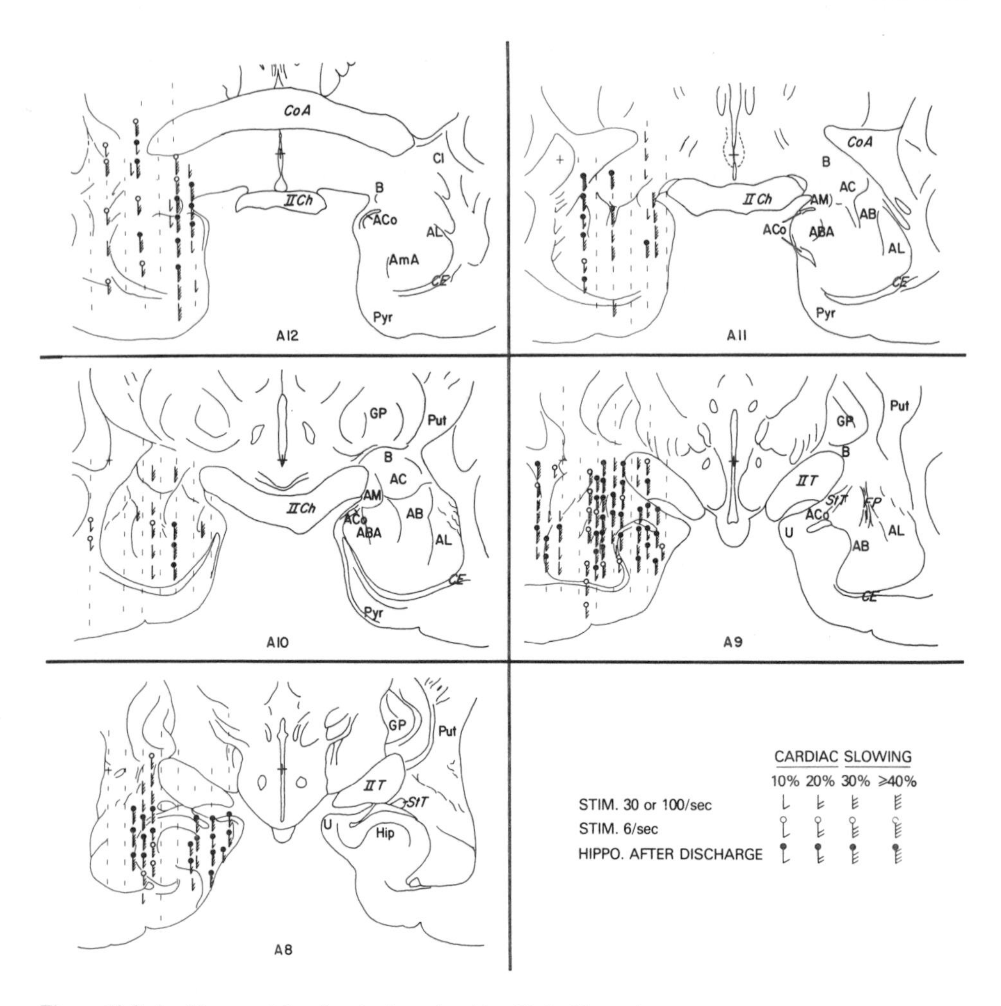

Figure 19-9. Loci in amygdala of squirrel monkey identified with maximum cardiac slowing under conditions stipulated in lower right-hand corner. Vertical dashes (spaced 1 mm apart) identify loci at which stimulation failed to elicit cardiac deceleration. Same atlas levels and abbreviations as in Figure 19-8.

Oral–Genital Responses

When the results of stimulating various parts of the septal and thalamocingulate divisions of the limbic system are reviewed, it will be emphasized that stimulation at certain loci in these circuits regularly elicits penile erection at relatively fixed latencies. Since amygdalectomy may result in a "release" of sexual behavior and since in many species feeding may be transiently associated with genital tumescence, the question of the effect of amygdalar stimulation on genital function is of special interest. In a systematic exploration of the amygdalar region in eight squirrel monkeys (using the parameters of stimulation shown in Table 19-1), we found no loci at which stimulation consistently elicited penile erection at a relatively fixed latency.[52] In both these and the original studies to be described, the magnitude of erection was graded on a six-point scale ranging from ± to 5+, the maximal response seen under natural conditions. Of the 270 points shown in Figure 19-10, there were 57 at which stimulation elicited some degree of penile erection. The most effective locus for stimuli at 30 or 100 per sec was in the region of fasciculus profundus. The magnitude of the response did not exceed 3+, and latencies were highly variable. With 6 per sec stimulation or during hippocampal afterdischarges, the response never exceeded 3+. The heavy dots in Figure 19-10 indicate loci at which rebound erection of 1 to 2+ occurred following stimulation and without an afterdischarge. There were 37 loci at which stimulation at 6 per sec elicited erection. Tumescence appeared only after a long latency, averaging 48 sec, and was invariably preceded by one or more of the following facial and oral manifestations listed in order of incidence—namely, ipsilateral

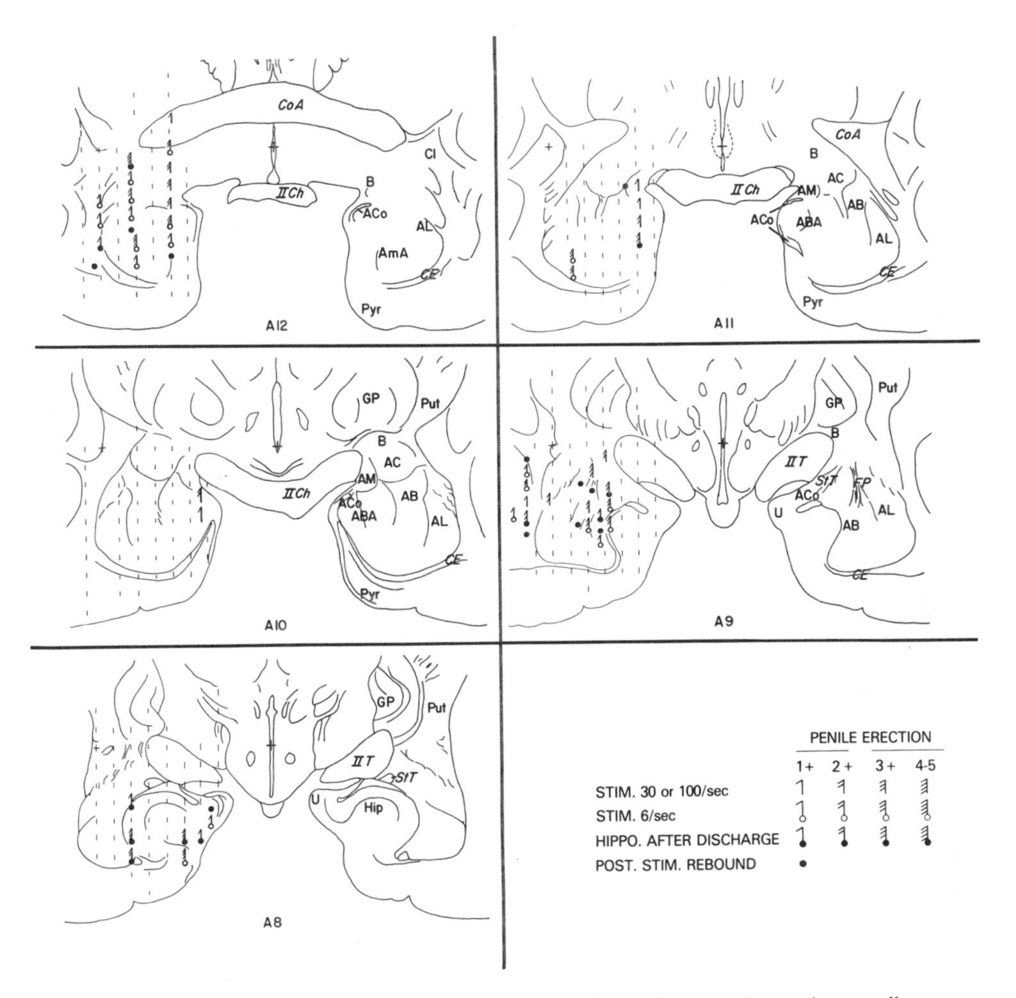

Figure 19-10. Loci in amygdala of squirrel monkey where stimulation elicited penile erection, usually occurring at varying and long latencies. Key to symbols in lower right-hand corner gives conditions under which erection was observed, as well as magnitude of the response. Same atlas planes and abbreviations as in Figure 19-8.

contractions of the face (78%), chewing (70%), salivation (60%), licking (5%), and retching (5%). Other associated effects were urination (46%), cardiac slowing (35%), respiratory changes (24%, including three instances of poststimulation gasps), and defecation (5%).

Neurohumoral Influences of Amygdalar Division

An extensive literature has accumulated with respect to the role of the amygdala and other limbic structures in neurohumoral mechanisms. Many of the studies are difficult to evaluate because of uncertainties about the extent of lesions or the sites of stimulation; differences in methods of stimulation, including the frequency and duration of electrical stimulation; failure to record electroencephalographic activity; and so on. Here I will attempt to do no more than to single out salient findings in certain studies.

Apropos of the role of the amygdalar division in agonistic forms of behavior, the original findings of Mason are of particular interest because they show that amygdalar stimulation (at unspecified loci) in macaques results in an elevation of plasma 17-hydroxycorticosteroid (17-OH-CS) levels comparable to those resulting from hypothalamic stimulation or the injection of an effective dose of adrenocorticotropic hormone (ACTH).[53] In this respect, McHugh and Smith made a significant contribution by showing that such an elevation did not occur unless stimulation produced a local afterdischarge.[54] Slusher found that cortisol implants in the ventral hippocampus in rats resulted in a significant morning increase of plasma corticosteroids.[55] Hayward demonstrated that amygdalar stimulation activates the neurohypophysis, having an antidiuretic effect comparable to what occurs following noxious stimulation, "emotional stress," and hypoxia.[56] There are also several reports that amygdalar stimulation triggers the release of thyrotropic hormone.[57]

Since the introduction of the combined use of autoradiography and tritiated gonadal hormones, it has been found in many species that the amygdala and other limbic structures possess cells having an affinity for estradiol[58] and testosterone.[59] As clearly demonstrated in rats,[60] cats,[61] and monkeys,[62] one finds labeling of amygdalar cells in the region associated with the stria terminalis, particularly in the medial nucleus and the posteromedial cortical nucleus adjacent to the hippocampus. There is also labeling of the bed nucleus of the stria terminalis in the septum. Amygdalectomy in young animals has been reported to result in atrophy of the endocrine glands,[63] including the gonads.[64] Based on a wide assortment of data, Sawyer suggests that the corticomedial amygdala of the female rat appears to project via the stria terminalis to two groups of neurons in the hypothalamus—one group that is "inhibitory to gonadotrophic function in general" and the other being "facilitatory to the ovulatory surge of pituitary LH release."[65] In this respect it is of interest that Elwers and Critchlow found that electrolytic destruction of the medial amygdala resulted in precocious ovarian development.[66] Stimulation of the corticomedial region in immature female rats, on the contrary, had the effect of delaying the onset of puberty.[67]

Finally, with respect to differences between male and female temperaments, it is of possible significance that the medial nucleus of the amygdala and the associated bed nucleus of the stria terminalis, which are involved in combative behavior, are the only two telencephalic structures showing an affinity for unaromatized androgen.[68]

Summarizing Comment

Overall, the behavioral responses elicited by stimulation of the amygdalar region in cats and monkeys fall into two categories: (1) In the first category are olfactory and investigatory responses, together with oral and alimentary automatisms, that one associates with the animal's search for food and its ingestion; (2) in the second category are components (or sequences) of behavior seen in defense or attack, including vocalization, biting, and defensive or combative postures and movements. Associated autonomic-dependent manifestations include lacrimation, salivation, pupillary effects, piloerection, cardiorespiratory changes, and sometimes urination and defecation. The findings that loci for chewing, swallowing, and the like are intermixed with those for searching, fighting, and self-defense indicate that the mechanisms for feeding are intimately geared in with those required for obtaining food. To a remarkable degree, the findings upon stimulation are the opposite of those following ablation, involving deficits in self-protection and feeding that would be prejudicial to survival in a natural environment. In a word, the amygdalar division of the limbic system appears to be "largely concerned with self-preservation as it pertains to feeding and to the behavior involved in the struggle to obtain food."[69] Several neuroendocrinological findings are also indicative that the amygdalar division plays an important role in the struggle for survival.

As will be mentioned in the next section, and further described and illustrated in the following chapter, it is probable that in the case of the oral–genital responses seen with 6 per sec stimulation, the tardy appearance of penile erection is due to recruitment of neural activity in amygdala-related structures involved in sexual functions.

THE SEPTAL DIVISION

By midcentury numerous studies had demonstrated the important role of the hypothalamus in neuroendocrinological aspects of sexual behavior. But, curiously enough, there was almost no information about the representation in the forebrain of such elemental sexual functions as genital tumescence and seminal discharge. "It was as though nature, a master playwright and producer, had staged a new production and inadvertently overlooked a part for one of the main actors."[70] Penfield, who had stimulated the greater part of the lateral and medial cortex in patients undergoing neurosurgery, had never elicited genital tumescence or symptoms of an erotic nature.[71]

The oral–genital responses that were described above in connection with the amygdalar stimulation raise the question as to what cerebral structures account for the late occurrences of penile erection. This question provides a natural point of departure for considering the role of the septal division in genital and other procreational functions. It should be said by way of introduction that on the basis of what was known about the hippocampus in the early 1950s, there was little to indicate that it was more than a useless appendix. For example, it was a matter of considerable interest and surprise that either during electrical stimulation of the hippocampus or during a subsequent high-voltage afterdischarge (commonly referred to as a seizure discharge) there appeared to be few, if any, changes in an animal's behavior.[72] Later in the decade, the use of conditioning techniques was to show that however little effect propagated hippocampal seizures might have on the behavior of an animal, they produced a profound change in its state of awareness. In the meantime, the emphasis placed on negative findings, together with the considerations outlined in Chapter 4, led me to undertake the experiments next to be described in which it was the purpose to learn whether or not chemical stimulation of the hippocampus with cholinergic agents might elicit behavioral changes that would not otherwise be apparent.[73] It was this study that suggested that the septal division of the limbic system is involved in procreational functions.

Effects of Stimulation

Chemical and Electrical Stimulation of Proximoseptal Hippocampus in Cats

The method used for either chemical or electrical stimulation of the same site in the brain and recording the locally induced bioelectrical changes was described in Chapter 4. As background for assessing the effects of chemical stimulation, I will describe first the behavioral changes occurring during electrical stimulation of the proximoseptal hippocampal segment, as well as during electrically induced hippocampal afterdischarges. In the cat, the behavioral response during hippocampal stimulation may amount to no more than a turning of the head to the opposite side, a pricking of the ears as though attentive, and a dilatation of the pupils.[74] If an afterdischarge is induced, any turning movements that were present during stimulation appear to reverse themselves, and the cat assumes a facial expression and posture suggestive of rapt attention or fearful alerting. The end of an afterdischarge is usually signaled by a meow or "yowl," followed by a series of such vocalizations.[74] For a brief period the cat looks as though confused or bewildered.

I conducted experiments on 18 cats in which cholinergic agents were deposited in the proximoseptal segment of the hippocampus, as well as the part transitional with the proximoamygdalar segment.[73] The findings upon depositing carbachol in the hippocampus were of particular interest. Since this agent with both the muscarinic and nicotinic action of acetylcholine is not destroyed by cholinesterase, it exerts a prolonged stimulating effect and allows ample time to test various reactions of the animal during the generation of, and recovery from, a fully developed hippocampal seizure. The behavioral manifestations correlate with three stages of the electroencephalographic changes illustrated in Figure 19-11 and referred to as (1) stage of development, (2) stage of culmination, and (3) stage of subsidence. As to be described, sexual manifestations occur during the last mentioned stage.

Stages of Development and Culmination

During the stage of development, the recording from the site of stimulation shows continuous rhythmic activity that becomes progressively greater in amplitude. Throughout this period, there is little or no evidence of any change taking place. The cat usually appears as though settling down to sleep, but arouses occasionally to sniff or to look around. Thereafter during the stage of culmination, continuous high-voltage spiking appears in the hippocampogram and persists for nearly an hour. While maintaining a resting posture, the cat sinks into a stupor, showing no spontaneous movement and seeming oblivious to any disturbance in the environment. If stimulated into movement, it will suddenly freeze and assume a statuesque posture. Although its extremities remain

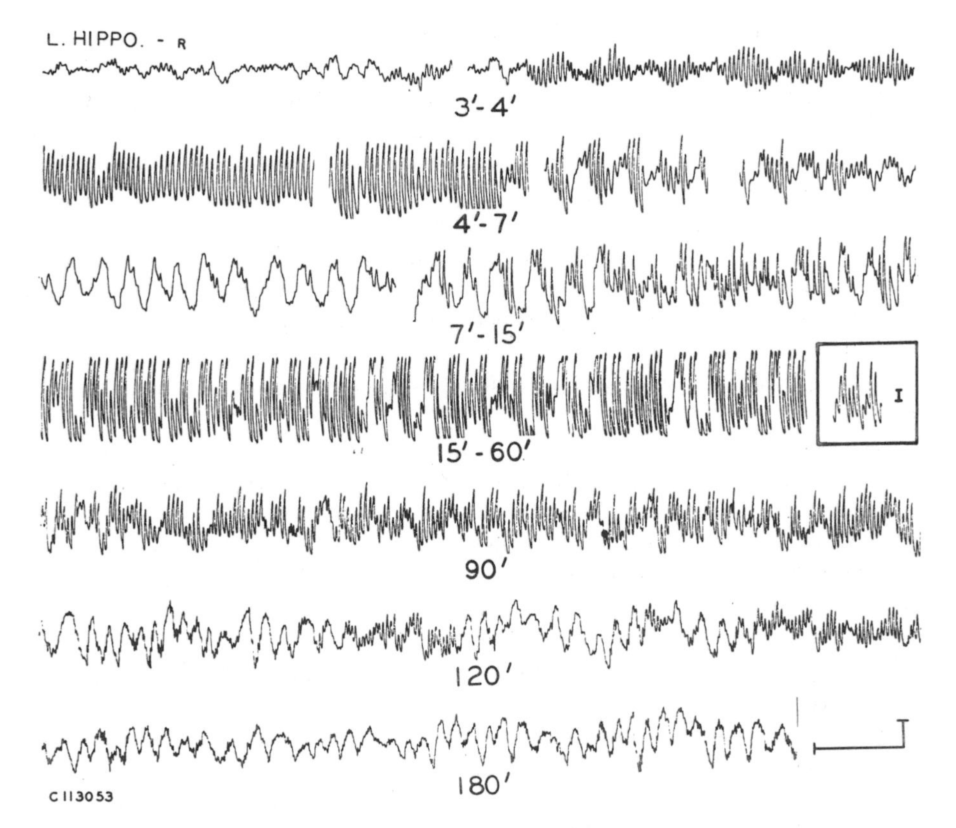

Figure 19-11. Excerpts showing stages of seizure activity induced in hippocampus by the deposit of crystalline carbachol. (1) Stage of development: upper three tracings, showing first 15 min. (2) Stage of culmination: middle tracing representative of maximal spiking activity, with sample in box showing activity at reduced gain. (3) Stage of subsidence: three lowermost tracings, recorded at 90, 120, and 180 min. Calibrations: 1 sec and 200 μV. From MacLean (1957a).

supple, the cat can be draped into postures that will be maintained for minutes at a time (see Figure 1 of Ref. 74). Righting reflexes are present; the cat can be dropped from a height and right itself. Generally, it is underreactive to noxious stimulation, but can be provoked into angry, *undirected* attack that terminates almost as abruptly as it appears. It was shown that animals trained in conditioned avoidance failed to respond appropriately to either the conditioned or unconditioned stimulus.[75]

Stage of Subsidence

It is in the second and third hour during the stage of subsidence that the sexual manifestations appear.[74] The cat undergoes a dramatic awakening from its stuporous state and displays enduring, enhanced pleasure reactions characterized by loud purring, strenuous rubbing of itself against every object; kneading and fanning of the forepaws; turning over on the back, rolling, twisting, stretching, and spreading of the forepaws. It will also seek out the observer, invite stroking, and engage in gentle biting and prolonged licking of a proffered hand. Attempts at copulatory grasping and sustained erections (see Figure 2 of

Ref. 74) have been observed during the expression of these enhanced pleasure reactions. The pleasure reactions may be interrupted by bouts of prolonged grooming.

Once the prolonged stimulatory effect of carbachol brought to light the enhanced pleasure and grooming reactions, as well as sexual manifestations, it thereafter became evident that these same kinds of behavioral changes occurred following electrically induced hippocampal seizures. The protracted, intensive grooming might, from the beginning, "be largely confined to the genital region or . . . involve an orderly progression from the head and forepaws, to body and hindlimbs, to stump of tail and region of anus and genitalia,[76] sometimes associated with penile erection (see Figure 3 of Ref. 74). The grooming was suggestive of what occurs during feline courtship. In ancillary comparative observations on rats, I also observed prolonged grooming activity and penile erection following hippocampal afterdischarges induced by electrical stimulation within the proximoseptal segment of the hippocampus.[74]

The enhanced pleasure and grooming reactions induced by cholinergic stimulation of the proximoseptal segment in cats are in striking contrast to the behavioral changes seen when acetylcholine or carbachol is either deposited directly into the ventricle or reaches the ventricle from the site of application in the hippocampus. Under these conditions the cat begins within a few minutes to growl and to hiss and to circle toward the contralateral side.[74] It will show defensive behavior and sometimes attack if a threatening gesture is made in its contralateral (but not the ipsilateral) visual field. The pleasure, grooming, and sexual manifestations are also in marked contrast to the "rage" states that have been observed in some instances during stimulation of the hippocampus near the amygdala or during a subsequent afterdischarge.[77] Pointing out that such "rage-like phenomena" may be due to propagation of the afterdischarge to the hypothalamus, Naquet, a French neurophysiologist, noted at the same time that "stimulation of the posterior portion of Ammon's horn produces an electrical after-discharge without 'a motor after-discharge' and without important behavioral changes."[78]

Chemical and Electrical Stimulation of Septum in Cats

While I was performing the experiments described above, Bruce Trembly, a medical student at Yale, chose for his medical thesis a similar study on the septum.[79] He performed chemical and electrical stimulation of the septal region in 22 cats at the loci shown in Figure 19-12. He observed enhanced pleasure reactions following cholinergic stimulation of the medial septal nuclei in 8 of 11 cats (Figure 19-12A), as opposed to only one such reaction in 11 cats stimulated elsewhere in the septum (Figure 19-12B). In some cases the pleasure reactions were punctuated by bouts of grooming and playful behavior. Spontaneous penile erection has been noted in the cat following a propagated afterdischarge elicited by electrical stimulation of the septum.[80] During electrical stimulation of the medial or lateral septum in the cat, there may be few behavioral changes except for some apparent alerting or staring or turning of the head usually to the ipsilateral side, together with an occasional meow.[81] If an afterdischarge is induced, there is usually little, if any, overt manifestation of its occurrence. Even a cat appearing somewhat dazed will usually respond appropriately to different stimuli. Trembly describes an unusual response, however, in a case in which an electrode in the lateral ventricle impinged on the lateral septum.[81] During electrical stimulation, the cat showed signs of alerting, raised its head, and developed dilatation of the pupils. Concurrent with the subsequent afterdischarge, it appeared alert and responsive, but when its tail was gently tapped, it suddenly leapt up and jumped around in "jackrabbit fashion, hissing, and spitting" for a period of 40 sec. This observation is to be kept in mind apropos of the activation of the system of structures connected with the bed nucleus of the stria terminalis that will be considered in the next chapter.

Following afterdischarges induced by septal stimulation, there may be enhanced pleasure and grooming reactions like those already described in the hippocampal experiments. In a retrospective analysis of their extensive material, Hess and Meyer found that the septal region (particularly the part near the columns of the fornix) was the place where electrical stimulation most commonly led to grooming (Fellreinigung).[82] Laursen observed that stimulation of the septum (loci not specified) was followed by sexual responses appearing within 1 to 2 min and persisting for 3 to 5 min.[83] The sexual response "consisted of purring, licking, rolling on the floor and in females, coital crouching with the tail held to one side."[84] In one cat Laursen observed that septal stimulation resulted in hissing. In my experience, I have never observed hissing with septal stimulation (exclusive of the bed nucleus of the stria terminalis) unless there was a propagated afterdischarge.[85] The hissing occurred simultaneously with the appearance of an afterdischarge in the part of the hippocampus adjoining the amygdala. The occurrence of hissing is of particular interest in the light of the copulatory afterreaction of a female cat in which she suddenly lunges forward, rolls on her back, hisses, and lashes out with bared claws toward the male. A reciprocal innervation of the septal and amygdala divisions that might account for the close relationship of sexual and agonistic behavior will be considered in the next chapter.

Comment

It has long been inferred that the sex steroids sensitize certain constellations of neurons for participation in patterning the experience and expression of sexual forms of behavior. In terms of computers it was as though these substances allowed the organism to switch programs for a special set of operations. Further along this line, it has recently piqued interest that some neuropeptides have the capacity to entrain certain behavioral sequences. Apropos of the experiments described above, the behavioral manifestations resulting from the injection of peptide hormones into the cerebrospinal fluid are particularly pertinent. In 1955, Ferrari *et al.* reported that the intracisternal injection of ACTH resulted in bouts of stretching and yawning.[86] Since then, not only this so-called stretching–yawning syndrome (SYS), but also penile erection has been observed in such diverse species as rats, rabbits, cats, dogs, and monkeys following the introduction of $\beta^{1\text{-}24}$-ACTH or α-melanophore stimulating hormone (MSH) (or certain fractions of them) into the ventricular system.[87] Gispen and Isaacson have drawn particular attention to the occurrence of excessive grooming, which in the rat may be the first manifestation to occur. Since the various behavioral changes do not depend on the presence of the pituitary or upon peripheral neuroendocrine mechanisms, it is assumed that the peptides have a direct action on the brain.[88]

Significantly, in regard to the role of the septal division in grooming and penile erection, Kinnard and I found in ancillary experiments on squirrel monkeys that the deposit of α-MSH in the diagonal band at the base of the septum resulted initially in excessive scratching (a prominent part of grooming in this species), followed by bouts of stretching, yawning, and penile erection.[89] The bouts were fully developed by $1\frac{1}{2}$ hr,

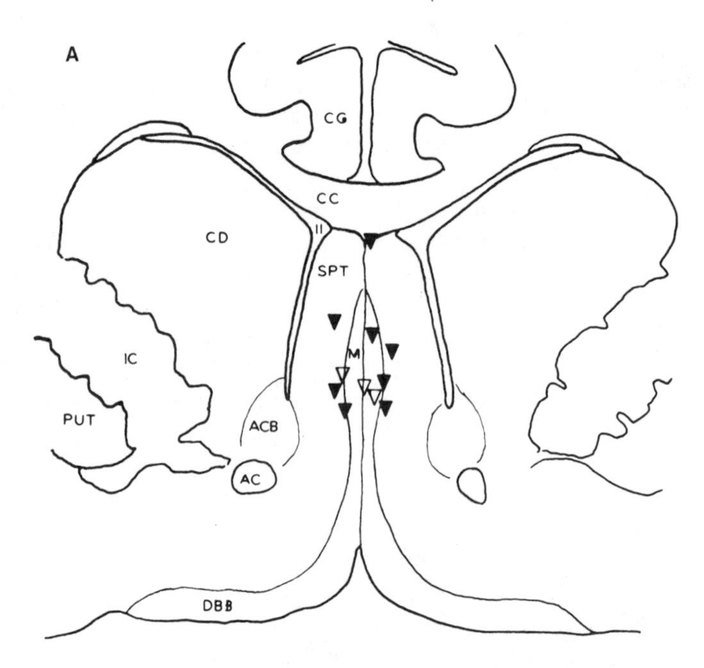

Figure 19-12. Illustrations from Trembly's experiments, showing loci at which chemical stimulation of the septum with acetylcholine (in conjunction with physostigmine) resulted in enhanced pleasure reactions in cat. A and B show sections through middle and caudal septum. Note that most positive responses (filled triangles) were located in or near the medial septal nucleus (M). Open triangles identify negative loci.

Abbreviations: AC, anterior commissure; ACB, nucleus accumbens; CC, corpus callosum; CD, caudate

occurred twice per min at 3 hr, and persisted until nearly 6 hr after treatment. After an interval of several days the same syndrome occurred after a deposit of α-MSH near the base of the medial preoptic region on the opposite side. The application of $\beta^{1\text{-}24}$-ACTH in the premammillary region or of α-MSH in the ventromedial nucleus of the hypothalamus failed to elicit the syndrome.

There would appear to be an opiate-sensitive component in ACTH-induced grooming, as evident by the observations that the amount of grooming is reduced by the opiate antagonist naloxone.[90] It is also notable that the intraventricular injection of small doses of morphine induces excessive grooming.[90] The hippocampus possesses opiate receptors and enkephalin fibers.[91] In view of the grooming responses following electrical or chemical stimulation of the hippocampus, it is of correlative interest that with respect to telencephalic structures, only large lesions of the hippocampus were effective in reducing excessive grooming induced by the intraventricular injection of ACTH.[92]

The mention of opiates raises the question as to whether or not the enhanced pleasure reactions induced by electrical or chemical stimulation of the hippocampus are partially dependent on neural mechanisms involving endorphins. Unlike most cerebral neurons hippocampal pyramids are excited, rather than depressed, by the local application of opiates.[93] Naloxone reverses this effect. How is the paradoxical opiate excitatory action to be explained? It has been suggested that opiates suppress the discharge of the Golgi type II hippocampal basket cells, which, through their short axons, ordinarily inhibit the firing of the type I pyramidal cells.[94]

The intraventricular injection of opiate substances produces prolonged hippocampal

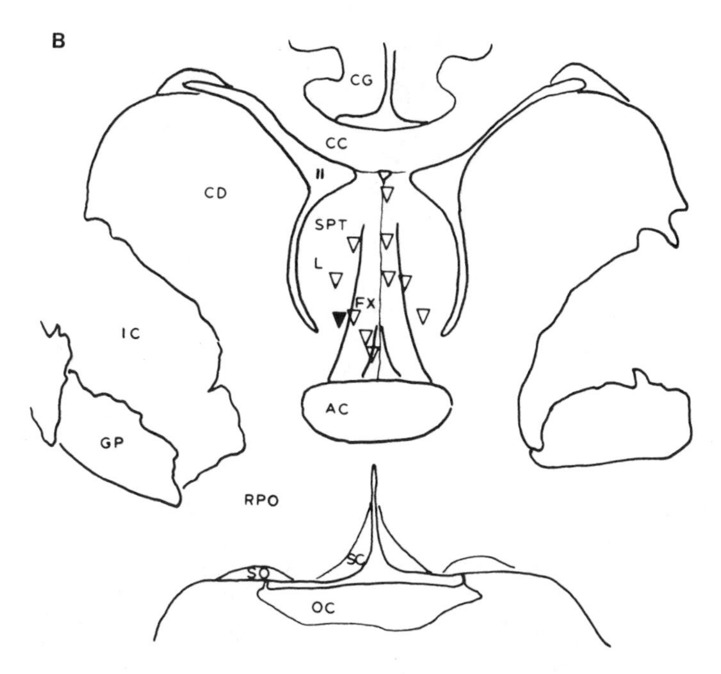

nucleus; CG, cingulate gyrus; DBB, diagonal band of Broca; Fx, fornix; GP, globus pallidus; IC, internal capsule; OC, optic chiasm; PUT, putamen; RPO, lateral preoptic area; SC, suprachiasmatic nucleus; SO, supraoptic nucleus; SPT L, lateral septum.

From Trembly (1956).

seizures[95] that compare in intensity to those induced by the local application of carbachol. At the height of a morphine-induced seizure, animals develop an immobility and other symptoms that have a partial resemblance to those seen during the stage of culmination of carbachol-induced seizures.[73] Although such immobility has been compared to the ''freezing'' of the ''death-feigning reflex,'' it is to be recalled that animals may also come to a complete standstill when stalking a prey.

Electrical Stimulation in Monkeys

Apart from their value as clues to functions of the septal division of the limbic system, the findings in the cat were of special interest because there appeared to be no previous evidence that brain stimulation elicited such sex-related forms of behavior. The findings suggested to me that it would be worthwhile as a next step to carry out a systematic exploration of the brain so as to identify structures involved in primal sexual functions. Because of the comparative and other reasons explained in Chapter 13, I selected squirrel monkeys as experimental subjects. In order to avoid the depressant effects of anesthesia, I devised the method for brain stimulation and recording already described in connection with the experiments on the amygdala. Although the focus of the study was on genital responses (penile erection, urination, and seminal discharge), we also systematically noted the occurrence of vocalization and movements of different parts of the body, as well as electroencephalographic and electrocardiographic changes. As detailed in the original paper, the magnitude of penile erection was graded on a scale of ± to 5+, with 5+ corresponding to full erection seen under natural conditions.[96] A positive response was defined as one that could be obtained repeatedly within a narrow range of latencies. In those instances in which stimulation elicited seminal discharge, the presence of spermatozoa was confirmed by microscopic examination.

In Figure 20-1 the diamonds and squares indicate loci in the telencephalon, diencephalon, and ventral midbrain, at which electrical stimulation elicited penile erection.[96] The squares give the added information that stimulation was followed by hippocampal afterdischarges. At this time I will summarize only the results of exploring the septal subdivision, leaving until the next chapter the findings on the third subdivision and the question of effector pathways in the brainstem. Stimulation of the medial and lateral septum (exclusive of the bed nucleus of the stria terminalis) (Figure 20-1, A13.5–A11) resulted in consistent 2 to 3+ responses occurring at regular latencies of about 6 sec. Stimulation of the lower part of the septum and contiguous medial preoptic area elicited full erection. With stimulation just below the anterior commissure, the monkey remains so quiet that one might assume that the electrode was in a silent area unless attention were directed to the genital region. Note also that stimulation of the posterior gyrus rectus, which receives direct hippocampal projections,[97] also elicited full erection (see also Chapter 20).

As indicated by the square symbols, stimulation in the septum commonly elicited hippocampal afterdischarges. Although genital tumescence did not depend on the occurrence of an afterdischarge, there was invariably hippocampal recruitment during stimulation. During afterdischarges elicited by stimulation at highly effective loci, the erection would become throbbing in character and reach maximum size. Ejaculation, however, was never seen under these conditions. Subsequently, there might be waxing and waning of a 1–3+ erection for several minutes, together with the appearance of random slow-wave or spiking activity in the hippocampus. Following a number of hippocampal afterdischarges, it was typical for a monkey to become placid and somnolent. This apparent change in disposition sometimes seemed to persist for a prolonged period of time. A consideration of other associated responses, including vocalization, will be deferred until the next chapter.

In exploring the brains of female squirrel monkeys, Maurus *et al.*[98] observed full enlargement of the clitoris upon electrical stimulation at sites in the septal region and medial preoptic area corresponding to those in the male. In male macaques Robinson and Mishkin elicited penile erection by electrical stimulation in the septopreoptic region.[99]

Stimulation of Hippocampus

Upon stimulation of the body of the hippocampus in squirrel monkeys, there may be few, if any, behavioral changes during the application of stimuli or during a succeeding afterdischarge. In some instances, however, we found that stimulation involving the dorsal psalterium or the fimbrial region of the proximoseptal segment (i.e., the part proximal to the splenium) resulted in penile erection.[100] As illustrated in Figure 19-13, there was invariably a recruitment of high-voltage hippocampal potentials. Genital tumescence would occur within a few seconds, but both the latency and magnitude of the erection were variable. In some cases, a throbbing 4 to 5+ erection developed during a subsequent afterdischarge. During, as well as following a prolonged afterdischarge, the monkey might scratch various parts of its body, including the anogenital region, and pull

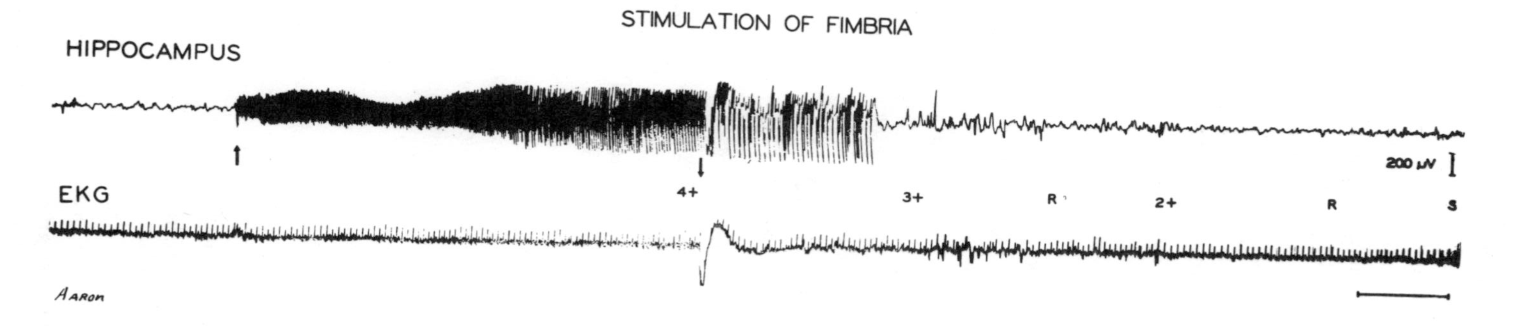

Figure 19-13. Recording from caudal hippocampus during stimulation of the adjacent fimbria that resulted in penile erection. Upward and downward arrows indicate onset and termination of stimulation with negative, constant-current pulses of 1 msec applied at 30/sec with an intensity of 0.7 mA. Penile erection of 4+, 3+, and 2+ appeared at times shown. Note afterdischarge upon cessation of stimulation. A comparison of this record with one previously published (see Figure 10 of MacLean *et al.*, 1962) illustrates the degree of variability in the hippocampal and genital responses elicited by a repetition of the same stimulation at this same site. Calibrations: 3 sec and 200 µV. Scale bar = 3 sec.

at its penis. As in the case of septal stimulations, the monkeys developed a state of quietude bordering on somnolence.

Comment

Heath and co-workers have made clinical observations that provide information about subjective feelings during septal stimulation. At a special meeting at Tulane University in June 1952, they reported on the effects of stimulating the septal region of patients for the avowed purpose of alleviating pain or psychotic symptoms.[101] "Frequently," they stated, "one noticed a pleasurable reaction to stimulation."[102] One patient said during stimulation, "I have a glowing feeling. I feel good."[102] In a later report, Heath described the symptoms of a patient who had been provided a push button for self-stimulation of the septum. The patient said that during stimulation he experienced a "good" feeling and that "it was as if he were building up to a sexual orgasm. . . . [but] that he was unable to achieve the orgastic end point."[103]

After the findings at the Tulane meetings were circulated by word of mouth, Olds and Milner tested the effects of septal stimulation in rats and reported that the subjects would repeatedly press a bar to obtain electrical stimulation in that region.[104] These striking findings were later confirmed by Brady in the cat[105] and by Lilly in the monkey.[106] In the discussion following a paper that Olds gave at a London Ciba Conference in 1957, I asked him if he had ever seen erections in his animals. He replied: "In about one-third of our animals we get erection; and almost always when we get erection we get self-stimulation. . . . We almost invariably get the enhanced grooming after the stimulus."[107]

Effects of Destruction of Septum or Proximoseptal Hippocampus

As illustrated in the case of the amygdala, the effects of brain stimulation and of destruction may be of an opposite nature. As will now be described, the effects of septal lesions were in an opposite direction to those of pleasure and grooming reactions elicited by electrical stimulation. In addition, the results of septal and dorsal hippocampal lesions have disclosed that the septal subdivision of the limbic system plays a role in maternal behavior.

Changes in "Emotionality"

In 1953, Brady and Nauta published an account of what they called increased "emotional reactivity" in rats immediately following surgical lesions of the septum.[108] For a period of several days the animals appeared to be hypersensitive to the slightest touch, reacting by jumping in the air, and attempting to bite the handler. In a later study it was demonstrated that recovery occurs within 60 days after the operation.[109] These transitory changes in "emotionality" in rats have been observed by several workers. Sodetz and Bunnell, however, reported that septal ablation in hamsters resulted in increased intraspecific aggression but no changes in "emotionality."[110] Slotnick and McMullen reported that septal lesions in mice induced increased emotionality characterized by vigorous biting and attempts to escape.[111] Unlike rats and hamsters the mice did not show intraspecific aggression. They would neither initiate attacks nor fight back. In the evaluation of the results of these different studies, it should be pointed out that it is not clear from the anatomical reconstructions what structures of the septum were eliminated. In most studies, it would appear that there was sparing of the bed nucleus of the stria terminalis, which, as noted above, has been implicated in combative behavior.

In the monkey I have never observed changes in emotionality following partial lesions of the septum involving its rostral, caudal, or ventral portions.[112] By a special manipulation interfering with the blood supply, I produced a virtually complete lesion of the septum, exclusive of the bed nuclei of the stria terminalis, in two squirrel monkeys (B-4 and E-4).[112] Figure 19-14B and C shows the extent of the ablations in these animals. In addition, there was a complicating bilateral infarction of the pericallosal anterior cingulate cortex. On the first postoperative day both of these animals showed what appeared to be a hyperirritable state characterized by brief, frenzied jumping in the plastic recovery box, as though the bottom had suddenly become red hot. In one of these animals (E-4) this behavior was associated with a piercing cry. This same animal behaved as though it were trying to remove something from its head and body. These rare and momentary occurrences were not observed after the first postoperative day. The behavior suggested a transient irritative hyperesthesia.

Effects on Genital Display

Partial lesions of the septum or bilateral preseptal section of the fornix have resulted in no interference with the genital display behavior of squirrel monkeys.[112] Figure 19-14A shows a bilateral lesion of the caudal septum and fornix produced in a squirrel monkey by electrocoagulation that gave a full display on the first postoperative day and performed regularly at criterion level thereafter. In the two monkeys described above (B-4 and E-4; Figure 19-14B and C), there was no sign of genital tumescence in the mirror display test until the 10th and 12th postoperative day, respectively, and full trump display did not occur until the 36th and 53rd postoperative day, respectively.

Since stimulation of the fimbria and dorsal psalterium at the level of the splenium elicited penile erection and hippocampal discharges in squirrel monkeys (see above), it is of interest to mention the outcome of surgical ablations in this region. In four monkeys (Xenia, Gump, Ian, and Hyde) the splenium, together with the fornix on each side, was aspirated.[112] The surgical approach also involved destruction of part of the posterior cingulate and parasplenial cortex. In one case (Hyde) the caudalmost part of the hippocampal formation on each side was also aspirated. The monkey (Gump) with the least extensive lesion showed partial erection when first tested on the second postoperative day and performed the full display 4 days later. The monkey (Hyde) with the most extensive lesion began to display on the 13th postoperative day and performed a trump display on the 15th day.

Effects on Maternal Behavior

Slotnick has observed marked deficits in maternal behavior of both mice and rats following septal lesions.[113] Quantitative measures of maternal behavior in rodents can be obtained on the basis of their nest building, nursing and care of the young, and retrieval of

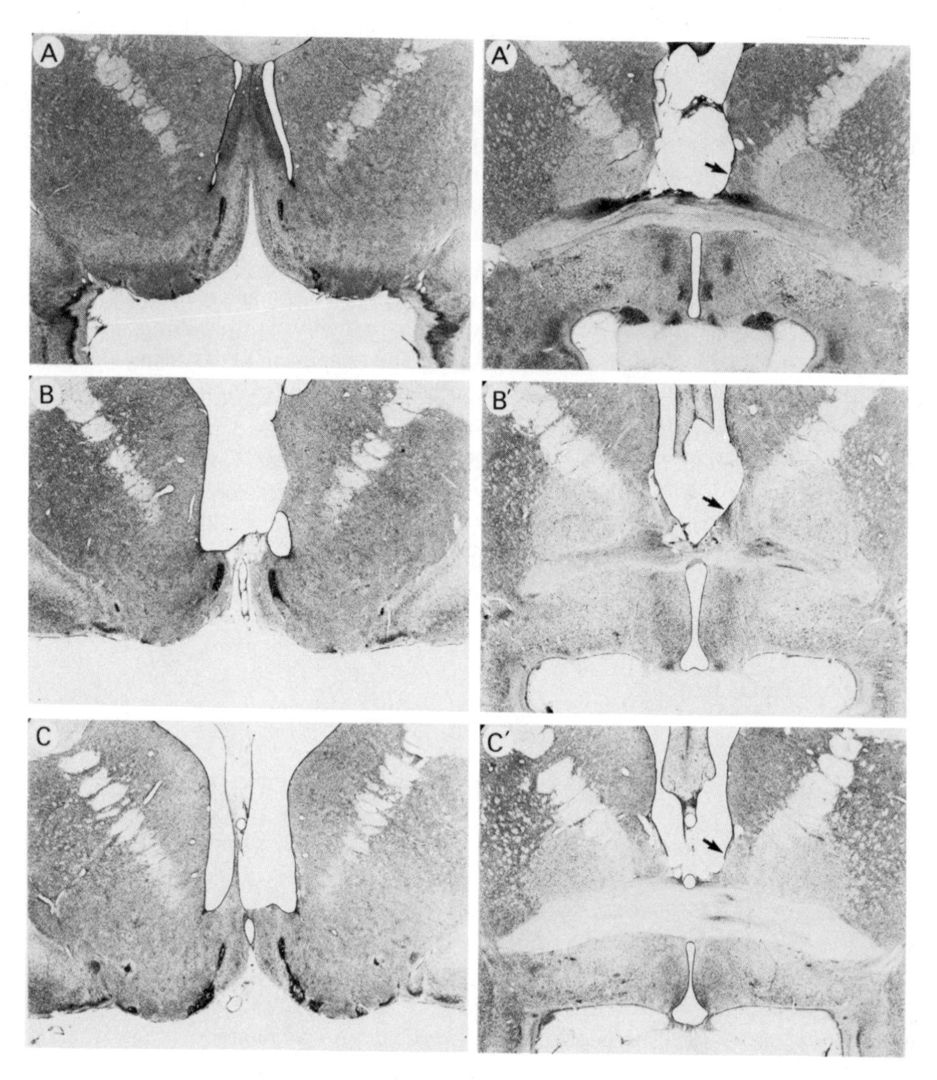

Figure 19-14. Representative histological sections showing extent of septal lesions in three squirrel monkeys. Arrows point to the bed nucleus of the stria terminalis.

(A and A′) Microscopic section on left shows intact rostral septum, while the one on right (A′) shows lesion of caudal septum (including bilateral destruction of the fornix) produced by electrocoagulation.

(B, B′, and C, C′) Disappearance of virtually entire septum except for the bed nucleus of the stria terminalis, subsequent to a procedure interfering with the blood supply.

pups. Since this matter will be a major consideration in describing the functions of the third subdivision of the limbic system, I will at this time simply refer to the experimental results and defer a description of the method of scoring nest building and other maternal performance until Chapter 21. Fleischer,[114] a graduate student working with Slotnick, found that septal lesions resulted in the following deficits in maternal behavior: After parturition, the dams did not clean all of their pups. They failed to build a nest and showed abnormal retrieving behavior, bringing the pups to the nest site and then carrying them away again. Only a few of the lesioned animals nursed their young or adopted a correct nursing posture. None of the pups survived longer than 48 hr. Slotnick is inclined to attribute the impaired maternal behavior of septal-lesioned rats to the "syndrome of hyperemotionality" described above, and he cites experiments that give support to such an inference.[115]

Consonant with the effects of septal lesions, Kimble *et al.* found deficits in both nest building and maternal behavior among rats with large lesions of the dorsal hippocampus, i.e., the proximoseptal hippocampal segment.[116] Their experiments were performed on primiparous rats. In contrast to the unoperated controls and a group with neocortical lesions, the hippocampal animals exhibited inferior nest building and inadequately nursed their young. As a consequence, there was statistically a highly significant reduction in the number of their offspring that survived. In the operative approach to the dorsal hippocampus, the overlying neocortex had also been aspirated. In a control group with destruction of 40 to 50% of the neocortex alone, there were no maternal deficits. The authors suggested that in earlier studies by Beach[117] and others in which deficits in maternal behavior had been attributed to neocortical lesions, there had probably been inadvertent damage to the underlying hippocampus. In Chapter 21, I will describe an experiment performed in our laboratory in which hamsters reared from birth without neocortex were able to breed and rear their young.

Nest Building and Other Changes in Males

It is relevant to the findings just discussed that Chul Kim found that large ablations of the hippocampus proximal to the septum in male rats resulted in inferior nest building.[118] Male rats with similar ablations showed increased mounting activity in the presence of females. Kimble *et al.* observed that ablations of the dorsal hippocampus in male rats did not interfere with their copulatory ability.[116]

Hoarding

Hamsters are noted for accumulating a hoard at their nest site. Bunnell and coworkers observed decreased hoarding activity in hamsters following septal lesions.[119]

Grooming

Reference has already been made to decreased grooming activity seen in rats following large ablations of the hippocampus.[92] Reymond and Battig have described decreased grooming in rats with septal lesions.[120] Since grooming occurs notably following feeding, soiling, wetting, after combat, various kinds of stress, and in association with mating and parental care, it would be of interest to learn whether or not ablations primarily of the proximoamygdalar or proximoseptal segments of the hippocampus would have a differential effect on these various forms of grooming.

Concluding Comment

Anatomical and functional considerations suggest that the limbic system consists mainly of three overlapping corticosubcortical subdivisions. For the two phylogenetically older subdivisions the amygdala and septum, respectively, serve as telencephalic internodes relating respective parts of the limbic cortex to the brainstem—hence the terms *amygdalar* and *septal* subdivisions. Unlike the thalamocingulate division to be considered in the following two chapters, the amygdala and septum contain nuclei that establish extensive connections with the hypothalamus via the medial forebrain bundle (Chapter 18). These two subdivisions have robust reciprocal connections that, in a communicative sense, reinforce their relationship to the olfactory apparatus.

The behavioral findings reviewed in the present chapter indicate that the amygdalar division is primarily concerned with self-preservation as it pertains to feeding, the search for food, and the fighting, attack, and defense that may be involved in obtaining food. The septal division, on the contrary, appears to be primarily implicated in procreation as evident by its role in primal sexual functions and in behavior conducive to mating and copulation. If viewed in the light of findings to be described in Chapter 21, there are indications that the septal division may have provided the initial potentiality for mammalian maternal behavior. As further discussed in the next chapter, the close linkage between oral and genital functions in the amygdalar and septal subdivisions of the limbic system appears to be owing to their close relationship to the olfactory apparatus.

Clinical and other findings related to the global limbic participation of these subdivisions in the experience of emotion, a sense of personal identity, and the memory of ongoing experience will be dealt with in Chapters 22–27.

References

1. Klüver and Bucy, 1939b
2. Klüver and Bucy, 1937, 1938, 1939a,b
3. Jackson and Stewart, 1899/1958
4. Bucy, 1975
5. Klüver, 1951, p. 150
6. Pribram and Bagshaw, 1953
7. Brown and Schäfer, 1888
8. Brown and Schäfer, 1888, pp. 310–311
9. Bucy and Klüver, 1940
10. Bucy and Klüver, 1955
11. Klüver and Bucy, 1939b, p. 991
12. Terzian and Ore, 1955
13. Terzian and Ore, 1955, p. 374
14. Terzian and Ore, 1955, p. 375
15. Akert *et al.*, 1961
16. Woods, 1956
17. Schreiner and Kling, 1953, see p. 656
18. Hagamen *et al.*, 1963
19. Schreiner and Kling, 1953, Fig. 5, p. 653
20. Phillips, 1964
21. Phillips, 1968
22. Sawa *et al.*, 1954; Narabayashi *et al.*, 1963
23. Pool, 1954
24. Scoville *et al.*, 1953
25. Rosvold, personal communication
26. Scoville, 1954
27. Rosvold *et al.*, 1954
28. Kling, 1968
29. Klüver, 1958
30. Franzen and Myers, 1973b
31. Bucher, 1970
32. Plotnik, 1968
33. Fuller *et al.*, 1957
34. Bunnell, 1966
35. Bunnell *et al.*, 1970
36. Dicks *et al.*, 1969
37. Myers and Swett, 1970
38. Kling *et al.*, 1970
39. Kling and Mass, 1974
40. Kaada, 1951
41. MacLean and Delgado, 1953
42. MacLean and Delgado, 1953, p. 94
43. Gastaut *et al.*, 1951, 1952
44. Magnus and Lammers, 1956; Fernandez de Molina and Hunsperger, 1959; Norris and Lanauze, 1960
45. Fernandez de Molina and Hunsperger, 1959
46. MacLean, 1967; MacLean and Ploog, 1962
47. Gergen and MacLean, 1962
48. MacLean *et al.*, unpublished experiments
49. Frost *et al.*, 1958
50. Reis and Oliphant, 1964
51. Hilton and Zbrozyna, 1963
52. MacLean *et al.*, unpublished experiments
53. Mason, 1959
54. McHugh and Smith, 1967
55. Slusher, 1966
56. Hayward, 1972
57. Eleftheriou and Zolovick, 1968; Kovacs *et al.*, 1965; Shizume *et al.*, 1962
58. Pfaff, 1968a
59. Pfaff, 1968a, p. 958
60. Pfaff and Keiner, 1973
61. Rees *et al.*, 1980
62. Keefer and Stumpf, 1975; Pfaff *et al.*, 1976
63. Koikegami *et al.*, 1955
64. Riss *et al.*, 1963
65. Sawyer, 1972
66. Elwers and Critchlow, 1960
67. Elwers and Critchlow, 1961
68. Sheridan, 1979
69. MacLean, 1962
70. MacLean, 1973a, p. 53
71. Penfield and Jasper, 1954

72. Andy and Akert, 1953; Liberson and Akert, 1955
73. MacLean, 1957a, b
74. MacLean, 1957b
75. MacLean *et al.*, 1955–56; MacLean, 1957b; Stevens *et al.*, 1961; Flynn *et al.*, 1961
76. MacLean, 1957b, p. 134
77. MacLean and Delgado, 1953; Naquet, 1954; Fernandez de Molina and Hunsperger, 1959
78. Naquet, 1954, p. 712
79. Trembly, 1956; MacLean 1957b
80. Kim, quoted by MacLean, 1957b, 1959
81. Trembly, 1956
82. Hess and Meyer, 1956
83. Laursen, 1962
84. Laursen, 1962, p. 181
85. MacLean, unpublished observations
86. Ferrari *et al.*, 1955
87. Gessa *et al.*, 1966, 1967
88. Gispen and Isaacson, 1981
89. MacLean and Kinnard, unpublished observations; MacLean, 1973d
90. Colbern *et al.*, 1977
91. Sar *et al.*, 1978; Atweh and Kuhar, 1977; Khachaturian *et al.*, 1983
92. Colbern *et al.*, 1977
93. Nicoll *et al.*, 1977; Hill *et al.*, 1977
94. Henriksen *et al.*, 1978
95. Zieglgansberger *et al.*, 1979
96. MacLean and Ploog, 1962
97. Poletti *et al.*, 1973; Poletti and Creswell, 1977
98. Maurus *et al.*, 1965
99. Robinson and Mishkin, 1968
100. MacLean *et al.*, 1962
101. Heath and Monroe, 1954
102. Heath and Monroe, 1954, p. 348
103. Heath, 1963, p. 573
104. Olds and Milner, 1954
105. Brady, 1958
106. Lilly, 1958
107. Olds, 1958, p. 146
108. Brady and Nauta, 1953
109. Brady and Nauta, 1955
110. Sodetz and Bunnell, 1970
111. Slotnick and McMullen, 1970
112. MacLean, unpublished observations
113. Slotnick, 1969
114. Fleischer, 1973
115. Slotnick, 1975
116. Kimble *et al.*, 1967
117. Beach, 1937
118. Kim, 1960
119. Bunnell, 1966
120. Reymond and Battig, 1964

20

Participation of Thalamocingulate Division in Limbic Sex-Related Functions

From the standpoint of mammalian evolution it is of unrivaled interest that three forms of behavior that distinguish the evolutionary transition from reptiles to mammals are represented in the cingulate gyrus. Before taking up this matter in the next chapter, I will first summarize experiments showing that parts of the thalamocingulate division are involved in elemental sexual functions. This will make it possible to complete a review of what is known about the role of the limbic system in sex-related functions while the data on the amygdalar and septal divisions are still fresh in mind. The review takes into account the question of effector pathways involved in genital tumescence; the uncovering of evidence in regard to somatosensory pathways involved in genital sensation and seminal discharge; and the close relationship of mechanisms underlying oral, sexual, and agonistic behavior.

Thalamocingulate Involvement in Genital Tumescence

Mammillothalamic Tract and Associated Nuclei

In describing genital effects elicited by electrical stimulation of certain parts of the thalamocingulate division, we will begin with findings on the mammillothalamic tract and associated nuclei shown schematically in Figure 19-1. Commenting in 1950 upon the enigmatic functions of the mammillothalamic tract, Clark and Meyer referred to a paper by Sigrist, "who made a careful study in Hess's laboratory of the effect of stimulation of the mamillo-thalamic tract in the cat [and] was forced to conclude that there is no definite physiological response which can be related to it."[1] Sigrist characterized it as a *stumme Zone* (silent zone).[2] In stimulating the corresponding structures in the monkey, our conclusions might have been like those of Sigrist had we not been looking specifically for genital responses. As part of the same study on primal sexual functions described in the preceding chapter, we found that partial erection occurred with stimulation of the mammillary bodies and at points along the mammillothalamic tract (Figure 20-1, A9 to A7).[3] It should be emphasized that the response obtained by stimulation of the mammillary body could not be attributed to a spread of current to the medial forebrain bundle because a small electrical coagulation at the site of the electrode almost completely eliminated the effect.

Stimulation of the mammillothalamic tract near its entry into the anterior group elicited partial erection and commonly resulted in hippocampal afterdischarges (square symbols in Figure 20-1). Following such stimulations abnormal spiking might occur in the

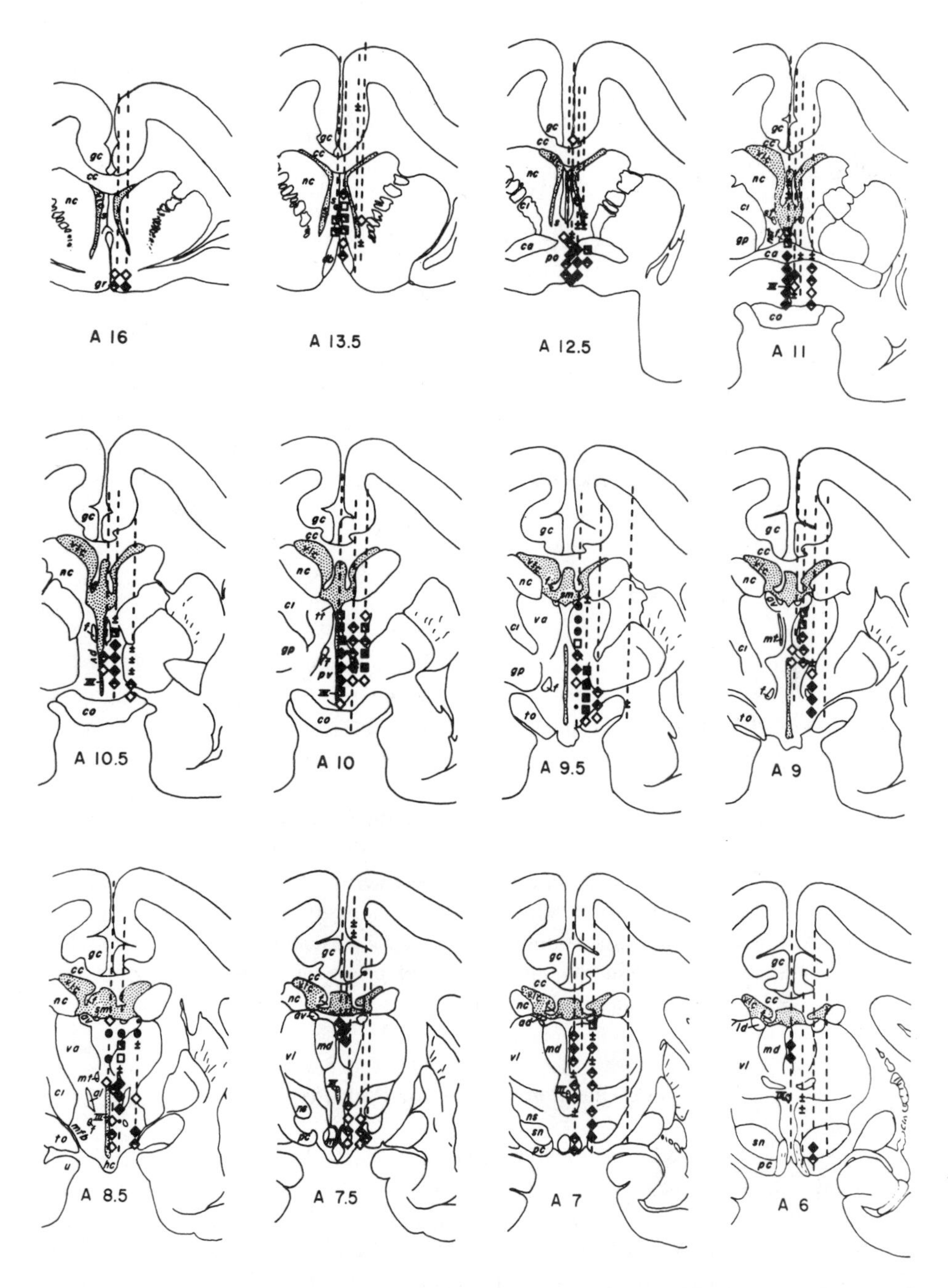

Figure 20-1. Drawings of frontal sections from brain atlas, showing loci at which stimulation elicited penile erection in squirrel monkeys. Region explored extends from the caudal level of the gyrus rectus to the midlevel of the medial dorsal nucleus. The letter "A" and numerals below each diagram give the distance (in millimeters) anterior to the interaural plane. White, half black, and solid black diamonds indicate loci at which stimulation resulted, respectively, in 1+, 2–3+, and 4–5+ erection. Correspondingly shaded squares give additional information that stimulation resulted in hippocampal afterdischarges. Large dots signify rebound erection associated with hippocampal afterdischarge; small dots indicate rebound erection. A ± denotes penile tumescence with separation of glans from prepuce. Vertical dashes at millimeter intervals identify negative points. Key to abbreviations in Appendix at end of chapter. From MacLean and Ploog (1962).

hippocampus and persist for as long as 10 min.[3] During such times there was waxing and waning of penile erection.

Medial Dorsal Nucleus

As reviewed in Chapter 18, the medial part of the medial dorsal nucleus (MD) projects to the caudal part of the gyrus rectus, and in the monkey, as in the rat, MD appears to be a source of projections overlapping with those of the anterior medial nucleus (AM) to the rostral cingulate cortex. A narrow, myelin-poor strip (containing cells belonging to the midline central nuclei) separates the medial dorsal nuclei of the two sides. The expression "medial part of MD" is to be understood as including this midline strip. Caudal to the level of the mammillothalamic tract, stimulation of the medial part of MD elicited full erection (Figure 20-1, A7.5 to A6). From this highly effective zone loci for partial erection were traced to the caudal pole of the thalamus (Figure 20-2, A6.5 through A4).[4]

Rostral Limbic Lobe

In view of the positive genital responses elicited by stimulation of the medial part of MD and in the region of AM, it was of correlative interest that full penile erection was produced by excitation of rostral limbic areas to which these nuclei project (Figure 20-3).[5] Figure 20-4 shows histologically the site of one such locus just forward of the knee of the corpus callosum. Stimulation here or at other positive sites shown in Figure 20-3 evoked long-latency slow potentials in the hippocampus.[5] With stimulation at the locus shown in Figure 20-4, penile erection appeared concurrently with the recruitment of high-amplitude potentials in the hippocampus (Figure 20-5). Self-sustained afterdischarges in the hippocampus were induced only by stimulation of the posterior part of gyrus rectus.[5] Since this cortex receives direct projections from the hippocampus,[6] such afterdischarges may have resulted from antidromic excitation of the cells of origin or from excitation of other hippocampal cells receiving their collaterals.

The Question of Effector Pathways

The findings reviewed here and in the preceding chapter indicate that the medial part of MD and the medial septopreoptic region are nodal zones for penile erection. The consistency and regular latency of the genital response observed upon stimulation within these highly effective zones are indicative that they serve to transmit excitation downwards from the archicortical and mesocortical areas. As will be described, these zones may, in turn, establish connections with effector neurons in respective parts of the hypothalamus.

Pathways for Septal Division

Traced caudalwards from the medial septopreoptic region, the distribution of positive loci for penile erection indicated a divergence of two effector pathways[3]—one veering laterally to join the medial forebrain bundle, the other turning medially to become associated with the periventricular fiber system. Positive loci for erection were followed along the course of the medial forebrain bundle to the ventral tegmental area at the junction of the midbrain.

In an experiment in which electrodes were implanted at highly effective loci in the septum and in the medial forebrain bundle (at midlevel of the hypothalamus), it was shown that coagulation at the hypothalamic locus almost eliminated the response elicited by septal stimulation.[7] This finding suggests that the medial forebrain bundle, rather than the periventricular fiber system, provides the major effector pathway from the medial septopreoptic region. Such an inference is further supported by a pilot experiment in which an attempt was made to induce chemical excitation of MD of the thalamus with tungstic acid.[8] Because of a complicating reflux of the chemical into the ventricular system, the brain matter surrounding the third ventricle and aqueduct of the midbrain was destroyed to a depth of 1 mm. Three weeks later this animal developed a unique form of seizures characterized by the occurrence every few minutes of 4+ erection and atypical vocalizations never heard in squirrel monkeys and sounding like cackling laughter. The seizures persisted until the time of sacrifice 48 hr later.[8] The results show that close to full erection can be achieved after elimination of the periventricular fiber system and its caudal continuation along the aqueduct.

Differences in latencies and parameters of stimulation were an indication that telencephalic and thalamic effector pathways establish connections with hypothalamic structures that in turn transmit excitation for the genital response.[3] The latencies observed with stimulation of telencephalic and dorsal thalamic structures ranged from 6 to 3 sec. Based on frequency–amperage curves, relatively long pulses of 1 msec applied at a frequency of 20–30 per sec are optimum for eliciting the genital response with telencephalic or dorsal thalamic stimulation. Figure 20-6A is illustrative of such a curve for the medial septopreoptic zone.[3] With stimulation farther caudally along the course of the medial forebrain bundle (beginning at about midlevel of the hypothalamus) the latency for erection becomes shorter (2.5–3 sec), and the response is elicited with considerably less total current and at optimum frequencies of 60–100 per sec and pulse durations of 0.01–0.05 msec. The curve shown in Figure 20-6B is illustrative of such observations. It deserves emphasis that the use of longer pulses at the hypothalamic level frequently resulted in aversive responses and that such stimulation would have obscured the genital response.

The dorsomedial hypothalamic area bounded rostroventrally by the fornix and dorsocaudally by the mammillothalamic tract[4] proved to be another hypothalamic locus where stimulation was highly effective in eliciting erection at latencies of 2.5 to 3 sec and with optimum parameters of stimulation comparable to those for the medial forebrain bundle.[3] There is some anatomical evidence that fibers from this region may course laterally through the hypothalamus and join the medial forebrain bundle.[9]

Pathways for Thalamocingulate Division

With autoradiographic techniques Müller-Preuss and Jürgens[10] have shown that the rostral limbic cortex under consideration projects to the rostral part of MD and to the dorsomedial hypothalamic area, and they have confirmed previous anatomical findings[10a] that other projections descend in the internal capsule and medial part of the cerebral peduncle. Since stimulation in the region of the medial peduncle elicits penile erection, it is possible that some effector fibers involved in this response descend in the internal

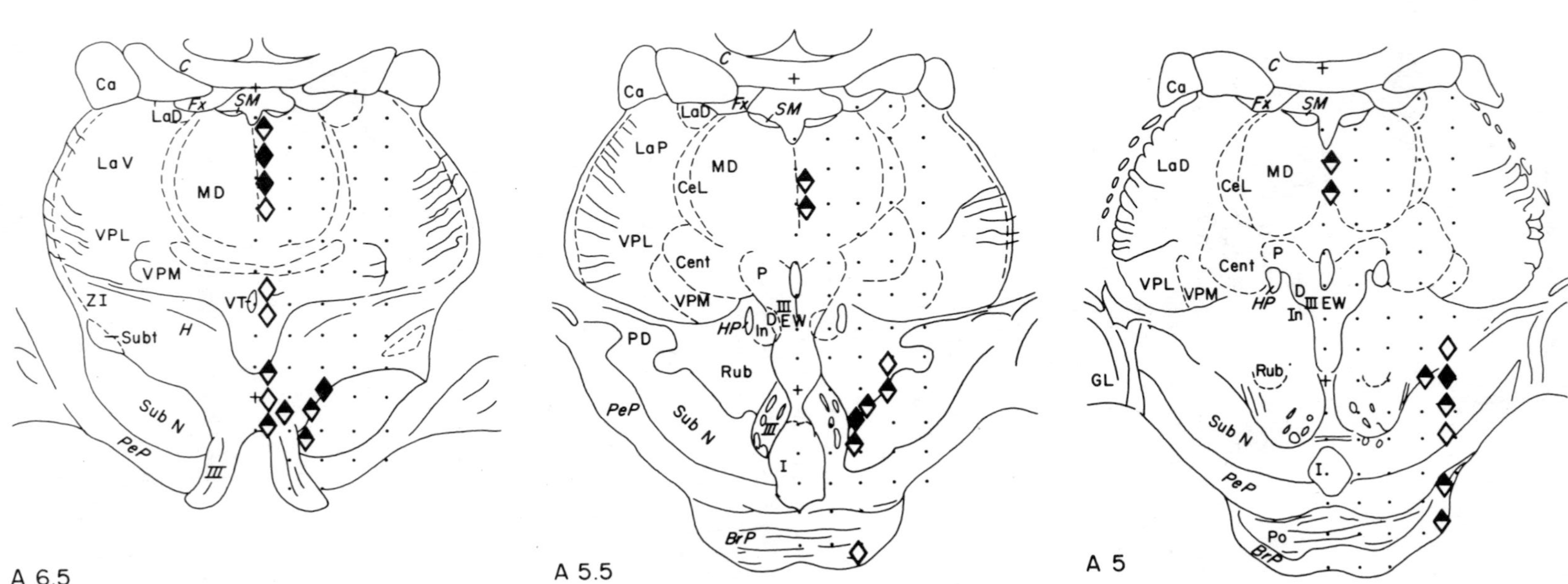
C
Ca
LaD
Fx
SM
LaV
MD
VPL
VPM
ZI
VT
Subt
H
Sub N
PeP
III
A 6.5
C
Ca
LaD
Fx
SM
LaP
MD
CeL
VPL
Cent
P
VPM
III
D
EW
HP
In
PD
Rub
PeP
Sub N
III
I
BrP
A 5.5
C
Ca
Fx
SM
LaD
MD
CeL
P
Cent
VPL
VPM
HP
D
III EW
In
GL
Rub
Sub N
I.
PeP
Po
BrP
A 5

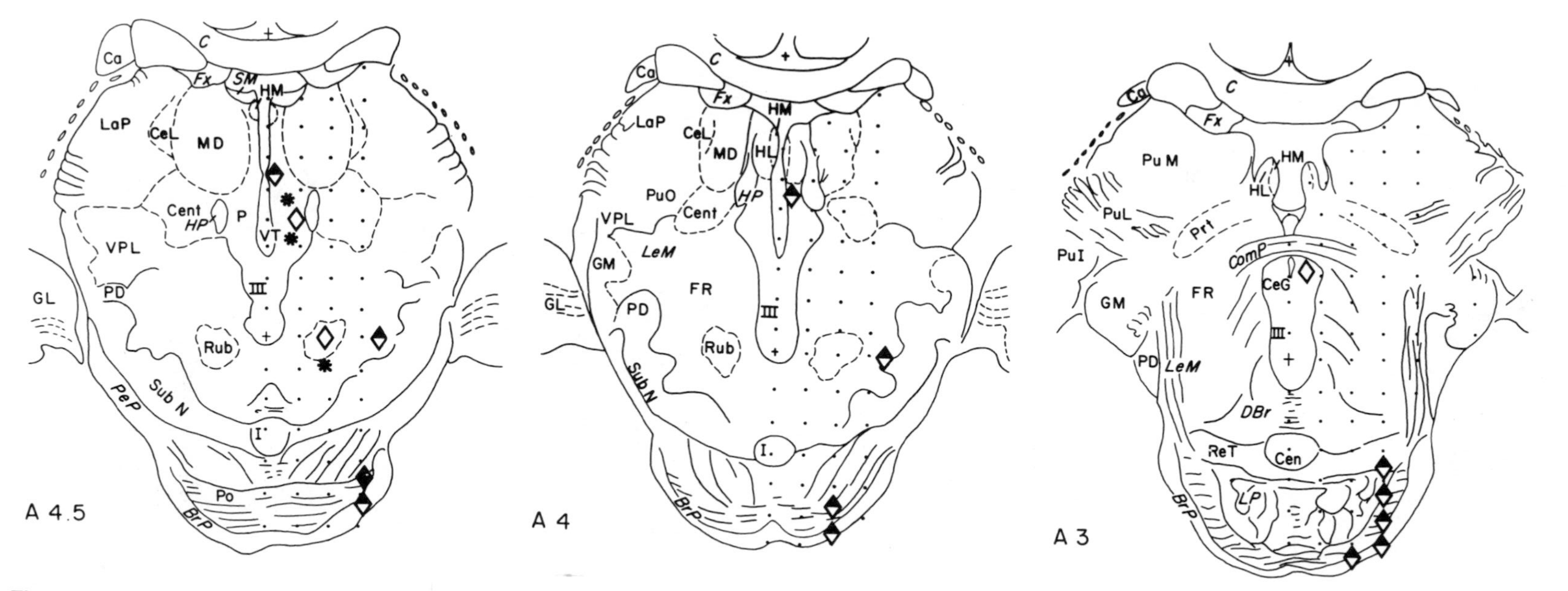

Figure 20-2. Loci for penile erection in squirrel monkey in the part of brainstem encompassing the caudal diencephalon and its junction with the midbrain. Symbols have same signification as in preceding figure, with exception of the use of the asterisk to indicate rebound erection. Black dots placed 1 mm apart overlie parts of brainstem explored and found to be negative with respect to penile erection. Key to abbreviations in Appendix at end of chapter. From MacLean *et al.* (1963a).

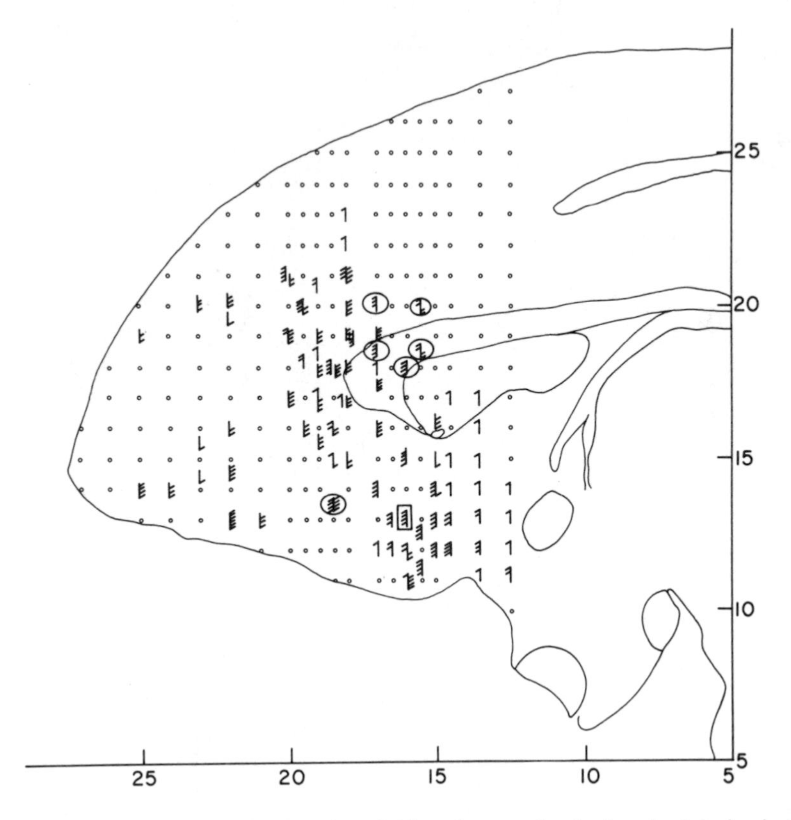

Figure 20-3. Drawing of sagittal section through medial frontal cortex of squirrel monkey's brain, showing loci at which stimulation elicited penile erection and/or slowing of the cardiac rate. Upward-directed arrows indicate loci for erection, with number of bars corresponding to 1, 2, 3, and 4–5+ erection. The number of bars on downward-pointing arrows gives percentage slowing of initial heart rate; each bar has a value of 10. Symbols combining bars on each side denote loci where stimulation elicited both genital and cardiac effects. Symbol in square denotes associated hippocampal afterdischarge; encircled symbols signify accompanying generalized seizure. Scales below and at right, respectively, give distance in millimeters forward of and above zero axes of the stereotaxic atlas. From Dua and MacLean (1964).

capsule. In regard to the nodal zone for penile erection in MD we were unable to trace any continuity of positive loci leading into the periaqueductal gray matter. Rather, we traced highly positive loci for erection in a rostroventral direction along the course of the inferior thalamic peduncle and the site of its convergence with the medial forebrain bundle. It is possible that another effector pathway extends ventrally from MD and establishes connections with the dorsal hypothalamic area.

Associated Manifestations

Upon stimulation at about 25% of the positive loci, penile erection was the only elicited manifestation.[3,5] In the remaining instances erection was variously associated with urination, cardiac and respiratory changes, vocalization, aversive somatic reactions, chewing and salivation, and defecation.[3–5]

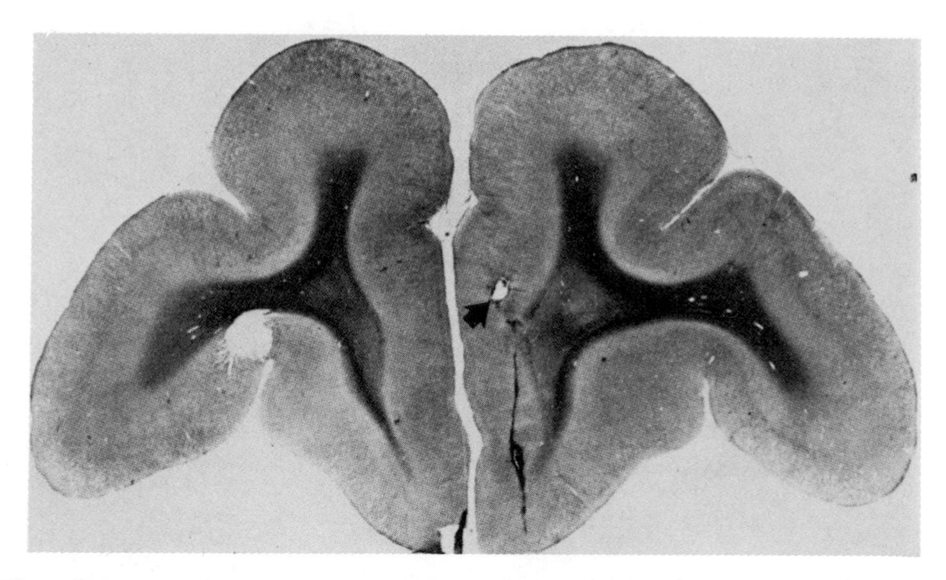

Figure 20-4. Brain section showing site of tip of implanted electrode (arrow) where stimulation elicited full erection. The locus was in cingulate cortex just ahead of the knee of the corpus callosum. The thin line of gliosis shows track of exploring electrode prior to being withdrawn to the level of implantation. Weil–Weigert stain. From Dua and MacLean (1964).

Urination was the most common associated response, occurring with stimulation at about 40% of the positive loci. As illustrated in Figure 20-3, stimulation at positive sites in the rostral limbic lobe frequently elicited a 10 to 40% slowing of the cardiac rate. Cardiac slowing, sometimes complicated by extrasystoles, occurred with stimulation at about 20% of the positive sites in the diencephalon.

In some instances stimulation in the rostral limbic lobe elicited both erection and acceleration of respiration.[5] In two experiments vocalization and erection were regularly obtained upon stimulation of the pregenual cingulate gyrus near the cingulate sulcus.[5]

In exploration of the dorsal thalamus, vocalization occurred with stimulation at 5 of 20 positive loci in MD.[4] Chirping vocalization (alias twitters) during or immediately following stimulation was frequently elicited by stimulation at positive sites in midline septal structures, medial preoptic region, and in the roof of the third ventricle.[3] It often occurred after termination of hippocampal afterdischarges induced by stimulation of the septum.[3] It also occurred as a "rebound" effect following stimulation at points along the course of the medial forebrain bundle and in the mammillary bodies.[3] Interpreted according to findings on spinal cord mechanisms,[11] such a rebound effect would be attributed to an induced excitatory state outlasting inhibition.

As described below, erection occurring in conjunction with aversive stimulation might be accompanied by cackling or piercing vocalization, baring of the fangs, and retraction of the corners of the mouth. Adduction of the knees in association with penile erection was a frequent aversive effect of stimulation at points in the region of the anterior thalamus and near the habenula.[3,4]

In one instance stimulation of the caudal part of gyrus rectus was followed by the appearance of a throbbing 4+ erection and "an alimentary automatism characterized by chewing, licking, and salivation."[5] Stimulation at some loci in the caudal thalamus

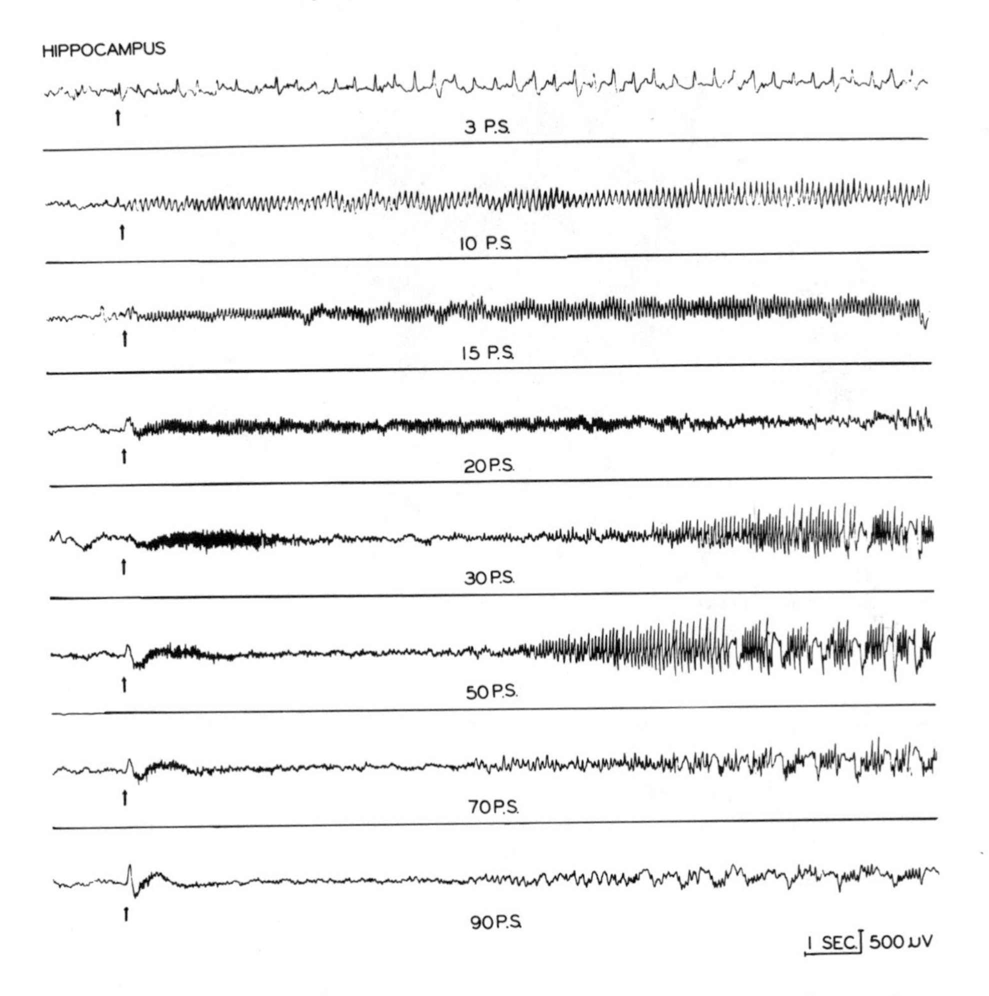

Figure 20-5. Recruited potentials in hippocampus associated with stimulation (arrows) of pregenual cortex (locus of cathodal electrode is shown in Figure 20-4). Upper four tracings show the gradual recruitment and ''decrementation'' occurring with application of negative, constant-current pulses of 1 msec at frequencies ranging from 3 to 20 per sec. In this experiment tumescence did not occur until the beginning of recruitment of high-amplitude slow potentials, typified in the tracings recorded with stimuli at 30 and 50 per sec. The magnitude of the response reached 4–5+. With stimulus frequencies above this range, there was a decline in the amplitude of recruited potentials, and the magnitude of the genital response decreased until at 100 per sec, no tumescence occurred. From Dua and MacLean (1964).

resulted in erection and the simultaneous or subsequent appearance of salivation, licking, and chewing.[4]

Descending Effector Pathway in Midbrain, Pons, and Medulla

As noted above, the diencephalic major effector pathway for penile erection follows the course of the medial forebrain bundle to the ventral tegmental area. At this level, the

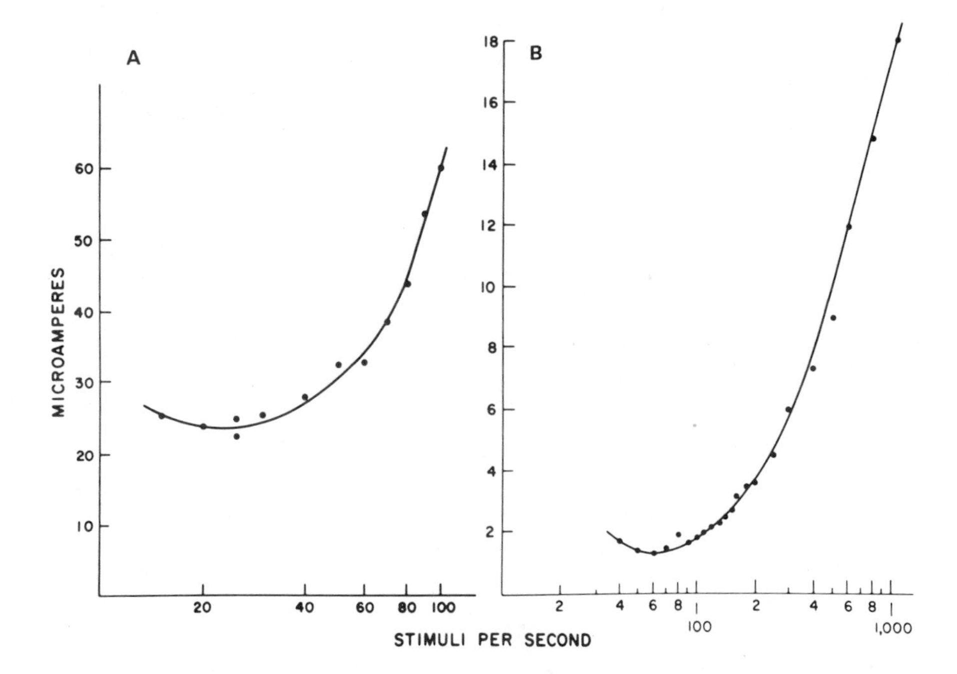

Figure 20-6. Frequency–amperage curves illustrating optimum parameters of stimulation for eliciting penile erection at loci rostral and caudal, respectively, to midlevel of the hypothalamus.

(A) Frequency–amperage curve obtained by stimulation in septopreoptic region. Semilog plotting is used to show nadir of curve to better advantage. As illustrated here, frequencies of 20 to 30 per sec with pulse durations of 1 msec were optimum for stimulating rostral limbic cortex, as well as anterior and midline dorsal thalamus.

(B) Frequency–amperage curve obtained by stimulating through an electrode impinging on medial margin of subthalamic nucleus, where current spread would also have involved fibers of medial forebrain bundle. Very short pulses of 0.01–0.05 msec applied at frequencies ranging from 60 to 100 per sec were optimum for eliciting penile erection when stimulating the medial forebrain bundle and its environs. Note that the ordinate values represent *total* current, not peak current of pulse. Value for peak current is obtained by dividing value for ordinate by product of the pulse repetition rate and the pulse duration.

From MacLean and Ploog (1962).

results of stimulation indicated that the pathway turns abruptly lateralwards and, as diagrammed in Figure 20-7, skirts the dorsal aspect of the substantia nigra before descending toward the pons.[5] In the pons the positive loci follow a course through the triangular area containing the lateral tegmental process that is bounded by the peduncular fascicles and brachium pontis. At one locus in this region, unlike any other encountered in the investigation, stimuli at a rate as slow as 2 per sec maintained full erection.[5]

The finding of positive loci in the region of the lateral tegmental process, together with the identification of some positive points in the brachium pontis, provides inferential evidence of involvement of cerebellar mechanisms in penile erection. In some instances stimulation at positive loci in the forebrain potentiated the genital response elicited by excitation of the brachium pontis. In view of the throbbing erections that have been described in conjunction with hippocampal afterdischarges, it was of particular interest that stimulation in the region of the dorsal psalterium potentiated the genital response elicited by stimulation of the lateral pontine nucleus and that 4+ throbbing erection persisted throughout the duration of the induced hippocampal afterdischarge.[4]

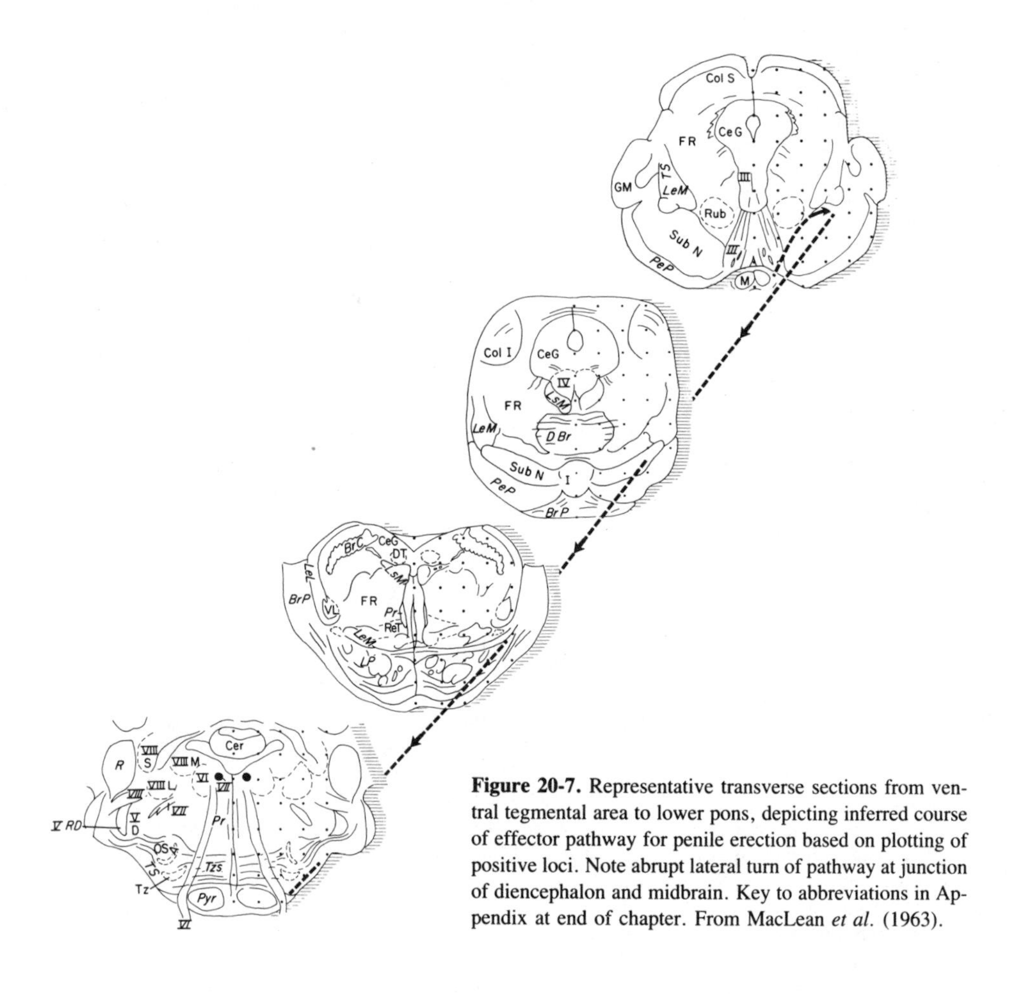

Figure 20-7. Representative transverse sections from ventral tegmental area to lower pons, depicting inferred course of effector pathway for penile erection based on plotting of positive loci. Note abrupt lateral turn of pathway at junction of diencephalon and midbrain. Key to abbreviations in Appendix at end of chapter. From MacLean *et al.* (1963).

In the lower part of the pons the plotting of positive loci for erection indicated that the effector pathway descends ventrolateral to the superior olive before turning medially to enter the medulla immediately lateral to the pyramids and just medial to the exit of the sixth nerve (Figure 20-7). We did not explore the brainstem below the level of the obex.[4]

Comparative Findings

Subsequent to the publication of our findings on the cerebral representation of penile erection in male squirrel monkeys, Maurus *et al.*[12] found a similar distribution of positive loci for clitoral tumescence in the upper brainstem of female squirrel monkeys. In the macaque, Robinson and Mishkin's findings on the two limbic divisions were essentially comparable to ours except for the failure to elicit penile erection by stimulation of the mammillary bodies.[13]

Findings Elsewhere in Forebrain

Striatal Complex

In the macaque, Robinson and Mishkin identified a few loci in the caudate and putamen at which electrical excitation induced partial erection.[13] We did not systematically explore the striatal complex, but stimulation at many points within the caudate nucleus, putamen, and globus pallidus elicited no genital responses.[3] Some results of stimulation at lower levels, however, would be compatible with the possibility that structures closely related to the striatal complex may be implicated in genital function. Penile erection, for example, was elicited by stimulation at a locus in the medial forebrain bundle just medial to the subthalamic nucleus[3] where current spread might have activated basal ganglia efferent and afferent fibers encapsulating the subthalamic nucleus. In regard to the substantia nigra, we concluded that the positive responses obtained by stimulating a narrow strip over its dorsal aspect (Figure 20-7) were owing to activation of the continuation of the medial forebrain bundle, which is known to turn lateralwards at the level of the ventral tegmental area.[4] It has been suggested that in higher forms the medial forebrain bundle and lateral forebrain bundle become temporarily separated in the nigral region because of the great development of the cerebral peduncle.[4]

Neocortex and Related Structures

Apropos of the role of the neocortex in genital function, it was mentioned in Chapter 19 that in operations for treatment of epileptic disorders, Penfield stimulated the greater part of the human neocortex, as well as the limbic cortex of the insulotemporal region and parts of the cingulate gyrus, but never elicited signs or symptoms of genital tumescence or feelings of an erotic nature. Penfield and Rasmussen reported one case in which electrical stimulation of the posterior postcentral gyrus "produced sensation in the contralateral side of the penis."[14] In the macaque, Woolsey *et al.* showed that tactile stimulation of the genitalia evoked potentials in the parietal cortex on the medial wall of the hemisphere.[15] In one instance, Lockhart observed that penile erection occurred in the cat following application of acetylcholine to the cortex of the superior bank of the splenial sulcus just above the posterior cingulate gyrus.[16] In our experiments on squirrel monkeys we explored the entire medial parietal[3,4,8] and medial frontal cortex.[5] Electrical stimulation of the parietal area was ineffective in eliciting penile erection, but as shown in Figure 20-3, positive loci for erection were found in the medial frontal neocortex just forward of and above the rostral limit of the cingulate cortex.[5] We did not explore the neocortex of the outer convexity, but it is significant in this respect that no genital responses were elicited by stimulation of the neothalamic nuclei or of the middle division of the pyramids that carries fibers from the sensorimotor area.[3,4]

Genital Sensation and Seminal Discharge

Spinothalamic Pathway

In regard to somatosensory aspects of genital function, it was of special interest to find that stimulation at loci along the course of the spinothalamic tract or in the region of termination of its medial thalamic division resulted in genital scratching and seminal discharge.[17] The dashed line in Figure 20-8 interconnects such loci found in the pons and midbrain, while Figure 20-9 gives the picture for the caudal thalamus. In each figure black stars indicate genital scratching, encircled black stars signify that seminal discharge with motile sperm also occurred. White stars identify loci for scratching of other parts of the body.

The lowermost locus shown in Figure 20-8 lies at a level near the junction of the medulla and pons. Here, as at higher levels, we were alerted to the possibility that we were approaching a locus for genital scratching because upon incremental lowering of the electrode the focus of the monkey's scratching shifted downwards from the upper part of the body to the abdomen. At the point where the electrode was chronically fixed in place, stimulation induced intensive scratching of the genitalia, pulling and kneading of the skin of the genitalia, after which, without further manipulation, there was ejaculation (findings documented by motion pictures[18]). With reapplied stimulation, the monkey immediately began to scratch and knead the genitalia, but upon periodic testing ejaculation did not occur again until 1 day later.

At the thalamic level, positive loci for genital scratching or other parts of the body, as well as for seminal discharge, were located primarily in the region of junction of MD with the central lateral, parafascicular, and centromedian nuclei.[17] By placing a partition in the special chair, a monkey could be prevented from scratching its genitalia. It was shown under these conditions that seminal discharge could occur independently of genital manipulation. In one case seminal discharge preceded the appearance of throbbing penile erection.

Other Findings

Within a millimeter of a positive point for seminal discharge stimulation might elicit other visceral-related effects such as salivation, vomiting, urination, or defecation.[17]

Isthmus Region

There was one other region in the brainstem where stimulation elicited genital scratching: As indicated in Figure 20-8, positive loci were found along the course of the longitudinal fiber system in the floor of the fourth ventricle just lateral to the sixth-nerve nucleus and as far forward as the dorsal tegmental nucleus of Gudden in the floor of the aqueduct.[17] Unlike the localized scratching responses elicited by stimulation along the spinothalamic tract and in the caudal thalamus, the monkeys behaved as though they were experiencing generalized formication: with the onset of stimulation they would begin to scratch in one area and then as stimulation continued, scratching would be performed with both hands and extend to every accessible part of the body. In some instances there might be partial penile erection, but seminal discharge was never observed.

Optimum Parameters and Latency

Stimulation with moderately long pulses of 1 msec at 30 per sec proved to be most effective for eliciting scratching responses. The same was true for seminal discharge. The

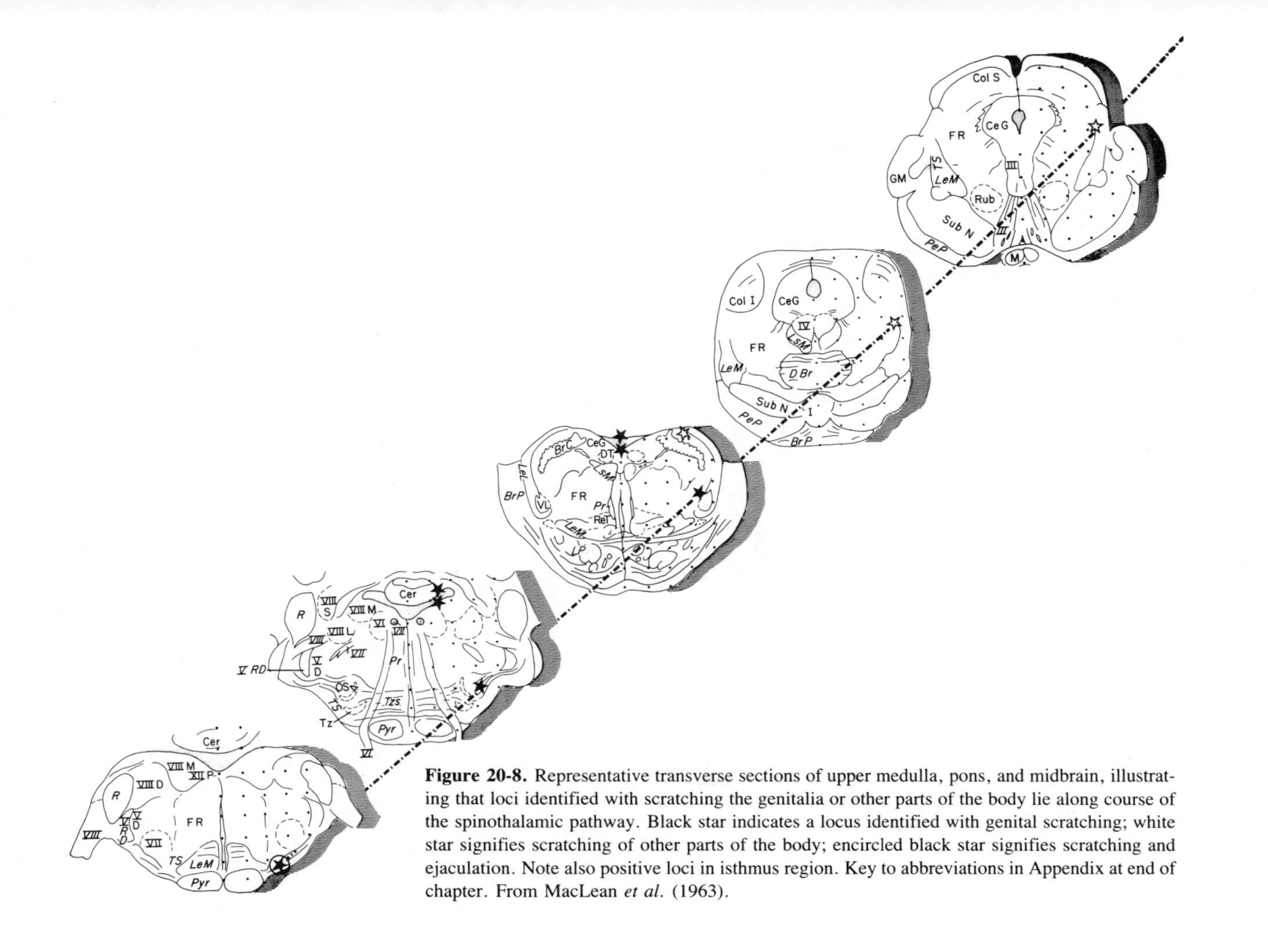

Figure 20-8. Representative transverse sections of upper medulla, pons, and midbrain, illustrating that loci identified with scratching the genitalia or other parts of the body lie along course of the spinothalamic pathway. Black star indicates a locus identified with genital scratching; white star signifies scratching of other parts of the body; encircled black star signifies scratching and ejaculation. Note also positive loci in isthmus region. Key to abbreviations in Appendix at end of chapter. From MacLean *et al.* (1963).

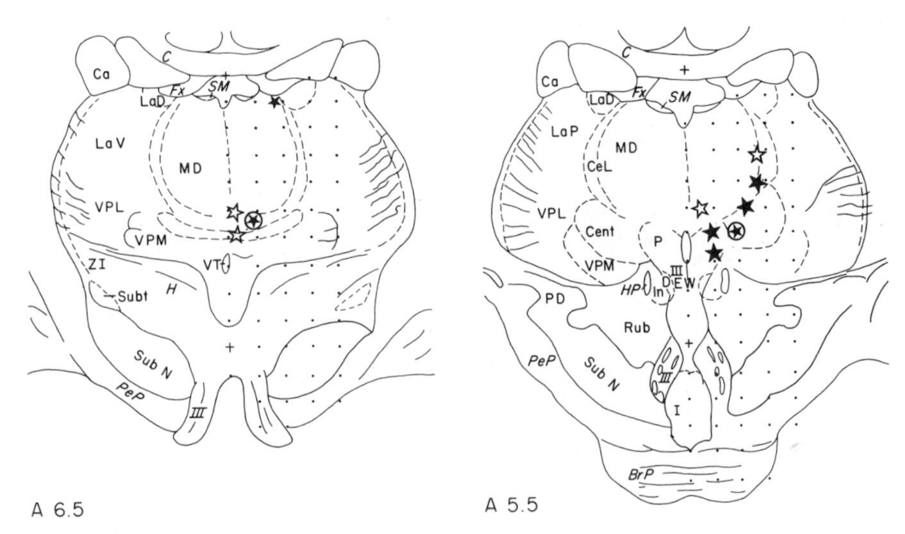

Figure 20-9. Drawings of sections from brain atlas, showing thalamic loci at which stimulation resulted in genital scratching (black star); scratching of other parts of the body (white star); and both genital scratching and seminal discharge (encircled black star). Small dots 1 mm apart overlie other parts of brainstem explored and found to be negative for the responses in question. Dot replaced by cross indicates 10 mm above horizontal axis 0 of the stereotaxic atlas. Key to abbreviations in Appendix at end of chapter. From MacLean *et al.* (1963).

higher peak currents required for pulses of short duration had the tendency to induce aversive effects that competed with the scratching response. By counting individual frames of motion pictures, one obtains a close estimate of the latencies for scratching. The latency range of 200 to 400 msec, together with several additional considerations listed in the original paper, provided inferential evidence that the scratching response was owing to afferent, rather than efferent, stimulation.[17] Of foremost significance were the observations that the animal would frequently alternate with its hands in scratching and would overcome an overlying obstruction in order to scratch the skin. The impression that the sensation was of a pruritic nature, rather than a paresthesia such as tingling, was supported by the finding that the positive loci fell along the course of the spinothalamic tract and that scratching was not elicited by excitation of the medial lemniscus. The kneading of the skin of the genitalia, such as occurs in kraurosis vulvae, was another indication of the pruritic nature of the symptom.

Comment

The foregoing findings provide a developing picture of somatosensory pathways and sensorimotor mechanisms involved in elemental sexual functions (see Figure 20-10). The caudal intralaminar structures shown to be implicated in genital scratching and seminal discharge are in juxtaposition to nuclei that project to the rostral limbic cortex and to the hypothalamus where stimulation is highly effective in eliciting penile erection.

Since genital scratching is also elicited by stimulation at points along the longitudinal fiber system in the floor of the fourth ventricle and aqueduct of the isthmus region, there are possibly more "primitive" somatosensory pathways for influencing structures of the

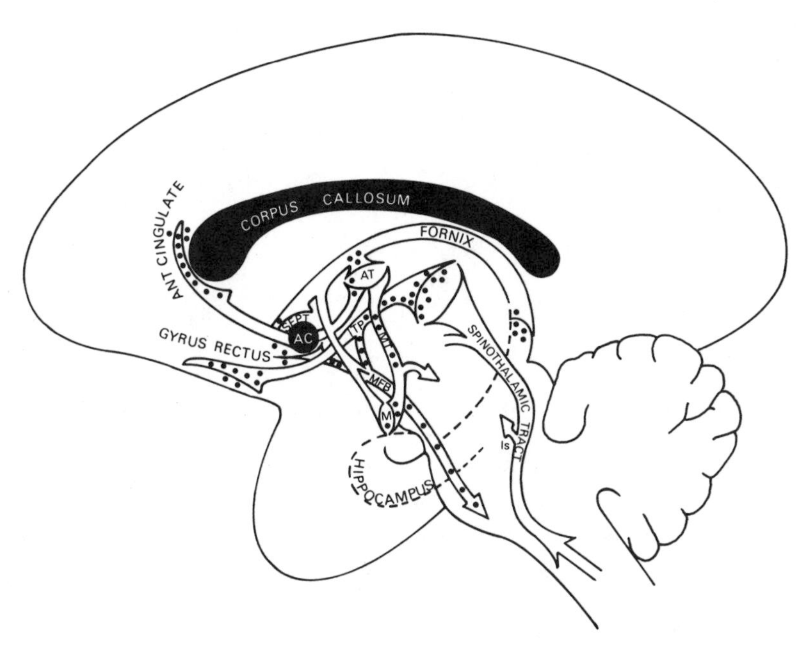

Figure 20-10. Sagittal diagram of squirrel monkey's brain, summarizing the results of mapping structures involved in primal sexual functions. The clumping of black dots indicates regions where stimulation is highly effective in eliciting erection, while linear placement of dots identifies pathways such as the mammillothalamic tract (MT), inferior thalamic peduncle (ITP), and medial forebrain bundle (MFB) along which there were positive loci for erection. Based on responses manifest by genital scratching and/or seminal discharge, the spinothalamic pathway and its medial distribution to the caudal intralaminar region of the thalamus are inferred to play an important role in genital sensation and function.

Abbreviations: AC, anterior commissure; AT, anterior thalamic nuclei; ITP, inferior thalamic peduncle; M, mammillary body; SEPT, septum. See Appendix for further abbreviations. After MacLean (1962).

forebrain involved in genital function. As noted in Chapter 18, there is a potential pathway from the genitalia that ascends via the ''pelvic vagus'' to the solitary nucleus, which in turn projects to the deep tegmental nucleus of Gudden within the caudal periaqueductal gray matter. From this latter nucleus the following three pathways ascend to forebrain nuclei involved in penile erection: (1) a mammillopeduncular path to the septum; (2) a mammillopeduncular path to nuclei of the mammillothalamic tract; and (3) the dorsal longitudinal bundle of Schütz conveying fibers to caudal intralaminar and midline nuclei. The septal projections possibly represent a ''primitive'' pathway for relating somatosensory and olfactory information. Finally, it should be noted that Mantyh has recently demonstrated in an autoradiographic study in the squirrel monkey that all parts of the central gray project via the periaqueductal fiber system to the dorsomedial area of the hypothalamus,[19] where stimulation elicits not only full erection, but also chirping vocalization. In a study on freely moving rhesus monkeys, Perachio *et al.* found that the dorsomedial nucleus was the only hypothalamic locus where radiostimulation led not only to mounting and thrusting but also to ejaculation.[20] Other areas, however, may be involved in ejaculation: Robinson and Mishkin reported the case of a rhesus monkey in which ejaculation resulted from self-stimulation at a locus in the medial preoptic region.[21] Herberg observed that seminal discharge occurred in rats following self-stimulation at loci said to be in the posterior hypothalamus.[22]

As noted in Chapter 18, Herrick long ago called attention to the isthmus region as a phylogenetically ancient neural substrate of basic importance for the integration of somatic and visceral functions. In ablation experiments on frogs, Aronson and Noble found that destruction of the isthmus region caudal to the trochlear nucleus interfered with spawning.[23] I am not aware of experiments in mammals that have pursued this lead in regard to ejaculation. In both macrosmatic and microsmatic mammals urination plays an important role in sexual and other social communication. Kuru traced ascending connections from the bladder to the region ventral to the superior cerebellar peduncle.[24] Barrington found that bilateral lesions involving the subpeduncular part of the isthmus region in cats interfered with complete emptying of the bladder and that in some cases the animals behaved as though unaware of impending urination or defecation.[25] It is most relevant to the present discussion that the so-called pneumotaxic center of Lumsden is located in this region and that the parabrachial nuclei have been shown to have reciprocal ascending and descending connections linking the vagal nuclei with the forebrain (see Chapter 18). On the basis of existing evidence it may be assumed that these connections are fundamental for the interplay of the forebrain and lower brainstem in the control of respiration, urination, copulation, and orgasm.

The Question of Hippocampal Influence

As described in the preceding chapter, in some instances stimulation of the caudal hippocampal formation (proximoseptal segment) resulted in 2 to 3+ tumescence at relatively long and variable latencies, with 4 to 5+ throbbing erection occurring during ensuing afterdischarges. During the afterdischarge, the monkey might scratch various parts of its body, including the anogenital region. As noted above, stimulation of the dorsal psalterium potentiated the genital response to stimulation of the lateral tegmental region of the pons, with full throbbing erection occurring during the ensuing hippocampal afterdischarge.

As also noted, with stimulation of the pregenual cingulate cortex erection appeared simultaneously with the recruitment of potentials in the hippocampus. Afterdischarges induced by stimulation of the inferior septal region, anterior thalamus, and posterior part of the gyrus rectus were associated with throbbing penile erection. Ejaculation, however, was never observed under such conditions. Since sneezing has been characterized as having an orgastic quality, it is of parallel interest to point out that in one experiment, stimulation of the olfactory bulb resulted in recruitment of high-voltage potentials in the hippocampus, followed by self-sustained afterdischarges, during which sneezing occurred (see text and Figure 11 of Ref. 5). In the few instances in which hippocampal activity was monitored during seminal discharge, there were no significant changes.

Comment on the role of the hippocampus in the neuroendocrine regulation of sexual functions will be deferred until Chapter 26 on electrophysiological studies on the limbic system.

Orosexual Mechanisms in Aggression

Observations made during exploration of the hypothalamus at rostral and intermediate levels were of particular interest in regard to the close association of brain mechanisms linking oral, sexual, and combative behavior. The nature of the findings may be illustrated by the results of exploring along the electrode track shown in Figure 20-11. The gliosis reveals the path of the electrode as it was lowered from the dorsal hypothalamic area past the medial side of the fornix into the ventromedial nucleus. The electrode was then withdrawn to the level of the roof of the third ventricle and permanently fixed in place. The clear area identifies the site of the electrode tip, which impinged upon the mammillothalamic tract. Stimulation at this point elicited full erection and chirping vocalization. In the initial exploration the same response was elicited down to the level indicated by the upper arrow. At this site just medial to where the ''pallidohypothalamic'' tract loops over the fornix, stimulation elicited agitated behavior, cackling vocalization, and only partial erection. From this level to that of the lower arrow, stimulation resulted in showing of the fangs, piercing vocalization, biting movements, and shrinking of the penis. Upon cessation of the stimulation, there was a ''rebound'' erection. Stimulation at loci along the rest of the track shown by the gliosis induced only chewing movements.

The same sequence of events occurs when an electrode is lowered at a more rostral level in the hypothalamus, with the angry manifestations appearing as the tip nears the supraoptic commissure of Ganser.

Along the course of the medial forebrain bundle caudal to midlevel of the hypothalamus, it is possible with optimum parameters of stimulation to elicit oral and genital responses, respectively, by stimulation at loci about 0.5 mm apart, an indication that effector pathways involved in these effects descend in close association.

The reciprocal functional relationship between oral and genital function that is observed upon stimulation within the amygdalar and septal divisions has already been commented upon in the preceding chapter. Here it should be noted that on the basis of anatomical connections (Chapter 18) and the results of electrical stimulation, a reciprocal functional relationship also probably exists between the amygdalar and thalamocingulate divisions: Stimulation at caudal levels of the medial part of MD may result in penile erection followed by salivation and chewing movements.

How does one account for the close relationship of oral and genital functions in combative behavior? This question brings us to the deferred discussion of the preceding chapter in regard to effector pathways accounting for angry vocalization and defensive behavior elicited by stimulation of the amygdala. As was illustrated in Figure 19-6, Fernandez de Molina and Hunsperger found that, starting from the medial amygdala, the defense reaction could be elicited by stimulation at points along the course of the stria terminalis to its bed nucleus in the lateral septum.[26] From the bed nucleus they traced positive loci for the defense reaction to that part of the anterior hypothalamus where Hess and Brugger had originally found that stimulation resulted in angry behavior.[27] In proposing that the stria terminalis constituted the major descending pathway from the amygdala, Fernandez de Molina and Hunsperger stated that this assumption was supported by their finding that the amygdala-induced reaction could no longer be obtained after coagulation of the stria.[26] Hilton and Zbrozyna found, however, that if one observed such animals beyond the acute 24-hr period, coagulation of the stria had no effect on the response induced by amygdalar stimulation.[28] On the basis of their own experiments, they argued

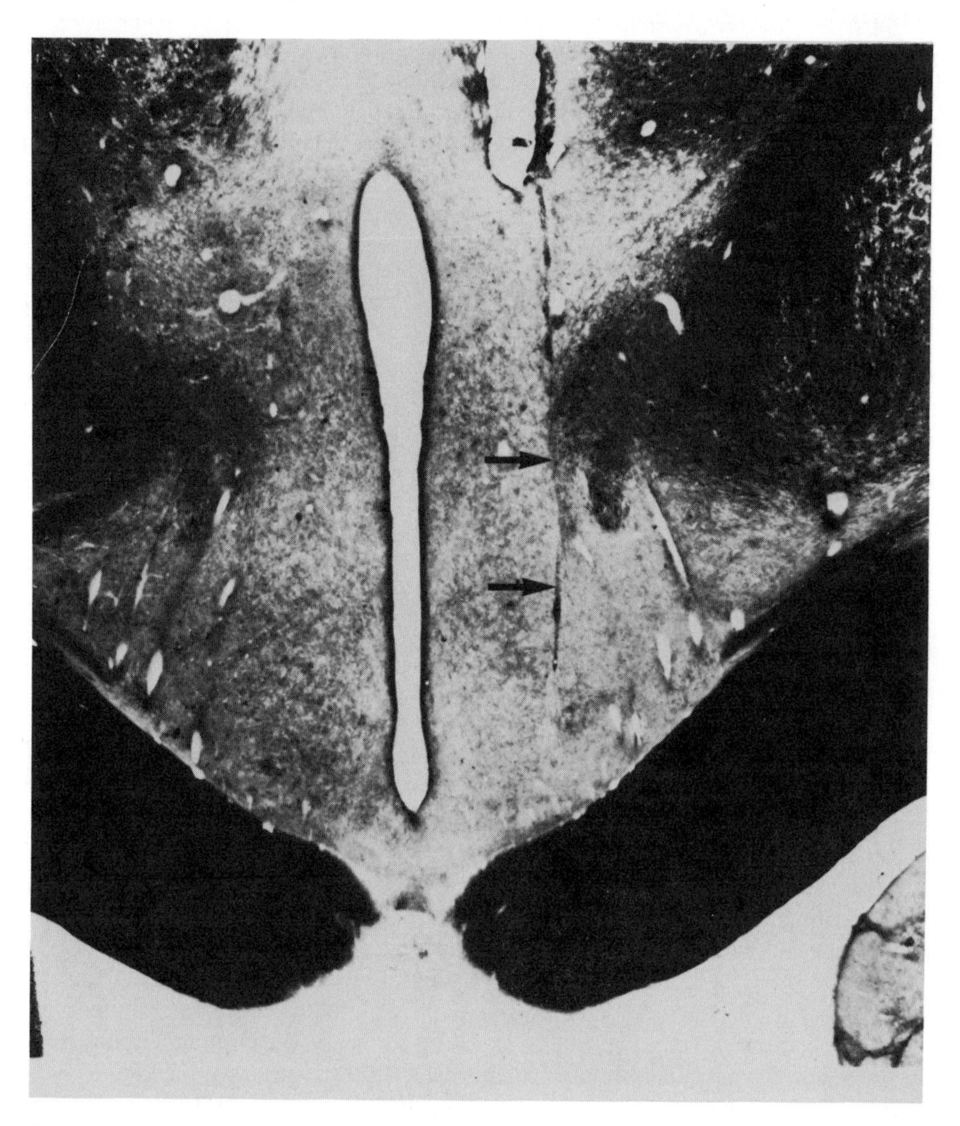

Figure 20-11. Brain section showing path of the electrode track along which stimulation elicited genital, agonistic, and oral manifestations. Line of gliosis shows path of exploring electrode prior to its being withdrawn and implanted at the site of the clear area where the tip impinged on the rostral aspect of the mammillothalamic tract. Stimulation both here and at a point halfway toward the upper arrow resulted in full erection and ''chirping'' vocalization. At the site of the upper arrow, stimulation elicited agitated behavior, cackling vocalization, and partial erection. As electrode descended toward the lower arrow, stimulation resulted in piercing vocalization and biting, followed by rebound erection. Note that part of track indicated by arrows lies medial to the ''pallidohypothalamic'' tract where it loops over and descends past the medial aspect of the fornix. At the lowest point of stimulation in the ventromedial nucleus, stimulation elicited primarily biting. Weil–Weigert stain. From MacLean (1973c).

that the response obtained by stimulation of the stria terminalis was the result of activating afferent fibers to the amygdala. Contrary to the conclusions of Fernandez de Molina and Hunsperger, they claimed that the pathway for the defensive reaction coincides with the amygdalofugal bundle of Nauta that fans out in the sublenticular region "to connect with the whole length of the hypothalamic center."[28] Subsequently, Hunsperger conducted an extensive series of experiments that failed to bear out their claim.[29] However this matter may be eventually resolved, there can be no doubt that pathways from the septum and amygdala converge in the hypothalamus in the region that appears to be nodal with respect to the expression of angry behavior. Hess referred to the critical locus as the perifornical region.[30] Although Hess and Brugger did not elicit angry behavior by stimulation of the body of the fornix, it may prove to be of functional significance that fibers from the fornix, as originally shown in rodents,[31] and more recently found in primates,[32] splay off into the perifornical nucleus.

Hunsperger found that stimulation of the lateral part of the periaqueductal gray matter lying between the level of the third and fourth cranial nerve nuclei elicits the same angry responses as excitation in the perifornical region, but with a shorter latency.[33] Since electrocoagulation of this particular central gray region eliminated the angry response to perifornical stimulation, he regarded it as a primary autonomous zone for the expression of combative behavior.

As illustrated in Figure 20-12, if one uses the shield of Mars for oral responses and his spear for genital responses and plots them on a sagittal diagram of the brain, the symbols for the mouth cluster in the region of the amygdala, while those for the genital appear in the region of the septum.[34] Followed caudally, one finds a reconstitution of the warrior Mars at the locus in the hypothalamus that is nodal for the expression of angry behavior. From this picture it would appear that the close functional relationship between the amygdala and the septal division and their descending pathways is owing to the olfactory sense, which, dating far back in evolution, plays a role in both feeding and mating, as well as in the fighting that frequently precedes.

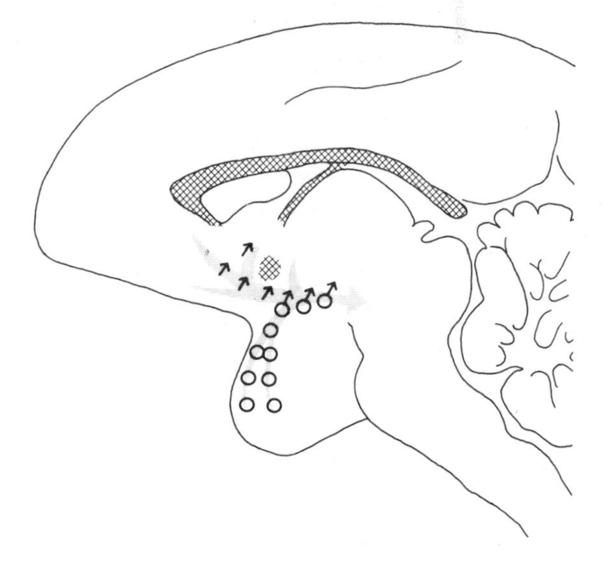

Figure 20-12. Parasagittal diagram of the brain, illustrating a hypothalamic convergence of neural mechanisms involved in oral, genital, and agonistic responses. The symbol for the shield of Mars represents loci at which electrical stimulation elicit oral and facial responses, while his spear signifies loci positive for genital responses. Symbols for oral responses cluster in the amygdala region and those for genital responses in the septopreoptic region. Traced caudally, the symbols join and show a reconstitution of the warrior Mars in a part of the hypothalamus involved in combative behavior. Since fighting may be a preliminary to both feeding and mating, it would appear that the same mechanisms for combat are involved in each situation. From MacLean (1964).

Appendix: Abbreviations

Except for abbreviations entirely in lowercase, roman letters and italics are used, respectively, to label nuclear and fiber structures. Initials for a number of generic terms (e.g., corpus, fasciculus, nucleus) are omitted. Except where noted in square brackets, the terminology is the same as that of Henry Alsop Riley's *An Atlas of the Basal Ganglia, Brain Stem, and Spinal Cord* published in 1943 and reprinted in 1960.

III	Nervus oculomotorius
IIIEW	N. nervi oculomotorii (Edinger–Westphal)
III	N. nervi oculomotorii or Ventriculus tertius
IV	N. nervi trochlearis
VD	N. radicis trigeminalis descendentis
VM	N. motorius nervi trigemini
VRD	Nervus trigeminus, radix descendens
VI	Nervus abducens
VI	N. nervi abducentis
VII	Nervus facialis
VIII	Nervus acusticus
VIIID	N. radicis vestibularis descendentis
VIIIL	N. vestibularis lateralis [Papez]
VIIIM	N. vestibularis medialis [Papez]
VIIIS	N. vestibularis superior [Papez]
XIIP	N. praepositus nuclei nervi hypoglossi
ad	N. antero-dorsalis thalami
al	Ansa lenticularis
av	N. antero-ventralis thalami
BCI	Brachium colliculi inferioris
BrC	Brachium conjunctivum
BrP	Brachium pontis
C	Corpus callosum
Ca	N. caudatus
ca	Commissura anterior
cc	Corpus callosum
CeG	Substantia centralis grisea
CeL	N. centralis lateralis [Olszewski]
Cen	N. centralis superior (Bechterew)
Cent	N. centri mediani thalami
Cer	Cerebellum
ci	Capsula interna
co	Chiasma opticum
ColI	Colliculus inferior
ColS	Colliculus superior
ComP	Commissura posterior

D	N. of Darkschewitsch
db	Diagonal band of Broca
DBr	Decussatio brachii conjunctivi
DT	N. dorsalis tegmenti (Gudden)
f	Fornix
FR	Formatio reticularis
Fx	Fornix
gc	Gyrus cinguli
GL	Corpus geniculatum laterale
GM	Corpus geniculatum mediale
gp	Globus pallidus
gr	Gyrus rectus
H	Area tegmentalis
hc	Hypophysis cerebri
HL	N. habenularis—pars lateralis
HM	N. habenularis—pars medialis
HP	Fasciculus habenulo-interpeduncularis
I	N. interpeduncularis
In	N. interstitialis (Cajal)
LaD or ld	N. lateralis dorsalis [Olszewski]
LaP	N. lateralis posterior thalami
LaV	N. lateralis ventralis thalami
ld	N. lateralis dorsalis thalami
LeL	Lemniscus lateralis
LeM	Lemniscus medialis
LoC	Locus ceruleus
LP	[Fasciculi longitudinales pontis]
LsM	Fasciculus longitudinalis medialis
m or M	Corpus mammillare
md or MD	N. medialis dorsalis thalami
mfb	Fasciculus medialis telencephali (medial forebrain bundle)
mt	Fasciculus mammillo-thalamicus
nc	N. caudatus
ns	N. subthalamicus
nst	N. stria terminalis
OS	N. olivaris superior
OSA	N. olivaris superior accessorius
P	N. parafascicularis thalami

p	Putamen
pc	Pedunculus cerebri
PD	N. peripeduncularis dorsalis
PeP	Pes pedunculi
Po	N. pontis
po	Area preoptica
Pr	Fasciculus praedorsalis
Prt	N. praetectalis
PuI	N. pulvinaris inferior thalami
PuL	N. pulvinaris lateralis thalami
PuM	N. pulvinaris medialis thalami
PuO	N. pulvinaris oralis thalami [Olszewski]
pv	N. paraventricularis hypothalami
Pyr	[Pyramis]
R	Corpus restiforme
ReT	N. reticularis tegmenti pontis
Reu	N. reuniens thalami
Rub	N. ruber tegmenti
s	Septum pellucidum
SM or sm	Stria medullaris thalami
sn	Substantia nigra
st	Stria terminalis
SubN	Substantia nigra
Subt	Corpus subthalamicum
to	Tractus opticus
TS	Tractus spino-thalamicus
tt	Tuberculum thalami
Tz	N. corporis trapezoides
Tzs	Corpus trapezoides
u	Gyrus uncinatus
va	N. ventralis anterior thalami
VL	N. ventralis lemnisci lateralis
vl	N. ventralis lateralis thalami
vlc	Ventriculus lateralis cerebri
VPL	N. ventralis postero-lateralis thalami
VPM	N. ventralis postero-medialis thalami
VT	Ventriculus tertius
ZI	Zona incerta

References

1. Clark and Meyer, 1950, p. 342
2. Sigrist, 1945
3. MacLean and Ploog, 1962
4. MacLean *et al.*, 1963a
5. Dua and MacLean, 1964
6. Poletti *et al.*, 1973; Poletti and Creswell, 1977
7. MacLean, 1959
8. MacLean, unpublished observations
9. Ariëns Kappers *et al.*, 1936, p. 1180
10. Müller-Preuss and Jürgens, 1976

10a. Showers, 1959

11. Sherrington, 1947
12. Maurus *et al.*, 1965
13. Robinson and Mishkin, 1968
14. Penfield and Rasmussen, 1952
15. Woolsey *et al.*, 1942
16. Lockhart; see MacLean, 1957b
17. MacLean *et al.*, 1963b
18. MacLean, 1966
19. Mantyh, 1983
20. Perachio *et al.*, 1979
21. Robinson and Mishkin, 1966
22. Herberg, 1963a, b
23. Aronson and Noble, 1945
24. Kuru, 1956
25. Barrington, 1925
26. Fernandez de Molina and Hunsperger, 1959
27. Hess and Brugger, 1943
28. Hilton and Zbrozyna, 1963
29. Hunsperger and Bucher, 1967
30. Hess *et al.*, 1945–46
31. Nauta, 1956
32. Poletti and Creswell, 1977
33. Hunsperger, 1956
34. MacLean, 1964b

21

Participation of Thalamocingulate Division in Family-Related Behavior

The term *thalamocingulate division* was introduced in Chapter 18 as a designation for the cingulate *mesocortical* areas innervated by the anterior thalamic nuclei and certain other thalamic structures (Figure 19-1). An important recent finding has been the demonstration of a co-innervation of the rostral cingulate cortex by the anterior medial nucleus and parts of the medial dorsal nucleus and intralaminar nuclei. As diagrammed in Figure 19-1, the mammillothalamic tract is the main source of afferents to the anterior thalamic nuclei. Le Gros Clark and Meyer[1] have emphasized that "there is nothing in the reptilian brain which corresponds to the medial mamillary nucleus, the mamillothalamic tract, or the anterior nucleus of the thalamus." "It follows," they concluded, "that the functions subserved by the circuit must have a relation to forms of cerebral activity which are not to be found in any vertebrates other than mammals."[1] Hence, they doubted that the circuit could be involved in functions "of an elementary character."

The foregoing considerations become of enhanced interest in the light of recent findings indicating that the thalamocingulate division is concerned with three forms of behavior that distinguish the evolutionary transition from reptiles to mammals—namely, (1) nursing, conjoined with maternal care, (2) audiovocal communication for maintaining maternal–offspring contact, and (3) play.[2] The present chapter deals with experimental findings on the cingulate representation of this family-related behavioral triad. From the standpoint of human evolution, no behavioral developments could have been more fundamental because they set the stage for a family way of life with its evolving responsibilities and affiliations that has led to worldwide acculturation.[2]

Before describing the neurobehavioral findings, it is relevant to ongoing and future research to consider some cytoarchitectural details that were not dealt with in Chapter 17 when emphasizing the transitional nature of the cingulate cortex. In comparative terms, the most pronounced cytoarchitectural differences apply to the posterior cingulate area of primates corresponding to area 23 in the maps of Brodmann[3] (Figure 17-7). In primates, this area gives the impression of having expanded at the expense of the retrosplenial area.

Aspects of Cingulate Comparative Cytoarchitecture

Attention has already been called to three features that characterize the transitional cortex (mesocortex; proisocortex) of the cingulate gyrus—namely, (1) the deeply stained layer II with its closely packed cells; (2) the "bandlike" layer V; and (3) the tufted cells of layer II with their dendrites extending into layer I. Based on a study of the cat, Vogt and

Peters concluded that the tufted cells (fusiform pyramids) are the only cells "unique to cingulate cortex," having "only one basally projecting dendrite."[4]

Four Main Cingulate Areas

The brain of the domestic rabbit *(Oryctolagus cuniculus),* belonging to the order Lagomorpha, serves to illustrate features of the four main areas of the prototypical cingulate cortex. For this purpose I draw upon the descriptions of Rose and Woolsey.[5] In their parcellation of the cingulate cortex shown in Figure 21-1A, they identified three main areas denoted as the anterior limbic, cingular, and retrosplenial (compare also brain of cat shown in Figure 18-9). Here I will also include their infralimbic area because of its demonstrated thalamic connections.

Infralimbic Area

The infralimbic area corresponds approximately to Brodmann's[3] area 25 and to M. Rose's[6] regio infraradiata a (Figure 21-1A,B,C). Lying rostrodorsal to the taenia tecta, it

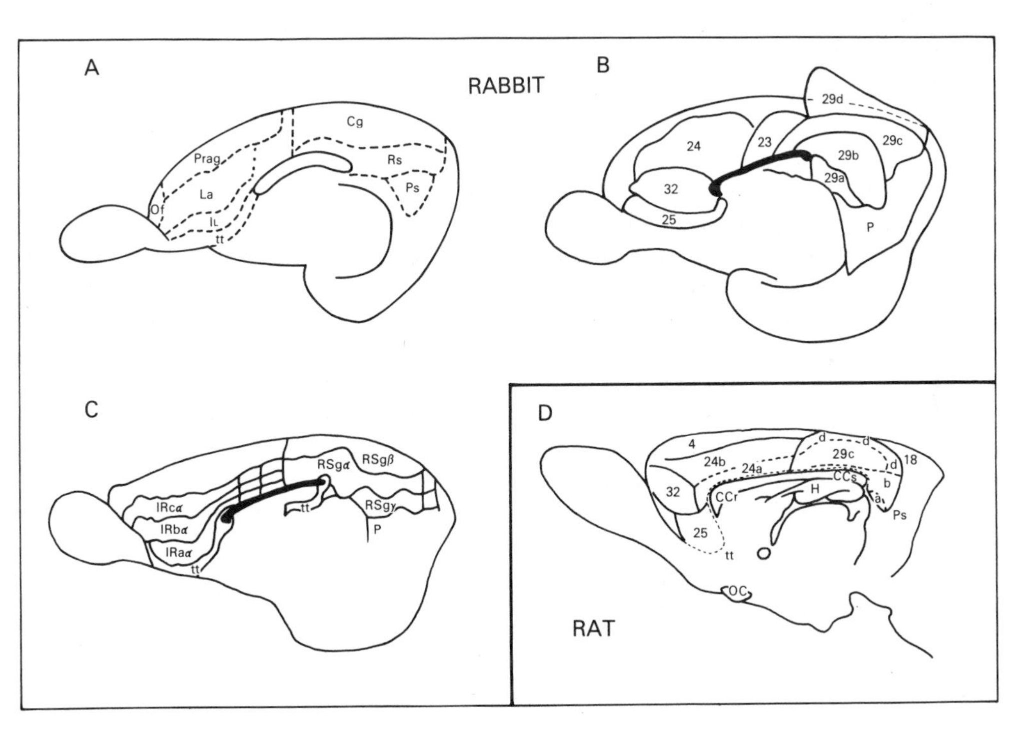

Figure 21-1. Parcellation, according to different authors, of cytoarchitectural areas of the cingulate cortex in the brain of a lagomorph and a rodent. A, B, and C represent, respectively, partially redrawn figures of rabbit brain from J. Rose and Woolsey (1948), Brodmann (1909), and M. Rose (1931). D shows one scheme for the rat (Vogt and Peters, 1981). See text for other details.

Abbreviations: CCr and CCs, rostral and splenial corpus callosum; Cg, cingular area; H, hippocampus; Il, infralimbic area; IR, regio infraradiata; La, anterior limbic region; OC, optic chiasm; Of, orbitofrontal area; P or Ps, postsubicular area; Prag, precentral agranular area; Rs, retrosplenial area; RSg, granular retrosplenial area; tt, taenia tecta.

extends from the level of the knee of the corpus callosum to almost the frontal pole, forming a strip that lies medial to the rostral extension of the lateral ventricle. Although the molecular layer (layer I) is as wide or wider than that of other cingulate areas, the outer margin of layer II is uneven and, as characteristic of cingulate cortex, the cells stain more deeply than those of the underlying layers.

It should be emphasized that this cortex is of special interest because of its close relationship to the olfactory apparatus and possible functional involvement in nursing and the separation call. In the rat, reciprocal connections have been found between nerve cells in this area and the vagal solitary nucleus (see Chapter 26). The same reciprocal relationship applies to the anterior limbic area next to be described.

Anterior Limbic Area

The anterior limbic area corresponds to areas 32 and 24 of Brodmann[3] and to the dorsal part of the infraradiate region (bα + cα) of M. Rose[6] (see Figure 21-1A,B,C). The superficial elements of layers II are dark staining. There is no distinct border between the cells of layers II and III, until dorsally the precentral agranular neocortex is approached. The typically small cells of these layers are distinctly larger than the so-called granular cells of the cingular area. There is no discernible granular layer IV. Layer V is about the thickness of layers II and III combined and contains larger cells that are distinctly pyramidal in shape. Layer VI is characterized by deeply staining fusiform cells. At the junction with the neocortex, the cells of this layer appear to pile up into multiple concentric rows.

The Cingular Area

Proceeding in a caudal direction, one finds that the cingulate cortex changes rather abruptly in character at about the level of the anterior thalamic nuclei. The superficial layers appear more prominent because the elements stain more deeply, and the cortex also assumes a dysgranular appearance because of an incipient intermingling of granulelike cells with the cells of layer III. Rose and Woolsey's cingular area occupies a considerable extent of the posterior limbic cortex, corresponding approximately to Brodmann's area 23, together with his area 29d and 29c (Figure 21-1A,B). The cortex of this region is distinguished by the presence of a thin but distinct granular layer IV separating the supragranular layers II and III from the infragranular layers V and VI.

Retrosplenial Area

The parasplenial and retrosplenial region comprises the retrosplenial area of Rose and Woolsey,[5] corresponding to Brodmann's areas 29a and 29b (Figure 21-1B). Examination in a rostrocaudal direction of the supracallosal, parasplenial cortex reveals a thinning of the second layer and a progressive piling up of granular cells in layer III. This is area 29b. Behind the level of the splenium such cells almost totally occupy the width of cortex usually identified as layers II and III. Immediately below this wide granular layer are layers V and VI. This area corresponds to Brodmann's area 29a. Inferiorly, and toward the presubiculum, superficial dark-staining cells practically disappear, and this part of the retrosplenial cortex acquires a distinctive appearance because of the lining up in horizontal rows of small, irregular cells that give the appearance of the warp and weft of a silken garment. According to Rose and Woolsey,[5] this distinctive area would correspond to Brodmann's area 29e, 27b, and 48 that he referred to as *area retrosubicularis* (area 48) in other mammals.[3]

Cingulate Cortex of Rodents

Because of the neurobehavioral work on rodents to be described, and in prospect of future research, it should be noted how the cytoarchitectural picture differs significantly in one respect from that of the lagomorph brain. Except for the cingular area, Rose and Woolsey's[5] description of the cingulate cortex of the lagomorph brain would apply almost equally well to such a rodent as the rat (Figure 21-1D). As opposed to the cingular area (area 23) in the rabbit that has a thin granular layer IV below layers II and III, area 23 in the rat is, in Vogt and Peters's[4] words, "a transitional region between areas 24 and 29, where the granularity is forming in layers II–III, and not in layer IV." Moreover, area 29d included in Rose and Woolsey's cingular area is dysgranular in the rat. Accordingly, Vogt and Peters contend that area 23 does not exist in the rat. Apropos of a recent cytoarchitectural map for the rat prepared by Krettek and Price,[7] it may be noted that their prelimbic area corresponds to Brodmann's[3] area 32.

As noted in Chapter 17, it is characteristic of the limbic cortex that the main afferents terminate in the outer layers. Whether or not there is the same parallel in regard to layer IA and IB as exists in regard to the secondary and tertiary afferents of the olfactory system (Chapter 18) deserves investigation. In sections well stained for myelin, the outer half of layer I of the anterior limbic area reveals a darker appearance because of the presence of distinctly myelinated fibers. Such a difference does not appear in Ramón y Cajal's Golgi preparations of the rodent brain; rather, in his studies on the interhemispheric cortex he places particular emphasis on a well-developed plexus in the third layer.[8] Some of Krieg's[9] illustrations of the myelin picture in rats show a darker outer half of layer I. At the transition between the anterior cingulate and posterior cingulate cortex (cingular area of Rose and Woolsey), the outer half of layer I continues to show a darker appearance because of the presence of distinctly myelinated fibers. As the splenium is approached and the typical retrosplenial cortex appears, the myelinated fibers congregate in a strand between the outer and inner halves of layer I. Finally as both the superficial and deep layers of the retrosplenial area thin out toward the hippocampal formation, the myelin strand of layer I surfaces from its intermediate position to become superficial in the presubicular zone.

Gyrencephalic Animals

In mammals with well-formed gyri, it may be said as a generalization that the distinctive features of the cingulate cortex are found within the limiting borders of the cingulate sulci. As Ariëns Kappers, Huber, and Crosby[10] have commented, "It is remarkable that the sagittal fissures, which bound this region dorsally, remain constant to a considerable extent throughout the mammalian series, not only morphologically but also in their relation to the underlying cytoarchitectonic fields." Taking into account both the Nissl and myelin picture, Campbell[11] observed that all "the facts" indicate that, like the hippocampal gyrus, the cingulate gyrus is "invested by cortex of great phylogenic age."

Carnivores

Rose and Woolsey's[5] areal parcellation of the cingulate cortex in the cat was referred to when describing cingulate connections with the thalamus (Chapter 18). A comparison of Figure 18-9 with Figure 21-1A shows that these authors identified equivalent areas in the cat as in the rabbit. In his comparative studies, Campbell commented that the cingulate areas in the dog have "a similar distribution" and exhibit "variations similar to those observed in Felis."[12]

Based on their studies of both the cytoarchitecture and thalamic connections of the cat and rabbit, Rose and Woolsey[13] concluded: "There is no reason to assume that the retrosplenial and cingular fields should be separate regions, as is widely accepted. They are certainly related and it seems reasonable to combine them both in one posterior limbic region." These comments are to be kept in mind in view of the possible visual functions of the posterior limbic region that will be touched upon next and again under the heading of primates.

Splenial Cortex of an Ungulate

As reviewed in Chapter 26, there is both anatomical and electrophysiological evidence of visual inputs to the retrosplenial area. In this connection it is of interest to cite an original observation by Campbell,[14] who, in his comparative studies, selected the common pig *(Sus communis)* as an example of an ungulate. According to Campbell, the retrosplenial area in the pig is unusually extensive. Apropos of the visual question, he pointed out that in the depth of the stem of the parasplenial sulcus (corresponding to the rostral stem of the calcarine in primates) the retrosplenial cortex in the pig can be easily confused with the abutting primary visual cortex. In his words, "[C]are must be exercised not to confound it with the visual, for in sections stained for nerve fibres, a linear arrangement occurs, which a superficial observer might mistake for a line of Gennari."[15] Even Ramón y Cajal,[8] in his earlier observations, mistook the retrosplenial cortex in rodents for the visual cortex. In such visually deficient animals as the ground shrew and the bat, the granular retrosplenial area is small.[16]

Cetacea

Because of the present focus on the role of the cingulate gyrus in maternal behavior, vocalization, and play, it is of interest to consider the development and structure of this convolution in whales. The behavioral triad in question is remarkably expressed in this order of mammals.[17] The cingulate part of the limbic lobe is well developed. The Nissl picture of the cingulate cortex has been described and illustrated by Morgane *et al.*[18] Given a familiarity with the rodent brain, it is surprising to find that in these animals that separated from land mammals some 60 to 90 million years ago,[19] one can detect in the main cingulate areas many of the same diagnostic features. In Golgi studies, Morgane *et al.*[20] have observed that cells of layer II generally are characterized by the presence of "extraverted neurons" (Figure 21-2) resembling the primitive type in the hedgehog by having four or more dendritic processes extending into the molecular layer and possessing no basilar dendrites. Such cells are indicative of the primitive pattern of afferent inputs to the outer cortical layers discussed in Chapter 17.

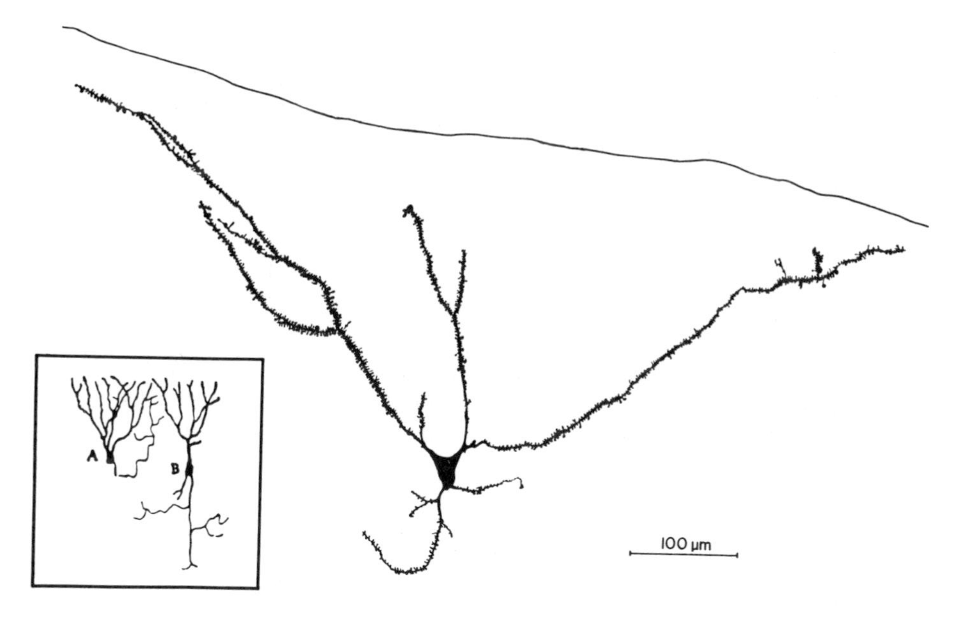

Figure 21-2. "Extraverted" neuron characteristic of layer II in the cortex of whale. In most mammals such "tufted" cells are found only in the archicortical and mesocortical areas, reflecting the primitive condition that their dendrites reach out in layer I to receive their main innervation. Inset shows for comparison Ramón y Cajal's (1893) representation of a pallial neuron of a frog (A) and lizard (B). Adapted from Morgane *et al.* (1986).

Cingulate Cortex in Primates

The question is raised as to whether or not the cingular area (area 23) exists in primates. In his well-known areal scheme for subdividing the cingulate cortex, M. Rose[6] did not include the posterior cingulate cortex in primates as a mesocortical area. Rather he regarded this cortex as fully developed cortex with the basic seven layers (septemstratificatus) (Chapter 17). Other authors, however, on the basis of both cytoarchitecture and thalamic connections, regard the posterior cingulate cortex as transitional in character.

Human Cingulate

Figure 21-3 shows Campbell's[14] depiction of the different cytoarchitectural areas he delineated in the limbic lobe of the human brain. The concentric shading overlies most of the cingulate gyrus that spans the corpus callosum like a great arc. He noted first of all that, on the basis of fiber-stained material, "it will be granted immediately that the [cingulate cortex] is deserving of separate representation on a cerebral map."[21] Specifically, he explains, "Perhaps the chief distinguishing feature of this type of cortex is that it contains absolutely no large evenly-medullated [sic] fibres like those seen in the central convolutions and calcarine region, and indeed practically none of medium calibre; they are all fine and wavy, and in most cases varicose, hence the pallor of the cortex when compared with that of adjoining fields."[21]

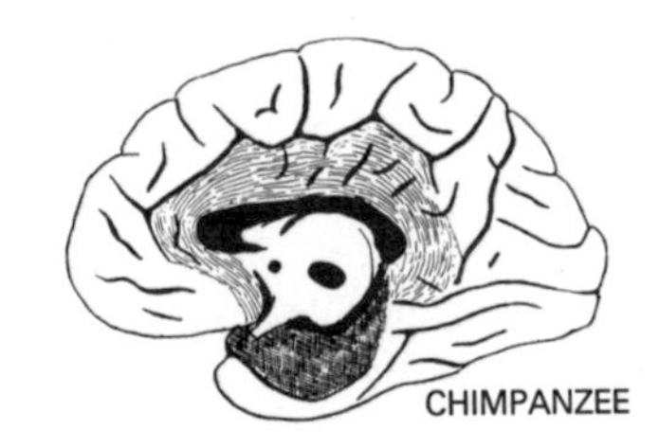

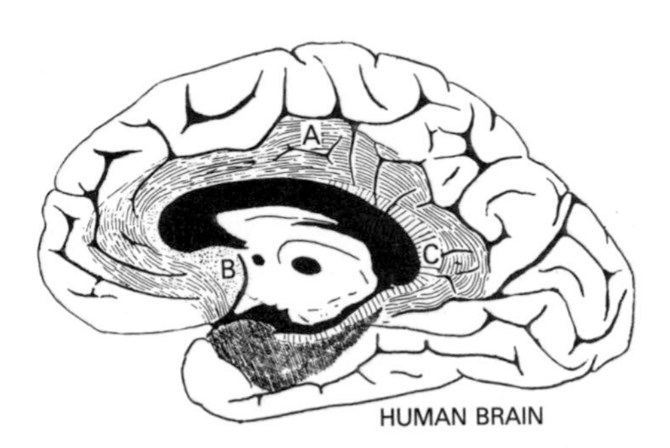

Figure 21-3. Campbell's drawings of the medial surface of the brain in a chimpanzee and a human being, showing his demarcation of the limbic lobe. From his classical study of 1905, this redrawn figure shows only the shading he used for the great limbic lobe of Broca. Concentric lines (A) identify most of the exposed cingulate cortex; stipple (B) and radiate shading (C), respectively, overlie paragenual and retrosplenial cingulate cortex. Radiate shading also identifies location of the lamina medullaris superficialis, with its widest part in the retrosplenial region. The hippocampal gyrus is shown in crosshatch. Redrawn after Campbell (1905).

In his book on the connections of the primate brain, Krieg[22] states, "The cortex of the cingulate gyrus is regarded by all close students of cortical architectonics as a distinct type, set off from all other cortex. . . ." At the same time he points out that posterior cingulate cortex "so nearly resembles parietal cortex that M. Rose has excluded it from the cingulate type."[23] Nevertheless, Krieg insists, it retains "the cingular stamp."[24] Presumably, he had reference to the same kind of observation made by Campbell on the basis of fiber-stained sections—namely, that not only the anterior, but also the posterior cingulate gyrus appears pale when compared with the supracingulate cortex. This situation seems to apply generally to the primate brain, as one can satisfy oneself by examining the brains of such animals as the pygmy marmoset, squirrel monkey, and on upwards.

As evident in Figure 21-3, Campbell recognized only three areal divisions of the cingulate gyrus. B and C label two distinctive subareas in the human brain. He defined area B as a 5- to 10-mm strip in the paragenual region that *extends inferiorly as far as the medial root of the olfactory nerve.*[14] He described this area as being characterized by deeply staining elongated pyramidal cells in the deep layers. He delineated the parasplenial region labeled C as a 5- to 10-mm strip of cortex covered over in part by a band of myelinated fibers that represent the continuation of the lamina medullaris superficialis overlying the subicular cortex of the temporal region, i.e., the same band illustrated in Figure 18-18. The so-called indusium griseum extending the length of the corpus callosum represents a rudimentary continuation of the hippocampus. It is covered by the thin part of the lamina medullaris, while the narrow strip of gray matter lateral to it, and overlain by the thick part of the band, corresponds to the subiculum and its perforating fibers. The classically recognized retrosplenial cortex lies adjacent to the subicular cortex in the supracallosal and parasplenial areas within the region labeled C. According to Flechsig[25] (1847–1929), the entire region is one of the first cortical areas to become myelinated.

As evident by the region labeled A, Campbell did not draw a distinction between the anterior and posterior cingulate areas that were subsequently identified by Brodmann[3] as areas 24 and 23. As noted in Chapter 18, these two areas are clearly distinguished by the presence in the latter of a distinct granular layer IV, and the border between them is marked roughly by a line extending from the central sulcus to the corpus callosum. Brodmann's area 24 and 23 correspond approximately to the regions labeled LA and LC by von Economo and Koskinas.[26] The latter authors also recognized a retrosplenial area, corresponding to Brodmann's area 29, that they labeled LD.

The numbers used by different authors to refer to the granule layer of the retrosplenial areas are confusing, as is also the terminology for differences in the degree of granularity. Rose[6] used the numbers II–IV to identify the granular layer, while von Economo and Koskinas[26] refer to the same layer as III (IV).

Nonhuman Primates

In cytoarchitectonic studies, Lorente de Nó[27] emphasized that reliance on the Nissl and myelin picture alone may lead to faulty conclusions, and that Golgi-stained material for identifying plexi, types of cells, and cellular connections is also necessary. Brent Vogt[28] has noted that except for Ramón y Cajal's[8] studies on the interhemispheric cortex of the rodent, there exist no extensive Golgi studies on the cingulate cortex of other species. In pursuing such a study in the macaque, Vogt[28] focused on the retrosplenial cortex (area 29), including both its so-called dysgranular and granular areas. Vogt concludes that the neurons in the granular layer of the latter area are the antecedents of the pyramidal cells of the lower part of layer III, as well as of the small, star pyramidal cells of Lorente de Nó[29] in layer IV. Vogt's diagrammatic portrayal of the location of the granular and dysgranular areas shows them extending along the posterior, superior part of the corpus callosum, and around the splenium in the same relative position as depicted by Campbell in the human brain (see Figure 21-3).

In the primate, as in the lagomorph brain, the lineup of the superficial granule cells in rows appears most distinctively just behind the splenium. Rostrally, this layer becomes transitional with the retrosubicular cortex. Followed in the opposite direction toward the striate cortex, it is found to lie in the superior bank of the rostral stem of the calcarine sulcus that corresponds to the retrosplenial stem of the splenial sulcus in advanced macrosmatic animals. Here the superficial granular layer becomes transitional with a thin granular layer that in turn meets the butt of the thick layer IV of the striate cortex. In the squirrel monkey, the rapid transition of these three distinctive cortical areas occurs within a distance of 1.5 mm and is best seen in horizontal brain sections.[30] As described in Chapter 26, there is both electrophysiological and anatomical evidence that the retrosplenial cortex is involved in visual functions.

Finally, it is to be noted that although area 23 of the posterior cingulate cortex has a resemblance to the medial parietal neocortex, it shows a distinctive difference insofar as the granular layer is thinner and lies nearer the surface. Stated otherwise, the posterior cingulate cortex on the medial wall (and most notably that in the lower half of the gyrus) shows a poorer development not only of the granular layer, but also of the supragranular layers.

Summarizing Comment

In summary, the cingulate cortex of primates has the diagnostic features of four regions seen in other mammals and located, respectively, in the infragenual, anterior, posterior, and retrosplenial parts of the cingulate gyrus. The posterior cingulate cortex shows the most marked divergence from the other areas, possessing supragranular and granular layers that in thickness and complexity approach those of the medial parietal neocortex. In addition to the cell layer features just considered, there is the notable difference that the cortex of the posterior cingulate gyrus is distinctly paler than the supracingulate cortex because of the lesser myelination of the radiate fibers. The expansion of the posterior cingulate cortex peripheral to the retrosplenial area possibly reflects the progressive importance of visual functions in the evolution of primates.

Neurobehavioral Findings

Introductory Considerations

In modern times neurobehavioral work on the cingulate gyrus can be said to have started with the studies of Wilbur Smith[31] and his observations on the effects of stimulating area 24 in the monkey. Performing electrical stimulation in lightly anesthetized animals, he obtained not only cardiovascular, respiratory, and pupillary changes, but also vocalization.[31] Subsequently, Arthur Ward,[32] in addition to performing similar experiments, ablated ''all of area 24'' in four monkeys and observed their behavior for as long as 2 months. Following surgery the animals were described as having decreased mimetic behavior and a loss of fear of man. In a large cage with other monkeys they ''showed no grooming or acts of affection'' and would walk over their cage mates ''as though they were inanimate.''[33] There was no mention as to whether or not they vocalized.

In the 1940s following the introduction of ''psychosurgery'' for the relief of mental disorders, a number of surgeons sought less maiming procedures than the standard frontal lobotomy for alleviating symptoms. Since it was known that the anterior cingulate is involved in autonomic functions, and since such changes are a common accompaniment of emotional experience, some neurosurgeons reasoned that ablation of parts of the anterior cingulate cortex alone might be of therapeutic value in treating patients with incapacitating anxiety or obsessive–compulsive behavior.[34] Presumably, such reasoning was reinforced by Ward's[32] observations that monkeys appeared less fearful following anterior cingulate ablations. The reported outcome of such surgery was that a substantial, but ill-defined, number of patients had a beneficial relief of symptoms and that there were no alterations of psychological or other functions of an undesirable nature.[34]

It is evident that observations under the above conditions could hardly have revealed that the cingulate gyrus is involved in the three forms of behavior that are the focus of the present chapter. In 1955, John Stamm[35] reported that cingulate ablations resulted in the disruption of maternal behavior of rats. In view of the pivotal role of nursing and maternal care in mammalian evolution, it is curious that these findings attracted little attention. Stamm had been led into this line of research by the suggestion that the limbic system was involved in instinctual behavior such as hoarding investigated in his original experiments.[36] A decade later, Burton Slotnick[37] confirmed and extended Stamm's findings on maternal behavior in rats.

We ourselves,[38] employing a different experimental approach in hamsters, observed similar deficits in maternal behavior in animals in which the lack of the cingulate convolution appeared to be the key factor and noted, in addition, that in the case of young animals, play behavior failed to develop.

Our ongoing experiments in squirrel monkeys indicate that the rostral cingulate cortex is involved in the production of the separation cry (isolation call), which, as has been emphasized, may represent the most basic mammalian vocalization, serving originally to maintain maternal–offspring contact.

Altogether the results of these diverse studies provide evidence that the thalamocingulate division of the limbic system, for which there is no evident counterpart in the reptilian brain, is implicated in three forms of behavior that characterize the evolutionary transition from reptiles to mammals—namely, (1) nursing, in conjunction with maternal care, (2) audiovocal communication for maintaining maternal–offspring contact, and (3) play. Because of the original order in which the experimental findings unfolded, I will deal successively with maternal behavior, play, and the separation cry.

Maternal Behavior

Although it has been shown in the rat that parental behavior in both females and males may be activated by a several-day exposure to sucklings and may occur in preparations devoid of the pituitary gland and gonads, fully developed maternal behavior is dependent on the glands of internal secretion (see Slotnick, 1975, for review[39]). Consequently, in assessing the effects of cingulate ablations on maternal behavior, it is essential to keep in mind the adjuvant role of neuroendocrine factors in the maintenance and termination of pregnancy; the preparation for and initiation of lactation; nest building; and maternal care. The following summary provides background for the ablation studies in rats.

Neuroendocrine Factors

The main neuroendocrine mechanisms requiring consideration are those that regulate the ovarian secretion of estrogen and progesterone via the anterior pituitary and those that account for the production of prolactin and oxytocin via the anterior and posterior pituitary, respectively.

The infundibulum (funnel) is a site of convergence of nerve fibers conveying gonadotropin releasing hormones that enter the portal circulation of the pituitary supplying the adenohypophysis (anterior pituitary). These releasing hormones liberate the pituitary follicle-stimulating and luteinizing hormones. The vital junction between terminating nerve fibers and the gateway to the portal system is in a specialized part of the neurohypophysis called the median eminence that shares a vascularization with the adenohypophysis. It was so named by Tilney (1936)[40] to distinguish it from the *lateral* eminences of the tuber cinereum. It forms the part of the "funnel" adjoining the hypothalamus.

It has been shown by immunohistochemical techniques that luteinizing hormone-releasing hormone (LHRH), a neuropeptide, is present within the perikarya and axonal processes of neurons in the medial septal nucleus, medial preoptic area, and anterior hypothalamus.[41] Since the axons of these cells have been traced to the median eminence, this means that not only diencephalic, but also telencephalic structures are directly involved in the release of luteinizing hormone. One group of fibers has been traced medially along the periventricular fiber system, while a lateral group follows the medial forebrain bundle before descending toward the median eminence.[41] Some fibers also lead to the organum vasculosum of the lamina terminalis adjoining the third ventricle. Those headed for the median eminence are described as reaching the portal vessels in the superficial capillary plexus of this structure.

Prolactin, the lactating hormone, is produced by acidophilic cells in the anterior pituitary gland. During the greater part of pregnancy its release is prevented by the so-called prolactin inhibitory factor (PIF) in the hypothalamus.[39] Dopamine is now regarded as being the probable inhibitory agent.[42]

Maintenance of Pregnancy

In the preovulatory phase of the estrous cycle, follicle-stimulating hormone promotes the release of estrogen from ovarian follicles in various stages of maturation. In feeding back to the brain through the circulating blood, a progressive increase in estrogen acts to suppress the neural output of the releasing hormone for follicle-stimulating hormone. Ovulation occurs following a surge of luteinizing hormone. After conception and ovarian implantation, secretion from the corpus luteum of ruptured follicles maintains a high level of circulating progesterone, which, together with adjuvant amounts of estrogen, serves to maintain pregnancy. Near term, and for reasons unexplained, there is a precipitous decline in the secretion of progesterone, after which the combined action of estrogen and the release of prolactin set the stage for parturition.

Preparation for, and Initiation of, Lactation

Mammary growth and the prelactational phase depend not only on prolactin, but also on the interacting effects of other hormones, including estrogen, progesterone, adrenal steroids, and growth hormone.[39] Lactogenic agents are also derived from the placenta. In association with the development of the mammary glands, the female engages in self-licking along the nipple line, a form of stimulation believed to contribute to the development of the milk glands. As already noted, at the end of pregnancy there is an abrupt fall of progesterone, permitting estrogen to suppress the hypothalamic production of PIF and thereby the release of prolactin.

At parturition, the hormone oxytocin serves the dual capacity of promoting parturition and lactation by inducing contractions of the uterine muscle and the myoepithelial cells of the mammary glands. Oxytocin is a product of neurons in the paraventricular nucleus, the axons of which terminate in the neurohypophysis. After birth, milk secretion is reinforced by a neural reflex originating with stimulation of the maternal nipples by the sucklings. The reflex involves a multisynaptic pathway conveying impulses to the paraventricular nucleus.

Nest Building

Unlike the mouse, nest building in the rat is inhibited by progesterone. Maternal nest building occurs 24 to 48 hr before parturition and appears to be associated with a secretory decline in progesterone and an increase in estrogen.

Nursing and Maternal Care

Slotnick[39] has reviewed the evidence indicating that exposure to pups is sufficient to induce maternal care and that such care need not depend on the action of reproductive hormones. At the same time he describes cross-transfusion experiments by Terkel and Rosenblatt[43] providing evidence that substances in the blood of postpartum rats are capable of inducing maternal behavior in otherwise unresponsive animals. It has been shown that once maternal behavior is well established, attempts to abolish it by altering hormonal conditions are unsuccessful.

Scoring Maternal Behavior

Since most neurobehavioral work on maternal behavior in mammals has involved rodents, it is necessary to give a brief description of the quantifiable aspects of such behavior.[35,37] Maternal behavior in these animals involves not only the act of parturition, nursing, retrieving, and other maternal care, but also the preliminary building and maintenance of a nest. In the wild, nests are usually made with a greater assortment of material than is provided in the laboratory. The lining of a nest, as well as its architecture, will depend on the nature of the available materials. Insufficient attention has been given to nidal architecture. The provision of only one variety of material such as strips of paper limits what can be learned about an animal's ability to structure a nest. However, even with a single material such as string pulled into the nesting site from a ball, there exists the opportunity to obtain information not only about the structure of a nest, but also about the amount of material the animal is motivated to use.

Different authors have different criteria for scoring the structure of a nest. A simple example of scoring would be that for a string nest of a mouse rated on a scale of 0 to 4, with 0 referring to no nest and the other values applying, respectively, to nests with the shape of a saucer, a cup, a three-quarter dome, and a full dome. In addition to such a rating, some studies include an evaluation of a dam's capability of restoring a destroyed nest, the site chosen for it, and the amount of time spent in rebuilding it.

Besides nest building, one judges the maternal behavior of rodents according to the mother's ability to (1) free the pups from the fetal membranes and to clean them; (2) gather them together in the nest; (3) adopt the correct nursing position; (4) retrieve the young when they leave the nest or when the nest is destroyed; and (5) sustain them through to the time of weaning.[35,37]

Effects of Cingulate Ablations

In his study showing the deleterious effects of cingulate lesions on maternal behavior, Stamm[35] made observations on three groups of pregnant rats, one of which served as a control. In two of these groups, he compared the effects of bilateral ablations of the cingulate cortex with those following equally large lesions of the neighboring neocortex. He found that unlike the control animals and those with neocortical lesions, the animals with ablations of the cingulate cortex showed deficits both in nest building and in maternal care. Prior to parturition, the cingulate animals failed to construct a preparatory nest. The dams in each of the three groups produced litters of comparable size, the average being about nine. As opposed, however, to a survival rate of 93% of the pups for the unoperated animals and 79% for the neocortical group, there was only a 12% survival of the pups reared by the cingulate animals. In tests of retrieval, the cingulate animals would run aimlessly around, picking up and dropping the pups in all parts of the observation box.

In 1967, Slotnick[37] reported a confirmation and extension of the study by Stamm in which he tested the effects of regional, as well as extensive, ablations of the cingulate cortex. It should be noted that the midline lesions in the rats of Slotnick's study did not extend as far laterally as those in Stamm's animals. Slotnick's findings were similar to those described by Stamm, with the exception that there was a 24%, as opposed to a 12%, survival of the pups belonging to the mothers with cingulate ablations.

Of Slotnick's 16 cingulate animals there were 4 with predominantly anterior lesions; 3 with posterior cingulate lesions; and 9 with involvement of both regions. In "several" instances the anterior ablations included the subgenual cingulate cortex. The posterior lesions largely spared the retrosplenial cingulate cortex. The maternal deficits of animals with lesions of both the anterior and posterior cingulate cortex were greater than those with destruction of one region alone. The animals with both regions involved failed to construct nests and showed marked impairment in the retrieval of the pups. Although all of the motor patterns seen with normal pup retrieval were present, they were performed in an "irregular and confused manner."[37] The retrieving behavior, documented by Slotnick's motion pictures, appeared completely disorganized; the pups would be repeatedly carried in and out of the nest site or dropped randomly about the cage. Contrary to normal retrieval, the dam might first lick the pups before picking them up and carrying them to a new place. When nursing, rather than assembling all the pups together, she might place herself over one or two and do so either in or outside the nesting area. Although most of the dams kept their litters alive during the 5-day postpartum test, the weight gain of their pups was less than half that of the control animals.

Neurohumoral Factors

Both Stamm[35] and Slotnick[37] showed that cingulate ablated animals had an adequate supply of milk. Apropos of a possible interference with neurohumoral mechanisms, Slotnick found that pre- and postpartum injections of prolactin resulted in no improvement of maternal behavior in cingulectomized rats.[37] In addition, he showed that maternal retrieval behavior induced in *castrated* male rats by exposure to 1- to 3-day-old pups was disrupted by cingulate ablations, but not by neocortical lesions.[37]

Comment

As noted in Chapter 19, Slotnick and co-workers found what they interpreted as a species difference in limbic structures accounting for maternal behavior in mice and rats. Mice with septal lesions showed severe deficits in all aspects of maternal care, whereas those with cingulate lesions primarily showed a slowness in retrieval of their pups as compared with the controls. In retrospect, it appears possible that in guarding against injury to the critical septal region, the investigators did not extend the cingulate lesions into the preseptal cingulate cortex. Rather the reconstructed lesions indicate that there was primarily destruction of the supracallosal cingulate cortex, as well as part of the retrosplenial cortex.[44] Significantly, with respect to nursing, Aulsebrook and Holland[45] elicited release of oxytocin in the rabbit by stimulation of "area 32." They referred to unpublished observations by Woods suggesting that this cortex projects indirectly to the paraventricular nucleus.[45] Powell[46] had previously reported that in the cat a lesion near the dorsolateral septum resulted in degeneration leading to the paraventricular nucleus. Using cats, Beyer *et al.*[47] elicited milk ejection and uterine contractions by electrical stimulation of the pregenual cingulate cortex.

In addition to the rodent studies on maternal behavior, mention should be made of the experiments on monkeys by Franzen and Myers[48] cited in Chapter 19. It will be recalled that their investigations included observations on the effects of bilateral prefrontal, anterior cingulate, and lateral temporal ablations on maternal behavior. There were two monkeys with anterior cingulate lesions. The findings in one animal were of no account because it exhibited no maternal behavior either before or after surgery. It was noted that the other monkey continued to nurse its infant. It was stated that the ablation of the cingulate cortex "included the major part of this structure but failed to include the gyrus subcallosus and the retrosplenial region."[49]

The application of improved techniques has revealed a high concentration of opiate receptors throughout the "rostro-caudal extent" of the cingulate cortex of the rhesus monkey.[50] The cingulate cortex of the rat is also rich in such receptors.[51] It may prove to be relevant to this that the administration of morphine disrupts maternal behavior in mice and rats.[52]

Selective Interference with Maternal Behavior and Play

Using a different experimental approach, we have conducted an investigation in hamsters that supports the previous findings in rats that the cingulate cortex plays an important role in maternal behavior. Since our experiments also provide evidence that the cingulate convolution is implicated in play behavior, I will briefly describe the background of this particular investigation and the methodology. The purpose of our experiments was to test in rodents the hypothesis that the two older evolutionary formations of the brain—namely, the striatal complex and the limbic system—are sufficient, along with the remaining nervous system, for the expression of most forms of species-typical behavior.[53]

Methods

Our subjects were Syrian golden hamsters derived from feral stock captured by Murphy[54] in 1971. Instead of using adolescent or mature animals, it was the purpose in these experiments to eliminate the neocortex soon after birth and then observe the development of hamster-typical forms of behavior under laboratory conditions. In pups 1 to 2 days old under cryoanesthesia, the neocortex was eliminated either by heat applied to the skull or by aspiration.[53] Littermates used for controls were treated in the same manner except that there was no destruction of brain tissue. Since a number of animals had complicating destruction of the midline cortex, there was the opportunity to observe how the additional loss of the cingulate cortex altered behavior.

A checklist based on an extensive ethogram was used to record the day-to-day behavioral development of the experimental animals and their littermates used as controls.[55] Many of the routine observations were made on animals housed individually in clear plastic cages (18″ × 10″ × 8″) with water and food always available and cotton and wood chips provided, respectively, for nesting and bedding. The light–dark cycle consisted of 16 hr of white light and 8 hr of dim red light. A special viewing apparatus was used for obtaining quantitative measurements of mating and of species preference behavior. The occurrence of various behaviors was registered by means of an event recorder and the data stored on a magnetic disc memory for later analysis. In the case of mating, tests were conducted for 30 min. For observations on tunnel blocking, hamsters were placed in an artificial tunnel system for 5 days. In order to obtain information on day-to-day routine and circadian activity, animals were observed in a large structured environment in which they could be monitored continuously by time-lapse television. The methods used to test scent marking, territorial aggression, and species preference are described, respectively, in Refs. 56, 57, and 58.

Results

Preservation of Hamster-Typical Behavior

The maturation of the hamsters without neocortex resembled that of their littermates with respect to weight gain, physical development, and the time of appearance of different hamster-typical forms of behavior. In addition to maternal behavior and playful behavior, the hamsters without neocortex developed the following kinds of activity at the expected times: thermotaxis, nest building, digging, seed cracking, food pouching, hoarding, tunnel blocking, and scent marking. Circadian activity rhythms were like those of the normal animals. The animals without neocortex showed the typical territorial defense and aggression when another hamster was introduced into its cage. Most significantly in regard to conspecific recognition discussed in Chapters 13 and 16, when Syrian hamsters devoid of neocortex were exposed to a member of their own or another species *(Mesocricetus brandti),* they showed a strong sexual preference for their own species. Although males without neocortex appeared, during mating, to have some difficulty in mounting and required twice the usual number of intromissions to achieve ejaculation, they displayed normal sexual arousal, performed the complete copulatory act, and successfully impregnated females. The females devoid of neocortex had a regular estrous cycle, were normally receptive to males, conceived, and gave birth to young.

Maternal Behavior

Figure 21-4 shows a dorsal view of the brain of a normal and two experimental animals, with representative sections of the thalamus underneath. The brain shown in B is that of a female in which virtually all of the neocortex except for a far frontal remnant on the right side was lacking. By comparison with the normal brain in A, it might be judged that some of the paramedian neocortex was still present, but it should be noted that some of the midline limbic cortex extends over onto the lateral surface and can be readily distinguished both by the myelinated fibers in layer I and by the cellular layers. The subject in this case gave birth to a litter of five pups, provided them with an adequate nest, and nursed them until weaning. Except for some clumsiness, her retrieval behavior was like that of normal hamsters.

In these experiments it was the purpose to eliminate the neocortex without damage to the limbic cortex. It was therefore of special interest that the behavioral findings proved to be an indication of damage to the medial limbic cortex. Figure 21-4C shows the brain of one of these animals (LC3) in which there was a complete absence of the retrosplenial and supracallosal cingulate cortex, as well as extensive damage to the left preseptal cingulate cortex, including some invasion of the taenia tecta. After mating on the 100th day of life,

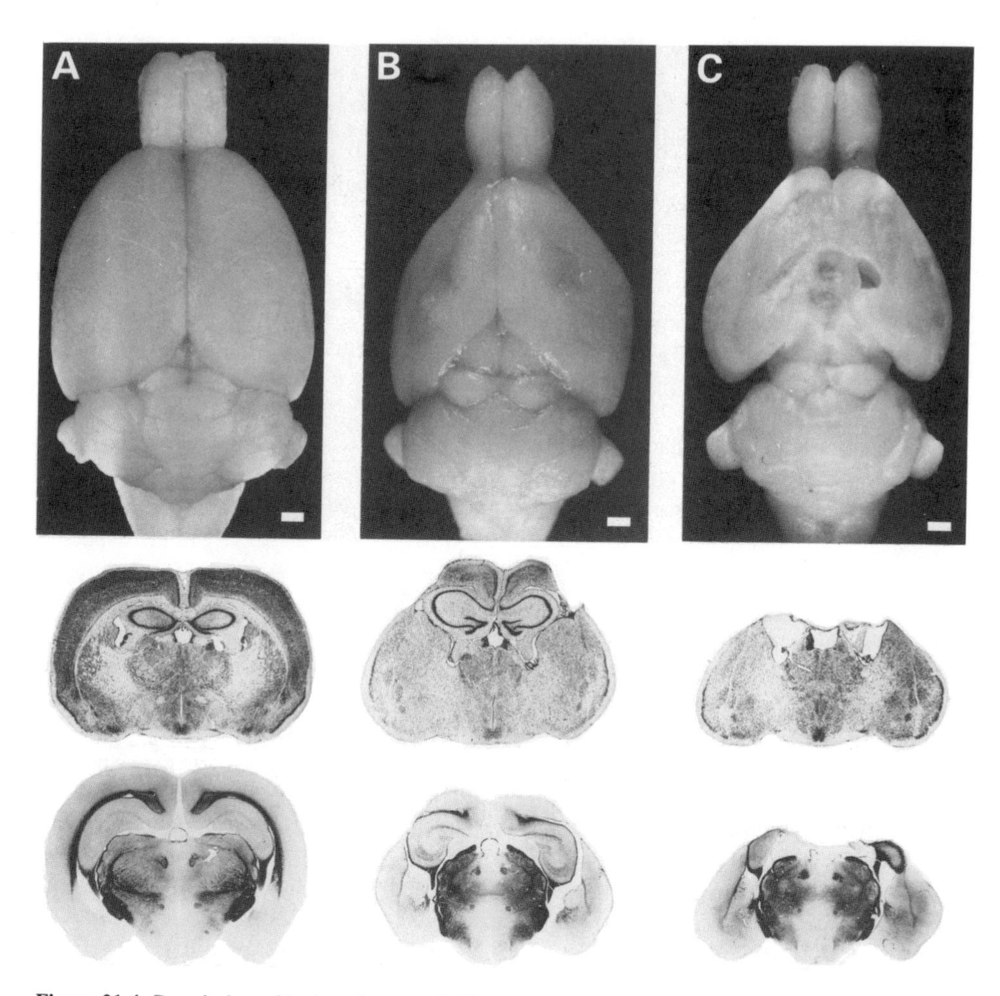

Figure 21-4. Dorsal view of brains of a control (A) and two experimental hamsters (B), (C). Representative sections through diencephalon at rostral (cell stain) and caudal levels (fiber stain) are shown underneath.

(B) Brain of a female deprived postnatally of all the neocortex except a remnant at right frontal pole. The preserved cingulate cortex is seen above the hippocampus. The fibrous zonal layer (not visible at this magnification) helps to identify the lateral border of the cingulate cortex in rodents.

(C) Brain of another female with additional loss of cingulate cortex and part of the underlying hippocampus. White bars correspond to 1 mm.

From Murphy *et al.* (1981).

she delivered 10 pups 16 days later. On the first day, seven of the pups were found in a nest and three were scattered. She was observed to pick up a pup and carry it around "aimlessly." On the second day only one pup remained. Other more active foster pups were placed in her cage, but she did not show any maternal interest in them. The findings in another female (LC2) suggest that preservation of the medial frontal cortex, including the paragenual and preseptal cingulate, may be necessary for nursing. This female with this area intact continued to nurse, but like LC3 with no supracallosal and retrosplenial cortex, showed no retrieval behavior.

Play Behavior

In view of the studies by Stamm and by Slotnick described above, the changes in maternal behavior of the hamsters with complicating loss of the midline cingulate cortex were to be expected. What was of new and particular interest was the failure of play behavior to develop. In rodents, play behavior is often referred to as "play fighting." In the hamster, play fighting can be expected to occur in the nest as early as the 13th postnatal day.[55]

In the protocol for the female LC3 described above (see also Figure 21-4C), it was noted that her male sibling control "often approached her with playfighting maneuvers, but she only fended him off, never playing with him." Two male hamsters with play deficits are of comparative interest because in one (LC5) not only the cingulate cortex, but also the proximoseptal hippocampus underneath, was absent. It was noted in the case of this animal that, in addition to not playing, it was repeatedly observed walking on top of the other pups "as though they did not exist."

Comment

As pointed out in Chapter 9, except for rudimentary aspects in some species, maternal behavior appears to be nonexistent in reptiles. It was also stated that there is nothing resembling play in reptiles. Consequently, one might say in regard to the deficits in maternal behavior and play observed in the foregoing experiments, it was as though the animals had reverted toward the condition of reptiles.

Given the evidence of these particular deficits in rodents, it remains now to learn whether or not the same kind of manifestations will occur in different orders of mammals. The findings on maternal behavior also indicate the need to investigate the functional role of areal subdivisions of the cingulate cortex, as, for example, the possible role of retrosplenial cortex in visual aspects of maternal behavior (Chapter 26).

In view of the prominence of play among mammals and its civilizing influence in human evolution, it is curious that it has received so little attention in neurobehavioral research. In one handbook of experimental psychology, for example, the subject of play is dealt with in less than a page,[59] and in a three-volume handbook of neurophysiology, there is no reference to play.[60] Play, admittedly, is a "fragile" behavior being subject to suppression by many ill-defined conditions. It is also difficult to deal with experimentally because of the problem of obtaining quantitative measures of play. As a quantitative measure of play in neuropharmacological studies on rats, Panksepp scores the number of times one animal pins another in wrestling.[61] Pinning is also part of play wrestling in immature squirrel monkeys. In our observations on the play behavior of these animals we score the incidence of (1) the "invitation" bounce; (2) chasing; (3) grabbing at different parts of the body (e.g., ears, tail); (4) mock biting; (5) wrestling; (6) pinning; (7) pseudocopulatory activity; and (8) associated vocalizations. At present, we are attempting to learn whether or not the scoring of play in the squirrel monkey, as well as the predictability of play, is sufficiently reliable to merit formal neurobehavioral studies.

Two neurobehavioral studies on monkeys deserve mention in regard to changes in play behavior. For his dissertation, Theodore Cadell[62] (a student of the late Harry Harlow) observed the effects of section of the fornix on the behavior of rhesus monkeys about $2\frac{1}{2}$ to 3 years in age. There were two groups of four monkeys in which the fornix had been sectioned by different surgical approaches, as well as a control group of four unoperated animals. It was found that the animals with section of the fornix were "markedly retarded in the onset of mutual play behaviors."[63] The retardation was ascribed to a failure on the part of subordinate fornix animals to "reciprocate in the initiation of play."

In a study of the effects of prefrontal lesions on social behavior in rhesus monkeys, Franzen and Myers[64] noted that there was decreased "play activity." They did not specify their method for scoring play.

As regards the question of function, Groos's[65] book *The Play of Animals,* published in 1896, helped to popularize the notion that play reflects an acting out of instincts before they are called upon for "serious" use by the adult animal. In modern times, play is usually interpreted as affording the animal a means of learning adult behavior and developing the strength and skills required in the struggle for survival. Another popular suggestion is that play serves as a means of dissipating excess energy (see Beach and Fagen for reviews[66]). Contrary to (but not excluding) other interpretations, I have suggested that the development of play behavior may have served originally to promote harmony in the nest, and then, later in life, affiliation among members of social groups.[67] The peacekeeping function of play at the anthropoid level may be illustrated by a playful chimpanzee mother described by Jane Goodall, which would take to playing with her son Frodo, "almost always when he got too rough with his small sister."[68]

Other considerations pertaining to maternal behavior and to play can be more adequately discussed after having dealt with the close anatomical relationship to the thalamocingulate division with parts of the prefrontal neocortex (Chapter 28).

Audiovocal Communication, with Special Reference to the Separation Cry

As noted in Chapter 17, the first mammals are presumed to have been tiny animals that lived in the dark floor of the forest and were possibly nocturnal. If so, it is evident that audiovocal communication would have provided a valuable additional means of maintaining maternal–offspring contact. Since any prolonged separation of the young from a nursing mother is of fatal consequence, the possibility suggests itself that the so-called *separation cry* may be the oldest and most basic mammalian vocalization (Chapter 17). As Romer[69] points out, some "supposed" Mesozoic mammals may have been semitherapsid because of retention of one or more of the small "reptilian" bones in the dentary–squamosal jaw articulation, whereas in "proper mammals" the quadrate and articular bones have become part of the middle ear. This latter condition reflects an improved sense of hearing. But whether or not the mammal-like reptiles or early mammals were capable of vocalization is a moot question.

On the basis of accumulating data, it would appear that separation cries (alias isolation or distress calls) are typical of most infant mammals. It has been recognized only since 1954 that rodent pups emit ultrasonic cries when isolated or subjected to uncomfortable conditions.[70] Such sounds have been shown to arise from the larynx.[71] The vocalization that occurs when pups are taken from the nest is attributed to decline in body temperature.[72] Noirot[73] found that the cries of isolated pups are effective stimuli for inducing retrieval behavior of the dam and guide her in finding the pup. The ultrasonic nature of the call possibly serves as a protection against predators such as owls with insensitivity to sounds above 20 kHz.[74] Noirot has pointed out that primiparous mice cease to retrieve when the pups are about 13 days old, and that this change appears to be correlated with a decline in the production of ultrasonic calls in pups of this age.[75]

Unfortunately, little is known about forebrain structures involved in the vocalization of rodents. Thanks particularly to the investigations of Ploog and his colleagues at the Max-Planck Institut für Psychiatrie at Munich, more is known about the vocal repertoire of squirrel monkeys and the underlying forebrain mechanisms than that of other mammals, including higher primates (see below). Because of this and because of the focus of our own work on the separation cry, most of the neurobehavioral studies to be described involve this one genus. Preparatory to describing investigations on the role of cingulate and related neural mechanisms in the separation cry of squirrel monkeys, I will summarize what is known about the vocal repertoire of these animals, as well as the cerebral representation of certain vocalizations.

Vocal Repertoire of Squirrel Monkeys

Winter *et al.*[76] originally identified six main groups of calls with a total of about 30 individual calls. Newman[77] has arrived at a somewhat different classification of calls, the major difference being his identification of a group of calls referred to as "chucks" (see below). In an accompanying table he lists equivalent names for different calls and their sources. For present purposes I am amalgamating Newman's classification with the original one of Winter *et al.* The six different groups of calls are identified as (1) peeps, (2) chucks, (3) twitters, (4) errs, (5) cackles, and (6) noisy vocalizations. The following summary gives a definition of the calls and conditions under which they occur.

1. Peeps are tonal in character and are sounded in such diverse situations as those associated with separation distress, alarm, play, and genital display. The separation cry (originally called the *isolation peep*) will be described below in connection with the neurobehavioral studies on its cerebral representation.

2. Chucks, which are characterized by a rapidly descending frequency-modulated element, have a sound like that of sucking lips and tongue breaking contact with the nipple ('tsik). Variations of this sound are emitted by the mother when encouraging the infant to resume nursing and by the infant when searching for the nipple. Later on when the infant begins to wander afield, the mother makes the call as a means of bringing it back. The "location trill" is a double or triple chuck that an infant emits upon regaining contact with its mother. Err chucks are sounded in sexual encounters by breeding animals. Keckers signal excitement during disputes, while yaps are an indication of alarm or something novel.

3. Twitters, characterized by a periodic changing tone, have the sound of a chirping bird, and are commonly heard among squirrel monkeys in feeding situations. They are also sounded during exploration and at times when monkeys find themselves partially out of sight of one another. In keeping with such situations, twitters also serve to signal recognition and greeting.

4. Errs (purrs) are pulsed sounds heard during extended nursing (milk purr), or when a female in estrus is soliciting contact with a male (purr call). Growls are produced only by "romans" and occur in situations involved in directed aggression as when a dominant male displays to a subordinate.

5. Cackles, having a fundamental below 1 kHz, are produced when members of a

group are in competitive situations or when something in the environment becomes disturbing.

6. Noisy calls, as typified by screams and shrieks, manifest the highest form of excitement and distress and are heard when animals are losing a fight or being captured, or when members of a group see one of their own molested, particularly an infant monkey. Prior to 1 year, the squirrel monkey's isolation call may contain noisy segments, while some such calls may "appear to be fricative, consisting of continuous wide-band noise, with tonal components restricted to the beginning and end of the call."[78] The *display* "peeps" of adult squirrel monkeys may show similar noisy features.

Cerebral Representation of Different Calls

In awake, sitting squirrel monkeys, Jürgens and Ploog[79] explored the brain from the level of the cortex to the medulla, using various parameters of electrical stimulation to identify structures involved in various vocalizations. Altogether, they specified five vocalizations that, according to the sonographic illustrations, are to be classified in the following listing with the call group shown in parentheses: (1) "chirping" (peeps); (2) "trilling" (twitters); (3) "growling" (errs); (4) "cackling" (cackles); and (5) "shrieking" (noisy calls). In view of the neurobehavioral studies on the separation cry to be described, it should be emphasized that their designation "chirping" applied entirely to the *peeps* of Winter *et al.*,[76] including the isolation peep. It should also be pointed out that in reporting their findings they did *not distinguish between vocalizations that occurred during stimulation and those emitted immediately afterwards*. The cortical and subcortical structures identified with the five specified vocalizations were as follows.[79]

1. "Chirping" (peeps) was elicited by stimulation of the subcallosal gyrus and caudal part of the gyrus rectus; the part of the nucleus accumbens adjoining the gyrus rectus; the midline part of the thalamus contiguous with the medial dorsal nucleus; the caudal part of the periaqueductal gray matter; and the pontine portion of the spinothalamic pathway.
2. "Trilling" (twitters) resulted from stimulation of the subcallosal gyrus; the region of the precommissural fornix; and at a few loci along the pontine portion of the spinothalamic pathway.
3. "Growling" (errs) was elicited by stimulation of the dorsal bank of the cingulate gyrus at the level of the genu of the corpus callosum; the temporal polar region and along the uncinate fasciculus toward the amygdala; parts of the amygdala; the stria terminalis and its bed nucleus; the preoptic region; the periventricular gray matter; posterior hypothalamus; the periaqueductal gray matter and central tegmental tract; and the parabrachial region.
4. "Cackling" (cackles) was obtained by stimulation of the pregenual cingulate cortex; ventromedial orbital cortex; temporal polar area and caudally along the uncinate fasciculus; basal and central nuclei of the amygdala; inferior thalamic peduncle and anterior thalamic radiation; anterior medial nucleus of the thalamus; periventricular fiber system; caudal part of the periaqueductal gray matter; and parabrachial nuclei.
5. "Shrieking" (noisy calls) was elicited by stimulation of the same circuit of structures described for the cat in Chapter 19, including the medial part of the amygdala; the stria terminalis and its bed nucleus; the ventromedial part of the hypothalamus; the

rostral part of the central gray matter and tegmental structures lateral to it; and the region of divergence of the spinothalamic pathway and medial lemniscus.

Comment

It should be pointed out that special difficulties arise in performing and interpreting the above kinds of experiments because the type of vocalization may vary according to the intensity of stimulation. Nevertheless, there is the persuasive finding that stimulation of moderate intensity in one particular region such as the anterior hypothalamus may elicit an agonistic form of vocalization, whereas the same stimulation applied elsewhere, such as in the dorsal hypothalamic area, may evoke a nonstressful type of call.

In view of the present focus on the separation call, it is to be emphasized that in Jürgens and Ploog's study[79] peeping vocalizations were primarily identified with parts of the thalamocingulate division as well as with certain loci in the septum and hippocampus. As evident from Figure 21-5, our own findings on vocalization differed from those of Jürgens and Ploog insofar as it was elicited by stimulation of the pregenual cingulate cortex, but not the subcallosal gyrus and gyrus rectus.[80] This discrepancy may possibly be explained by the fact that Jürgens and Ploog did not differentiate in their report between vocalizations induced during stimulation and those occurring when stimulation was terminated. In this respect, it is of interest that in his study on rhesus monkeys, Robinson[81] shows loci in the gyrus rectus, as well as some in the subcallosal gyrus, identified with vocalization occurring *after* stimulation. Apropos of the separation cry, Robinson noted that all "koo" calls, which he describes as being similar to the "separation call," were elicited only by stimulation of the "anterior cingulate gyrus or its immediate environs."[81]

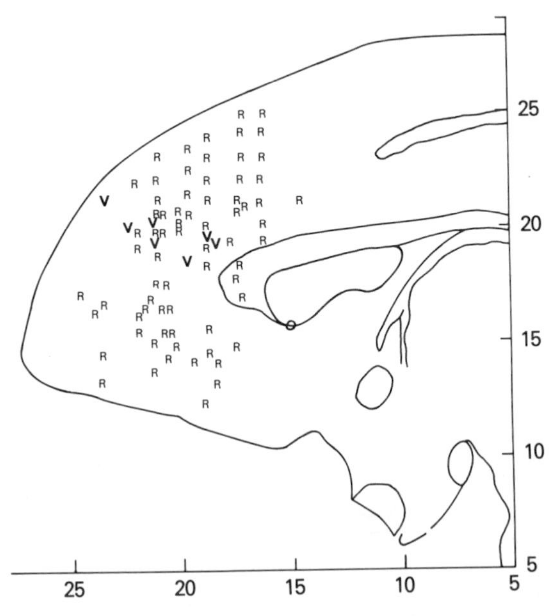

Figure 21-5. Diagram of sagittal section through medial frontal cortex of squirrel monkey, showing loci at which stimulation resulted in vocalization (V) and acceleration of respiration (R). Abscissa and ordinate, respectively, show distance in millimeters rostral to and above zero axis of brain atlas. From Dua and MacLean (1964).

Cerebral Substrate of the Separation Cry

The reasons have been given for the inference that the separation cry may represent the earliest and most basic mammalian vocalization. Judged by available evidence, the separation cries of primates have certain commonalities in all species ranging from the marmoset to human beings. As illustrated by the sound spectrograms in Figure 21-6 of the squirrel monkey, macaque, and human infant, the separation cries of primates are characterized by a slowly changing tone. This commonality, Newman points out, suggests that mechanisms controlling infant cry patterns have a "conservative evolutionary history."[82]

Because of the inferred basic nature of the separation cry, it is evident that a knowledge of the brain mechanisms underlying the production of the call might be helpful in reconstructing its evolutionary history.[83] Based on the original findings of Magoun *et al.*[84] regarding the role of the periaqueductal gray matter and adjoining tegmental structures in vocalization, one might expect that the mechanisms accounting for the structure and production of the separation cry would be located in the midbrain. They identified loci in rhesus monkeys where stimulation resulted in "faint" vocalizations, whines, and

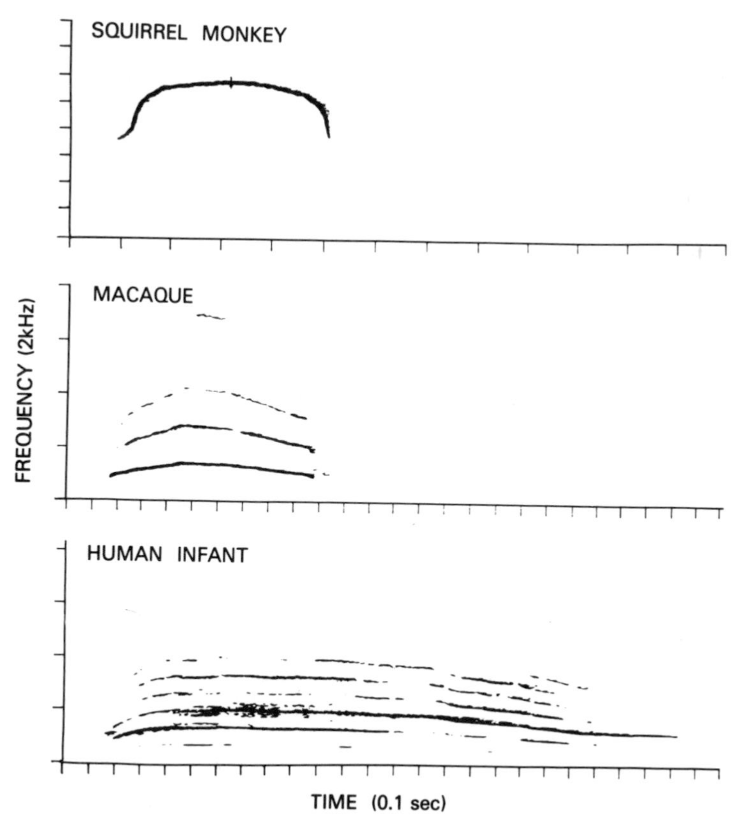

Figure 21-6. Sound spectrograms of separation cries of a squirrel monkey, a macaque, and a human infant. In primates such cries have the basic pattern of a slowly changing tone. Recordings by Newman from MacLean (1985a).

barks, while in cats, they elicited mews, growls, and hissing.[84] Given this information, they demonstrated by placing various lesions in the cat that the integrity of these structures is necessary for the production of most of an animal's vocal repertoire.[85]

In regard to the separation cry, however, fortuitous observations on two squirrel monkeys used in the display studies described in Chapter 14, drew attention to structures bridging the thalamomidbrain junction. The impression that these two animals produced abnormal vocalizations was confirmed by spectrographic analysis undertaken by my colleague John Newman of the Laboratory of Comparative Ethology, NIH, at Poolesville, Maryland. Newman regarded the findings in these two animals to be unique insofar as he was unaware of any experimental studies in which cerebral lesions resulted in an altered structure of a particular vocalization. Accordingly, we decided to undertake an investigation of the cerebral representation of the separation cry in squirrel monkeys.

In squirrel monkeys the separation cry, as in many species, serves to maintain not only maternal–offspring contact,[86] but also contact of individuals within a group.[87] As explained in Chapter 13, there are two main varieties of squirrel monkeys—gothic and roman—that were originally distinguished not only by their circumocular markings, but also by behavioral differences in certain aspects of their displays and vocalizations. The sound spectrograms shown in Figure 13-3 illustrate that the gothic and roman types can be readily distinguished by their separation cries, which respectively have a downward and upward deflection at the termination of the cry. The karyotypic differences depicted in Figure 13-4 attest, along with other evidence, that these vocal differences are inherited.[88] The separation cry lends itself to investigation in squirrel monkeys because it is readily elicited by temporarily isolating a subject from other monkeys. By 1 year of age, noisy components in the call disappear, and the cry thereafter remains stable in structure over a period of several years.[89] Its characteristic pattern makes it easy to distinguish it from other calls of the vocal repertoire. The cry may be sounded individually or in bursts of two to six, with the intervals between each vocalization being shorter than the cry itself.

Methods

In our present investigations of the cerebral representation of the separation cry, adult male and female *Saimiri* monkeys are tested for their ability to produce such vocalizations before and after ablation of different parts of the brain. Criterion performance is the production of 20 spontaneous cries during a 15-min period of isolation in a sound-reducing chamber. Cries made under these isolated conditions are hereafter referred to as isolation calls. Recordings are made with a standard Uher microphone, and sound spectrograms are obtained with a VII model 700 sound spectrograph.[83] Postoperative testing is conducted weekly during the first month and at longer intervals thereafter. The method for producing lesions of the brainstem by electrical coagulation are the same as described in Chapter 13. Ablation of various parts of the frontal cortex is usually accomplished surgically by subpial aspiration of the gray matter. Electrocoagulation was used to make isolated lesions of the inaccessible subcallosal cortex. The methods for histological examination of the brain are the same as described in Chapter 13.

Findings on Brainstem

Here it will be sufficient to give but a brief summary of the results of coagulating parts of the thalamic tegmentum and periventricular gray matter, as well as the tegmentum and central gray of the midbrain. The reconstructed lesions in eight animals are shown in Figure 8 of the original paper.[83] By the process of elimination it was concluded that lesions involving the periventricular gray matter and underlying tegmentum at the thalamomidbrain junction resulted in alterations of structure of the call. Figures 21-7 and 21-8 show the nature of the alterations in structure observed in the two animals mentioned in the introduction of this section. The locations of the respective brainstem lesions are shown in Figures 14-10 and 14-11. The locus of the effective lesion can further be defined insofar as the alteration in structure of the call was associated with destruction of the greater part of the interstitial nucleus of Cajal and nucleus Darkschewitsch. Lesions of the midbrain involving the ventral central gray matter and contiguous parts of the central tegmental tract were followed by a marked reduction in the production of the call, but no alteration of its structure.

Findings on Midline Frontal Cortex

Background Considerations

Given the evidence that structures near the core of the thalamomidbrain junction are essential for the structure and emission of the isolation call, what parts of the telencephalon might be expected to participate in influencing the production of this particular vocalization?

In the monkey, it has been the experience of all investigators that electrical stimulation of the neocortex is ineffective in eliciting vocalization. As early as 1876, Ferrier[90] had noted that electrical excitation in the region corresponding to Broca's area in the monkey, resulted in movement of the vocal cords, but no vocalization. In a review of their own findings in the squirrel monkey, Ploog[91] stated that "there was not a single site in the neocortex which yielded a vocal response in many thousands of stimulations in dozens of [animals]." Excitation of the motor area near the sylvian fissure resulted in movement of the vocal cords, but no vocalization.[92]

Wilbur Smith[31] at the University of Rochester appears to have been the first to demonstrate that stimulation of the anterior cingulate cortex produces vocalization in a primate species. Using rhesus monkeys, he was able to elicit vocalization only in lightly anesthetized animals.[31] Ward's[32] failure to duplicate his results was probably owing to deep anesthesia induced by allobarbital (Dial), because subsequent experimentation in monkeys, both in anesthetized and in chronically prepared waking animals, has confirmed Smith's findings. The V's in Figure 19-3C from a study by Kaada show loci in the pregenual cingulate cortex where stimulation elicited vocalization in lightly anesthetized macaques. As indicated in Figure 21-5, Dua and MacLean[80] found that stimulation in the corresponding region of awake, sitting squirrel monkeys also results in vocalization. Significantly, in regard to the thalamocingulate division, Jürgens and Ploog[79] concluded that the midline rostral limbic cortex, including the anterior cingulate gyrus, subcallosal gyrus, and posterior part of the gyrus rectus, was the only cortical area where stimulation in the awake, sitting squirrel monkey consistently produced vocalization.

It is to be noted, however, that in the squirrel monkey (see Figure 19-8), as in the macaque (Figure 19-3C), stimulation of the anterior hippocampal gyrus (piriform cortex) elicits vocalization. Additionally, Jürgens and Ploog[79] reported that stimulation of the temporal polar cortex rostral to the piriform area elicits cackling and growling. Their findings indicated (see above) that the implicated pathways passed through the amygdala.

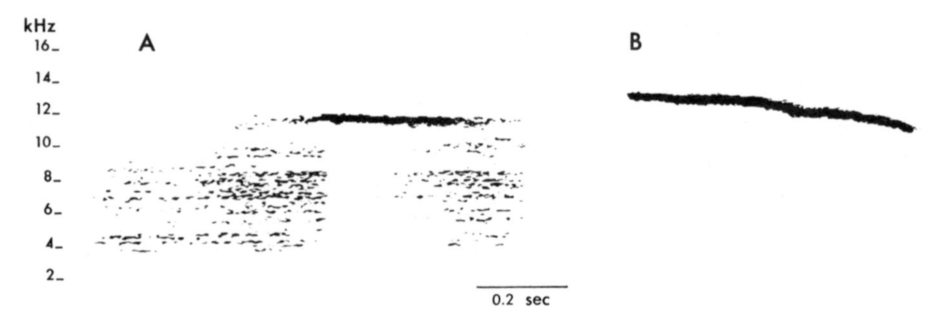

Figure 21-7. Sound spectrograms recorded from monkey I-5 with tegmental lesions shown in Figure 14-9A and B. (A) Record shows expiratory noise typical of this monkey's attempts to vocalize when isolated. Note predominant lack of tonal character. (B) 9 months later, showing spectrogram within normal limits. From Newman and MacLean (1982).

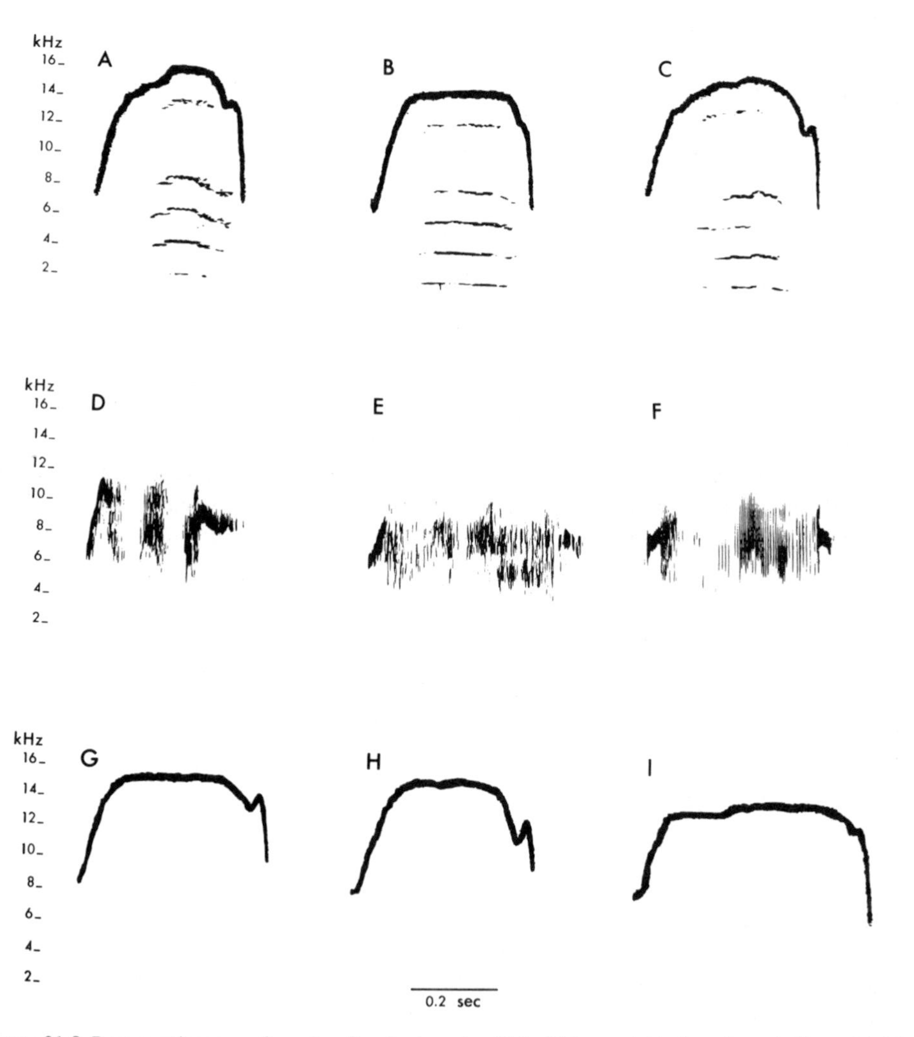

Figure 21-8. Postoperative separation cries of squirrel monkey H-5 with tegmental lesions shown in Figure 14-10.

(A–C) Separation cries showing abnormal, harmonically related components below the fundamental frequency.

(D–F) Examples of infantile nature of cries recorded during same sessions as in A–C, with characteristic fricative noise between short, tonal components.

(G–I) 12 months later, showing disappearance of abnormal subharmonics.

From Newman and MacLean (1982).

Consequently, our finding that two monkeys (Y-4, Z-4) with virtual bilateral destruction of the amygdala produced separation cries,[30] predisposed us to focus on the medial frontal cortex, rather than the temporal cortex. This inclination was reinforced by the anatomical and stimulation studies of the Munich workers[93] that indicated that the ''cingular vocalization pathway'' follows a course close to the thalamomidbrain junction that we had found to be implicated in the patterning and production of the isolation call.[83]

Results

By the process of elimination Newman and I[94] have narrowed down a continuous strip of limbic cortex responsible for the spontaneous production of the call. The findings in three monkeys provided information about the function of gross subdivisions of the frontal lobe in the isolation call. In monkey SC-7 bilateral prefrontal lobectomy rostral to the cingulate gyrus (Figure 21-9B) did not significantly affect the call during a 3-month period of testing. On the contrary, bifrontal lobectomy (SC-8), or bilateral lobotomy (X-5), just rostral to the knee of the corpus callosum and lateral ventricles (Figure 21-9B), resulted in an immediate and enduring failure in the production of spontaneous calls. Other forms of vocalization continued to be expressed.

The findings in a fourth monkey (R-5) indicated that the cortex essential for the call is located on the medial frontal surface. Aspiration of area 24 rostral to AP 14, together with area 25, most of area 12, and all of areas 8 and 9 (see Figure 21-9C) resulted in permanent elimination of the spontaneous production of the call during the period of 9 months that the subject was tested. Figure 21-11 shows his performance record. Other vocalizations including yaps, cackles, errs, and shrieks were emitted under appropriate conditions. If, however, in an otherwise comparable lesion there was sparing of the supragenual area 24 and posterior parts of areas 25 and 12 (subject SC-3, Figure 21-9D), there was an early recovery of the spontaneous production of the isolation call. The same was true of isolated lesions of parts of the band of limbic cortex contained in areas 12, 25, and paragenual 24. Figure 21-11 shows the performance record and early recovery of subject SC-9 with a symmetrical coagulation of the subcallosal cortex represented in Figure 21-10D.

If, however, there was bilateral aspiration of the entire limbic area in question, there was an enduring, almost complete failure to produce spontaneous isolation calls. Figure 21-11 shows the 5-month performance record of subject Z-5 in which the reconstructed bilateral lesion seen in Figure 21-10B was largely confined to areas 12, 25, and paragenual 24. Another subject SC-11 with an ablation of all the midline neocortex peripheral to area destroyed in the preceding case served as an excellent control (see Figure 21-10A). As shown by his performance record in Figure 21-11, recovery of the isolation call returned by the third weekly trial.

In human beings, it is known that stimulation of the supplementary area above the anterior cingulate cortex may induce a variety of vocal effects and that lesions of this area[95] may temporarily interfere with vocalization and speech[96] (see also Chapter 28). The presumed corresponding area—rostral area 6 and adjoining area 8—was included in the extensive midline neocortical lesion in the animal just described (SC-11) and in which there was recovery of the spontaneous isolation call. In another monkey in which an attempt was made to confine the lesion to the supplementary area in question, but in which there was invasion of the cingulate vocal area, there was also early recovery of the call.

In summary, the spontaneous production of the isolation call appears to depend upon the concerted action of a continuous band of limbic cortex contained in the supragenual,

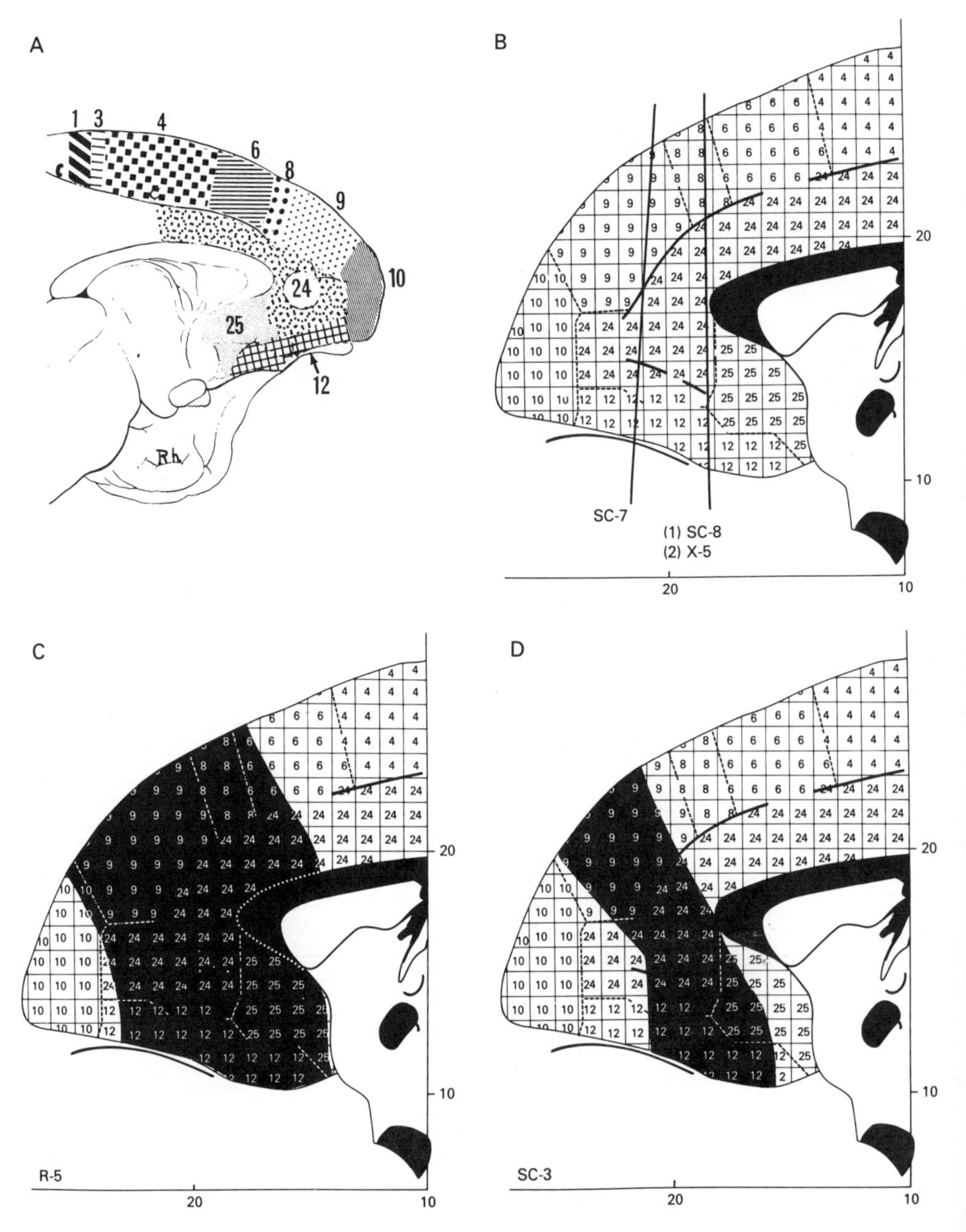

Figure 21-9. Sagittal representations of levels of frontal lobe sections (vertical lines) and cortical ablations (shading) in *Saimiri* monkeys, based on Rosabal's areal parcellation combined with coordinates of stereotaxic brain atlas.

(A) Rosabal's (1967) cytoarchitectural scheme. For facilitating comparison, other panels show numbers for Rosabal's areas inserted into mm squares conforming to planes of atlas (Gergen and MacLean, 1962).

(B) The off-vertical line corresponds to level of prefrontal lobectomy in subject SC-7. Single vertical line for subjects (1) SC-8 and (2) X-5 indicate approximate pregenual level of frontal lobectomy and lobotomy, respectively. Pregenual severance of connections, but not prefrontal, resulted in failure to produce spontaneous isolation calls. C and D show extent of midline cortical ablations in subjects R-5 and SC-3. Shading corresponds to cortical destruction common to both sides of the brain. Ablation shown in C, but not that in D, eliminated spontaneous calling. Note in D sparing of supragenual cingulate (part of area 24) and parts of subcallosal 25 and 12. From MacLean and Newman (1988).

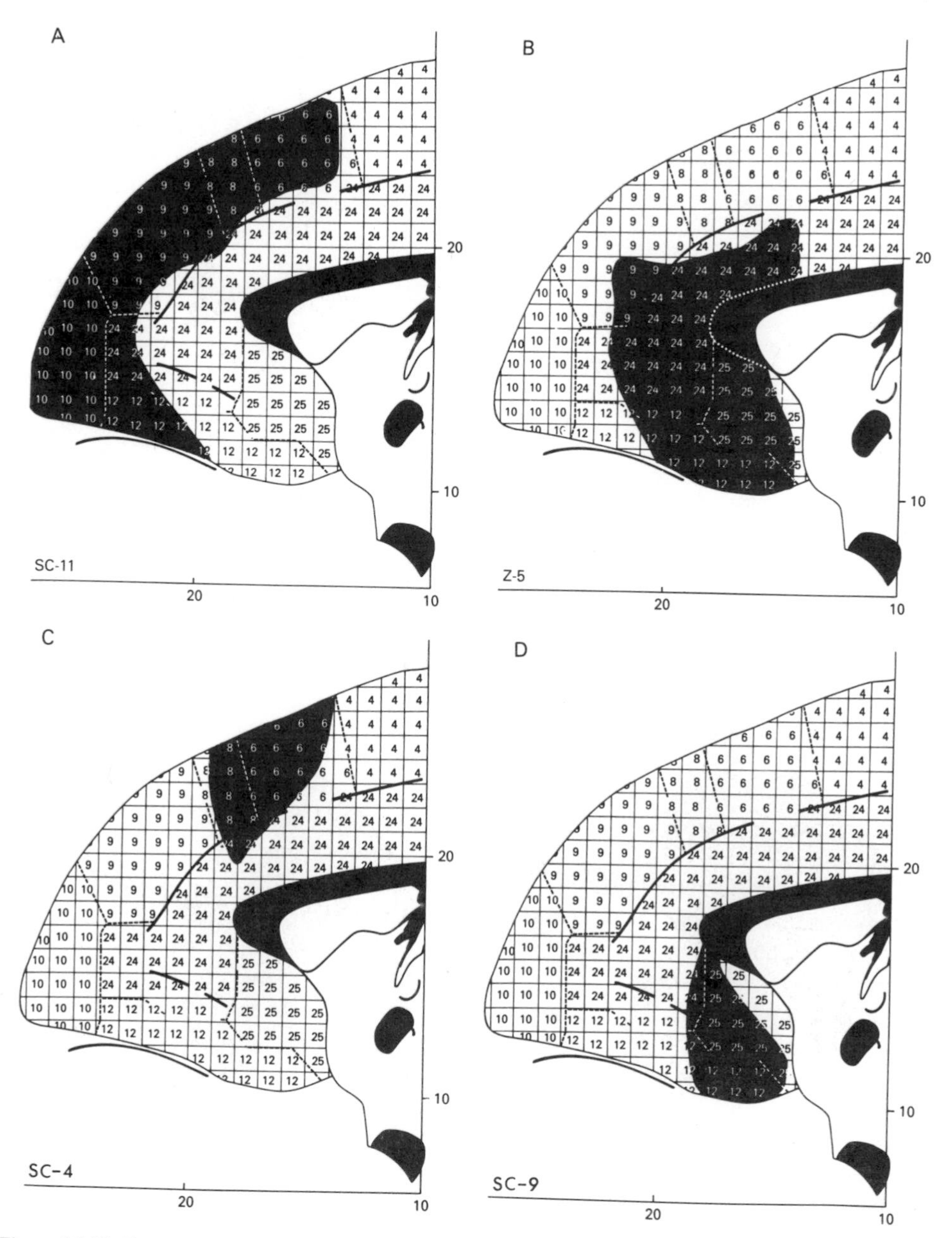

Figure 21-10. Representation of midline cortical ablations in four *Saimiri* monkeys. See legend of Figure 21-9 regarding numeration of mm squares corresponding to planes of brain atlas.

(A) Aspiration of all neocortex peripheral to limbic areas shown in B was followed by early recovery of spontaneous production of the isolation call in subject SC-11.

(B) Ablation largely limited to limbic areas shown here (parts of areas 24, 25, and 12) resulted in virtual elimination of the call in subject Z-5.

(C) Aspiration of "rostral supplementary area" with extension into "cingulate vocalization area" was followed by resumption of spontaneous calling by subject SC-4 within 3 weeks.

(D) Partial ablations of the limbic area shown in B (in this case most of the subcallosal cortex) failed to eliminate the spontaneous isolation call in subject SC-9.

From MacLean and Newman (1988).

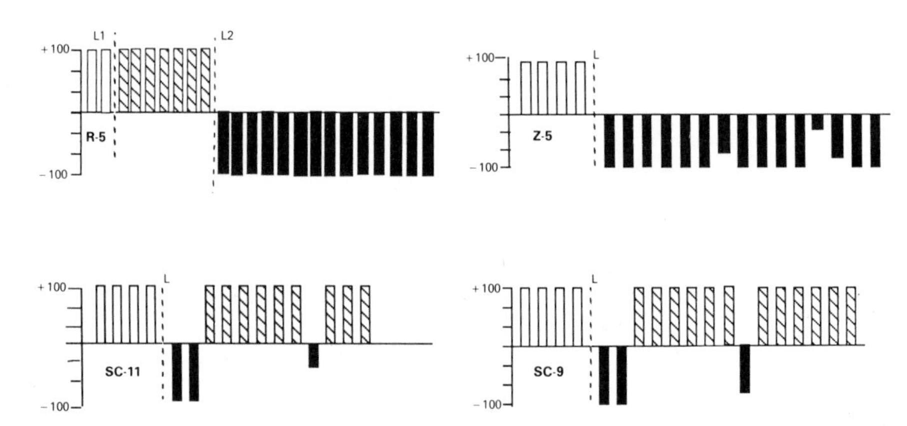

Figure 21-11. Bar graphs showing pre- and postoperative performance of four *Saimiri* monkeys with midline frontal ablations shown in Figures 21-9 and 21-10. Upward open and shaded bars represent, respectively, criterion performance in pre- and postoperative trials (20 or more spontaneous isolation calls within 15 min). Downward black bars show percent deficit short of criterion. As explained, ablation of coextensive limbic areas involved in subject Z-5 virtually eliminated spontaneous isolation calls, while aspiration of all of midline neocortex peripheral to this limbic zone (SC-11) was ineffective. Partial ablation of the critical limbic zone (e.g., SC-9) was also ineffective. L1, L2, and dashed lines indicate time of first and second lesions. L1 in subject R-5 proved to be a control operation. Excerpted illustrations from MacLean and Newman (1988).

pregenual, and infragenual cingulate cortex, together with that in the caudal part of gyrus rectus.

Comment

Kirzinger and Jürgens[97] have reported a study in which they investigated the effects of various frontal ablations on the spontaneous vocal repertoire of squirrel monkeys. They found that ablations of the anterior cingulate cortex, the neocortical face area, and the posterior supplementary area had no effect on the production or structure of vocalizations. They observed, however, that ablations of the "anterior supplementary" area resulted in reduced vocalization that was primarily attributable to a "drastic reduction of the isolation peep."[97] *It should be noted, however, that their animals were not individually isolated during recordings, but rather each was paired with a surgically matched subject.* It is our understanding that their postoperative observations did not extend beyond 4 weeks. As was mentioned, following aspiration of the corresponding area we observed an early resumption of the call with full recovery in 2 to 3 weeks.

What can be said about the relevance of the present findings to other primates? Similarities in the separation cries of primates were pointed out above when illustrating the spectrograms of a squirrel monkey, macaque, and human infant shown in Figure 21-6. Among macaques and the great apes, cooing occurs between mother and infant for assuring an acceptable distance of separation. Lawick-Goodall refers to such calls in the chimpanzee as the "hoowhimper."[98] In human infants comparable cries have been variously characterized as hunger cries, pain cries, and the like, but as Wolff has noted, the so-called hunger cry can be more predictably elicited by separating an infant from its mother.[99]

In experimental work, the so-called "koo" or "coo" call of rhesus monkeys is regarded as a separation call. Sutton *et al.*[100] have investigated the effects of various cortical lesions on the spontaneous and conditioned coo calls of rhesus monkeys. Initially they compared the effects of ablating (1) the homologue of Broca's area, (2) the transitional parieto-occipital cortex, (3) the temporal association cortex, and (4) the anterior cingulate/subcallosal cortex. They found no changes in vocal performance except in two animals with combined ablations of the anterior cingulate cortex and subcallosal cingulate cortex. As opposed to the monkeys with neocortical ablations, the postoperative calls in these two animals were weak and infrequent in the test situation. If, as in a subsequent study on four other animals, ablations were confined to the anterior cingulate and pregenual cingulate cortex, there was a reduction in the number of conditioned coo calls.[101] The spontaneous calling rate when the animals were in a group was said to be unchanged. That situation would have compared to the experimental conditions of Kirzinger and Jürgens[97] (see above).

In a long report on the excitable cortex of the orangutan, chimpanzee, and gorilla, Leyton and Sherrington[102] noted that they obtained the "emission of a sound" upon stimulation of Broca's area in the chimpanzee. In the great apes, however, I know of no observations on the effects of lesions or of stimulation of the cingulate area on vocalization. Clinically, a few cases have been described in which akinetic mutism resulted from occlusion of the anterior cerebral artery and an infarction largely confined to the cingulate gyrus.[103] Vocal alterations have not been reported in cases in which cingulate lesions were produced for the purpose of alleviating psychiatric illness. In such cases, however, it would appear that the lesions have been insufficient in extent, as well as too far caudal from the knee of the corpus callosum, to affect vocalization. In collaboration with Penfield and Jasper,[104] Kaada[105] had the opportunity to observe the effects of pregenual cingulate stimulation in patients undergoing neurosurgery. In his words: "An interesting feature was the observation that the patients were partly *able to overcome the respiratory arrest* when asked to count during the stimulation. But difficulties were encountered in talking, obviously due to interference with the respiratory mechanism."[106] It should be noted in this respect that in the experimental studies on monkeys that have been reviewed, the conditions were not appropriate for testing inhibition of vocalization. In the preceding chapter mention was made of the experiment in which stimulation just rostral to the knee of the corpus callosum suppressed vocalization in the squirrel monkey. I have observed that stimulation at the corresponding locus in the macaque elicits the same effect.[30]

It is of timely interest that Newman *et al.*[107] have shown in the squirrel monkey that doses of morphine insufficient to interfere with the general behavior of the animal, eliminate the production of separation cries (isolation calls), and that treatment with the antagonist naloxone reinstates the calls.[107] In the rhesus monkey Wise and Herkenham[50] have found a high concentration of opiate receptors in the cingulate cortex. Inspection of their autoradiograms reveals a distinct demarcation between the labeled area and the supracingulate cortex.

In conclusion, one other anatomical matter requires mention in regard to the cortical ablations of the present study. It has long been recognized that with the exception of the medial polar area, ablations of the medial frontal cortex do not result in clearly discernible retrograde thalamic degeneration. The use of improved neuroanatomical techniques has recently thrown a whole new light on connections of the midline frontal cortex with the thalamus. These findings will be dealt with in Chapter 28 when considering the subject of crying and laughter in connection with the close relationship of parts of the frontal neocortex with the thalamocingulate division.

Summary

Since running commentaries and summaries have been given for different sections of this chapter, only the main items of interest will be noted here. The thalamocingulate division of the limbic system comprises the cingulate mesocortical areas innervated by the anterior thalamic nuclei and certain other thalamic structures. Anatomical considerations may prove helpful in guiding the direction of future neurobehavioral research. Going beyond anatomy reviewed in Chapter 18, the present chapter includes a description of certain constancies and changes in the cytoarchitecture of the principal cingulate areas in the evolution of mammals, with attention called particularly to altered development of the retrosplenial and posterior cingulate areas of primates that possibly reflects an increased behavioral reliance on vision.

There appears to be no clearly defined rudimentary counterpart of the thalamocingulate division in the reptilian brain. In the light of this, it is of unusual interest that recent research points to a representation of three forms of behavior that characterize the evolutionary transition from reptiles to mammals—namely, nursing, in conjunction with maternal care; audiovocal communication for maintaining maternal–offspring contact; and play. Evidence of the role of the cingulate gyrus in maternal behavior and in play is derived from experimental work on rodents, the salient findings of which are described. The rest of the chapter deals with audiovocal communication and focuses on the cerebral representation of the separation cry, which perhaps ranks as the earliest and most basic mammalian vocalization, serving originally to maintain maternal–offspring contact.

Experiments in squirrel monkeys indicate that the rostral cingulate cortex (including that of the infragenual part) is selectively implicated in the spontaneous production of the separation cry. Further implications of these findings will be discussed in Chapter 28 when dealing with the relationship of the frontal neocortex to the thalamocingulate division.

References

1. Clark and Meyer, 1950, p. 342
2. MacLean, 1985a
3. Brodmann, 1909
4. Vogt and Peters, 1981
5. Rose and Woolsey, 1948
6. Rose, 1927
7. Krettek and Price, 1977a
8. Ramón y Cajal, 1901–02/1955
9. Krieg, 1946
10. Ariëns Kappers *et al.*, 1936, p. 1592
11. Campbell, 1905, p. 282
12. Campbell, 1905, p. 232
13. Rose and Woolsey, 1948, p. 332
14. Campbell, 1905
15. Campbell, 1905, p. 275
16. Rose, 1912
17. Herman, 1980
18. Morgane *et al.*, 1982
19. Kesarev *et al.*, 1977; Gingerich *et al.*, 1983
20. Morgane *et al.*, 1986
21. Campbell, 1905, p. 184
22. Krieg, 1963, p. 219
23. Krieg, 1963, p. 220
24. Krieg, 1963, p. 222
25. Flechsig, 1921
26. von Economo and Koskinas, 1925
27. Lorente de Nó, 1933
28. Vogt, 1976
29. Lorente de Nó, 1949
30. MacLean, unpublished observations
31. Smith, 1945
32. Ward, 1948
33. Ward, 1948, p. 15
34. Le Beau, 1952; Livingston, 1953; Tow and Whitty, 1953
35. Stamm, 1955
36. Stamm, 1954
37. Slotnick, 1967
38. Murphy *et al.*, 1978, 1981; MacLean, 1978c
39. Slotnick, 1975
40. Tilney, 1936
41. Bennett-Clarke and Joseph, 1982; King *et al.*, 1982
42. Gill, 1985
43. Terkel and Rosenblatt, 1972
44. Slotnick and Nigrosh, 1975
45. Aulsebrook and Holland, 1969
46. Powell, 1966
47. Beyer *et al.*, 1961
48. Franzen and Myers, 1973b
49. Franzen and Myers, 1973b, p. 143
50. Wise and Herkenham, 1982
51. Lewis *et al.*, 1983
52. Nettles *et al.*, 1982; Bridges and Grimm, 1982
53. Murphy *et al.*, 1981

54. Murphy, 1971
55. Dieterlen, 1959; Murphy, 1977
56. Murphy, 1970
57. Murphy, 1976
58. Murphy, 1973, 1977
59. Stevens, 1951, pp. 358–359
60. Field *et al.*, 1959–60
61. Panksepp, 1981
62. Cadell, 1963
63. Cadell, 1963, p. 134
64. Franzen and Myers, 1973a
65. Groos, 1896/1915
66. Beach, 1945; Fagen, 1981
67. MacLean, 1978c, 1981, 1982
68. Goodall, 1981–82
69. Romer, 1966
70. Anderson, 1954
71. Roberts, 1975
72. Okon, 1972
73. Noirot, 1972
74. See Smith and Sales, 1980; see also Okon, 1972
75. Noirot, 1968
76. Winter *et al.*, 1966
77. Newman, 1985a
78. Newman and MacLean, 1982, p. 319
79. Jürgens and Ploog, 1970
80. Dua and MacLean, 1964
81. Robinson, 1967
82. Newman, 1985a, p. 308
83. Newman and MacLean, 1982
84. Magoun *et al.*, 1937
85. Kelly *et al.*, 1946
86. Baldwin, 1967, 1968; Newman and Symmes, 1983
87. Thorington, 1968
88. Newman and Symmes, 1982; Newman, 1985b
89. Symmes *et al.*, 1979
90. Ferrier, 1876
91. Ploog, 1981, p. 37
92. Jürgens, 1976
93. Jürgens and Pratt, 1979a,b
94. Newman and MacLean, 1985; MacLean, 1987; MacLean and Newman, 1988
95. Brickner, 1940; Penfield and Rasmussen, 1952; Penfield and Welch, 1951; Penfield and Jasper, 1954; Penfield and Roberts, 1959
96. Penfield and Jasper, 1954
97. Kirzinger and Jürgens, 1982
98. Lawick-Goodall, 1969
99. Wolff, 1969
100. Sutton *et al.*, 1974
101. Trachy *et al.*, 1981
102. Leyton and Sherrington, 1917
103. Barris and Schuman, 1953; Buge *et al.*, 1975; Nielsen and Jacobs, 1951
104. Penfield and Jasper, 1954, p. 108
105. Kaada, 1951
106. Kaada, 1951, p. 64
107. Newman *et al.*, 1982

22

Phenomenology of Psychomotor Epilepsy

Pathogenic Aspects

There is probably no clinical condition that provides more windows for viewing the neural substrate of the human psyche than psychomotor epilepsy. Since clinical findings in this disease provide the crucial evidence that the limbic system plays a basic role in the subjective experience of emotion, we deal in this chapter with the pathogenesis of this cruel affliction and then in the following three chapters with the symptomatology. During the aura at the beginning of the seizure, patients experience, according to the individual case, a wide variety of vivid emotional feelings that range from intense fear to ecstasy. In addition to such conventionally recognized emotional feelings, they also experience other affective feelings identified with the basic needs and with the special senses. As to be explained (Chapter 23) the latter two forms of feelings will be classified, respectively, as "basic" and "specific" affects, while the conventionally recognized emotional feelings will be categorized as "general" affects. The word *affect* will be used to apply to the subjective aspect of the emotional experience, whereas Descartes's word *emotion* will be reserved for denoting the behavioral expression of affect.

Introductory Considerations

Terminology

The word *epilepsy* commonly evokes a picture of someone undergoing a generalized convulsion. In that sense the word is poorly descriptive of psychomotor epilepsy because in the majority of attacks there are no *grand mal* convulsions.[1] Rather the attacks are characterized initially by an assortment of subjective states after which the patient may engage in simple or complex automatisms for which there is no memory. Since the time of Hippocrates, such episodic, nonconvulsive attacks were suspected of being a form of epilepsy, but it remained for the introduction of clinical electroencephalography to demonstrate that they were associated with abnormal bioelectrical discharges in the brain. In 1938, nine years after Hans Berger's[2] first report on recording human brain waves, Gibbs *et al.*[3] introduced the term *psychomotor* to distinguish the epilepsy in question from the grand mal type and the so-called petit mal triad in which there are brief lapses of consciousness that usually occur without any subjective symptoms. In the last century, the French neurologist P. Janet had used the shorter expression *psycho-lepsy*.[4]

The term *psychomotor* refers to both the psychic and motor aspects of the epilepsy. With improvements in electroencephalographic diagnosis, the term *temporal lobe epilep-*

sy became a substitute label for *psychomotor epilepsy* because of the recognition that the manifestations are commonly the result of abnormal discharges in the temporal lobe.[5] In 1953, Fulton (1899–1960) commented that "temporal lobe seizures . . . might more properly be designated as seizures involving the limbic lobe, for such attacks involve not only temporal lobe structures, but usually also the posterior orbital gyrus, the hippocampus, and cingulate gyrus."[6] In line with such reasoning, Glaser introduced the expression *limbic epilepsy*.[7] Given the historic principle of adhering to original names, *psychomotor* will be used here as the preferred term, but in some contexts the other expressions will be applied interchangeably. Since 1970, the noncommittal expression *complex partial seizures* has been officially adopted by electroencephalographers as a term for the kind of epilepsy under consideration.[8]

Limbic Propagation of Seizure Discharges

Apropos of Fulton's comments quoted above, it is to be emphasized that the epileptic discharges arising in or near the limbic cortex have the tendency to spread in and be confined to the limbic system. Kaada, working in Fulton's laboratory, was one of the first to draw attention to this striking phenomenon. In regard to the hippocampus he states that afterdischarges initiated there "readily spread into the limbic . . . cortex on both sides, and further to the cortex just lateral to the rhinal fissure. . . ."[9] His findings on anesthetized animals were later confirmed in awake preparations.[10] Limited observations made during neurosurgery indicate that the same predilection of spread of afterdischarges to limbic structures applies to the human brain.[11] As will unfold in subsequent chapters, the importance of these findings in regard to epistemics cannot be overemphasized.

Etiology and Pathogenesis

In human beings, it is the limbic subdivision associated with the amygdala that is most often the site of epileptic activity. There are several conditions that account for this, one being that the location of this division at the base of the brain makes it subject to contusions resulting from blows to the head incurred in falls, fighting, athletic contests, automobile accidents, and other mishaps.[12] As Rasmussen has commented, "In my opinion . . . the principal reason that temporal lobe epilepsy is the most common of all epilepsies is due to the fact that the temporal lobe is in a relatively enclosed bony and ligamentous box so that any abnormal movement or swelling of the brain inside the skull is peculiarly apt to produce damage to the temporal pole and medial temporal structures."[13] Middle ear infections, with a complicating meningitis involving the temporal lobes, may be another cause of temporal lobe epilepsy.[14] The herpes simplex has predilection for limbic structures, and acute encephalitis attributable to that agent may be complicated by epilepsy (see below).[15] An encephalitis involving primarily the limbic lobe may also be associated with oat-cell carcinoma of the lung.[16] Tumors comprised of glial elements are more often a cause of temporal lobe epilepsy than is generally recognized. Since such tumors may be astrocytomas of microscopic size, they may not be evident upon x-ray examination and for the same reason may be overlooked in the pathological examination of tissue obtained at operation or after death.[17]

However, as will now be described, there have been two chief contending views regarding the main cause of psychomotor epilepsy—the first view implicating birth injury and the second, febrile convulsions.

A Favored First View

The well-known neurosurgeon Wilder Penfield (1891–1975) of the Montreal Neurological Institute, believed that injury at the time of birth to medial, basal, limbic structures was the most common cause of psychomotor epilepsy. As he noted, "That portion of the temporal lobe which rests upon the incisura of the tentorium as it encircles the midbrain is the most frequent site of pathological alteration."[18] Although the tentorium was so named because this particular thickening of the dura gave the impression of a tent over the cerebellum, it is more comparable to a support holding up the pendulous occipital lobes. The incisura refers to the slit (incision, notch) between the tent flap and anteromedial aspect of the pontine fossa. Using stillborn infants, Earle *et al.*[19] showed that the overall compression of the head at the time of birth could result in incisural herniation of this part of the brain on one or both sides, indicating not only how injurious deformation could occur, but also impairment of the blood supply, particularly by pressure against the anterior choroidal and posterior cerebral arteries where they cross the free edge of the tentorium (see Figure 17-8). Their demonstration required freezing of the brain in place after applying compression; otherwise there was spontaneous reduction of the herniation. The blood supply to the hippocampus reflects a primitive condition insofar as the terminal vessels to part of it branch off from the main vessel somewhat like the prongs of a rake (Figure 22-1A).[20] These vessels are somewhat like end-arterials of the opossum brain because of the few anastomoses between them. If the blood pressure falls in the part of the vessel corresponding to the handle of the rake, there results a progressive falloff of pressure in the pronglike vessels farther and farther from the handle. Such embarrassment of the circulation may lead to death of nerve cells and their replacement by glia. The glial tissue amounts to a scar that upon aging becomes shrunken and hard. It is presumed that a retracting scar pulls upon surrounding tissue and at the same time pinches off capillaries supplying adjacent gray matter.[14] The resulting impairment of metabolism of nerve cells may be conducive to an abnormal discharge of nerve cells. Prolonged discharges may lead to cellular exhaustion and death. Under such circumstances one may imagine a vicious circle in which there occurs more scarring and a further death of nerve cells. Within a number of years, the slowly creeping "reparative" process may spread to involve the parahippocampal gyrus and neighboring deep cortex of the first temporal convolution, resulting in an extensive scar that looks yellow and feels rubbery. As Penfield has observed, "Local destruction in the brain, adjacent to a scar does continue and we have found incontrovertible evidence of it at the border of a 28-year-old scar."[21]

Penfield and his group referred to the hardening in the region of the hippocampus as *incisural sclerosis*.[19] Upon reviewing 157 cases of temporal lobe epilepsy, they concluded that the pathological findings in 100 (63%) of them "suggested that compression or anoxia during birth or infancy was the cause."[22] In support of this thesis, it is to be noted that X-ray examination in some cases reveals that the middle temporal fossa at the base of the skull may be smaller and the petrous bone higher on the side of the brain on which there is incisural sclerosis. This finding would indicate that damage occurred at

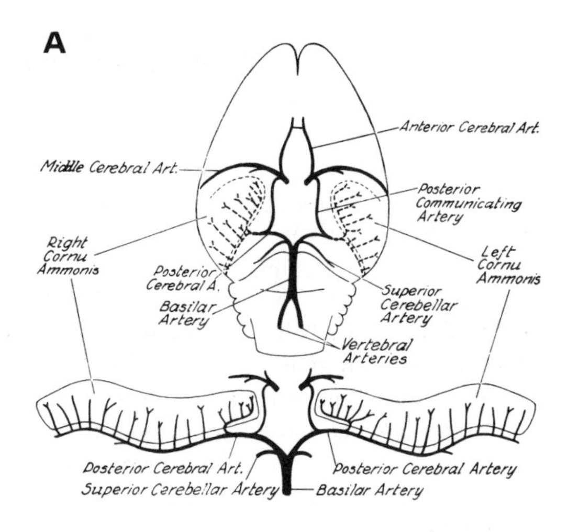

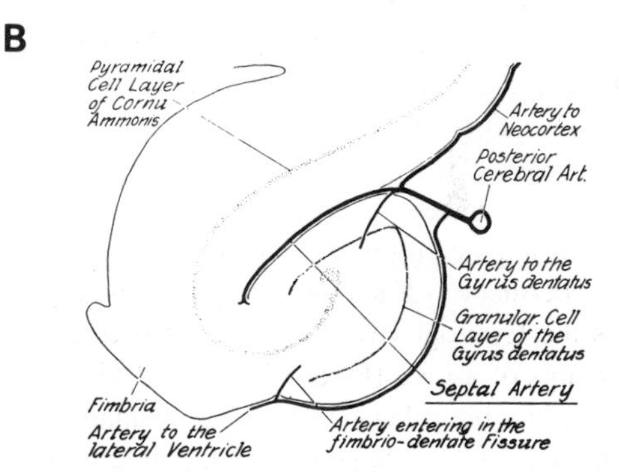

Figure 22-1. Illustrations from Nilges, showing arterial blood supply to the hippocampus in a typical mammal.

(A) Ventral surface of brain and the typical rakelike pattern of blood vessels derived from the posterior cerebral artery.

(B) Transverse section of hippocampus, focusing on the origin and course of the septal arteries supplying inter alia, Sommer's sector (see text).

Slightly modified after Nilges (1944).

least before the age of 2.[23] Such X-ray changes have also been observed by others, including Falconer and his group,[24] whose views are next to be described.

A Favored Second View

The late Mr. Murray Falconer (1910–1977), a British neurosurgeon who resected the temporal lobe in more than 150 cases of psychomotor epilepsy, believed that childhood convulsions are a much more common cause of hippocampal injury than is birth trauma.[25] Instead of *incisural sclerosis,* Falconer and his colleagues[26] preferred the expression *mesial temporal sclerosis* because of the recognition, as already explained, that the hardening may spread to the neighboring medial temporal cortex. Falconer believed that mesial temporal sclerosis will prove to be "the single, most common lesion underlying temporal lobe epilepsy."[27] As is well known, young children are subject to febrile convulsions occurring in conjunction with teething and a variety of bacterial and viral infections. Complicating illnesses include "colds," tonsillitis, measles, whooping cough, and illness following inoculation of triple antigen.[24]

In proposing that febrile convulsion is the most common cause of mesial temporal sclerosis, Falconer admitted to an inability to explain why the lesion so often occurs only on one side of the brain. In 80% of this first 100 cases, for example, only one side was involved.[25] This question leads to a historical reconstruction of the slowly developing interest in peculiarities of the hippocampus, with respect both to its susceptibility to damage during generalized convulsions, and its being a site of epileptogenic foci. After a brief historical survey, we will return to the question of unilateral temporal sclerosis and Falconer's suggested explanation of its occurrence.

Historical Perspective

As Falconer and others have pointed out, the association of hardening of the hippocampus in cases of epilepsy has been known for 150 years, but, curiously enough, it was only 50 years ago that such disease was regarded seriously as a causative factor in epilepsy. In 1825, two French physicians, Bouchet and Cazauvieilh[28] reported that in 9 out of 18 cases with a history of epilepsy, they detected changes in the region of the hippocampus, with "induration" noted in at least 4 cases. Since these perceptive observations were made well before the introduction of histological methods (see Chapter 18), the findings were based entirely on gross examination. Because the brains were unfixed, it was possible to feel a hardening of the tissue in the region that looked diseased.

During the next 75 years, several other papers appeared in which disease of the hippocampus was described in cases of epilepsy. In 1847, Bergmann[29] called attention to an asymmetry of Ammon's horn that he was inclined to regard as a change resulting from, rather than a cause of, epilepsy. In 1880, Pfleger[30] of Vienna, stimulated by Meynert's[31] findings of atrophy and sclerosis of Ammon's horn in epilepsy, reviewed the brains of 300 individuals and found that of 43 known to have had epilepsy, atrophy and sclerosis of Ammon's horn were present in 25 (58%). The disease was homolateral in 88%. He includes a useful table comparing his findings with those of Meynert, Holler, Snell, and Hemkes. He raised the question of whether or not the hippocampal change might be attributable to peculiarities of its blood supply.

The most widely quoted paper of the 19th century is that of Sommer,[32] who in 1880 gave the first description of microscopic changes of the hippocampus seen in epilepsy. He found evidence of hippocampal disease in 30% of the 90 cases examined. In his detailed description, he noted that there was one area of the hippocampus that was particularly involved—the so-called area CAl, that has since become known as Sommer's sector (see Figure 17-8).[33] Sommer did not speculate about a causal relationship between hippocampal disease and epilepsy. Bratz[34] confirmed and extended Sommer's findings, emphasizing changes also found in the "end blade" (Figure 22-1B).

Spielmeyer (1879–1935),[35] a renowned German neuropathologist, sought to establish once and for all whether or not the hardening of the hippocampus was a result or a cause of generalized convulsions. He finally concluded that during convulsions there was a reduced blood supply and that this condition made the small hippocampal vessels susceptible to spasm that resulted in a still greater deprivation of oxygen transported to this structure. Uchimura,[36] one of Spielmeyer's students, demonstrated that the so-called septal vessels (Sektorgefäss) were exceptionally long and narrow, with relatively few interconnections between their terminals. These are vessels that supply Sommer's sector, as well as the "end blade" (Figure 22-1B), the parts of the hippocampus where death of the nerve cells most commonly occurs following convulsions. In diameter, they are hardly more than two to three times the width of a red blood cell (16–20 μm).[38] The parent vessels (12 to 15 in number[37]) have the rakelike arrangement described above. Scholz, who succeeded Spielmeyer, also adhered to the view that hardening of the hippocampus was the result, rather than a cause, of epilepsy, and his authoritative monograph (1959)[39] further solidified thinking along these lines.

In 1935, the suggestion that hippocampal sclerosis might be a cause of epilepsy cropped up again when Stauder,[40] a German pathologist, described such changes in 36 of 53 epileptics who had died a natural death. In the face of prevailing views, however, and in the absence of electroencephalographic evidence, his suggestion that *Ammon's horn sclerosis* played a positive role in epilepsy failed to attract attention. A few years later, on the other side of the Atlantic, cases of epilepsy were beginning to be studied intensively at the Montreal Neurological Institute. In a paper on the classification of epilepsies, Jasper and Kershman (1941)[41] pointed out that in some cases the electroencephalographic findings were suggestive of disturbances deep in the temporal lobes, "probably in the archipallium." At that same time, Fred and Erna Gibbs were accumulating electroencephalographic records on a great number of patients with various forms of epilepsy. With William Lennox, they reported that in cases of psychomotor epilepsy, the electroencephalogram was frequently characterized by a series of flat-topped waves (otherwise called saw-toothed waves) appearing in all leads from the scalp.[3] It was their practice to use electrodes applied to one or both ears as an "indifferent reference." It subsequently turned out that the saw-toothed waves were "inverted spikes" and that one of the ear electrodes was nearest to the spike focus. Consequently, their method of "monopolar recording" showed the abnormal recording in all leads from the scalp. In 1948, Gibbs *et al.*[42] reported that in psychomotor epilepsy the spike focus usually reached its greatest amplitude in the anterior temporal region on one or both sides. They also made the important observation that in some cases the abnormality could be demonstrated only when the patient was in a state of light sleep. "Sleep activation" thereafter became routine in some laboratories as a means of arriving at a diagnosis of psychomotor epilepsy.

During part of the period under consideration (1947–1949), I was a U.S. Public Health Fellow under the aegis of Dr. Stanley Cobb (1887–1968) at Massachusetts General Hospital and was conducting research in the electroencephalographic laboratory of Dr. Robert Schwab (1903–1972). I had devised an improved nasopharyngeal electrode for the purpose of recording the activity of structures at the base of the brain.[43] With Arellano, who had subsequently designed a soft electrode that could be placed next to the eardrum just underneath the temporal lobe, we examined a series of patients with psychomotor epilepsy whose previous electroencephalograms had shown either no epileptogenic focus or no abnormality at all.[44] In using sleep activation, we found that a spike discharge showed up in the recording from the basal electroencephalogram. In 7 of 12 cases (58%), the origin of the spike activity appeared to be nearer one of the electrodes at the base than at the scalp. Since of all the electrodes applied to the outer cranium, the nasopharyngeal is located nearest the hippocampus, it was suggested that in those cases in which the spike showed maximum amplitude in the nasopharyngeal leads, the disturbance was owing to disease originating in the hippocampus.[44]

In 1953, Sano and Malamud[45] reported their finding of hippocampal sclerosis in more than half of 50 cases with a history of epilepsy. Thirteen years later, Margerison and Corsellis[46] reported a comparable incidence in 55 cases of epilepsy and provided the additional information that in 80% the lesion was unilateral. In a series of papers since then, Falconer and his colleagues have described their findings in about 150 cases in which the temporal lobe, including the hippocampal formation, had been resected on one side for the treatment of psychomotor epilepsy (see Ref. 25). What is unique about their case material is the availability in each case of the operated lobe for neuropathological study. Case reports from other clinics lacked valuable information about the neuropathology because the neurosurgical technique involved aspiration of the medial parts of the temporal lobe. On the basis of the Falconer material, it can be concluded that a creeping sclerosis originating in the hippocampus and spreading to nearby convolutions is a common cause of psychomotor epilepsy.

The Question Regarding Unilaterality of Ammon's Horn Sclerosis

How is one to explain the apparent high incidence of unilateral mesial temporal sclerosis? As further background for considering this question, other factors that may contribute to the ''selective vulnerability'' of the hippocampus should be mentioned. In the first place, it has long been recognized that the hippocampus has the lowest seizure threshold of any structure in the brain.[47] Mechanical injury alone is sufficient to induce a seizure discharge.[48] These factors together with the peculiar nature of the blood supply of the hippocampus and its special metabolic needs (Chapter 17) might all be expected to contribute to the vulnerability of hippocampal neurons. One metabolic requirement in particular deserves emphasis. In an autoradiographic study, we observed an especially high uptake of radioactive methionine in the hippocampal cortex (Figure 17-10).[49] This observation provides indirect evidence of a high protein turnover in the hippocampus. Relevant to febrile infectious illnesses, we observed that an induced renal infection in the rat was associated with a greatly depressed uptake of methionine in the hippocampus. Moreover, it was found that insulin-induced hypoglycemia had a similar effect on methionine uptake. All things considered, it would appear that the hippocampus, more than other forebrain structures, would be susceptible to an exhaustion of its metabolic resources by seizure activity.

With their method for producing temporal lobe herniation, Earle *et al.*[19] showed that it was more usual for one side or the other to herniate rather than for both to do so. Unilateral mesotemporal sclerosis might therefore be explained on this basis. But in regard to Falconer's proposal, what might account for unilateral sclerosis following febrile convulsions? It is well known that increased intracranial pressure or edematous swelling of the brain may result in temporal herniation. Since swelling of the brain may occur as a complication of febrile convulsions, it might be argued that such a condition may induce herniation on one side or the other. Falconer,[25] however, suggested that unilateral involvement was the result of the tendency of one side of the head to be in a lower position during protracted seizures. Such an interpretation was suggested by McLardy's[50] observations on the dependent hemispheres of convulsing guinea pigs. If a prolonged convulsion were to result in a marked fall of blood pressure, it would set the stage for a pooling of blood on the lower side of the brain that would be conducive to a slower venous drainage. Venous stasis or venous congestion, just as a reduced blood supply through arterial channels, may be damaging to nervous tissue. In a study of the effects of repeated generalized convulsions in adolescent baboons, Meldrum and Brierley[51] noted that in a few instances a lower position of the hemisphere on the right side may have contributed to ischemic changes in the hippocampus on that side.

In connection with unilateral sclerosis (and also the question of febrile convulsions) there is the possibility that an infectious agent might reach the medial temporal region by neural or perineural channels on one side. In 1890, Karoly Schaffer[52] (for whom the Schaffer collaterals are named; see Chapter 18) proposed that the rabies virus may invade the nervous system by traveling along nerve fibers. In the 1920s it was postulated that the ubiquitous herpes simplex virus might also reach the central nervous system via nerve axons or perineural spaces. Since 1955, it has become evident that this virus has a predilection for the limbic cortex.[53] In most cases, however, the inflammation, inclusion bodies, and necrosis tend to be localized in the limbic and adjacent cortex of the fronto-temporal region.[54] As reviewed by Johnson,[54] the possibility exists that the spread of infection to these areas might take place via olfactory nerves from the nose or via nerve fibers leading from the gasserian ganglion to the basal meninges of the anterior and middle fossae.[54] As Johnson[54] has also noted, studies of pathology have shown that the fronto-temporal localization is ''*often strikingly unilateral*'' (italics added). Most reported cases of herpes simplex virus encephalitis indicate that the disease is severe, fulminating, and often fatal. Convulsions are a frequent complication.[55] It is my impression, however, that pathologists do not exclude the possibility that less severe cases may occur and that the extent of the infection might be sufficiently restricted so as to result in only the manifestations of febrile convulsions. In some such cases one might postulate that a deviated nasal septum or other asymmetric condition of the nasal passage might favor the occurrence of infection on one side, with unilateral spread along the olfactory pathways.

In various surveys dealing with etiological factors in epilepsy, one finds that the listed incidence of childhood febrile convulsions is usually less than 6%,[56] but ranges as high as 27%.[57] One study involving 666 patients with temporal lobe epilepsy is particularly relevant because it provides statistics on a representative population and thereby lends assurance that a relatively low incidence of febrile convulsions cannot be ascribed to the exclusion of individuals in the older age groups. In this study the incidence of febrile convulsions was 5%.[58] Over 50% of the patients experienced their first epileptic attack after the age of 25, the mean age being 28. Regarding Falconer's claims (see above), another possible bias is to be considered. He pointed out that a history of febrile convulsions was relatively rare except in cases of mesial temporal sclerosis.[59] Moreover, with respect to the above hypothesis regarding a milder form of viral disease, febrile convulsions need not be considered as a necessary precondition for the later development of epilepsy. At the present time hybridization, immunochemical, and other techniques are being developed for the detection of virus material in brain tissue[60] that may help to resolve questions about viral factors in the etiology of psychomotor epilepsy.[60]

Problems of Localization and Treatment

It has been estimated that in the United States about 60% of the million people with epilepsy suffer from ''partial seizures'' and that in one-third of these medical treatment fails to provide an adequate alleviation of symptoms.[61] In refractory cases of psychomotor epilepsy with a unilateral epileptogenic focus, surgical excision of the diseased tissue may be a highly effective form of treatment. In their book of 1954, Penfield and Jasper wrote, ''The recent increase in understanding of temporal lobe epilepsy has led us to lobe ablation and to the discovery of what may be called *incisural sclerosis*. In case after case we found the cortex to be tough in the anterior and deep portion of the first temporal convolution. This abnormality extended into, and grew more marked in, the uncus and hippocampal gyrus. The tissue was tough, rubbery, and slightly yellow. . . . We have gradually realized the importance of this discovery, as the epileptogenic focus was often shown by electrography to be situated here, and further more patients return with continuing seizures when we had made anterior temporal removals without excision of this area.''[62] In other words, they were saying that it not infrequently happened that a presumed epileptogenic focus on the lateral surface of the temporal lobe led the surgeon to excise the lateral cortex, only to discover later that the pathological examination showed no lesion. The reason for the mistaken localization may be explained by the extensive interconnections of the frontotemporal cortex that comprises part of the amygdalar circuit. As Pribram and I demonstrated[63] by strychnine neuronography in cats and monkeys, under Dial (allobarbital) anesthesia, the application of strychnine to any part of the frontotemporal cortex diagrammed in Figure 18-12 will ''fire'' all of the interrelated cortex of that region. Based on such findings, it is evident how a discharge at the site of an irritative lesion in the buried medial part of the temporal lobe might be transmitted to the exposed lateral surface and thereby mislead the neurosurgeon and electroencephalographer as to the location of the epileptogenic focus.

In addition to the indispensable use of electroencephalography for demonstrating epileptogenic foci, the introduction of positron-emission tomography (PET), and other such noninvasive methods, promises to be increasingly helpful in identifying pathofunctioning cerebral loci in epilepsy. It has been shown by PET scans, in conjunction with the use of fluorine-18-labeled fluorodeoxyglucose, that areas of ictal onset in psychomotor epilepsy show increased metabolism, whereas hypometabolic regions may reflect postictal or interictal depression of function.[64] Studies of blood flow by means of single photon emission computerized tomography in conjunction with [^{123}I]iodoamphetamine have been advocated as a readily available substitute for costly and complex PET procedures.[65] According to one study, high-resolution cerebrospinal fluid enhanced computerized scanning in 25 cases revealed chronic incisural herniation of medial temporal structures in 12 patients that was subsequently found at operation and confirmed by histological examination.[66]

Summarizing Comment

In relation to the goals of the present investigation on brain mechanisms underlying paleopsychic processes, the study of psychomotor epilepsy is essential because no other clinical condition provides such extensive information regarding cerebral structures involved in the subjective experience of emotional feelings.

Commonly considered etiological factors include birth injury, febrile convulsions of childhood, head injury, infections, and tumors. The medial limbic structures located in the hippocampal formation are particularly vulnerable with respect to most of these factors. Although hippocampal sclerosis has been frequently observed in cases of epilepsy since the early 1800s, its recognition as a major cause of psychomotor epilepsy began to be appreciated only within the last 50 years. Incisural herniation at the time of birth has been proposed as a major cause of hippocampal sclerosis (incisural sclerosis). Complications of febrile convulsions have also been suggested as another major cause of hippocampal injury leading to a slowly spreading mesial temporal sclerosis. Because of the frequent unilaterality of hippocampal sclerosis, the possibility is discussed here that a nonfulminating viral infection of medial temporal structures might occur as a result of an invasion of the infectious agent via the olfactory or trigeminal pathways. The slowly spreading nature of mesotemporal sclerosis may explain the delayed onset of psychomotor epilepsy until late adolescence or adulthood.

References

1. Gastaut, 1954; Penfield and Jasper, 1954
2. Berger, 1929
3. Gibbs *et al.*, 1938
4. Quoted by Lennox, 1960
5. Lennox, 1951
6. Fulton, 1953, p. 77
7. Glaser, 1967
8. Gastaut, 1970; Penry, 1975; Merlis, 1970
9. Kaada, 1951, p. 226
10. Creutzfeldt, 1956; Creutzfeldt and Meyer-Mickeleit, 1953; MacLean, 1957a
11. Jasper, 1964; Pagni, 1963; Buser *et al.* 1968
12. Courville, 1958
13. Rasmussen, personal communication
14. Penfield and Jasper, 1954
15. Bennett *et al.* 1962; Drachman and Adams, 1962
16. Brierly *et al.*, 1960; Corsellis *et al.*, 1968
17. Gonzalez and Elvidge, 1962
18. Penfield and Jasper, 1954, pp. 782–783
19. Earle *et al.*, 1953
20. Nilges, 1944
21. Penfield and Jasper, 1954, p. 314
22. Earle *et al.*, 1953, p. 27
23. Penfield, 1956, p. 81
24. Falconer, 1974
25. Falconer, 1971
26. Falconer *et al.*, 1964
27. Falconer, 1971, p. 14
28. Bouchet and Cazauvieilh, 1825
29. Bergmann, 1868
30. Pfleger, 1880
31. Meynert, 1868
32. Sommer, 1880
33. Bratz, 1899
34. Bratz, 1899
35. Spielmeyer, 1927, 1930
36. Uchimura, 1928
37. Altschul, 1938
38. Alexander and Putnam, 1940
39. Scholz, 1959
40. Stauder, 1935
41. Jasper and Kershman, 1941
42. Gibbs *et al.*, 1948
43. MacLean, 1949a
44. MacLean and Arellano, 1950
45. Sano and Malamud, 1953
46. Margerison and Corsellis, 1966
47. Gibbs and Gibbs, 1936; Jung, 1949
48. Renshaw *et al.*, 1940; Andy and Akert, 1953; Liberson and Akert, 1955; MacLean, 1957a
49. Flanigan *et al.*, 1957
50. McLardy, 1969
51. Meldrum and Brierly, 1973
52. Schaffer, 1890
53. van Bogaert *et al.*, 1955; Corsellis *et al.*, 1968; Glaser and Pincus, 1969
54. Johnson, 1982
55. Drachman and Adams, 1962
56. Friedrichsen and Melchior, 1954; Frantzen *et al.*, 1968; Nelson and Ellenberg, 1978; Lee *et al.*, 1981
57. Lindsay *et al.*, 1980
58. Currie *et al.*, 1971
59. Falconer, 1970
60. Sequiera *et al.*, 1979; Gannicliffe *et al.*, 1985
61. Spencer, 1981
62. Penfield and Jasper, 1954, p. 333
63. MacLean and Pribram, 1953; Pribram and MacLean, 1953
64. Theodore *et al.*, 1984; Engel 1984
65. Magistretti *et al.*, 1982
66. Wyler and Bolender, 1983

23

Phenomenology of Psychomotor Epilepsy

Basic and *Specific* Affects

In this and the next chapter we proceed to consider the subjective symptomatology of limbic epilepsy. Since this is the first time that the matter of subjectivity has come up since the introductory comments in Chapters 1 and 2, this might seem to be an appropriate place to deal with the question of whether or not subjective data have scientific, factual value. However, to tackle that question now would be premature because an explanation of what constitutes a scientific "fact" depends on a clarification of the subjective contribution of affect to mentation. In this respect the study of psychomotor epilepsy is an absolute necessity because no other condition provides such a range of information about cerebral structures involved in the *subjective* experience of affects. For the purpose of analyzing the symptomatology of psychomotor epilepsy it is desirable to provide a few operational definitions in regard to the psyche; subjectivity; kinds of psychological information; and the nature of affects.

Subjective Forms of Psychological Information

The "Psyche"

It will be recalled that the word *psyche* derives letter-for-letter from the Greek word for "breath" (ψυχή). In Greek times, air (Greek *pneuma;* Latin *anima*) was regarded as a universal, nonmaterial, and immortal life principle. Through its assimilation, organisms transform themselves into animated matter, as expressed by the Latin derived word *animal.* This interpretation found perpetuation in the meaning of the Anglo-Saxon word *soul.* Contemporary dictionaries variously define the psyche as "the human soul" or, more specifically, as "the mind, considered as an organic system . . . serving to adjust the total organism to the needs or demands of the environment."[1]

Recalling Wiener's statement (Chapter 2) that information is information, not matter or energy, one might say that the psyche is information and not matter or energy. Bertrand Russell[2] maintained that introspective data are scientifically inappropriate for investigation because they do not obey physical laws. If it were agreed, however, that introspective data represent psychological information, then such data obey at least one law that makes them amenable to scientific investigation. That law is the law of communication, which states that there can be no communication of information without the intermediary of what are recognized as physical, behaving entities (Chapter 2). It is the behaving entities of the brain that make known psychological information to the self, and it is the behaving entities of the nervous system that make possible the public communication of this information. One might say that entities and information are, in the present context, somewhat comparable to the particle and wave of quantum mechanics.

Subjectivity

Introspection suggests that it is the element of subjectivity that most clearly distinguishes psychological from other functions of the nervous system.[3] By subjectivity is meant the awareness that is attached to various forms of psychological information such as sensations and perceptions (see below). A philosopher such as Kant might have referred to subjectivity as an *a priori* "form of consciousness." It is doubtful that the condition of sleep offers any exception, because introspection reveals that in the stage of sleep associated with dreaming there coexists an element of subjectivity. It has yet to be shown that psychological processes transpire in the deep stage of sleep correlated with an electroencephalogram hardly distinguishable from that of a comatose state.

Somewhat like a lamp under the control of a potentiometer, subjectivity covers a wide range of brightness. The lowest level would correspond to the faintest glimmer of consciousness. Verbal descriptions attempting to define such a state tend to become tautologous like Wiener's definition of information. The words become meaningful only in terms of the behavioral correlates. When another human being or an animal reveals vocal, ocular, facial, or bodily evidence of being the least bit responsive or attentive, one concludes on the basis of subjective identification that a state of consciousness exists.[3] In recent years, it has become evident that even the subjective awareness of dreaming has its behavioral correlates in eye movements, muscle twitches, and autonomic changes.

It is sometimes argued that subjectivity is an unessential epiphenomenon—that the brain could perform everything it does without the need of subjectivity. It is pointed out, for example, that the brain takes action to prevent the body from falling "before there has been time to think about it." Or it may be stated that some public speakers may be able to give a lecture "with their minds being some place else." Such "balloonous" arguments are pricked by the realization that the mere existence of subjectivity means that it is an additional source of information that may be drawn upon for adapting to the environment.[3] In the case of human beings, language would hardly exist without the need for words to express subjective experience.

In comparing one's own experience with what has been stated in various psychological writings, one recognizes, in addition to the global subjective state, five main kinds of psychological information. In making allowance for varying terminology, these can be characterized under the headings of (1) sensations, (2) perceptions, (3) compulsions (including proclivities), (4) affects, and (5) conceptions (thoughts).

Sensations and Perceptions

Sensations represent the "raw" feelings associated with activation of "intero-ceptive and extero-ceptive fields."[4] In such Sherringtonian terms they fall into two broad classes of interoceptions and exteroceptions. They are distinguished in terms of quality (modality) and intensity. Individually or in combination, sensations become more informative as they are appreciated in terms of time and space. In such cerebral transformation they are introspectively recognized as perceptions. It may be presumed that sensations and perceptions are basic to the original generation of compulsions, affects, and conceptions, which, paralleling the triune development of the brain, would appear evolutionarily to represent a hierarchic order of information.[3]

Sensing and Perceiving vis-à-vis Mentation

In considering compulsions, affects, and conceptions, it is important to emphasize how they differ generally and fundamentally from sensations and perceptions. With the exception of such phenomena as afterimages, as well as certain pathological conditions such as causalgia, it is characteristic of sensations and perceptions that they occur contemporaneously with signals to the brain that originate from activation of receptors associated with specific afferent systems. Such conditions are not necessary for compulsions, affects, and thoughts. Although the latter forms of information may arise in association with specific incoming signals, they need not be contingent upon them. (This is not out of harmony with current assumptions that a state of consciousness depends on a continuing diffuse bombardment of the brain with signals from extero- and interoceptive systems.) Hence, compulsions, emotions, and thoughts can be subjectively distinguished from sensations and perceptions by their capacity to occur and persist "after-the-fact."[5] The unexplained process that makes this possible will hereafter be referred to as *mentation.* In terms of a behaving nervous system, one might say that mentation involves self-regenerating neural replica of events either as they first occurred or in some rearrangement. How the original ordering of the events is preserved (i.e., memorized) or reordered (i.e., imagined or conceived) remains a mystery. A further general distinction pertains to what Descartes[6] qualified as "lively" and Hume[7] otherwise called "vivacity." Sensations and perceptions are characterized by a sense of immediacy and vivacity. However, when compulsions, affects, and thoughts occur independently of signals from receptors, they characteristically lack these subjective attributes.

Analysis of Affects

The words *affects* and *emotions* are commonly used interchangeably. For reasons now to be explained, we shall use them to refer, respectively, to emotional experience and expression.

Descartes[6] introduced the word *emotion* in his 1649 essay on the "Passions of the Soul." He used it for the purpose of distinguishing bodily sensations associated with specific sensory systems from feelings identified with the so-called passions such as joy, sadness, love, and hate. Descartes regarded the pineal body as the seat of the immaterial soul because it was the only unpaired organ in the brain and as such was in a position to fuse impressions from the "double organs" such as the eyes "before arriving at the soul." Suspended like a little rudder above the aqueduct through which the animal spirits flowed from the anterior into the posterior ventricles, the pineal was in a position to be deflected in various directions. Descartes attributed the passions to "feelings" of the soul produced by unusual turbulence of the animal spirits. As he explained, "We may . . .

call them feelings because they are received into the soul in the same way as are the objects of our outside senses . . . ; but we can yet more accurately call them emotions of the soul . . . because of all the kinds of thought which it may have, there are no others which so powerfully agitate and disturb it as do these passions.''[8]

According to Galenic tradition, the animal spirits (the ancient counterpart of axonal transport, as well as nerve impulses) were manufactured and supplied to the brain by the heart. The spirits were believed to pass back and forth between the brain and all parts of the body through the tubular nerve filaments that opened into the ventricles of the brain through little pores. Sensations were conveyed by the animal spirits from the sense organs to the ventricles of the brain where they passed through pores into nerves leading to the muscles and organs of the body to produce appropriate responses. Reminiscent of the James–Lange theory of recent times, Descartes[6] conceived of the emotions as being aroused by excitement in the viscera (particularly the heart) and the somatic musculature. In situations conducive to passions there was a greater than usual excitement of the sensations, with a corresponding increase in the flow of animal spirits to the heart and other parts of the body. Such excitement, in turn, resulted in an increased feedback of the animal spirits to the anterior ventricle and enhancement of their flow past the pineal body into the posterior ventricle. The direction of flow and the degree of turbulence around the pineal body determined the nature of the emotion of the soul.

In the Cartesian sense, the word *emotion* covers both the subjective and expressive aspects of the so-called passions. Because of its mechanistic implications, it is particularly suitable for referring to the behavioral manifestations. Another word, however, is needed for identifying purely subjective aspects of emotion. The word *affect,* as usually defined, satisfies this purpose. In the strict sense, we can speak of ''affect'' only with respect to ourselves, because only we as individuals have direct access to the private signals necessary for subjectivity. The existence of affect in another person must be inferred solely on the basis of behavior (i.e., emotion) conveyed by publicly shared signals.

The Nature of Affects

The affects impart subjective information that is instrumental in guiding behavior required for self-preservation and preservation of the species.[9] The subjective awareness of affects is characterized by a sense of bodily pervasiveness or by feelings localized to certain parts of the body. Because of their apparent sensory qualities, affects are commonly referred to as feelings. Introspection reveals, however, that as opposed to simple sensations, affects always have the distinction of being either *subjectively agreeable* or *disagreeable.*[3] This negative and positive polarity of affects is schematized by the hemispheric labeling in Figure 23-1. There are no neutral affects, because emotionally speaking, it is impossible to feel unemotionally.[10] On the contrary, it is easy ''to think'' of null-affective states or other conditions of nothingness. Such considerations illustrate a fundamental distinction between affects and thoughts because it is evident that, in contrast to propositional thinking, there is no provision in affective mentation for reckoning with ''zero.'' There is the rejoinder to this argument that some people complain that they no longer have emotional feelings, but the complaint itself testifies to the disagreeable nature of this condition.

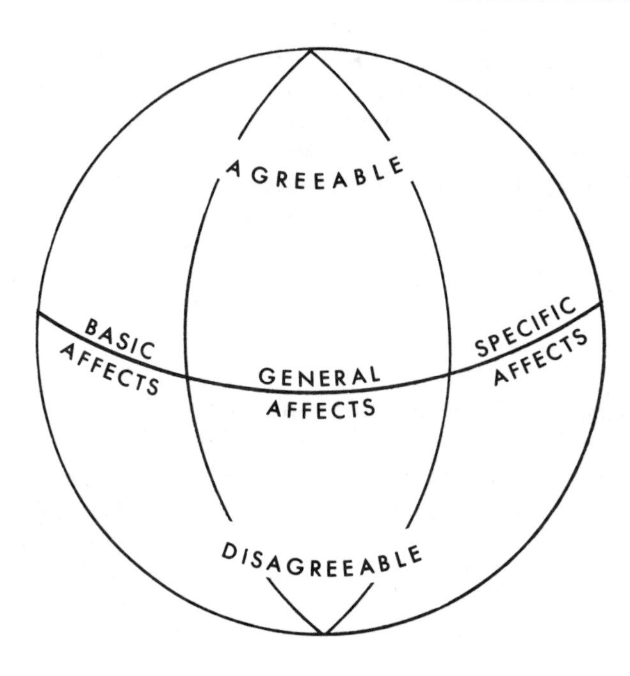

Figure 23-1. A scheme for analyzing affects. See text. From MacLean (1970a).

Three Kinds of Affects: Basic, Specific, and General

Given the two main classes of affects—agreeable and disagreeable—how can they be further qualified with respect to subjective information? All of the other main forms of psychological information—sensations, perceptions, compulsions, and thoughts—may be conjoined with affects of an agreeable or disagreeable nature. An attempt to specify the different varieties of affects along these lines, however, would amount to little more than compiling a thesaurus (see Chapter 24). What is desired is a workable classification for use in various investigations, including an identification of brain mechanisms underlying psychological experience.

One approach to this problem begins with an analysis of affects associated with the intero- and exteroceptive systems.[10] This approach has the appeal of lending itself to self-testing and self-analysis and making a comparison with what is experienced by other people. In this way one may arrive at an identification of three main forms of affects that are identified in Figure 23-1 as "basic," "specific," and "general."[10] The basic and specific affects depend on concurrent activation of interoceptive and exteroceptive systems. The basic affects derive from interoceptions signaling different kinds of internal states identified with basic bodily needs—namely, the needs for food, water, air, sexual outlet, sleep, and needs associated with various emunctories. The specific affects are so called because they apply to exteroceptions and perceptions immediately generated by activity in specific sensory systems. Some specific affects are unlearned, while others are conditioned. The former are illustrated by agreeable and disagreeable sensations such as fragrant or foul odors, harmonious or noisy sounds, while the latter include esthetic affects identified with agreeable and disagreeable aspects of music and various forms of art.

The general affects are distinctive feeling states that do not depend on immediate experience, since, owing to mentational processes, they may persist or recur "after-the-fact." They rank as general affects because they may apply to feelings aroused by other individuals, situations, or things. All of the general affects may be considered from the standpoint of self-preservation and the preservation of the species. Those general affects that are informative of threats to the self or species are disagreeable, whereas those that signal the removal of threats and gratification of needs are agreeable.

In this chapter attention will be confined to symptomatology of patients with limbic epilepsy that are relevant to the basic and specific affects.

Feelings Related to Basic Needs (Basic Affects)

Feelings immediately dependent on interoceptions are identified with the basic bodily needs. The basic needs are traditionally referred to in Freudian terms[11] as *instinctual drives*. In striving to maintain a purely objective attitude, some psychologists have variously defined drives as "strong stimuli" or "physiological states." Others define them as the amount of work an animal will do to obtain or avoid something, failing to suggest an equally important measure for the "drive" to *desist* from work in order to obtain or avoid something. In attempts at further clarification, some writers equate the driving stimuli with subjective states. In one text, for example, drives are defined as strong stimuli impelling action, as illustrated by "parching thirst," "pangs of hunger," and "stabbing pain."[12]

Since the word *drive* implies a force and has been used interchangeably to refer to behaving entities or subjective states, it is desirable to use another term for a purely informational condition that has no energy of itself, and about which only we as individuals have firsthand experience. Since the subjective awareness of a basic need is typically disagreeable (unpleasant) or agreeable (pleasant), it would be appropriate to refer to the subjective aspects of the basic needs as basic affects.[10]

For purposes of public comparison, a major difficulty in analyzing the basic affects is that most parts of the interoceptive systems are not directly accessible for observable, selective stimulation. Regardless of such limitations, there are indirect methods for obtaining a public comparison of most basic needs. By deprivation or by varying the amount of intake of food, water, air, and the like, observers may compare their subjective feelings resulting from different kinds of interoceptive stimulation. They may also compare their subjective symptoms following the administration of drugs known to affect particular interoceptive systems.

In psychological texts, the basic needs for food, water, and sexual expression have received primary consideration because the behaviors associated with them have proved to be particularly amenable to study. But also basic are the need to breathe, the need to rid the body of excreta and noxious agents, the need to move, and the need to rest and to sleep.

Curiously enough, the English language has only a few words for qualifying the feelings informative of basic needs. The familiar utterances are "hungry," "thirsty," "breathless," "restless" (for the need to move), "tired," and "sleepy" (for the needs to rest and to sleep). But there are no socially acceptable words strictly applicable to the urgent feelings associated with the need for sexual expression and with the emunctories such as clearing the respiratory passages, urinating, and defecating. For these and for most subjective states signaling the urges or the satisfaction of the basic needs, whole phrases are necessary.

Ictal Feelings Akin to Basic Affects

In psychomotor epilepsy, the ictal symptoms like those of basic affects are most commonly identified with the alimentary, cardiovascular, respiratory, and genitourinary systems. It should be emphasized that the symptoms may be the immediate result of the epileptic discharge in the brain and need not depend on secondary activation of various organs or systems. Some epileptics, for example, have undergone abdominal operations for the removal of an appendix[13] or the treatment of some other condition when there was actually no disease to account for the particular symptoms.

Alimentary Symptoms

The most common visceral symptom generated by an epileptogenic discharge in or near the limbic cortex associated with the amygdala is what neurologists have long referred to as an epigastric aura. In the last century the vagus nerve (great wandering nerve of Galen) was alternately known as the pneumogastric nerve. As Gowers commented in his textbook of 1888, "In many epileptic fits the central representations of the nerve are the parts through which the consciousness is first affected, and hence the so-called 'epigastric aura.' "[14] The term covers a multitude of symptoms that patients feel in the pit of the stomach or, seemingly, in the stomach itself. In one report such symptoms constituted part of the aura in nearly 40% of 155 cases.[15] As will be noted again in Chapter 24, an epigastric aura is commonly associated with intense emotional feelings identified as fear or terror. The epigastric sensations themselves are usually unpleasant and are only rarely described as having a pleasant quality. For example, the fluttering feeling of "butterflies" is usually associated with some degree of tension or apprehension, but one patient described her epigastric sensation as a "warm butterfly feeling."[13] When at a loss for words, some patients characterize the peculiar feeling as funny,[16] foolish,[16] or ticklish.[17] A feeling of nausea[18] (otherwise called a bilious[19] or sick[20] feeling) is common. Some patients experience a feeling of movement in the gastric region, describing it as a nervous,[21] shivering,[22] rumbling,[23] rolling,[24] or twitching[25] sensation. In other cases the sensation conveys a feeling of pain.[26] One patient complained of an abrupt, intense pain "like a knife."[27] Another compared the feeling to a pressure of "crushing intensity."[27] Sometimes the aura is a hollow feeling.[28]

Most relevant to the basic affects, patients may experience hunger or thirst as one of the aural manifestations. Daly[29] describes a patient who experienced a warm epigastric sensation, followed by a feeling of hunger. As opposed to the initial symptom in this case, a patient whom I treated experienced a feeling of sadness, followed by a feeling of hunger.[30] Penfield and Jasper,[31] noting that it is an uncommon symptom, describe two patients with thirst as part of the aura.

Oftentimes, the epigastric aura gives the impression of rising,[29] with the sensation reaching the chest, throat, mouth, or head. Upon reaching the throat, it may change into a sensation of taste[27] or choking.[27] Rarely, the sensation, instead of "rising," gives the impression of descending into the abdomen and sometimes into the genital region.[32]

Cardiac and Respiratory Symptoms

The next most common visceral symptoms identified with the aura are those that seem to involve the heart. Patients report that at the beginning of their seizures they may have a fluttering sensation in the region of the heart, a feeling of palpitation,[33] pounding,[34] or throbbing.[35] There may be a feeling that the heart is racing, although there may be no objective sign of it.[36] Ictal symptoms related to respiration are invariably unpleasant. They are described as sensations of choking,[37] smothering,[38] or suffocation.[39] As in the case of the epigastric aura, the cardiac and respiratory symptoms are commonly associated with unpleasant emotional feelings such as a feeling of fear[28] or depression.[28]

Genitourinary Symptoms

The urge to urinate[40] (as well as the urge to defecate[40]) is among the less common visceral symptoms experienced during the aura. Based on the experience of attending a seizure clinic, one would gain the impression that sexual symptoms during the aura are rare. Nevertheless, several cases characterized by epileptic sexual experiences have been reported.[41] Blumer[42] described a patient who experienced frequent trances characterized by a sensation of tightening of his body and a brief feeling of "sexual climax." The attacks were usually associated with a feeling of fear. The spells lasted about 10 to 20 sec and recurred as often as 10 times a day, occurring, apparently, without erection or ejaculation. Stevens[43] reported the case of a woman who experienced a visual illusion followed by palpitation and then vomiting or orgasm. A 38-year-old man experienced seizures beginning with a pleasant orgastic sensation triggered (either deliberately or accidentally) by the sight of a safety pin.[44] A 46-year-old woman experienced an aura that began with "a feeling in the stomach" and descended into the genital region; there was a feeling of intromission and orgasm, followed by a loss of consciousness.[45] Another woman described her aura as a feeling of itching in the perineum and a sensation "like a red hot poker being inserted into her vagina."[46] One woman experienced a "sexual feeling" on only one side of her genitalia.[47]

Symptoms of Fatigue

Finally, apropos of the basic need for rest and sleep, a patient at the beginning of a seizure may experience a tired[48] or a sleepy[25] feeling.

Comment

Because of an important symbolic parallel that will be discussed in connection with the general affects, it should be pointed out that the basic affects in general are informative either of the needs (1) *to assimilate* or (2) to *eliminate* (expel), as would be exemplified by eating or defecating. In this respect, the whole organism mimics what have been the fundamental processes of individual cells since earliest evolutionary times. These bipolar functions might not be immediately apparent in regard to sexual needs if it were not pointed out that ejaculation is equivalent to expelling, whereas the feminine role in copulation is comparable to assimilation. The needs for rest and sleep are correlated with the "elimination" of catabolites and the "assimilation" of anabolites.

Ictal Symptoms Often Akin to Specific Affects

There are many cases in which the outstanding symptoms during the aura are identified with one or more of the special senses.[49] The symptoms range from crude sensations to complex hallucinatory experiences that are often affective in nature. Malamud described an unusual case in which, at one time or another, there were symptoms referable to all five of the sensory systems—olfactory, gustatory, visual, auditory, and somatic.[50] The patient was a 26-year-old man who was found to have a small tumor in the amygdalohippocampal region.

Olfactory Symptoms

Olfactory sensations are almost invariably described as disagreeable,[51] unpleasant,[52] or terrible.[53] Patients commonly associate the phantom odor with something burning—like burning meat,[54] rubber,[55] or paint.[56] But there are reports of many other types of odors as well, such as the smell of blood,[29] fish,[57] chloride of lime,[53] phosphorus,[53] or acid fumes.[58] One patient compared the odor to that of a chicken coop,[59] while another said that the smell was like that of "stuffed cabbage in a dirty outhouse."[60] Occasionally, odors seem to be confused with other kinds of sensations: in trying to describe the abnormal symptoms, for example, a patient may say that "it smells sour"[61] or that it smells like "something cold."[61] Williams cites the case of a man who "smelled a smell like the sound of blue thunder."[28] Rarely, one finds a report of pleasant olfactory experiences: in conjunction with visual hallucinations a young physician experienced a pleasant odor of incense.[29] Sometimes the experienced odor gives the impression of strangeness[62] or greater awareness.[63]

Gustatory Symptoms

Gustatory sensations are also usually referred to as disagreeable,[29] unpleasant,[64] or terrible.[65] Tastes are variously characterized as bad,[66] bitter,[67] metallic,[29] neutral,[29] rotten,[13] rough,[67] and sour.[68] Rarely, as in the case of olfactory symptoms, a taste will be described as pleasant: A 44-year-old man with a calcified lesion deep in the temporal lobe experienced a dreamlike "faraway feeling" associated with a peculiar smell and taste that reminded him of pines.[29]

Auditory and Vestibular Symptoms

In regard to auditory sensations, a patient may report that at the beginning of a seizure sounds seem to get louder or fainter, nearer or more distant.[69] A 36-year-old man "was driving in his automobile with his children and stopped at a traffic light. Suddenly the sunlight appeared to be more intense, and he felt that his perception was unusually acute."[70] Then *sounds became distant* and "he was nauseated, sweated, and had an urge to defecate.''[70] There may be actual hallucinations such as hearing ''an inside voice,''[49] ''a ringing bell,''[71] a piece of ''music.''[72]

Symptoms associated with balancing functions of the vestibular apparatus are described as dizziness,[49] floating,[66] sinking,[66] and falling.[73] Here again, as with other sensory experiences, the sensation may be elaborated into a complex feeling state. One patient experienced a sinking feeling like falling into a tunnel[66]; another had the sense of falling into a mountain.[74] L. F. Chapman *et al.*[75] reported the case of a man in whom electrical stimulation of the left mid-hippocampal gyrus re-created the feeling that he experienced as a child upon falling into a hole.

Visual Sensations

As with auditory sensations, a patient may have visual experiences in which things seem to be near or far, large or small.[76] A man who experienced a psychomotor seizure while watching a motion picture had the impression that the image on the screen seemed ''to be coming very close to him.''[77] There may be hallucinations. An English woman described by Foster Kennedy[78] always saw at the beginning of an attack a room filled with Negroes. Since she claimed that she had never seen a black person, it was difficult to explain this hallucination. Further inquiry about her past history revealed that she had been born in Jamaica and had lived there during the first 2 years of her life. As will be noted in the next chapter, visual experiences at the beginning of a seizure may be compounded with elaborate mental states.

Bodily Sensations

In records of patients with psychomotor epilepsy, one finds that virtually every part of the body is mentioned in connection with unnatural sensations occurring during the aura. Feelings referred to the body as a whole have been described as a sense of pressure,[63] heaviness,[66] numbness,[50] tenseness,[70] or fatigue.[79] Somewhat like the distortions of perception mentioned in connection with hearing and vision, the body may seem to be becoming smaller or taller. ''My body,'' said one patient, ''felt much longer than it is,''[80] recalling the familiar example of delirium in *Alice in Wonderland*.

Symptoms involving the head (the so-called cephalic auras) are the most common somatic manifestations in psychomotor epilepsy.[81] The sensation may be compared to a tight band around the head,[82] an empty feeling,[83] a funny feeling,[83] a cold feeling,[84] a heavy feeling,[66] a feeling of pressure,[85] a tingling feeling,[25] or a feeling of movement across the forehead.[25] Sometimes there is actually a feeling of pain[86] that may be one-sided.[80] Burning,[29] tickling,[87] or tingling[88] feelings may be perceived as either just outside or inside the nose. Pulling sensations in the jaw have been described.[66]

Mention has already been made of some of the sensations referring to the throat. Other symptoms include a feeling of a lump in the throat[28]; a burning[29] or painful[89] feeling; or a sensation of dryness.[28]

A sensation of pain[89] or a tight[90] feeling may be experienced in the chest. One patient described the sequences of his feelings, involving pain in his chest, as follows: ''I begin to tremble. It feels like a fear spasm. Something seems about to happen, with the next step uncertain. A pain goes from my chest into my throat. Every fiber in my body feels like screaming, like I'm torn apart.''[91]

In some cases sensations are felt in one or more extremities and are usually characterized as tingling[92] or painful[89] or like electricity.[93] A hand and arm may feel as though bandaged[57] or there may be a spreading feeling of pins and needles[94] of the kind sensed when a limb goes "asleep." As in delirium, part of an extremity such as the hand may have the feeling of becoming enlarged.[95] One patient reported the feeling that a Coca-Cola bottle in his hand seemed to be growing bigger.[63] Another had the impression of something in his hand that felt like the surface of burnt wood.[80]

Thermoregulatory Manifestations

As periodically stated like a recurring theme, it has been a curious evolutionary development that many of the signs and feelings of emotion are the same as those that accompany changes induced by thermoregulatory mechanisms. During the period of the aura some patients complain of feeling warm or cold all over.[83] Others may experience local changes such as flushing of the face,[96] a sense of cold in the forehead,[84] or a shivery feeling in the region of the stomach.[97] Since vasomotor changes during the aura commonly result in flushing or pallor of the face and other parts of the body, it is possible that symptoms of the above kind might be explained by reflex sensations arising from autonomic changes. "Gooseskin,"[28] shivering,[98] and sweating[99] may be observed in such cases. But as mentioned earlier, the subjective symptoms might also be the direct result of the cerebral discharge, being experienced before the peripheral manifestations had time to appear. The feeling of chills-going-up-and-down-the-spine[29] that some patients experience during an aura are often equated with fear. In other cases such symptoms are suggestive of a suddenly aroused pleasant state of excitement such as accompanies the feeling of patriotism triggered by the sight of a flag passing by in procession or by hearing the national anthem. Such feelings are usually attributable to reflex sensations resulting from the action of piloerector muscles along the upper part of the back and which in furry animals cause the hair to stand on end. One patient, in describing his aura to me, said the feeling was like pigeons walking on the back of his neck.[25] Others compare it to the sensation of having something crawling up the spine.[29] Such feelings have been elicited at operation by stimulation of the insular region.[100]

In tense situations some men are aware of contraction of cremasteric muscles that raise the testicles into a protected position, an action that also occurs under conditions of cold. In an analysis of 50 cases with psychomotor epilepsy, Feindel and Penfield cited 2 cases (4%) with scrotal sensations during the aura.[101] In one of these patients (Case 9) who experienced a "drawing up of the scrotum as the warning of his seizure," there was a yellowing and atrophy of the uncus and hippocampus and first temporal convolution.

Comment

Of all the affects, those associated with the exteroceptions and perceptions are most easily subject to public demonstration and comparison. This is because appropriate stimuli can be publicly applied to specific receptors of the special senses.

The question arises as to whether or not *all* forms of sensation and perception are imbued with affect. Sherrington's chapter on "Cutaneous Sensations" in Schäfer's classical *Text-Book of Physiology* published in 1900 provides an interesting historical context in which to consider this and related questions.[102] Sherrington commented,

"Mind rarely, probably never, perceives any object with absolute indifference, that is, without 'feeling.' In other words, affective tone is the constant accompaniment of sensation. . . . All are linked closely to emotion."[103] A few pages before, however, he had said, "With the minimal sensation the agreeableness is *nil*. . . ."[104] The possibility of drawing a clear distinction between sensation and affect is illustrated under certain pathological conditions. von Economo,[105] for example, described a patient who, following encephalitis lethargica, could recognize objects as cold, but denied experiencing any of the disagreeable feelings of cold.

Sherrington reviewed the findings of Weber, Fechner, and others on the relations of the intensity of sensation to the strength of a stimulus. When combined, the laws of Weber and Fechner state that for a particular sensation to increase arithmetically by equal amounts, the strength of the stimulus must increase geometrically.

Sherrington pointed out that sensations of every kind "become unpleasant when their intensity exceeds a certain degree."[106] As examples, he mentions an "intense light" and a "piercing note." "Some species of sensations," he explains, "*e.g.* certain tastes and odours, seem 'unpleasant' even at the threshold of the sensible intensity of their stimuli; and, similarly, others seem in other individuals to possess from very limen up to maximum a pleasurable, certainly no painful, character, *e.g.* the taste of 'sweet.'"[107]

Sherrington drew a distinction between the "projicient" and "non-projicient senses."[102] By the former he meant the teleceptive auditory and visual senses, each with their so-called distance receptors. Sensations of the "non-projicient" senses that are often associated with affects are fragrant and acrid odors; sweet and bitter tastes; somatic feelings of pain and tickle, cold and warmth, smooth and rough. The vertiginous and floating feelings associated with the vestibular sense have distinctive affective qualities. Many of the affects connected with the nonprojicient senses appear to be unlearned.

Sherrington noted that the sensations "belonging to the projicient senses are less strongly [emotionally] coloured than those of the non-projicient."[108] But he seems to have underestimated music. In connection with the projicient senses are affective feelings associated with loud, soft, harmonious, and discordant sounds; colors, variations of light and shade, utter darkness, shape and pattern. Because of the vague character of many of these conditions Sherrington referred to projicient senses as "wordless"; they must be experienced in order to be appreciated. But it may be pointed out that some people are prone to wince when hearing the two words "discordant sound" and others may grimace upon hearing the word "bitterness." Such examples illustrate that a conditioned, as well as unconditioned specific affect can be evoked by the symbolic representation.

In anticipation of the consideration of general affects, it should be noted that the affects of exteroception and perception are basic to the so-called aesthetic or cultural affects because they signal an agreeable sense of harmony, rhythm, equilibration, focus, form, taste, and so forth, that are essential to the appreciation of music, art, and other cultural enjoyments. Considered as a group, and characterized in simplest terms, the affects of exteroception and perception might be regarded as the affects of agreeable and disagreeable recognition.

Summarizing Statement

For the purpose of analyzing the symptomatology of psychomotor epilepsy, it has been necessary to provide some operational definitions with respect to the psyche, subjectivity, five main kinds of psychological information, and the nature of emotional feelings (affects).

Conjoined with the element of subjectivity, the main forms of psychological information may be identified as sensations, perceptions, compulsions, affects, and conceptions. As opposed to sensations and perceptions, the other three main classes of psychic information have the capacity to occur and persist after-the-fact.

Affects have the subjective distinction of being either pleasant or unpleasant. By definition there are no neutral affects. The term *affect* is reserved for the subjective experience of emotional feelings, while Descartes's word, *emotion,* is used to denote emotional expression. Affects may occur in conjunction with all other forms of psychic information. Through the introspective process, three main forms of affects are recognized—namely, (1) those identified with basic needs incited by interoceptions and called basic affects; (2) those associated with specific sensory systems and referred to as specific affects; and (3) those that rank as general affects because they may apply to individuals, situations, and things. In the present chapter, a description is given of the ictal phenomenology indicative of the involvement of the limbic system in the elaboration of basic and specific affects.

References

1. *Webster's New World Dictionary,* 1968, p. 1175
2. Russell, 1921
3. MacLean, 1960a
4. Sherrington, 1906/1947, p. 317
5. MacLean, 1975d
6. Descartes, 1649
7. Hume, 1739/1888
8. Descartes, 1649, p. 344
9. MacLean, 1960a, 1969a, 1970a
10. MacLean, 1969a, 1970a
11. Freud, 1900/1953
12. Dollard and Miller, 1950
13. Mulder and Daly, 1952
14. Gowers, 1888, Vol. 2, p. 255
15. Feindel and Penfield, 1954
16. MacLean and Arellano, 1950; Chessick and Bolin, 1962; Ferguson *et al.*, 1969
17. Hunter *et al.*, 1963
18. Jackson, 1888/1958; Shenkin and Lewey, 1943; Kershman, 1949; Penfield and Jasper, 1954, p. 107; Williams, 1956; Glaser, 1957; Daly, 1958; Van Reeth, 1959; Maspes and Pagni, 1964; Kamrin, 1966
19. Kershman, 1949
20. Penfield and Jasper, 1954, p. 431; Bennett, 1965
21. Penfield and Jasper, 1954, p. 430
22. MacLean and Arellano, 1950; Williams, 1956
23. Penfield and Jasper, 1954, pp. 427, 429
24. Glaser, 1957; Daly, 1958
25. MacLean and Arellano, 1950
26. Jackson, 1888/1958; Penfield and Jasper, 1954, p. 427; Glaser, 1957, 1967
27. Mulder *et al.*, 1954
28. Williams, 1956
29. Daly, 1958
30. MacLean, 1952
31. Penfield and Jasper, 1954, pp. 423, 457
32. Van Reeth, 1959
33. Penfield and Jasper, 1954; Rodin *et al.*, 1955; Stevens, 1957; Daly, 1958; Glaser, 1967; Richardson and Winokur, 1967
34. Williams, 1956; Van Buren and Ajmone Marsan, 1960; Wadeson, 1965
35. Rodin *et al.*, 1955
36. Jasper, personal communication
37. Hamilton, 1882; Rodin *et al.*, 1955; Anastasopoulos *et al.*, 1959
38. Rodin *et al.*, 1955; Serafetinides and Falconer, 1962
39. Hamilton, 1882; Anderson, 1886; Jackson and Beevor, 1889
40. Penfield and Jasper, 1954; Feindel and Penfield, 1954; Williams, 1956; Daly, 1958
41. Stevens, 1957; Daly, 1958; Van Reeth, 1959; Ehret and Schneider, 1961; Serafetinides and Falconer, 1962; Heath, 1963; Blumer, 1970
42. Blumer, 1970
43. Stevens, 1957
44. Serafetinides and Falconer, 1962
45. Van Reeth, 1959
46. Freemon and Nevis, 1969
47. Erickson, 1945
48. Penfield and Jasper, 1954; Crandall *et al.*, 1963
49. Penfield and Jasper, 1954
50. Malamud, 1966
51. Hamilton, 1882
52. Henry, 1963; Hierons and Saunders, 1966; Kamrin, 1966
53. Jackson, 1888/1958
54. Ironside, 1956; Kamrin, 1966
55. Penfield and Jasper, 1954, p. 109
56. Penfield and Perot, 1963
57. Ferguson *et al.*, 1969
58. Maspes and Pagni, 1964
59. Jackson and Beevor, 1889
60. Weil, 1959
61. Chitanondh, 1966
62. Jasper, 1964
63. Mullan and Penfield, 1959
64. Williams, 1956; Kamrin, 1966; Hierons and Saunders, 1966
65. Jackson, 1888/1958
66. Crandall *et al.*, 1963

67. Anderson, 1886
68. Shenkin and Lewey, 1943
69. Penfield and Jasper, 1954; Williams, 1956; Daly, 1958
70. Daly, 1958, p. 102
71. Wadeson, 1965
72. Rodin *et al.*, 1955; Penfield and Perot, 1963; Chapman *et al.*, 1967
73. Jackson, 1880/1958; Chapman *et al.*, 1967
74. Hallen, 1954
75. Chapman *et al.*, 1967
76. Jackson and Stewart, 1899/1958; Penfield and Jasper, 1954; Williams, 1956; Lehtinen and Kivalo, 1965; Ferguson *et al.*, 1969
77. Penfield and Jasper, 1954, pp. 440–441
78. Kennedy, 1932
79. Penfield and Jasper, 1954, p. 363; Crandall *et al.*, 1963
80. Lehtinen and Kivalo, 1965
81. Janz, 1955; Penfield and Perot, 1963; Ajmone Marsan and Goldhammer, 1973; King and Ajmone Marsan, 1977
82. MacLean, unpublished case material
83. Rodin *et al.*, 1955
84. Ferguson, 1962
85. Remillard *et al.*, 1974
86. Penfield and Jasper, 1954, p. 427
87. Hunter *et al.*, 1963
88. MacLean and Arellano, 1950; Maspes and Pagni, 1964; Daly, 1958; Hierons and Saunders, 1966
89. Chessick and Bolin, 1962
90. Anastasopoulos *et al.*, 1959; Kamrin, 1966
91. Chessick and Bolin, 1962, pp. 74–75
92. MacLean and Arellano, 1950; Penfield and Jasper, 1954, p. 429; Daly, 1958; Maspes and Pagni, 1964; Hierons and Saunders, 1966
93. Pagni and Marossero, 1965
94. Penfield and Perot, 1963, pp. 625–626
95. Anderson, 1886
96. Golub *et al.*, 1951; Rodin *et al.*, 1955
97. Glaser, 1957; Daly, 1958
98. MacLean and Arellano, 1950; Williams, 1956
99. Van Buren and Ajmone Marsan, 1960
100. Penfield and Jasper, 1954, p. 429
101. Feindel and Penfield, 1954, Case 9
102. Sherrington, 1900
103. Sherrington, 1900, p. 974
104. Sherrington, 1900, p. 967
105. von Economo, 1931
106. Sherrington, 1900, p. 966
107. Sherrington, 1900, p. 968
108. Sherrington, 1900, p. 969

24

Phenomenology of Psychomotor Epilepsy

General Affects

The ictal symptoms next to be considered pertain to affects that are conventionally regarded as emotional feelings. In the present scheme illustrated in Figure 23-1, such feelings are referred to as general affects.

The Nature of General Affects

Inclusive of the two distinctions explained in the preceding chapter, the general affects differ from the basic and specific affects in four main respects:

1. They merit the term *general* because they may apply to individuals, situations, and things.
2. They may be incited directly by either exteroceptive or interoceptive stimulation, and, furthermore, owing to mentation, they may occur and persist "after-the-fact." The enduring anger of Achilles offers a classic example.
3. The general affects transcend the self, giving goal-directing information as to how near or how far other individuals or social groups are from succeeding or failing in various activities.
4. Symbolically, the general affects are like the basic affects with respect to the bipolarity of assimilation and elimination. There is the significant difference that because of a one-to-many relationship (i.e., an individual reacting with individuals) the assimilative and eliminative aspects must be considered in both the active and passive sense. In other words, the combinations of possibilities are such that the individual or a group of individuals may be actively or passively involved in the assimilative process or actively or passively involved in the eliminative process. Marriage, for example, may be an active or passive process with respect to two individuals, whereas the assimilative and eliminative process existing in groups would be illustrated by a person's being accepted into, or rejected by, a social organization.

For purposes of self-testing, self-analysis, and public comparison, the general affects present special difficulties because unlike the basic and specific affects, there are no specific gateways to the sensorium for inciting different affects. Given such limitations, how does one affix a handle to the general affects for dealing with them scientifically? The

This chapter provided the substance of an article entitled "Ictal Symptoms Relating to the Nature of Affects and Their Cerebral Substrate" that appeared in a volume on emotion (MacLean, 1986b).

historical approach provides a means for identifying the kinds of feelings in question, after which one may attempt to devise a formula for their categorization.

From time to time philosophers have attempted to arrive at a basic number of affects, out of which, like the hues obtained by mixing primary colors, all others may be derived. Spinoza[1] recognized three basic "emotions," which he identified as desire, pleasure, and pain. According to our scheme portrayed in Figure 23-1, his "pleasure" and "pain" would apply not to a particular affect, but rather to the two main subclasses of affects denoted as "agreeable" or "disagreeable."

Descartes[2] identified six "simple passions" and listed them as wonder, love, hatred, desire, joy, and sadness, concluding that "all the others are composed of some of these six, or are species of them."[3] He considered anger to be a species of hatred.

Hume[4] drew a vague line of distinction between what he called *direct* and *indirect* passions, but attributed to all of them the attributes of "good or evil" or "pain or pleasure." William James[5] recognized two categories of "emotions," saying, "I shall limit myself in the first instance to what may be called the *coarser* emotions, grief, fear, rage, love, in which everyone recognizes a strong organic reverberation, and afterwards speak of the *subtler* emotions, or of those whose organic reverberation is less obvious and strong."[6] Among the subtler emotions he included "the moral, intellectual, and aesthetic feelings."[7]

In tackling a classification of affective states in 1913, Jaspers commented, "Extensive analysis of every different kind of a feeling would only end in a vast array of trivialities."[8] In reviewing recent contributions, Plutchik[9] cites a number of authors who have compiled lists of several score of emotions. Plutchik himself concludes that there are eight primary emotions oppositely paired as anticipation–surprise, anger–fear, disgust–acceptance, and sadness–joy.[9]

It helps to narrow down and to categorize the general affects if they are considered broadly in terms of self-preservation and the preservation of the species.[10] The general affects generated by threats to the self or species (or by a decline in well-being) fall into the category of disagreeable affects, as for example, affective states identified as anger, fear, and sorrow. In the opposite category are affects informative of the removal of threats, the promotion of well-being, and the achievement of goals. The words *joy* and *affection* would apply to such affective states.

Six Main General Affects

Since there are no specific gateways to the sensorium for arousing general affects, is there any experimental approach for identifying a basic number of affects? When it is recalled that the communication of subjective states must be manifest by some form of behavior, it widens the possibilities for dealing with this kind of problem. Although as human beings we derive the greatest amount of information through the verbal communication of subjective states, it is evident that in the realm of emotion, prosematic (nonverbal) behavior is a highly effective conveyor of information. This granted, the investigation of emotions and their inferred underlying subjective states (affects) may be widened to include animals as well as human beings. Drawing upon ethological descriptions of mammalian behavior, one can identify six main forms of behavior that we identify with the six well-recognized general affects. These six forms of behavior are (1) searching, (2) aggressive, (3) protective, (4) dejected, (5) gratulant, and (6) caressive.[10] Corresponding words that would be broadly descriptive of the associated affective states are (1) desire,

(2) anger, (3) fear, (4) sorrow, (5) joy, and (6) affection.[11] Symbolic language and the introspective process make it possible to identify many variations of these affects, but in ethological investigations of prosematic behavior in both animals and human beings, inferences about the existence in them of any particular emotional state must be based largely on the six stipulated general types of behavior.

Other Questions of Clarification

Of other questions relating to the classification of general affects, consideration may be limited here to a few remarks about the particular nature of (1) feelings of familiarity and strangeness, (2) anxiety, and (3) mood.

Familiarity and Strangeness

It is of interest to note that with the exception of wonder, the above-mentioned affects are the same as the six "simple passions" listed by Descartes. Descartes's inclusion of wonder calls one's attention to two truant affects that do not lend themselves to any classification thus far considered and which may be identified as a "feeling of familiarity" and a "feeling of strangeness." Both of these affects may be associated with the three main classes of affects—basic, specific, and general—and each may on occasion be either agreeable or disagreeable. In this latter respect (and viewed in reference to Figure 23-1) they compare to switches that may occur in the earth's magnetic polarity. Judged by prosematic behavior alone, there are many occasions when one might confidently infer that an animal or person had happened upon something familiar or strange, but there would be other instances when the behavior would suggest feelings of fear or gratification. A special point is made of this here because during the auras of psychomotor epilepsy a feeling of familiarity or strangeness (or an alternation of the two states) is a rather common symptom. As human beings we recognize that these affects themselves are quite distinctive from any that have been mentioned. Because of their ambiguous nature, they will be dealt with here as though they could not be partitioned in any part of the globe shown in Figure 23-1, but occupied a position comparable to the earth's atmosphere.

A distinction is commonly made between anxiety and other forms of emotional feeling. Anxiety may be defined as the unpleasant feeling that accompanies alerting for, and anticipation of, future events (see Chapter 28). As such, it would rank as a general affect akin to fear.

A distinction is also made between various moods and emotional feelings. Introspective analysis suggests, however, that moods are equivalant to various forms of general affects extended in time.[10] Chronic depression (dejection) would serve as an example.

Ictal Varieties of General Affects

For the purpose of considering the complex mental states experienced during the auras of psychomotor epilepsy, it will be convenient to group them according to the six main types of general affects that have been discussed. Accordingly, the symptomatology will be described under six successive headings—namely, (1) feelings of desire, (2) feelings of a fearful nature, (3) feelings akin to anger, (4) dejected feelings, (5) gratulant feelings, and (6) feelings of affection.

Feelings of Desire

The automatisms of psychomotor epilepsy are frequently of short duration and sometimes of such a trivial nature (e.g., fingering one's tie) as not to attract notice. During such episodes, however, the patient commonly gives the impression of looking for something. In the light of such behavior, it might be presumed that an aura, characterized by the feeling of wanting something, would be quite common. That such has proved not to be the case in reviewing numerous case histories may be owing in part to the failure of the interviewer to educe such information. It must be remembered that the main goal in history taking in a seizure clinic is to learn whether or not a patient has epilepsy and in the case of positive findings to establish the type of epilepsy. It is therefore understandable that many subtleties, in regard to both the subjective and objective manifestations of psychomotor epilepsy, have not received the attention they deserve. In his 1901 textbook on epilepsy, Gowers[12] had reference to 1000 cases of convulsive disorders, in 505 of which there was a warning. In describing the symptoms he stated, "A sudden intense desire to be alone is sometimes described. Another of the infinitely various states that occur was a vague but strong feeling that something was wanted, and that an effort must be made to obtain it. . . ."[13] Such symptoms have been described in the recent literature in which the diagnosis of psychomotor epilepsy has been confirmed by electroencephalography. Of cases of this type, I recall a patient J.L., a machinist, who referred to his epigastrium a feeling of "wanting somebody near him."[14] This desire contrasted with that of another patient, a young housewife, who described a feeling inside her head of "wanting to be alone,"[14] followed by an automatism in which she went to the bathroom and urinated or defecated, sometimes on the floor. Other instances of the feeling of desire are illustrated by a school teacher with the feeling that "I must get to the bottom of it"[15] and by a man experiencing the sense that "I want to go through with it."[16]

Fearful Feelings

Most neurologists agree that a feeling of some kind of fear is one of the most common aural symptoms of psychomotor epilepsy. In five representative papers,[17] for example, in which an analysis was made of the incidence of different types of auras, fear was listed as the most common emotional symptom, ranging from 10 to 37%. Anyone who has not treated patients suffering from psychomotor epilepsy, might suppose that their feelings of fear were a natural response to the warning of an impending seizure. Hughlings Jackson, the noted English neurologist (1835–1911), discussed the commonly held view that "the patient is naturally frightened because experience tells him that a fit is coming on."[18] He argued that there is no support for such a supposition since the patients themselves "usually repudiate this interpretation."

In a study on the nature of fear in epilepsy, Macrae[19] cites the case of a 23-year-old medical student who had been subject to psychomotor attacks for a year and a half. It is stated that "he invariably had an aura consisting of a peculiar sensation in the epigastrium and an intense feeling of fear."[20] He described the fear as having an "abrupt onset, more intense and unpleasant than any fear he otherwise ever experienced," emphasizing that it was "different from any normal fear, inasmuch as it arose for no apparent reason and irrespective of present mood or thought or situation."[20] "Yet," he explained, "it had a quality of peculiar familiarity 'as though I had previously experienced it, even though at the time I know that this is not so.' " He had no problem in distinguishing this unaccountable fear from "the anxiety he might feel concerning the imminence of the fit. . . ." The strange fear might last only 15 to 20 sec and be followed by a feeling that he was hearing voices and that the radio sounded strangely different. At this stage he went into a cold sweat and showed pallor of the face.

A 42-year-old newspaper editor described his aura as "a feeling of impending disaster. It's a horrible sensation; I can't describe it; a feeling that I'm going stark, raving mad."[21] In citing this, the first case in his perceptive paper on "ictal affect," Daly comments that "fear is the emotion most commonly described."[22]

In reviewing the records of 2000 epileptics, Dennis Williams[23] identified 100 cases (5%) in which some kind of emotion was experienced as part of the epileptic attack. Fear was the initial symptom in 14 cases (14%). In all these cases, there was evidence of disease in the temporal lobe. Other neurologists have reported occasional cases with lesions in other lobes and in which discharge could have involved the limbic cortex and in which there were aural feelings of fear.[24] In a study of 40 cases with an aura of fear, Macrae[19] found that epigastric sensations anteceded the feeling of fear in 10 patients (25%). In various case histories the words used to describe the fear add up to an extensive list. Some patients refer to the fear as awful,[22] horrible,[22] indescribable,[25] intense,[22] while others say they have a feeling of panic[26] or terror.[27] Others simply say they feel afraid,[28] anxious,[29] frightened,[30] or that they have a scared feeling,[31] a sinking feeling,[32] or a feeling of hollowness.[23] For others the feeling is one of apprehension,[33] foreboding,[34] a feeling of impending disaster,[22] and a feeling of impending death.[35] Along with such fearful feelings there may be the compulsion or urge to flee, to run, or to seek someone's help.[22,23]

During the removal of part of the temporal lobe in one case, Green and Scheetz found that upon "stimulation of the anterior portion of the hippocampus . . . the patient immediately cried out, 'That's the feeling! It's an indescribable fear. I'm going to have an attack.' "[36] This response was elicited on four occasions. Jasper[37] describes a case in which stimulation of either the hippocampus or the amygdala elicited a "scared feeling" that the patient associated with the spontaneous attacks. As indicated in Figure 24-1, the feeling occurred when the afterdischarge in the hippocampus spread to the amygdala. Upon stimulation of the amygdala (Figure 24-1), the patient experienced the scared feeling at the moment an afterdischarge appeared in the amygdala. This finding suggests that the amygdala is farther downstream along the lines of communication leading to wherever the "viewer" resides in the brain. It should be noted that in this case, as in several others reported in the literature, the limbic discharge did not propagate to the neocortex.

Paranoid Feelings

There are several other complex mental states experienced during the aura that may be regarded as having shadings of fear. Paranoid feelings fall into this category. Daly[22] describes a number of such cases: One patient had attacks of inexplicable fear accompanied by the feeling that people were saying derogatory things about him. A 10-year-old boy experienced a feeling as if some men were after him. The aura of a young girl began with a sense of fear and a feeling that "a man is going to grab me." Another of Daly's patients had attacks that began with a feeling of fear followed by a conviction that someone was standing behind him. Several cases of this last kind have been described.

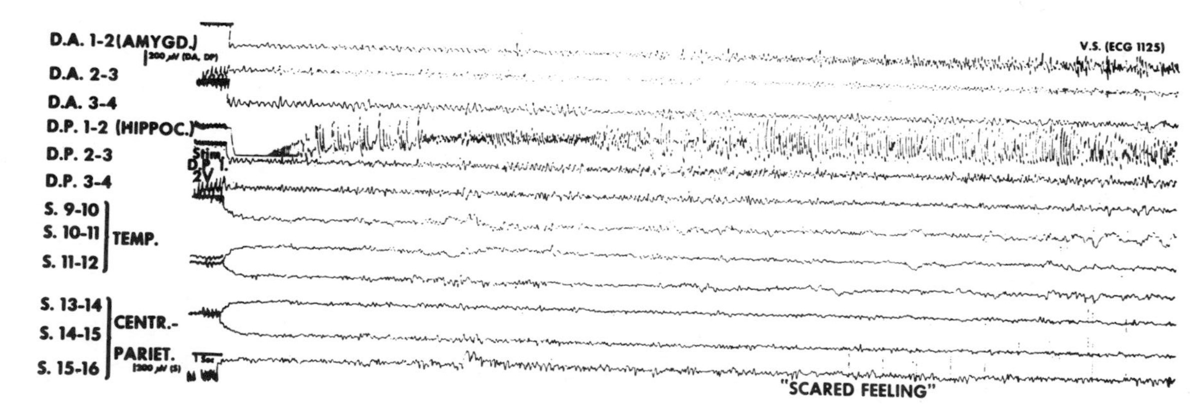

Figure 24-1. Depth and cortical recordings from case illustrated by Jasper, showing buildup of hippocampal afterdischarge induced by hippocampal stimulation. Coincident with buildup of high-voltage hippocampal potentials and a propagation of discharge to amygdala, patient experienced a "scared feeling." Abbreviations: D.A.1-2, depth anterior electrodes in amygdala; D.P.1-2, depth posterior electrodes in hippocampus; S. refers to cortical surface electrodes in temporal, central, and parietal regions. Calibrations: vertical bar, 200 μV; horizontal bar: 1 sec. From Jasper (1964).

Williams,[23] for example, gives an account of one man (Case 7) who experienced an intense feeling that there was someone behind his right shoulder. A woman (Case 24) had seizures beginning with a "hollow feeling" followed by a feeling of futility ("is it all worthwhile?"), and then a feeling as though she were standing behind her own right shoulder and could see herself from the back.[23] I, myself, treated a young man who experienced during the aura a feeling that someone was standing behind him.[38] If he turned to see who it was, the feeling became more intense and he would go into a generalized convulsion. He discovered that by resisting the impulse to see who was standing behind him, he could prevent a generalized seizure. A 32-year-old woman described by Mullan and Penfield[39] had attacks in which she saw herself as a little girl walking across the field. "Suddenly she felt as though someone from behind was going to smother her. . . , and she felt frightened all over."[40] At the time of operation, stimulation deep within the temporal lobe re-created the feeling of fear. "She seemed disturbed, wept, was terribly afraid, and looked afraid."[40] In most of such cases the patient realizes that the feelings are not real.[41] A number of patients experience paranoid feelings between seizures, and their symptomatology may bear such a close resemblance to schizophrenia as to be misdiagnosed as such.

Other Feelings of a Fearful Nature

Other aural symptoms with shadings of fear are feelings of unreality,[42] depersonalization,[43] and strangeness.[44] Penfield and Jasper[45] described a patient with an epileptogenic focus in the undersurface of the right temporal lobe who during the aura felt terrified and as though in another world.[46] In most cases it is difficult to obtain good subjective reporting of what the patient means by a feeling of unreality. They may use such expressions as "far off"[47] or "out of this world."[48] Sometimes a feeling of strangeness attaches itself to voices or to a radio being heard at the time.[20] Williams describes a 35-year-old woman who experienced the same kind of attack over a period of 5 years: "Suddenly she would feel outside her body, looking at what she was doing. . . . Then there is an unnatural fear without cause. She is terrified of something unknown. . . . She involuntarily holds her breath, is nauseated, retches, then recovers in a few moments."[49] Hughlings Jackson[50] referred to such autoscopic experiences as a mental diplopia—literally, a double vision of the mind. Penfield and Perot[51] describe five patients with autoscopic hallucinations. One of the patients, a 32-year-old woman (Case 51), had attacks beginning with a sensation in her throat, followed by an abdominal sensation and a visual hallucination in which she could see herself in different situations. She had scarring in the hippocampal region on the left side. Another of their patients was a woman (Case 42) who experienced an auditory hallucination about which she said, "It was as though I were two persons, one watching, and the other having this happen to them."[52] She felt that she was the one watching! In Penfield's words, "It was as though the patient were attending a familiar play and was both actor and audience."[53]

Anger and Related Feelings

As was noted when reviewing experimental findings (Chapter 19), angry behavior is evoked by stimulation within certain parts of the amygdalar subdivision, whereas ablations of this same subdivision may result in tameness and placidity of wild animals.

Reference was also made to parallel findings in clinical cases, with the additional information that stimulation within the amygdala elicited angry feelings. In the literature one finds wide disagreement about the incidence of anger as an aural symptom in psychomotor epilepsy. For example, in their book on epilepsy published in 1954, Penfield and Jasper[54] stated that in their experience neither localized epileptic discharge nor electrical stimulation is capable of eliciting feelings of anger. Dr. Peter Gloor,[55] who succeeded Jasper at the Montreal Neurological Institute, stated in 1972 that there had been no cases in which stimulation had elicited ''rage.'' As opposed to these negative findings, Bingley[56] reported that of 90 noninstitutionalized patients, 17% manifested ''anger and/or expression of aggressivity.'' Among 50 institutionalized patients with symptoms of psychomotor epilepsy, Sano and Malamud[57] found that ''ictal psychic phenomena'' characterized by ''confusion and rage'' were described in most cases with sclerosis of Ammon's horn, occurring in 14 of the 16 cases without complicating malformations of the brain.

The discrepancy in different reports about the incidence of anger may be partly owing to misinterpretations of what the patient or the patient's family stated. Some patients, for example, may report that anger is one of the symptoms of their attacks because they have learned from individuals seeing them during a seizure that they became assaultive. If left alone during their automatisms, patients commonly engage in harmless activities, but if anyone attempts to restrain them, they may become suddenly combative (see Chapter 25). Consequently, electroencephalographers try to avoid interfering with a patient who develops a seizure and begins to show automatic behavior.

At the same time, it must be remembered that in everyday life the white rage precipitating assault and battery may occur so suddenly that in retrospect the individual has difficulty in reconstructing his or her feelings. Since epileptic patients have complete amnesia for happenings that occur once the automatism begins, they might be expected to have even greater difficulty in recollecting a momentary subjective state immediately prior to the automatism. As noted above, Williams analyzed 100 cases in which some form of emotion was experienced during the aura. In regard to anger he said, ''It is difficult to divide the emotion anger from motor aggression, particularly in the stories of epileptic patients. Altogether 17 examples of ictal aggression were encountered in [our] series and in all of them the lesion was in the anterior half of the temporal lobe or the inferior part of the frontal. . . . There was one patient who described fury as well as a number of other feelings in his attacks. . . .''[58] The patient (Case 40), a 37-year-old businessman, described his aura this way: ''I feel most uncomfortable. I want to get away. I feel furious and have an unpleasant abdominal sensation, then a sense of terrific release, of elation.''[59]

In my own experience, I have encountered three patients, all men, in whom assaultive or ''rage'' behavior was a major complication. Two of them periodically required commitment for periods of a month or more because of continued episodes of violent behavior. One of them had been admitted to the medical service because of an undiagnosed comatose condition. The electroencephalogram was like that seen after a hypnotic dose of a barbiturate. It turned out that the patient had become so guilt-ridden and depressed by his unpredictable attacks on other persons that he had attempted to commit suicide. Upon recovery he impressed everyone as a mild-mannered, middle-aged person who seemed especially endearing because of his eccentric use of an ear trumpet to aid his deafness. Then suddenly out of the blue, he tried to stab a nurse with scissors and found himself again committed to a state hospital. In both this and one of the other two mentioned patients (Case P.R.[60]), the basal electroencephalogram revealed a unilateral epileptogenic focus with the greatest spike amplitude recorded by the left nasopharyngeal electrodes. As in many other cases with psychomotor epilepsy, the disturbance in the medial temporal region was not reflected in the routine electroencephalogram.

Neither of the two patients just mentioned was able to recall an angry feeling in association with his aura. In the third case[61] an interview under Amytal sedation helped him recall a feeling of getting "mad" that had not been expressed on previous occasions when he had been encouraged to describe his feelings, no matter how trivial, at the time of a seizure. His seizures might occur during the day or night. In one nighttime attack, his mother found him choking his brother. As to his aura during daytime attacks, he was unable to recall anything more than a feeling of "electricity" throughout his body. During the first Amytal interview, he further remembered that during the aura he had "dreams" of animals chasing him and that there also might be a "tension feeling," "sort of a darkness closing in on me, and I wanted to grab on to something, and I would just black out right there."[62] In a second Amytal interview, he very significantly added, "I just get the electrical feeling, and it goes all the way through me; it starts in my head (I'd say both the stomach and head) and then it makes me do things I don't want to do—*I get mad.*"[62]

Feelings Akin to Anger

In other cases the presenting symptoms of the aura may be interpreted as having some shading of anger. One patient, for example, described by Mullan and Penfield,[39] experienced a feeling of disgust that could be reproduced by stimulation of the temporal lobe. At operation, this patient was found to have sclerosis of the uncus and hippocampus on the left side. Ferguson *et al.*[63] describe the case of a 14-year-old girl who before a seizure became "unusually irritable, uncooperative, and nasty."[64] In her aura she felt a "funny feeling in my stomach of seeing something dead, something disgusting, something red."[64] Another example of a nuance of anger would be their Case 3—a patient quoted as saying, "Some feelings come over me. I feel very morose. . . ."[65] The question concerning the ambiguity of the subjective awareness of anger and its expression will come up again in connection with epileptic automatisms that follow the initial aura.

Dejected Feelings

Dejected feelings are associated with accidental or enforced separation from something to which one feels attached, as, for example, another person or group of people; an animal; an object; a place of abode; or something for which one was striving to obtain. Ostracism and imprisonment are examples of socially enforced separation.

In cases of psychomotor epilepsy, feelings of sadness or sorrow may occur during the aura.[66] One patient whom I treated had a feeling in the pit of the stomach that he characterized as a feeling of sadness and wanting to cry.[67] This feeling was accompanied by welling up of tears and a sensation of hunger. Ferguson *et al.*[63] describe a patient whose attacks were precipitated by the mention of, or reading about, death. With the beginning of the seizure there was a feeling of sadness.

An 18-year-old girl (Case 2) mentioned by Daly[22] suffered from attacks in which there was a sudden welling up of intense fear and a feeling of loneliness. Another of his patients (Case 5) experienced an unpleasant emotion like the feeling of being alone. A third patient (Case 4) had a homesick feeling in the epigastrium associated with a sense of sadness. Crying and tearing as limbic epileptic manifestations will be dealt with in Chapter 28.

Guilty Feelings

Feelings of guilt are also appropriate to consider in the present context. Daly[22] describes the case of an 18-year-old girl (Case 6) who was subject to attacks in which she might be looking at an object when it would suddenly assume a frightening quality and induce in her a feeling of being "ashamed and guilty." "If she looked away, the intensity of the emotion decreased, but if she returned her gaze to the object, the emotion returned to its former intensity and then slowly increased until she lost consciousness."[68] This case is reminiscent of the one mentioned earlier in which the patient had a generalized convulsion if he looked to see who was standing behind him. Stevens[69] cites two cases in which a "troubled conscience" ("conscience bothers") constituted part of the developing aura. In an analysis of 90 cases of temporal lobe epilepsy, Bingley[56] alludes to a few patients in whom a feeling of anxiety was accompanied by "pangs of conscience and self-accusatory delusions."

Depression

Bingley[56] also refers to a few cases in which anxious feelings during the aura were associated with feelings of profound depression. In one of these cases there were suicidal ideas. Several other authors have commented upon the occurrence of feelings of depression during the aura.[70] Weil[71] analyzed the ictal emotions of 28 subjects with temporal lobe disease and ictal emotions. He stated that half of them had ictal depression "characterized by rather sudden let-down of mood and psychomotor retardation," with shades of melancholia ranging "from simple listlessness and apathy to agitated depression with suicidal attempts."[72] Among his 100 cases, Williams[23] described 21 patients who experienced some degree of ictal depression. A woman aged 34 (Case 23) had attacks in which there was "a dramatic onset" of depression and a compulsive urge to suicide.[23] There were at least two instances in which she had to be restrained from killing herself, including one occasion when she tried to jump from a moving train. Altogether, there were five patients in Williams's series who had experienced suicidal ideas.

Gratulant (Gratifying, Triumphal, Successful, Ecstatic) Feelings

Of the other complex mental states, we have yet to consider those pertaining to gratulant and to caressive behavior. All of the ictal feelings considered under the heading of "gratulant" can be characterized as having a pleasurable feeling tone. The word *gratulant* is used here to refer to all gradations of pleasure and joy, as well as gratifying feelings associated with enhanced awareness, satisfying recognition, achievement, success, and discovery. The feelings may range all the way from mild satisfaction, to exultation, to ecstasy. If one were to attend a seizure clinic or to ask a neurologist, one would gain the impression that the occurrence of pleasurable feelings during an aura is extremely rare. The rareness of such cases may be explained in terms of the different functions ascribed to the main subdivisions of the limbic system. As already noted, it is the amygdalar subdivision that is particularly prone to injury or disease. And it is this subdivision that appears to be primarily concerned with activities involving self-preservation, including those activities that provide sustenance of the self and protection from harm.

Nevertheless, there are numerous papers in which auras of a pleasant kind have been described. The feelings are variously characterized as those of completeness,[23] contentment,[73] elation,[74] exhilaration,[23] fascination,[23] gladness,[23] satisfaction,[23] and security.[23] Other expressions include feelings of "eternal harmony,"[75] "immense joy,"[75] "intense happiness,"[76] "paradisiacal happiness,"[77] "a well being of all the senses,"[23] "a feeling like whiskey taking effect,"[22] "relaxed feeling,"[22] "a wonderful feeling."[56]

In his analysis of 100 cases in which some emotion comprised part of the aura, Williams[23] identified 9 in which there was the experience of some form of pleasure variously described as "elation," "satisfaction," "gladness," or "exhilaration." Based on the 2000 cases of epilepsy that he reviewed, such findings would suggest that about 0.5% of patients with epilepsy have some form of pleasant feeling during an aura. One such case (Case 32) was that of an accountant who, at the beginning of an attack, had the compulsion to look to the left side, after which he experienced a "mood of pleasure and satisfaction about the moment." A schoolboy (Case 33), aged 15, had the impression that things became small and indistinct; almost simultaneously he became aware of "a very pleasant feeling—a sense of pleasure."[23] An electrical fitter (Case 24) experienced a pleasant emotion that he identified with a realization of "half-knowing, half seeing that a man is doing a job properly . . . a pleasant sense of satisfaction, of completeness."[78] At the same time, he was aware of "a warm sinking feeling in the abdomen." Then he went into a generalized convulsion.

Williams pointed out that "all the patients who experienced ictal pleasure had visceral sensations" of an abdominal, respiratory, cardiac, or vasomotor type.[79] In one such case (No. 28) there appeared to be a mixture of unpleasant and pleasant feelings; the patient first experienced an unpleasant taste and a dry sensation in his throat, along with a feeling of depression and a smothering sensation. At the same time, there was a curious "pleasureable fascination—a sense that 'I want to go through with it.'"[80] In another example, Williams referred to a "Polish intellectual" (Case 36) who suffered from attacks that suddenly began with "a pleasant epigastric sensation." In the patient's words, there was at the same time "a sudden feeling of extreme well-being involving all my senses. I see a curtain of beautiful colors before my eyes and experience a pleasant but indescribable taste in my mouth. Objects feel pleasureably warm. The room assumes vast proportions, and I feel as if in another world."[81]

Daly[22] reported that 12 of his 52 patients who experienced some form of emotion during the aura described pleasurable feelings. Here again, visceral symptoms were prominent. One such patient (Case 5), a 36-year-old man, had his first attack while driving with his children in an automobile. At a traffic light, he suddenly felt the sunlight as more intense, and his perceptions generally seemed more acute. Then sounds became distant. He felt sick to his stomach, sweated, and had the urge to defecate. He felt disassociated from his body and as though he were looking down upon the scene. Yet despite all this, he experienced a pleasurable feeling, illustrating it by saying that it was like "a sunny day when your friends are all around you."[82] Another of Daly's patients experienced in the epigastrium a "warm butterfly feeling" and a sense of hunger,[22] whereas in a third case there was an epigastric "burning sensation," followed by a pleasant feeling "like whiskey taking effect."[22]

In reviewing Penfield's cases, I turned up only one (Case 58[83]) in which there was a feeling of happiness and well-being during the aura. The feeling of happiness was followed by a rising epigastric sensation and an indescribable smell or taste.

An illustration has already been given (one of Williams's cases) of the curious epileptic condition in which the patient seems to experience pleasant and unpleasant feelings at the same time. As another example, we may mention one of Daly's patients, a 24-year-old man who experienced a "feeling of sadness which at the same time was somehow pleasant, like Juliet's 'sweet sorrow.' "[84] It is possible that in these cases there is actually an unrecognized seesaw of feelings. Several years ago, I pointed out that the study of limbic epilepsy suggests that there is a reciprocal innervation for opposite feeling states that compares to the reciprocal innervation of muscles.[85] In the present context one is reminded of a case report by Hughlings Jackson (1899)[86] in which he commented that a pleasurable aura may be replaced by a feeling of depression. In a moment, reference will be made to a rather famous literary illustration of the reverse situation.

Ecstatic feelings during the aura appear to be extremely rare. In his extensive practice, the epileptologist William G. Lennox encountered only one such case. The patient, a 36-year-old woman, described her feelings this way: "My fingertips began to vibrate thrillingly, and then the sensation passed to my head, giving me the most ecstatic physical pleasure. Over and over, either by clenching my hands or by an effort of will, I reproduced this exciting, pleasureable stimulus in my head."[87] Had there not been electroencephalographic evidence of epilepsy, Lennox would have been inclined to interpret her symptoms as a form of hysteria.

If there is a popular tendency to associate epilepsy with ecstasy, it can probably be attributed to Dostoyevsky's writings. Alajouanine[88] points out that in *The Idiot*[89] the character Prince Mishkin describes his ecstatic feelings in almost the same words that Dostoyevsky privately recounted his own fits. Prince Mishkin describes his spiraling elation as a "direct sensation of existence in the most intense degree."[90] "Yes," he concludes, "for this moment one might give one's whole life."

Subirana,[77] a Spanish neurosurgeon, briefly described two patients whose epileptic attacks were characterized by ecstatic "paradisiacal" feelings. One, a man aged 44, experienced feelings of "extraordinary beatitude," "feelings completely out of this world," while the other, a 45-year-old man, had an aura that made him feel as though he knew what it was like to be in heaven. Subirana[91] told me that the second patient felt reluctant to undergo an operation lest it deprive him of his "wonderful feeling which gave the impression of lasting for hours."

Laughter has been reported as one of the manifestations of temporal lobe epilepsy.[92] Laughter and crying as limbic epileptic manifestations will be dealt with in Chapter 28. In case reports, it is usually stated that the patient experiences no feeling of laughter or the sense of joy and surprise that may accompany laughter.[71] An exception with which I am familiar is the case of a patient who was interviewed under Amytal sedation. When he first described his symptoms to us, he had referred to his feeling during the aura simply as a laughing sensation. Under Amytal sedation, he was able to elaborate by saying, "The feeling in the heart is as though you need extra air—and then when the air comes out—that's when I get the laughing sensation. The climax of anything is the top—the finale, isn't that right, doctor? The end before the finale? Then this laughing comes about . . ."[93] (see Chapter 28 regarding other evidence).

Other Kinds of Gratulant Feelings

Other kinds of ictal gratulant feelings do not convey quite so clearly the feeling of pleasure or happiness of those already dealt with. Nevertheless, they represent feelings with definitely agreeable qualities. They include feelings of greater awareness[39]; the feeling that "things are more real"[22]; the feeling that things are "crystal clear," "frightfully clear"[88]; the feeling of clairvoyance[94]; the feeling that what is happening or what one is thinking at the moment is all important[95]; feelings of certainty and conviction[96]; and feelings of revelation of the truth.[96] The last two kinds of feelings are of the kind associated with mystical experiences and discovery. Apropos of play, I have uncovered no cases with allusions to "playful feelings" (see, however, Chapter 25 on automatisms).

In reviewing papers in which the number of cases totaled about 3000, I found that in about 0.5% of them there was mention of the occurrence of one of the above kind of gratulant feelings during the aura. The experience of "recognition" carries along with it a feeling of gratification provided that there is nothing threatening in what is recognized. If one were to add cases in which positive feelings of familiarity or *déjà vu* occurred during the aura, the recorded incidence of gratulant aural feelings would be many times greater. Feindel and Penfield,[97] for example, reported that a feeling of *déjà vu* occurred in 18% of their 155 cases of temporal lobe epilepsy.

The feeling of enhanced awareness or the feeling of clairvoyance creates an agreeable sense of expected recognition. Mullan and Penfield[39] describe two such cases. One patient (Case 38) had a feeling of greater awareness that applied to smells, sounds, visible objects, and pressures. He called it a "new awareness." Stimulation within the temporal lobe re-created the feeling of increased awareness. In the other case (Case 24) the patient reported the effects of electrical stimulation in the left temporal region as follows: "I knew everything that was going to happen in the near future. . . . As though I had been through all this before, and . . . I knew exactly what you were going to do next."[98] Stimulation near the uncus induced familiar memories. Chapman *et al.*[99] describe a case in which stimulation of the anterior hippocampus elicited visual images and a feeling that the patient could read the doctor's mind. Williams[23] describes the case of a housewife (Case 35) who during the aura experienced "a sudden feeling of being lifted up, of elation, with satisfaction, a most pleasant sense. With it there is the feeling, 'I am just about to find out knowledge no one else shares—something to do with the line between life and death.' "[100]

From the standpoint of epistemics and epistemology, the study of the phenomenology of psychomotor epilepsy provides profound insights into the neuropsychology of feelings underlying beliefs and a sense of conviction about what is real, true, and important. Such feelings accompany not only everyday experiences of the moment, but also, in a more intense form, the realization of discovery (the so-called eureka feeling) and the revelation of a mystical experience. Here again, we may turn to the writings of Dostoyevsky for examples of the verbal expression of such feelings. In *The Demons,* Kirilov is "plagiarizing" the author's own ictal feelings when he says to Chatov, "There are some instances, they last for five or six seconds, when you suddenly embrace the entire creation and you say, well, it is like that, it is true. . . . The most terrible thing is that it is so frightfully clear and such an immense joy at the same time."[101] Or again, to quote from the little-known work *The Mirage* by the same author: "It seemed to me that I understood at that moment something of which I already had a presentiment, without having ever expressed it. . . . I believe that my real existence dates from that moment."[102]

In shifting the scene from Russia to North America and the time from the 19th to the 20th century, we now consider another example of a case in which ictal feelings are expressed in somewhat similar words. There is, however, the crucial additional information of a confirmed epileptogenic focus in the left medial temporal region. The patient was a 22-year-old art instructor whose symptoms had resulted in his undergoing psychoanalysis for 3 years. His case came to my attention when his lay analyst explained the nature of the symptoms to me. I suggested that psychomotor epilepsy was a probable diagnosis, and the patient subsequently was admitted to the Neurological Clinic (Yale University School of Medicine) for a diagnostic work-up. The patient's history revealed an ill-defined head injury about 10 years previously. Sometime afterwards, he began to have occasional attacks that he presumed to be a kind of fainting spell. Then the attacks began to occur as often as four times a day and were characterized by a "bursting feeling in the stomach" accompanied by "clear, bright thoughts." He had learned from others that during these spells he became flushed, stared, gripped his belly, and showed stiffening of the right leg. Not surprisingly, the routine waking electroencephalogram appeared to be normal, but a recording during "sleep activation" showed spiking in the anterior temporal region. In a subsequent basal electroencephalogram, the spike was recorded at maximum amplitude with the left nasopharyngeal electrode, signifying that the temporal focus was more medial than lateral.

The following account of the patient's aura was recorded on tape by Dr. Janice Stevens on the same day that he had had an attack while driving his car in the company of his children; he had had sufficient warning to pull up alongside of the road and stop. Describing his thoughts that accompany the bursting sensation in his stomach, he said: "Each time this thing happens, these thoughts occur very clear and bright to me. They seem as if 'this is what the world is all about—this is the absolute truth.'"[93]

As already pointed out, it is characteristic that under the abnormal conditions of a limbic discharge, the feelings that light up in the patient's mind are free-floating and out of context—being attached to no particular person, situation, or thing. In the present case, the patient said of his thoughts: "It was as though they were just everyday things going on as usual, only they seem so much more important and vital than they do in ordinary living."[93]

Feelings of Affection

In reviewing several hundred case histories, I have encountered none suggestive of feelings or sentiments related to parental behavior. In regard to feelings of affection or feelings related to caressive or attachment behavior generally, the symptomatology is so ambiguous as to merit little more than a listing. In all cases in which such symptoms occurred, the epileptogenic focus appeared to be in the temporal lobe. Wadeson[103] mentioned one patient who, during the aura, experienced a feeling of cephalic warmth and micropsia and, in some attacks, "feelings of love." Penfield and Perot[104] described a case of a 44-year-old woman (Case 9) whose attacks began with a flushing of the face and neck and were followed by automatisms in which she was apt to say, "I'm all right," and would walk around her room showing marked affection to anyone who happened to be present. In her case there was demonstrated incisural sclerosis. Mention has already been made of Daly's patient (Case 5), who during the aura had feelings "like on a sunny day when your friends are all around you."[82] In reporting cases of ictal and postictal sexual intercourse, Blumer[105] does not allude to any feelings of affection. Penfield and Perot[51] refer to a 34-year-old veteran (Case 53), who experienced the hallucination of seeing himself having intercourse with his fiancee. One patient is described as "verbalizing" her sexual needs while at the same time spreading her legs and placing her hand over her perineum.[106]

Indeterminate Affects

Familiarity and Strangeness

The terms *familiar* and *strange* may apply to something observed or to one's subjective feelings. Recalling the discussion in Chapter 10, one might say that familiarity and strangeness represent opposite sides of the same coin. They are referred to here as indeterminate affects because (1) they may be associated with the experiencing of basic, special, and general affects, and (2) depending on the situation, the feelings of familiarity and strangeness may be subjectively agreeable or disagreeable. It was noted above that the alternation of opposite affects experienced by some patients during the aura suggests a reciprocal innervation of affects that compares to the reciprocal innervation of muscles. The same question arises in regard to familiarity and strangeness. For example, as illustrated by three patients in Stevens's[69] study on the ictal "march," a feeling of strangeness was replaced by a feeling of familiarity, and vice versa. L. F. Chapman *et al.*[99] reported that unilateral stimulation of the amygdala or hippocampus elicited feelings of familiarity in 7 of 17 patients. Penfield and Perot[51] cite a case (No. 23) of a patient with "incisural sclerosis" who upon brain stimulation experienced a feeling of strangeness like that at the beginning of his usual spells. In their case study, Mullan and Penfield[39] found that a feeling of familiarity occurred twice as often when the nondominant hemisphere was the primary site of epilepsy.

Time and Space

Kant[107] regarded space and time as "transcendental aesthetic" because they represent "pure *a priori* intuitions" that provide the *form* for experiencing the temporal succession in which things occur and their spatial dimensions and relationships. One might otherwise say that time and space are predetermined derivatives of the subjective brain that do not exist *per se* but are purely informational, being of themselves neither matter nor energy. Under many conditions, temporal and spatial information are affectively sensed as pleasant or unpleasant, leading one to question whether or not there is a neural apparatus that generates affects pertaining particularly to time and space. As has been described, things seen or heard may give the impression of speeding up or slowing down. A patient may describe the feeling as though words took a whole minute to be said.[103] Another may experience the feeling of living his life backwards.[108] Things may seem to come to a standstill[39] or, as in delirium, seem to either speed up or slow down. There may be a disorientation with respect to time.[109] As was described, in regard to space, things may seem farther or nearer, larger or smaller. As one patient described the aural experience, space "seems to open up,"[15] while another had a feeling of being in a room of vast proportions.[79]

"Interictal" Symptomatology

In regard to behavioral disorders and the psychoses, it has long been a matter of contention as to whether or not there is an unusually high incidence of such forms of psychopathy in patients with psychomotor epilepsy. Stevens and co-workers[110] were among the first to undertake a systematic study in regard to this question. They concluded that there was no significant difference in this respect between patients with psychomotor epilepsy and those with other types of convulsive disorders. Stevens[111] has since concluded that although the overall incidence of psychiatric disorders was comparable in the two groups, "dysphoric and schizoid" manifestations were somewhat more common in the psychomotor group. In 1963, Slater *et al.*[112] reported findings on 69 patients that indicated that epilepsy preceded the appearance of schizophrenic-like psychoses by 10 to 20 years. Stevens, however, has cited and discussed evidence that if age alone is taken into account, the incidence of schizophrenia in patients with epilepsy is not significantly greater than that for the general population.[111]

In 1969, Flor-Henry[113] published his findings (based on 50 epileptic patients with psychotic symptoms and 50 epileptic patients without such manifestations) that temporal lobe epilepsy involving the dominant hemisphere results in a predisposition to a schizophrenic-like psychosis, whereas with signs of a predominant disturbance in the right hemisphere there is a significant inclination to a manic–depressive condition. He also called attention to the finding in his analysis that there is an inverse correlation between frequent psychomotor seizures and the presence of a psychosis. As Stevens[111] points out, epileptologists have long been aware that the reduction of seizures associated with pharmacological treatment appears in some cases to precipitate a psychotic condition.

Some of the uncertainties in regard to the questions that have been raised could be clarified if it were possible for professional observers, as in a hospital, to accompany patients in their day-to-day and nighttime activities. But even under such conditions there would remain the question as to what constitutes "interictal" symptomatology, an expression that implies a condition exclusive of outright seizure activity in which electroencephalographic abnormalities may or may not be present. Such a state would leave open the possibility of postseizure depression or enhancement (potentiation) of excitation in structures under the influence of epileptogenic foci. But as Monroe[114] points out in a discussion of episodic behavioral disorders ("dyscontrol syndrome") even with surface monitoring of the electroencephalogram, seizure activity in deep limbic structures could be overlooked without the use of depth electrodes. This, he states, will continue to be the case until some "nonintrusive measure" is developed.

In regard to indisputable ictal episodes, there can be no doubt that patients experience at the time symptoms associated with psychoses. As I have discussed elsewhere,[115] one can single out four main kinds of psychotic symptoms associated with psychomotor seizures—namely, (1) disturbances in emotion and mood, (2) feelings of depersonalization, (3) distortions of perception, and (4) paranoid symptoms. Such manifestations may carry over into the so-called interictal periods. Drawing upon cases I have seen myself, I recall, for example, a patient with paranoid symptoms who was persistently obsessed by a feeling that God was punishing her for overeating. While she was expressing such thoughts, the basal electroencephalogram showed random spiking at a site near the tympanic lead just underneath the temporal lobe.[116] In regard to combined paranoid and hallucinatory symptoms, I am reminded of a patient who claimed to hear clicking sounds and imagined that people were snapping pictures of her.[116] It is appealing to cite this case because in schizophrenia, as opposed to psychomotor epilepsy, auditory illusions and hallucinations are more frequent than those in the visual sphere. Picturing the first-mentioned case in metaphorical terms, one might imagine the spiking subtemporal focus as a cerebral "itch," generating in turn a "paranoid itch."

As Trimble[117] has commented, a difficulty in dealing with the extensive literature on interictal symptomatology is that most authors have had to rely on "clinical impressions." For further information on the subject the reader is referred to two symposia dealing with various aspects of the problem.[118]

Concluding Discussion

The study of psychomotor epilepsy provides the best evidence that the limbic system is basically implicated in the generation of basic, specific, and general affects (Chapter 23). In the present chapter the focus of attention has been on the auras of psychomotor epilepsy that are identified with the *general* affects that represent what are traditionally regarded as emotional feelings. Such affects are referred to as "general" because of the generality of their application to individuals, situations, and things. As defined in this and the preceding chapter, they also have other characteristics that distinguish them from basic and specific affects.

The existence of affects in another individual must be gauged solely on the basis of behavior, either prosematic or verbal. In animals and human beings, there are six main forms of prosematic behavior that through the introspective process are subjectively identified with the general affects—namely, searching, aggressive, protective, dejected, gratulant, and caressive behavior. In connection with these six forms of behavior, one can subjectively identify variations of six main affects, characterized, respectively, as desire, anger, fear, sadness, joy, and affection. In the present chapter it turns out that the various *general* affects experienced by patients as part of the aura of psychomotor epilepsy can be categorized under the headings of these six forms of affective feelings.

At this point it should be emphasized that the study of psychomotor epilepsy also provides the invaluable information that affective feelings are genetically constituted. Although there does not appear to be any paper dealing specifically with this question, case reports from various countries in different parts of the world indicate that the same kind of subjective phenomenology presents itself in psychomotor epilepsy regardless of the person's race or ethnic background.

In addition to the basic, specific, and general affects, the subjective phenomenology of psychomotor epilepsy raises the question as to the existence of two forms of affects that were described under the tentative heading "indeterminate affects"—namely, (1) feelings of familiarity or strangeness that may be variously associated with the basic, specific, and general affects, and (2) affects identified with space and time. The latter question will require further consideration in the final chapter.

In Chapter 2 it was remarked that whatever else the brain may be, it serves as both a detecting and an amplifying device. The subjective phenomenology of psychomotor epilepsy suggests that the limbic system (corresponding to the paleomammalian formation) has the capacity to act as an amplifying device for turning up or down the "volume" of affective feelings that guide behavior required for self-preservation and the preservation of the species. Analysis of ictal auras, for example, indicates that *general affects* such as fear may occur in all gradations, ranging from "butterfly" feelings to terror, while changes in intensity of *specific affects* may be illustrated by sounds that seem to become louder or fainter, nearer or farther. Fewer details are available regarding aural gradations in *basic affects*.

In everyday life the affective feelings that guide our behavior have relevance to something in particular, but under ictal conditions the affects are usually free-floating feelings unattached to specific individuals, situations, or things. Hence, the phenomenology of psychomotor epilepsy provides a strong argument against the contention of those who would claim that it is inadmissible to make a sharp distinction between "emotion" and "reason" (see Chapter 2).

Several illustrations were given of the alternation of opposite feeling states during the aura, as, for example, the alternation of feeling of cold and warmth, displeasure and pleasure, fear and anger. Such a subjective phenomenology raises the question as to the possibility of reciprocal innervation of feeling states that compares to the reciprocal innervation of muscles.[85] In Chapter 28 the question of the reciprocal innervation of laughing and crying will come up in connection with the prefrontal cortex and its relation to the thalamocingulate division. At that time additional case material on limbic epilepsy relevant to crying and laughter will be presented. That will also be an appropriate place to comment upon the question of differential effects of irritative lesions in the dominant and nondominant hemispheres in generating various affects.

In some writings the limbic system seems to be regarded as though it functioned globally in generating diffuse emotional feelings. It is evident, however, from the experimental and clinical findings reviewed thus far that the three main limbic corticosubcortical divisions generate emotional feeling-states and expression germane to their respective functions. The predominance of agnostic affects associated with psychomotor epilepsy can be attributed to the circumstance that the medial temporal structures are more susceptible to injury and disease than those in other lobes. The temporal lobe contains most of the amygdalar subdivision that is involved in self-preservation, including the agonistic behavior incidental to the search for food and struggle to obtain food. Although the fearful and other aural symptoms may be "free-floating," the subsequent automatisms are not infrequently in keeping with the subjective state, as when, for example, a patient runs and screams, as if in fear, or fights and struggles, as if angry (see Chapter 25).

In Chapters 26 and 27 the phenomenology of psychomotor epilepsy will be used, in part, to develop the argument that the integrity of function of the limbic system is essential for a feeling of personal identity, which in turn is requisite for the memory of ongoing experiences.

As pointed out in the preceding section, the matter pertaining to the part played by the "interictal" symptomatology of psychomotor epilepsy in psychological disorders is fraught with uncertainties. Nevertheless, it cannot be overemphasized that the study of psychomotor epilepsy, in general, has greater potential for providing insights into the neural substrate of affect-related disorders than any other clinical condition.

Finally, the study of psychomotor epilepsy poses profound questions pertaining to epistemics and epistemology. As was illustrated under gratulant affects, a patient may experience during the aura free-floating, affective feelings of conviction of what is real, true, and important. Does this mean that this primitive part of the brain with an incapacity for verbal communication generates the feelings of conviction that we attach to our beliefs, regardless of whether they are true or false? It is one thing to have the anciently derived limbic system to assure us of the authenticity of such things as food or a mate, but where do we stand if we must depend on the mental emanations of this same system for belief in our ideas, concepts, and theories? In the intellectual sphere, it would be as though we are continually tried by a jury that cannot read or write. These are questions that will become the focus of attention at the end of the final chapter.

References

1. Spinoza, 1677/1955, Vol. 2, p. 138
2. Descartes, 1649
3. Descartes, 1649, p. 362
4. Hume, 1739/1888
5. James, 1890
6. James, 1890, p. 449
7. James, 1890, p. 468
8. Jaspers, 1913/1963, p. 108
9. Plutchik, 1980
10. MacLean, 1960a
11. MacLean, 1969a, 1970a, 1976
12. Gowers, 1901
13. Gowers, 1901, p. 72
14. MacLean, 1952
15. Williams, 1956, Case 32
16. Williams, 1956, Case 28
17. Williams, 1956; Bingley, 1958; Daly, 1958; Mullan and Penfield, 1959; Weil, 1959
18. Jackson, 1958, p. 301
19. Macrae, 1954
20. Macrae, 1954, p. 387
21. Daly, 1958, p. 99
22. Daly, 1958
23. Williams, 1956
24. Macrae, 1954; Ludwig and Ajmone Marsan, 1975; Ajmone Marsan and Goldhammer, 1973
25. MacLean and Arellano, 1950; Green and Scheetz, 1964
26. Penfield and Jasper, 1954, p. 134; Williams, 1956; Daly, 1958
27. Daly, 1958; Ferguson, 1962
28. Daly, 1958; Mullan and Penfield, 1959
29. Bingley, 1958; Daly, 1958
30. Penfield and Jasper, 1954, pp. 88, 374, 434; Williams, 1956; Chapman *et al.*, 1967; Remillard *et al.*, 1974
31. Penfield and Perot, 1963; Jasper, 1964; Gloor, 1977
32. Williams, 1956; Crandall *et al.*, 1963
33. Lennox and Cobb, 1933; Daly and Mulder, 1957; Crandall *et al.*, 1963
34. Penfield and Jasper, 1954, p. 411; Caveness, 1955; Daly, 1958; Chessick and Bolin, 1962
35. Jackson and Stewart, 1899/1958
36. Green and Scheetz, 1964, p. 139
37. Jasper, 1964
38. MacLean, 1952
39. Mullan and Penfield, 1959
40. Mullan and Penfield, 1959, pp. 278–279
41. Daly, 1958; Penfield and Jasper, 1954; Penfield and Perot, 1963; Wadeson, 1965
42. Daly, 1958; Penfield and Perot, 1963; Wadeson, 1965; Hierons and Saunders, 1966
43. Williams, 1956; Bingley, 1958
44. Jackson, 1880/1958; Macrae, 1954; Stevens, 1957
45. Penfield and Jasper, 1954, p. 133
46. See also Penfield and Perot, 1963, p. 641
47. Penfield and Jasper, 1954, p. 434
48. Green and Scheetz, 1964
49. Williams, 1956, p. 45
50. Jackson and Stewart, 1899/1958
51. Penfield and Perot, 1963
52. Penfield and Perot, 1963, p. 658
53. Penfield, 1952, p. 183
54. Penfield and Jasper, 1954, p. 451
55. Gloor, 1972
56. Bingley, 1958
57. Sano and Malamud, 1953
58. Williams, 1956, p. 38
59. Williams, 1956, p. 58
60. MacLean and Arellano, 1950
61. Stevens *et al.*, 1955
62. Stevens *et al.*, 1955, unpublished quote from patient's record
63. Ferguson *et al.*, 1969, p. 490
64. Ferguson *et al.*, 1969, p. 491
65. Ferguson *et al.*, 1969, p. 487
66. Penfield and Jasper, 1954, pp. 398, 440; Williams, 1956; Stevens, 1957; Daly, 1958; Ferguson *et al.*, 1969
67. MacLean, 1952; Stevens, 1957
68. Daly, 1958, p. 103
69. Stevens, 1957
70. Mulder and Daly, 1952; Williams, 1956; Daly, 1958; Weil, 1959
71. Weil, 1959
72. Weil, 1959, p. 89
73. Heath and Guerrero-Figueroa, 1965
74. Williams, 1956; Dostoyevsky, 1962
75. Alajouanine, 1963
76. Van Reeth, 1959
77. Subirana, 1953
78. Williams, 1956, pp. 56–57
79. Williams, 1956, p. 57
80. Williams, 1956, p. 54
81. Williams, 1956, p. 57
82. Daly, 1958, p. 102
83. Penfield and Perot, 1963, p. 662
84. Daly, 1958, p. 101
85. MacLean, 1955, 1958a
86. Jackson and Stewart, 1899/1958, p. 468
87. Lennox, 1960, Vol. 1, pp. 272–273
88. Alajouanine, 1963
89. Dostoyevsky, 1962
90. Dostoyevsky, 1962, p. 214
91. Subirana, personal communication
92. Daly and Mulder, 1957; Weil *et al.*, 1958
93. Stevens and MacLean, unpublished records
94. Mullan and Penfield, 1959; Kamrin, 1966
95. Williams, 1956; Stevens, 1957
96. Subirana and Oller-Daurelia, 1953; Subirana, 1953; Alajouanine, 1963
97. Feindel and Penfield, 1954
98. Mullan and Penfield, 1959, p. 277
99. Chapman *et al.*, 1967
100. Williams, 1956, p. 57
101. Quoted by Alajouanine, 1963, p. 214
102. Quoted by Alajouanine, 1963, p. 217
103. Wadeson, 1965
104. Penfield and Perot, 1963, p. 623
105. Blumer, 1970
106. Freemon and Nevis, 1969
107. Kant, 1899
108. Williams, 1956, Case 27
109. e.g., Ferguson *et al.*, 1969
110. Small *et al.*, 1962
111. Stevens, 1975
112. Slater *et al.*, 1963
113. Flor-Henry, 1969
114. Monroe, 1986
115. MacLean, 1970b, 1973a
116. MacLean, 1969c
117. Trimble, 1986
118. Girgis and Kiloh, 1980; Doane and Livingston, 1986

25

Phenomenology of Psychomotor Epilepsy

Prototypical and Emotional Behavior

In addition to being the only condition that provides extensive information about cerebral structures involved in the subjective experience of affects, psychomotor epilepsy affords continuing opportunities for enlarging the knowledge of brain mechanisms implicated in prototypical behavior. Subsequent to the experience of the aura, a patient characteristically engages in automatic forms of behavior (automatisms) that initially are usually stereotyped, but later may involve a complex sequence of actions. Both the stereotyped and complex automatisms may entail emotional forms of expression that in some instances may be in keeping with the kind of feeling experienced during the aura. For example, Gloor *et al.*[1] describe the case of a 19-year-old woman whose seizures began with a feeling of "intense fear" followed by an "automatism in which she acted as though she were in the grips of the most intense terror." As further related, "She let out a terrifying scream and her facial expression and bodily gestures were those of someone having a horrifying experience." In this case, an epileptogenic focus was demonstrated by depth recordings to be in the region of the right amygdala where electrical stimulation elicited her aura of fear.

In regard to the elucidation of cerebral functions, there is another most notable aspect of the automatisms of psychomotor epilepsy. Clinically, it has long been recognized that following the onset of an automatism the patient is totally unable to recall anything that happens during the period of automatic behavior. Owing to the reproducibility of these phenomena during diagnostic procedures used at the time of therapeutic neurosurgery, Penfield and Jasper[2] and others have been able to contribute to the knowledge of structures involved in the registration and memory of ongoing experience. This and related matters will be the subject of the next two chapters. In the present chapter it is the purpose to describe the nature of automatisms occurring in psychomotor epilepsy and then, based on electroencephalographic recordings and diagnostic cerebral stimulation, summarize what has been learned about the identification of structures involved in automatisms.

Automatisms of Psychomotor Epilepsy

On the basis of continuous monitoring with depth electrodes, Gloor *et al.*[3] point out that psychomotor epilepsy may manifest itself in three main forms—namely, (1) electroencephalographic seizure activity without subjective or objective symptoms; (2) subjective symptoms and/or barely notable automatisms; and (3) subjective symptoms followed by "socially disabling" automatisms. Generalized convulsions are an unpredictable complication of psychomotor seizures.

Automatisms are commonly classified as "simple" and "complex." Adhering to this broad classification, but otherwise using different descriptive labels, the present account will deal with automatisms under the following four headings: (1) simple somatomotor and somatovisceral automatisms; (2) simple pseudomimetic automatisms; (3) simple and complex pseudoemotive automatisms; and (4) complex disorganized or quasi-organized automatisms. As background, a brief statement will be made regarding the incidence of automatisms and what has been learned about the associated electroencephalographic changes.

Incidence

In an analysis of 500 extensively studied cases of epilepsy, King and Ajmone Marsan[4] show a tabulation of the incidence of aural symptoms and automatisms in patients with temporal lobe epileptogenic foci as opposed to those with extratemporal foci. In 270 cases of temporal lobe epilepsy, automatisms occurred in 95%. This incidence is a little more than twice that for patients with extratemporal foci (43%). The aural symptomatology in many of the extratemporal cases would suggest that foci in other lobes may have involved either limbic or perilimbic structures. In an earlier study involving a series of 155 patients with psychomotor epilepsy, Feindel and Penfield[5] reported that 121 (78%) "showed some evidence of behavioral automatism."

Electrophysiological Observations

In Penfield and Jasper's[2] book on epilepsy, Penfield describes automatisms induced by brain stimulation at the time of surgery. He depicted, as follows, the sequence of events in one patient in whom the olfactory aura of an attack was reproduced by electrical stimulation of the left uncal region: "There was an initial period of low-voltage rapid local discharge, which would certainly not be detected in the usual E.E.G. . . . [T]he electrical activity from the entire temporal region was suppressed. . . . The patient became unresponsive . . . . and began aimless automatic movements . . . as in his habitual seizures. . . . When the attack was over [after 90 seconds], the patient remembered the olfactory aura but was amnesic for all the rest of the episode."[6] This patient proved to have induration and sclerosis of the uncus, neighboring insular cortex, and hippocampal gyrus. A decade later, Jasper,[7] basing his comments on experience with depth and surface recordings such as those illustrated in Figure 24-1, pointed out that symptoms of the aura may occur when stimulation of medial temporal structures elicits local afterdischarges in the amygdala and the hippocampus, but that automatisms do not develop until electroencephalographic changes occur elsewhere. Under these conditions, there is typically a "suppression" of activity recorded in the surface electrocorticogram (Figure 24-1), together with signs of propagation of the discharge beyond the amygdalohippocampal region. As Jasper noted in the legend explaining Figure 24-1, concurrently with distant spread of the seizure "there was immediate amnesia and automatism." Conditioning experiments described in Chapter 27 indicate that loss of awareness is attributable to propagation of the seizure discharge to corresponding structures of the opposite side, thereby resulting in bilateral disruption of function.

There is electrophysiological evidence that the initial automatisms develop concurrently with propagation of the epileptic discharge to structures of the brainstem, whereas a subsequent complex automatism, for example, in which a patient continues to carry out (though somewhat deficiently) ongoing activities, may be the joint manifestation of exhaustion of neural structures involved in the seizure and the continued function of noninvolved structures. Clinically, the phenomenon of ''exhaustion'' in seizure-involved structures has been recognized at least since 1840, when Todd[8] (1809–1860), an English neurologist, described transient ''epileptic hemiplegia.''

Some Relevant Experimental Findings

Experimentally, low-frequency stimulation (stimuli ~ 6–12/sec) may be used to mimic the recruitment of neural activity within distant structures that is seen during a propagating seizure discharge. In awake, unrestrained animals under such conditions, there may be the development of stereotypic manifestations such as would occur during the propagation of an electrically induced seizure discharge (afterdischarge). A masticatory automatism such as chewing and salivation would be an example (see Chapter 19).

Our findings during the recording of single nerve cells in awake sitting monkeys provide an illustration of an ''exhaustion'' phenomenon that may occur in the wake of a propagated hippocampal afterdischarge elicited by electrical stimulation.[9] When one to three electric shocks are applied to the hippocampus or fornix, the majority of nerve cells in the basal forebrain and hypothalamus respond with a discharge of one or a few spikes. Following an electrically induced hippocampal afterdischarge, however, we observed that some units in the hypothalamus did not recover the ability to respond to stimuli for as long as 11 min.

Simple Somatomotor and Somatovisceral Automatisms

In instances of complex automatisms, patients themselves can sometimes reconstruct the nature of their behavior during the automatism. Because of the total amnesia attending simple and complex automatisms, however, most information about ictal behavior has derived from family or friends who have witnessed a patient's attack. In recent years, increasing use has been made of telemetric and video/audio techniques for obtaining continuous recording of patients in a hospital, providing the invaluable diagnostic aid of a simultaneous comparison of the development of behavioral and EEG changes during spontaneous seizures.

In their original paper of 1948 on psychomotor epilepsy, Gibbs *et al.*[10] reviewed the symptomatology in 270 cases and listed 18 predominant forms of behavior in automatisms—namely, staring, searching, groping, chewing, salivation, smacking lips, spitting, rubbing, plucking, pushing, undressing, laughing, crying, confused talking, shouting, screaming, ''incoordination,'' and ''negativism.''

In a study employing simultaneous video and EEG recording in 79 patients with automatisms, Escueta *et al.*[11] listed essentially the same kinds of behavior, specifying, in addition, standing, walking, running, and other automatisms that will be considered here under the headings ''pseudomimetic'' and ''pseudoemotive.'' The analysis of King and Ajmone Marsan,[4] referred to above, gives a comparable listing, adding confirmation to what has been observed over and over again in previous studies of psychomotor epilepsy.

Simple Somatomotor Automatisms

Escueta *et al.*[11] emphasized that in 76% of their cases, the automatism began with a motionless stare. Otherwise the automatisms in this group that were designated as Type I were essentially of the same kind as those in the rest of the patients that were denoted as Type II.

In some instances somatomotor manifestations may be hardly noticeable, as, for example, making plucking movements, fumbling with one's tie.[2] There may be simple grimacing or pursing of the lips. Pawing, picking, rubbing movements may occur, or the simple act of circling a table. The video recordings of Escueta *et al.* afford a detailed temporal analysis of various tonic posturing of the head, limbs, and body, as well as adversive turning of the eyes, head, and/or body, that may occur at the beginning of automatism. Other forms of simple automatisms will be dealt with under descriptive headings.

Somatovisceral Manifestations

It will be evident that many of the autonomic and somatovisceral manifestations to be described are of the kind that accompany the basic, specific, and general affects. Although dealt with separately here, it deserves emphasis that many of the autonomically related manifestations occur in conjunction with somatomotor components of automatisms.

Masticatory Automatisms

In 50 "well-documented" cases of psychomotor epilepsy Feindel and Penfield[5] found that masticatory automatisms were observed in half of them. In addition to chewing,[12] salivation,[13] and swallowing,[14] there may be lip smacking[15] or licking.[16] Spitting,[17] as though to rid the mouth of a foul-tasting substance, may also occur.

Instances of actual eating[18] or drinking[19] during the automatism have been observed. One patient with an aura of thirst would call for water.[20] Williams[21] describes a patient who as a young girl would say "please nurse me" and after repeated swallowing would, if any food were around, cram it into her mouth.

Other Alimentary Manifestations

Other manifestations involving the alimentary tract include belching,[22] hiccoughing,[23] retching,[24] and vomiting.[25] Loud borborygmi[26] are frequently heard during automatisms. There may be repeated noisy passing of flatus.[27] There may be incontinence of feces or partial-to-adequate preparation for defecation.[28]

Cardiovascular Manifestations

The unpredictability of automatisms has made it difficult to document cardiac and vasomotor changes concurrently with, and following, a patient's various cardiovascular symptoms during the aura. Van Buren[29] conducted a systematic study of 13 patients in whom he recorded cardiovascular and other autonomic changes during 20 automatisms.

The findings were quite uniform in all patients, with tachycardia, peripheral vasoconstriction, and a rise in blood pressure developing during the automatism.

Respiratory and Vocal Manifestations

Respiratory manifestations include breath holding,[30] deep breathing,[21] hyperpnea,[21] gasping,[31] sighing,[21] and coughing.[32] The automatism of one of our patients was characterized by yawning and stretching.[33]

Vocalizations include crying,[34] groaning,[35] hissing,[36] snorting,[37] shouting,[38] shrieking,[39] screaming,[40] giggling,[41] cackling laughter,[42] and laughter.[43] In some patients, muttering,[44] cursing,[45] uttering nonsense words,[46] and incoherent speech[47] characterize the beginning of an automatism.

Genitourinary Manifestations

Incontinence of urine[48] or urination with the assumption of the appropriate posture[49] has been frequently reported in connection with automatisms. Sexual automatisms have been described, including exposure of the genitals,[50] masturbatory activity, pelvic thrusting,[50,51] and receptive spreading of the thighs.[52]

Thermoregulatory Manifestations

Vasomotor and piloerector signs that under ordinary conditions are associated with thermoregulation or emotional reactions commonly occur in automatisms and may be unusually pronounced. Pallor,[53] goose flesh,[21] piloerection,[21] shivering,[54] flushing,[55] sweating,[56] and contraction of the cremasteric muscles[57] are observed. In the analysis of their 199 cases, King and Ajmone Marsan[4] listed an incidence of 39% for blushing and 24% for pallor.

Pupillary Changes

It is obviously difficult to obtain a quantitative assessment of pupillary changes during automatisms. Pupillary dilation has been frequently observed, as witness its being noted in 51% of the 199 cases reviewed by King and Ajmone Marsan.[4]

Simple Pseudomimetic Automatisms

Pseudomimetic automatisms refer to simple acts in which the patient, as in a charade, appears to be performing some familiar gesture, maneuver, or activity. As yet, there have been no persuasive suggestions as to why automatisms take this form. Nor is there any indication that they reflect epileptic activation of particular structures. With the increasing use of video monitoring, it may be expected that a greater variety and increased incidence of pseudomimetic automatisms will be observed. In their video study of 79 patients, Escueta *et al.*[11] reported that automatisms suggestive of "karate postures," "shuffling cards," and pulling in a fish were among stereotyped forms of behavior that were "frequently" observed.

Simple and Complex Pseudoemotive Automatisms

In conformance with the categorization of emotional behavior used in the preceding chapter, pseudoemotive automatisms will be described under the same six main headings, beginning with searching behavior.

Searching Behavior

The occurrence of searching behavior during automatisms has already been referred to in the preceding chapter. In brief attacks a patient may seem to do no more than perform picking or plucking movements or to peer under a sheet so as to uncover something; or the behavior may seem more suggestive of groping.[58] If, as in some prolonged complex automatisms described below, a patient succeeds in achieving a specific goal or in arriving at a particular destination, it is evident that neural mechanisms ordinarily used in searching or in finding one's way have adequately responded to some original intention (i.e., desire, purpose; see below).

Aggressive Behavior

The aggressive behavior manifest in some automatisms provides further evidence beyond the subjective symptomatology (as well as the experimental findings described in Chapter 19) that limbic temporal lobe structures are involved in agonistic behavior. One of the most persuasive cases with respect to localization is that of a girl described by Vonderahe[59] who at the age of 17 developed epilepsy complicated by outbursts of rage. The postmortem examination revealed in the left temporal lobe a small tumor, "the size of a cherry," that "encroached upon the amygdaloid nucleus and adjacent portion of the hippocampus." Prior to her development of epilepsy, the patient had been characterized as having "a happy, even-tempered disposition." She was a gifted musician and had played at the age of 11 as a soloist with the Cincinnati Symphony Orchestra. Her ictal rages were precipitated by minor situations and made her family fearful of being harmed. For example, if someone at the dinner table failed to pass her something immediately, she "would fly into a fury of excited rage." A more extreme case is that of a 14-year-old girl cited by Falconer and Pond[60] who exhibited rage behavior "for hours on end" and who was found to have a calcified hemangioma in one temporal lobe.

In many case summaries, the nature of angry behavior during automatisms is not described beyond saying that the patient was subject to episodic rage or engaged in assaultive, destructive, maniacal, and violent behavior. One uncovers, however, in different reports mention of specific aggressive gestures and actions. Some of the vocalizations listed above were of the aggressive type—hissing, cursing, spitting, snorting, shouting, screaming, and shrieking. Other agonistic manifestations are fierce grimacing,[61] gritting of the teeth,[62] gnashing of teeth,[63] clenching of the fist(s),[64] shaking the fist,[65] punching with the fist,[66] stamping,[67] beating of the head against a wall,[68] or throwing oneself against a wall.[69]

There are also instances of chest beating.[70] Ferguson describes a patient who had an aura of terror followed by automatisms during which he grimaced fiercely, paced up and down beating his chest and shrieking. It is of comparative interest that, as in the case of the gorilla, it is specifically mentioned in one case report that the person beat her chest with her hands (not the fists).[52]

Other manifestations of aggression and violence include throwing of objects,[71] destruction of furniture or other things,[72] breaking through doors,[11] pushing,[4] flailing with the upper extremities,[11] and actively striking or otherwise assaulting someone.[73]

Under clinical conditions, it is usually observed that assaultive behavior occurs only when an attempt is made to restrain protectively a patient during an automatism (see Chapter 24). Seven of the nine episodes of violent behavior listed in King and Ajmone Marsan's study[4] were of this nature. In Stevens's study[19] of the "epileptic march," four patients fought restraint, while two were spontaneously assaultive. Epileptologists are sensitive to the probability that countless individuals have been tried and convicted because of violent acts carried out during an automatism. I am unaware of reports of surreptitious violent acts during automatisms. I remember a patient who "came to" from an automatism just as he observed his hand at the throat of his baby lying in a crib.

Protective Behavior

Examples have already been given in Chapter 24, and in the introduction of the present chapter, of fearful and protective behavior. Williams refers to cases of children who during an attack "look afraid, run for comfort, and call out in fear in a period of amnesia."[74] "In them," he continues, "the sequence of events suggests that fear is being experienced, but not recalled. . . ." Green and Scheetz[75] describe the case of a 22-year-old woman who at the age of 11 experienced spells in which she would run in fear to her mother. At the time of a right temporal lobectomy, stimulation of the hippocampus near the uncus resulted in her immediately crying out and saying, "That's the feeling . . . an indescribable fear" (see Chapter 24).

A 31-year-old man described by Penfield and Perot[76] had seizures characterized by a scared feeling, *scrotal movement*, and visual hallucinatory experiences, followed by automatisms. At the time of surgery, depth stimulation in the region of "incisural sclerosis" on the left side induced him to say, "I feel the beginning of a spell." There was trembling of his legs, and he said, "I saw a man coming towards me."

Such as the ones already given, most examples of fear-expressing automatisms are of a protection-seeking nature—fleeing as if in terror, running to someone for protection, and the like. In their video studies of automatisms, Escueta *et al.*,[11] without going into other details, refer to the occurrence of *defensive* flailing and kicking.

Dejected Behavior

Crying,[77] or alternate crying and laughter,[78] has been described as part of automatisms. This matter will be one of the main topics in Chapter 28. In the preceding chapter, reference was made to a patient who experienced a feeling of sadness and wanting to cry that was associated with a welling up of tears. Other than that, I am unaware of any cases in which crying or weeping was linked to the feeling of the preceding aura. Nor have I encountered in case reports descriptive accounts of behavior that mimics states of depression or guilt. Mention was made in the preceding chapter of the patient described by Williams, who, following a sudden onset of ictal depression, experienced a compulsive urge to commit suicide and, upon two occasions at least, had to be restrained from killing herself.

Gratulant Behavior

Except for the vocal manifestations of giggling and laughter,[79] a review of case records reveals only scant evidence of gratulant behavior during automatisms. One patient is described as euphoric and talkative during a seizure. When asked how she felt, she replied, "Wonderful," but had no memory afterwards of what had occurred.[79]

One of Hughlings Jackson's[80] notes on one of his patients would seem pertinent to gratulant automatisms of a playful nature. During an office visit, he observed the following automatism of a young doctor identified as Case Z. In Jackson's words, the young physician broke off in conversation and

> leaned over one arm of his chair and felt about on the floor as if searching for something: . . . Shortly, having a pin in his hand . . . he made a feint of pricking my hand; the action was as if in fun, for he stopped well short of my hand and was smiling. This little affair was exactly after the manner of joking with a child. . . .[81]

The next day Z told Jackson that he had remembered nothing until he returned home.

The remaining cases to be mentioned under the present heading will appear again in Chapter 28. Apropos of the inferred role of the cingulate gyrus in play (Chapter 21), a case referred to by Geier *et al.*[82] is of interest. The patient was a 19-year-old woman who was shown by depth recording to have a medial frontal epileptogenic focus involving the cortex of the anterior cingulate gyrus at the level of the genu. She experienced a seizure in which she suddenly sat up and moaned for about 30 sec. There were some left-sided clonic movements, predominantly of the left face and arm. Afterwards she said that she had just dreamed of "*playing* with a friend." It is also of interest with respect to the question of play and its cerebral representation that Talairach *et al.*[83] elicited a "playful gesture" in one patient by stimulation of the anterior supracallosal cingulate gyrus.

There are several reports of laughter as a seizure manifestation of temporal lobe epilepsy.[84] In a remarkable case reported by Van Buren, electrical stimulation with electrodes in the amygdalohippocampal region elicited laughing, followed by a laughing automatism.[85] In some studies, the electroencephalographic findings have indicated a medial frontal or orbitofrontal focus.[86] In one case, retraction of the left side of the mouth and a "smile" like that of the automatism were induced by stimulation applied with an electrode near the medial orbital surface.[87]

Caressive Behavior

Mention was already made in the preceding chapter of a few instances illustrating caressive kinds of behavior, including that of a woman who during an automatism would walk around her room showing marked affection to anyone present and that of another woman who verbalized her sexual needs in conjunction with invitational behavior. Gowers[88] cited the case of a 20-year-old woman in whom "each slight seizure was followed by a paroxysm of kissing." There are some reports in which the case histories were indicative that sexual intercourse constituted part of the automatism.[89]

Complex Disorganized or Quasi-Organized Automatisms

In complex disorganized automatisms, a patient may behave in a confused, disoriented manner for a prolonged time. Although such automatisms are frequently referred to in case reports, there is usually little description of them beyond such statements as the patient "would find himself blocks away"[90] in some unfamiliar place. Cases of traffic violations may provide witnesses and more details, such as the man described by Penfield and Jasper,[91] who drove his automobile through a red light, proceeded two blocks, and came to a stop in the midst of traffic.

As opposed to complex disorganized automatisms, some patients will continue during an attack to perform an ongoing job or other activity in a quasi-appropriate manner. In making this point, Forster and Liske[92] cite cases such as those of a dishwasher and a laundry worker who, except for a certain loss of agility, continued their jobs without interruption.

Historically, perhaps the best-documented complex quasi-organized automatism was the one described by Jackson's patient (Case Z) mentioned above.[80] As already noted, the patient was a physician. Having suffered from attacks since the age of 12, he became depressed by the hopelessness of his condition and took an overdose of chloral at the age of 43. At autopsy, a small cavity was found in the uncinate gyrus (Figure 25-1). According to the report, the cavity was "$\frac{5}{8}$ inch below the surface just in front of the recurved tip of the uncus . . . [and] was like those seen long after softening from thrombosis or embolism; there was no surrounding inflammation nor any indication that it was of recent origin."[93] The patient experienced the automatism in question while in his office examining a patient. In describing the occurrence, he wrote:

> I was attending a young patient whom his mother had brought me with some history of lung symptoms. I wished to examine the chest, and asked him to undress on a couch. I thought he looked ill, but have no recollection of any intention to recommend him to take to his bed at once, or of any diagnosis. Whilst he was undressing I felt the onset of a *petit-mal*. I remember taking out my stethoscope and turning away a little to avoid conversation. The next thing I recollect is that I was sitting at a writing-table in the same room, speaking to another person, and as my consciousness became more complete, recollected my patient, but saw he was not in the room. I was interested to ascertain what had happened, and had an opportunity an hour later of seeing him in bed, with the note of a diagnosis I had made of 'pneumonia of the left base.' I gathered indirectly from conversation that I had made a physical examination, written these words, and advised him to take to bed at once. I reexamined him with some curiosity, and found that my conscious diagnosis was the same as my unconscious—or perhaps I should say, unremembered diagnosis had been.[94]

This case recalls the complex automatisms of two other physicians. Lennox,[95] the noted epileptologist, cites the case of an obstetrician who, during an automatism, successfully carried out a difficult podalic version (feet-first delivery) and perineal repair. There is a report of another physician, an accomplished organist, who was accompanying carol singers on Christmas eve when he suddenly "switched" to jazz.[92] At the termination of his seizure, he returned to the "exact bar" of the carol that he had been playing.

To emphasize again that complex automatous behavior may be not only well organized, but also appropriate, I cite the case of an engineer for the New York, New Haven, and Hartford Railroad who I saw in consultation.[96] It was reconstructed that he had suffered a psychomotor attack at the 125th Street Station in New York City, but had appropriately taken his train through a series of red and green lights to Grand Central

588 ORIGINAL ARTICLES AND CLINICAL CASES

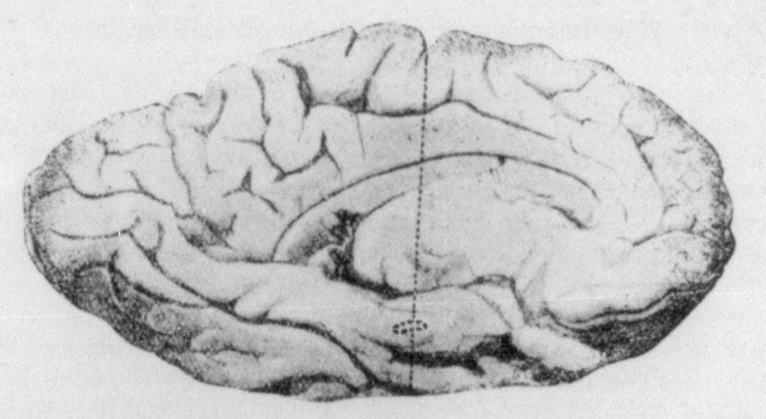

FIG. 1.

Drawing of internal surface of left hemisphere (case of Z). The black ring corresponds with the position of the cavity found, which is shown in section in Fig. 2.

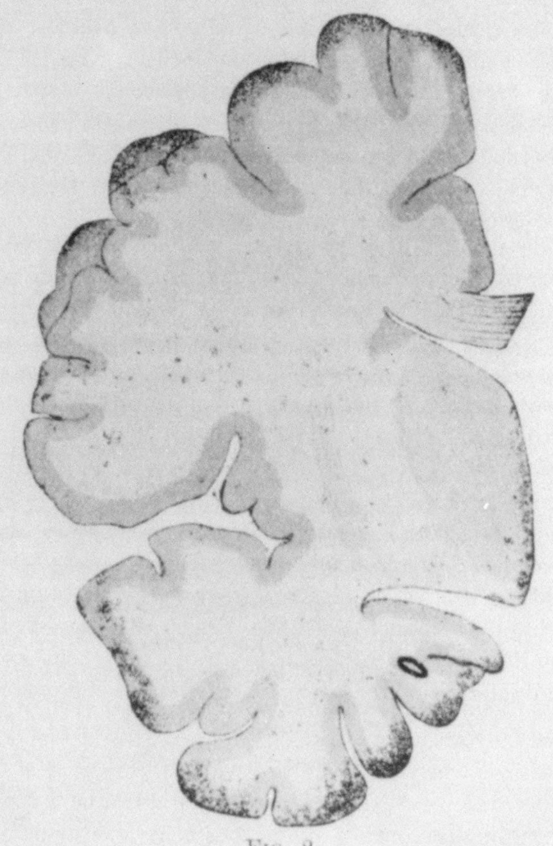

FIG. 2.

Vertical section of left hemisphere (case of Z, from a tracing). Site of small cavity apparently due to softening, shown by black ring.

Figure 25-1. Illustration from Jackson and Colman's (1898, p. 588) paper describing the cerebral lesion in Jackson's famous case Z.

Station. When he stepped down from the cab, he realized that he had experienced one of his seizures.

Concluding Comment

It is evident that the nature of the many automatisms observed in psychomotor epilepsy lends support to the introspective data on affects showing that the limbic system is fundamentally involved in the experience and expression of emotion.

Granted that many of the acts carried out during automatisms reflect learned forms of behavior, the question arises as to whether or not a more detailed analysis of automatisms might provide insights into the nature and evolution of unlearned forms of human behavior. Take, for example, the question about chest beating sometimes seen during epileptic automatisms. This is a form of behavior that few people would have the opportunity to learn in the course of day-to-day living. Chest beating can probably be considered a natural human proclivity because, in addition to epileptic automatisms, it may also occur *de novo* in states of delirium. In human beings, is it the natural tendency to strike the chest with the open hand as in the case of gorillas,[97] or is it performed with the clenched fist that is sometimes portrayed by actors? Mention has been made here of a case in which it is clearly stated that the woman beat her chest with her *hands*.

To ask a related question, what may have been the evolution of the use of the clenched fist, both as a threatening, aggressive posture or as an instrument to be used in fighting? Cases have been cited in which the patient either showed a clenched fist, shook the fist, or displayed actual fisticuffs during automatisms. As far as I have been able to ascertain, none of the pugnacious behavior observed during automatisms resembles the formalized boxing used in either ancient or modern times. The rare assaultive behavior that I have observed during automatisms has involved flailing of the upper extremities and striking with the flat of the hand that has been documented in the video recordings of Escueta *et al.*[11] The behavior is reminiscent of what has been observed in chimpanzees.[98]

One wonders whether or not the throwing of the body against the wall as seen in some automatisms does not reflect an innate behavior used in cases of entrapment, as might be the situation in a cave in which earth fell against an outlet. A more puzzling question is that regarding the frequent disrobing observed during automatisms. Such disrobing has been observed in panic-stricken members of primitive societies (e.g., Ref. 99). Might this ridding of clothes reflect an innate form of behavior used by a primitive people to avoid ensnarement while escaping from some critical situation, or perhaps burning in the case of fire? Perhaps a continuing refinement in the analysis of automatisms will not only give further insights into innate forms of human behavior, but also provide additional diagnostic information regarding the localization of brain lesions.

References

1. Gloor *et al.*, 1982, p. 130
2. Penfield and Jasper, 1954
3. Gloor *et al.*, 1980
4. King and Ajmone Marsan, 1977
5. Feindel and Penfield, 1954
6. Penfield and Jasper, 1954, pp. 525–526
7. Jasper, 1962
8. Todd, 1861
9. Poletti *et al.*, 1973
10. Gibbs *et al.*, 1948
11. Escueta *et al.*, 1982
12. Jackson and Stewart, 1899/1958; MacLean and Arellano, 1950; Mulder and Daly, 1952; Feindel and Penfield, 1954; Van Buren and

Ajmone Marsan, 1960; Pagni and Marossero, 1965; Wadeson, 1965
13. Penfield and Jasper, 1954; Penfield and Perot, 1963; Wadeson, 1965; Stevens *et al.*, 1969
14. Penfield and Jasper, 1954; Feindel and Penfield, 1954; Williams, 1956; Hunter *et al.*, 1963; Maspes and Pagni, 1964; Stevens *et al.*, 1969; Remillard *et al.*, 1974
15. MacLean and Arellano, 1950; Golub *et al.*, 1951; Caveness, 1955; Green and Scheetz, 1964; Glaser, 1967; Remillard *et al.*, 1974
16. Stevens *et al.*, 1969
17. Gibbs *et al.*, 1948; Anastasopoulos *et al.*, 1959; Glaser, 1967; Stevens, 1957
18. Williams, 1956; Stevens, 1957
19. Stevens, 1957
20. Penfield and Jasper, 1954, p. 424
21. Williams, 1956
22. Penfield and Jasper, 1954; Caveness, 1955; Williams, 1956; Daly, 1958
23. MacLean and Arellano, 1950; Anastasopoulos *et al.*, 1959
24. Williams, 1956; Glaser, 1957
25. Coats, 1876; Kershman, 1949; MacLean and Arellano, 1950; Stevens, 1957; Henry, 1963; Glaser, 1957, 1967
26. Penfield and Jasper, 1954; Williams, 1956; Penfield and Perot, 1963; Maspes and Pagni, 1964
27. MacLean and Arellano, 1950; Mulder and Daly, 1952; Williams, 1956
28. MacLean and Arellano, 1950; Caveness, 1955
29. Van Buren, 1961
30. Williams, 1956; Bennett, 1965
31. Hunter *et al.*, 1963
32. Mulder and Daly, 1952
33. MacLean and Arellano, 1950
34. Lehtinen and Kivalo, 1965; King and Ajmone Marsan, 1977; Escueta *et al.*, 1982
35. Bennett, 1965; Geier *et al.*, 1977
36. Glaser, 1967
37. Stevens, 1957
38. MacLean, unpublished observations
39. Ferguson, 1962
40. Gibbs *et al.*, 1948; Chessick and Bolin, 1962
41. Ironside, 1956; Druckman and Chao, 1957
42. Ironside, 1956; Lehtinen and Kivalo, 1965
43. Daly and Mulder, 1957; Druckman and Chao, 1957; Weil *et al.*, 1958; Van Buren, 1961; Ajmone Marsan and Goldhammer, 1973; Ludwig *et al.*, 1975; Jandola *et al.*, 1977; Bachman *et al.*, 1981
44. Glaser, 1967
45. Forster and Liske, 1963
46. Gibbs *et al.*, 1948; Penfield and Jasper, 1954; Bennett, 1965
47. Daly, 1958
48. Ironside, 1956; Williams, 1956; Freemon and Nevis, 1969
49. MacLean and Arellano, 1950; MacLean, 1952; Penfield and Jasper, 1954
50. Hooshmand and Brawley, 1969
51. Spencer *et al.*, 1983
52. Freemon and Nevis, 1969
53. Macrae, 1954; Rodin *et al.*, 1955; Williams, 1956; Daly, 1958; Maspes and Pagni, 1964
54. Anderson, 1886; MacLean and Arellano, 1950; Penfield and Jasper, 1954
55. MacLean and Arellano, 1950; Penfield and Jasper, 1954; Rodin *et al.*, 1955; Penfield and Perot, 1963; Lehtinen and Kivalo, 1965
56. Mulder and Daly, 1952; Macrae, 1954; Williams, 1956; Daly, 1958; Maspes and Pagni, 1964; Wadeson, 1965
57. Feindel and Penfield, 1954; Penfield and Perot, 1963
58. Gibbs *et al.*, 1948; MacLean and Arellano, 1950; Green and Scheetz, 1964
59. Vonderahe, 1944
60. Falconer and Pond, 1953
61. Ferguson, 1962
62. MacLean, unpublished
63. Chessick and Bolin, 1962
64. Coats, 1876; Hill, 1949; Feindel and Penfield, 1954; Lennox, 1960; Crandall *et al.*, 1963
65. Wilson, 1930
66. Lennox, 1960; Crandall *et al.*, 1963
67. Jackson and Colman, 1898/1958; Stevens, 1957
68. R. G. Heath, 1962
69. Lennox, 1960; Delgado *et al.*, 1968
70. Ferguson, 1962; Jasper, 1962; Freemon and Nevis, 1969; King and Ajmone Marsan, 1977
71. King and Ajmone Marsan, 1977; Escueta *et al.*, 1982
72. Mulder and Daly, 1952; Sano and Malamud, 1953; Chessick and Bolin, 1962
73. Stevens, 1957; Ferguson, 1962; Delgado *et al.*, 1968; King and Ajmone Marsan, 1977
74. Williams, 1956, p. 49
75. Green and Scheetz, 1964, p. 139
76. Penfield and Perot, 1963, p. 636
77. Gibbs *et al.*, 1948; Lehtinen and Kivalo, 1965
78. Lehtinen and Kivalo, 1965
79. Mulder and Daly, 1952
80. Jackson and Colman, 1898/1958
81. Jackson and Colman, 1898/1958, p. 460
82. Geier *et al.*, 1977, p. 956
83. Talairach *et al.*, 1973
84. Daly and Mulder, 1957; Weil *et al.*, 1958
85. Van Buren, 1961
86. Lehtinen and Kivalo, 1965; Loiseau *et al.*, 1971; Ludwig *et al.*, 1975
87. Ludwig *et al.*, 1975
88. Gowers, 1881
89. Van Reeth *et al.*, 1958; Stevens, 1957; Blumer, 1970
90. Green and Scheetz, 1964, p. 139
91. Penfield and Jasper, 1954, p. 429
92. Forster and Liske, 1963
93. Jackson and Colman, 1898/1958, p. 462
94. Jackson, 1888/1958, pp. 404–405
95. Lennox, 1960
96. MacLean, unpublished observations
97. Schaller, 1963
98. Lawick-Goodall, 1968
99. Yap, 1952

26

Microelectrode Study of Limbic Inputs Relevant to Ontology and Memory

In addition to providing the most definite evidence that the limbic system is implicated in the experience and expression of emotional feelings, the study of psychomotor epilepsy supplies invaluable leads in regard to two related matters, indicating that the limbic cortex may process information derived from inputs from the various intero- and exteroceptive systems requisite for (1) feelings of being an individual and (2) memorative functions. The present chapter will deal with the question of inputs, while the next chapter will focus on subject matter pertaining to individuation and memory.

So as to have a single frame of reference for describing various findings on limbic inputs, I will summarize a series of our microelectrode studies dealing with this problem, noting in each case relevant anatomical findings and how the results supplement or differ from existing information. In our microelectrode studies, we have tested the responsiveness of more than 14,000 units, of which nearly 40% were located in various limbic cortical and subcortical structures.

As noted in the chapters on phenomenology, crude or elaborate sensory experiences related to virtually all of the intero- and exteroceptive systems have been reported in psychomotor epilepsy. It was partly on the basis of such evidence that I suggested in the visceral brain paper of 1949[1] that all the intero- and exteroceptive systems have inputs into the hippocampal formation. The argument could be made, of course, that such symptoms do not arise because of a representation of sensory systems in the limbic cortex, but rather are the result of propagated discharges from the epileptogenic focus to parts of sensory systems in other regions of the brain. Such an argument, however, is tempered by the realization that there appears to be a reciprocal innervation of most structures of the brain involved in any particular function.

Antecedent Macroelectrode Findings

Except for olfaction,[2] there existed prior to 1950 hardly any experimental neuroanatomic or electrophysiological evidence that other sensory systems projected to the limbic lobe. Many of the earlier comparative neurologists took it for granted that the basal forebrain areas were a site of integration of the visceral with the olfactory and gustatory

Parts of this chapter adhere closely to an article entitled "An Ongoing Analysis of Hippocampal Inputs and Outputs: Microelectrode and Neuroanatomical Findings in Squirrel Monkeys" that appeared in a volume on the hippocampus (MacLean, 1975e).

senses. Thus, Edinger[3] referred to this general region as the site of integration of the "oral senses." Herrick (1921)[4] concluded that the anterior olfactory nucleus receives an innervation from ascending sensory systems. Johnston (1923) referred to the amygdala as "a complex in which olfactory, gustatory, and general somatic sense impressions are brought into correlation."[5]

In 1938, Bailey and Bremer[6] reported that vagal stimulation in the cat elicited synchronized activity in the cortex of the orbital gyrus, an area corresponding to the part of the insular cortex overlying the claustrum. In 1949, Pribram and I, while working in the Laboratory of Physiology at the Yale University School of Medicine, conducted experiments on cats and monkeys in which we explored the medial and lateral cortical surface of the cerebral hemispheres with recording bipolar electrodes while applying shocks to the vagus nerve. The animals were anesthetized with sodium pentobarbital (Nembutal), which had become the standard agent used in studies on evoked slow potentials. Except for the occurrence in the monkey of inconsistent slow waves of variable latency recorded from the lateral frontal cortex near the caudal end of the principal sulcus, the findings were entirely negative.[7] In 1951, Dell and Olson[8] reported that in the encephale isolé preparation of Bremer (animals with section of the cord at the high cervical level and allowed to waken from anesthesia following surgery), vagal shocks evoked a slow-wave response in the buried cortex within the anterior rhinal sulcus as well as in the amygdala.

The following year, MacLean *et al.*[9] reported that in the rabbit gustatory or noxious stimulation resulted in rhythmically recurring olfactorylike potentials in the piriform area, as well as in the "insular" cortex above the rhinal sulcus. It is important to emphasize that such responses could be obtained only when the animal was in the transitional phase from deep to light Nembutal anesthesia. Stimuli of noxious intensity applied to the midline nasal area or to the base of the tail were more effective than such stimulation elsewhere in eliciting a train of olfactorylike potentials. Figure 26-1A and B illustrates that a brief clamp-induced pressure of the stump of the tail was as effective as a prolonged pressure in inducing a train of rhythmic potentials, an observation recalling one's subjective experience upon feeling a brief pinch of the skin.

In the same study, olfactory and gustatory stimulation sometimes elicited rhythmic potentials in the hippocampus. Attention was called to a report by Jung and Kornmuller[10] (1938) in which they included a figure illustrating that pinching of the hind leg or other parts of the body in unanesthetized rabbits resulted in the occurrence of rhythmic potentials at about 5 to 6 per sec in the hippocampus. In following up on these various leads, Green and Arduini[11] found in experiments on unanesthetized, curarized rabbits that different forms of sensory stimulation resulted in the appearance of theta activity in the hippocampus. Such changes were also evident in cats but not in monkeys.[11] MacLean observed that in rats, rhythmic activity at about 6 per sec "characteristically appeared when an animal seemed to focus its attention or was exploring, and disappeared when it was eating or drinking."[12] Altogether, the occurrence of theta activity under so many various conditions was conducive to the view that hippocampal theta activity reflected nonspecific responses resulting from activation of the reticular system, rather than of specific sensory systems. On the basis of ablation studies, Green and Adey[13] concluded that theta activity depends on afferent connections from the septum. In this regard, it is to be noted that in the cat, the proximoamygdalar segment of the hippocampus shows low-voltage fast activity at the same time as theta activity appears in the proximoseptal segment.[14] In macrosmatic animals, the theta rhythm appears to be generated in the pyramidal and radiate layers of the hippocampus.[15]

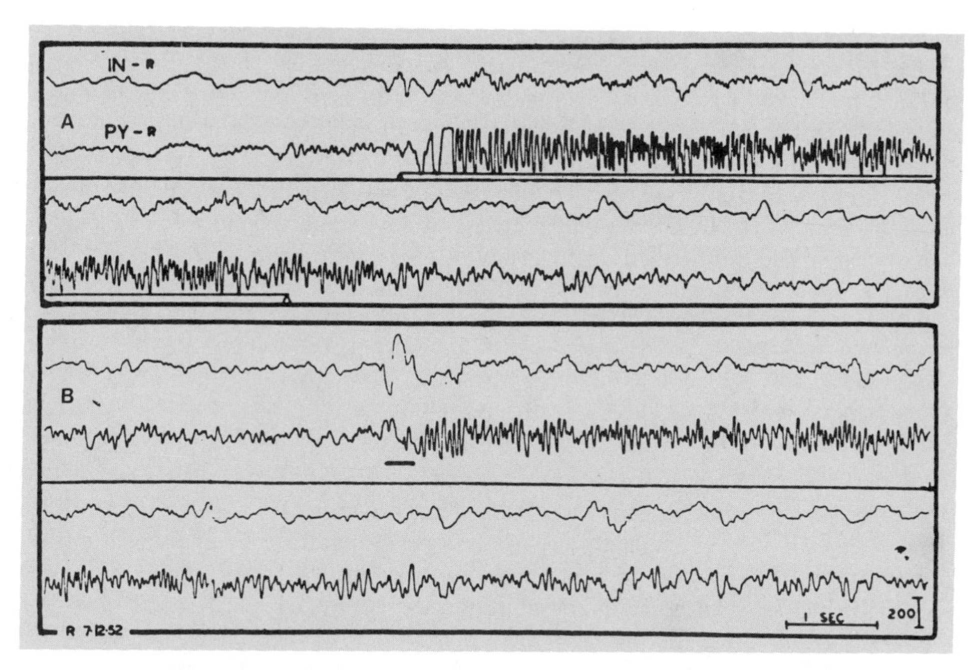

Figure 26-1. Comparison of rhythmically recurring olfactory-like potentials recorded from piriform cortex of a rabbit in response to a momentary and 9-sec application of pressure to stump of tail (Nembutal anesthesia). Note that the brief pressure (bar under tracing in B) induced almost as great and long a response as the one applied for 9 sec (A). From MacLean *et al.* (1952).

Over the years, various authors had reported that visual, auditory, and somatic stimulation evoked slow-wave potentials in the hippocampus or other areas of limbic cortex,[16] but in such experiments, there remains doubt about the authenticity of such responses because of the possibility of volume-conducted potentials from neighboring structures. For example, when averaging techniques became available, one could show that photic stimulation resulted in a buildup of a slow wave in hippocampal recordings that had the same latency as that occurring in the overlying lateral geniculate body.

For our purposes, there was another major disadvantage of the evoked slow potential technique. The electrophysiological work of Woolsey, Marshall, and others[17] had demonstrated that moderately deep Nembutal anesthesia provides stable conditions for studying the configuration and locus of evoked responses in neocortical areas receiving projections from *exteroceptive* systems. Thereafter, it became evident that anesthesia had a vitiating effect on transsynaptic conduction in interoceptive systems.[18] Mention was made above of Dell and Olson's positive findings on vagal projections to the forebrain. Years later, in commenting on these results, Dell[19] remarked that in their initial experiments on animals anesthetized with chloralose or Nembutal, they "never found one vagal projection."

MICROELECTRODE FINDINGS ON LIMBIC INPUTS

Because of the limitations of the evoked, slow-wave potential technique for demonstrating the occurrence and locus of the kind of responses in question, we developed microelectrode techniques for recording evoked unit responses in chronically prepared, awake, sitting squirrel monkeys. Such experimentation not only makes it possible to pinpoint the locus of a response, but also circumvents the depressive effects of anesthesia on neural transmission. It is to be emphasized that the main purpose of these investigations was to learn whether or not limbic cortical cells, like those of neocortical sensory areas, can be specifically and regularly activated by stimulation of various sensory systems. In using extracellular recording we were aware of the possibility of overlooking inputs that induce only partial polarizing effects on single neurons. Subsequently, it proved possible in some structures to obtain intracellular recordings in awake, sitting monkeys.

Methodology

For the microelectrode experiments, we used the same stereotaxic platform that was described and illustrated (Figure 19-7) in connection with the stimulation studies dealt with in Chapter 19. The method provides a closed system for exploration with microelectrodes and among other advantages makes it possible to obtain serial histological sections in the same plane as the electrode tracks. Heretofore, such experiments had commonly required the use of anesthesia, open surgery on the brain, paralysis of the animal, and other measures to prevent movement of brain tissue by vascular pulsation and respiration. In the original experiments, platinum microelectrodes tipped with platinum black were used,[20] but subsequently glass microelectrodes made from theta glass were used for either extracellular or intracellular recording. No more than two experiments were performed in one week, and no more than eight tracks were explored in one animal. The other conditions pertaining to the animal's comfort and freedom of movement in the home cage between experiments were the same as described in Chapter 19. By leaving the electrodes implanted, both the shank and tip were clearly outlined by gliosis. Histological sections were projected and drawn at a magnification of 10× on millimeter graph paper. Because the frozen sections stained for cells conform closely to the original size of the brain (fiber-stained sections may shrink by as much as 10%), there was usually a good fit of the location of units within cellular areas along the reconstructed track.

Exteroceptive Inputs

The findings on respective limbic inputs will usually be presented in the following order: (1) a brief description of the methods; (2) a statement of the principal microelectrode findings; (3) relevant findings by other workers; (4) neuroanatomical correlations; and (5) comment on significance.

Visual Inputs

Methods

Unit responses to photic stimulation were tested by means of a stroboscopic flash of 50 μsec and an on–off presentation of light from a 2-W tungsten lamp for various durations.[21] For these experiments, it was not practicable to plot receptive fields. As a gross means of testing a response to moving edges, an eight-spoked figure was projected on a translucent screen 60 cm in front of the monkey and swept back and forth at various frequencies.

Principal Findings

Altogether, units tested under the above conditions totaled 4600, including 1199 in limbic cortical areas.[21] For comparative purposes, units were recorded in structures known to be photically responsive, including the lateral geniculate body, the superior colliculus, and neocortical visual areas. In the limbic cortex, photically responsive units were encountered only in the posterior part of the parahippocampal gyrus, the parahippocampal part of the lingual gyrus, and the retrosplenial cortex. Their topological distribution is illustrated in Figure 26-2, with the letters H, L, and R signifying, respectively,

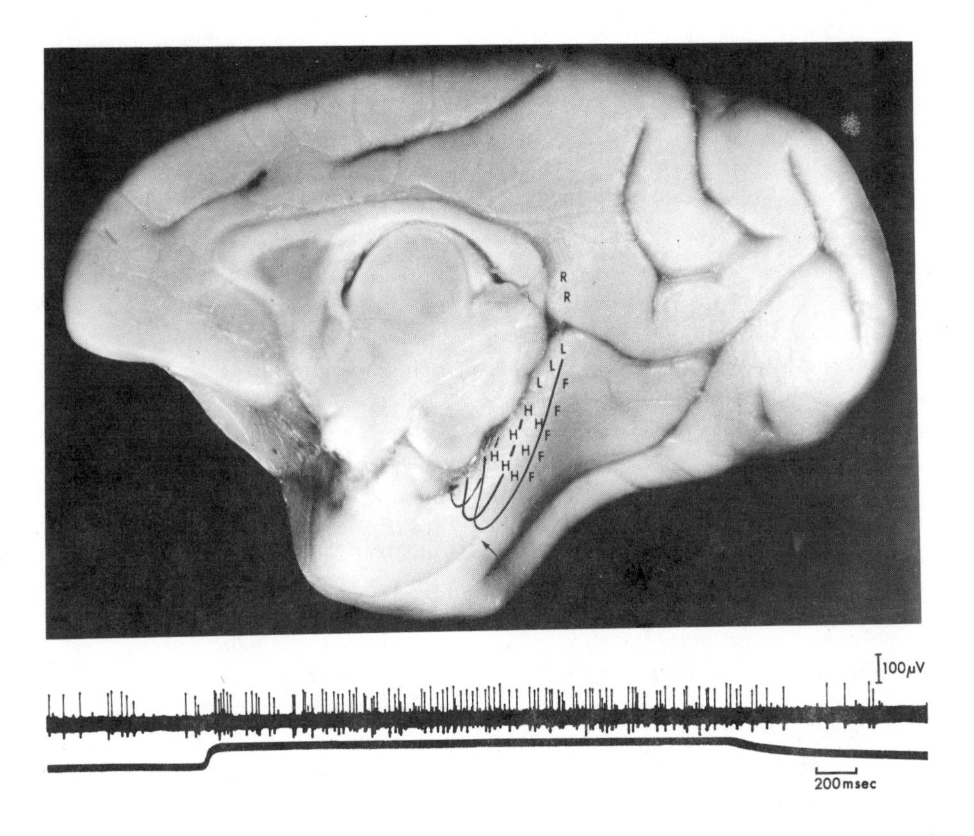

Figure 26-2. Topographical location of photically responsive units in temporo-occipital region of squirrel monkey (top), together with illustration of the photically sustained on-responses typical of half of the responsive cells in the posterior hippocampal gyrus (H). Ls and Rs, respectively, overlie the parahippocampal part of the lingual gyrus and the retrosplenial cortex, where phasic responses were evoked by photic stimulation. Fs indicate the location of the neocortex in the fusiform gyrus in which photically responsive units were recorded.

(Top) The curved lines schematize the location and course of the part of the optic radiations forming the temporal loop.

(Bottom) The electroencephalographic tracing illustrates a sustained on-response of a unit located in the posterior parahippocampal gyrus induced by ocular illumination with a tungsten light. The tonic discharge endures throughout the application of the stimulus. For locations of such units see Figure 26-3.

From MacLean *et al.* (1968).

the three areas. In an earlier study on squirrel monkeys anesthetized with α-chloralose, Cuénod *et al.*[22] had found photically responsive units in these same areas. Of 279 units recorded in the posterior part of the hippocampal gyrus,[21] 18 (6.5%) were responsive, with latencies ranging from 42 to 95 msec. All gave "on" responses. It was of special interest that half of these gave a sustained on-response during ocular illumination (Figure 26-2). Representative sites of electrodes recording such units are shown in Figure 26-3. In the adjacent neocortex of the fusiform area (F in Figure 26-2), 10% of 210 units responded to photic stimulation, but none showed a sustained on-response. Twenty-nine percent of 42 cells in the parahippocampal part of the lingual gyrus and 21% of 68 in the retrosplenial area were photically responsive. Most of the units in these areas gave phasic on-responses, but none gave a tonic, sustained on-response. Some cells in the retrosplenial cortex were characteristic insofar as they responded only to stimulation of the contralateral eye, suggesting the possibility that impulses originated in the primitive temporal monocular crescent.

Of 600 units recorded in the hippocampus, none responded to photic stimulation. The same was true of 90 units recorded during exploration of the amygdala. One hundred and one units tested while exploring the supracallosal, pregenual, and subcallosal cingulate cortex were photically unresponsive except for one that showed a decrease in its firing rate upon ocular illumination with a tungsten light.[23]

The photically activated limbic units in these studies appeared to be modality specific insofar as they were unresponsive to auditory and somatic stimulation. Those giving a sustained on-response showed an increased firing rate in the range of 9 to 33 per sec (mean 18 per sec) throughout the period of ocular illumination. Elsewhere, units of this particular type were recorded only in the lateral geniculate body and lateral tegmental process of the pons.[21]

Other Forms of Visual Stimulation

As illustrated in Figure 26-4, during exploration of the insula, a few units were encountered that responded to any approaching object with an accelerating discharge.[24] Such responses raise the question as to whether or not neural activity of this kind might be relevant to symptoms of macropsia experienced by patients with epileptic discharges involving the insular cortex.

In a study testing the effects of vagal volleys on unit activity in the amygdala, we by chance encountered a unit that responded with a rapid discharge only when there was a moving stimulus in either temporal field.[25] Moving objects elsewhere were without effect. Photic stimulation was ineffective. The electrode was fixed at the site where the unit was recorded, and it proved to be in the medial basal nucleus of the amygdala just beneath the capsule surrounding the central nucleus.

Findings of Other Workers

In patients with epilepsy, Babb *et al.*[26] recorded from the "hippocampal formation" units with response characteristics apparently similar to those giving sustained "on" responses in the monkey. They stated that the neurons may respond "as a geniculate X-cell by sustained firing to sustained illumination" and that "the rate of firing may be directly related to the light intensity."[26] They recorded from some neurons with "discrete receptive fields similar to those plotted in animal striate cortex."[26] The latencies of the photically responsive units were as short as 27 msec. On the basis of their findings, the authors inferred that there are "lateral geniculate axons that project to the hippocampal formation of man, representing a source of visual information to the limbic system significantly different . . . [from] the cortico-cortical projections from peri-striate" cortex.[26]

Evidence of visual anatomico-physiological correlations in regard to the retrosplenial cortex is suggested by experiments on lissencephalic animals. In 1950, Thompson *et al.*[27] reported that in the rabbit, contralateral photic stimulation of the eye evoked potentials in the limbic cortex medial to the splenial sulcus. O'Leary and Bishop[28] elicited potentials in this same area by stimulation of the optic chiasm. Rose and Malis[29] found in the rabbit that the "peristriate" cortex of this area medial to the splenial sulcus receives projections from the "dorsal lateral geniculate body." Karamian *et al.*[30] have presented evidence of a convergence of retinogeniculate and retinotectothalamic projections within area 29 of the retrosplenial cortex.

Neuroanatomical Findings

What is the origin of the visual input to the posterior parahippocampal gyrus, the parahippocampal cortex of the lingual gyrus, and the retrosplenial region? In one of our earlier studies on squirrel monkeys anesthetized with chloralose, it was found that bipolar shocks applied to the lateral geniculate body evoked short-latency unit responses in the posterior hippocampal gyrus and adjoining part of the lingual gyrus.[31] These and the findings described above led to a neuroanatomical study in which an electrocoagulator was used to produce lesions in the lateral geniculate body and surrounding structures, and the resulting degeneration was traced by improved Nauta techniques.[32] As a control for damage produced by the inserted electrode, it proved possible in two cases to introduce an electrode into the lateral geniculate body by an extracerebral approach along the course of the optic tract.

Following a lesion largely confined to its ventrolateral part (Figure 26-5), a continuous band of degeneration extended from the lateral geniculate body into the core of the hippocampal gyrus. Some fibers entered the cortex of the posterior hippocampal gyrus, as well as contiguous areas in the lingual gyrus and in the fusiform cortex. Traced caudally, the medial degenerating fibers in the optic radiation intermingled with those of the cingulum, and, in addition to degeneration in the primary visual cortex within the rostral calcarine sulcus, some fibers were traced into the adjoining retrosplenial cortex. In keeping with the electrophysiological findings, degeneration did not extend rostrally into the entorhinal cortex of the hippocampal gyrus. Lesions involving the inferior pulvinar resulted in a coarser type of degeneration that was traced to the same limbic and perilimbic areas. Such pulvinar projections were contained in a band of fibers just lateral to the optic radiations.

As originally shown by Myers[33] and confirmed by Mishkin,[34] silver degeneration studies in the monkey reveal a strong projection from the superior colliculus to the inferior pulvinar. Snyder and Diamond[35] have demonstrated a comparable pathway in the tree shrew in which the collicular fibers project to the lateral dorsal nucleus believed to be the counterpart of the inferior pulvinar in primates. Hence, it is evident that a collicular pulvinar pathway may be a source of one visual input to the posterior hippocampal gyrus. However, it is unlikely that it would be a sole source of visual afferents because there were no units in either the superior colliculus or pulvinar that gave sustained on-responses such as those in the posterior hippocampal gyrus.[21]

Experiments by Conrad and Stumpf[36] and others indicate an extrageniculate visual

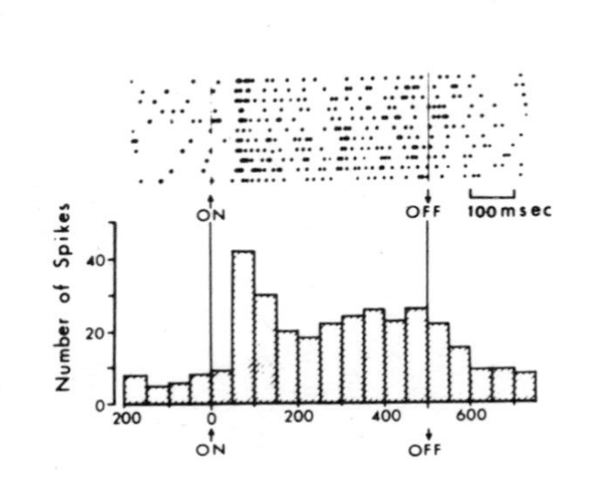
A
Number of Spikes
40
20
0
200
0
200
400
600
ON
OFF
100 msec

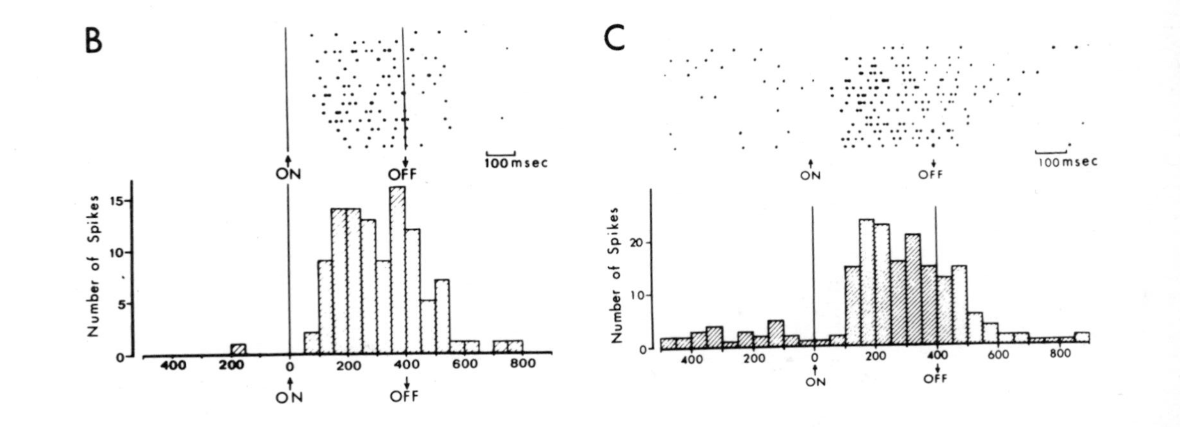
B
Number of Spikes
15
10
5
0
400
200
0
200
400
600
800
ON
OFF
100 msec
C
Number of Spikes
20
10
0
400
200
0
200
400
600
800
ON
OFF
100 msec

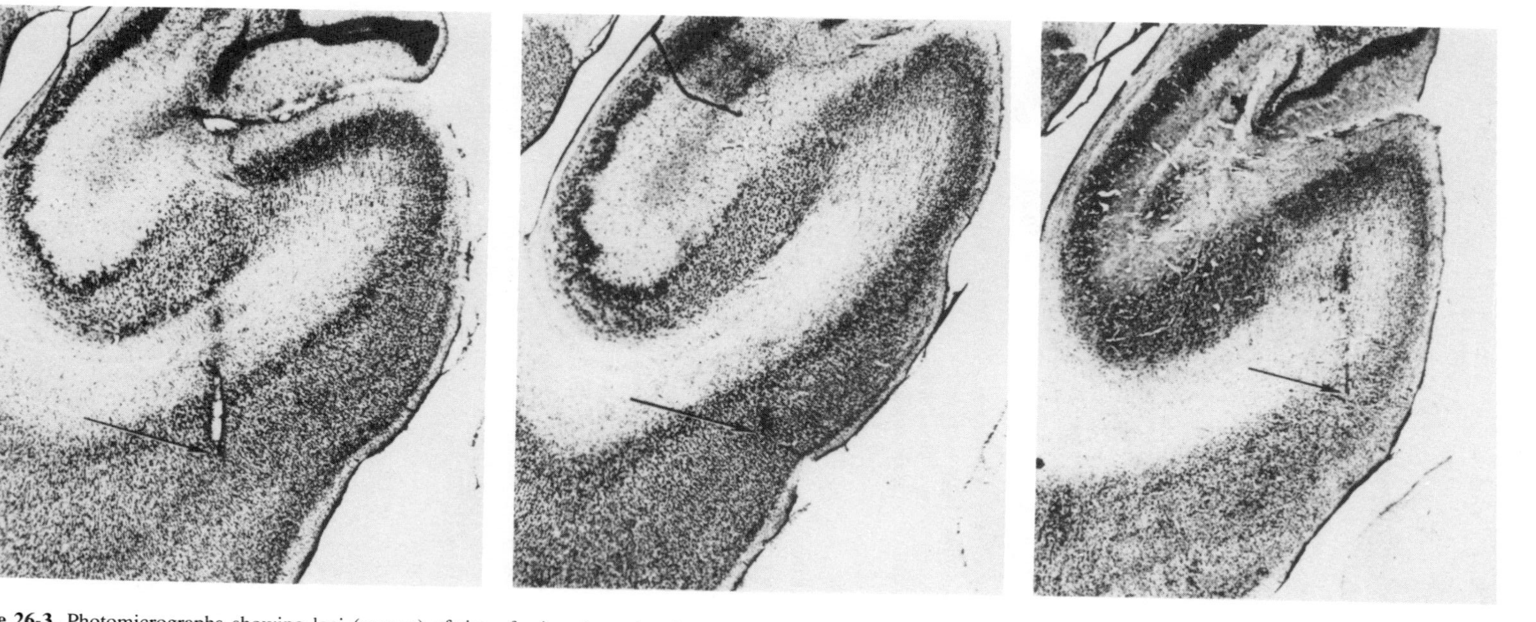

Figure 26-3. Photomicrographs showing loci (arrows) of tips of microelectrodes that recorded photically sustained on-responses of units in the posterior part of the hippocampal gyrus. In each case, the electrolytic marks involve the thin granular layer. The oscillographic dot displays of pre- and poststimulus time interval histograms for units recorded at these loci are shown above. From MacLean *et al.* (1968).

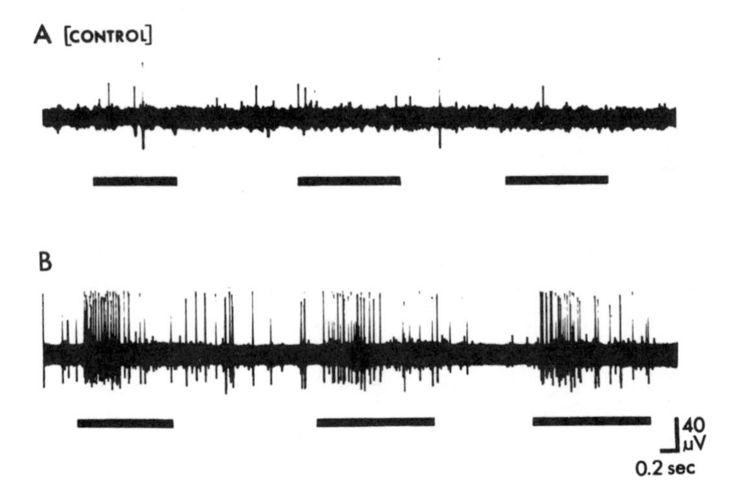

Figure 26-4. Response of an "approach-type" unit in insular cortex of squirrel monkey. Units of this type (lower tracing) showed an increase in the rate of discharge whenever the subject looked at an approaching object. As shown in upper control tracing, such movements performed in the dark were without effect, demonstrating that response was not owing to tactile stimulation from air currents. From Sudakov *et al.* (1971).

input to the retrosplenial cortex. In the tree shrew, Conrad and Stumpf[36] injected tritiated leucine into the eye and traced a contingent of optic nerve fibers to the contralateral anterior dorsal nucleus of the thalamus. Using HRP for anterograde labeling of fibers, Itaya *et al.*[37] described comparable findings in the rat.

Robertson *et al.*[38] have described another potential pathway by which visual information might reach the retrosplenial cortex. As reviewed in Chapter 18, the lateral dorsal nucleus has extensive projections to the cingulate cortex in both primates and subprimates. Following injections of HRP into the lateral dorsal nucleus, Robertson *et al.* found labeled neurons in several pretectal cell groups (olivary, anterior and posterior pretectal nuclei of the optic tract, and medial pretectal region), as well as anterograde labeling in layers I and III of the midline posterior cingulate and retrosplenial cortex (including areas 29a, b, and c; see Figure 18-9). As reviewed by these same authors, the pretectal region receives direct projections from the retina.

In agreement with the described evidence of four different visual inputs to the retrosplenial cortex, Akopyan[39] found that following application of dry HRP to area 29 of the rat, labeled cells appeared in the dorsal part of the lateral geniculate body, and in the anterior dorsal, lateral posterior, and pretectal nuclei.

In regard to the amygdala responses, it is to be noted that Pickard and Silverman,[40] using HRP ocular injections for anterograde labeling of retinal projections, traced some fibers to the piriform cortex and region of the septum.

Some investigators have described visually evoked slow-wave[41] and unit[42] responses in the claustrocortex of the cat. Bignall *et al.* observed that the slow-wave response was eliminated by destruction of the lateral geniculate body.[43] Reference has been made in Chapters 4 and 18 to the anomalous nature of the claustrum. Some workers have reported elicitation of photically evoked slow-wave[44] and unit responses in the claustrum,[45] while Olson and Graybiel[46] have described units that respond to moving bars and spots of light. Reciprocal connections between the claustrum and primary visual cortex have been demonstrated by the application of recent neuroanatomical techniques (Chapter 18).

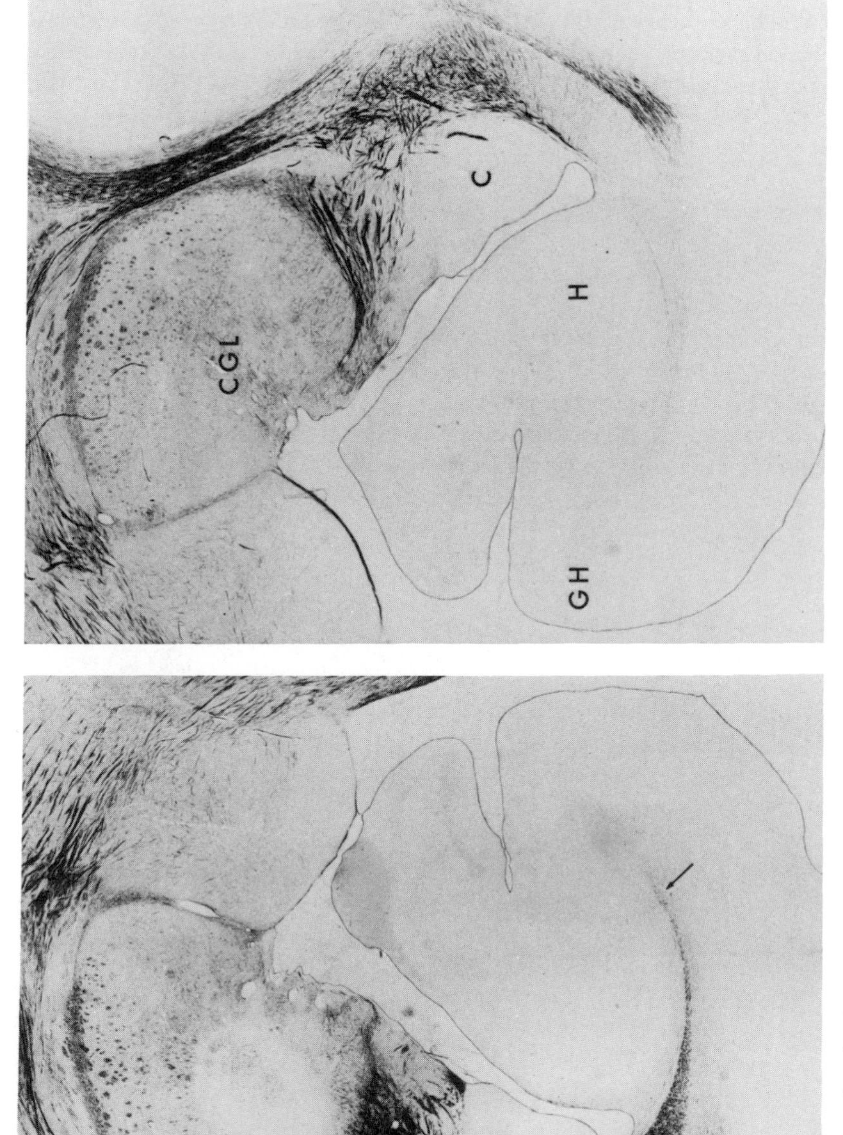

Figure 26-5. Histological section showing degeneration in Meyer's loop extending into the core of the hippocampal gyrus (small arrow). The lesion in the ventrolateral part of the lateral geniculate body shows as a clear area (large arrow). The nonlesioned side from this same section is shown on the right. See text for description of degeneration seen under higher power. Modified Nauta stain; ×10 magnification. From MacLean and Creswell (1970).

Comment

On the basis of the foregoing electrophysiological and anatomical findings, it is evident that limbic cortical and subcortical structures are not solely dependent on transcortical association pathways for visual information, but are also directly connected with visual centers via subcortical pathways. The main visually responsive limbic cortical areas are the posterior hippocampal gyrus, the parahippocampal part of the lingual gyrus, and the retrosplenial cortex. In retrospect, it appears that the traditional emphasis on the connections of the hippocampal formation with the olfactory apparatus created a predisposition to overlook the possibility of connections of the limbic cortex with other sensory systems.[47] As evident from Figure 26-6, a sagittal section through the human hippocampal formation shows a folding of the hippocampal formation in the region of the calcarine sulcus that almost compares to that in the region of the rhinal fissure. In 1902, Elliot Smith[48] pointed out that the calcarine sulcus ranks along with the hippocampal and rhinal ''fissures'' as being one of the three most constant furrows in the brain. In the first figure of that paper, it is shown as corresponding to the retrosplenial stem of the splenial sulcus of the infraprimate brain. Campbell[49] confirmed Smith's observation[50] (Figure 26-7) that the depth of this furrow marks the boundary between the limbic and striate cortex (Chapter 21). In primates, the abrupt transition between the limbic and striate

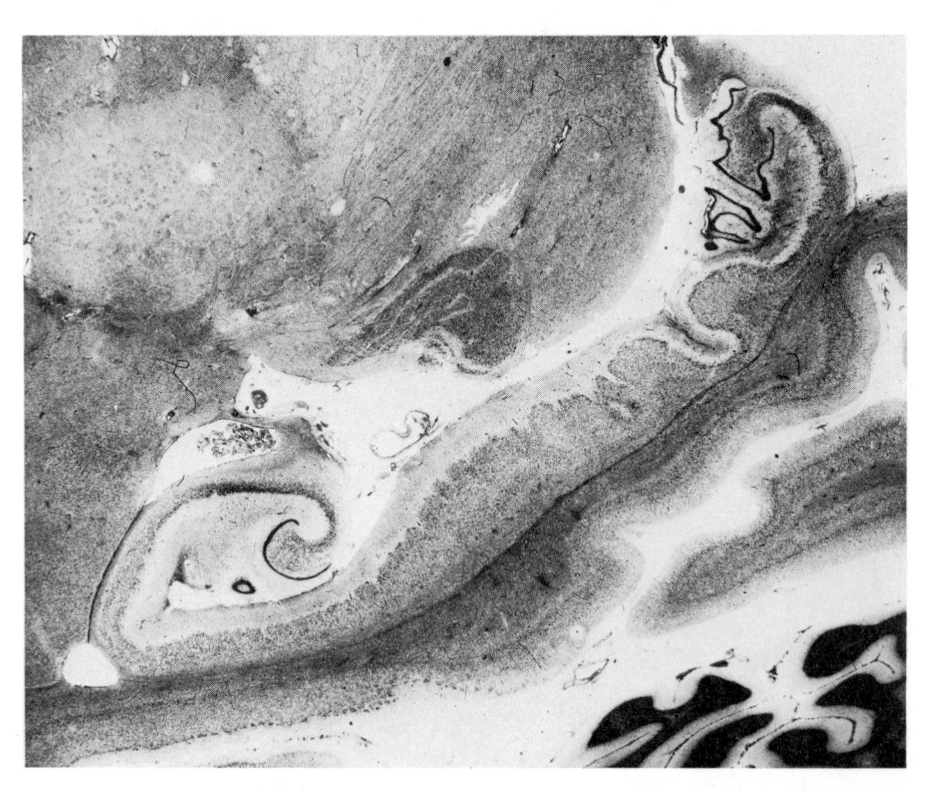

Figure 26-6. Longitudinal section illustrating the long axis of the hippocampus in the human brain. Note that there is almost as much folding of the hippocampus in the calcarine region (right side) as there is in the region of the uncus. From MacLean (1972b).

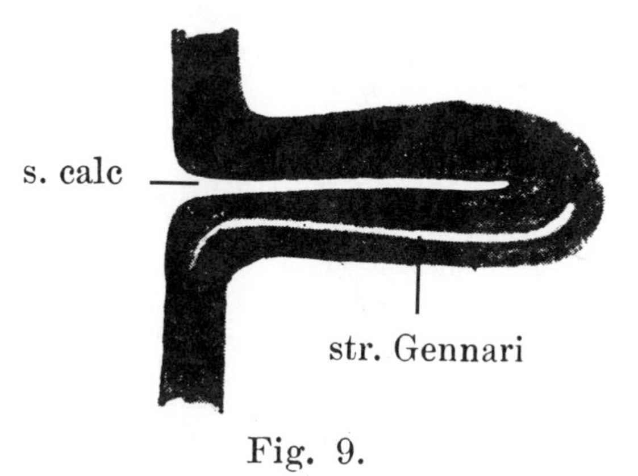

Figure 26-7. Figure from Elliot Smith's paper of 1903–04, illustrating his observation that the visual cortex, as identified by the line of Gennari, terminates abruptly in the depth of the anterior calcarine sulcus. The limbic cortex is immediately adjacent. Hence, the border between the striate and limbic cortex is as abrupt as the border between the striate cortex and the peristriate visual cortex. From Smith (1903–04).

cortex occurs near the junction of the anterior and posterior calcarine sulci. In primates generally, and in the human brain in particular, there is a great expansion of the perilimbic and perivisual cortex in the medial parietal and medial occipitotemporal regions. The expansion of the hippocampal formation, including the hippocampal gyrus, was illustrated in Figure 18-18. A fusiform gyrus marked off by a sulcus between it and the hippocampal gyrus represents a new development in the brains of higher primates.[51]

It has long been puzzling why the optic radiations in the inferior longitudinal fasciculus make a detour around the lateral ventricle. As Polyak noted, this portion of the radiation "thins out into a narrow fiber sheet, which slips medially underneath the ventricle and becomes the core of the hippocampal gyrus."[52] "It is very probable," he adds, "that the most inferior and medial bundles of this fiber lamina of the temporal lobe are in no way concerned with the visual function." The temporal detour of the optic radiation was originally known as Flechsig's "temporal knee."[53] Thanks to Adolf Meyer, Cushing,[54] one of the early neurosurgeons, became aware that temporal lobe lesions frequently involved the detouring fibers and resulted in a defect in the upper quadrants of the visual fields. In the legend of the first figure of the 1911 paper by Cushing and Heuer, the roundabout projection was referred to as "Meyer's loop."[54] Subsequently, this designation, rather than "Flechsig's knee," became the familiar expression. In a comparative study, Putnam[55] made reconstructions of the optic radiation and concluded that a "temporal knee" existed not only in primates but also in carnivores. On the basis of the electrophysiological and anatomical studies described above, it would appear that the temporal loop owes its existence to a contribution from the optic radiations to the cortex of the posterior hippocampal gyrus and, beyond that, to the parahippocampal part of the lingual gyrus and retrosplenial cortex.[56]

In our original paper describing the tonic on-units in the posterior hippocampal gyrus,[21] it was suggested that these units possibly reflect activity of mechanisms involved in maintaining a state of wakefulness or inducing alerting and attention.

Since the parahippocampal cortex of the posterior hippocampal and lingual gyri and of the retrosplenial area are a major source of afferents to the hippocampus, the question arises as to why none of 600 units recorded from the hippocampus responded to visual stimulation. A possible explanation is suggested by intracellular findings to be described later, indicating that excitation induced by impulses conducted by the perforant pathway may be subliminal for generating spikes. Such subliminal excitation would suggest mechanisms by which visual influences on hippocampal cells could sum with those of other origin and through projections to the hypothalamus and other parts of the brainstem affect neurovegetative, memorative, and affective functions. In view of the hippocampal changes that occur in association with the rapid eye movement (REM) phase of sleep,[57] the new findings are of interest in regard to the visual aspects of dreaming and the autonomic manifestations identified with REM sleep.

As noted in the introduction, clinical observations on psychomotor epilepsy suggest that limbic structures are involved in seizure discharges during which patients may experience visual hallucinations, *déjà vu,* macropsia, and paranoid visual feelings. Although the experimental findings provide evidence of several different routes by which visual information might reach, and be processed in, the limbic system, such data are insufficient for clarifying mechanisms by which subjective visual symptoms would become manifest in psychomotor epilepsy. In clinical observations on the effects of stimulation of deep occipital temporal structures, Horowitz *et al.*[58] reported that "the posterior hippocampus was the site of greatest interest in terms of visual events." In half of the 26 reports of fully formed imagery, the stimulating electrodes were believed to be in the posterior hippocampus. In some of these instances, "the visual imagery occurred in conjunction with an electrically evoked aura or seizure."[58] Apropos of visual symptomatology in cases of temporal lobe tumors, Cushing commented, "One would naturally expect visual hallucinations to be a feature of occipital rather than of temporal lobe tumors, but the former are less common in my series. . . ."[59] Some patients experienced visual hallucinations in the region of the field defect. Penfield and Jasper[60] demonstrated at operation that stimulation of the hippocampal gyrus of some patients with epilepsy might evoke visual recollections or feelings of *déjà vu*. The approach-type units found in the insular region are of interest to consider in the light of macropsia. In Chapter 23, mention was made of the case of a man described by Penfield and Jasper who, during an aura while seeing a motion picture, had the feeling that the screen was coming closer to him.

Objects entering the peripheral (i.e., temporal) field of vision commonly induce feelings of alarm or startle. Some retrosplenial units responded only to stimulation of the contralateral eye, suggesting that they are activated by retinal projections from the primitive temporal monocular crescent.[21] Such units are possibly involved not only in mechanisms of alarm and startle, but also in the generation of paranoid visual feelings of the kind described in Chapter 23.

Auditory Inputs

In exploring the cingulate gyrus[23] and the hippocampal formation,[21] we encountered no units that responded to auditory stimulation. Such units were found only in the insular cortex overlying the claustrum.[24] This cortex, which Brockhaus[61] referred to as claustrocortex, is by definition limbic because it borders upon the brainstem. It also represents a transitional type of cortex (see Chapter 18). For auditory stimulation, we applied clicks as well as pure tones presented in the form of 5-msec triangular pulses at frequencies ranging from 500 Hz to 20 kHz.[24] We tested 1828 units while exploring the insula and the surrounding areas. Of nearly 500 units recorded in the claustrocortex, 11% responded to auditory stimulation. They appeared modality specific insofar as they were unaffected by visual or somatic stimulation. There were two main types of auditory units. One type responded with a discharge of one to six spikes at short latencies ranging from 7 to 15 msec and occurred regularly if the stimulus frequency did not exceed 10 per sec. The other type responded at longer latencies and with a discharge persisting as long as 250 msec. There was a low probability of firing at stimulus rates greater than 1 to 2 per sec. It was observed by chance that one unit responded only to the vocalization of another monkey.[24]

Comment

In the cat, Desmedt and Mechelse[62] identified a fourth auditory area (A IV) in the "insular" cortex overlying the claustrum. Their observed latencies of evoked slow potentials fell in the same range as the shortest latency of 7 msec for responsive units in our study. The finding of regularly responding units at such short latencies is indicative of a direct orthodromic pathway. This inference is supported by anterograde[63] and retrograde[64] degeneration studies in the cat showing that the cortex overlying the claustrum receives projections from the medial geniculate body. Rasmussen[65] found that the degenerating fibers from the medial geniculate body passed rostrolaterally, following a course just medial to the optic tract.

In summary, the recent electrophysiological findings, together with anatomical studies, provide evidence of direct ascending auditory inputs to the limbic cortex of the insula. In addition, findings with improved anatomical techniques reveal evidence of inputs from neocortical auditory areas into the insula.[66] Outputs from the insula could, in turn, affect viscerosomatic functions of the hypothalamus via pathways to the amygdala and hippocampus mentioned in Chapter 18.

Somatic Inputs

The claustrocortex also proved to be the only limbic cortex in which cells responded to somatic stimulation.[24] Somatic stimulation was performed by light touch with a camel's hair brush and by application of blunt, sharp pressure of a probe attached to a strain gauge for signaling the onset, duration, and strength of the stimulus. About 9% of 497 units responded to somatic stimulation with latencies ranging from 30 to 300 msec. Most of these units were located rostral to those responsive to auditory stimulation. Only three units responded solely to noxious stimulation, the rest being activated also by light touch. The receptive fields were large and usually bilateral. Most of the fields involved the head or upper extremities. As in the case of the insular auditory units and the visual temporo-occipital units, the somatic units appeared to be modality specific.

Since ablation experiments had indicated that the cingulate cortex was implicated in the alleviation of symptoms of morphine addiction,[67] as well as in the relief of pain,[68] it was anticipated that exploration of the cingulate gyrus might reveal units responding to somatic stimulation. None of 84 units tested in the subcallosal, pregenual, or supracingulate cortex responded to light touch, pinprick, or pressure.[23] In contrast, 14% of 130 units in the supracingulate cortex were responsive to somatic stimulation.[23] The receptive fields were large and usually bilateral and always involved the lower extremities.

Comment

The region of the claustrocortex with units responding to somatic stimulation lies in proximity to the so-called somatic areas I and II of the neocortex.[69] The recent anatomical demonstration of insular connections with these somatic areas was referred to in Chapter 18. It may be expected that current improved neuroanatomical techniques will resolve the question as to whether or not the claustrocortex receives direct connections from ascending somatosensory pathways.

The electrophysiological findings, together with neurosurgical observations, would suggest the likelihood that the insular cortex, more than any other limbic area, is implicated in the numerous variations of somatic symptoms experienced in cases of psychomotor epilepsy (Chapter 23). In their clinical paper on the function of the insula, Penfield and Faulk[70] reported that in the accessible part of the insula caudal to the level of the central sulcus, stimulation resulted in subjective feelings that were experienced in the contralateral extremities. The somatic sensations associated with stimulation here and elsewhere in the insula were those of tingling, numbness, warmth, and movement.

Gustatory Inputs

In exploring the insular cortex overlying the claustrum, we tested the effects of gustatory stimulation on 437 units.[24] Ten percent saline, 10% sucrose, and cow's milk were applied as stimulants. Acid and bitter solutions were not used. As a control for tactile and pressure stimulation, the lingual, buccal, and pharyngeal surfaces were mechanically stimulated. Fourteen (3.2%) units responded only to gustatory stimulation.

The majority of the gustatory units fell within the same area of the rostral insula and frontal operculum from which Benjamin and Burton[71] recorded evoked slow potentials upon stimulation of the chorda tympani in the squirrel monkey. In our study, the number of units in the frontal operculum responding to gustatory stimulation was comparable to the value for that of the insula. In their analysis of taste deficits in rhesus monkeys following cortical ablations, Bagshaw and Pribram[72] concluded that the focal cortical area for gustation is located at the junction of the anterior insula and frontal operculum. On the basis of retrograde degeneration, they inferred that the "nucleus medialis ventralis" (alias nucleus reuniens) of the thalamus projects to the rostral part of the insula. Roberts and Akert,[73] on the contrary, reported that anterior insular lesions result in retrograde degeneration in the nucleus ventralis posteromedialis, pars parvocellularis. As reviewed in Chapter 18, Norgren[74] and co-workers found combined electrophysiological and anatomical evidence of an ascending gustatory pathway from the solitary nucleus to the dorsolateral parabrachial region and from there to the ventral posteromedial thalamus and to the central nucleus of the amygdala.

Olfactory Inputs

Electrophysiologically, it has long been recognized that olfactory stimulation elicits rhythmically recurring potentials in the piriform cortex.[75] For many years, however, it has been a matter of contention as to whether or not the hippocampus receives any direct connections from the olfactory apparatus. As mentioned in Chapter 17, Brodal,[76] in his review of 1947, argued that there was no direct evidence of olfactory connections with the hippocampus. In a number of species, however, it has now been conclusively demonstrated by improved techniques that the main olfactory bulb projects to the taenia tecta (Chapter 18).

In our microelectrode studies on the squirrel monkey, we made no systematic attempt to test the effects of natural olfactory stimulation on units of the limbic cortex. We did, however, investigate the differential effects of electrical stimulation of the olfactory bulb and of the septum on both intracellular and extracellular potentials of hippocampal neurons in awake, sitting squirrel monkeys.[77] We also tested the effects of such stimulation on neurons in the entorhinal cortex. Upon extracellular recording of hippocampal neurons with glass micropipettes, we found that double or triple shocks applied to the olfactory bulb failed to elicit unit spikes in either hippocampal or entorhinal cells. Such stimuli, however, elicited excitatory postsynaptic potentials (EPSPs) in 26% of 54 hippocampal cells that were successfully penetrated. Similar EPSPs without spikes were also induced in entorhinal cells.[77,78] The response latencies for the hippocampal cells ranged from 15 to 17.5 msec. The latencies for the entorhinal units were shorter by 2 to 2.5 msec,[77,78] a finding consistent with other evidence that the entorhinal area transmits olfactory impulses to the hippocampus (Chapter 18).

Interoceptive Inputs

Vagal Inputs

In our initial attempts to test the effects of vagal stimulation on forebrain units in awake, sitting squirrel monkeys, there developed the complication that because these animals flex and turn their necks to such an extreme degree, the vagal electrodes tended to produce mechanical injury of the nerve. Finally, a modification entailing fixation of the electrodes next to the nerve at the entrance of the jugular foramen made it possible for these animals between experiments to engage in all their usual cage activities without risk of vagal injury.[79]

Cingulate Cortex

In the first study, we tested the effects of single, double, and triple shocks to the vagus on the responsiveness of 518 cingulate units.[79] One hundred (19.3%) units were responsive to vagal volleys, with about half of them showing initial excitation and half initial inhibition. For comparative purposes, we also explored the supracingulate cortex and found that 110 (22%) of 498 units were responsive, with 82 showing initial excitation and 28 initial inhibition. As a control for adventitious activation by extravagal somatic afferents, vagally responsive units were tested during shock-induced facial and cervical twitches. In agreement with a preceding study mentioned above, somatic stimulation failed to excite cingulate units.[23] For further evidence of vagal involvement, microamounts of serotonin were injected through a catheter in the superior vena cava as a means of exciting pulmonary receptors.[79] Of 80 cingulate units thus examined, 18 showed excitatory or inhibitory effects, with a ratio of excited to inhibited units being 3:1.

An analysis of latencies to vagal volleys showed that the cingulate units responded in a significantly shorter time (12–20 msec for the majority) than did those of the supracingulate cortex. There were two main types of initially excited units: Type 1 units were characterized by a discharge of one to three spikes at relatively short and constant latencies, while type 2 units responded with a burst of three or more spikes at longer and more variable latencies. The mean latency of type 1 units of the cingulate cortex was significantly shorter than for units of the supracingulate cortex.

Thalamus

The response latencies of cingulate units suggested the possibility of both oligosynaptic and polysynaptic pathways. What thalamic nuclei might participate in the transmission of vagal impulses to the cingulate and supracingulate cortex? Exploration of the thalamus revealed that vagal shocks elicited responses of a large percentage of units in the anterior medial, paracentral, lateral dorsal, central lateral, and medial dorsal nuclei.[80] As many as 27% of 367 units in the medial dorsal nucleus responded to vagal volleys. As in the case of the cingulate and supracingulate cortex, the units could be categorized as either type 1 or type 2. The range of latencies of the type 1 units of the anterior ventral, paracentral, medial dorsal, and lateral dorsal nuclei was such as to be compatible with their participation in the transmission of vagal impulses to the respective parts of the cingulate and supracingulate cortex.

Basal Limbic Structures

In subsequent investigations of the vagal projections to the forebrain, the use of improved computer methods[81] led to the discovery of forebrain units that discharged in phase with respiration.[82] While investigating the striopallidum, we observed that application of triple shocks to the left cervical vagus nerve every 4 sec resulted in the entrainment of respiration[82] and that in the case of 6% of the units, there was an associated periodic discharge commensurate with the respiratory rhythm (Figure 26-8).[82] In a subsequent study on basal limbic structures,[83] respiration-related units were found in the limen insulae, anterior amygdala, hippocampus, and hippocampal gyrus. Figure 26-9 shows the records for two such units in the amygdala (see also hippocampal unit response in Figure 26-10A).

Altogether, 16% of a population of more than 200 limbic units responded to vagal volleys. The ratio of initially excited to initially inhibited was about 2:3. Figure 26-10A and B shows examples of initial inhibition and excitation in the case of two hippocampal units. The response latencies of respective units ranged from 15 to 280 msec. In the amygdala, the largest percentage of responsive units was located in the central nucleus, none of which was of the respiratory type.

Subsequently, we undertook a study in which we used the peak of the inspiratory phase to trigger recording of units.[84] In these experiments, we encountered units discharging in phase with respiration in such structures as the corpus striatum, insula, amygdala, hippocampal formation, cingulate cortex, and cortex of the outer convexity. Such patterns were most easily recognized in recordings from the hippocampus. By elimination of inputs from the olfactory bulb,[83,84] it was shown that phasic discharge of hippocampal units was not dependent upon impulses from the olfactory apparatus.

Comment

The finding that some forebrain units show a phasic discharge correlated with either vagally entrained or spontaneous respiration has implications recalling the Greek use of the word *psyche*—meaning air, breath—to refer to the mind or spirit. The findings raise such questions as: (1) What is the possible role of respiration-related units in contributing to the central excitatory state required for consciousness? (2) How might the activity of such units influence the occurrence of sleep, as well as dreaming in which there may be marked cardiorespiratory changes? (3) What is the possible role of such units in psychological functions such as recognition (''breath of recognition''); motivation (e.g., respiratory changes in feats requiring unusual strength); emotional states (e.g., startle, joy, depression); memory and creative thought (''inspiration'')? (4) How is the activity of such units possibly related to the development of hypnotic or hypnagogic states?

Differential Effects of Extero- and Interoceptive Inputs on Hippocampus

Of all the telencephalic limbic structures, the septum, amygdala, and hippocampus are the ones that have been shown anatomically to be most closely connected with the hypothalamus via the pathways described in Chapter 18, including the medial forebrain bundle, stria terminalis, and fornix. As emphasized in Chapter 19, the hippocampus may be pictured as an arciform body anchored medially in the septum and laterally in the amygdala. Through its afferent and efferent connections with these structures and through its own direct connections, it is in a position to exert a widespread influence on the viscerosomatic and neuroendocrine functions of the hypothalamus. It would therefore appear that the adduced evidence of fairly direct inputs to the hippocampus from all exteroceptive and interoceptive systems is of fundamental significance in regard to viscerosomatic and neuroendocrine functions. In this connection, and particularly with respect to neural mechanisms involved in the sense of personal identity, memory, and conditional learning to be considered in the next chapter, it is of special relevance to consider next the differential effects of interoceptive and exteroceptive inputs on hippocampal neurons. Vagal stimulation proved to be the only form of sensory excitation to evoke unit discharge in the hippocampus.

Introductory Considerations

In a study reported in 1964, Gergen and MacLean[85] applied shocks to the olfactory tract and to the septum as a means of comparing the effects of representative inputs, respectively, from exteroceptive and interoceptive systems on hippocampal activity. As diagrammed in Figure 26-11 (see p. 492), the typical hippocampal pyramid is characterized by a main apical dendrite (in stratum radiatum) and several basal dendrites (in stratum oriens). In monkeys anesthetized by α-chloralose, we found that septal volleys evoked a negative slow wave in stratum oriens that was associated with a discharge of cells in the pyramidal layer, whereas an olfactory volley applied to the olfactory tract elicited a high-amplitude negative wave in stratum radiatum that was poorly correlated with pyramidal cell discharge. In seeking an anatomical explanation, it was found that medial septal lesions in the squirrel monkey resulted in degeneration in stratum oriens,[85] which contains the basal dendrites of the pyramidal cells as well as a number of interneurons. This finding contrasted with what was known about olfactory inputs, which, according to both anatomical and electrophysiological evidence, follow the perforant pathway that terminates on the distal portions of the apical dendrites of hippocampal

A

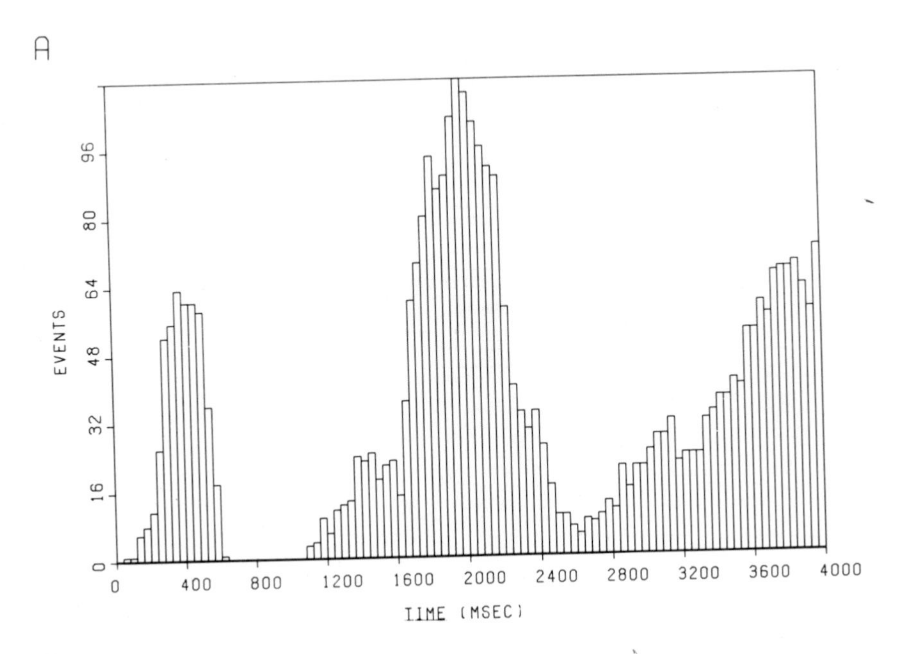

B

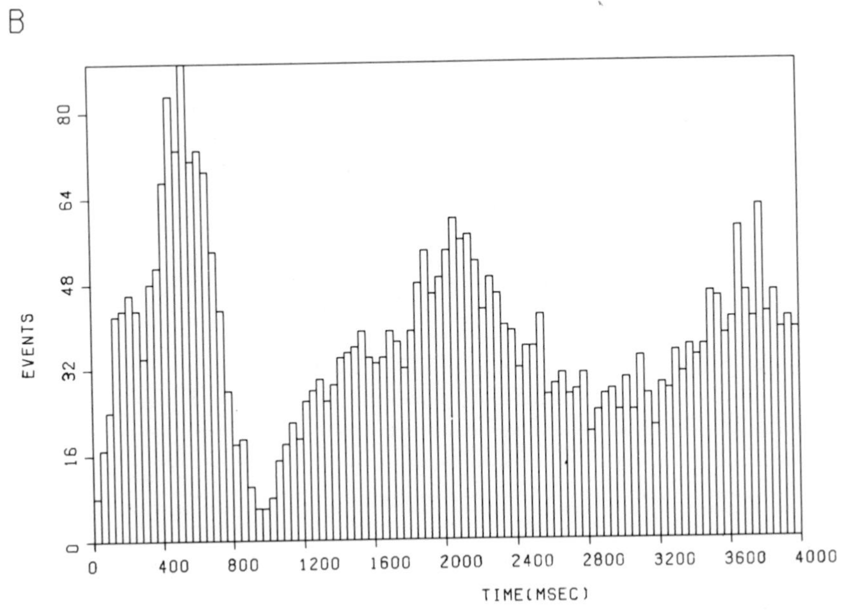

Figure 26-8. Comparison of the firing pattern of a pharyngeal muscle unit and of a unit in the putamen associated with entrainment of respiration by vagal volleys.

(A) Poststimulus time interval histogram for pharyngeal unit representing 50 uninterrupted trials in a monkey anesthetized with pentobarbital.

(B) Histogram of a unit recorded in putamen in an awake, sitting squirrel monkey (76 uninterrupted trials).

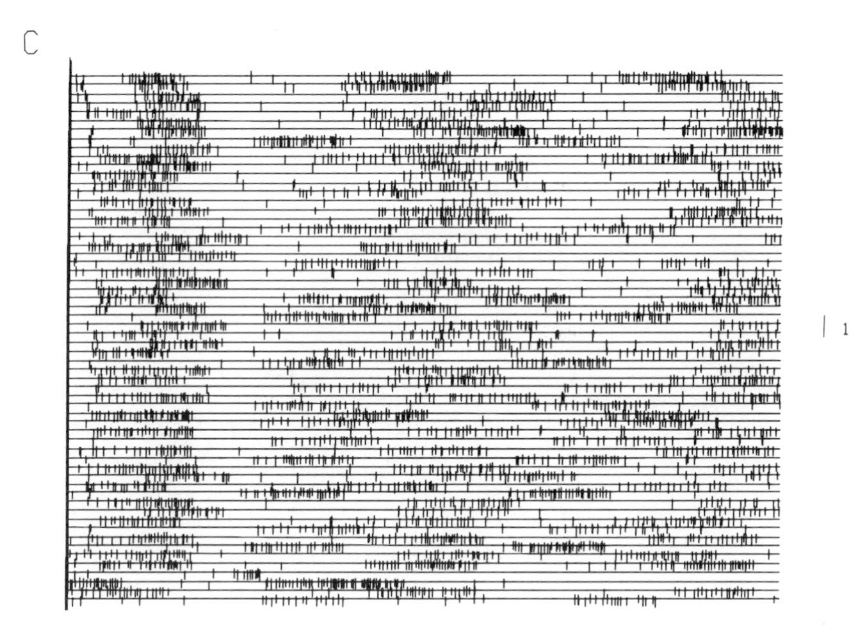

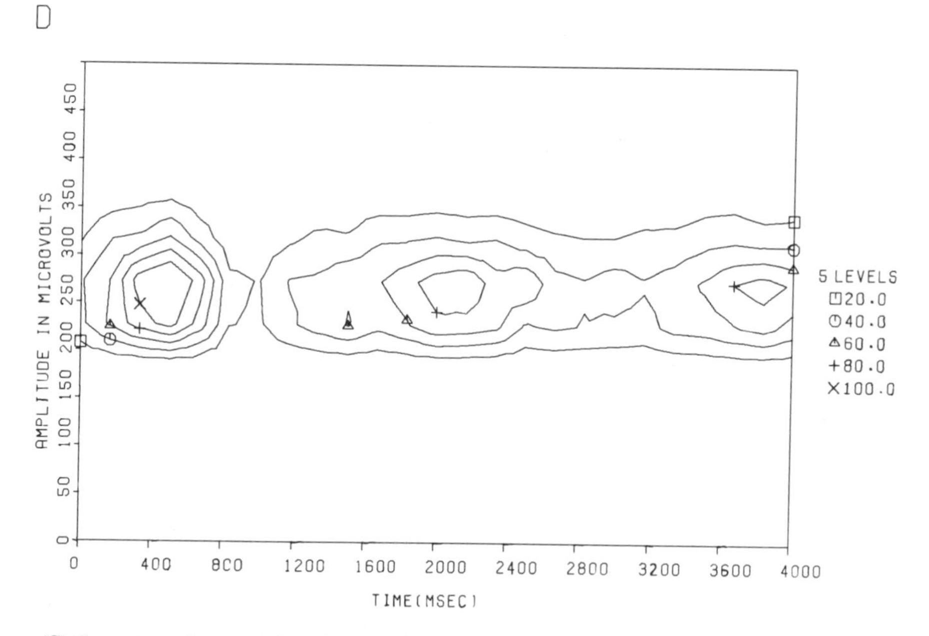

(C) Computer write-out of the microelectrode recording of the unit described in B.
(D) Surface plot of same unit, representing firing pattern as a function of amplitude and latency. From Radna and MacLean (1981a).

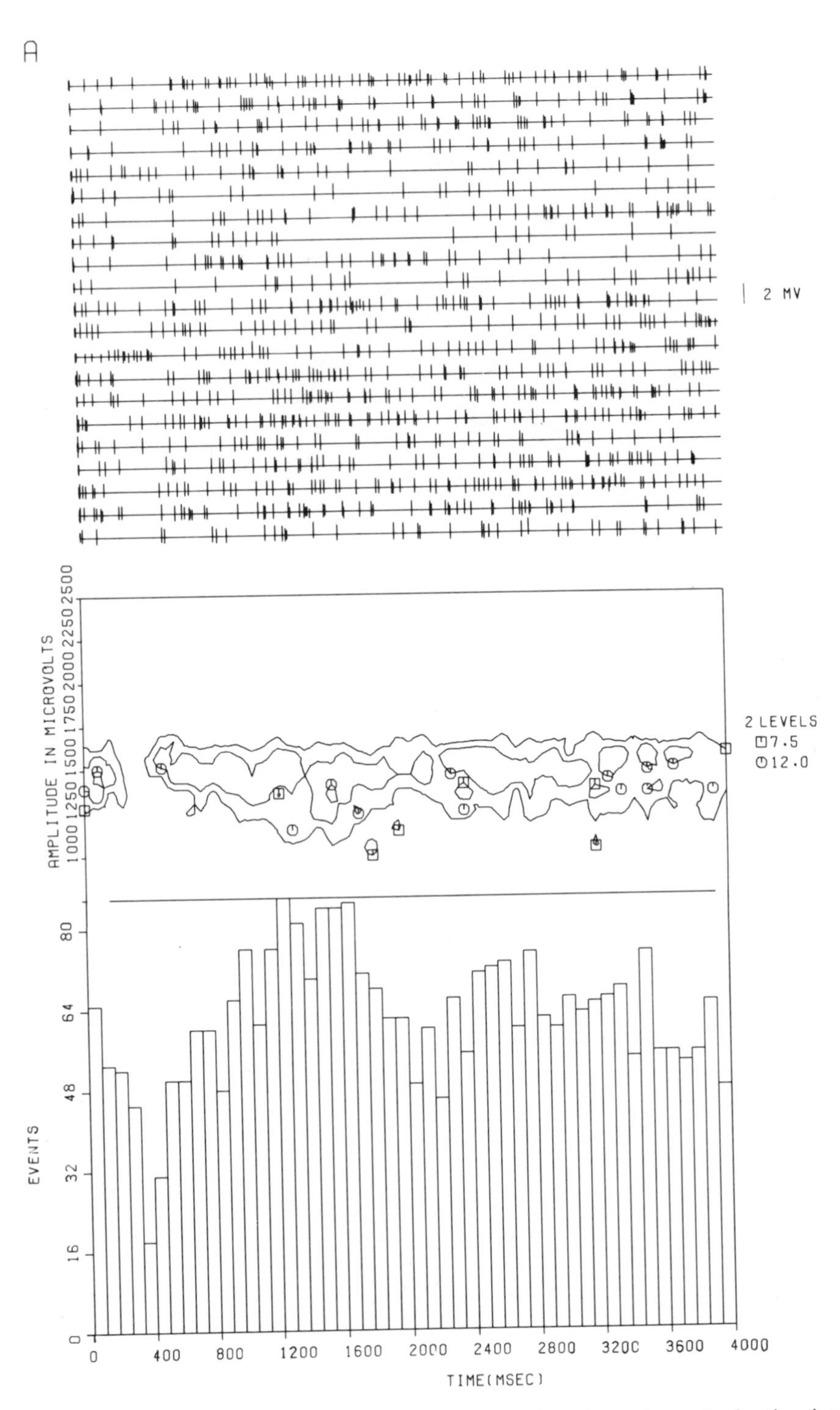

Figure 26-9. Computer write-out of microelectrode recording, surface plot, and poststimulus time interval histogram of two units in the anterior amygdala associated with entrainment of respiration by vagal volleys. The unit responses in A show initial inhibition at 320 msec followed by two main undulating discharges. The unit in

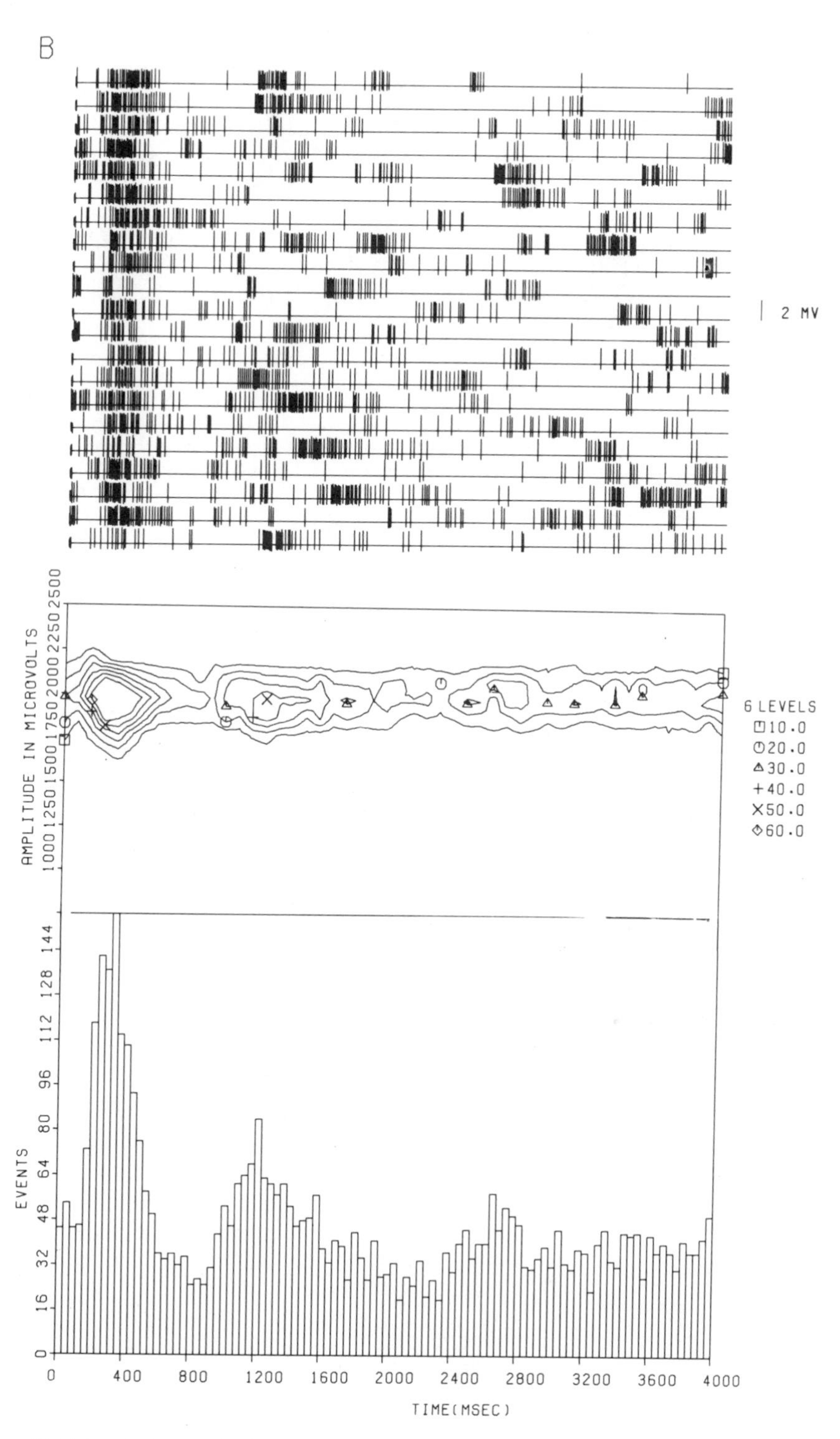

B showed initial excitation maximal at 320 msec, followed by some inhibition and then two phasic discharges. From Radna and MacLean (1981b).

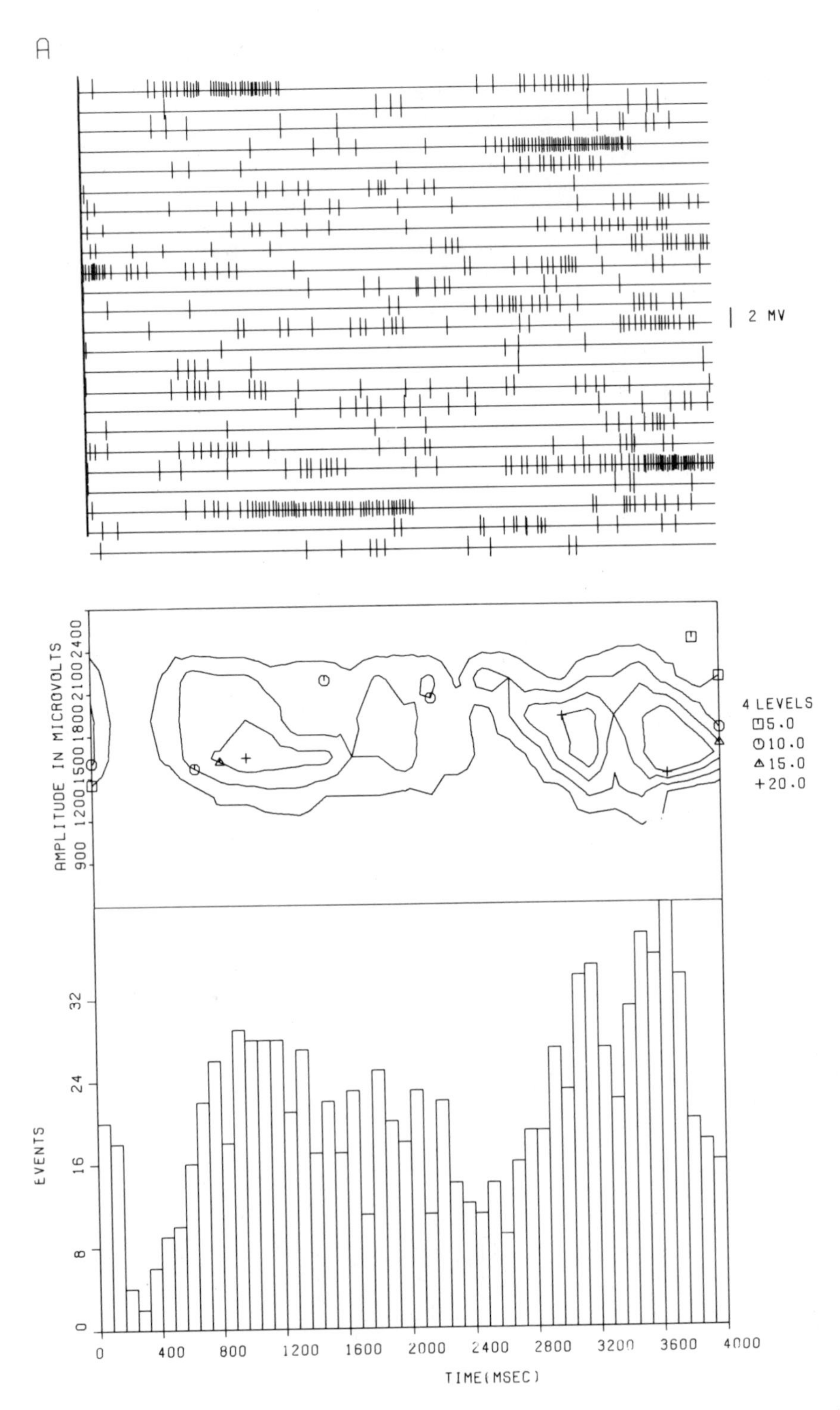

Figure 26-10. Records of responses to vagal volleys of two units in area CA1 of the hippocampus, illustrating initial inhibition in one case (A) and initial excitation in the other (B). Records for each unit show computer write-out of unit recording; contour plot of unit's firing rate, amplitude, and response latency; and poststimulus

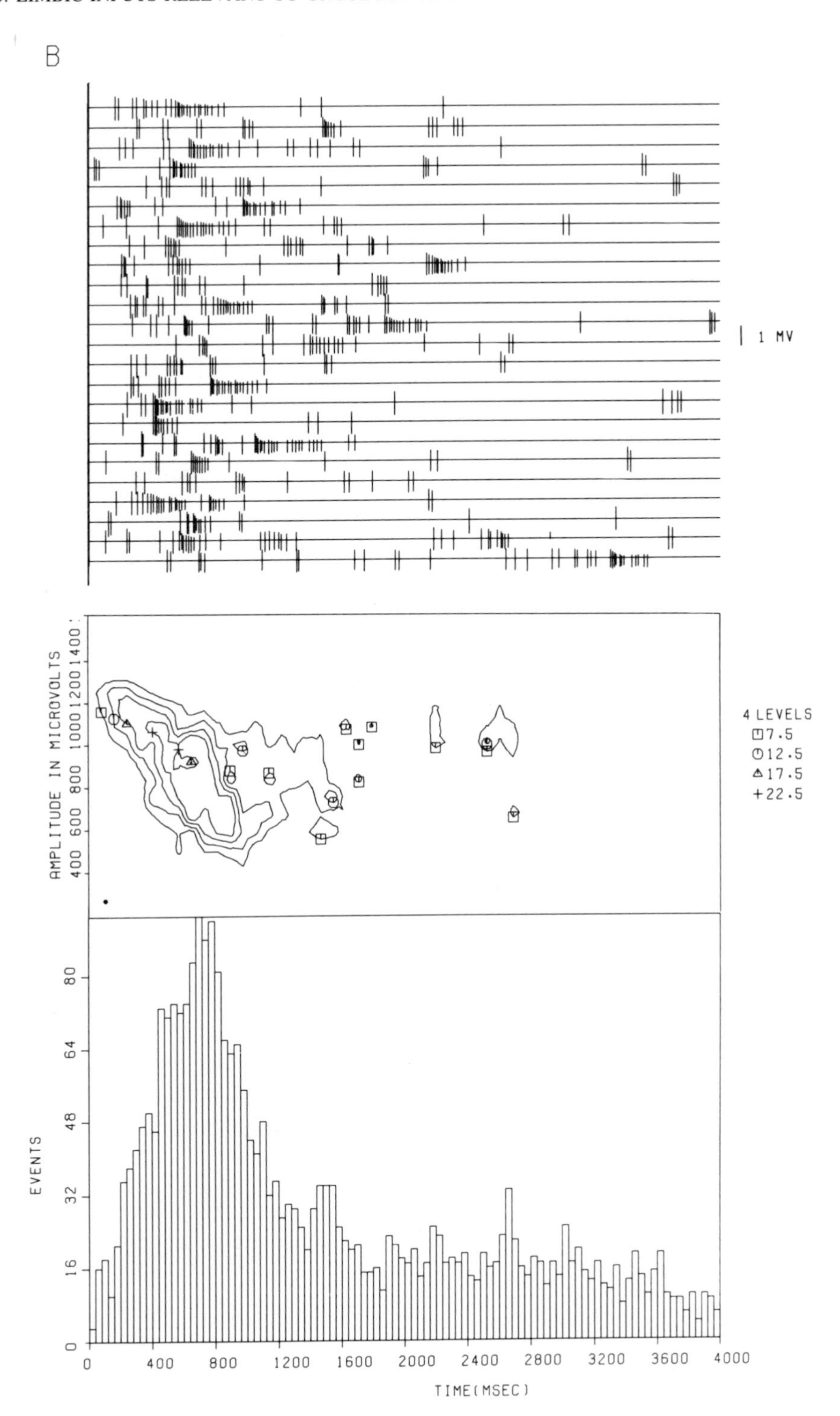

time interval histogram. Note in A the periodic nature of the discharge and the ''rebound'' character of the second undulation (69 trials). In B, the contour plot reflects the bursting discharge and falloff in amplitude of the spike potentials. Histograms represent 126 uninterrupted 4-sec trials. From Radna and MacLean (1981b).

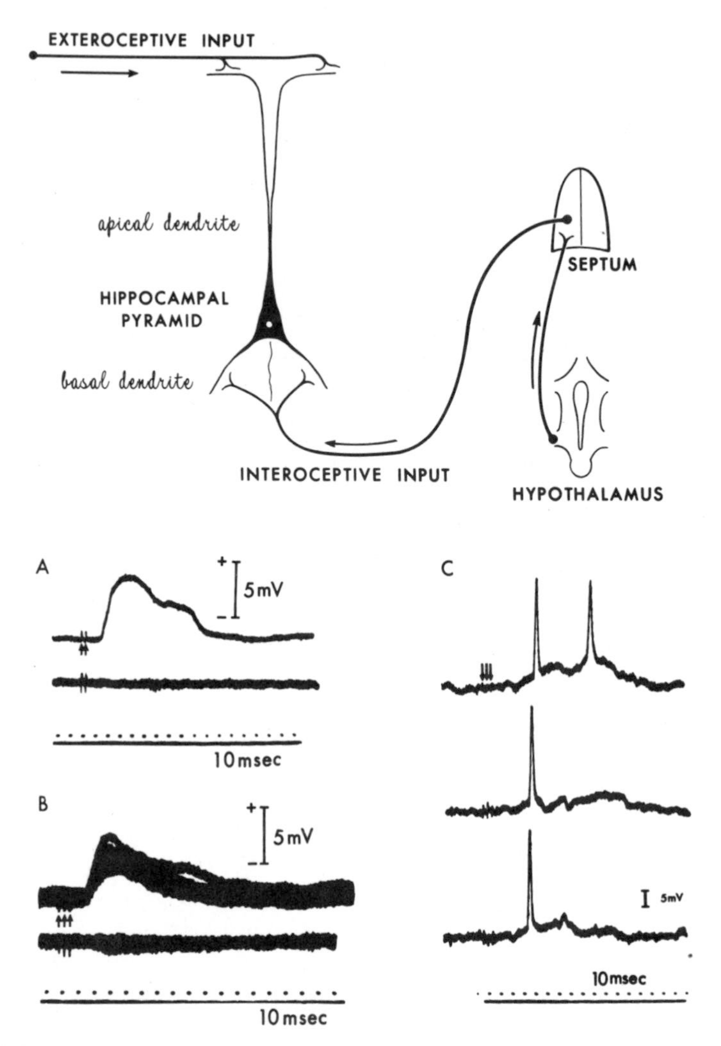

Figure 26-11. Upper diagram illustrates circuitry described in text regarding differential effects of stimulating intero- and exteroceptive inputs on a hippocampal pyramidal cell. As shown in A and B (multiple sweeps), olfactory volleys generated only EPSPs without spikes, whereas septal volleys elicited EPSPs that reached threshold for neuronal discharge (C). Lower recordings in A and B show potential changes just outside cell. Resting potentials were stable, measuring between −40 and −50 mV. Diagram from MacLean (1969b) and records from Yokota *et al.* (1970).

pyramids. In line with current concepts of neuronal excitation, it was suggested that septal impulses act predominantly on the basal dendritic and perisomatic regions of hippocampal pyramids where they are highly effective in triggering neuronal discharge, whereas those of olfactory origin are relatively ineffective because they play upon the distal portions of the apical dendrites.[85]

Findings with Intracellular Recording

Extracellular recording with microelectrodes is useful for detecting the discharge of single nerve cells, but it is inadequate for revealing excitatory or inhibitory postsynaptic potentials developing in an individual cell. As a further means of testing the above hypothesis, Yokota *et al.*[77] obtained both intra- and extracellular recordings in awake, sitting squirrel monkeys. Stimuli were one to three negative, constant-current 0.5-msec pulses with a 3.2-msec interval. Shocks applied to the septum and fornix were of sufficient strength to elicit hippocampal afterdischarges if applied at 30 per sec for 10 sec. Stimuli to the bulb were at threshold strength for eliciting sneezing if applied in long trains.

Septal volleys induced postsynaptic potentials in 60 of 80 impaled hippocampal units that were characterized as (1) simple EPSPs, (2) inhibitory postsynaptic potentials (IPSPs), and (3) combined IPSP–EPSPs. As illustrated in Figure 26-11C, it is to be emphasized that in all cells showing pure EPSPs there was a superposition of one to two spikes. Evidence was presented that IPSPs were attributable to antidromic activation of fornix efferent fibers and the consequent excitation of recurrent fornix collaterals having an inhibitory effect on hippocampal pyramids. Such antidromic effects seen with stimulation of the fornix[86] would tend to counteract excitation of hippocampal neurons induced by afferent stimulation. In two units showing combined IPSP–EPSPs, spike discharges were superposed on the EPSPs.

In contrast to septal stimulation, paired or triple shocks applied to the olfactory bulb elicited only EPSPs. Figure 26-11A, B shows typical responses. Fourteen of fifty-four units showed such EPSPs and in no instance were they associated with spike discharges.

With respect to the occurrence of unit discharge, the extracellular recordings correlated with the intracellular findings. During extracellular recording, 25 of 102 hippocampal units responded to septal stimulation with spike discharges, whereas none of 77 units discharged in response to stimulation of the olfactory bulb.

Comment

The findings with intracellular recording provide additional evidence compatible with the paradigm suggested above regarding the differential action of interoceptive and exteroceptive systems on hippocampal neurons.[85] As originally stated, if the hippocampal circuitry pictured in Figure 26-11 were viewed analogously in terms of classical behavioral conditioning, "the interoceptive impulses conducted by the septal–fornix system would be comparable to unconditional stimuli, since they are capable by themselves of discharging units," whereas those of exteroceptive origin relayed by the perforant pathway from the entorhinal area "would be analogous to conditional stimuli, lacking the capacity when at first acting alone to bring about a consummated response."[87]

Hippocampal Influence on Diencephalon and Basal Telencephalon

Because of the evidence that exteroceptive and interoceptive systems converge either indirectly or directly on the hippocampus and exert a differential action on hippocampal neurons, the present account will conclude with a summary of the results of our observations on the influence of hippocampal volleys on unit activity in neurons of the diencephalon and basal telencephalon. The effects of hippocampal afterdischarges were also observed. Stimulating electrodes were bilaterally implanted in the anterior (proximoamygdalar) and posterior (proximoseptal) segments of the hippocampus, as well as in the subcommissural fornix.

Hypothalamus and Basal Telencephalon

As a prefatory note, it should be stated that in the awake, sitting, unmedicated monkey, we found that, contrary to what had been reported in the cat,[88] no hypothalamic units responded to photic or somatic stimulation (including noxious stimuli) and only a few were affected by auditory stimulation.[89] Such findings would indicate that exteroceptive information affecting the hypothalamus is first integrated and processed in related structures. This inference should be qualified by noting that we were unable to obtain satisfactory intracellular recordings from hypothalamic units and that it is possible that, as in the case of hippocampal and entorhinal units, exteroceptive stimuli may induce depolarizing changes that are subthreshold for eliciting unit discharge.

The effects of hippocampal volleys were tested on 666 units located in the main cellular groups of the hypothalamus and preoptic area, and in the part of the basal telencephalon encompassing the nuclei of the septum (including the nuclei of the diagonal band and stria terminalis), the nucleus accumbens, olfactory tubercle, and caudal part of the gyrus rectus.[88] Altogether 22% of the units of the combined areas were responsive. Units responding with short (9.5–12.5 msec) and constant latencies suggestive of a direct pathway were found in the nucleus accumbens, nucleus of the diagonal band, lateral septum, bed nucleus of the stria terminalis, medial and lateral preoptic areas, anterior hypothalamus, dorsal hypothalamus, perifornical area, medial mammillary nucleus, and posterolateral hypothalamus. Apropos of function (see below), it is to be emphasized that of all of the responsive units in the basal telencephalon, preoptic region, and hypothalamus, more than 83% showed initial excitation, while the rest were initially inhibited.

As in the case of hippocampal volleys, hippocampal afterdischarges more commonly elicited unit excitation than inhibition, with the ratio being 3:1. Units that were excited during an afterdischarge usually became silent upon its termination, whereas those that were inhibited showed a post-afterdischarge increase in firing rate. These post-afterdischarge changes were manifest for periods *ranging from 50 sec to 11 min.*

Neuroanatomical Correlations

In a companion neuroanatomical study,[90] it was found that unilateral section of the fornix resulted in degeneration leading to the structures listed above with units responsive at short and constant latencies. Degeneration was also traced to the posterior part of gyrus rectus and olfactory tubercle. Heretofore there had been no reports of fornix projections to the medial preoptic and perifornical areas of the monkey.

Hippocampal Projections via Amygdala

In an extension of the above studies, Poletti and co-workers have tested the effects of anterior and posterior hippocampal stimulation on (1) unit activity of the amygdala[91] as well as on (2) unit activity of basal telencephalic and hypothalamic structures subsequent to bilateral section of the subcommissural fornix.[92] Hippocampal stimulation elicited responses in 20% of the 476 units in the amygdala.[91] The probability of eliciting a response was almost ten times greater with anterior than with posterior hippocampal stimulation. The amygdaloid nuclei with 20% or more responsive units were the basomedial (39%), accessory basolateral (33%), central (22%), and basolateral (21%).

Most of the responsive units in the central (90%) and basomedial (82%) showed initial excitation, whereas the majority of responsive units in the basolateral (63%) and accessory basolateral (67%) showed initial inhibition.

Subsequent to bilateral section of the fornix, hippocampal stimulation elicited responses in 80 (13%) of 603 units tested in the basal telencephalon and hypothalamus.[92] It is of special hodological interest that posterior hippocampal stimulation elicited a response in only one unit.

Comment

On the basis of the findings in these studies, together with an assortment of other evidence, Poletti and his co-workers have proposed that the influence of the posterior hippocampus is mediated largely through the fornix system, whereas the amygdala probably provides a major nonfornix pathway from the anterior hippocampus.[92] As will be discussed in the next chapter, this interpretation may be relevant to an explanation of the outcome of experiments in monkeys on the role of the amygdala and hippocampus in memorative functions.

Hippocampal Influence on Medial Thalamus and Intralaminar Structures

Clinically, it has long been recognized that during automatisms of psychomotor epilepsy, patients may burn or otherwise injure themselves because of an apparent insensitivity to noxious stimuli. The midline and intralaminar nuclei of the thalamus might in some respects be regarded as representing somatosensory nuclei for the striatal complex and for the hypothalamus. In awake, sitting squirrel monkeys, we tested the effects of fornix and hippocampal volleys on the responsivity of cells of the medial thalamus and intralaminar nuclei to a potentially noxious stimulus.[93] We used the application of triple shocks to the Gasserian ganglion as a potentially noxious stimulus because with repetitive stimulation, the monkey responded with vocalization and agitation of the kind associated with nociception. Application of the triple shocks by themselves was judged not to be painful, because the monkey gave no signs of distress but, on the contrary, would tend to become somnolent. We found that hippocampal or fornix volleys inhibited, but not did augment, responses of caudal intralaminar units to fifth-nerve stimulation. Hippocampal afterdischarges had a similar effect, except that the signs of inhibition persisted for many seconds. Stimulation of the fifth nerve had no effect on 609 units recorded in the medial dorsal nucleus, whereas fornix volleys activated 51 units and inhibited the spontaneous firing of 2.

Comment

The large percentage of initially excited units induced in the basal telencephalon and hypothalamus by hippocampal volleys calls into question the speculations based on behavioral studies that the hippocampus "operates primarily through mechanisms of inhibition."[94] To be sure, a few physiological studies provide support for such an inference. It has been reported, for example, that hippocampal stimulation suppresses the release of ACTH[95] and that such stimulation also inhibits cortically induced extensor reflexes in cats.[96] There is, however, some evidence to indicate that facilitatory or inhibitory effects with respect to ACTH release,[97] cardiovascular reflexes,[98] or to visceral responsiveness[99] may depend on the physiological state of the animal at the time of stimulation. The microelectrode findings in the awake, sitting monkey suggest that if the hippocampus exerted a predominantly inhibitory influence on structures of the brainstem, it would do so largely through the agency of neurons that are initially excited.[100]

The findings on hippocampal afterdischarges add to the evidence that the hippocampal formation induces excitation more often than inhibition of individual units. The prolonged aftereffects of afterdischarges on unit activity may help to explain the prolonged "rebound" autonomic and behavioral changes that may occur subsequent to hippocampal seizures. As mentioned in Chapter 19, the manifestations seen in different species include prolonged eating and drinking; agitation; vocalization; scratching and grooming of the body; pleasure reactions; taming; quietude; somnolence; cardiac slowing and irregularities; waxing and waning of penile erection. Sherrington's work on spinal reflexes indicated that rebound is the result of an excitatory state outlasting that of inhibition subsequent to the simultaneous stimulation of excitatory and inhibitory fibers.[101] The rebound effects seen following hippocampal afterdischarges possibly reflect a similar mechanism. Similarly, some of the alternation of opposite feeling states that occurs in psychomotor epilepsy might be interpretable along such lines.

The new anatomical findings on the distribution of fornix fibers to the preoptic and perifornical regions of the monkey were of particular interest because it was the first time that direct hippocampal projections had been traced to these particular structures in a primate. The microelectrode findings revealed that the hippocampus exerts primarily excitatory effects on cells in these areas.[89] In 1943, Hess and Brugger presented evidence that the perifornical region plays an important role in the expression of angry behavior.[102] The medial preoptic area has proved to be implicated in the control of thermoregulation, water balance, food intake, cardiovascular functions, sexual behavior, and sleep.[103] This area has also been subject to extensive research because of the finding in rodents that circulating testosterone during the first 2 weeks of life has the capacity to determine sexual differentiation by its action on this area and adjoining parts of the septum and hypothalamus.[104] Autoradiographic studies have shown that cells of this region have a special affinity for L-testosterone[105] and estradiol.[106] Electrical stimulation of the medial preoptic area and adjoining structures results in ovulation[107] and genital tumescence,[108] whereas the direct application of estrogen induces changes in sexual receptivity.[109] In males, the preoptic region appears to exert a tonic effect on the secretion of gonadotropin, whereas in the female, the influence is cyclical. It had heretofore been the belief that limbic influences on the medial preoptic area are mediated primarily by the amygdala.[110] On the basis of the microelectrode and neuroanatomical findings reviewed here, however, the hippocampus also exerts a direct and potent influence on the preoptic area. If, as the phenomenology of psychomotor epilepsy indicates, the hippocampus is involved in affective processes, "the new findings would suggest neural mechanisms by which either the agreeable or disagreeable aspects of affective experience could influence genital and gonadal function."[100]

Summarizing Comment

The phenomenology of psychomotor epilepsy suggests the possibility that the limbic cortex receives information from all the intero- and exteroceptive systems. This is a question of basic relevance in regard to forebrain mechanisms underlying the feeling of individuality and memorative processes. The experimental findings reviewed in this chapter are indicative of limbic inputs from all the exteroceptive systems and at least a number of interoceptive systems depending on vagal transmission. The microelectrode studies in awake, sitting squirrel monkeys indicated that in almost every instance the unit responses were modality specific.

Elsewhere, I have referred to the parcellations of sensory thalamocortical projections innervating the neocortex as first-order, second-order, etc., "biocones." The projecting nucleus would represent the apex of the cone and the recipient cortex the face of the cone. (Corticocortical associations between cone faces, whether ipsilateral or commissural, would be analogous to hourglass cones.) If various sensory projections to the limbic cortex were viewed in a like manner, the biocones would appear as diagrammed in Figure 26-12: The olfactory cones would be the only ones not having the apical part of the cone in the brainstem. The biocones for the gustatory, auditory, and somatic senses are pictured as having the faces of the cones represented in an overlapping rostrocaudal sequence in the claustrocortex. The cone face for most of the visual projections overlies the parahippocampal part of the posterior hippocampal gyrus, lingual gyrus, and retrosplenial cortex. Cone faces for vagal projections to the hippocampus, limen insulae, and cingulate cortex are not depicted.

The hippocampus is afferently related to all the limbic areas just mentioned. Findings with intracellular recordings suggest that interoceptive systems may be effective in discharging hippocampal units because of afferents terminating on the basal dendrites and

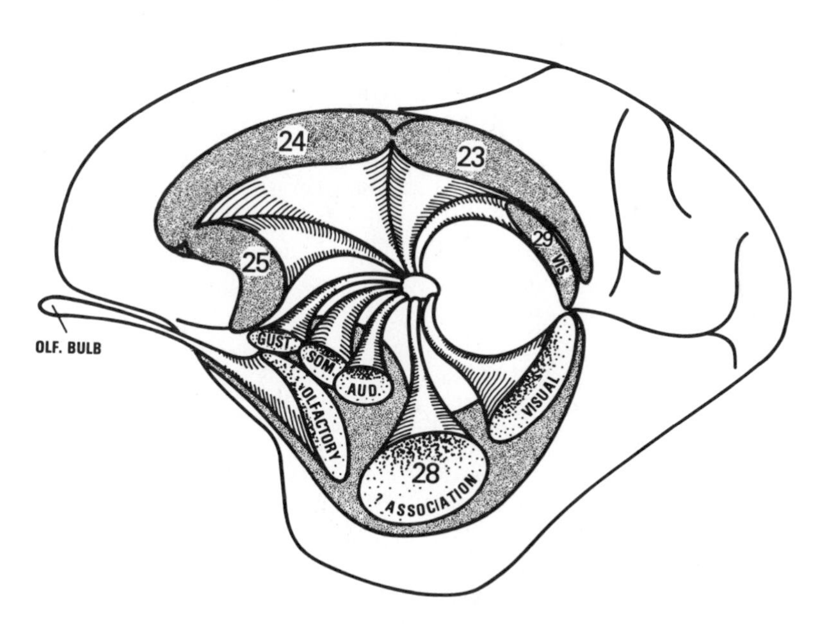

Figure 26-12. Diagram of medial aspect of squirrel monkey brain schematizing the location of limbic "sensory biocones." The buried insular cortex overlying the claustrum is externalized to show the partially overlapping areas containing units specifically activated by gustatory, somatic, and auditory stimulation. The areas marked olfactory and visual overlie respectively, the piriform cortex and the posterior parahippocampal cortex of the posterior hippocampal gyrus, lingual gyrus, and retrosplenial area (area 29). Cone faces for the areas in which units were activated by vagal volleys (hippocampal, entorhinal, limen insula, and cingulate area 24 and 23) are not delineated. Redrawn after MacLean (1975e).

perisomatic parts of hippocampal pyramids, whereas exteroceptive inputs reaching the hippocampus by the perforant pathway may induce only partial depolarizing effects because of terminal endings on the distal parts of the apical dendrites (Figure 26-11). Viewed in terms of classical conditioning, the degree of pyramidal excitation induced by the interoceptive and exteroceptive inputs would compare, respectively, to unconditional and conditional stimuli.

As dealt with in the next chapter, the findings that the hippocampus is in a position to integrate information from both extero- and interoceptive systems and thereafter influence neurovegetative, somatovisceral, and emotional functions are of special significance in regard to mechanisms underlying a feeling of individuality that, in turn, is requisite for certain aspects of memorization.

References

1. MacLean, 1949
2. Ramón y Cajal, 1901–02/1955; Penfield and Erickson, 1941; Adrian, 1942; Fox *et al.*, 1944; Clark and Meyer, 1947; Meyer and Allison, 1949
3. Edinger, 1899
4. Herrick, 1921
5. Johnston, 1923
6. Bailey and Bremer, 1938
7. Pribram and MacLean, unpublished observations
8. Dell and Olson, 1951b
9. MacLean *et al.*, 1952
10. Jung and Kornmuller, 1938
11. Green and Arduini, 1954
12. MacLean, 1957b, p. 135
13. Green and Adey, 1956
14. MacLean, 1959, pp. 111–112, Figure 32
15. Green *et al.*, 1960
16. Gerard *et al.*, 1936; O'Leary and Bishop, 1938; Robinson and Lennox, 1951; Harman and Berry, 1956; Desmedt and Mechelse, 1959; Gergen and MacLean, 1962; Brazier, 1964
17. Woolsey and Walzl, 1942; Marshall *et al.*, 1943
18. e.g., Stuart *et al.*, 1964
19. Dell, 1972, p. 169
20. Kinnard and MacLean, 1967
21. MacLean *et al.*, 1968
22. Cuénod *et al.*, 1965
23. Bachman and MacLean, 1971
24. Sudakov *et al.*, 1971
25. Orr and MacLean, unpublished findings
26. Babb *et al.*, 1980, p. 838
27. Thompson *et al.*, 1950
28. O'Leary and Bishop, 1938
29. Rose and Malis, 1965
30. Karamian *et al.*, 1984
31. Casey *et al.*, 1965
32. MacLean and Creswell, 1970
33. Myers, 1963
34. Mishkin, 1972
35. Snyder and Diamond, 1968
36. Conrad and Stumpf, 1975
37. Itaya *et al.*, 1981
38. Robertson *et al.*, 1980
39. Akopyan, 1982
40. Pickard and Silverman, 1981
41. Bignall *et al.*, 1966; Avanzini *et al.*, 1969
42. Loe and Benevento, 1969
43. Bignall *et al.*, 1966
44. Spector *et al.*, 1970
45. Segundo and Machne, 1956
46. Olson and Graybiel, 1980; see also Sherk and LeVay, 1981
47. MacLean, 1966, 1972b
48. Smith, 1902
49. Campbell, 1905
50. Smith, 1903–04
51. Ariëns Kappers *et al.*, 1936
52. Polyak, 1957, p. 400
53. See Polyak, 1957, p. 178
54. Cushing and Heuer, 1911, see legend of Figure 1; see also Cushing, 1922
55. Putnam, 1926
56. MacLean, 1966, 1968
57. Jouvet *et al.*, 1960; Passouant and Cadilhac, 1962
58. Horowitz *et al.*, 1968
59. Cushing, 1922, p. 393
60. Penfield and Jasper, 1954
61. Brockhaus, 1940
62. Desmedt and Mechelse, 1959; see also Avanzini *et al.*, 1969
63. Woollard and Harpman, 1939; Wilson and Cragg, 1969; Rasmussen, personal communication
64. Diamond *et al.*, 1958; Rose and Woolsey, 1958
65. Rasmussen, personal communication
66. Mesulam and Mufson, 1982
67. Foltz and White, 1957
68. Foltz and White, 1962; Ballantine *et al.*, 1967
69. Woolsey *et al.*, 1942
70. Penfield and Faulk, 1955
71. Benjamin and Burton, 1968
72. Bagshaw and Pribram, 1953
73. Roberts and Akert, 1963
74. Norgren, 1976
75. Adrian, 1942
76. Brodal, 1947
77. Yokota *et al.*, 1970
78. Yokota *et al.*, 1967
79. Bachman *et al.*, 1977
80. Hallowitz and MacLean, 1977
81. Radna and Vaughn, 1977, 1978
82. Radna and MacLean, 1981a
83. Radna and MacLean, 1981b
84. Radna *et al.*, 1980; Orr *et al.*, 1981
85. Gergen and MacLean, 1964
86. Kandel *et al.*, 1961
87. Gergen and MacLean, 1964, p. 85
88. Dafny *et al.*, 1965; Stuart *et al.*, 1964
89. Poletti *et al.*, 1973
90. Poletti and Creswell, 1977
91. Morrison and Poletti, 1980
92. Poletti and Sujatanond, 1980
93. Yokota and MacLean, 1968
94. Douglas, 1967; Grastyan, 1959; Isaacson and Wickelgren, 1962; Pribram, 1967

95. Endroczi and Lissak, 1959; Mason, 1958; Porter, 1954
96. Vanegas and Flynn, 1968
97. Kawakami *et al.*, 1968
98. Hockman *et al.*, 1969
99. MacLean, 1957b
100. MacLean, 1975e, p. 204
101. Sherrington, 1906
102. Hess and Brugger, 1943
103. For review see Haymaker *et al.*, 1969
104. Arnold and Gorski, 1984
105. Pfaff, 1968a
106. Michael, 1965; Pfaff, 1968b; Stumpf, 1968
107. Everett, 1965
108. MacLean and Ploog, 1962
109. Lisk, 1962
110. Nauta, 1962

27

Question of Limbic Mechanisms Linking a Sense of Individuality to Memory of Ongoing Experience

In this chapter a main focus of attention is on a condition that has been traditionally referred to as a loss of recent memory. But this would seem to be a mistaken characterization because it is not possible to lose what is not possessed. Hence, it is necessary to substitute another expression. Clinically, the term *amnesia*, the Greek word for forgetfulness, refers to a memory loss, but also means a *lack* of memory. Clinicians recognize two main forms of amnesia—anterograde and retrograde amnesia. Anterograde amnesia pertains to a person's inability to recall ongoing experiences *subsequent to* some disease, trauma, or other affliction, whereas a retrograde amnesia applies to a failure to remember experiences throughout a particular period (or periods) *prior to* such an affliction. The term *affliction* covers a wide range of conditions, such as a blow to the head, carbon monoxide poisoning, a viral encephalitis, an epileptic seizure.

Depending on the nature of the affliction, the duration of a retrograde amnesia may involve periods of minutes, hours, days, months, or years. Without documentation it may be impossible to conclude whether or not a retrograde amnesia is actually a failure to recall something that was once remembered or whether the memory deficit is attributable to a preexisting anterograde amnesia that made impossible the acquisition of memories. When, as may be the case in anterograde amnesia, there is an inability to recall information derived via all the sensory systems, the condition is referred to as a global amnesia.

For present purposes, it is necessary to take into account another kind of memory variously characterized as "immediate" or "short-term," usually implying a fleeting (evanescent) memory.

Anterograde amnesia is of primary interest here because it is a condition that has led to the recognition of cerebral structures accounting for the memory of ongoing experiences that, in turn, is requisite for learning. Hence, it has proved to be a matter of utmost significance and unweaning interest that destruction of certain parts of the limbic system may result in a profound anterograde amnesia. The following account deals successively with (1) the history of clinical findings relevant to anterograde amnesia; (2) attempts of animal experimentation to duplicate the clinical picture; and (3) insights provided by the analysis of the amnesia of epileptic automatisms. This assorted material is then discussed in the light of the electrophysiological and anatomic findings on intero- and exteroceptive

This chapter provides the substance of an article entitled "A Reinterpretation of Memorative Functions of the Limbic System" that is to appear in *Festschrift for A. R. Luria* (The IRBN Press, New York).

inputs to the limbic cortex described in the preceding chapter. The analysis of the various lines of evidence leads to the proposal that the limbic integration of information derived from intero- and exteroceptive systems is a prerequisite for a sense of self-identity and a memory of self-involved experience.

Anterograde Amnesia in Historical Perspective

A chronological account of the revelation of brain structures involved in anterograde amnesia requires one to deal successively with the mammillary bodies, hippocampal formation, and medial dorsal nucleus of the thalamus.

In a textbook on diseases of the brain, Wernicke[1] in 1881 described a condition that he called "polio-encephalitis superior hemorrhagica" characterized by diplopia, ataxia, delirium, and mental confusion. Six years later, Korsakoff[2] published a report of cases of alcoholism in which, in addition to a multiple neuritis, there was a mental disorder most strikingly manifest by a disturbance of "recent memory." He later referred to it as a "psychic disorder in conjunction with multiple neuritis."[3] The condition subsequently became known as Korsakoff's psychosis. In cases showing both the memory deficit and the other neurological signs, the condition is referred to as the Wernicke–Korsakoff syndrome. Many, if not all, of the symptoms have been ascribed to a deficiency of thiamine (vitamin B_1).[4] The pathology is characterized by a noninflammatory type of reaction primarily affecting the vasculature of certain structures of the diencephalon and lower brainstem. The ocular paralysis of Wernicke's encephalopathy is attributable to disease of parts of the nuclei of the third and sixth nerves, while the ataxic condition can be ascribed to loss of Purkinje cells in the vermis and uvula of the cerebellum.[5] The structures believed to be involved in the memory deficit will be dealt with after giving a clinical picture of the amnestic syndrome, beginning with a description in Korsakoff's own words:

> . . . the disorder of memory manifests itself in an extraordinarily peculiar amnesia, in which the memory of recent events, those which just happened, is chiefly disturbed, whereas the remote past is remembered fairly well. . . . At first, during conversation with such a patient, it is difficult to note the presence of psychic disorder; the patient gives the impression of a person in complete possession of his faculties; he reasons about everything perfectly well, draws correct deductions from given premises, makes witty remarks, plays chess or a game of cards; in a word, comports himself as a mentally sound person.[6]

Yet later, the author continues, one discovers that the patient "remembers absolutely nothing of what goes on around him." In conversation the same thing may be repeated "twenty times." "Patients of this type," he points out, "may read the same page over and over again sometimes for hours, because they are absolutely unable to remember what they have read."

This last statement recalls the description of the behavior of a monkey (Chapter 19) that, after bilateral temporal lobectomy, may pick up and examine a nail "a hundred times" as though it had never been seen before. Clinically, what may be the most surprising contrast is that the patient's apparently normal social comportment and intellectual functioning are interlarded with glaring memory lapses. This is illustrated by the conversation of a former naval person to whom I offered a ride as he seemed to be rushing for shelter from an approaching thunderstorm. He was so polite and well-spoken that I might have left him at his destination without having been aware of his mental derangement, had not the tower of the naval hospital kept coming in and out of view as we drove along the highway. Every time his eye caught the tower of the hospital, he would repeat, almost word for word, an experience that he had had there many years before.

Based on a study of 245 patients, Victor, Adams, and Collins[5] found that *retrograde* amnesia occurs in conjunction with *anterograde* amnesia in individuals suffering from Korsakoff's psychosis. They point out that the retrograde amnesia is "variable in extent and degree," but usually extends back several years prior to the present illness. They cite one extreme case, that of a 40-year-old woman who could remember nothing since leaving Ireland at the age of 19. In most cases, the retrograde amnesia is not complete and, comparable to a partially destroyed videotape randomly spliced together, the patients recount remembered periods as though they had occurred successively. Such ill-remembered past experiences may contribute to the clinical impression that patients with Korsakoff's psychosis are inclined to confabulation.

Victor[7] observes that in patients who recover from Korsakoff's psychosis, the anterograde and retrograde amnesia "always recover together." It seems quite probable that when there appears to be a permanent gap in memory for a period prior to the onset of the present illness, there may have been a preexisting anterograde amnesia.

In anticipating the account of animal experimentation, it is to be emphasized that "the capacity for immediate recall or so-called short-term memory" is unimpaired, as evident by the patient's ability to repeat a series of six to nine numbers.[7]

Clinicopathological Correlations

In 1896, in an article illustrating the peripheral nerve findings in alcoholic polyneuritis, Hans Gudden[8] (not to be confused with Bernard von Gudden) also described partial to extensive degeneration in the medial mammillary nucleus in three of the four cases (sections not available in fourth case). Some 30 years later, Gamper,[9] an Austrian neurologist, called attention to these observations when reporting findings on nine men and six women with Korsakoff's psychosis. Since lesions of the mammillary bodies were the "constant" finding in all of his cases, Gamper regarded damage of this structure as a crucial factor accounting for the psychological changes characteristic of Korsakoff's psychosis. He referred to current views that the mammillary bodies constituted a nodal structure in the cerebral regulation of autonomic function.

Since then, there have been repeated reports of the high incidence of lesions of the mammillary bodies in Korsakoff's psychosis. In two of the most extensive case studies, however, the conclusions of the respective authors were in disagreement regarding the significance of the mammillary lesions. In 1956, Malamud and Skillikorn[10] described the cerebral pathology in 70 institutionalized cases of Korsakoff's disease in which the memory deficit was documented. Their finding of the presence of disease of the mammillary bodies in 65 of the cases (95.7%) was a clear indication to them that Gamper was correct in his conclusions that damage to these structures was primarily responsible for the manifestations of Korsakoff's disease. The periventricular and periaqueductal gray matter was variously involved in all 70 cases. Contrary to the case study next to be mentioned, the medial dorsal nucleus of the thalamus was involved in only 37 cases (52.8%), while the pulvinar was damaged in 3 (4.2%).

In their monograph on the Wernicke–Korsakoff syndrome, Victor *et al.*[5] report the cerebral pathology in 62 cases. They found that the mammillary bodies were grossly involved in 46 of the cases (74%). The lesions were almost always symmetrical in distribution and occupied the central part of the medial nuclei. Hemorrhages were present in 10% of the cases. The walls of the third ventricle were grossly involved in 24% of the cases. It should be emphasized that a zone 0.5 to 1.0 mm in width immediately next to the ependyma was always spared. Parts of the thalamus were available for examination in 45 cases. Material from 43 cases revealed disease of the medial dorsal nucleus in 38 (88.4%). The medial pulvinar was examined in 20 cases and showed involvement in 17 (85%). Of the thalamic structures in general, the authors noted that "the degree of cell loss was always greatest in the medial dorsal nucleus and medial pulvinar."[11] Of interest in the light of recent findings on frontal lobe and thalamic connections, there might also be extensive cell loss in the anterior and lateral dorsal nuclei and in the submedius, medial ventral, and parataenial nuclei. There was loss of myelinated fibers in the ventral anterior and medial dorsal nuclei and in the medial pulvinar.

In discussing clinicopathological correlations, Victor *et al.*[5] point out that with respect to the memory defect, analysis showed a high degree of correlation with disease of the mammillary bodies, the medial dorsal nucleus, and the medial part of the pulvinar. In regard to the question as to which structures are crucial, they emphasized that five patients in their series who had shown no memory defect were found to have disease of the mammillary bodies and that in none of them could it be described as "minimal." Significantly, in all five the medial dorsal nucleus was "free of disease." There were an additional five cases in which there had been a serious defect in memory and in which the medial dorsal nucleus "was virtually the only thalamic nucleus affected." On the basis of these findings the authors concluded that "the mammillary bodies may be significantly affected in the absence of a memory defect,"[12] whereas there was strong evidence that disease of the medial dorsal nucleus may be the critical factor.

As opposed to cases of Korsakoff's psychosis with its multiple lesions, Delay and Brion[13] have reported a case in which there was anterograde amnesia and in which the postmortem findings revealed a hamartoma that resulted in an almost exclusive destruction of the mammillary bodies. Other evidence, next to be considered, indicates that the mammillary bodies, regardless of the nature of their specific functions, lie within a circuit of structures that are requisite for the retention and recall of ongoing experiences.

Hippocampal Formation

Bechterew,[14] the renowned Russian physiologist and neurologist, is credited with being the first to call attention to the involvement of the hippocampal formation in memorative functions. As briefly described in an abstract, the author presented before the Moscow Society the case of a 60-year-old patient with alcoholic polyneuritis and a psychosis who, for 20 years, suffered from a memory disturbance. Examination of the brain revealed bilateral softening of the uncinate gyrus and hippocampus.

In 1951, Conrad and Ule[15] described a case of Korsakoff's psychosis with bilateral lesions of the hippocampal formation that were ascribed to a toxoplasma infection. There was transneuronal degeneration of the mammillary bodies. In discussing these findings, Ule[16] referred to a case described by Grünthal[17] in 1947 in which a bilateral destruction of the hippocampus subsequent to insulin coma was associated with advanced intellectual and emotional deterioration. In 1952, Glees and Griffith[18] described a case in which there was bilateral destruction of the hippocampus in a patient in whom there had been a long-standing loss of memory. Although the fornix is said to have had only about 25% of "the normal population" of fibers, the mammillary bodies presented a normal appearance. In the light of these findings, the authors suggested that the hippocampal formation is "essential for recent memory."

The possible role of the hippocampal formation in memorative functions began to attract wider interest in 1958 when Penfield and Milner[19] reported two cases of severe anterograde amnesia following unilateral extirpation of a large part of the hippocampal formation for the treatment of psychomotor epilepsy. It was believed that the memory impairment in these two cases resulted because, after partial destruction of the hippocampal formation in the dominant hemisphere, there remained a "more or less completely destructive" lesion of the right hippocampal zone.[20] The outcome in the case of these two patients was regarded as remarkable because the memory loss "appeared in isolation from any disturbance of reasoning, attention, or concentration."[21] The first patient was a glove cutter and the second a civil engineer. The surgery did not produce any significant alteration in the IQ, as witness the engineer's postoperative score of 125, versus a preoperative score of 120. The outstanding finding in each case was the inability to retain the memory of any ongoing experience for longer than 5 min, particularly if there occurred any interruption. Extensive psychological testing revealed that "the difficulty with recent memory was a general one, affecting both verbal and nonverbal material. There was no corresponding loss of attention, concentration, or reasoning ability."[22] As was noted in regard to Korsakoff's psychosis, there was in these two patients a retrograde amnesia, extending over the preceding 4 years for the glover and for the preceding 2 months for the civil engineer.

In 1951, when visiting the Montreal Neurological Institute, I saw the civil engineer (patient P.B.) about 2 months after his operation.[23] At that time he was totally dependent on a little notebook that he carried in his hand for finding his way around and keeping appointments. His keeping it in his hand served as a continual tactile and visual reminder of why it was there to refer to. Tasks to which he had been long accustomed were performed, so to speak, automatically. He had, for example, been in the habit of following the weather and recording the daily temperature and barometric pressure. He continued to carry out this routine following surgery. In this respect, it is of interest to mention that prior to surgery he would sometimes have epileptic automatisms during which he would observe and record the temperature and barometric pressure, having no memory of it afterwards.

In response to a question about his capacity to experience emotion, he said to me that generally things did not bother him so much because he so quickly forgot them.

Support for the inference that the memory deficit in the two cases under consideration was owing to disease and malfunction of the remaining hippocampus was obtained when P.B. died 15 years later. It is said that the pathological examination by Dr. Gordon Mathieson disclosed that the "right hippocampus was atrophic, the extent of destruction actually exceeding that on the left caused by the surgical removal."[24]

The most extensively studied case of a patient subjected to a bilateral surgical ablation of the medial temporal region is that of a man who has become widely known in the literature as H.M. On September 1, 1953, H.M. underwent a "bilateral medial temporal-lobe resection" by the late William B. Scoville, who explained that "this frankly experimental operation" was undertaken with the hope that it would remedy the patient's incapacitation by major and minor seizures uncontrolled by various forms of maximal medication.[25] The patient was a 29-year-old motor winder with a history of epilepsy since the age of 16. The bilateral medial temporal lobe resection is said to have extended posteriorly for a distance of 8 cm from the tips of the temporal lobes and that the temporal horns constituted the lateral edges of the resection. Although it is more than risky to assume that the extent of the ablations conforms to the operative description, the patient's symptomatology will be briefly described because he has now been followed for 30 years and has become better known than any other case of this kind. After operation, the patient "could no longer recognize the hospital staff nor find his way to the bathroom, and he seemed to recall nothing of the day-to-day events of his hospital life."[26] There was also a partial retrograde amnesia extending back 3 years prior to surgery, as dated by his failure to remember the death of a favorite uncle 3 years earlier. Conforming to the description given above by Korsakoff, he would "read the same magazines over and over again without finding their contents familiar."[26] A half an hour after luncheon, he could not remember a single item of food that he had eaten. "Yet," it is said, "to a casual observer this man seems like a relatively normal individual, since his understanding and reasoning are undiminished."[26]

In 1968, Milner *et al.*[27] described the condition of this same patient 14 years after his surgery. Throughout that period he continued to have a "profound amnesia for most ongoing events." H.M., they explained, "still fails to recognize people who are close neighbors or family friends"[28] with whom he became acquainted subsequent to his operation. He had not lost any of his social graces, and kept himself neat. "It is characteristic," the authors state, "that he cannot give us any description of his place of work, the nature of his job, or the route along which he is driven each day. . . . "[29] Yet he was able to draw an accurate floor plan of the place in which he lived, and he seemed familiar with the neighborhood within, but not beyond, three blocks from home base. He would describe his condition as like that of "waking from a dream," apparently because of a failure to remember what went on before. H.M. had been given a wide assortment of tests, all of which were indicative of a memory disability rather than one of intelligence and perception.[27] For example, on an alternate form of the Wechsler memory scale, he achieved a quotient of only 64, which is "grossly abnormal" and in striking contrast to his full-scale IQ of 118. He proved superior to controls on Mooney's facial perception task, but showed marked impairment in the delayed matching of photographed faces. Given a shortened version of a visual or tactual maze, he proved capable of some learning and partial retention.

Some other notable items are the authors' observations that H.M. rarely mentions being hungry; does not report physical pain such as headache or stomach ache; and has lack of sexual interest.[27] They do not refer to his subjective response to feelings associated with urination and defecation. Speaking of his "emotional tone," they state that it is instructive that a feeling of uneasiness associated with the vague knowledge of his mother's illness, appeared to fade away rapidly.

In addition to these and other case reports of anterograde amnesia with bilateral lesions of the hippocampus, there is some evidence that unilateral destruction of the hippocampus, together with atrophy of the fornix and anterograde transneuronal degeneration, may contribute to the same kind of impairment.[30] For example, Torch *et al.*[31] have since described a case of impairment of "recent and short-term memory" in which severe unilateral atrophy of the fornix, mammillary body, mammillothalamic tract, and anterior thalamic nucleus followed an infarction of the posterior hippocampal and fusiform gyrus. Boudin *et al.*[32] have cited cases, including one of their own, in which asymmetrical lesions of the limbic system are associated with a Korsakoff-type deficit in memorization.

In their own case there was an occipitotemporal softening on the left side including the posterior hippocampus and fornix, while on the right side there was a softening of the anterior cingulate gyrus and other structures fed by the right anterior cerebral artery.

Finally, mention should be made of two cases in which large parts of the limbic lobe were destroyed by a viral infection and in which, in addition to cognitive and emotional deficits, there was a severe anterograde amnesia.[33] The case described by Friedman and Allen,[34] for example, was that of a 50-year-old automobile mechanic who survived 11 years following the onset of his symptoms. The "amnestic syndrome was characterized by complete loss of memory for the period of the acute illness and total impairment of recording of new memoranda. Instantaneous retention of sequential perceptions . . . was intact, but nothing could be retained from one minute to the next."[35] Consequently, although he retained the ability to spot a mechanical difficulty in a car and to use tools correctly, he could no longer carry out repairs.

In view of the experimental work to be described and the question of the role of the amygdala in memorative functions, mention should be made of a case of anterograde amnesia following bilateral infarction of the hippocampal formation, in which it is specifically stated that the amygdaloid complex was "normal."[36]

Investigation of the Amnestic Syndrome in Animals

Since the recognition that the limbic structures of the temporal lobe are implicated in memorative functions, numerous experiments have been performed on rats, cats, dogs, and monkeys in an attempt to define the nature of the deficit and to identify specific structures involved. As a generality, it may be said that the results of such experimentation are difficult to evaluate because of (1) the variation in the location of the brain lesions; (2) the failure in most cases to provide *volumetric* measurements of the amount of particular brain structures destroyed; and (3) lack of expert neuropathological analysis of the nature and extent of changes in surrounding tissue as a result of surgical probing and interference with the arterial blood supply and venous drainage. One might also regret the lack of information regarding the pre- and postoperative bioelectrical activity of the brain in the region of the surgery, as well as elsewhere. For example, the discrepancies in the findings of some experiments might be attributable to discharges occurring in injured brain tissue. It should be noted, however, that the placement of electrodes themselves may result in damage that will give rise to an irritable focus.[37] The implantation of an electrode in the hippocampus, for example, may occlude a small vessel, with nerve cell death and gliosis leading to a discharging focus.[37] The numerous hippocampal ablation studies that have been performed on rats present special difficulties in their interpretation, not only because of the great variation in the location and size of the lesions, but also because of the uneven experience of the workers performing the surgery and assessing the anatomical extent of the lesions.

Apart from such considerations, the experimental work on animals has been generally frustrating for investigators because of the failure to produce the kind of amnestic deficit observed clinically in patients with lesions of the hippocampus and neighboring structures. In the research on nonprimates it is noteworthy for comparative purposes to call attention to experiments such as those conducted by Olton and co-workers.[38] They use a radial arm maze that they believe provides a test of spatial memory. In a maze with bait at the end of the arms, a fasted rat will usually obtain the bait in one arm and then go on to another. Rats with hippocampal lesions, however, may repeatedly enter an arm in which they had already obtained the bait. They are reminiscent of the Korsakoff patient who will keep rereading the same page, or the Klüver–Bucy monkey that keeps examining the same object as though it had not been seen before. Since reviews by Isaacson[39] and others[40] have dealt with the outcome of hippocampal experimentation on nonprimates, the following account will deal with the results of experiments on monkeys.

Tests Used for Experiments on Monkeys

The tests usually employed in work with primates are (1) the delayed response test, (2) the delayed alternation test, and (3) the delayed matching-from-sample tests. The delayed response test involves baiting one of two cups to the right and left of a monkey enclosed in a Wisconsin-type apparatus. One of the cups is baited in a random manner and then an opaque screen is lowered between the monkey and the cups for a period usually ranging from 5 to 10 sec. A correct response requires that the monkey go directly to the baited cup and uncover it to obtain food. A variation of this test is to bait alternately the right- and left-hand cups. Interestingly, this delayed alternation test has proved to be a more difficult task for the monkey than the delayed response test.[41]

Tests involving *delayed matching-from-sample* or *non-matching-to-sample* have been advocated as preferable methods for assessing deficits in retention. In such tests, a left right orientation can be minimized. A delayed matching-from-sample test, for example, may simply involve the presentation of a red or green light as the cue whereupon, after a delay of 5 sec, the monkey is required to indicate the matching color upon the simultaneous appearance of green and red lights. The color to be matched is randomly presented as is the position of the two lights. The delayed non-matching-to-sample test will be described when the reason is explained for its introduction in recent experimentation.

Since tests involving repeated trials lead to a "learning set" (namely, the animal's learning an approach to the test that facilitates its accomplishment), advantage may be taken of this factor for assessing retention of memory. Once an animal has reached criterion in the performance of a test, one can, after a lapse of several days or weeks, administer the test again and obtain a measure of how many trials it requires for the animal again to reach criterion. Rapid relearning is indicative of retention of the learning set. In this way, one obtains a measure of an animal's long-term retention before and after surgery.

All of the tests that have been described thus far involve the use of vision. Consequently, it is necessary after brain surgery to confirm that a deficit in performing a test is not owing to a loss of vision. For this purpose, animals are trained in a task in which they must visually differentiate between two objects.

In tests of memorization, it is also important to show whether or not the loss is attributable to other sensory deficits. For this purpose, one has access to tests involving retention of information conveyed via auditory, somesthetic, and other channels.

Combined Amygdalar and Hippocampal Ablations

In 1960, Orbach *et al.*[42] reported the first extensive study in monkeys in which an attempt was made to duplicate the kind of memory deficit attributed to damage of the hippocampal formation in human beings. Eleven of their macaques survived lesions of various medial temporal structures. In seven subjects the amygdala, hippocampus, and hippocampal gyrus were aspirated via a frontal approach. In addition there was one animal with an amygdalectomy and transection of the fornix; a ninth with an amygdalectomy; a tenth with the hippocampus aspirated via a posterior approach; and an eleventh subject with a partial bilateral lesion of the amygdala and hippocampus. For comparison, there were two animals with ablations of the inferior temporal cortex; one with a transection of the olfactory tracts; two with electrolytic lesions of the cingulate gyrus; and two with electrolytic lesions of the pulvinar.

As tests of ''short-term memory,'' the investigators used the delayed response and delayed alternation tests. For assessing long-term retention, they administered tests of visual and tactile discrimination. Finally, they observed the effects of distraction on the performance of tests involving short-term and long-term retention.

The main findings were these: None of the monkeys with extensive removals of the amygdala and hippocampus showed savings in relearning the delayed alternation test. The delayed response test had not been administered prior to surgery. Following surgery, the animals with extensive medial temporal lesions learned to perform both the delayed response and delayed alternation tests with a 5-sec delay as readily as the controls. The introduction of a distraction during trials did not interfere with their performance. The seven monkeys with extensive medial temporal lesions showed impaired retention, as well as retardation in relearning the visual discrimination problem. A similar impairment in a tactile, size-discrimination problem indicated that the retention and relearning deficits were not modality specific.

Comment

It is evident that the deficits observed in this study seemed to correspond only in part to those that have been described in connection with hippocampal destruction in human beings. The results would indicate that medial temporal ablations involving a large part of the hippocampus and the amygdala affected long-term memory, but had no apparent effect on the so-called ''short-term memory.''

Delayed Matching-from-Sample Tests

Lawicka and Konorski[43] were among those who called into question the use of the delayed response and delayed alternation tests for assessing impairment of ''short-term memory.'' It was pointed out that animals might bridge the delay by maintaining orientation with respect to one baited cup or the other. Drachman and Ommaya[44] used a delayed matching-from-sample test that was designed to avoid this objection. First, they trained eight macaque monkeys in a matching-from-sample test and then in a delayed matching-from-sample test, using an apparatus in which the use of display of white, green, and red lights avoided spatial cues. After reaching criterion, the subjects were tested again 7 to 10 weeks later so as to obtain a measure of their retention. Five monkeys were subjected to surgery, and three served as controls. There were three monkeys in which the amygdala and the greater part of the body of the hippocampus were aspirated. In matching-from-sample, these three subjects required more trials to attain criterion than in the preoperative assessment, but there were ''considerable savings.'' In the delayed matching-from-sample (5-sec delay), these same three animals required more trials to relearn this task than did two monkeys with small medial temporal ablations and the three unoperated animals. However, they were as readily trained to delay for 12 sec as the two animals with small lesions. It was also shown that the introduction of a distraction did not interfere with performance in the delayed matching-from-sample test.

In the early 1960s, Correll and Scoville[45] had also been investigating the effects of medial temporal lesions on memory and learning. In one study utilizing the delayed matching-from-sample test, they compared the performance of 13 rhesus monkeys—four that had amygdala–hippocampal ablations; four with amygdala–uncal ablations; three with combined hippocampus and hippocampal gyrus ablations; and two that were sham-operated. The preoperative retention scores showed ''a high degree of stability in performance.'' Postoperatively, however, the animals with extensive bilateral amygdala and hippocampal ablations showed a statistically significant deficit in both the matching-from-sample and the delayed matching-from-sample tests with delays of 5 sec. The authors concluded, however, that the deficit was attributable to the conditional structure of the problem rather than to a ''rapid decay of the memory trace.''

Comment

In each of the studies involving delayed matching-from-sample, the investigators were in agreement that the ability to bridge a delay is not impaired after extensive bilateral temporal lesions. Drachman and Ommaya concluded that the ''loss of pre-operative retention and impairment of acquisition are the important factors'' that explain the retarded reacquisition of delayed matching-from-sample.[46] According to their interpretation, the ''studies of short-term memory in human beings and animals are in reality testing different properties.''[47] Since the terms *short-term* or *immediate* now tend to be used clinically for evanescent memory, the term *recent,* in Korsakoff's sense, would more appropriately refer to what the authors had in mind. In other words, they were referring to anterograde amnesia in which there is a failure to acquire and store memories of ongoing experiences, whereas the animal experiments seem to involve the testing of short-term (immediate) retention, of which a patient with anterograde amnesia is capable.

Delayed Non-Matching-to-Sample

Mishkin[48] contends that none of the tests heretofore used on animals are comparable to those used in assessing human amnesia. Consequently, he has employed a test of the kind that appeared to be effective in disclosing the memory impairment of the patient H.M. described above. Referred to as the delayed non-matching-to-sample test, it involves a one-trial object recognition for which the animal receives a peanut. A monkey is tested in a Wisconsin general testing apparatus in which it is required to show that a new object just presented is not the same as the one to which it had been exposed 10 sec before. Since no object is seen more than twice, the experimenter draws upon a collection of 300 dissimilar objects. The new object presented in each test becomes the familiar item of the next test. The procedure thus ''exploits'' the monkey's natural tendency to focus upon the novel object. After reaching criterion of 90 correct choices in 100 trials, monkeys are tested with longer delays between the presentation of the sample and nonmatching objects.

Mishkin[48] undertook to learn whether or not this test would distinguish monkeys with bilateral ablations of the amygdala and the greater part of the hippocampus from subjects that had been submitted to either an amygdalectomy or a hippocampectomy (including much of the fusiform–hippocampal gyrus). There were three animals in each group, as well as three controls. Two weeks following surgery, the operated animals were retrained to criterion and then tested under conditions in which the delay was increased from 10 to 30 to 60 to 120 sec. All of the monkeys achieved criterion, except those with the combined lesions in which the average score was 60% correct, ''or just above chance.''

Mishkin interpreted the results of this study as demonstrating that combined ablation of the amygdala and hippocampus resulted in a loss of memory comparable to that seen in patients. It was proposed that the amygdala and hippocampus have overlapping projections to the diencephalon that allow one structure to compensate for the loss of the other.

On the basis of experiments conducted by Horel,[49] however, there was the question of whether or not Mishkin's findings were owing to an interruption of the ''temporal stem'' that corresponds to the white matter above the lateral ventricle and which contains connections going to and from the neocortex of the temporal lobe. Accordingly, Zola-Morgan *et al.*[50] carried out a study in which they compared the effects of transecting the temporal stem, as opposed to a bilateral removal of the amygdala and hippocampus by an approach that left the stem intact. The performance of the delayed non-matching-to-sample test of the animals with the temporal stem lesions was essentially no different from that of the controls, whereas the results with respect to the combined amygdala–hippocampal ablations were practically the same as in the previous study.

Medial Thalamic Lesions

As was noted under clinical findings, lesions of the medial dorsal nucleus of the thalamus may be responsible for the anterograde amnesia of Korsadoff's psychosis. In proposing an explanation for the apparent deficit resulting from combined amygdala and hippocampal lesions, Mishkin[48] noted that the amygdala projects to the medial part of the medial dorsal nucleus whereas the hippocampal formation projects to the anterior group. Aggleton and Mishkin[51] found that lesions of the medial thalamus involving both the anterior group and the medial dorsal nucleus resulted in the same kind of impairment in the delayed non-matching-to-sample test as that following combined amygdala and hippocampal ablations.

Role of Different Sensory Modalities

Using a test comparable to the delayed non-matching-to-sample, Murray and Mishkin[52] found that monkeys with combined amygdala and hippocampal lesions also showed a memory deficit when performance required tactile recognition. They inferred from these findings that combined amygdala–hippocampal ablations probably resulted in a global type of amnesia. These observations recall those of Stepien *et al.*[53] reported in 1960. Using Konorski's tests involving compound stimuli, they compared the effects of various medial temporal lesions on the ability of green monkeys to bridge a delay when required to indicate whether or not one of two sounds (clicks at 5 and 20 per sec) or intermittent photic stimuli (also at 5 and 20 per sec) were presented. Although anatomical confirmation was not available, it was said that combined amygdala and hippocampal ablations impaired performance in both tests. Posterior inferior temporal ablations interfered only with the visual test, whereas ablation of the cortex rostral to the primary auditory area interfered only with the auditory test. In testing these same animals 2 years later, Cordeau and Mahut[54] concluded that the deficits resulting from medial temporal ablations were transitory.

Unresolved Role of Hippocampus

Based on the studies that have been reviewed, there is evidence that combined ablations of the amygdala and hippocampus result in some impairment of retention of past experience, as well as the registration and recall of ongoing experience. But there remain opposing views as to whether or not hippocampal ablations alone impair memorative functions. Kimble and Pribram[55] conducted experiments on monkeys that led them to conclude that hippocampal lesions interfere with the sequential performance of a task, rather than ''short-term'' memory *per se.* Waxler and Rosvold[56] called into question an existing view that impairment of delayed alternation was ''an inevitable consequence of removing the hippocampus.'' In a study involving eight monkeys with hippocampal lesions and eight controls they found that failure in performing the delayed alternation test might or might not occur regardless of the size or locus of hippocampal destruction. It was their impression that the test itself presented the inherent difficulty of requiring the animal to withhold responding to the just-baited cup, and that some monkeys were able to bridge the delay by learning to orient in a manner not evident to the experimenter. The authors, however, failed to take into account the possibility that an irritative focus near the site of hippocampal destruction may have, in some cases, resulted in hippocampal discharges or cerebral dysrhythmias that might have had a disrupting effect on performance.

Recently, Mahut *et al.*[57] reported their findings that ablations primarily involving the hippocampus and adjacent hippocampal–fusiform gyrus significantly interfere with non-spatial, ''short-term'' memory. They used the delayed non-matching-to-sample test, as well as a test involving recollection of a list of objects. A ''concurrent discrimination task'' was also employed.

In seeking an explanation of the discrepant findings on the hippocampus, Squire and Zola-Morgan[58] reviewed the descriptions of the postmortem examination of the brain in 20 different studies. In those cases in which there was a frontotemporal approach, they thought it likely that the amygdala and the greater part of the hippocampus back to the level of the lateral geniculate body had been ablated. In the cases, however, in which there had been a caudal approach to the hippocampus, it seemed probable, as Orbach *et al.*[42] had found, that a considerable segment of the hippocampus proximal to the amygdala was left intact. They attributed this situation to the operator's caution against not extending the ablation rostrally into the amygdala. It was estimated that in many cases 20 to 30% of the body of the hippocampus toward the amygdala might be left intact.

The Mammillary Bodies

As was noted, Victor *et al.*[12] were inclined to believe that disease of the mammillary bodies does not contribute significantly to the amnestic syndrome of Korsakoff's psychosis. There are only a few experimental studies that bear on this question. Thus far, it has not been possible to induce lesions of the mammillary bodies in monkeys[59] by giving diets deficient in thiamine, and damage of these structures is only an irregular occurrence in carnivores subjected to thiamine deficiency.[60] In a study on squirrel monkeys, Ploog and MacLean[61] found that electrocoagulations of the mammillary bodies did not interfere with a conditioned avoidance test in which a short delay was introduced between the conditional stimulus and the called-for response. Dahl *et al.*[62] obtained essentially the same results in experiments on cats.

A Comparison of Experimental and Clinical Findings

As was illustrated, verbal ability and verbal communication may be intact in cases in which lesions of parts of the limbic system are associated with an anterograde type of amnesia. The same applies to forms of reasoning and perception that are dependent on verbal ability. Therefore, it would appear that discrepancies between clinical and experimental findings cannot necessarily be ascribed to unbridgeable differences between animals and human beings because of verbal factors.

A major difficulty in comparing the experimental and clinical findings on memorative functions arises in part because of differing interpretations with respect to what is designated as recent memory. As was noted, Drachman and Ommaya[44] put their finger on one source of the problem by their emphasis on the distinction between an evanescent retention of an experience, as opposed to the registration and storage of an experience that makes it possible for it to be recalled at a later time. The capability of monkeys with medial temporal ablations to bridge a short delay in delayed response tests would correspond to the patient's ability with a Korsakoff deficit to recall temporarily a list of items, followed by an evanescence of this memory. The delayed non-matching-to-sample test used by Mishkin[48] was calculated to show whether or not animals with medial temporal lesions were able to recall a novel object after much longer delays than routinely used with delayed response tests. It will be recalled that monkeys with combined amygdala and hippocampal lesions tested at just above chance level with 2-min delays. This type of test, however, still seems to leave in doubt whether or not an animal with such lesions is capable of meaningful assimilation of new forms of experiences into its day-to-day existence. Perhaps situations might be devised in a quasi-natural environment with one or more conspecifics or unlike species that would more adequately demonstrate whether or not an experimental subject had the capacity to register, store, and recall ongoing experiences.

A Parallel to the Amnesia of Ictal Automatisms

As described in Chapter 25 subsequent to the initial aura, patients with psychomotor epilepsy may engage in simple or complex automatisms for which they have no memory. Clinically, there appears to be general agreement that such patients have a complete absence of recall of anything that happens after the onset of the automatism. In this respect, it is to be noted that interviews of patients under sodium amytal sedation have failed to elicit any evidence of awareness during the automatism.[63] Yet, as was illustrated, patients may be capable of performing such complex sequential acts as driving a train, delivering a baby, and playing an organ. Performances of this kind would be indicative of competent neocortical function. The preservation of cognitive functions for dealing with new situations is remarkably illustrated by Jackson's case Z, the young physician who, as described in Chapter 25, proved at autopsy to have a small cystic lesion near the junction of the left hippocampus and amygdala. During one automatism he continued examination of a patient, made a diagnosis, and wrote an appropriate prescription.

Since patients are unable to recall anything that happens after the onset of an automatism, the memory lapse would compare to an anterograde type of amnesia. Starting with the onset of the automatism, there appears to be an interference with cerebral processes that account for the registration, storage, and recall of ongoing experience. At the same time, it is evident that because of the preservation of motor skills and verbal capacities that have been illustrated, neocortical functions must be relatively intact. Experimentally, it has been observed that during propagating hippocampal seizures, sensory stimulation may evoke potentials in the neocortex like those recorded under normal conditions.[64]

Since it has been shown clinically and experimentally that seizures induced in the limbic cortex by electrical stimulation propagate preferentially in the limbic system, it has been inferred that such propagation may result in a ''functional ablation'' of the involved structures.[65] An inference of this kind is partly based on neurosurgical observations that seizures induced in the region of Broca's area, for example, may result in a postictal paresis of speech for several minutes, or that seizures induced in the medial occipital region may be followed by a bilateral hemianopsia lasting many minutes. Because of the regularity with which automatisms and the associated amnesia occur with afterdischarges initiated by electrical stimulation in the medial temporal region, Feindel and Penfield[66] were inclined to regard the amygdala and the hippocampal formation as requisite for the registration of memories.

Ingredients of a Sense of Individuality

What might be unique about the role of these structures in memory? This leads to the question of what accounts for a feeling of individuality—a sense of self—that is basic to a consideration of neural mechanisms of memory. Without a sense of individuality there is, so to speak, no place to deposit a memory. Philosophical and other writings have skirted the issue of what it is that accounts for a feeling of individuality, as well as the uniqueness of being an individual. Through introspection it becomes evident that the condition that psychologically most clearly distinguishes us as individuals is our twofold source of information from the internal private world and the external public world. Signals to the brain from the world within are entirely private, being self-contained, whereas those from the outside world can be publicly shared and lend themselves to a comparison among individuals.[67]

Where in the forebrain might one look for structures strategically located for effecting a fusion of internally and externally derived experience? Although the neocortex together with its related subcortical structures must obviously exert control over respiration and other somatovisceral functions, it serves primarily to promote the organism's orientation and survival with respect to the external environment. The evolution of the neocortex goes hand in hand with the elaboration of the visual, auditory, and somatic systems, which, more than any other of the ''sensory'' systems, provide a refined differentiation and discrimination of happenings in the external environment. Interestingly, the signals to which these systems are receptive are the only ones that lend themselves to electronic amplification and radiotransmission.[68] Smells, tastes, and interoceptions have no such avenue for communication.

As noted in Chapters 17 and 28, until relatively recent times the cortex of the great limbic lobe of Broca was regarded as part of the rhinencephalon, depending primarily on information received from the olfactory apparatus and the "oral senses." An analysis of the phenomenology of psychomotor epilepsy, however, leads to the recognition that symptoms experienced during the aura may be of the kind associated with any one of the intero- and exteroceptive systems. It was this realization that provided the impetus for the experimentation described in the preceding chapter in which we attempted to clarify the nature of inputs to the limbic cortex. Combined with the findings of others, the results provided evidence of inputs to respective areas of the limbic cortex from all exteroceptive systems and a number of interoceptive inputs depending on transmission via the vagus nerve. Significantly in regard to the following discussion, only vagal stimuli were effective in eliciting complete excitation of hippocampal units.

Concluding Discussion

Since the hippocampal formation receives connections from all the limbic areas innervated by extero- and interoceptive systems, it is in a position to achieve a synthesis of internally and externally derived information and, in turn, to influence the activity of the hypothalamus and other structures involved in somatovisceral functions variously called upon in self-preservation, procreation, and family-related behavior. In addition to its direct projections to the mammillary bodies and anterior thalamic nuclei, its connections with the amygdala and septum provide it the capacity to exert an extensive influence on the medial dorsal nucleus, which, as was noted, is one of the structures primarily implicated in Korsakoff's disease. Relevant to the matter of internal inputs, it will be recalled that 27% of 367 units tested in the medial dorsal nucleus were responsive to vagal volleys (Chapter 26). Luria and Homskaya observed that frontal lobe lesions interfered with regulation of the vegetative components of the orienting response.[69]

As reviewed in the preceding chapter, we suggested a hypothesis regarding a circuitry and mechanism of action that might account for the differential effects of intero- and exteroceptive systems on hippocampal pyramids. Since the hypothesis bears directly on the present question of memory, it is restated as follows: Impulses conducted by interoceptive inputs are highly effective in producing neuronal discharge because of their excitatory effects near the cell bodies, whereas impulses of external origin induce only partial depolarization because excitation occurs primarily at the distal portion of the apical dendrites. In terms of classical conditioning, interoceptive impulses transmitted by the septum would be comparable to unconditional stimuli since they are capable by themselves of inducing unit discharge, whereas olfactory and other exteroceptive impulses conducted by the perforant pathway would compare to conditional stimuli, being incapable, when at first acting alone, of bringing about a discharge. It is evident how such a physiology might be conducive not only to memory, but also to learning dependent on memory.

A study on cats in which we observed the effects of propagated hippocampal seizures on conditioned visceral and viscerosomatic responses is relevant to the questioned role of the hippocampus in memory.[70] For this study, we used classical *trace* conditioning in which the conditional stimulus was the sound of a buzzer followed 15 sec later by an unconditional stimulus consisting of a shock to the foreleg. Under these conditions, cardiac and respiratory changes are regularly elicited after about 40 trials. The animal, however, never develops an anticipated leg response. The visceral and viscerosomatic responses are characterized by cardiac slowing and accelerated shallow respiration. During the electrically induced hippocampal afterdischarges there is an elimination of the autonomic responses if the seizure propagates to the opposite side. The neocortical activity appears unaffected, and evoked auditory responses are unaltered. Moreover, if single muscle units are recorded in the foreleg during the propagated afterdischarge, there is no change in the rate of their discharge.

Subsequent to our original study, Flynn and Wasman[71] attempted to show whether or not the cat was capable of *learning* during hippocampal seizures. For this purpose they used a classical conditioning method in which the termination of a conditional stimulus (the sound of a buzzer) coincided with a shock to the foreleg. Under these conditions, they observed an anticipatory leg response. The outcome in this experiment is perhaps relevant to evidence that patients with a Korsakoff memory deficit appear to manifest some degree of learning when performing a simple maze test involving the use of a stylus.[27] Such a test might be regarded as comparing to the experimental situation in the cat where there is an *overlap* of the conditional and unconditional stimuli.

According to what has been proposed, either a temporary "functional ablation" during an attack of psychomotor epilepsy or bilateral disease or destruction of the hippocampal formation would interfere with the integration of internal and external experience and thereby tend to eliminate a sense of self. During the aura of psychomotor epilepsy, and prior to the automatism, some patients may have exaggerated feelings of self-awareness or, on the contrary, there may be feelings of bodily detachment and depersonalization, which Jackson characterized as "mental diplopia."[72] In the light of what was said about extrapersonal information derived by the neocortex, it is of interest to recall Penfield's patient who, during the aura, experienced "mental diplopia," having the feeling as though she were two persons, with one watching the other (see Chapter 24). When asked with which person she seemed identified, she replied, "I felt as though I were the one watching."[73]

This last observation recalls the clinical condition known as paramnesia, in which a patient may ascribe a personal experience to someone else. Talland[74] reported such a disturbance of self-reference in a patient who developed a Korsakoff type of memory deficit following an acute attack of "inclusion body encephalitis." When the patient described what he remembered about his brother's fatal automobile accident, Talland reminded him that this was a good example of the preservation of his memory prior to his present illness. To that the patient countered, "But was it not *you* who has just told me about [it]?"[75] It further reflects on this patient's disruption of self-reference that his family was in the habit of holding up a mirror to him so that he could observe and identify himself. Talland suggests that in this case, as has also been observed in Korsakoff patients with an alcoholic etiology, there were deficits in olfactory and gustatory discrimination and that this may have contributed to his inability to maintain a feeling of self-reference. Patients with a Korsakoff type of memory disturbance secondary to strokes or other disease may also have problems with self-reference. For example, when asked if they have a need to go to the bathroom, they may reply, "*You* have to go."

To sum up, it is proposed that the anterograde type of amnesia associated with ictal automatisms results from a interference with the integration of internal and external experience by a bilateral propagation of a seizure discharge within the hippocampal formation. It is suggested that a like failure of integration would occur in cases of anterograde amnesia in which, instead of a "functional ablation," there is an actual destruction of hippocampal tissue. Squire[76] has recently reported a case of severe 5-year anterograde amnesia secondary to an anoxic episode. At autopsy the major finding was a disappearance of the pyramidal cells of CA1 throughout the hippocampus. One might contend that a "clasp" of internal and external experience is as essential for memory as the antigen–antibody union in the anamnestic immune reaction. With no self-reference to the internal world there is, so to speak, no place to deposit the recollection of an experience.

In the final chapter, attention will be given to another limbic factor that would seem to be highly significant in regard to not only ontology, but also memory. There is a saying that "something does not exist until you give it a name." The study of the affective nature of the auras in psychomotor epilepsy suggests that there may be an essential precondition and that something does not exist unless it is imbued by an affective feeling, no matter how slight.

References

1. Wernicke, 1881
2. Korsakoff, 1887
3. Korsakoff, see Victor and Yakovlev, 1955
4. Jolliffe *et al.*, 1936; Phillips *et al.*, 1952
5. Victor *et al.*, 1971
6. Korsakoff, see Victor and Yakovlev, 1955, pp. 397–398
7. Victor, 1981, p. 10
8. Gudden, 1896
9. Gamper, 1928
10. Malamud and Skillikorn, 1956
11. Victor *et al.*, 1971, p. 112
12. Victor *et al.*, 1971, p. 132
13. Delay and Brion, 1954
14. Bechterew, 1900
15. Conrad and Ule, 1951
16. Ule, 1951
17. Grünthal, 1947
18. Glees and Griffith, 1952
19. Penfield and Milner, 1958
20. Penfield and Milner, 1958, p. 488
21. Penfield and Milner, 1958, p. 476
22. Penfield and Milner, 1958, p. 493
23. MacLean, personal observations
24. Mathieson, quoted by Milner *et al.*, 1968, p. 231
25. Scoville and Milner, 1957
26. Scoville and Milner, 1957, p. 14
27. Milner *et al.*, 1968
28. Milner *et al.*, 1968, p. 216
29. Milner *et al.*, 1968, p. 217
30. Schenk, 1959
31. Torch *et al.*, 1977
32. Boudin *et al.*, 1968
33. Friedman and Allen, 1969; Gascon and Gilles, 1973
34. Friedman and Allen, 1969
35. Friedman and Allen, 1969, p. 680
36. DeJong *et al.*, 1969
37. MacLean, personal observations
38. See Olton, 1978, for review
39. Isaacson, 1974
40. Douglas, 1967; Weiskrantz and Warrington, 1975
41. Orbach *et al.*, 1960; Correll and Scoville, 1967
42. Orbach *et al.*, 1960
43. Lawicka and Konorski, 1959
44. Drachman and Ommaya, 1964
45. Correll and Scoville, 1967
46. Drachman and Ommaya, 1964, p. 423
47. Drachman and Ommaya, 1964, p. 412
48. Mishkin, 1978, 1982
49. Horel, 1978
50. Zola-Morgan *et al.*, 1982
51. Aggleton and Mishkin, 1983
52. Murray and Mishkin, 1983
53. Stepien *et al.*, 1960
54. Cordeau and Mahut, 1964
55. Kimble and Pribram, 1963
56. Waxler and Rosvold, 1970
57. Mahut *et al.*, 1982
58. Squire and Zola-Morgan, 1983
59. Witt and Goldman-Rakic, 1983
60. Jubb *et al.*, 1956
61. Ploog and MacLean, 1963a
62. Dahl *et al.*, 1962
63. Stevens *et al.*, 1955
64. Flynn *et al.*, 1961
65. MacLean *et al.*, 1955–1956
66. Feindel and Penfield, 1954
67. MacLean, 1969b
68. MacLean, 1972b
69. Luria and Homskaya, 1964
70. MacLean *et al.*, 1955–1956
71. Flynn and Wasman, 1960
72. Jackson and Stewart, 1899
73. Penfield and Perot, 1963, p. 658
74. Talland, 1965
75. Talland, 1965, p. 69
76. Squire, 1986

IV

Neo-Encephalon with Regard to Paleocerebral Functions

28

Neocortex, with Special Reference to the Frontal Granular Cortex

In relation to epistemics the focus of the present investigation has, for the given reasons, been an analysis of the role of different parts of the forebrain in paleopsychic processes and prosematic communication. Paleopsychic processes are subsumed under what was defined as paleomentation (Chapter 2), a term covering protomentation and emotional mentation. Protomentation applies to cerebration accounting for the performance of activities involved in the daily master routine and subroutines, and in the rudiments of social communication. Emotional mentation pertains to cerebration dependent upon emotional feelings that guide behavior required for self preservation and the procreation of the species.

The question next arises as to the role of the neocortex in paleopsychic processes, particularly those indicative of innate forms of paleomentation. As mentioned in Chapter 2, there is a modern tradition largely traceable to Locke that at the time of birth the human brain compares to a clean slate, as witness the many psychologists and anthropologists who claim that all human behavior is learned. Since the greatly expanded human cortex is usually cited as being responsible for the uniqueness of human intelligence, it is implicit that the clean slate is largely represented by the neocortex.

In this age of computers it would be quite consistent with respect to the clean-slate hypothesis to regard the neocortex as an expanded central processor especially adapted to serve the protoreptilian and paleomammalian formations in performing calculations, making discriminations, and solving problems beyond their capabilities. The situation would be analogous to our own use of supercomputers to perform numerical calculations that otherwise would be impossible. Nevertheless, there are accumulating bits of evidence that the neocortex has built-in mechanisms that make possible an innate recognition of species-typical kinds of information. For example, there appear to be neural networks in neocortical auditory areas that are so tuned as to permit some cells to respond only to vocalizations of conspecifics.[1] At a much more complicated level there are grounds for the revival of inferences by Chomsky[2] and others[3] that the rules of grammar are laid down in the cerebral cortex. This would probably apply to the grammar of numerical usage (Hogben's "language of size"[4]) as well as to what we usually understand by language itself.

In the introductory chapter, the brain was compared to a detecting, amplifying, and analyzing device for making it possible for the organism to survive in its internal and external environment. In the chapters on the phenomenology of psychomotor epilepsy, evidence was presented for the inference that the limbic system serves in part as an amplifying device for turning up or down the intensity (volume) of affective feelings that guide behavior in self-preservation and the preservation of the species. This inference would apply not only to the *general affects,* but also to the *basic* and *specific affects* (Chapters 23 and 24). Examples of volume control of *specific affects* by limbic mechanisms were given in describing the auras of psychomotor epilepsy, as evident by the subjective accounts that sounds may become louder or fainter or that things may seem larger or smaller, and in a parallel manner, nearer or farther. In everyday life, it is recognized that mildly foul-smelling or bad-tasting substances may be subjectively amplified to the point of inducing retching and vomiting. Sounds that may be tolerable for one individual may for another be greatly distressing. In like manner, one can cite corresponding examples for both the basic and general affects.

Apart from the question of innate species-typical functions mentioned above, there is the formidable problem of explaining how thoughts or words arouse emotional feelings and exert control of the process by which such feelings may be amplified or diminished. On the basis of present knowledge, one might infer that the anatomical and functional connections discussed in Chapters 17, 18, and 26 provide some of the pathways by which the association areas of the neocortex influence limbic mechanisms involved in the experiencing of the various affects. Given also reciprocal connections, the opportunity would exist not only for thought to arouse affects, but also for affects to arouse thought. Clinical intimations of the existence of such mechanisms with respect to the occipitotemporal cortex are provided by cases of psychomotor epilepsy (Chapters 23 and 24) and by isolated cases manifesting the Klüver–Bucy syndrome (Chapter 19).

At the present time, however, the only neocortex that appears to be definitively known to be involved in the interplay of intellection and affective feelings is located in frontal areas. The same might be said in regard to neocortical functions influencing protomentation with respect to ongoing and future variations in routine activities and ritual. Accordingly, the following sections on neocortical function will largely focus on the frontal cortex. We begin with a historical résumé explaining how the frontal "association" areas became recognized as playing a role in emotional experience and behavior. This leads to considering the question of the role of the frontal cortex in laughing and crying, as well as the neural substrate of these manifestations suggested in part by new anatomical findings. This topic, in turn, calls for a brief consideration of the evolution of language and speech. Finally, we deal with the question of frontocerebellar relationships in predictive capacities (including those of a mathematical nature) and a "memory of the future." All such functions are to be regarded in the light of neofrontal relationship to the three main subdivisions of the limbic system and their respective roles in self-survival and survival of the species. Particularly noteworthy is the strong interrelationship with the thalamocingulate division that recent comparative studies have implicated in the functions of parental care, play, and social bonding—functions that would seem to have favored the evolution of the human sense of empathy and altruism.

AFFECT-RELATED FUNCTIONS OF FRONTAL "ASSOCIATION" AREAS

Areal Definition

The frontal association areas are identified with granular cortex lying rostral to the agranular motor cortex located in areas 4 and 6 (see Figure 28-1A and B). In the monkey, the granular cortex is characterized by a "clear-cut, but narrow, layer IV with granular

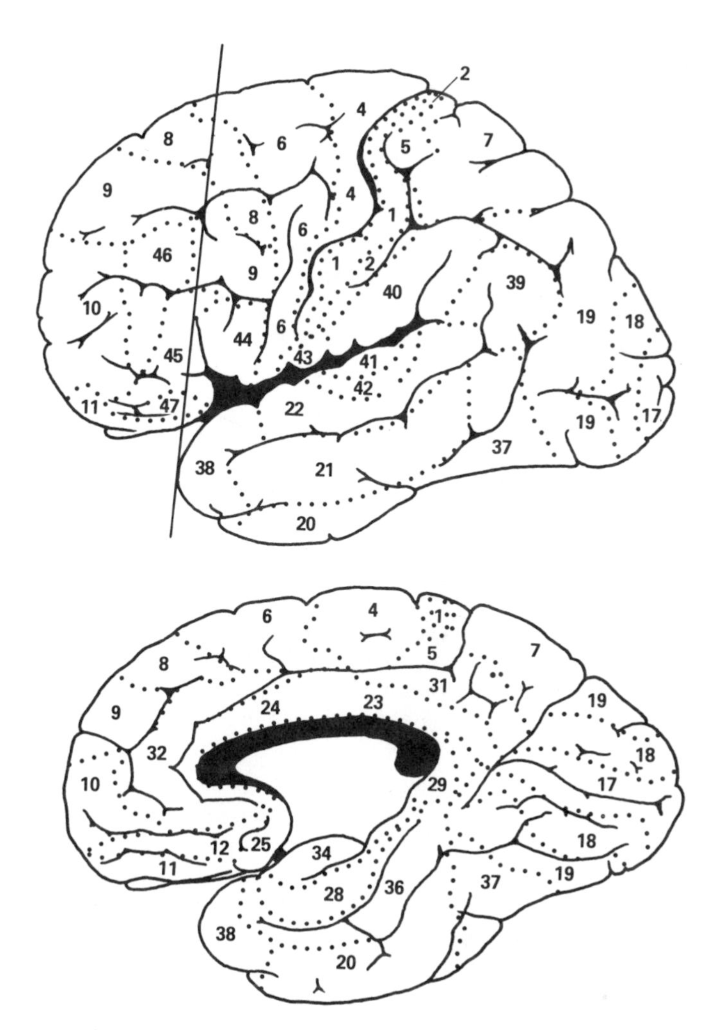

Figure 28-1. Simplified version of Brodmann's (1907) cortical cytoarchitectural scheme for the human brain. Dotted lines with enclosed arabic numerals show the limits of various areas depicted in Brodmann's original drawings by ornate shading. The vertical line in A shows approximate plane of section of the once-used standard frontal lobotomy to be discussed later. Slightly redrawn after Denny-Brown (1951).

cells."[5] If, in the rhesus monkey, one imagines the arcuate sulcus as a bow with the principal sulcus aimed like an arrow toward the posterior of the brain (Figure 28-2), then the boundary between the granular and agranular cortex could be defined as a line formed by the bow, extending above from the upper tip to the midline and from the lower tip to the forward end of the lateral (Sylvian) fissure. Transitional dysgranular cortex showing graded changes in the granular and pyramidal layers is located along the boundary line and partly encompasses part of the so-called frontal eye fields[5] of area 8 where stimulation elicits ocular movements.

On the lateral surface of the human brain (Figure 28-1A), the frontal granular cortex would include in a counterclockwise direction, areas 8, 9, 10, 11, 47, 45, 44, and 46, and on the medial surface (Figure 28-1B), areas 8, 9, 10, 11, and forward part of 12. The entire granular area is often referred to as the prefrontal cortex, but since this term means "at the

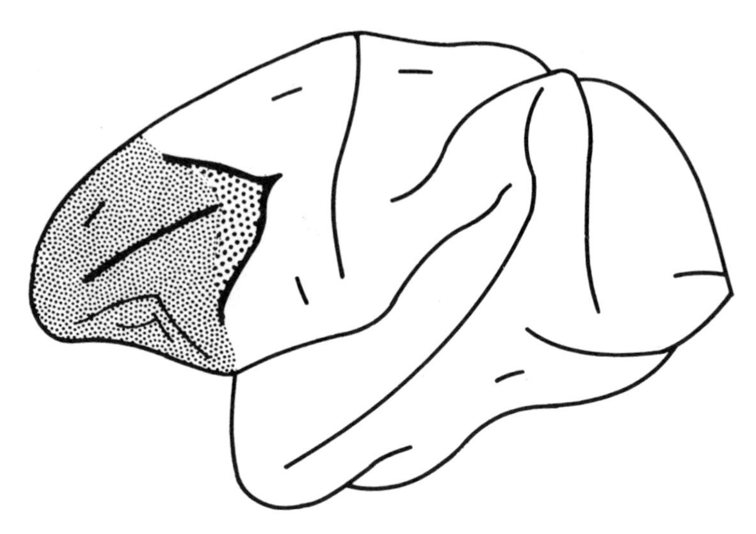

Figure 28-2. Distribution of granular and transitional frontal cortex of the lateral surface in macaque monkey. For purposes of definition, arcuate and principal sulci have been accented, giving the appearance of a bow and arrow. The caudal border of the cortex in question is delimited by the bow-shaped arcuate sulcus and imaginary lines extending upwards and downwards from the tips of the bow as shown here. The granular frontal cortex (small stipple) is characterized by a definite, but thin granular layer (see Akert, 1964). The large stipple identifies the dysgranular cortex of area 8 encompassing the so-called frontal eye-fields. Characterized by a well-developed granular layer IV and large pyramidal cells in layers III and V, it represents a transitional type of cortex lying between the granular and agranular cortex. Brain drawing after MacLean (1952) with stippled areas based on studies of Scollo-Lavizzari and Akert (1963) and Akert (1964).

tip,'' the latter would apply more appropriately to parts of areas 9 and 10. The so-called orbitofrontal cortex of Rose and Woolsey[6] includes not only the granular cortex but also the limbic agranular cortex on the inferior and medial frontal surface. The medial dorsal nucleus supplies the major source of thalamic inputs to the granular cortex. Except for parts of areas 9 and 10 near the frontal pole where pupillary responses are obtained, the granular frontal cortex is regarded as one of the electrically ''silent areas.''[7] Penfield and Jasper[8] found that spreading afterdischarges in the frontal granular cortex were associated with no subjective symptoms.

Clinical Insights from Historically Noteworthy Cases

Lawrence Kolb[9] points out that modern studies of the functions of the frontal lobes date back as early as 1805 when Cuvier in his ''Lessons of Comparative Anatomy'' concluded that the frontal lobes are concerned with intellectual functions. But perhaps the first insights into function based on injury of the frontal lobe were provided by the following case report by J. M. Harlow in 1848[10] and 1868.[11]

Harlow's Case

The case was that of a 25-year-old railroad foreman, Phineas Gage, who, in preparing to blast, was tamping a charge of powder that exploded and drove the tamping iron upwards through his left cheek and out the top of his skull (Figure 28-3). The iron was over $3\frac{1}{2}$ feet long, $1\frac{1}{4}$ inches in diameter, and pointed. Miraculously, he was able to walk afterwards and explain what happened. After a stormy 3-month recovery, he displayed a distressing change in character, described by Harlow as follows:

> He is fitful, irreverent, indulging at times in the grossest profanity . . . , manifesting but little deference for his fellows, impatient of restraint or advice . . . , at times pertinaciously obstinate, yet capricious and vacillating, devising many plans of future operation, which are no sooner arranged than they are abandoned. . . . A child in his intellectual capacity and manifestations, he has the animal passions of a strong man. Previous to his injury, though untrained in the schools, he possessed a well-balanced mind, and was looked upon by those who knew him as a shrewd, smart business man, very energetic and persistent in executing all his plans of operation. In this regard his mind was radically changed, so decidedly that his friends and acquaintances said he was 'no longer Gage.'[12]

Gage thereafter took to a roaming way of life, going to South America and then to California, where, 12 years after the original injury, he developed uncontrolled seizures and died. Throughout the intervening time he worked at a variety of jobs, including one in Barnum's circus, always taking his "iron" with him. Through persistence, Harlow was finally able to recover Gage's skull and the tamping bar. Figure 28-3 is Cobb's adaptation of Harlow's illustration, suggesting the parts of Gage's frontal lobes that must have been completely destroyed, as well as those partially damaged.

Because of the infection complicating Gage's injury, and the uncertainty as to the extent of the brain lesions, many neurologists were reluctant to cite this so-called "crowbar case" as evidence that the frontal lobes play a foremost role in the control of temperament, social interactions, and planned activities.

Prior to the introduction of frontal lobotomy (see below) there were three other cases in American neurology that have been prominently cited in regard to frontal lobe func-

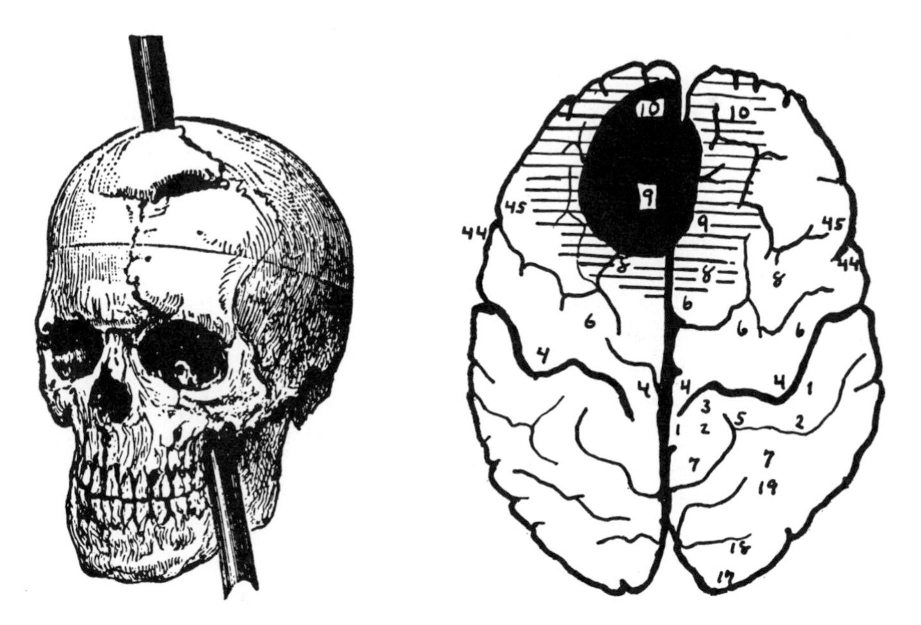

Figure 28-3. Harlow's (1868) reconstruction of the course of the tamping iron through the skull of Phineas Gage. For this illustration, Cobb (1943) added Brodmann's numbers for the cytoarchitectural areas of the frontal lobe presumed to have been destroyed (black) or injured (shading). From Cobb (1943).

tion—namely, Ackerly and Benton's Case J.P.,[14] Brickner's Case A.,[15] and Hebb and Penfield's Case K.M.[16]

Ackerly and Benton's Case

The case described by Ackerly and Benton[14] (identified in a later report as J.P.[17]) is of particular interest because it sheds light on what the personality might be like in an individual growing up devoid of the greater part of the premotor frontal cortex. The senior author first saw the patient when he was 20 years old and thereafter followed him for 14 years. Upon hospitalization for a neuropsychiatric evaluation, a pneumoencephalogram showed vacuities in the frontal region, and a neurosurgical exploration revealed a "cystic degeneration of the left frontal lobe [and] absence of prefrontal lobe on the right."[18] The history of a fall from a bed at the age of 4, followed by peculiar behavior and convulsions on the left side, was suggestive of a traumatic etiology, but a congenital defect or brain damage incidental to a prolonged instrumental delivery could not be excluded.

The patient's preschool childhood was characterized by two main manifestations: (1) ingratiating manner toward adults that earned him the title of "little Lord Chesterfield" and (2) incorrigible wanderings to distant places. Both these manifestations are to be kept in mind in view of subsequent interpretations. The patient's mother was both protective and scolding, while the father, a businessman, saw himself as the one to maintain discipline by administration of whippings.

Upon entering school, the patient's courteous, ingratiating manner helped him with adults, but rankled his classmates who were also put off by his irrepressible braggadocio. He is described as being heartily disliked by boys with whom "there were many fights."

At all stages in school it is said of him over and over again that he never had a single friend and was never accepted into any social group, nor did he seem to seek or feel the need of friends. Except for sexual encounters, there was no seeking of feminine companionship or expressions of love. He was never seen to be "daydreaming."

There is a paucity of history about play. It is only stated that "quick to flare up, he refused to play if he found himself at a disadvantage."[19] There is the added comment that "he had more than the usual play equipment—ice and roller skates, tennis racquet, baseball equipment, etc."[19] He studied piano for 2 years, was regarded as having a good voice, and enjoyed a year of dancing lessons.

Barred from the public school, he received private tutoring for a while and was evaluated by his mentors as having scholastic promise, particularly in the languages. Apropos of the question of memory that comes up in connection with the analysis of frontal lobe function, it is to be noted that beginning in childhood, he was recognized to have an unusually good memory for details in a story or what took place in a movie.

Because of truancy, he was unable to remain in a Catholic parochial high school and was sent to military school in another state where, after 2 months, he stole a teacher's car and was committed to a reform school for 2 years. Afterwards, he took to further and further wanderings, traveling hundreds of miles to distant cities either by hitchhiking or filching someone's car.

It was the pressing of charges because of stealing a car that forced his admission to the Louisville General Hospital where he was seen by Dr. Ackerly. Impressed by the atypical clinical picture, Ackerly advised the diagnostic work-up that disclosed the frontal lobe destruction described above.

The routine personality study revealed an "over-polite, boastful, somewhat over-talkative individual, with no evidence of anxiety''[20] or symptoms of a psychotic nature. There was ''no memory difficulty.'' His manifest inability to take the future into account was measurably borne out by his performance of the Porteus Mazes that reflected a specific defect in ''planning ability and foresight.''

Subsequent to his hospitalization, there were again misdemeanors and one commitment, but because of his pleasing manners, he was able to obtain and, for a time, hold down jobs such as salesman, factory worker, and more responsible positions such as watchman, bus driver, and interstate truck driver.

In the discussion the authors were at pains to explain why the patient did not deserve the diagnostic tag of ''psychopathic personality,'' the thrust of their argument being that the patient was incapable of wrongdoing requiring premeditation (i.e., planning and remembering the actions to be taken in a proposed crime). Rather, they were inclined to regard his lying, cheating, and wrongdoing as forms of protective behavior, or else ''indulged in to suit the prestige needs of the moment.'' In summing up, they placed emphasis on two primary deficits, one of which they called a ''primary social defect,'' attributable, in turn, to an incapacity to integrate past and present experience for the realization of the ''social-self'' in future personal interactions.

In conclusion, I would add a few comments about the patient's original most striking manifestations—namely, his ingratiating manners and his wanderings. I would suggest that the ingratiating manners might be protective in nature, being equivalent to the appeasement displays that young lizards may protectively employ in the presence of adults (Chapters 8 and 9). Since, in the patient's case, the striatal complex would have been lacking many of its afferent inputs from the frontal cortex, there is the possibility that frontal ''release'' might have allowed an almost reflex expression of youthful, human appeasement displays.

The wandering in his infancy and his aimless driving of hundreds of miles as an adult introduce for consideration the apparent lack of anxiety manifest in this patient, as well as in other individuals with frontal lobe lesions. As stated in Chapter 24, ''Anxiety may be defined as the unpleasant feeling that accompanies alerting for, and anticipation of, future events. As such, it would rank as a general affect akin to fear.'' Although the absence of granulofrontal cortex may deprive a person of fearful anticipation, this does not presuppose an incapacity to experience immediate fear or tension of a less defined nature. The early childhood wanderings of the patient, as well as his aimless driving as an adult, might be regarded as an expression of tension given release because of the frontal lobe deficit. One recalls the hyperactivity of rhesus monkeys with disconnected orbitofrontostriate pathways, which suggests the pacing of caged animals. As Grinker[21] commented in discussing this case, ''[B]oth physical symptoms and conscious anxiety may be absent when direct action is initiated . . .'', and he wondered whether or not ''the positive symptoms or the release'' in cases of frontal lobe lesions represented ''an acting out of impulses which cannot be delayed or dealt with in terms of consequences in the future.''[21]

Finally, one might raise the question as to whether or not frontal release of striatal mechanisms might have contributed to the patient's inveterate use of deception. As one examiner observed, ''His whole attitude is one of deception in order to gain his point.''[22]

Brickner's Case A.

Brickner's Case A.[15] was that of a man who had had the greater part of his frontal lobes excised because of bilateral involvement by a ''cellular meningioma.'' The operation was performed by Dandy at the Johns Hopkins Hospital. The case was of particular neurological interest because it was the first of its kind in which the frontal lobes had been surgically excised. The patient had been originally referred to Brickner (a neurologist at the College of Physicians and Surgeons, Columbia University) because of the complaint of "impaired memory" and absentminded behavior. In the preceding year, the patient had acquired a seat on the New York Stock Exchange and had continued his work as a broker. Following surgery, Brickner listed the outstanding behavioral manifestations as an apparent lack of initiative, apathy, inattentiveness, distractability, impairment of recent memory, and reduced learning capacity. Nevertheless, the patient continued to be facile with words, being readily able to supply synonyms when the word he wanted did not come to mind. Disturbances in the emotional sphere were manifest by boastfulness, a "slaphappy" demeanor, outbursts of temper, and a lack of insight into the serious nature of his affliction. Brickner commented that there was a "puerile quality" to all his actions. In his sexual life there was a regression to masturbation and to the solicitation of abdominal coitus. Over the years, other traits that characterized him were a lack of concern to be gainfully employed and the tendency to perform little rituals such as in the way he would tie his shoes. As has been reported repeatedly in cases of frontal lobe damage, the family and friends of this patient found him to be a "changed person." Finally, it deserves emphasis that in this case, as in others with either unilateral or bilateral lesions of the frontal lobes, most of the commonly used psychological tests failed to uncover a significant intellectual deficit.

Hebb and Penfield's Case K.M.

Hebb and Penfield's Case K.M.[16] is historically worth noting because it reflects one rather prevalent view about the equipotentiality of cortex that may have helped to promote the use of frontal lobotomy for the treatment of mental illness. In 1929, Karl Lashley's book *Brain Mechanisms and Intelligence* appeared,[23] and one of its most persuasive arguments was that the cerebral cortex is characterized by an equipotentiality of function, having the capacity, however, for some degree of localization of function. More than 30 years later, Hebb, in writing the introduction for republication of the Lashley monograph, said, "It is not far-fetched to date the beginning of the modern period in psychology from the publication of this book in 1929."[24] In his own book, *The Organization of Behavior* published in 1949, Hebb had himself stated in regard to the human brain that "no other localization of function" had been established other than for the "so-called speech area."[25] As part of the evidence for this statement, he could refer to the case now reported by Penfield and himself in 1940[16] and a follow-up report by Hebb in 1945.

The patient was a 27-year-old Nova Scotian on whom Penfield had performed a bilateral amputation of the frontal lobes as treatment for an epileptic condition resulting from a depressed fracture sustained in an accident in a sawmill when the patient was 16. Prior to surgery, the patient's interictal behavior had been characterized as "childish, violent, stubborn, and destructive," along with manifestations of gross defects in judgment. At operation, the forward half of the frontal lobes were excised beyond the plane defined by the rostral tip of the lateral ventricles and the lesser wing of the sphenoid bone, being comparable to the plane of section in a standard frontal lobotomy (see below).

In reporting the outcome of this case a year later, the authors concluded that the patient showed none of the symptoms generally attributed to loss of the frontal lobes.

Consequently, they commented that their findings were "opposed to the view that bilateral removal of the frontal poles, uncomplicated by pathological processes, has a special effect on human intelligence or personality."[26]

Five years later based on information from the patient's relatives and acquaintances in his hometown, Hebb said that K.M. appeared to be "normal in every way." Hebb adhered to this view despite the brother's saying that he was concerned about the patient's frequent changes of job and Hebb's own observation that K.M. showed no great concern about finding a job with permanence or future. Of K.M's half-formed intention of going to Toronto, a thousand miles away, where there was no particular job prospect, Hebb remarked that it suggested "an unexpected degree of initiative." He also commented that although K.M. probably was making no effort to save money for the future, "his provision for at least a few days ahead showed no lack of foresight."[27]

Hebb's appraisal was considerably different from that of Ackerly,[17] who saw the patient at the age of 49. The sister volunteered the information that "K.M. had never grown up since the accident, had always been a teen-age boy in his interest and behavior except that he had never been interested in girls."[28] K.M.'s sister, brother, and former employer all agreed in saying that he needed to be taken care of: "If he were alone for very long, he would not feed himself properly or even change his clothes. He certainly would not bathe."[28]

In summary, the thrust of Hebb's presentation of this case was that other cortical areas could fully compensate for the frontal destruction, a conclusion of no surprise to those who subscribed to the view of equipotentiality of cortical function. It seems probable that this quite prevalent view made it easier for many physicians to recommend lobotomy because the operation would not deprive the patient of some particular personality trait and, most important, run no risk of affecting a person's "soul." Whatever the explanation, the age of lobotomy that saw several thousand operations performed within two decades promises to be one of the bleakest periods in medical annals.

Insights from Frontal Lobotomy

Historical Aspects

The therapeutic procedure of frontal lobotomy is said to have been suggested by experiments on two chimpanzees in the Department of Physiology at the Yale University School of Medicine.[29] In an investigation of the functions of the premotor frontal areas, Fulton, head of the department, excised the rostral parts of the frontal lobes in two chimpanzees, Becky and Lucy, undergoing psychological testing by Jacobsen.[30] There were two outstanding findings. First, it was observed that the lobectomy resulted in a deterioration in problem-solving behavior. It interfered particularly with the solution of a problem that put the tools for a solution temporarily out of sight. If the food was put on one side of the cage, and the sticks for pulling it in on the other side, the two subjects appeared as though they were unable to keep the sticks and food in mind at the same time. The deficit was attributed to a lack of attention rather than a loss of memory.

The other finding that became of great medical consequence was the report about the emotional changes in the chimpanzee called Becky. In the preoperative testing period, this subject became so frustrated when she failed the test and went unrewarded by food that she developed temper tantrums in which she would lie down in the cage and kick, scream, urinate, and defecate. The tantrums became so frequent and violent that testing had to be discontinued. Nevertheless, it was decided to proceed with the bilateral frontal lobectomy. Much to Jacobsen's surprise, Becky no longer developed temper tantrums, but when unrewarded appeared entirely unruffled.

When Fulton described these findings at the Second International Neurological Congress in London in 1935, Egas Moniz, a Portuguese neurologist, was persuaded that disconnection of the frontal areas might alleviate certain psychotic illnesses. According to Freeman and Watts,[31] Moniz "sought out Fulton and asked him about the possibility of applying such an operation to human sufferers. This was too much for Fulton, but Egas Moniz . . . finally persuaded Almeida Lima to operate upon certain patients who had proven refractory to other methods of treatment." The first lobotomies were performed in Lisbon in the autumn of 1935,[32] and a year later Freeman and Watts[33] introduced a modification of the procedure in the United States. In 1949, Moniz became a Nobel laureate.

The Watts–Freeman modification[31] became known as the standard lobotomy and where otherwise not stated, will apply to the case material referred to hereafter in describing postlobotomy changes. The frontal plane of the standard lobotomy (see Figure 28-1) is on a line with the lower end of the coronal suture and the sphenoidal ridge. The four quadrants of the white matter of the frontal lobe were sectioned by the sweeping movement of a leukotome inserted on each side through a burr hole at the lower end of the coronal suture. Cognizant of the inexactness of the line of incision, some neurosurgeons devised special instruments for guiding the depth and plane of section of the leukotome.[34]

Manifestations of Lobotomy

Symptoms during Surgery

Freeman and Watts[35] describe the manifestations of patients during section of the four quadrants of the frontal lobes under local anesthesia. They note that discussion with a patient can be continued with perfect cooperation after incisions of the frontal lobe are completed on one side or after symmetrical section of either the upper halves or lower halves of the two frontal lobes. When the third quadrant is sectioned, however, there is a diminution in the length of the patient's replies, but the voice continues to show a certain liveliness. But when the fourth quadrant is sectioned, the patient usually becomes unresponsive except to urgent questions, when the answers may be monosyllabic. The face becomes expressionless, and there is a loss of orientation. It is said that in some cases it appeared as though there were a small bundle of fibers preserving the patient's contact with reality, and as far as could be judged, the bundle is located close to the midline at about the level of the genu of the corpus callosum, possibly involving the cingulum. With the severance of the white matter in this region, the patient might drift off into a state of confusion and unresponsiveness.

Early Postoperative Manifestations

For 2 to 5 days, there may be a state of lethargy and disorientation, during which time the patient has "to be tended like a baby"—having to be fed, watered, and have his bed linen changed because of incontinence.[35] Then there may follow a period of several days during which the patient is characterized as being in a semidazed condition, showing a placid acceptance of relatives and making playful gestures and remarks. There is a prolongation of eating and of toilet.

The authors[35] state that upon returning home after 2 weeks, the patient presents a picture best described as immaturity, since they are frequently likened to children by their relatives. Their interest span is short, and they are easily distracted. In some, indolence is outstanding, whereas others show lack of self-control as evidenced by explosiveness, petulance, talkativeness, and laughter.

Long-Term Changes

Rylander,[36] a Swedish neurologist, said of his patients that they all showed "changes of the well-known type, namely, tactlessness, emotional lability with tendencies to outbursts, extravertness, and slight euphoric traits." Freeman and Watts[35] also comment on the characteristic distractibility, pointing out that a patient may complain of forgetfulness that actually should be attributed to distraction.

Freeman and Watts[35] note that some manifestations such as tactlessness become less evident after a period of several months. So do the signs of indolence, but loss of usual interests and of ambition (so dependent on a sense of futurity) continue to be noted in most long-term case reports. Patients may be aware of the need for employment, but cannot come to grips with looking for a job. It was as though they were "drifting."[35]

Rylander found that the more introspective patients may be aware that they are "unable to feel as before," that something has "died within them," and that they "can feel neither real happiness nor deep sorrow."[37] An operating room nurse commented upon her loss of sympathy with patients. According to Freeman and Watts,[35] the lobotomized individual laughs easily, but seldom cries (see next section).

Based on the assessment of relatives, there appears to be general agreement that lobotomy results in an indescribable change in a patient's personality. Rylander reports the mother of one patient as saying, "She is my daughter but yet a different person. She is with me in body but her soul is in some way lost. Those deep feelings, the tendernesses are gone. She is hard, somehow."[36] The wife of a schoolteacher said, "I have lost my husband. I am alone. I must take over all responsibilities." A friend of a lobotomized patient remarked, "I am living now with another person. She is shallow in some way." Commenting upon similar reports by other workers, Rylander[36] cites the wife of one of Hutton's patients who said of her husband, "His soul appears to be destroyed." Rylander then offers some of his own observations, emphasizing such changes as a loss of ambition and a loss of interest in books, in social affairs, in politics. "Some of my patients," he adds, "have lost the ability to dream."[37]

Modifications of Standard Lobotomy

Because of changes of the kind just described following a standard lobotomy, neurosurgeons began to look for modifications of the procedure that would be less maiming of the personality, but at the same time alleviate the psychological disturbance. Gyrectomies and topectomies[38] of various areas of Brodmann were attempted and subsequently abandoned. Penfield and co-workers[38] at the Montreal Neurological Institute, for example, undertook to compare the effects of lobotomy and gyrectomies. They reported the results of seven cases of gyrectomy involving different frontal areas and observed that these operations resulted in more complications such as epileptic seizures and recurrence of old symptoms, than did the cases of lobotomy. Resections involving the "superior midfrontal zone" appeared to result in greater deficits than ventral lesions, being followed in three cases by "confusion, disorientation, stupor, apathy, repetitive automatic activity, and incontinence." It was concluded that there was no merit in continuing the investigation "in its present form." In the psychological testing, it should be noted that Malmo found no memory deficit in any of the seven patients, but five showed a decline in the performance of the Porteus Maze Test used to assess "planfulness."[38]

As reviewed by Fulton[39] the results of combined clinical experience and animal experimentation, suggested that the dorsolateral areas of the frontal lobe were primarily implicated in intellectual functions, whereas the ventral and medial surfaces played an important role in emotional and visceral activities. Accordingly, various modifications of the lobotomy procedure were devised for interrupting the white matter limited to the ventral quadrants. McIntyre *et al.*[40] described the results on 30 patients treated with the Grantham procedure[41] in which the ventromedial quadrants are coagulated. This series included only patients suffering from anxiety and tension. They cite the case of one of the most anxious patients, who immediately after the second quadrant had been coagulated, volunteered the remark, "That is strange, I am no longer afraid." Twenty-seven of these subjects are said to have been relieved of their anxiety.

It was noted above that severance of the cingulum was suspected as accounting for the release of tension in patients submitting to a standard lobotomy. Reference was made in Chapter 21 to the ambiguous outcome of various forms of cingulectomy for conditions ranging from obsessive, compulsive states to intractable pain.

Frontal Lobotomy for the Relief of Pain

In the classification of affects discussed in Chapter 23, pain was categorized as a *specific* affect. After introducing their laboratory procedure to this country in 1936,[33] Freeman and Watts became aware that several of their patients who complained of pain before the operation no longer do so afterwards. For example, Freeman and Watts[42] observed that a woman who had been confined to her bed for 2 years was ambulatory after 4 days and no longer complained spontaneously of pain. In follow-up visits, she reported that she "felt fine" but if asked specifically about her back, she would say, "This back of mine hurts so I can hardly walk."[43] It was observations of this kind that eventually led neurosurgeons to perform lobotomies specifically for the relief of intractable pain such as incurable causalgia and the pain endured in terminal cancer. On the basis of preliminary findings, Koskoff *et al.*[44] suggested that, initially, unilateral lobotomy be performed. Subsequently, Scarff[45] advocated unilateral frontal lobotomy on the same side as the pain, or unilateral lobotomy of the dominant hemisphere when the pain was experienced on both sides. Based on observations in 27 cases, White and Sweet[46] concluded that the side selected for unilateral lobotomy is of no consequence. They commented that the disfigurement from cutaneous ulcerations and the foul smell that arises from infected, sloughing tissue are no longer dreaded. They noted a remarkable absence of withdrawal symptoms or dependence on narcotics following surgery. They concluded that lobotomy has its greatest effectiveness in patients suffering from "fear, depression, and agitation." They make one notable observation that neither they nor other workers have singled out as particularly significant from the standpoint of perhaps the most universally dreaded human experience—namely, the "spectre of imminent death." They describe the case of Roy T., a 26-year-old naval veteran with a massive tumor in his right upper chest pressing on his trachea and esophagus that filled him with terror of "imminent death from asphyxia." After a left frontal lobotomy, he is said to have had a release from pain and all concern about his condition.

Summarizing Comment

The alteration in a patient's psychological reactions to pain following frontal lobotomy provides a frame of reference for a general assessment of frontal lobe functions. Although such an individual may give the impression of being less responsive to pain, there is the paradoxical finding, as Chapman *et al.*[47] demonstrated, that there is actually a lowered threshold to pain. And, as was illustrated by the case material, a patient who no longer gives the impression of being bothered by pain may upon questioning attest that the pain is worse than ever. Of various explanations, there were two in particular that were of special interest. One explanation might be characterized as anatomicophysiological and the other as psychological. To the first would now be added neurochemical considerations, while the second could hardly be considered apart from the first.

In regard to the anatomicophysiological explanation, Fulton[39] suggested that the relief of pain was possibly attributable to a dampening of frontal excitation because of a disconnection of inputs conveying impulses from visceroceptive systems accounting for the appreciation of "deep pain." Relevant to mechanisms, it has been already described in Chapter 26 that 27% of the cells located in the medial dorsal nucleus were responsive to vagal stimulation. Other frontal inputs from structures believed to be involved in the appreciation of somatic pain will be dealt with in a later section when describing recent neuroanatomical findings.

The *psychological explanation* in question is based on the inferred special capacity of the granular frontal cortex to anticipate the nature of future experience on the basis of past and ongoing experience. In such terms, it was proposed that the apparent postlobotomy alleviation of pain is owing to a nullification of the anticipation of pain. This statement is not enough in itself because it requires the qualification of the word *anxiety*, which, as noted in Chapter 24, may be categorized as the unpleasant *general* affect that accompanies alerting for, and anticipation of, potentially harmful future events. As such, anxiety would rank as a general affect akin to fear. Hence, a proponent of the psychological explanation would contend that lobotomy affords an attenuation of the experience of pain because of the relief from the anxiety of facing a continuation of pain. Operationally, it was as though the patient were in a state in which the continuous pain of the moment is of passing concern because it has no future. Further support for the present argument derives from the physician's experience upon making morning rounds. Whereas the usual hospital patient afflicted by pain seeks to be reassured that an order has been written for an analgesic "as required," the lobotomized patient may express no such concern.

To sum up, the relief of psychological suffering that physicians hoped for in recommending a lobotomy was a relief from the psychic "tension" generated by anxiety. For example, McIntyre *et al.*[48] commented: "The beneficial effects of any type of prefrontal lobotomy are to be explained solely in terms of release of tension. . . . The chief symptom of such tension is anxiety. . . ."

The inability to project oneself into the future has also been suggested as an explanation for the lobotomized patient's tactlessness and other manifestations of asocial behavior. Thus, the inability to foresee the consequences of one's own behavior might explain the seeming readiness "to speak out of turn," to use foul language, to tell smutty jokes, to make inappropriate sexual advances, and the like. A similar explanation has been given in cases in which there is an unconcern about financial matters as evident by careless spending and no apparent worry about future sources of income.

In regard to both pain and anxiety the psychological changes following lobotomy might also be considered in the light of a sense of individuality dealt with in the preceding two chapters. On the basis of electrophysiological findings and the manifestations of epileptic automatisms, it was proposed that a sense of individuality depends on an integration of both internally and externally derived experience. In the case of frontal lobotomy, one might speculate that a blunting of the sense of pain or of anxiety occurs because of a failed synthesis of interoceptive and exteroceptive information reaching the frontal cortex, and a consequent partial lack of realization of a sense of self. As Hutton[49] observed, lobotomized patients appear to be able to be stimulated from without, but not from within.

While considering frontal lobe function with respect to anxiety, it is relevant to contrast the behavior under stressful conditions of an animal with a poorly developed prefrontal cortex with one in which the prefrontal cortex is quite well developed. As explained in Chapter 27, we used cats for recording hippocampal and neocortical activity while testing the effects of hippocampal seizures on trace conditioning in which a shock was used as the unconditional stimulus. Since we were mainly interested in the changes in heart rate and respiration, it was necessary for each test to have stable, baseline recordings of these functions. Within 5 min after a test, the heart and respiratory rates would return to a baseline level. Upon attempting to duplicate this study with rhesus monkeys, however, we were unable to carry out the experiments, because once these animals experienced one or two shocks, their apparent anxiety was such that the heart and respiratory rates would not return to baseline levels for the rest of the day.

Attempts to Define Deficits by Psychological Testing

Since it had been repeatedly observed that a patient's score on the intelligence test following standard lobotomy failed to show significant changes either in an upward or downward direction, there was an evident need to devise tests that might show a particular deficit. Various tests of memory usually showed no deficits and when this was not the case, the deficit was usually ascribed to distractability. In presenting Ackerly's case, reference has already been made to the usefulness of the Porteus Mazes in disclosing deficits presumably owing to a defect in planning ability. Milner,[50] in testing the effects of lesions in various parts of the hemispheres, found that patients with ablations of the dorsolateral frontal cortex showed significant deficits in a card sorting test that was attributed to perseveration. She also observed a significant reduction in spontaneous speech, noting that Bonner *et al.*[51] had demonstrated a like deficit after left-sided leukotomy. Her analysis suggested that destruction of the inferolateral cortex rostral to Broca's area accounted for the reduced fluency.

Halstead[52] cites Forster and von Monakow as stating that head-injury cases "were wholly unsuitable for studies of localization of function." In his book *Brain and Intel-*

ligence, Halstead[53] reports his findings on 237 cases, including 9 pre- and postlobotomy patients and 50 cases with cerebral lobectomies involving various parts of the hemispheres. Utilizing 27 neuropsychological tests, and applying factor analyses, Halstead[53] arrived at a four-factor description of performance on these tests. The factors were identified as (1) a *c*entral integrative field factor, C; (2) a factor of *a*bstraction, A; (3) a *p*ower factor, P; and (4) a *d*irectional factor, D. The measures of the different factors were combined so as to obtain an impairment index score. He found that patients with frontal lobectomies had an impairment index score about six times that for control subjects and about three times that for patients with nonfrontal lobectomies. Impairment could not be attributed to any sensory or motor defects. On the basis of pre- and postoperative scores, however, patients following frontal lobotomy did not show a significant difference in the impairment index. The lack of change in such cases was ascribed to the preservation of association fibers relating the frontal cortex to other cortical areas.

Summarizing Comment

The explanation just given for the differences in the psychological findings in frontal lobectomy and in frontal lobotomy points up another variable that must be taken into consideration in evaluating the effects of lobotomy: the standard frontal lobotomy that, according to the Freeman and Watts method, involves a temporal approach would transect a different assortment of association fibers than would a lobotomy in the same standard plane, but in which the leukotome was introduced through a dorsal approach. Aside from these variables (as well as differences in the plane of section and the number of frontal quadrants involved) an analysis of the manifestations following frontal lobotomy indicates that the granular frontal cortex is implicated generally in functions related to an anticipation of, and preparation for, future experience in the light of past and ongoing experience. As would be expected, the changes are of the same genre as observed after frontal lobe injuries and frontal lobectomies. Hence, the psychological alterations can be credited for the various clinical impressions leading to the characterization of such patients as ''puerile,'' lacking in planfulness, deficiency in motivation, and the like.

But as was also brought out, there is missing in such individuals another ingredient that is crucial for both ongoing and future social relationships, which may be denoted as ''insight'' requisite for projecting oneself into, and identifying with, the feelings of others. As noted, this deficit may be owing to interruption of pathways conveying interoceptions not only necessary for appreciation of pain, but also for a sense of self. Given impaired insight in conjunction with impaired foresight, there would exist a combination of factors contributing to the impression that frontal-deficient individuals are asocial, lack empathy, and have blunted anxiety about the outcome of events affecting themselves and others.

As discussed in Chapter 15, many of the ''negative symptoms'' of neuropsychiatry might just as appropriately be referred to as nonevident symptoms. Unfortunately (and in many instances attributable to the psychological background of patients), there appears to be almost a total lack of information in the case histories about the effects of frontal lobe lesions on parental and play behavior—two forms of behavior of unparalleled importance in the evolution of mammals that would be of focal interest here because of the close relationship of the parts of the frontal neocortex with the thalamocingulate division of the limbic system. The absence of such information is all the more lamentable because it may be hoped that the practice of medicine will never again be involved in such maiming ''therapies'' as frontal lobotomy.

The Question of Relationships to Crying and Laughter*

As noted in Chapter 21, a further consideration of neural mechanisms underlying crying and laughter was deferred until this time because of the necessity to consider new anatomical findings, not only in relation to the midline frontolimbic cortex, but also the frontal neocortex.

The question as to specific brain structures involved in crying and laughter continues to be one of the unresolved problems of clinical neurology. This lack of information is significant because the manifestations of crying and laughter would rank closely to language in reflecting the human condition. Related questions include those regarding mechanisms of emotion and mood; the human propensity to tearing in connection with altruistic acts; and alternating waves of laughing and crying. Since the last-mentioned bouts are suggestive of the prolonged alternating movements of cerebellar seizures, how may the cerebellum be involved? Although chimpanzees and gorillas display elements of the facial, vocal, and postural elements of crying and laughter, why are human beings the only creatures known to shed tears with crying?

As evident by the work reviewed in Chapter 21, comparative neurobehavioral studies may be helpful in identifying brain structures involved in certain aspects of crying. And it is possible that some information might be gained regarding the cerebral representation of equivalents of laughter such as ''squeals'' associated with play. But in an attempt to obtain an all-around picture of mechanisms of crying and laughter in human beings, clinical histories and the associated findings provide the most valuable resource.

Introductory Comment on Clinicopathological Findings

As pointed out in reviews,[54] pathological laughter and crying may be a complication of pseudobulbar palsy resulting from multiple lesions of the cerebrum secondary to such conditions as cerebral arteriosclerosis and multiple sclerosis. It is emphasized that almost without exception the uncontrolled episodes of laughter and/or crying are not in keeping with the patients' feelings and are often a great embarrassment to them. Contrary to the view that unilateral lesions are insufficient to induce pathological crying and laughter, Poeck[55] gave emphasis to the finding by himself and Pilleri[56] that out of 30 cases in their study, only one hemisphere was involved in a third of them. They pointed out that lesions of the anterior limb of the internal capsule were a regular finding. This and the frequent complication of one or more lesions in the corpus striatum, globus pallidus, nucleus ventralis anterior, and nucleus ventralis lateralis of the thalamus led them to conclude that disease of this system of structures was conducive to pathological crying and laughter. It should be noted, however, that in over half of the cases of pseudobulbar palsy reviewed by Tilney and Morrison,[57] there was no apparent involvement of these structures. The relevance of this question will be further discussed after reference to additional clinical

*Material from this section was used for an article in a volume on the frontal lobes (MacLean, 1987a).

data and the consideration of new anatomical findings on the connections of the frontal lobes with the thalamus.

Involvement of the Limbic System

A review of cases of psychomotor (limbic) epilepsy suggests that symptoms of laughing and/or crying may result from disease of parts of the limbic system—particularly disease involving the amygdalar and thalamocingulate divisions. The parts of these two divisions linked by the fornix are sometimes referred to as "the Papez circuit" (Figure 28-4).[58] The recent anatomical evidence reviewed in Chapter 18 indicates that it is the prosubicular and subicular parts of the hippocampal formation that project via the fornix and uncrossed tract of Gudden directly to the anterior thalamus and, via the fornix, to the medial mammillary nucleus.

Amygdalar Division

Since lacrimation may be a prominent feature of crying and laughter, it is curious that in case reports on pathological crying and laughter, there may be no mention of the occurrence of this autonomic manifestation, or, at most, it may be alluded to without details as to the amount and duration of tearing. Consequently, in regard to case reports of laughing and/or crying in temporal lobe epilepsy, it is of great interest that during one

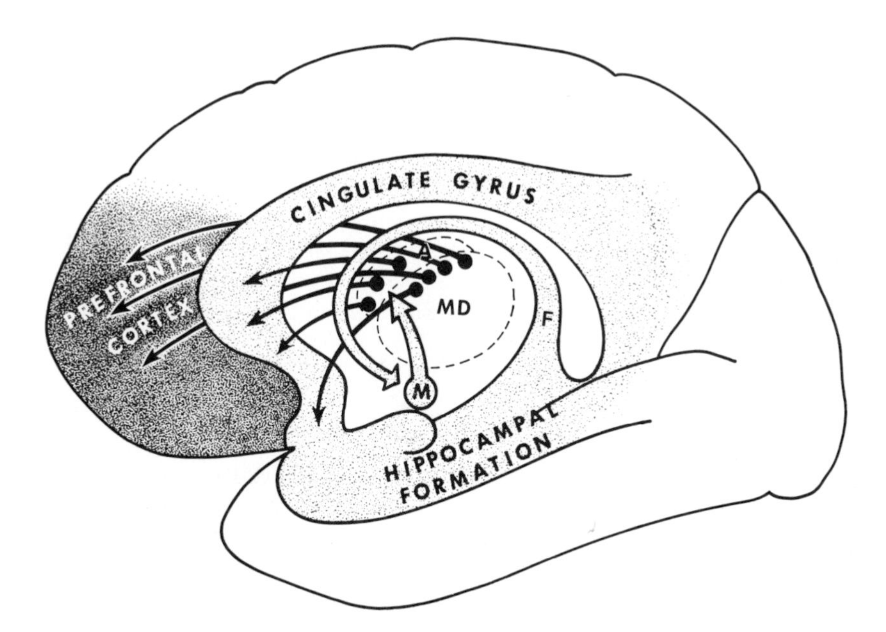

Figure 28-4. Drawing of human limbic lobe (light shading), together with diagrammatic representation of its relation to the prefrontal cortex (darker shading). Clinical findings indicate that both the frontotemporal region and midline frontal limbic and neocortex, as well as the mammillary bodies (M) and anterior thalamic nuclei (A) of the "Papez circuit," are involved in crying and laughter (see text). Other abbreviations: F, fornix; MD, nucleus medialis dorsalis. From MacLean (1973a).

operative procedure involving exposure of the hippocampus under general anesthesia, Pool[59] observed "profuse lacrimation" upon "stimulation of its anterior lateral extent" (see his Figure 4, p. 56). Penfield and Jasper cite the case of a patient (Case M.Pn.)[60] with a large astrocytoma of the temporal lobe extending into the corpus striatum in whom tearing of the eyes occurred during a seizure. In another patient (Case L.J.)[61] seizures were characterized by yawning and a flow of tears.

Although feelings of depression, sorrow, or sadness are a fairly frequent symptom during the aura in psychomotor epilepsy, there are relatively few descriptions of the manifestations of crying or weeping. Offen *et al.*[62] report one case, and cite six others, describing a condition that they refer to as dacrystic epilepsy (Gr. *dakry,* a tear). In Chapter 24, I mentioned under affects of dejection an unusual case of weeping in a patient I treated. As briefly reported in 1952, it was that of a factory worker who experienced an epigastric aura associated with a feeling of sadness and wanting to cry. This feeling was followed by a welling up of tears and a sensation of hunger.

One finds crying incidentally referred to in papers reporting cases of epileptic laughter. It is said that epileptic laughter was first described by Trousseau[64] in 1872. Since the introduction of electroencephalography, there have been several reports of such cases in which paroxysmal discharges were localized either in the temporal or frontal regions. In 1957, Daly and Mulder[65] introduced the expression *gelastic epilepsy* for seizures characterized by laughter. *Gelastic* refers to something provoking laughter. They describe two cases of ictal laughing in which there was spiking or slow-wave activity localized in the left temporal region. A year later Weil *et al.*[66] reported four additional cases in which the temporal lobe was implicated. Loiseau *et al.*[67] described one case (Case 4) with a left temporal focus in which paroxysms of laughter persisted for *10 to 20 min.*

Almost no information exists about the subjective feelings associated with laughter that might be expected during the aura. Stevens and I[68] treated one patient who, in the course of interviews that avoided leading questions, spontaneously characterized his aura as a "laughing sensation" (see Chapter 24). In this respect it is of parallel interest that Swash[69] described a woman who "was subject to sustained involuntary laughter" subsequent to an infarction of the left temporal lobe. The author stated that "she was adamant that she was actually amused during the act of laughing."[70]

In a case involving diagnostic brain stimulation, Van Buren[71] obtained convincing evidence of a neural link to laughter in the limbic part of the temporal lobe. On different occasions, he stimulated in the region of the amygdala of a 46-year-old man and elicited laughter as a part of the automatism.

The Thalamocingulate Division

Upon advancing an additional step in the Papez circuit (Figure 28-4), one uncovers several cases in which the mammillary bodies have been implicated in laughter and crying. Dott[72] cites the case of a patient who had been subject to fits of laughter or screaming. At autopsy an astrocytoma was found "replacing the floor of the third ventricle between . . . the tuber cinereum and the posterior margins of the mamillary bodies,"[73] which were destroyed. Martin[74] reported the case of a 25-year-old man who became seized by uncontrollable laughter at the graveside of his mother during the interment service. He subsequently died of a ruptured aneurysm that was found compressing the mammillary bodies. List *et al.*[75] reported two cases in which a hamartoma was discretely attached to the mammillary bodies. One was that of a 15-year-old girl who was subject to attacks that began either as laughing or as crying. Microscopic examination revealed a hamartoma attached only to the left mammillary body, being completely free and separated from the surrounding brain. The other case was that of an 8-year-old girl who had been subject to attacks of laughter since the age of 6 months. The hamartoma appeared to arise from the mammillary bodies and the tuber cinereum.

Further along the Papez circuit (Figure 28-4), it is of great interest that Penfield and Jasper described a case (J.Hl.)[76] suggestive that the thalamic part of the thalamocingulate division is involved in lacrimation. As detailed by the nurse's notes, the patient, a 41-year-old woman, was subject to seizures in which tearing of both eyes was a manifestation. At autopsy, a small tumor, "the size of a hazelnut," was found wedged in the foramen of Monro, where it exerted pressure on the anterior thalamus.

Finally, cases are to be cited in which the clinical findings either indicated or demonstrated the presence of a lesion affecting the medial frontolimbic cortex of the thalamocingulate division. Loiseau *et al.*[67] described the case of a 20-year-old woman (Case 1) who had been subject to seizures in which she burst out crying or laughing. The episodes lasted for 20 to 60 sec. One of the diagnostic procedures indicated the presence of "a left subfrontal space-occupying lesion."

Geier *et al.*[77] have provided an analysis of the symptomatology of 22 patients with frontal lobe epilepsy. There were phonatory manifestations in 19 cases. In the authors' words, the "vocalizations were "brief, single, or repeated," and sounded like "cries, moans, or grunts." The findings in one case (Case 1) were of particular interest. The patient was a 19-year-old woman who had been subject to seizures since the age of 2. During a depth recording that revealed an epileptic zone in the genual part of the cingulate gyrus and another on the medial surface of the frontal pole, the patient suffered one of her seizures. At the beginning, she "suddenly lifted herself up, turned to the left, . . . and *moaned* for about 30 seconds."[78] Shortly afterwards, she said that "she had just dreamed she was *playing* with a friend" (italics added). In regard to her dream of playing with a friend, it is of parallel interest that Talairach *et al.*[79] reported an incident in which stimulation of the supracallosal cingulate gyrus elicited in one patient a playful gesture of the kind communicating an invitation to play.

In the case study of frontal lobe epilepsy already mentioned, Geier *et al.*[77] reported that 2 of the 22 patients had motor manifestations of laughter. Direct evidence of the involvement of the medial frontal cortex in the expression of laughter has been provided by Ludwig, Ajmone Marsan, and Van Buren.[80] They described the case of a 13-year-old boy (Case 4) who was subject to seizures in which he felt himself "smile" and unable to speak. Members of his family gave the added information that an upward movement of either side of the mouth was followed by a "silly laugh." During a diagnostic procedure by Van Buren it was observed that stimulation of the medial orbital area elicited retraction of the left side of the patient's mouth and a smiling expression. Following such stimulation, rhythmical spike and waves were recorded by the most medial of four subdural electrodes on the orbital surface.

Loiseau *et al.*[67] described the case of a 5-year-old girl (Case 3) whose seizures, beginning at the age of 30 months, were characterized by laughing "in a queer way." At operation, a cystic tumor was found that appeared to originate near the junction of the corpus callosum and cingulate gyrus and extended rostrally along the medial frontal cortex.

A case recorded by Lehtinen and Kivalo[81] provides an illustration of an ictal switch from laughing to crying. Their patient was a 12-year-old girl (Case 1) who was subject to laughing attacks occurring as many as 20 times a day. The laughter "lasted from a few seconds to a couple of minutes," during which time she became very red in the face. Toward the end of the attack "*the laughter might change into crying*" (italics added). The electroencephalographic examination indicated a midline frontal disturbance.

These cases of ictal laughter complicating frontal lobe lesions recall Kramer's[82] report of "laughing spells" in patients following bilateral prefrontal lobotomy—laughing variously characterized as loud and noisy; childish laughter; incessant laughter followed by bouts of screaming; and laughter of the kind that neurologists associate with "witzelsucht."

In concluding this section on clinicopathological findings, it is pertinent to call attention to an inquiry by Sackeim *et al.*[83] regarding hemispheric representation of laughing and crying. They state that they uncovered reports of 91 cases of predominantly epileptic laughter, but only 6 of predominantly epileptic crying. Of the 91 cases of epileptic laughter, the incidence of a left-sided focus was twice that of a right-sided focus. Only one case of dacrystic epilepsy had a left-sided focus. For comparison, there were reports of 119 cases of brain damage associated with pathological laughing and crying, 62 of which were manifest primarily by laughing and 28 by crying. As opposed to the epileptic (and presumed "excitatory") condition, laughing was predominantly associated with damage to the right hemisphere (ratio of 3 to 1 in 33 cases) and crying with damage to the left hemisphere (ratio of 2 to 1 in 23 cases).

Question of Role of Neofrontal Cortex

In the frontal case histories that have been summarized, the diagnostic and operative findings indicated a midline epileptogenic disturbance. Such indications immediately present a special problem in explaining the neural circuitry involved in the manifestations of crying and/or laughter. It has long been recognized that with the exception of the midline polar area, ablations of the midline frontal cortex fail to result in clearly defined retrograde degeneration in the thalamus. For example, Pribram *et al.*,[84] upon analyzing the brains of 20 macaques with various frontal cortical ablations, concluded that thalamic retrograde degeneration did not occur following midline aspirations unless there was involvement of the medial polar cortex. Some 10 years later, Akert[85] stated, "The existence in rhesus monkey of *athalamic* frontal areas is already suggested by the work of Walker[86] (1983, Fig. 39). . . . [F]rontal granular cortex consists of two principal regions: one (lateral-ventral) which receives essential projections from the medial dorsal nucleus, and another (dorsal-medial) which receives no essential projections from the thalamus and at most may be supplied by sustaining ones."

New Anatomical Findings

The use of improved neuroanatomical techniques has thrown a whole new light on connections of the frontal cortex with the thalamus. Using macaques and autoradiographic techniques, Tobias[87] appears to have been the first to show that the medial dorsal nucleus (caudal part) does indeed project to the granular cortex of the dorsomedial wall of the frontal lobe, as well as to the anterior cingulate cortex (see Figure 3, p. 197 of his report). Also in the monkey, Carmel[88] found that lesions in the ventral anterior nucleus resulted in widespread sparse degeneration in the frontal cortex, being most evident in the lateral orbital region. Since then, with the utilization of retrograde transport of HRP, it has been demonstrated in primates that both the midline frontal limbic cortex and neocortex receive connections from several thalamic nuclei, including the anterior, medial dorsal, ventral anterior, and several intralaminar nuclei.[89]

Because of additional questions raised by our studies on the separation cry, I have employed cytochemical tracing techniques for obtaining clarification of the connections of the subcallosal cortex and adjacent areas. The findings to date have been obtained with a modification of a technique using wheat germ agglutinin conjugated to HRP (WGA-HRP). The results using this substance support, in general, what others have found upon the application of HRP alone to the anterior cingulate cortex in the squirrel monkey and the macaque.[90] With a few exceptions in regard to loci and number of labeled cells, the findings on subcallosal cortex were quite parallel to those of Jürgens[91] upon applying HRP to the anterior cingulate cortex of the squirrel monkey. On the basis of my own additional findings on the subcallosal cortex it would appear as a generalization that the area of limbic cortex implicated in the separation cry receives connections from the following thalamic nuclei: nucleus reuniens; nucleus anterior medialis et ventralis; nucleus densocellularis; nucleus parataenialis; nucleus centralis superior lateralis; dorsal and caudal parts of nucleus medialis dorsalis; medial part of nucleus pulvinaris medialis; nucleus limitans; nucleus parafascicularis; nucleus ventralis anterior, pars magnocellularis (VAmc); and nucleus ventralis lateralis, pars medialis (VLm).[90]

With respect to the ventral anterior nuclei, the retrograde labeling seen after subcallosal injections is located mainly in the part of the magnocellular division (VAmc) surrounding the mammillothalamic tract, being so conspicuous as to suggest the picture of a bed nucleus. With HRP injections progressively more rostral toward the frontal pole, and then backwards in the dorsomedial midline frontal neocortex to the supplementary area, the retrograde labeling in VA extends more and more laterally into the parvocellular portion (VApc).[90]

Of the frontal neocortical areas, it is particularly pertinent from the standpoint of functional correlations about to be considered, to refer to anatomical findings on the supplementary area. On the basis of findings in the macaque, it was believed that the supplementary area received all of its thalamic connections from the oral part of the ventral lateral nucleus (VLo)[92] that receives projections from the caudal part of the internal segment of the globus pallidus, [93] but not from the cerebellar nuclei.[94] Wiesendanger and Wiesendanger,[95] however, found that if WGA-HRP is applied more rostrally in the supplementary area (presumably the face area), there is, in addition, labeling in area X (of Olszewski) and VLc, which receive cerebellar projections,[94] as well as in VApc, a recipient of fibers from the rostral part of the pallidal internal segment.[93] Jürgens's [96] earlier findings reported in the squirrel monkey would appear to be quite parallel, and the somewhat heavier labeling seen in our material with the use of WGA-HRP[90] involves areas corresponding to those shown in Jürgens's maps.

Some Possible Anatomico-Functional Correlations

Before discussing anatomico-functional correlations, it should be explained why the supplementary area has been singled out over other neocortical frontal areas. This is not to deny that the other areas may also be involved in either the affect or expression of crying and laughter. Here we deal first with the question of expression.

As already noted in Chapter 21, the midline frontal supplementary area in human beings has been shown to be involved in vocalization. In the squirrel monkey, as was described, ablation of the rostral supplementary area was followed by early recovery of the spontaneous isolation cry in 2 to 3 weeks. Clinical findings indicate that at least this part of the human frontal neocortex may be necessary for the phonation in crying, as well as in laughter. Rubens[97] describes a patient with a vascular lesion involving the left supplementary area who was aphonic for about 10 days and in whom "weeping and hearty laughter were not accompanied by any sounds." In regard to function, however, there is the reservation in this case that the brain scan suggested that some portions of the frontoparietal cingulate gyrus were probably involved.

Striopallidonigral Linkup

Given this clinical background, it is to be noted that in line with our original neuronographic findings,[98] neuroanatomical studies have demonstrated reciprocal connections between the rostral cingulate cortex and the supplementary area (area 6 and adjacent area 8).[99] As just reviewed, the supplementary area is innervated by the parvocellular part of VA and by VLo, two thalamic areas receiving projections, respectively, from the rostral and caudal parts of the pallidal internal segment. The rostral limbic cortex involved in the separation cry, on the contrary, has connections with the magnocellular part of VA and the medial part of VL—structures that are innervated by pars reticulata of the substantia nigra (see Chapter 4). Given these connections, there is clinical evidence of how frontolimbic structures are geared in with parts of the striatal complex shown to be involved in laughing and crying. When neurosurgical procedures were being used for the treatment of Parkinson's disease, it was found that stimulation of the inner pallidum or of VLo might elicit laughing,[100] whereas coagulations of the pallidum on one side and VLo on the other might result in pathological crying.[101]

Cerebellar Linkup

It is to be assumed that the cerebellum participates in mechanisms of crying and laughter, including the alternating waves of these manifestations. Based on the new neuroanatomical findings, there is now evidence that the supplementary area is under the direct influence of cerebellothalamic projections. But there are no reported cerebellothalamic connections that overlap with the parts of VA and VL innervating the rostral limbic cortex.[94] It is to be noted, however, that Jones,[102] in his parcellation of the thalamus, shows the central lateral nucleus (CL) as including the cellular group medially adjacent to the laterodorsal nucleus that Olszewski labeled *nucleus centralis superior lateralis* (CSL). Asanuma *et al.*[103] present evidence that this particular part of Jones's CL receives projections from the deep cerebellar nuclei. Hence, if CSL, which shows consistent heavy labeling following HRP injections of the midline frontolimbic cortex, proves to correspond to CL, it would account for some participation of the cerebellum in the separation cry.

It should be emphasized that the anterograde labeling seen with WGA-HRP indicates that respective frontal areas project back upon the same thalamic nuclei by which they are innervated. Other efferent projections, particularly those from the frontal neocortex to the pons, will be discussed in a following section on other frontocerebellar functions. But here in regard to the limbic areas involved in crying, it is to be recalled from Chapter 18 that one autoradiographic study[104] has provided evidence of projections from areas 32,

24, and 25 to the peripheral medial border of the pontine nuclei. These latter findings suggest how a flow of frontopontine impulses to the cerebellum might not only have a playback influence on the midline frontolimbic cortex, but also via projections of the deep cerebellar nuclei modulate the neuronal activity in the red nucleus and its descending pathways.

Connections Relevant to Affective Aspects

The question raised above as to whether or not parts of CSL and CL represent the same nucleus is also relevant to the distressful (or what otherwise might be called "painful") nature of the separation cry, both as it applies to infants and to adults. CL receives spinothalamic[105] as well as cerebellar[103] connections. The application of HRP to the anterior cingulate cortex of the squirrel monkey results in labeling in both CL and CSL.[91] It is also relevant to the affective aspects of crying that the parafascicular nucleus (Pfc) is labeled after application of HRP to the paragenual and subcallosal limbic cortex: In our laboratory, Casey[106] found that this nucleus is one of the thalamic structures that contain cells responding with greater changes to noxious than to innocuous somatic stimuli.

Finally, in regard to the frontal neocortex, it has been shown in both the squirrel monkey[96] and the macaque[95] that the supplementary area receives connections from both CL and Pfc. Since the tickling conducive to laughter may represent an intermediate sensation associated with spinothalamic activation, the same neural mechanisms considered here may apply to risus. The tearing of crying and laughter would depend largely on nervous excitation via the facial autonomic nerve supplying the sphenopalatine ganglion.

Based on the new anatomical findings and the case material reviewed above, there are indications of a linkup of thalamofrontocingulate, striopallidonigral, and cerebellar mechanisms implicated in both the affective feeling and the expression of crying and laughter. Although the emphasis has been placed on the frontal midline limbic and supplementary areas, case histories of patients with frontal lobotomy provide evidence that the frontal granular cortex is also involved in the experience and expression of crying and laughter. Questions regarding the genesis of tearing in connection with crying and laughter will be considered in the final discussion.

The Evolution of Handedness and Speech

Relevant to the identification of neural mechanisms involved in paleopsychic processes and prosematic communication, it is timely at this juncture to ask how a jump was made from limbic affective vocalization to neocortical, propositional speech. This is a question that requires a consideration of how vocalization becomes linked to handedness and speech. According to various estimates, about 95% of people are right-handed,[107] while clinical evidence indicates that almost every right-handed individual has the faculties for speech represented in the left, or so-called dominant, hemisphere.[108] Although there are strong cultural influences favoring the use of the right hand, there are clear indications that handedness itself has a genetic basis.[109]

Since the present argument in regard to the development of speech and cerebral dominance hinges on factors leading to handedness, the latter question will be dealt with first. As a caveat concerning the complexity of the problem, it is to be noted that the condition of "rightness" exists in some snails, flatfish, and other animals. Several explanations have been given for right-handedness, including an inequality of left- and right-hand side of the body affecting the center of gravity that favors use of the right upper extremity for manipulation[110]; conditions promoting use of the right eye in sighting[111]; the ''warrior'' hypothesis[112]; and so on. Since most of the explanations are not relevant to the present question, I am singling out the ''warrior'' hypothesis (hereafter referred to as the ''weapon'' hypothesis) and am introducing the ''infant carrying'' hypothesis because, in the light of contemporaneous findings, these two proposals have special interest.

Weapon Hypothesis

First it is to be noted that analysis has shown that the early bronze tools and weapons were made for right-handed individuals[113] and that dextrals produced the oldest human artwork.[114] Going back further in time, the examination of stone implements has provided quite convincing evidence that they were made by right-handed individuals.[115] Toth[116] contends that as far back as 2 million years ago hominids were preferentially right-handed. Examination of both the tools and the flakes at dated sites at Lake Turkana, Kenya, indicates that the striking stone was held by the right hand, while the left hand held the fabricated object (core), turning it in a clockwise direction. Hence, Toth suggests that a genetic basis for right-handedness has existed for at least a period of 2 million years.

It has been part of accepted lore that the use of tools by human progenitors did not amount to anything until there was a discovery of how to fashion implements out of stone. This argument seems quite specious in the light of the availability of bones, wood, and materials for twine that could be made into a wide assortment of implements for which stone would have been no substitute. Raymond Arthur Dart, who described *Australopithecus africanus*,[117] was one of the first to give emphasis to findings indicative that the australopithecines employed bones, teeth, and horn for various kinds of utensils, tools, and weaponry.[118] His arguments to this effect fell on such deaf ears that, as he related it to me, he felt it necessary to use the difficult, unheard-of expression ''osteodontokeratic culture'' in the titles of his presentations so as to attract interest. Thereafter he was obliged to respond to the counterargument that the australopithecine artifacts were no more than cave collections of bones broken and left there by hyenas or nibbled by porcupines.[118] Since then there has accumulated considerable evidence that bones and flakes of bones were indeed used for various instruments. Kitching,[119] in examining A. L. Armstrong's collection of bones from the Pin Hole Cave in Derbyshire, England, has illustrated refinements of Mousterian (Neanderthal) people in the fashioning of bony instruments, including a Mammoth vertebral dagger with a natural haft smoothed by wear (see his Plate 11). Writing in 1987, Irving[120] describes an ''Old Crow bone industry'' involving the use of Mammoth long bones that with a new radiocarbon dating technique, range in age from 22,000 to 43,000 years old. He points out that the paleolithic period ''cannot be understood by paying attention only to implements made of stone.''[121] The American Indians used bones for utensils and weapons. Apple-corers made of bone are still used in New England.

Among artifacts associated with the australopithecines are variously shaped skulls that could have lent themselves for use as cups and bowls. Dart[118] provides some evidence that australopithecines used bony weapons, and he suggests that the dagger-shaped artifacts made from long bones may have been used for committing murder, or, as might be added, in combat against opposing groups. He describes the demonstration by his colleague Kitching that when a fresh long bone is hit at midpoint and then twisted, it can be broken into the form of long, sharp daggers.[118]

Given the possibility of such weapons, the Carlylean weapon hypothesis[112] might provide an explanation of how right-handedness could have developed in the predawn of human evolution. For example, had there been a chance favoring of the left or right manus (as occurs in some nonhuman primates[122] and other animals[123]) an inclination to right-handedness might have resulted in the survival of a greater number of individuals: According to the weapon hypothesis, in the case of a right-hander's fighting another armed individual, the left upper extremity would have been free to flex as a shield to protect the heart against a sharp, penetrating object. Under such conditions, the right-hander would have had a better chance than the left-hander to survive and reproduce. In the course of time, actual shields were to take the place of the flexed left arm. The custom of driving on the left-hand side of the road in England is said to be a carryover from the days when knights rode horseback with the shield held on the left side and the sword or lance carried on the right.

Given an inherited predisposition to right-handedness, one can offer an explanation of cerebral dominance of speech, first noting how it would be neurologically advantageous for a midline organ of speech such as the tongue with a bilateral innervation to receive its commands from a single hemisphere.[124] Before elaborating, it will be recalled that controlled vocalization in monkeys appears to depend on limbic-related structures and to have an emotional overlay (Chapter 21). Among human beings, prosematic vocalization commonly occurs in the form of expletives expressive of pain (ouch), surprise (shriek), triumph (whoop), and the like. On the basis of clinical experience it would appear that expletives may erupt regardless of cerebral dominance[125] and perhaps do not depend on the *neo*cortex. Moreover, unlike slurred speech, expletives may erupt without much loss of meaning.

This situation is to be contrasted with what is required when enunciating words with precise meanings. Since the tongue is a midline organ, there must be synchronized action of both sides if there is not to be slurring of speech.[124] Since the cerebral hemispheres are mirror images of one another, and since delay is involved in relaying information from one side to the other, it is unlikely that both could cerebrate exactly alike and "speak as one voice." Each side of the tongue might receive impulses for the same word at slightly different times, or, worse, receive the neural command for two different words. The result would be stammering or stuttering. Penfield and Welch[126] have pointed out that upon stimulation of the motor cortex in human beings, the responsive movement is always on the contralateral side *except* for such structures as the tongue and pharynx that straddle the midline.

Granted the desirability of a unified command for midline organs, why has it turned out that the left, rather than the right hemisphere has been singled out in most individuals? Recognizing the uncertainties and prolixities that apply to all such explanations, we limit ourselves here to the consideration of two possibilities. The first relates to the possibility that speech originated as the result of the cooperation of members of hunting bands mainly comprised of males. The second will be dealt with under the infant carrying hypothesis.

It has been proposed that the teaming up of individuals would result in both greater safety and productivity than would be the case if only one or two were hunting.[127] It is also proposed that such teaming up would be conducive to socialization and to the use of vocal sounds as a way of communicating. Since emotional expletives are deficient for transmitting precise information, it might be postulated that the neocortex, rather than the less educable limbic cortex, was called upon to produce an assortment of sounds with specific meanings. Since the right hand had already become the final effector organ for directing the use of a weapon, the representation of speech in the left hemisphere would have provided the quickest and most effective means of coordinating speech and action. It is evident how under conditions of group hunting or of internecine strife, split-second timing in enunciating a directional signal might make the difference between life and death.[124]

The same neurological economy with respect to dominance would apply not only to speech, but also to a written language, because whenever the idea first occurred to jot things down, "the right hand was ready and waiting."[124]

It does not vitiate the foregoing argument to point out that findings on "crossed aphasia" have shown that about half of individuals favoring the left hand have left hemispheric dominance with respect to speech.[128] It would serve to support the argument, however, if such individuals were found to have a slower reaction of the left hand in response to verbal commands than those with handedness and speech represented in the same hemisphere.

The Infant-Carrying Hypothesis

As in the case of handedness and speech, it is commonly assumed that agriculture also originated with the male. Campbell,[129] however, is one who has suggested that it was women, rather than men, who discovered agriculture. The women, staying at home while the male bands were off after game, had the opportunity to observe how seeds take root and grow, and thus they learned to plant them. One might likewise argue that handedness originated with women. In 1960, Lee Salk[130] published his observations that most women carry their babies with the left arm so that the infant's head lies against the left breast. In seeking an explanation, he undertook studies that suggested that they unwittingly held the baby this way because the sound of the beating heart has a soothing effect on the infant. Given these conditions, one can see how, with the baby cradled by the left arm, the mother's right hand would be left free for the dextrous performance of ongoing tasks.[131]

But how, as proposed in the case of the weapon hypothesis, might the original chance inclination of some mothers to hold the baby in the left arm and to use the right hand for manipulation have a genetic-selective value for favoring the survival of right-handed persons? One may imagine conditions under which absolute silence would be essential for preventing detection by a human enemy or a dangerous predator. At such times a soothed and quiet infant might mean the difference between life and death, not only for itself and mother, but also, if present, members of an affiliated group.

Origin of Vowel–Consonant Combinations

Certain comparative considerations suggest that the beginnings of language may have had origins in mother–infant relationships, and, if so, the refinement of language meant, as in the case of the weapon hypothesis, a linkup of lingual and manual mechanisms in the same hemisphere as the one issuing commands to the favored hand.

It is said that among human infants (regardless of race, ethnic groups, or geographic location), spontaneous babbling involving vowel–consonant combinations begins to occur at about the age of 8 weeks.[132] Based on personal inquiries of workers observing anthropoid apes, it is my understanding that such babbling does not occur among the young of these animals. According to Lieberman and co-workers,[133] the shape of the nasopharynx in Neanderthal people would not allow the production of a vowel sound. How might babbling, a presumed harbinger of speech, have developed? In this respect it is tempting to consider the separation cry and the so-called chuck, respectively, as prototypes of vowel and consonant sounds. As noted in Chapter 21, the separation cry of higher primates has the character of a slowly changing tone that, in the human, has predominantly the vowel sound *aaah*. It is an innate sound that results from what can be one of the most distressful of mammalian conditions, namely, separation. Hence, it is a sound associated with great motivation to communicate.[131]

Perhaps the sound next to occur in ontogeny is one made during nursing and characterized as chucklike (see also Chapter 21).[134] The sound compares to that heard when the sucking lips of the infant suddenly break contact with the nipple ('tsik'). Its consonant quality is self-evident. It is otherwise familiar to us as the sound made by someone who is encouraging a horse to get moving. Its use in maternal–offspring communication may be illustrated by the squirrel monkey.[134] When the sound is made by the mother, it is an encouragement to the infant to resume nursing, whereas the infant emits the sound when searching for the nipple. (The human mother may make the sound when weaning an infant and encouraging it to take milk from a bottle.) Later on, when the infant squirrel monkey begins to wander afield, its mother makes the chuck sound as a means of calling it back. One might speculate that there exists in the separation cry and the chuck two basic sounds later incorporated as vowels and consonants in speech.[131]

Given this maternal–offspring background as to the possible origin of phonemes used in speech, we not only have a competing argument for the hunting hypothesis regarding speech, but also, as stated above, for handedness.[131]

The Question of Neofrontocerebellar Functions

Since the focus of the present study is on the aspect of epistemics concerned with neural mechanisms underlying paleopsychic processes, why, at this point, are we to delve even further into mechanisms of neopsychic processes? We consider now the matter of neofrontocerebellar functions because it raises questions about the possible role of the involved circuitry in anticipation and prediction, including numerical computation for estimating the probability of various happenings. Since the human capacity to anticipate and to predict so greatly enlarges the realm of experience, it multiplies interminably the number of prospects that, at the affective level, can induce concern and anxiety or joyful expectation.

Anatomical Considerations

Regarding function of the neocerebellum, Robert Dow[135] comments, "One must . . . ask why this system reached such extraordinary size in man if it is not concerned with the traditional control of motor activities. It completely overshadows the older parts of the cerebellum present in all subhuman forms." (Figure 1 of Chapter 2 shows schematically, by the concentric shading, that the cerebellum can, in evolutionary terms, be subdivided into archi-, paleo-, and neocerebellar parts.) Dow further notes that the "extreme phylogenic development is not related to the cerebral cortex as a whole," and he raises the question in particular as to why frontal association areas project to the pons if the cerebellum is purely motor in function.[135] He cites his electrophysiological studies in 1942[136] that indicated that the frontal association areas project to Crus II of the ansiform lobule of rhesus monkeys that corresponds to the part of the neocerebellum that "reaches such enormous size in man."[135]

As illustrated in Figure 28-5, the ventral pons associated with the cerebellum reaches such great proportions in the human brain that at first glance it appears in sagittal section like a large tumor compressing and choking off the overlying brainstem. The red nucleus must also be considered when taking into account the great development of the pontocerebellar system in the human brain because, like the pons, it, too, reaches tumorous proportions (Figure 28-5). This nucleus receives most of its afferent supply from the roof nuclei of the cerebellum[137] and from the motor and sensory neocortex.[138] It used to be thought that the red nucleus served to relay impulses from the cerebellum to parts of the thalamus supplying the frontal motor and premotor cortex, but this has proved not to be the case.[139] Rather, its main projections are from its parvocellular part to the inferior olive and from the magnocellular division that gives rise to the rubrotegmentospinal tracts.[138]

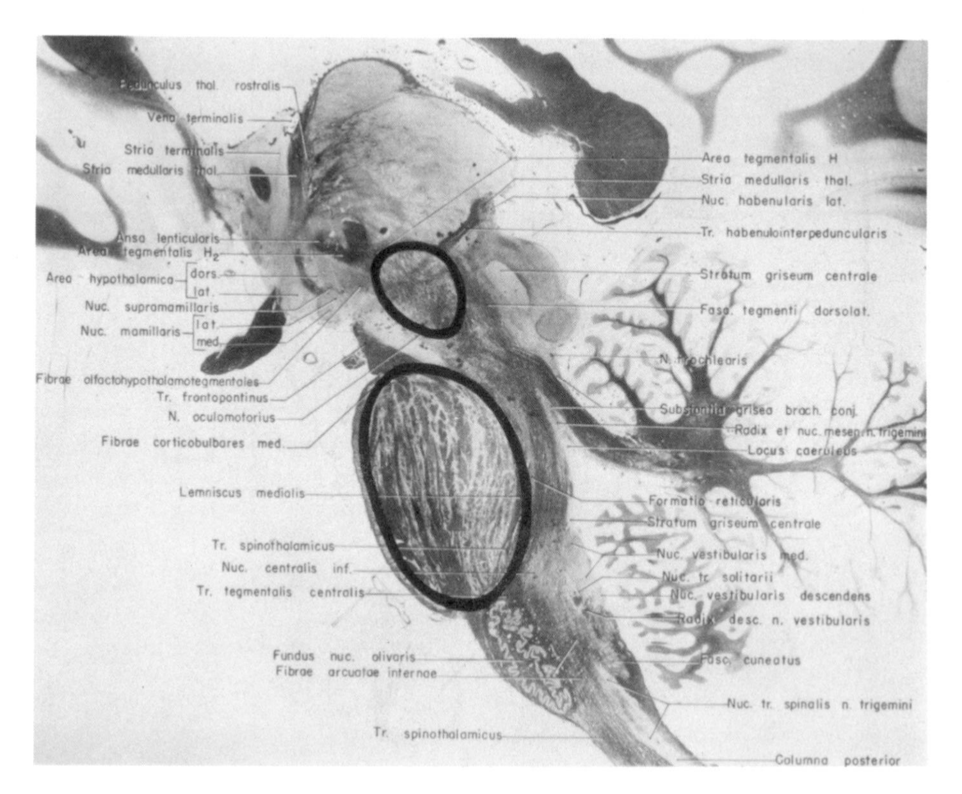

Figure 28-5. Sagittal section of the human brain with the greatly developed pons and red nucleus outlined in black. The pons gives the impression of compressing the overlying brainstem. From Singer and Yakovlev. (1954).

Dow[140] also points out that the dentate nucleus in the human brain, as well as in the brain of the great apes, shows a distinctive difference from that in less advanced primates. In the latter forms, the nucleus consists only of a microgyral, magnocellular part, whereas in the human brain and that of the great apes, there is in addition a ventrolateral, macrogyral, parvicellular part. In Figure 28-6 in which the upper arrow points to the macrogyral part of the dentate nucleus in the human brain, one should note also the great development of the inferior olive (lower arrow). In 1924, Gans[140a] reported his histochemical findings that the macrogyral part had a greater amount of iron than the microgyral component.

Existing Hypotheses as to Motor Functions

Although the pons, cerebellum, dentate nucleus, red nucleus, and inferior olive in the great apes show a much greater development than in the monkey, they fall far short in comparing to the size of these structures in the human brain. What then might account for the disproportionate development of these parts of the human brain? Various explanations

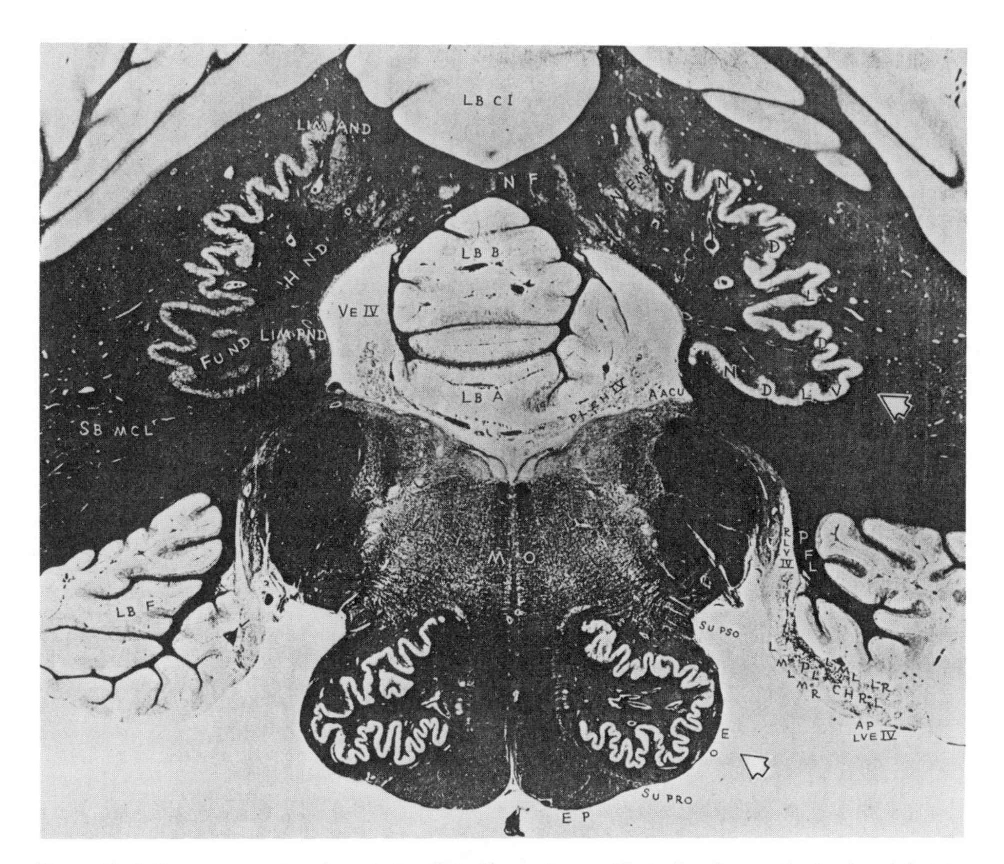

Figure 28-6. Transverse section through the cerebellum and pons, illustrating the greatly developed dentate nucleus (upper arrow on right) and olive (lower arrow). See text regarding comparative and clinical implications. From Riley (1960).

have been offered, including the human assumption of an upright posture; human dexterity with its involvement of the very fine movements of the fingers; and the unique development of speech in human beings. After commenting on these factors, I want to suggest the possibility that the neocerebellum cofunctions with the frontal neocortex in calculation and prediction.

The noted English neurologist Gordon Holmes was categorical in stating that the "main function of the cerebellum is the control of muscular contractions."[141] He furthermore suggested that the "chief function of the cerebellum as a whole is to concentrate and organize the different forms of afferent impulses outside consciousness for the service of other parts of the nervous system."[142] Holmes's sole emphasis on the role of cerebellum in the control of muscular contractions must be weighed in the light of several negative experimental and clinical findings in regard to neocerebellar functions. For example, Dow[140] points out that in monkeys with lesions carefully restricted to the lateral cerebellum (i.e., neocerebellum) the classic signs of cerebellar ataxia associated with damage to other parts of the cerebellum fail to develop. Then he observes that clinically, when stereotaxic surgery was performed on the dentate systems for therapeutic purposes, there were no classic signs of cerebellar deficits following "purely dentate lesions."[140]

Mechanical and Mnemonic Demands of Speech

The past arguments that have been proposed regarding neocerebellar expansion in conjunction with the assumption of upright posture and of manual dexterity are not altogether persuasive in the light of the capacity of many monkeys to perform arboreal acrobatic feats far surpassing in timing and precision the bodily capabilities of anthropoid apes and human beings. From a mechanistic point of view the requirements of human speech, along with the mnemonic demands of an extensive vocabulary, would suggest a plausible explanation of the great enlargement of the neocerebellum and related structures. Lieberman[143] has commented that the motor control patterns required for speech "are probably the most complex maneuvers that a human being can master." He elsewhere points out that speech allows the transmission of vocally phonetic segments "at an extremely rapid rate, up to 25 per second."[144] On the contrary, it is not possible to identify nonspeech sounds at rates exceeding 7 to 9 items per second. Hence, he notes, the high transmission rate of speech allows the communication of complex thoughts "within the constraints of short-term memory." In like vein, Dingwall[145] has commented that speech represents "the ultimate in the tendency towards elaboration of neural structures controlling fine motor movements and their coordination which Tilney and Riley (1928) termed *neokinesis*." He refers to the calculations by Darley *et al.*[146] that no less than 140,000 neuromuscular events per second are required for speech production.

Disturbances of speech, particularly slurring of speech such as occurs following the drinking of alcohol, have been associated with lesions of the paleocerebellum, but as Fulton observes, "It would be more logical to believe that speech was integrated by the newer parts of the cerebellum. . . ."[147] However, he points out that "from clinical experience it is clear that speech is not seriously affected except with very large lesions of the cerebellum. . . ."[147] Speaking to this point, Holmes says, "[T]he delicately coordinated mechanism of speech is rarely affected by even extensive unilateral lesions of the cerebellum, but its disturbance may be a prominent symptom when the lesion involves both sides. . . ."[148] In his experience, speech deficits were most often associated with midline lesions of the vermis, but could occur with "disease strictly limited to one lateral lobe."[149]

Holmes describes the speech disturbance as characterized by the "slow, drawling, and monotonous character of the voice, the unnatural separation of the syllables, and the slurred, jerky, and often explosive manner in which they are uttered."[150] He comments that patients with recent and severe lesions complain that they cannot use the proper word if it is a long and difficult one. During speech, there may be grimacing or other facial contortions, while the accompanying gestures are often irregular and inappropriate. And then he adds, there may be a tendency to unnatural explosive laughter.[150] (He makes no comment about crying.)

Holmes's paper of 1907 describing his observations on members of a family with apparent primary involvement of the cerebellar cortex provides significant details regarding the underlying neuropathology.[151] Four of eight siblings developed in their 30s locomotor signs of cerebellar disease; and then later on, abnormalities in speech of the kind just mentioned. In the one case (No. 4) in which there was a postmortem examination, the cerebellar cortex was diffusely atrophied. The pons and inferior olive were also much reduced in size. Significantly, the roof nuclei, including the dentate, appeared about normal in size and "were well provided with cells." Most relevant also to the present topic was brief mention of the cerebral gyri. All appeared normal except those of the medial and lateral prefrontal areas. These were described as "rather small and worm-like, suggesting a certain amount of general atrophy."

Given the evidence of cerebellar involvement in speech, it should be noted that there is no evidence of a dominance factor with respect to laterality as in cerebral dominance.[152]

Question of Role in Calculation and Prediction

Apart from purely motor functions, is it possible that the great development of neocerebellar systems in the human brain reflects capacities transcending those of a mechanical nature? In speculating about this matter, Dow[153] asks, "Is the cerebellum, in particular its hemispheral parts, a place where we store . . . motor patterns once they are learned?" In other words, he is asking, "Is the cerebellum requisite for the memory of past performance?" With Leiner and Leiner,[154] he has since drawn upon their own clinical findings in conjunction with new anatomical data to support their suggestion that the frontal association areas enlist cerebellar participation in the "anticipatory use of cues" and "skillful manipulation of ideas." Here I wish to focus on another possibility—namely, that parts of the granular frontal cortex, together with the agranular motor and premotor cortex, are implicated in memorized, planned activity that may involve both intuitive and formalized mathematical operations for making predictions.

Additional Anatomical Considerations

In providing a little more anatomical background for this discussion, it is to be recalled that the cerebellothalamic projections terminate in patchy cell-sparse areas that are coextensive with the caudal and "postrema" parts of the ventral lateral nucleus (VLc and ps); paralamellar part of VL denoted as "area X"; and oral part of the ventral posterolateral nucleus (VPLo). Asanuma *et al.*[155] regard the entire "zone" as one nucleus that embraces the region to which Cecile Vogt[156] traced the cerebellar projections and which she identified as the "Noyau ventral intermediaire." If so, it deserves in her honor to be referred to as the ventral intermediate nucleus of Cecile Vogt. The irregular limits of this nucleus are best seen in horizontal sections. Cytochemical tracing techniques have provided further clarification of the cerebellar influx to the frontal motor cortex (area 4), showing that it derives mainly from the main body (VPLo) of the "intermediate" nucleus.[157] The innervation of the premotor cortex including the eye fields of area 8 appears to derive from the paralamellar part of VL labeled area X. The innervation of the midline frontal supplementary cortex (area 6 and part of area 8) was described in a preceding section.

In regard to the frontal granular cortex, it is primarily noteworthy that (1) the posterior part of area 9 appears to have connections with the paralamellar part of area X[158] and (2) anatomical studies involving both anterograde and retrograde labeling[159] in the macaque have also shown that the caudal part of area 9 on the lateral convexity projects to the pontine nuclei, but in much less profusion than the motor and premotor areas. It is possible that a greater area of granular cortex will be found to be involved, if in view of the small caliber of the projecting fibers and their collaterals, the experimental protocols allow for a longer survival time of the animals.

Clinical Observations Relevant to Extrapolation and Calculation

In the 1940s, problems came up in connection with communication and with the control of antiaircraft weapons that led to the collaboration of physicists, physiologists, and mathematicians. Norbert Wiener[160] introduced the term *cybernetics* (derived from the Greek word for steersman) to designate the specialty dealing with "control and communication in animals and machines." In this respect, *negative feedback* became a popular expression, and in the discussion of animal behavior, the cerebellum was a focus of interest because of its recognized role in the smooth operation of the body and its extremities in goal-directed activity. An oft-repeated illustration was that of a frog with but a meager cerebellum that would snap directly at a fly, as opposed to a cat with a well-developed cerebellum that would extrapolate the direction and speed of its prey and pounce accordingly. How much superior was the extrapolative ability of a baseball player in catching a high and long fly ball! It was as though the brain had a built-in calculus.

Since Bruns'[161] original observations in 1892, there have been several reports that lesions of the frontal lobe may be accompanied by cerebellar symptoms, usually manifest by an ataxia of gait. There has been an inclination to attribute these symptoms to increased intracranial pressure with a resulting disturbance of function of the pons and cerebellum. Meyer and Barron,[162] however, conducted a study on seven patients (including one with an atrophic lesion of the prefrontal areas) in whom an increase of intracranial pressure was unlikely, and in whom there was a disturbance of gait. Since these authors attribute the symptoms to a disconnection of cortical association fibers, rather than an involvement of cerebellar circuits *per se,* they prefer to characterize the symptoms as an "apraxia of gait" rather than as an ataxia. They suggest that the motor deficit is owing to a disturbance of abstraction. For example, a patient proves capable of kicking a ball, but finds it impossible to execute the same movements *if asked to kick an imaginary ball.* It is not clear, however, why these same authors discount involvement of frontopontine connections. Meyer *et al.*[163] found degeneration in frontopontine pathways in patients who had undergone frontal lobotomy.

Some neurophysiologists[164] use the expressions *long-term* and *short-term planning* to distinguish between the premovement role of the cortical association areas and the corrective control attributable to cerebellar nuclei once a movement has started. But in terms of function, there is in addition the obvious consideration that human beings forsee their actions not just in the immediate future, as when anticipating where to run when catching a fly ball, but also visualize their movements days, months, and years into the future. In conjunction with such anticipation, either intuitive or arithmetic calculation may be employed as a means of increasing the accuracy of prediction. Intuitive calculation would be illustrated by a South Seas islander who walks to the shore, raises a wet finger to the wind, and then navigates a boat to a pinpoint island 50 miles away; whereas the formal logistics in preparing for D day would be an example of an opposite extreme.

In either case, for a planned feat to be successful, there must be a registration and memory of the order and timing of each calculated event. In other words, there must be a memory of what is to transpire in the future, or, as one might say for short, a "memory of the future."

The present argument requires clarification regarding the frequent clinical impression that patients with prefrontal lesions suffer from a failure of memory. Examination shows that in actuality the patient is capable of remembering, but in reporting what is remembered, may relate it inappropriately to other happenings and thereby gives the impression of memory failure.[165] As a consequence, such patients might be expected to be incapacitated in utilizing the memory of a past experience for integrating what must be remembered in carrying out future actions. In addition, there is evidence that a person, say, in maintaining a checking account and planning future expenditures might be further handicapped by an impaired ability to perform calculations. As Luria[166] has commented, "Perhaps the disturbances of intellectual activity in patients with a 'frontal lobe syndrome' are seen most clearly when they try to solve arithmetical problems." As an example, he cites a patient's response to a question about how many of 18 books would be on two shelves if there where twice as many on one shelf as the other. The patient immediately hears the word *twice* and gives the answer 36 + 18 = 54.

In view of what has been said about the extrapolatory role of the cerebellum in goal-directed behavior, it would not be surprising if it eventually proves to play a role in numerical calculations ranging from the simplest type to the most complex of which the human being is capable. At the present time, almost everything that can be said on this score is speculative. Unfortunately, there exists little information about the role of the cerebrum in the language of numeration that represents one of the greatest human intellectual accomplishments. The condition acalculia (a term that Henschen[167] introduced for the inability to do simple arithmetic calculations) is treated like an afterthought in most neurological textbooks and handbooks. Unlike the acalculias associated with occipital, temporal, and parietal lesions owing respectively to visual, auditory, and spatial factors, the acalculia associated with frontal lobe lesions appears to be secondary to inabilities in abstraction. But the deficits in simple calculation of the kind described would hardly seem to be exacting for the kind of mechanisms that allow some individuals to perform in a matter of seconds (and with the accuracy of a calculator) computations involving large numbers.

Cases are described of patients with olivopontocerebellar atrophy in whom dementia or "intellectual deterioration" developed, but it remains ambiguous as to what extent the disease of cerebellar circuits, as opposed to complicating lesions elsewhere, contributed to the mental disability. Perhaps cases of agenesis of the cerebellum would provide the least complicated picture of human potential in the absence of the greater part of the cerebellum. Rubinstein and Freeman[168] describe a remarkable case of cerebellar agenesis in a 72-year-old man who had shown no sign of motor disability prior to a vascular lesion in the left parietal region. The postmortem examination revealed that the cerebellum consisted of two small nubbins measuring no more than $8 \times 7 \times 5$ mm. The patient had worked all his adult life either as a handyman, repairman, or gardener. As a boy he had apparently participated in games without attracting attention because of disability. But most notably he never went to school and was described by his brother as always being "mentally subnormal." His ability to perform calculations is not mentioned, but it is said that he understood the value of money and "did not lose out" in handling his financial affairs.

Comment

Answers as to whether or not the frontal lobes and cerebellum play a special role in calculation and prediction will require an unusual degree of alertness on the part of neurologists for potentially revealing cases. In the past, incomplete case histories and only partial examination of the brain have curtailed knowledge of brain correlates of psychological processes. In the future the ideal goal with respect to problems of special interest will be the availability of case histories obtained by neurologists sophisticated in psychodynamics and human ethology, combined with computer-assisted neuropathological examination of serial sections of the entire brain.

If indeed the cerebellum should turn out to be important for mathematical calculation and prediction, it is evident how the subject of fractals (see Chapter 2) would have special relevance. For example, it would seem important in numerical computation that both axoplasmic flow and the flow of nerve impulses in the branching of the ever finer and finer terminals would be crucial to the timing leading either to a correct or incorrect solution. Since it is likely that computers have been designed, so to speak, "in our own image," it is possible that they might, mirrorwise, reflect insight into the circuitry and mechanics of neural computation. At the present time, it simply must be admitted that there exist no compelling explanations of how the avalanche flow of nerve impulses affords a repeatably exact outcome of calculations involving multiple steps of addition, multiplication, and division even when "performed in the head" without benefit of pencil and paper. If it should turn out that "idiot savants" are able to carry out complicated calculations with the kind of cerebellar damage reported in some autistic individuals,[169] then perhaps one should look elsewhere in the brain for structures underlying the *mechanics* of computation.

Concluding Discussion

Recalling that in this chapter our primary concern is with paleopsychic processes, we focus in this final discussion on human evolution as it pertains to (1) the shedding of tears with crying; (2) the role of play in acculturation and creativity; and (3) "the memory of the future." All of these topics reflect in some degree the concurrent evolution of a sense of empathy and altruism. Since they are conditions that appear to depend particularly on the linkage of the frontal neocortex with the thalamocingulate division of the limbic system, it is relevant as background to summarize what is known about the time course of the evolution of the present-day human cranium with its distinctive elevation of the brow overlying the prefrontal region of the brain.

Evolution of the Human Cranium

In considering the evolution of mammals, we described evidence of "directional evolution" among several lines of the mammal-like (Chapters 5 and 17). In regard to mammals themselves, the comparative findings on 20 orders of existing mammals would indicate that the development of the limbic system represents another example of directional evolution. A like interpretation would seem to apply to the progressive development of the neocortex in advanced mammals. And based on the *de novo* appearance of the ventrolateral part of the dentate nucleus of the burgeoning neocerebellar system, there would appear to be a relatively recent directional evolution occurring in the brains of the gorilla, chimpanzee, and human being.

Historically, there has been a traditional effort to establish a line of continuity leading from a common simian ancestor to the appearance of human beings. As noted in Chapter 17, there existed in the Cretaceous Period, insectivore-like animals that may have been antecedents of the oriental tree shrew, which has many features (but not the nails) typical of primates. In tracing human descent, one main interest has been a search for fossil forms reflecting a pivotal condition between monkeys and apes. Such forms turned up in Egyptian Fayum sediments geologically identified with the early Oligocene.[170] One rather advanced type (*Aegyptopithecus zeuxis*) had certain skeletal features suggestive of some bipedal capability. Since the New World monkeys are seldom discussed in terms of human ancestral stock (possibly because of their divergence dating back to the breakup of Pangaea), it is of parenthetical interest that the howler monkey (*Alouatta*) has a forelimb similarity to *Aegyptopithecus* and that this feature, together with its locomotion in the trees and on the ground, suggests a model of locomotion for the Egyptian ape. Schön Ybarra[171] has noted that if this model were to hold, *Aegyptopithecus* would have been an arboreal quadruped that could have used a bipedal stance while reaching for food, and may also have walked bipedally for short distances on the ground.

In a review, Cartmill *et al.*[172] point out that, during the 1950s, findings of Miocene fossils "began to pose intractable problems for theorists." Early Miocene apes such as *Proconsul africanus* and *Dendropithecus macinnesi* that had been regarded as possible ancestors of gibbons and chimpanzees had limb bones that "turned out to resemble those of quadrupedal monkeys" rather than those of brachiating apes. Then in 1961, they continue, Elwyn Simons[173] reported that one of the late Miocene apes (*Ramapithecus brevirostris*) was characterized by small canine teeth, a lofty palate, and rows of upper teeth diverging posteriorly. Because of these human features, he proposed that it represented the long-sought Miocene hominid with features marking it as the earliest member of the human family (Hominidae) (see Ref. 172). But thereafter immunological studies pointed to a close linkage of human beings and modern apes, and based on Sarich and Wilson's[174] "molecular clock" (1967) and the estimated rate of change of serum albumin, the lineage of the chimpanzee, gorilla, and human being would have diverged about 5 million, rather than 10 million years ago.

As of today, the gracile form of *Australopithecus africanus* first described by Raymond Dart[175] in 1925 still stands as the best example of an intermediate form between apes and human beings (see Figure 28-7A). The cranial shape, brain endocast, and teeth all suggest a human trend. Subsequent findings of Broom *et al.*[176] established that the Australopithecinae were bipedal. In Ethiopia, Johanson and colleagues[177] uncovered a smaller variety (including their famous "Lucy") that they called *Australopithecus afaren-*

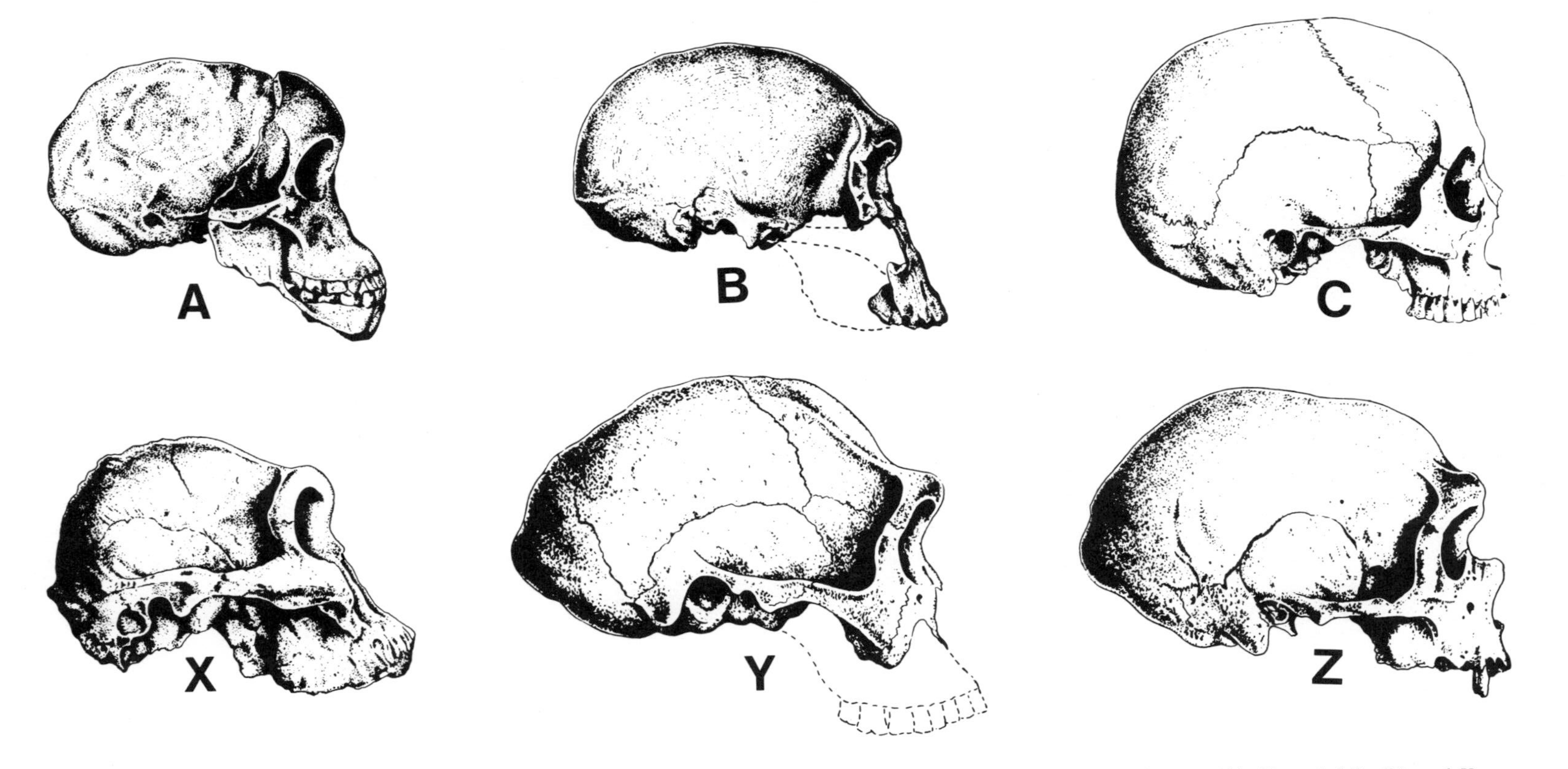

Figure 28-7. Representative skulls for discussing the evolution of the human cranium. The cranial contours of *Australopithecus africanus* (A), *Homo habilis* (B), and *Homo sapiens sapiens* (C) all show a rounding in the frontal and occipital region. In contradistinction, the craniums of *Australopithecus robustus* (X), *Homo erectus* (Y), and *Homo sapiens neanderthalensis* (Z) have the following similarities: heavy brow ridge and periorbital buttressing in all; sagittal ridges in *A. robustus* and *H. erectus;* and the long low cranium together with an occipital protuberance in *H. erectus* and Neanderthal. See text regarding justification of using the juvenile Taung skull for the present comparison. Selection from original drawings by Nicholas Amorosi in Eldredge and Tattersall (1982).

sis, which existed 4 to 3 million years ago and which also walked upright. In addition to the gracile australopithecines, there was contemporaneously a robust type with an accentuated brow and sagittal ridge, the South African type called *A. robustus* and the East African *A. boisei* (Figure 28-7X).

In 1964, Louis Leakey *et al.* had described a fossil with definite human characteristics that lived about 2 million years ago and that they identified as *Homo habilis*,[178] the handy man, a name, they said, suggested by Dart.[178] The skull capacity was estimated as being 673.5 cm^3 or more.[179] The Oldowan Beds in which it was found revealed stone tools.

Later, skull fragments were found from another specimen of *H. habilis* (KNM-ER 1470)[180] (Figure 28-7B) that, when reconstructed, afforded the preparation of an entire endocast.[181] Measurements showed a cranial capacity within the range of 750–775 cm^3. To her surprise, Falk[181] found that the sulcal markings in the basal frontal regions were indicative of gyri corresponding to, but smaller than, those of Broca's speech area in the human brain. In contrast, the smaller australopithecine brains of about 450 cm^3 had only a single fronto-orbital sulcus characteristic of the ape brain.[182] As Falk[181] points out, her findings naturally raise the question as to whether or not *H. habilis* was capable of speech. But this question becomes muted in the light of Lieberman's[133] statements cited earlier in regard to evolutionary changes in the nasopharynx required for speech (for contrary view see Dingwall[145]).

A mere inspection of the brow, face, and rounded skull of the famous Taung skull (*A. africanus*) and of *H. habilis* (ER 1470) (Figure 28-7A and B) explains why the question arises as to whether or not the latter represents an advanced form of *A. africanus*, as L. S. B. Leakey,[183] the discoverer, suggested. One might take exception to the use of the Taung skull for the present comparison because it is that of a juvenile ("a mere boy," as Dart informally expresses it), but one could have substituted for it the skull of an adult gracile form. As also illustrated in Figure 28-7, one might say that the cranium of *H. erectus* with its reduced sagittal ridge and heavy periorbital bones seems to reflect a carryover of these exaggerated features in the robust australopithecine (see Figure 28-7X and Y). Crania of the type of *H. erectus* were first discovered in Java and in China and were respectively identified as *Pithecanthropus erectus* and *Sinanthropus pekinensis* and dated as about 500,000 years in age. The cranial capacities of four Java skulls were in the range of 815 to 1059 cm^3, while five of Peking type measured from 915 to 1225 cm^3.[184] It is now assumed that the Asian forms of *H. erectus* originated in Africa where exploration on both the east and west side of Lake Turkana (formerly Lake Rudolf) has turned up the remains of this species in strata as old as 1.6 million years (Figure 28-7Y).[185] The calculated cranial capacity of this early human form ranges from 800 to 900 cm^3. Most significantly, a robust australopithecine (*Australopithecus boisei*) has been found in strata of the same age, indicating that *H. erectus* was contemporaneous.[186] A specimen of the boise variety (formerly known as *Zinjanthropus boisei*, discovered by Mary Leakey[187]) has been found in deposits with an age of 2.5 million years and counter to earlier belief, is suggested to be a more primitive, rather than later, form of the South African robust variety.[188] (The species name "boise" honors Charles Boise, a supporter of the Leakeys' work.[183]) Ranging in age from 500,000 to 200,000 years, human remains have been found in Europe, western Asia, and Africa that are identified as those of archaic "*Homo sapiens*" because the skull and cranial capacity begin to approach that of a modern human (1380–1400 cm^3).[189] Then from about 150,000 to 30,000 years ago there existed the well-known Neanderthal people with distinctive type skulls that were characterized by a great expansion in the occipital region. There are "very few" Neanderthal cranial endocasts in existence.[190] Based on his examination of three and the reported measurements of three others of the "classic" western European type, Holloway[190] calculated a mean cranial capacity of "1504 ml with S.D. of 148.8." He expresses his belief that on the average, Neanderthals had larger brains than human beings of today and that the difference may have been owing to a greater amount of body musculature.

Figure 28-7Z shows the familiar profile of the Neanderthal cranium. Its elongated somewhat flattened appearance, together with the rather low brow and elongated occipital region, agrees with the platycephalic form of endocasts. According to Trinkaus and LeMay[191] the occipital bunning involving lambdoid flattening might be expected as the result of normal growth and expansion of the occipital lobes up to the age of 7. It is a feature that may also show up in the craniums of archiac *H. sapiens* and occurs in modern human beings.

For comparison with the Neanderthal skull, Figure 28-7C portrays the high brow and rounded occiput typical of the cranium of Cro-Magnons, who appeared as though overnight about 40,000 to 30,000 years ago in Europe and whose presence was associated with the demise of the Neanderthals. The apparent suddenness of their appearance has remained a mystery, but one chink into the mystery may have been opened by a study employing thermoluminescence dating that has led the authors[192] to claim that "Proto-Cro-Magnons" existed contemporaneously with Neanderthals living in the Middle East 100,000 to 90,000 years ago.

It would seem that ever since the time of the australopithecines there persisted two main kinds of cranial features in the evolution of hominids that are respectively illustrated by A, B, C, and X, Y, Z, in Figure 28-7. The distinctive features of the gracile australopithecines (A) are represented by the rounded forehead and occiput that appear to be preserved in the expanded crania of *H. habilis* (B) and *H. sapiens sapiens* (C). The other distinctive cranium is typified by the accentuated features of *A. robustus* that have already been commented upon. As was noted, the prominent supraorbital ridges of *H. erectus* together with the attenuated sagittal ridge and the elongated flattened cranium suggest a stretched form of *A. robustus*. In turn, the elongated low Neanderthal cranium with its great supraorbital ridges appears somewhat like a stretched form of *H. erectus*. Indeed, several of the reconstructed crania of *H. erectus* appear to have an incipient Neanderthal bun.

In most current writings it seems to be taken for granted that *H. erectus* is the direct antecedent of human beings, but the fossil record itself leaves open the possibility that, as the cranial features might suggest, there evolved contemporaneously two different human species or subspecies. However the question may be resolved, it is the high forehead of Cro-Magnon people that survives today, and it may be assumed that it is the expanded frontal lobes underneath that have afforded the development of a sense of empathy and altruism that have the potential of inducing a 180° turnaround in attitudes toward the life-and-death struggle for existence. We consider next the possible relationship of the expanded brain to the long hominid association with fire and the honing of feelings associated with tears.

The Question Regarding the Evolution of Tearing

It is assumed that human ingenuity received additional stretching by various uses of fire that made possible migration and survival under the cold conditions met with in

Eurasia. It is the subject of fire that reverts our attention to the earlier section in this chapter dealing with the cerebral representation of crying and laughter, particularly as they are manifest by the highly developed human capacity to identify affectively with the feelings of other living beings.

Physiologically, both the somatic and autonomic manifestations of crying and laughter are of the kind that would rid the body of something noxious. When these manifestations occur, the *affect* is one of dejection with crying, and one of relief with laughing. Although chimpanzees and gorillas display elements of crying and laughter,[193] human beings are the only creatures known to shed tears with crying.[194] In regard to this distinction, I have pointed out that human beings, and perhaps their immediate predecessors, are the only creatures known to have used fire.[195] Because of this, I have suggested that in the course of time there may have arisen some connection between smoke and tears and activities surrounding fire, including ceremonies involved in disposing of departed ones.[195]

What is the time required for a new manifestation such as emotional tearing to evolve? If the frontal lobes increased substantially in size within 100,000 years (see above), it is a reasonable assumption that a manifestation such as tearing could have developed within an equal or shorter length of time. The hominid use of fire has a history of more than ten times that period. It is said that the use of fire has been documented as early as 1.4 million years ago in East Africa. With newer methods of analysis, the earliest use of fire at Yuanmou, Xihoudu, and Lantian in China has been dated as between 1 million and 700,000 years. The age of the hearths at Terra Amata (Nice), France, and at Vértesszölös, Hungary, would fall into the lower part of that range.[196] Desmond Clark[197] has cited evidence that hominids may have had an association with fire for 2.5 million years and perhaps as long ago as the beginning of bipedalism. Clark and Harris[196] have reported the results of the application of new techniques in archeometry (magnetic analysis, thermoluminescence, electron spin resonance, spectroscopy) at suspected sites of fire located at Middle Awash and Gadeb in Ethiopia and at Koobi Fora and Chesowanja in Kenya, where fossils of hominids have been identified. The best evidence of use of fire was found at Chesowanja, where there was uncovered a hearthlike arrangement of baked clay clasts not far from a spot where a hominid specimen was found and tentatively identified as belonging to *A. boisei*. However, the authors conclude that not only at Chesowanja, but also at Gadeb and Koobi Fora, the evidence suggests that "lower Pleistocene hominids may have been making use of, though probably not manufacturing, fire as early as 1.5–1.0 myr on the high plains of eastern Ethiopia and in the East African Rift Valley."[198]

Clark and Harris[196] emphasize that the knowledge of making fire would have been unnecessary because of its availability from natural sources such as volcanic eruptions or trees and savanna grasses ignited by lightning. They point out that to this day, the people of the Andaman Islands in the Bay of Bengal do not know how to make fire and are in the habit of carrying it from place to place. Like a number of opportunistic feeders including birds and carnivores, early hominids would have had the opportunity to observe how a grass fire drives insects and fleshy animals before it and concentrates them as edible targets. Once such an association had been recognized, the next step would have been to set fire purposely when hunting for game, just as has been the practice of hunter-gatherers up to the present time.

Clark and Harris[196] also suggest that the early hominids may have used burning brands not only as a means of transporting fire from one place to another, but also for driving carnivores away from their kills. I think that it is clearly implied in their discussion that perhaps the hand carrying of fire may have contributed to the assumption of a bipedal posture. I would also suggest that the use of fire for hunting game by burning over patches of grass would have favored a bipedal posture so as to keep the nose and face farther away from the burning ground. The assumption of an upright posture would also have been a natural reflex in backing away from the blazing of a hearth fire when new fuel was added.

Given this background, we return to the question of the association of fire and smoke with lacrimation. With the northward migration of early humans into Eurasia, the need for warmth would have been conducive to making campfires, and regardless of smoke, promoted a close mingling of individuals around the fire. Both for warmth, and as a protection against predators, a campfire would have had the effect of extending natural daylight and favoring sociability into the evening hours. Hearth fires in caves would have provided a more pungent atmosphere of smoke. Confined by the surrounding darkness and drawn toward the fire for warmth or the cooking of food, might the individuals of a family or other small social group have felt the compulsion through signs, sounds, and ultimately words to share their experiences of the day just spent or of times past?

Yet none of the conjectures provide a hint as to the connection of fire, smoke, and emotional tearing—least of all an explanation in neo-Darwinian terms of how emotional tearing might have originated. In an attempt to address this question, one might begin by noting that there is a degree of equivalence of somatic pain and psychological pain. Given the cerebral capacity to react to a painful experience as an unconditional stimulus, the possibility suggests itself that individuals who quickly learned a respect for fire (and attendant smoke) after accidental burning might have had a better chance of surviving and reproducing themselves than those who did not. Accidental burning could occur when groups were using grass fires as a means of hunting; during the process of transporting fire from one place to another; or as a result of some mishap around a campfire or hearth fire. The knowledge of present stone age customs suggests that there may have been a long history of an association of fire and smoke incidental to the treatment of various ailments; the keloid decoration of the body by both children and adults; the purposeful infliction of burns during frenzied, ceremonial dancing; and various mourning rites.[199] In regard to mourning, it has been a stone age practice in New Guinea to roast and eat the dead as a means of maintaining a ''feeling of closeness'' with the departed one.[200] Hunger in connection with separation and grief is a well-known human experience. The connection between hunger and tears might not be immediately apparent, but there is at least evidence of a close functional relationship in the brain: Reference was made in Chapter 24 to a patient with psychomotor epilepsy who, during the aura, experienced an epigastric feeling of sadness and wanting to cry that was followed by a welling up of tears and a sensation of hunger. There are no indications of how long the practice of cremation has occurred as a means of disposing of dead bodies, rather than burial as by the Neanderthals. Cremation as practiced in India may involve large public gatherings during which substances are thrown on the fire that exaggerate the intensity of smoke. Some might speculate that cremation accounts for the few remains that have been found of Cro-Magnon people.

As for a connection between laughter and smoke and tears, one might suggest that guffawing amusement around a hearth fire in a cave might be conducive to increased smoke inhalation and an aggravation of tearing.

With that muted observation, we consider next the question of influence of the frontal neocortex on play and laughter.

Question of Role of Neofrontal Areas in Play and Laughter

As pointed out in Chapters 2 and 21, play represents one of three cardinal forms of behavior that characterized the evolutionary transition from reptiles to mammals. There is no persuasive evidence that reptiles play, and the alleged play of feather tossing that occurs among some birds appears almost accidental and of short duration. Hence, it might be argued that individual play, and most particularly long bouts of social play, represents a uniquely mammalian trait. In commenting upon the function of play in Chapter 21, it was suggested that originally play may have served to promote harmony in the nest, and then later on affiliation of members among a group. The experimental findings that were described suggest that the function of play is identified with the most recently evolved division of the limbic system—namely, the thalamocingulate division. Because of the neural associations of the thalamocingulate division with parts of the frontal neocortex, it is a matter of great interest to consider the possible role of the frontal lobes in the regulation of play and laughter and in the intellectual, creative process as it is influenced by wit and humor. In *The Act of Creation* Koestler[201] uses the joke as a paradigm of the creative process, stating, "The pattern underlying all varieties of humour is 'bisociative'—perceiving a situation or event in two habitually incompatible associative contexts."

There is but little experimental information bearing on the role of the frontal cortex in play. In describing a female monkey with a frontal lobectomy on the right side and part of the frontal cortex of the outer convexity ablated on the other side, Bianchi[202] remarked that she no longer played with the other two monkeys that were her companions. Franzen and Myers[203] reported that three macaque monkeys, 2 to 3 years in age, with prefrontal lobectomy rostral to the eye fields, but including the orbitofrontal region, showed a marked reduction in play behavior.

Earlier in describing Ackerly and Benton's case, it was noted that their patient who grew up devoid of the frontal part of the brain had much sport equipment. Although it was not reported whether or not he used it, his virtual exclusion from groups his own age would indicate that he would not have had much opportunity for social play. Given the paucity of clinical attention to play in cases of frontal lobe deficits, there are abundant illustrations of a play on words, punning, and off-color jokes. In discussing frontal lobotomy mention was made of the uninhibited, inappropriate, and sometime queer laughter that may occur following that procedure. The laughter frequently gives the impression that such patients behave as though not identifying with the feelings of those in their company. These assorted observations, and the epileptic material that was reviewed, are indicative that the frontal neocortex is implicated in the regulation of playfulness and laughter.

For examples of how a play on words, punning, and wit have contributed to creative thought in literature and in the theatrical world, one hardly needs to look beyond Shakespeare. For the interplay of such factors in science, art, and music, Hofstadter's[204] book *Gödel, Escher, Bach* provides abundant illustrations. O. B. Hardison[205] points out that Friedrich Schiller, poet and friend of Goethe, "popularized the game metaphor at the end of the 18th century," tracing the human urge to create to the *Spieltrieb*, the play impulse. Hardison goes on to say that the play of modern science is "serious" but that "its games are so exhilarating, and the rules often so strange, that the play becomes overtly playful. The playfulness spills over into mathematical and logical puzzles and into language that is intentionally paradoxical, whimsical, and absurd.''[206] He cites for example the origins of such words as ''quarks'' and ''gluons.'' Then turning to quite a different kind of game, he refers to Mandelbrot's book *The Fractal Geometry of Nature* that ''offers picture after picture of breathtaking playfulness,'' commenting that ''part of the playfulness . . . comes from its unpredictability.''[207]

With the soaring of the world's population, it is of interest to consider the role of play in coping with crowding. Apart from the limits of the mother in giving birth to viviparous young, the number of offspring in a mammlian family is limited by the number of nipples. It may be partially for this reason that the optimum number in mammalian social groups tends not to exceed 12.[208] Whatever the reason, it would appear that, except for matters concerning family and language, the limbic cortex and neocortex have few inborn programs for regulating behavior, and none in particular for coping with large numbers of individuals.[131] In other words, it would appear that the adoption of a family way of life has made it awkward for most mammals to adapt to crowds. Even herd animals tend to group as families. When human beings meet in large numbers, they seem to do best in situations in which they are feeding together, as at feasts and music festivals; or, taking advantage of the mammalian trait of play, are engaged in local, national, or international games, including the Olympic Games. But even in the case of international games, there appears to be a primitive, childlike, fine line between enjoying fun at play and getting mad and fighting. Hence, we have seen international games become a political means of displaying national will in showdown situations.[131]

Role of Neofrontal Areas in Planning and a ''Memory of the Future''

This last topic regarding planning and a ''memory of the future'' recalls the earlier discussion of frontocerebellar mechanisms in terms of calculation and prediction. Here it is also relevant to recall Ackerly and Benton's patient, the young man with bilateral vacuities in the prefrontal region. It was argued in his case that it was impossible for him to commit a premeditated crime because of his inability to plan and remember the sequential acts involved. This view was expressed despite abundant evidence of the patient's ability to remember in considerable detail current and past happenings. It was as though premeditation required not only the ability to plan, but also the step-by-step memory of what is planned, or as one might otherwise say, a ''memory of the future.''

For clinical evaluation it would be difficult, if indeed possible, to devise a test comparable to one that would embody the motivation and affect involved, say, in making detailed travel plans for holiday in a foreign country; deciding on alternatives for a child's education at home or out of town; in undertaking some business enterprise; in making provisions for a will; and the like. Indeed, to fashion such tests would require special tailoring with respect to a person's past history and current background. A Porteus maze test, to be sure, may be indicative of deficits in short-term planning, but could in no way compare to the psychological challenges involved in the items just listed.

In regard to the motivational and affective aspects of planning, it is to be emphasized that because of its respective linkage to the amygdalar and thalamocingulate divisions of the limbic system, the frontal association cortex might be expected, in the former case, to function ''selfishly'' in terms of self-survival or, in the latter case, altruistically on behalf of others. Planning and saving for the day when one would no longer be earning a living would serve as an example of a concern for self-survival that might entail the weighing of many alternatives, financial calculation, and memory of the timing and nature of steps to be taken in the future. Performing a like exercise on behalf of one's family would exemplify an altruistic (affective) concern in long-range planning, as illustrated by all of the alternatives that must be taken into account in making provisions for a child's advanced education. An altruistic example in terms of the social, extended family would be provided by the planning and memory entailed in establishing a foundation for the promotion of some charitable endeavor. All such actions by an individual would pale compared with the planning, calculation, and memorization of statistics and logistics required of those responsible for preparing and justifying a national budget, or, say, of those involved in a multinational economic enterprise.

When the expression "memory of the future" is used, it may be asked, "How can there be a memory of something that has not yet happened? Wasn't it the intention to say *planning* instead of *memory?*" Planning, however, represents a developing, ongoing process. It is the blueprint in the *drawing stage,* not the blueprint itself. In a competitive sport such as football, it would generally be self-defeating to plan a play while the game is in progress. What is important is to remember the "blueprint" already available that provides a foreknowledge of where each player is supposed to be. This requires, one might say, a frontocerebellar "mind-set" of the number and direction of turns and running distances. Accordingly, it has been implicit in all the illustrations previously given that it is not only the planning, calculations, and predictions involved, but also the memory of what is planned, as, say, the restrictions in a will, the conditions of a national trust, and the like.

Cerebral Mechanisms

In regard to mechanisms of prediction and "memory of the future," it is of signal interest that the new neuroanatomical findings of Carmel[88] and others[89,90] have revealed extensive connections of the frontal granular cortex with the ventral anterior nuclei (VA, pc, mc) of the thalamus. These findings help to clarify the observations by Starzl and Magoun[209] and Hanberry and Jasper[210] in the early 1950s that this nuclear complex appears to be central to a short-latency activation by the diffuse "thalamic projection system,"[211] being most pronounced in the frontal cortex, but also accounting for widespread cortical responses in the other "association areas." Subsequently, Starzl and Whitlock[212] found that the frontal response (including cingulate areas) was considerably greater in the monkey than the cat. Such widespread effects on the association areas would seem relevant to the global functions of the frontal lobe, including functions depending on an integration of past memory and the memory of ongoing experience, as they would relate to planning and a "memory of the future."

It holds particular interest to consider how the new functional neuroanatomy relates particularly to the linkup of the frontal cortex with the thalamocingulate division and family-related behavior as it may have affected the development of a sense of altruism and empathy. Not considered under "tearing" was the question as to why misting of the eyes is such a common experience when people see an altruistic act. It has been suggested that this condition may reflect "a high order of generalization owing to a close association of neural mechanisms" conditioned, say, by childhood crying for help and the relief of parental rescue.[213]

Altruism (Auguste Comte, 1853), *empathy* (Theodor Lipps, 1903): these are almost new words, reflecting the acceleration of the humanitarian movement. In considering the linkup of intero- and exteroceptive systems requisite for a feeling of personal identity (Chapters 26 and 27), it was pointed out that microelectrode studies provide evidence that through its connections with the medial dorsal nucleus, the prefrontal cortex is the only neocortex that receives a strong projection from the great visceral nerve. Presumably such internally derived experience is necessary for an individual's identification with the feelings of others. And it may be imagined that such an interiorized sense is also necessary for the "insight" requisite for the foresight to feel a concern for the future of others, as well as the self. The possibility is suggested that through the neofrontal connections with the thalamocingulate division, a parental concern for the young generalizes to other members of the species, a psychological development that amounts to an evolution from a sense of responsibility to what we call conscience.[131] What is substantially new in the known history of biology is that this concern extends not only to the human family, but to all living things—an evolutionary turnabout that could affect a turnabout in what has heretofore seemed a vicious life–death struggle long recognized as the struggle between good and evil.

References

1. Newman, 1979
2. Chomsky, 1968
3. Lenneberg, 1967
4. Hogben, 1937/1946
5. Akert, 1964
6. Rose and Woolsey, 1948
7. Fulton, 1949a
8. Penfield and Jasper, 1954
9. Kolb, 1949
10. Harlow, 1848
11. Harlow, 1868
12. Harlow, 1868, pp. 339–340
13. Cobb, 1943
14. Ackerly and Benton, 1948
15. Brickner, 1934, 1952
16. Hebb and Penfield, 1940
17. Ackerly, 1964
18. Ackerly and Benton, 1948, p. 480
19. Ackerly and Benton, 1948, p. 487
20. Ackerly and Benton, 1948, p. 489
21. Grinker, 1948, pp. 502–503
22. Ackerly and Benton, 1948, p. 488
23. Lashley, 1929/1963
24. Hebb, 1963, p. v
25. Hebb, 1949, p. 284
26. Hebb and Penfield, 1940, p. 436
27. Hebb, 1949, p. 11
28. Ackerly, 1964, p. 211
29. Fulton, 1949a, see p. 459
30. Fulton and Jacobsen, 1935
31. Freeman and Watts, 1950
32. Moniz, 1936
33. Freeman and Watts, 1936
34. Davidoff and Brenner, 1948
35. Freeman and Watts, 1944
36. Rylander, 1948, p. 695
37. Rylander, 1948, p. 696
38. Penfield, 1948; Cameron and Prados, 1948; Malmo, 1948
39. Fulton, 1949b, 1951
40. McIntyre *et al.*, 1954
41. Grantham, 1951
42. Freeman and Watts, 1950
43. Freeman and Watts, 1950, p. 356
44. Koskoff *et al.*, 1948
45. Scarff, 1948
46. White and Sweet, 1969
47. Chapman *et al.*, 1948
48. McIntyre *et al.*, 1954, p. 119
49. Hutton, 1943
50. Milner, 1964
51. Bonner *et al.*, 1953
52. Halstead, 1948
53. Halstead, 1947
54. Tilney and Morrison, 1912; Wilson, 1924; Davison and Kelman, 1939; Ironside, 1956; Sackeim *et al.*, 1982
55. Poeck, 1969
56. Poeck and Pilleri, 1963
57. Tilney and Morrison, 1912
58. Papez, 1937
59. Pool, 1954
60. Penfield and Jasper, 1954, Case M.Pn., p. 416
61. Penfield and Jasper, 1954, Case L.J., p. 418
62. Offen *et al.*, 1976
63. MacLean, 1952
64. Trousseau, 1872, quoted by Loiseau *et al.*, 1971
65. Daly and Mulder, 1957
66. Weil *et al.*, 1958
67. Loiseau *et al.*, 1971
68. Stevens and MacLean, unpublished observations
69. Swash, 1972
70. Swash, 1972, p. 109
71. Van Buren, 1961
72. Dott, 1938
73. Dott, 1938, p. 179
74. Martin, 1950
75. List *et al.*, 1958
76. Penfield and Jasper, 1954, Case J.Hl., p. 414
77. Geier *et al.*, 1977
78. Geier *et al.*, 1977, p. 956
79. Talairach *et al.*, 1973
80. Ludwig *et al.*, 1975
81. Lehtinen and Kivalo, 1965
82. Kramer, 1954
83. Sackeim *et al.*, 1982
84. Pribram *et al.*, 1953
85. Akert, 1964
86. Walker, 1938
87. Tobias, 1975
88. Carmel, 1970
89. Baleydier and Mauguiere, 1980; Goldman-Rakic and Porrino, 1985; Jürgens, 1983; Vogt *et al.*, 1987
90. MacLean, 1987a, 1988
91. Jürgens, 1983
92. Schell and Strick, 1984
93. Nauta and Mehler, 1966; Kuo and Carpenter, 1973
94. Asanuma *et al.*, 1983b
95. Wiesendanger and Wiesendanger, 1985
96. Jürgens, 1984
97. Rubens, 1975

98. Pribram and MacLean, 1953
99. Pandya and Barnes, 1987
100. Hassler, 1961
101. Krayenbühl *et al.*, 1961
102. Jones, 1985
103. Asanuma *et al.*, 1983b
104. Vilensky and Van Hoesen, 1981
105. Mehler *et al.*, 1960
106. Casey, 1966
107. Stier, 1911; Chamberlain, 1928; Rife, 1940; Trankell, 1950; Bingley, 1958; Benson and Geschwind, 1987
108. Broca, 1861a,b, 1865; Jackson, 1874; Nielsen, 1946; Penfield and Jasper, 1954; Wada and Rasmussen, 1960; Benson and Geschwind, 1987
109. Trankell, 1950; Bingley, 1958; Warren, 1980
110. Chambers's Encyclopedia, 1892b; Parson, 1924
111. Parson, 1924
112. Carlyle, 1871; Pye-Smith, 1871; Parson, 1924
113. Chambers's Encyclopedia, 1892b; Evans, cited by Parson, 1924; Rust, 1974
114. Wilson, 1891; Coren and Porac, 1977; Marshak, 1976
115. Evans, cited by Parson, 1924; Rust, 1974
116. Toth, 1985
117. Dart, 1925
118. Dart, 1959
119. Kitching, 1963
120. Irving, 1987
121. Irving, 1987, p. 14
122. Finch, 1941; Warren, 1953
123. Peterson, 1931; Warren *et al.*, 1967; *see also* MacNeilage, 1987
124. MacLean, 1978c
125. Jackson, 1879–80/1958
126. Penfield and Welch, 1951
127. Campbell, 1979; Ardrey, 1970; Tiger, 1970
128. See Bingley, 1958, pp. 58–64
129. Campbell, 1979, pp. 207–210
130. Salk, 1960
131. MacLean, 1985a
132. Koehler, 1954; Ploog, 1979
133. Lieberman and Crelin, 1971; Lieberman *et al.*, 1972
134. Newman, 1985a
135. Dow, 1974, p. 110
136. Dow, 1942
137. Asanuma *et al.*, 1983c
138. Carpenter, 1981
139. Hopkins and Lawrence, 1975
140. Dow, 1974
140a. Ga s, 1924
141. Holmes, 1922a, p. 117
142. Holmes, 1907, p. 486
143. Lieberman, 1985, p. 659
144. Lieberman, in press
145. Dingwall, 1988
146. Darley *et al.*, 1975
147. Fulton, 1949a, p. 535
148. Holmes, 1907, p. 487
149. Holmes, 1922b, p. 63
150. Holmes, 1922b, p. 63; *see also* Holmes, 1907
151. Holmes, 1907
152. Holmes, 1922b; Fulton, 1949a
153. Dow, 1974, p. 111
154. Leiner *et al.*, 1986
155. Asanuma *et al.*, 1983a, see p. 231
156. Vogt, 1909
157. Strick, 1976; Jones *et al.*, 1979
158. Miyata and Sasaki, 1983
159. Brodal, 1978; Glickenstein *et al.*, 1985
160. Wiener, 1948
161. Bruns, 1892
162. Meyer and Barron, 1960
163. Meyer *et al.*, 1947
164. Allen and Tsukahara, 1974
165. Luria, 1969
166. Luria, 1969, p. 750
167. Henschen, 1920/1922
168. Rubinstein and Freeman, 1940
169. Bauman and Kemper, 1984; Ritvo *et al.*, 1986; Courchesne *et al.*, 1987
170. Simons, 1965; Romer, 1966
171. Schön Ybarra, 1984
172. Cartmill *et al.*, 1986
173. Simons, 1965
174. Sarich and Wilson, 1967
175. Dart, 1925
176. Broom *et al.*, 1949
177. Johanson and Taieb, 1976; Johanson and White, 1979
178. Leakey *et al.*, 1964
179. Tobias, 1964
180. Leakey, 1973
181. Falk, 1983
182. Falk, 1980
183. Leakey, 1961
184. Falk, 1987
185. Brown *et al.*, 1985
186. Leakey and Walker, 1976
187. M. Leakey, (1959), reported by L. S. B. Leakey, 1959, 1961
188. Walker *et al.*, 1986
189. Eldredge and Tattersall, 1982
190. Holloway, 1981
191. Trinkaus and LeMay, 1982
192. Valladas *et al.*, 1988; *see also* Gould, 1988
193. see e.g., illustration by Kohts in Campbell, 1979
194. Collins, 1932
195. MacLean, 1985a,b, 1987a
196. Clark and Harris, 1985
197. Clark, 1984
198. Clark and Harris, 1985, p. 17
199. Gajdusek, 1970
200. Sorenson and Gajdusek, 1969
201. Koestler, 1964
202. Bianchi, 1895
203. Franzen and Myers, 1973a
204. Hofstadter, 1979
205. Hardison, 1986
206. Hardison, 1986, p. 392
207. Hardison, 1986, p. 402
208. Calhoun, 1971
209. Starzl and Magoun, 1951
210. Hanberry and Jasper, 1953
211. Morison and Dempsey, 1942; Dempsey and Morison, 1943
212. Starzl and Whitlock, 1952
213. MacLean, 1985a, 1987

V

Conclusion

29

Implications for Future Thinking in Regard to Epistemics and Epistemology

The greater part of this final chapter will deal with epistemic questions raised by studies on the limbic system. But, first, I will consider some "reptilian" aspects of mammalian behavior that now, given the background on limbic function, can be more meaningfully discussed here than previously in the context of Chapter 16.

Further Comments on Comparative Behavioral Studies on the R-complex

The comparative neurobehavioral studies on the R-complex have been of value in calling into question the traditional view that this phylogenetically ancient part of the forebrain is *primarily* part of the motor system under the control of the neocortex. The studies have also been useful in directing attention to the neglected question concerning the nature of neural mechanisms required for the complicated functions of orchestrating the daily master routine and subroutines. The results of neurobehavioral experiments on animals as diverse as lizards and monkeys are indicative that the R-complex is basically implicated in integrating the somatic and autonomic components of displays used in prosematic, social communication. The associated findings would also suggest that the R-complex is implicated in the cognitive factors leading to the evocation of displays. If so, this and other considerations would indicate that more than just being part of the motor apparatus, the R-complex "has a mind of its own."

Nativism

In describing Llewellyn Evans's observations on the group of Mexican black lizards inhabiting the nooks and cranies of a cemetery wall (Chapter 6), reference was made to their "nativistic" response to the intrusion of strangers. In regard to nativistic responses, it is also of comparative interest that if one removes a turkey, say, from a confined flock, or indeed takes a young dog from its littermates in an enclosure, it will be attacked as if a stranger when reintroduced to the group (even after a period as short as 5 min). As witness many veterans of the Vietnam War, soldiers may find that upon returning home they are treated like unwanted strangers.

Blemishes

The element of strangeness is also symbolically related to blemishes, not only as that word applies to physical and behavioral blemishes but also as to what one might call ideological blemishes such as differences in religious or political views. Since there does not seem to be a comparative ethological study dealing with the role of blemishes in social behavior, I will describe an incident that I observed that will serve as an illustration. On our farm we had a white Muscovy duck that was unusually capable in rearing a large brood. She would lay so many eggs in her nest in the horse's stall that they would extend beyond her body surface. Nevertheless, most eggs would hatch, and she would be followed by as many as 21 ducklings. After one of the little ducklings developed a blood spot on the right side of its head from a slight injury, it began to suffer harassment by the mother, which, during the next few days, persistently drove it from the brood. One evening, at the fall of darkness, she drove it away, and this time the duckling swam to the opposite side of the pond and into the woods where surely it would have been snatched up by a predator. Consequently I rescued it, and we reared it separately. I was curious to learn whether some other factor that I did not recognize, had contributed to this "child abuse." Perhaps the blood spot on the head that I regarded as a blemish, had resulted from the mother's own pecking for some other reason. Consequently, I took another duckling and made a like spot on its head with a red marking pen. Two days went by, and it was treated no differently from the rest of the brood. Then on the third day, the mother began a relentless harassment, and we found ourselves with a second duckling to rear with the first.

It is a familiar observation among those who are pigeon fanciers or who raise poultry that a fowl with an open injury may be relentlessly pecked to death by members of the flock, an observation illustrating that blemishes also precipitate group harassment. Hence, the occurrence of harassment of strangers and of individuals with blemishes by reptiles and birds indicates that this form of behavior is phylogenetically deeply ingrained and, in the light of the neurobehavioral studies, might be inferred to have its neural basis in structures of the forebrain belonging to the R-complex.

The inferred built-in propensity to nativism and intolerance of blemishes invites special attention because the increasing world population will be conducive to an increasing number of human contacts. Such a situation is calculated to aggravate the needs of each individual for a certain amount of territorial "surround" (See Chapters 6, 9, and 16). Moreover, the population pressure conducive to both *intra*national and *inter*national migration will multiply the opportunities for nativistic reactions and the intolerance of what are physically, behaviorally, or ideologically regarded as blemishes, such as, for example, differences in color, race, ethnic forms of behavior, religion, and political beliefs. An innate propensity to an abhorrence of blemishes would perhaps help to explain why, even in the absence of severe crowding, there are now so many ethnic and racial groups locked in prolonged and bloody struggle because of physical, religious, or ideological differences.

Educational Considerations

Education is affected by similar factors, as witness, for example, the problem of racially integrating schools. But it is possible that the burgeoning population has another profound effect on education. As noted in the preceding chapter, the social group size of mammals appears to have been originally determined by the dependence on a nursing mother and the available number of nipples. Moreover, it was pointed out that the limbic system and neomammalian brain appear to have few "built-in" mechanisms for dealing with crowded conditions. With the increasing number of children in classrooms because of urban crowding, each child will receive less personal attention, and it may be expected that not only will the quality of education deteriorate, but there will also be a rise in disorderliness and truancy.

Legal Implications

In addition to what was said in Chapter 16 regarding some of the legal implications of the role of the R-complex in the acquisition of power and the exercise of power, including the power of law, it now merits additional comment on how limbic and neocortical functions contribute to the reification of the law and the administration of the law. When, in helping to draft the constitution of Massachusetts, John Adams[1] used the expression that we are a "government of laws, not of men," he was indulging in a reification because it was as though the body of the law had been made flesh through the agency of some external power. When discussing the *general affects* in Chapter 24, emphasis was given to the evidence that the limbic system has the capacity to generate out of context, affective feelings of conviction that we attach to our beliefs regardless of whether they are true or false. Accordingly, it must be presumed that it plays a fundamental role in hypostatizing the sovereignty (the supreme power) of the law. Without such a reification, the law could hardly be the powerful, pervasive force that is able to keep a society in line and to do so, for the most part, without even being seen.[2] And without the neocortex, we may rightfully imagine that there could be no evenhanded meting out of the law. Nor, as a counter to "reptilian" vindictiveness (Chapter 9), could there be a merciful weighing of justice without the frontal neocortex.

I should emphasize that the wording of the last sentence should not be construed as a general denigration of reptilian behavior. The reptilian brain appears to have a built-in set of rules for maintaining mannerliness and avoidance of bodily assault under ordinary conditions. However, under certain conditions of crowding, as was illustrated by the marine iguanas in the Galapagos Islands (Chapter 9), infringement upon the small territorial "surround" of another individual may result in a sudden resort to combat.

Compulsions Weighed against Emotions

It is a commonly held Spinozistic view that "men are governed by their emotions."[3] On the contrary, one would gain the impression from comparative studies on animals ranging from reptiles to advanced mammals that in the most essential everyday forms of behavior comprising the daily routine (the "staff of life" itself) animals behave as though governed by some quiet compelling agency. At the human level, it would seem that even people who are driven to commit some crime or asocial act are commonly at a loss to describe their feelings at the time. How does one characterize the feeling of a young man who, when asked why he stabbed a perfect stranger in the back and killed him, said, "I had the feeling of wanting to hear the sound of a knife against bone."[4] What gifted author could dissect out the ingredients of the subjective state of the farmer cited by William James,[5] who, after seeing his father cut his thumb while paring an apple, went to a neighbor's farm and slit a horse's throat? Of such incidents, newspaper accounts will commonly state that "the individual was driven as though by a singular compulsion." One might as well say, "The reptile does what it has to do."

Thus, it can be argued that routine day-to-day activities upon which our lives depend are associated with no particular emotion unless the intended acts are frustrated and that the same may apply to compulsive asocial acts unless the individual is apprehended. One therefore might say that emotions are oftentimes merely reflectors, rather than determinants, of action.[6]

THE LIMBIC SYSTEM IN REGARD TO EPISTEMICS

In this section we consider a number of neglected questions regarding the role of the limbic system in the subjective sense of "being" and assessment of what is real, true, and important—all questions relevant to epistemics (see Chapter 1).

It is understandable that those working in the natural sciences feel obliged to view the world as it appears to them and to assume that a line can be drawn between what is subjective and what is objective. A person working in the brain-related sciences, however, cannot avoid the realization that in the final analysis, everything reduces to subjectivity and that there is no rigorous way of defining a boundary between the subjective and what is regarded as objective.

This same statement must be cast in the light of how we regard the substance of the brain itself. Ommaya,[7] a neurosurgeon investigating brain injury, states that prior to 1968 "an extensive search of the literature has revealed only three sources of information regarding mechanical properties of brain tissue," the earliest one in modern times appearing in 1954. Ommaya concludes that brain matter may be characterized as a viscoelastic substance that is not so stiff as a gel or as plastic as a paste. Comprised of closely packed elastic cells held together chiefly by colloidal forces, the brain has a density slightly greater than that of water and a viscosity comparable to glycerin. The surging blood flow through the brain imparts to it a firmness that helps to resist deformation. Placed lifeless in a container, the brain tends to spread because of its own weight, and if rotated under these conditions, it will "be distorted like a soft gel."[7]

Given these properties of the ultimate receiver and analyzer, it is curious that scientists and people generally place so much confidence in their metallic instruments of precision while at the same time so constantly calling into question the workings of the subjective brain. At the same time as he questions the human ability ever to grasp the true nature of things, Jeans[8] states that "physics gives us exact knowledge because it is based on exact measurements" (see Chapter 1). In regard not only to its substance but also to its transmission and coding of signals, the gelatinous brain is imprecise and infinitely slow when compared with the functioning of manufactured instruments.

Some physicists are of the opinion that if all matter were to "evaporate," there would still exist space and time.[9] Kant's[10] scheme speaks to the inconsistency of such claims. He concluded that there exists without experience (*a priori*) a "form of sensibility" that imposes a sense of space and time on what we perceive through our "senses." Space in his scheme belongs to the "outer sense" and time to the "inner sense.'' Similarly, he ascribed to the ''understanding'' the *a priori* functions that give form to what is intellectually experienced. Thus, he contended that both the ''form of sensibility'' (otherwise referred to as the *transcendental aesthetic*) and ''pure understanding'' are necessary for the appreciation of total experience. One might say that like a television tube without a screen, there could be no picture of experience without these *a priori* formal properties of the mind and senses. Hence, it can be argued that time and space do not exist *per se* but are purely informational constructs derived by the subjective brain.

Even if the brain could stand outside of itself or to multiply itself so as to stand on any platform—any frame of reference—and judge itself moving in relation to its other selves at any speed, it could not assure itself that its perceptions of time and space and of the objects within space bore any true resemblance to what was observed. This situation cannot be attributed solely to the restraints of processing and communicating information in the gellike brain: even if the brain could reconstruct itself of any material the outcome would be the same. The reason for this is because there is no known means of circumventing the problem of self-reference. As Bronowski[11] has commented, ''[A]ny reasonably rich system necessarily includes reference to itself. This creates an endless regress, an infinite hall of mirrors of self-reflection.'' He goes on to say that no mathematical or scientific statements can be wholly cleared of self-reference or of some equivalent recursive regress and that ''no logical machine can reach out of the difficulties and paradoxes created by self-reference.''[12] Since the subjective brain is solely reliant on the derivation of immaterial information, it can never establish an immutable yardstick of its own. Hence, for these reasons it is left with nondimensional space and nondimensional time for which it must arbitrarily set standards of its own. In this respect, it is saddled with a nonyielding relativity.

Procedure for the Analysis of Facts

Given the conditions that we have no basis for anything that we think or do except immaterial information, how do we put a handle on things for dealing with them scientifically? Here, the introspective method comes to our aid, and, short of the limbic aspects, anyone living in the intellectual climate of the last 300 years might arrive at a formulation not too unlike that which is to follow. Indeed, the outcome in many respects is not unlike that arrived at by the founders in 1847 of physiological reductionism—namely, Du Bois-Reymond, Brücke, Helmholtz, and Ludwig, who, in emulating the natural sciences, wished to establish a new physiology on a firm scientific basis.[13] One suspects that they arrived more or less independently at their various tenets by the method of introspection, but 30 and 40 years later Helmholtz and Du Bois-Reymond, respectively, acknowledged their being influenced by Kant's epistemological views.[13] Although the formulation that follows might seem to give one a somewhat firm footing in making a scientific approach to a subject, it seems slippery at best when considered in the light of the epistemological question raised by limbic function that will subsequently be discussed.

How can one pin down some invariant regularities (the usual requisite for laws or rules) that will provide the basis for making observations and analyzing their significance? To begin with, one may make the assumption that everything known about the external and internal worlds is no more than information, and that the communication of this information, including its subjective aspects, depends on the nervous system. In information theory, the word *information* is used in a strict sense to refer to a numerical quantity that is a measure of uncertainty in a communication system. In what is to follow the word *information* will be used in the broad sense to refer to anything meaningful.

As Wiener[14] was quoted as saying in Chapter 2, "Information is information, not matter or energy." The significance of this statement is illustrated by reading it aloud, first forwards and then backwards and comparing the meaning. Mirror images, whether of objects or molecular configurations, provide another illustration. In each case the amount of energy involved in the transmission of information is the same, but the amount of information conveyed is quite different. It can be weighed and measured only as information. In Berkeley's[15] words: "All our ideas, sensations, notions, or the things which we perceive . . . are visibly inactive, there is nothing of Power or Agency included in them." What Hume[16] says on this point is practically a paraphrase: "All ideas are derived from, and represent impressions. We never have any impression, that contains any power or efficacy."

In brief, we are left to conclude that an informational transformation provides the only approach to the "objective" world, and that our derivation of this information is a function of the brain. The solid object at the fingertips may indeed be "out there," but the perceived object represents a cerebral transformation. Whether or not the transformation bears any resemblance to the experienced object is beyond demonstration. Electrophysiology suggests an analogy: the cathode-ray oscilloscope provides the best available picture of the nerve impulse, but the tracing on the screen in no way resembles the electrochemical changes inferred to be occurring within the nerve. What is desired scientifically is to establish that whatever exists "out there" is sufficiently reliable and reproducible to allow *subjective agreement* of observers about its informational properties. In other words, what is desired is agreement about what are usually referred to scientifically as "facts."

Traditional Views of Facts

A traditional view of what is meant by "facts" is found in Locke's[17] exposition of the primary and secondary qualities. His analysis amounted to little more than a paraphrase of a Greek point of view that was clearly expressed by Aristotle and found renaissance notably in the writings of Galileo and Descartes. According to Locke, the primary qualities produce simple ideas of solidity, extension, form, and motion. These qualities belong to objects and provide true (i.e., factual) resemblances of them. The secondary qualities, on the contrary, such as smells, tastes, sounds, and colors, exist in the mind and not in the object itself. To use his words, "light, heat, whiteness, or coldness are no more in them [objects] than sickness or pain is in manna" (manna, a laxative from the Manna ash).[18]

Information as Opposed to Fact

Twenty years later George Berkeley,[15] then 25 years old, pointed out that there is no essential difference between primary and secondary qualities—that both are equally ideas existing only in the mind. "In short," he says, "let anyone consider those arguments which are thought manifestly to prove that colors and tastes exist *only* in the mind, and he shall find that they may with equal force be brought to prove the same thing of extension, figure, and motion.''[19]

A few decades later, Hume[16] gave renewed emphasis to the contradiction of distinguishing between primary and secondary qualities. ''[H]ow,'' he asks, ''can an impression represent a substance otherwise than by resembling it? And how can an impression resemble a substance, since, . . . it is not a substance . . . ?'' In these quotations from Berkeley and Hume one sees the 18th century counterpart of Wiener's tautology that ''information is information, not matter or energy.''

Information may be otherwise regarded as the order that emerges from a background of disorder. Order, as usually conceived, implies temporal and spatial relationships, but it should be emphasized again that neither time nor space exists *per se,* but represents only a form of information derived by the brain. Regardless of how it is defined, information as we deal with it amounts to cerebral emanations without substance.

How then does one put a handle on information so as to deal with it scientifically? One may begin where Descartes began by making some such statement as ''I think, therefore I am'' (*cogito ergo sum*).[20] Descartes arrived at this assertion because of a distrust of his sensations, as typified by mirages. But ''cogito'' by itself is insufficient to give an impression of a material existence because to state, simply, ''I think'' is the equivalent of saying, ''I derive information'' or ''I am information.'' This is comparable to what Descartes said when he referred to ''this 'me' '' as an immaterial soul (Chapter 1). It is significant that in later writings he qualified ''I think'' by stating, ''By the word thought I understand all that of which we are conscious as operating in us. And that is why not alone understanding, willing, imagining, but also *feeling* are here the same as thought''[21] (italics added).

His addition of ''feeling'' was essential, but insufficient. As the phenomenology of psychomotor epilepsy suggests (Chapter 23), it would seem necessary to qualify ''I feel'' by further stating ''I feel affectively.'' This is because the sense of a personal existence and of an ''out there'' existence seem to depend on a blend of affective feelings derived from a combination of information from the internal and external environment (Chapter 27). ''Thinking'' and ''affectively feeling'' have little in common except for representing immaterial information. It is the peculiar property of the affective feeling that it lends a feeling of conviction—a feeling of reality to what is deemed material.

Communication of Information

Through the empirical pairing of the ''I know'' and the *affective* ''I feel'' processes, we derive information that there is no communication of information without the intermediary of something that we directly or indirectly identify as substantial. One might take exception to this by arguing that absolute nothingness conveys information. But it would not do so if it did not stand in contrast to something substantial. Moreover, an appreciation of nothingness depends on what we believe to be a substantial, behaving nervous system, as would also be the case if something could be communicated by ''mental telepathy.''

The word *substantial* applies to anything in the environment that can make itself felt, including the inferred particles and forces of nuclear physics that make themselves ''felt'' through recordings obtained with instruments. Hereafter, the word *entity* will be used to apply to any detectable thing in the environment, no matter how large or how small. Rosenblueth, Wiener, and Bigelow[22] define behavior as ''any change of an entity with respect to its surroundings.'' It is empirically established that communication (the transmission of information) cannot occur without the intermediary of behaving entities. The statement of this invariance might be considered as a *law of communication* (Chapter 2). Because of this invariance, behavior is to be regarded as an unconditional correlate of the communication of information. Although this means that the communication of information involves some expenditure of energy, it does not deny the original tautologous proposition that ''information is information, not matter or energy.''

Measurement of Information

Wiener[23] stated that ''the amount of information in a system is the measure of its degree of organization.'' He further commented that since the entropy of a system is a measurement of its degree of disorganization, information represents negative entropy—or ''negentropy,'' to use the shortened expression of Brillouan.[24] The communication of information depends on entities that conform ''to some sort of statistical regularity.'' Rosenblueth *et al.*[25] have proposed that various forms of behavior of animals or machines may be classified according to a hierarchical system with first, second, third, . . . orders of behavior. Their scheme would imply that the greater amount of orderliness within a system, the greater is its potentiality for conveying information. When so understood, the amount of orderliness derived from a system would provide a measure of the amount of information. In this respect, the immateriality of information lends itself to measurement.

Information may be spun out into an infinite amount of information, as, for example, in the case of an infinite series. But there is a limit to what is information to a human subject. To derive information from an unlimited amount of information, it is necessary to focus upon parts, intervals, or epochs of an endless continuum. It would make no sense to speak of an infinitely large or small object because such an ''object'' would be limitless and hence without the necessary boundaries to make it an object (see Kant[10]). A period of time that was infinitely divided into shorter and shorter temporal durations would never complete itself. It is a power of the calculus that it allows one to approach limits along a continuum.

Brain and Communication of Information

On the basis of present-day knowledge, it is inferred that our derivation and communication of information depends on a *behaving* nervous system that owes its behavior to physicochemical processes of receptors, nerve cells, and effectors that are metabolically sustained by transports in the blood and the surrounding tissues. The considerable advances in the understanding of how information is communicated in this neural network by chemical substances and electrochemically generated impulses have left completely unanswered how changing patterns of neural activity result in the subjective aspects of information.

Subjectivity

The question of subjectivity would be of no concern in an analysis of ''facts'' if, as claimed by some who subscribe to the doctrine of psychophysical parallelism, it represents only a by-product or epiphenomenon of a behaving system. This would be tantamount to saying that our brains could perform everything they do without our being there as subjective participants. It is recognized, for example, that reflexes go into play to prevent a person from falling before there is time for the realization of being off balance. Many of the behaviorist school ascribed the same automaticity to psychological processes and accordingly conducted their experiments as though the subjective element could be neglected. In arguing to the contrary, it is not enough to say that subjectivity is an inseparable part of all waking behavior, no matter how automatic that behavior seems to be. This would not satisfy the ''parallelist'' who would contend that one could dress or repeat a familiar lecture with the mind being some place else. I have pointed out elsewhere that there is one telling argument against the contention that subjectivity is a useless epiphenomenon: *that it exists at all means that the brain has an additional source of information for influencing behavior.*[26] It was as though the brain could hold up before itself the mirror of subjectivity as a means of reflecting information.[26] The most persuasive illustrations are provided by human beings in conveying information about their subjective states through verbal communication, as witness how poor our vocabulary and literature would be without words expressive of subjective states. Were it not for the communication of subjective feelings, immobilizers, rather than anesthetics, might have been developed for the relief of pain.

The Subject as an Individual

Thus far in attempting to derive what is meant by ''facts,'' it has been necessary to introduce ourselves as sentient subjects who derive from immaterial information the realization of substantial, behaving entities. From this information we gain the additional information that no communication can occur without the intermediary of behaving entities. We further infer that as individuals, our ability to derive and communicate information depends on a behaving nervous system. In taking the next step, one must consider what constitutes a sense of individuality, because an answer to what is meant by ''facts'' depends upon reaching *subjective agreement among individuals.*

As was discussed in Chapter 27, the condition that most clearly distinguishes us as individuals is the *duality* of our source of information—namely, one source of signals is from the world within, the other from the world without. Signals to the brain from the world within are entirely private (being self-contained), whereas those from the outside world can be publicly experienced and lend themselves to a comparison among individuals. Individuality, therefore, depends on a privacy due to an inaccessibility of the internal signals to anyone but the individual person. Because of the infinite variation in the way individuals are assembled, it must be assumed that the sentient properties of any one person, like his or her fingerprints, could never be identical with those of another. It is probable, therefore, that there does not exist or ever will exist one person exactly like another. If uniqueness were an indispensable requirement for an evolving society, every person would be indispensable.

In making a distinction with respect to an individual's two sources of information, it is recognized that signals from the outside world become private as soon as they pass the threshold of the body's surface or its orifices and affect the receptors, as, for example, signals of light passing the eye's window to reach the retina; signals of odors carried through the nose to the olfactory epithelium; signals of vibration affecting mechanoreceptors. The crucial point of the present argument is that before these signals become private, they are publicly available for public assessment in a way that is never possible for signals transmitted from within the individual. As an illustration, one may imagine a group of passengers sitting in the same compartment of a speeding train. If asked to judge the amount of vibration and bumpiness of the train, there would doubtless be subjective agreement among the group, unless one of them was affected by arthritis. The inflamed joints of the arthritic would be a source of additional signals not affecting the others and resulting in the complaint that the train was painfully bumpy.

One may therefore characterize a subjectively aware individual as someone who derives both private and public signals, but who can share directly only the public signals with other individuals. We are now in a position to reintroduce the factual question and in doing so, it helps to narrow down what is meant by a fact if we first consider Northrop's[27] categorization of "first-order" and "second-order" facts. He considers first-order facts to be "non man-made naturalistic species of facts," e.g., the planets, which of themselves "merely are, and are therefore, incapable of being in error." He describes second-order facts as "man-made cultural artifacts" that may be true or false, such as moral, legal, and political rules of conduct.

But the public signals by which the first-order facts are known become private as soon as they enter a person and are processed by the nervous system. In this transformation—and they can only be known through such a transformation—they are just as "man-made" as the second-order "man-made cultural artifacts." Thus, both Northrop's first-order and second-order facts—just as Locke's primary and secondary qualities—are reduced to informational correlates of a behaving nervous system. There always exists the barrier of the nervous system between us and the so-called first-order facts. In other words, the brain always stands between us and what we observe.

The preceding arguments lead to the conclusion that facts are to be considered only as those forms of information that are publicly manifested as behaving entities. This conclusion entails no reversion to the position of the behaviorists by either denying or disregarding the informational value of subjectivity and introspection. Inasmuch as subjectivity and introspection find expression through vocal utterances, gesticulation, postural attitudes, and other bodily changes of the individual, they become available to public inspection, communication, and scientific investigation.

With the recognition of what is common to all facts, it remains for individuals to find subjective agreement on how to assess and categorize various kinds of facts in some such manner as Northrop has done in considering first-order and second-order facts. This is not an easy thing to do, presumably, because of the propensity to subjective confusion as to the origin of informational correlates of behavior. Time and space, for example, do not exist in the immediately perceived behaving entities, but rather represent information derived from behaving entities of the brain. When so understood, there is no resulting *contradiction with respect to the possibility of immaterial information having the capacity to generate immaterial information without the intermediary of "behaving entities."*

Validity

Early in the *Principia,* Newton[28] attempted to arrive at a scientific definition of time, space, place, and motion, observing "that common people conceived those *quantities* under no other notions but from the relation they bear to sensible objects" (italics added). Accordingly, he sought to distinguish such *relative apparent common* concepts from *absolute, true, and mathematical* concepts. In modern times, Northrop[27] has correspondingly distinguished "concepts by intuition" (i.e., experience) from "concepts by intellection" (*a priori*). But here again it must be emphasized that both kinds of concepts represent forms of information that are inferred to be correlates of a behaving nervous system. There is the informational distinction, however, that "concepts by intuition" (phenomena) contain a publicity not present in "concepts of intellection" (Kant's noumena).

If there is public agreement about a piece of information, it is usually regarded as "true," whereas the lack of agreement either implies that it is "false" or "uncertain." Because of their private nature, concepts by intellection cannot be considered true, false, or uncertain until they are publicly communicated and evaluated on the basis of public experience. If a subject matter is beyond the realm of public experience, as in the case of Kant's antinomies,[10] they must remain forever uncertain. This is the motif that runs through Kant's *Critique of Pure Reason,* in which there appears again and again the refrain typified in the following sentence: "I maintain, then, that the possibility of things is not derived from . . . [*a priori*] conceptions *per se,* but only when considered as formal and objective conditions of an experience in general."[29] But in so saying, he was speaking of "objective conditions of an experience" as being as absolute in a Euclidean sense as the absolutes listed by Newton. Since then, times have seen a shift in scientific subjective agreement as to what is absolute and what is relative.

In summary, *facts* are to be considered as those forms of information that are *publicly* manifest as behaving entities. The term *validity* does not apply to the *facts* themselves, which *per se* are neither true nor false, but rather to what is agreed upon as true by subjective individuals after public assessment of the *facts*. What is subjectively agreed upon as "true" or "false" or "uncertain" by one group may be quite contrary to what is agreed upon by another—regardless of whether the groups are comprised of scientists or Newton's common man.

An Epistemological Impasse?

We now consider a psychological impasse that defies circumvention because of mechanisms underlying that variety of affective feelings described in Chapter 24 that occur during the aura of psychomotor seizures and range from an enhanced sense of reality to strong feelings of conviction as to what is real, true, and important. As was pointed out in Chapter 24, the latter feelings appear as if "free-floating" because they are not identified with any particular thing, situation, or individual. The electroencephalographic findings generally point to an epileptogenic focus in or adjacent to the limbic cortex of the temporal lobe. As will be recalled, one patient was quoted as saying, "I had the feeling that this is the truth and the whole truth; this is what the world is all about" (Chapter 24).

Feelings or utterances of this kind are encountered under other conditions. Jaspers stated that the scientist whom he admired most was Max Weber because he was "incurably torn within himself throughout his life in the search for truth."[30] It is reported that on his deathbed Weber's last words in his delirium were, "The true is the truth." William James[31] speaks of the "sense of reality and truth" experienced during the inhalation of nitrous oxide, saying, "this goes to a fully unutterable extreme in the nitrous oxide intoxication, in which a man's very soul will sweat with conviction, and he be all the while unable to tell what he is convinced of. . . .'' (Experimentally, it is of correlative interest that spiking activity may occur in the limbic structures of the frontotemporal region during nitrous oxide anesthesia.[32]) In present times, it is hardly necessary to elaborate on like states of mind experienced by persons consuming psychedelic (*delic* = clear) drugs such as LSD or when, without medication, persons experience mystical revelation during meditation or religious conversion. In *The Sleepwalkers,* Koestler[33] describes in detail Kepler's ''this is it'' revelation of the five perfectly fitting solids and explains how his mistaken notion became an *idée fixe* that led to the discovery of what have since become known as Kepler's three laws. In Koestler's words, Kepler was drawing an illustration for his class ''when an idea suddenly struck him with such force that he felt he was holding the key to the secret of creation in his hand.''[34]

As should be emphasized again, in psychomotor epilepsy the feelings of the kind just described that occur during the aura are free-floating, being unattached to any particular thing, situation, person, idea, or theory. In Chapter 25, cases were illustrated in which during the automatism following the aura, patients were able to carry out complicated forms of behavior either of a routine kind or of an appropriate nature under novel situations. The retention of an adequate neocortical capacity to deal with new situations during a limbic ictal episode was illustrated by a physician who examined a patient, arrived at a correct diagnosis, and wrote an appropriate prescription. In Chapter 27, the argument was presented that the complete lack of memory for what happened during automatisms is primarily owing to a disturbance of limbic function and a resulting failure of the integration of internally and externally derived experience, upon which a sense of being and reality of the self depends. Without an integrated sense of self, there is, so to speak, no place to deposit a memory of ongoing experience. The evidence was reviewed in Chapter 26 for inferring that the limbic cortex receives more extensive input from interoceptive systems than the neocortex and, also of crucial importance, receives input from the various exteroceptive systems.

It should be recalled at this point that psychomotor epilepsy provides evidence that the limbic system is involved in self-realization, as evident by such experiences during the aura as feelings of enhanced feelings of reality, increased awareness, or self-duplication (''mental diplopia''). And as pointed out in the preceding section, it is of great consequence apropos of a sense of being that the phenomenology of psychomotor epilepsy reveals that even the least obtrusive feelings generated by limbic activity are tinged with some degree of affect. As expressed at the end of Chapter 27, there may be an essential precondition to the saying that ''something does not exist until you give it a name'': Something does not exist unless it is imbued with an *affective* feeling, no matter how slight.

Given the previously described propagation of limbic seizures and the manifestations of psychomotor epilepsy, one is led to infer a dichotomy in function of neocortical and limbic systems that may account for a dissociation in intellectual and emotional mentation. Moreover (and this cannot be overemphasized), the phenomenology of psychomotor epilepsy suggests that without a co-functioning limbic system, the neocortex lacks not only the requisite neural substrate for a sense of self, of reality, and the memory of ongoing experience, but also a feeling of conviction as to what is true or false. This presents a problem of crucial epistemological significance because there is no evidence that the limbic structures of the temporal lobe are capable of comprehending speech, nor is there any basis for inferring a capacity to communicate in verbal terms. Hence, it would appear that the manufacture of belief in the reality, importance, and truth or falsity of what is conceived depends on a mentality incapable of verbal comprehension and communication. To revert to a comment in Chapter 24, it is one thing to have a primitive, illiterate mind for judging the authenticity of food or a mate, but where do we stand if we must depend on that same mind for belief in our ideas, concepts, and theories?

I have encountered no one who has been able to suggest a satisfactory circumvention of this seemingly delusional impasse. Because of the disparity of the two types of intelligence in question, it seems to be a more disturbing impasse than the one concerning "self-reference," even if, because of the triune nature of the brain's intelligence, one were to multiply Bronowski's "hall of mirrors" by three.

Perhaps it is premature for human beings to be concerned about these questions, as might be illustrated by imagining a new kind of pact that Faust made with the devil. One day when Faust was despairing of life, Mephistopheles (possibly recalling the mammal-like reptiles) tempted him by asking, "Faust, would you be willing to go on living *in your present condition* for 250 million years if at the end of that time, I told you the secret of the universe and the meaning of life?" Faust agreed, but like a prisoner with a life sentence, he began to lose track of time and, after 250 million years, had to be reminded of the bargain. Mephistopheles asked him if he was no longer interested. Faust looked up and after a slight gasp, answered, "Heavens, yes," and Mephistopheles explained the secret of the universe and the meaning of life. Faust looked puzzled and said, "But I still don't understand." And Mephistopheles replied, "Of course not, how could you? You agreed to remain just as you are, so your brain has not evolved in the meantime."

And so, one might conclude, we are left with the question as to whether or not there can ever evolve an intelligence that will be intelligent enough to take measure of itself and at the same time discover a braille for reading the blind message of evolution.

References

1. Adams, 1780
2. MacLean, 1982
3. MacLean, 1975a
4. Newspaper report
5. James, 1890, Vol. II, p. 412 (footnote)
6. MacLean, 1975a, p. 323
7. Ommaya, 1968
8. Jeans, 1943, p. 7
9. Schrodinger, 1951
10. Kant, 1781/1899
11. Bronowski, 1966, p. 7
12. Bronowski, 1966, pp. 9 and 12
13. Galaty, 1974
14. Wiener, 1948
15. Berkeley, 1710/1937, p. 76
16. Hume, 1739/1888, p. 233
17. Locke, 1690/1894
18. Locke, 1690/1894, p. 50
19. Berkeley, 1710/1937, p. 71
20. Descartes, 1637/1967
21. Descartes, 1644/1967, p. 222
22. Rosenblueth *et al.*, 1943
23. Wiener, 1948, p. 18
24. Brillouan, 1956
25. Rosenblueth *et al.*, 1943
26. MacLean, 1960
27. Northrop, 1963
28. Newton, 1729, p. 6
29. Kant, 1781/1899, p. 145
30. Stierlin, 1974, p. 214
31. James, 1890, p. 214
32. MacLean, unpublished observations
33. Koestler, 1963
34. Koestler, 1963, p. 247

Bibliography

Abbie, A. A. (1940). Cortical lamination in the Monotremata, *J. Comp. Neurol. 72,* 429–467.

Abbie, A. A. (1942). Cortical lamination in a polyprotodont marsupial, Parameles nasuta, *J. Comp. Neurol. 76,* 509–536.

Ackerknecht, E. H. (1959). *A Short History of Psychiatry,* Hafner, New York.

Ackerly, S. S. (1964). A case of paranatal bilateral frontal lobe defect observed for thirty years, in: *The Frontal Granular Cortex and Behavior* (J. M. Warren and K. Akert, eds.), McGraw-Hill, New York, pp. 192–218.

Ackerly, S. S., and Benton, A. L. (1948). Report of case of bilateral frontal lobe defect, in: *The Frontal Lobes, Res. Publ. Assoc. Res. Nerv. Ment. Dis. 27,* 479–518.

Adams, J. (1780). Novangus Papers (pseud), *Boston Gazette 7,* 1779.

Adey, W. R. (1952a). An experimental study of hippocampal afferent pathways from prefrontal and cingulate areas in the monkey, *J. Anat. 86,* 58–74.

Adey, W. R. (1952b). Hippocampal and hypothalamic connexions of the temporal lobe in the monkey, *Brain 75,* 358–384.

Adey, W. R. (1953). An experimental study of the central olfactory connexions in a marsupial (*Trichosurus vulpecula*), *Brain 76,* 311–330.

Adinolfi, A. M., and Pappas, G. D. (1968). The fine structure of the caudate nucleus of the cat, *J. Comp. Neurol. 133,* 167–184.

Adrian, E. D. (1942). Olfactory reactions in the brain of the hedgehog, *J. Physiol. (London) 100,* 459–473.

Aggleton, J. P., and Mishkin, M. (1983). Memory impairments following restricted medial thalamic lesions in monkeys, *Exp. Brain Res. 52,* 199–209.

Aghajanian, G. K., and Bunney, B. S. (1977). Dopamine "autoreceptor": Pharmacological characterization by microiontophoretic single cell recording studies, *Naunyn-Schmiedeberg's Arch. Pharmacol. 297,* 1–7.

Ajmone Marsan, C., and Goldhammer, L. (1973). Clinical ictal patterns and electrographic data in cases of partial seizures of frontal–central–parietal origin, in: *Epilepsy: Its Phenomena in Man* (M. A. B. Brazier, ed.), Academic Press, New York, pp. 236–260.

Åkerman, B. (1966). Behavioural effects of electrical stimulation in the forebrain of the pigeon. I. Reproductive behaviour, *Behaviour 26,* 323–338.

Akert, K. (1964). Comparative anatomy of frontal cortex and thalamofrontal connections, in: *The Frontal Granular Cortex and Behavior* (J. M. Warren and K. Akert, eds.), McGraw-Hill, New York, pp. 372–396.

Akert, K., and Andersson, B. (1951). Experimenteller Beitrag zur Physiologie des Nucleus caudatus, *Acta Physiol. Scand. 22,* 281–298.

Akert, K., and Andy, O. J. (1955). Experimental studies on corpus mammillare and tegmentomammillary system in the cat, *Am. J. Physiol. 183,* 591.

Akert, K., Gruesen, R. A., Woolsey, C. N., and Meyer, D. R. (1961). Klüver–Bucy syndrome in monkeys with neocortical ablations of temporal lobe, *Brain 84,* 480–498.

Akopyan, E. V. (1982). Visual thalamic afferents to field 29 of the rat limbic cortex, *Neurophysiology (USSR) 14,* 135–139.

Alajouanine, T. (1963). Dostoiewski's epilepsy, *Brain 86,* 209–218.

Albe-Fessard, D., and Buser, P. (1979). Is a theory on neostriatal functions now possible?, in: *The Neostriatum* (I. Divac and R. G. E. Oberg, eds.), Pergamon Press, New York, pp. 315–319.

Albe-Fessard, D., Oswaldo-Cruz, E., and Rocha-Miranda, C. (1960a). Activités évoquées dans le noyau caudé du chat en réponse à des types divers d'afférences. I. Étude macrophysiologique, *Electroencephalogr. Clin. Neurophysiol. 12,* 405–420.

Albe-Fessard, D., Rocha-Miranda, C. E., and Oswaldo-Cruz, E. (1960b). Activitiés évoquées dans le noyau caudé du chat en réponse à des types divers d'afférences. II. Étude microphysiologique, *Electroencephalogr. Clin. Neurophysiol. 12,* 649–661.

Alexander, L. (1942). The fundamental types of histopathologic changes encountered in cases of athetosis and paralysis agitans, in: *The Diseases of the Basal Ganglia, Res. Publ. Assoc. Res. Nerv. Ment. Dis. 21,* 334–493.

Alexander, L., and Putnam, T. J. (1938). Pathological alterations of cerebral vascular patterns, *Res. Publ. Ass. Nerv. Ment. Dis. 18,* 471–543.

Allen, G. I., and Tsukahara, N. (1974). Cerebrocerebellar communication system, *Physiol. Rev. 54,* 957–1006.

Allen, W. F. (1944). Degeneration in the dog's mammillary body and Ammon's horn following transection of the fornix, *J. Comp. Neurol. 80,* 283–291.

Allin, E. F. (1975). Evolution of the mammalian middle ear, *J. Morphol. 147,* 403–438.

Allin, E. F. (1986). The auditory apparatus of advanced mammal-like reptiles and early mammals, in: *The Ecology and Biology of Mammal-like Reptiles* (N. Hotton, P. D. MacLean, J. J. Roth, and E. C. Roth, eds), Smithsonian Institution Press, Washington, D.C., pp. 283–294.

Alonso, A., and Kohler, C. (1984). A study of the reciprocal connections between the septum and the entorhinal area using anterograde and retrograde axonal transport methods in the rat brain, *J. Comp. Neurol. 225,* 327–343.

Altschul, R. Von (1938). Die Blutgefässverteilung im Ammonshorn, *Z. Gesamte Neurol. Psychiatr. 163,* 634–642.

Amaral, D. G., and Cowan, W. M. (1980). Subcortical afferents to the hippocampal formation in the monkey, *J. Comp. Neurol. 189,* 573–591.

Amin, A. H., Crawford, T. B. B., and Gaddum, J. H. (1954). The distribution of substance P and 5-hydroxytryptamine in the central nervous system of the dog, *J. Physiol. (London) 126,* 596–618.

Anastasopoulos, G., Diakoyiannis, A., and Routsonis, K. (1959). Three cases of temporal lobe epilepsy with endocrinopathy, *J. Neuropsychiatry 1,* 65–76.

Anden, N. E., Carlsson, A., Dahlström, A., Fuxe, K., Hillarp, N. A., and Larsson, K. (1964). Demonstration and mapping out of nigroneostriatal dopamine neurons, *Life Sci. 3,* 523–530.

Andersen, H., Braestrup, C., and Randrup, A. (1975). Apomorphine-induced stereotyped biting in the tortoise in relation to dopaminergic mechanisms, *Brain Behav. Evol. 11,* 365–373.

Anderson, J. (1886). On sensory epilepsy. A case of basal cerebral tumour, affecting the left temporo-sphenoidal lobe, and giving rise to a paroxysmal taste-sensation and dreamy state, *Brain 9,* 385–395.

Anderson, J. W. (1954). The production of ultrasonic sounds by laboratory rats and other mammals, *Science 119,* 808–809.

Andy, O. J., and Akert, K. (1953). Electroencephalographic and behavioral changes during seizures induced by stimulation of Ammon's formation in the cat and monkey, *Electroencephalogr. Clin. Neurophysiol. 3,* 48.

Angevine, J. B., Jr. (1965). Time of neuron origin in the hippocampal region. An autoradiographic study in the mouse, *Exp. Neurol. Suppl. 2,* 70.

Angevine, J. B., and McConnell, J. A. (1974). Time of origin of striatal neurons in the mouse: An autoradiographic study, *Anat. Rec. 178,* 300.

Anschel, S. (1977). Functional specificity of vocalizations elicited by electrical brain stimulation in the turkey (*Meleagris gallopavo*), *Brain Behav. Evol. 14,* 399–417.

Ardrey, R. (1966). *The Territorial Imperative,* Atheneum, New York.

Ardrey, R. (1970) *The Social Contract,* Atheneum, New York.

Ardrey, R. (1972). Four-dimensional man, *Encounter (London)* 1–15.

Ariëns Kappers, C. U. (1908). Weitere Mitteilungen über die Phylogenese des Corpus striatum und des Thalamus, *Anat. Anz. 33,* 321–336.

Ariëns Kappers, C. U. (1909). The phylogenesis of the palaeo-cortex and archi-cortex compared with the evolution of the visual neo-cortex, *Arch. Neurol. Psychiatry 4,* 161–173.

Ariëns Kappers, C. U. (1922). The ontogenetic development of the corpus striatum in birds and a comparison with mammals and man, *K. Akad. Wet. Amsterdam Proc. Sect. Sc. 26,* 135.

Ariëns Kappers, C. U. (1928). The development of the cortex and the functions of its different layers, *Acta Psychiatr. Neurol. Scand. 3,* 115–132.

Ariëns Kappers, C. U., Huber, G. C., and Crosby, E. C. (1936). *The Comparative Anatomy of the Nervous System of Vertebrates, Including Man,* two volumes, Macmillan Co., New York.

Aristotle (1952). Nichomachean ethics, in: *The Works of Aristotle*, vol. 9 (W. D. Ross, ed.), Translated into English under the editorship of W. D. Ross, Oxford University Press, London, in irregular order.

Armstrong, E. (1982). Mosaic evolution in the primate brain: Differences and similarities in the hominoid thalamus, in: *Primate Brain Evolution. Methods and Concepts* (E. Armstrong and D. Falk, eds.), Plenum Press, New York, pp. 131–161.

Arnold, A. P., and Gorski, R. A. (1984). Gonadal steroid induction of structural sex differences in the central nervous system, *Annu. Rev. Neurosci. 7,* 413–442.

Aronson, L. R. (1970). Functional evolution of the forebrain in lower vertebrates, in: *Development and Evolution of Behavior: Essays in Memory of T. C. Schneirla* (L. R. Aronson, E. Tobach, D. S. Lehrman, and J. S. Rosenblatt, eds.), Freeman, San Francisco, pp. 75–107.

Aronson, L. R., and Noble, G. K. (1945). The sexual behavior of Anura. 2. Neural mechanisms controlling mating in the male leopard frog, Rana pipiens, *Bull. Am. Mus. Nat. Hist. 86,* 86–139.

Asanuma, C., Thach, W. T., and Jones, E. G. (1983a). Cytoarchitectonic delineation of the ventral lateral thalamic region in the monkey, *Brain Res. Rev. 5,* 219–235.

Asanuma, C., Thach, W. T., and Jones, E. G. (1983b). Distribution of cerebellar terminations and their relation to other afferent terminations in the ventral lateral thalamic region of the monkey, *Brain Res. Rev. 5,* 237–265.

Asanuma, C., Thach, W. T., and Jones, E. G. (1983c). Anatomical evidence for segregated focal groupings of efferent cells and their terminal ramifications in the cerebellothalamic pathway of the monkey, *Brain Res. Rev. 5,* 267–297.

Atweh, S. F., and Kuhar, M. J. (1977). Autoradiographic localization of opiate receptors in rat brain. III. The telencephalon, *Brain Res. 134,* 393–405.

Auffenberg, W. (1970). A day with number 19: Report on a study of the Komodo monitor, *Anim. Kingdom 73,* 18–23.

Auffenberg, W. (1972). Komodo dragons, *Nat. Hist. 81,* 52–59.

Auffenberg, W. (1978). Social and feeding behavior in *Varanus komodoensis,* in: *The Behavior and Neurology of Lizards* (N. Greenberg and P. D. MacLean, eds.), U.S. Government Printing Office, Washington, D.C., DHEW Publication No. (ADM) 77-491, pp. 301–331.

Auffenberg, W. (1981). *The Behavioral Ecology of the Komodo Monitor,* University Presses of Florida, Gainesville.

Aulsebrook, L. H., and Holland, R. C. (1969). Central regulation of oxytocin release with and without vasopressin release, *Am. J. Physiol. 216,* 818–829.

Avanzini, G., Mancia, D., and Pelliccioli, G. (1969). Ascending and descending connections of the insular cortex of the cat, *Arch. Ital. Biol. 107,* 696–714.

Axelrod, J. (1965). The metabolism, storage, and release of catecholamines, *Recent Prog. Horm. Res. 21,* 597–622.

Axelrod, J. (1971). Noradrenaline: Fate and control of its biosynthesis, *Science 173,* 598–606.

Axelrod, J. (1974). Neurotransmitters, *Sci. Am. 230,* 59–71.

Babb, T. L., Wilson, C. L., Halgren, E., and Crandall, P. H. (1980). Evidence for direct lateral geniculate projections to hippocampal formation in man, *Soc. Neurosci. Abstr. 6,* 838.

Bachman, D. S., and Katz, H. M. (1977). Physiologic responses to intravenous serotonin in the awake squirrel monkey, *Physiol. Behav. 18,* 987–990.

Bachman, D. S., and MacLean, P. D. (1971). Unit analysis of inputs to cingulate cortex in awake, sitting squirrel monkeys. I. Exteroceptive systems, *Int. J. Neurosci. 2,* 109–113.

Bachman, D. S., Hallowitz, R. A., and MacLean, P. D. (1977). Effects of vagal volleys and serotonin on units of cingulate cortex in monkeys, *Brain Res. 130,* 253–269.

Bachman, D. S., Shultz, J., and Cooper, R. (1981). Cursive and gelastic epilepsy: Case report, *Clin. Electroencephalogr. 12,* 32–34.

Bagshaw, M. H., and Pribram, K. H. (1953). Cortical organization in gustation (*Macaca mulatta*), *J. Neurophysiol. 16,* 499–508.

Bailey, P., and Davis, E. W. (1942). The syndrome of obstinate progression in the cat, *Proc. Soc. Exp. Biol. Med. 51,* 307.

Bailey, P. and Bremer, F. (1938). A sensory cortical representation of the vagus nerve, *J. Neurophysiol. 1,* 405–412.

Bakker, R. T. (1975). Dinosaur renaissance, *Sci. Am. 232,* 58–78.

Baldwin, J. D. (1967). A study of the social behavior of a semifree-ranging colony of squirrel monkeys (*Saimiri sciureus*), Ph.D. dissertation, Johns Hopkins University, Baltimore.

Baldwin, J. D. (1968). The social behavior of adult male squirrel monkeys (*Saimiri sciureus*) in a seminatural environment, *Folia Primatol. 9,* 281–314.

Baleydier, C., and Mauguiere, F. (1980). The duality of the cingulate gyrus in monkey. Neuroanatomical study and functional hypothesis, *Brain 103,* 525–554.

Ballantine, H. T., Jr., Cassidy, W. L., Flanagan, N. B., and Marino, R., Jr. (1967). Stereotaxic anterior cingulotomy for neuropsychiatric illness and intractable pain, *J. Neurosurg. 26,* 488–495.

Ballard, P. A., Tetrud, J. W., and Langston, J. W. (1985). Permanent human parkinsonism due to 1-methyl-4-phenyl-1,2,3,6-tetrahydropyridine (MPTP): Seven cases, *Neurology 35,* 949–956.

Bard, P. (1928). A diencephalic mechanism for the expression of rage with special reference to the sympathetic nervous system, *Am. J. Physiol. 84,* 490–513.

Bard, P., and Rioch, D. M. (1937). A study of four cats deprived of neocortex and additional portions of the forebrain, *Johns Hopkins Med. J. 60,* 73–153.

Barghoorn, E. S., and Schopf, J. W. (1967). Alga-like fossils from the Early Precambrian of South Africa, *Science 156,* 508–512.

Barnett, S. A. (1963). *A Study in Behaviour,* Methuen, London.

Barrington, F. J. F. (1925). The effect of lesions of the hind- and midbrain on micturition in the cat, *Q. J. Exp. Physiol. 15,* 81–102.

Barris, R. W., and Schuman, H. R. (1953). Bilateral anterior cingulate gyrus lesions, *Neurology 3,* 44–52.

Bauman, M. L., and Kemper, T. L. (1984). The brain in infantile autism: A histoanatomic case report, *Neurology 34,* Suppl. 1, 275.

Beach, F. A. (1937). The neural basis of innate behavior. I. Effects of cortical lesions upon the maternal behavior pattern in the rat, *J. Comp. Psychol. 24,* 393–440.

Beach, F. A. (1945). Current concepts of play in animals, *Am. Nat. 79,* 523–541.

Beach, F. A. (1951). Effects of forebrain injury upon mating behaviour in male pigeons, *Behaviour 4,* 36–59.

Bechterew, W. von (1900). Demonstration eines Gehirns mit Zerstorung der vorderen und inneren Theile der Hirnrinde beider Schlafenlappen, *Neurol. Zentralbl. 19,* 990–991.

Beck, E. (1934). Der Occipitallappen des Affen (Macacus rhesus) und des Menschen in seiner cytoarchitektonischen Structur, *J. Psychol. Neurol. 46,* 193–323.

Beck, E., and Bignami, A. (1968). Some neuro-anatomical observations in cases with stereotactic lesions for the relief of parkinsonism, *Brain 91,* 589–618.

Bedard, T. P., Larochelle, A. L., Parent, A., and Poirier, L. J. (1969). The nigrostriatal pathway: A correlative study based on neuroanatomical and neurochemical criteria in the cat and the monkey, *Exp. Neurol. 25,* 365–377.

Bellairs, A. (1970). *The Life of Reptiles,* two volumes, Universe Books, New York.

Benjamin, R. M., and Burton, H. (1968). Projection of taste nerve afferents to anterior opercular-insular cortex in squirrel monkey (*Saimiri sciureus*), *Brain Res. 7,* 221–231.

Bennett, A. E. (1965). Mental disorders associated with temporal lobe epilepsy, *Dis. Nerv. Syst. 26,* 275–279.

Bennett, D. R., ZuRhein, G. M., and Roberts, T. S. (1962). Acute necrotizing encephalitis, *Arch. Neurol. 6,* 96–113.

Bennett-Clarke, C., and Joseph, S. A. (1982). Immunocytochemical distribution of LHRH neurons and processes in the rat: Hypothalamic and extrahypothalamic locations, *Cell Tissue Res. 221,* 493–504.

Benson, F., and Geschwind, N. (1987). The aphasias and related disturbances, in: *Clinical Neurology,* Volume 1 (A. B. Baker and R. J. Joynt, eds.), Harper & Row, Philadelphia, pp. 1–2.

Bentivoglio, M., Van Der Kooy, D., and Kuypers, H. (1979). The organization of the efferent projections of the substantia nigra in the rat. A retrograde fluorescent double labeling study, *Brain Res. 174,* 1–17.

Berger, H. (1929). Über das Elektrenkephalogramm des Menschen, *Arch. Psychiatr. Nervenkr. 87,* 527–570.

Bergmann, G. H. (1847). Vorläufige Bemerkungen über die Verrücktheit nebst pathologisch-anatomischen Erlaüterungen gewisser dabei leidender Functionen des Gehirns, *Allg. Z. Psychiatr. 4,* 361–384.

Berkeley, G. (1710/1937). *The Principles of Human Knowledge* (T. E. Jessop, ed.), A. Brown, London.

Bernstein, L. (1976). Obsessive ritual by a performer, *Washington Post* Style section, January 7.

Bertler, A., and Rosengren, E. (1959). Occurrence and distribution of dopamine in brain and other tissues, *Experientia 15,* 10–11.

Beyer, C. F., Anguiano, G. L., and Mena, F. J. (1961). Oxytocin release in response to stimulation of cingulate gyrus, *Am. J. Physiol. 200,* 625–627.

Bianchi, L. (1895). The function of the frontal lobes, *Brain 18,* 497–522.

Bichat, X. (1799/1800). *Recherches physiologiques sur la vie et la mort,* Volume 8, Brosson, Gabon, Paris.

Bichat, X. (1815/1955). *Recherches physiologiques sur la vie et la mort,* Gauthier–Villars, Paris.

Bielschowsky, M. (1904). Die Silberimpragnation der Neurofibrillen, *J. Psychol. Neurol. 3,* 169–198.

Biemond, A. (1970). *Brain Diseases,* Elsevier, Amsterdam.

Bignall, K. E., Imbert, M., and Buser, P. (1966). Optic projections to nonvisual cortex of the cat, *J. Neurophysiol. 29,* 396–409.

Bingley, T. (1958). Mental symptoms in temporal lobe epilepsy and temporal lobe gliomas, *Acta Psychiatr. Neurol. 33,* 1–151.

Björklund, A., and Lindvall, O. (1978). The meso-telencephalic dopamine neuron system. A review of its anatomy, in: *Limbic Mechanisms* (K. E. Livingston and O. Hornykiewicz, eds.), Plenum Press, New York, pp. 307–331.

Blackstad, T. W. (1956). Commissural connections of the hippocampal region in the rat, with special reference to their mode of termination, *J. Comp. Neurol. 105,* 417–537.

Bleier, R. (1969). Retrograde transsynaptic cellular degeneration in mammillary and ventral tegmental nuclei following limbic decortication in rabbits of various ages, *Brain Res. 15,* 365–393.

Blinkov, S. M., and Glezer, I. I. (1968). *The Human Brain in Figures and Tables: A Quantitative Handbook,* Plenum Press, New York.

Bloom, F. E., Costa, E., and Salmoiraghi, G. C. (1965). Anesthesia and the responsiveness of individual neurons in the caudate nucleus of the cat to acetylcholine, norepinephrine, and dopamine administered by microelectrophoresis, *J. Pharmacol. Exp. Ther. 150,* 244.

Blumer, D. (1970). Hypersexual episodes in temporal lobe epilepsy, *Am. J. Psychiatry 126,* 1099–1106.

Bobillier, P., Petitjean, F., Salvert, D., Ligier, M., and Seguin, S. (1975). Differential projections of the nucleus raphe dorsalis and nucleus raphe centralis as revealed by autoradiography, *Brain Res. 85,* 205–210.

Bobillier, P., Seguin, S., Petitjean, F., Salvert, D., Touret, M., and Jouvet, M. (1976). The raphe nuclei of the cat brain stem: A topographical atlas of their efferent projections as revealed by autoradiography, *Brain Res. 113,* 449–486.

Bogert, C. M. (1954). Lizards, in: *The Animal Kingdom,* Volume 2 (F. Drimmer, ed.), Graystone Press, New York, pp. 1282–1322.

Bohringer, R. C., and Rowe, M. J. (1977). The organization of the sensory and motor areas of cerebral cortex in the platypus (*Ornithorhynchus anatinus*), *J. Comp. Neurol. 174,* 1–14.

Bonner, F., Cobb, S., Sweet, W. H., and White, J. C. (1953). Frontal lobe surgery. Its value in the treatment of pain with consideration of postoperative psychological changes, *Res. Publ. Assoc. Res. Nerv. Ment. Dis. 31,* 393–421.

Bonvallet, M., Dell, P., and Hugelin, A. (1952). Projections olfactives, gustatives, viscerales, vagales, visuelles et auditives au niveau des formations grises du cerveau anterieur du chat, *J. Physiol. (Paris) 44,* 222–224.

Boole, G. (1854). *An Investigation of the Laws of Thought,* Dover, New York.

Bouchet and Cazauvieilh (1825). Epilepsie et alienation mentale, *Arch. Gen. Med. 9,* 510–542.

Boudin, G., Brion, S., Pepin, B., and Barbizet, J. (1968). Syndrome de Korsakoff d'étiologie artériopathique, par lésion bilatérale, asymétrique du système limbique, *Rev. Neurol. 119,* 341–348.

Bouvier, M. (1977). Dinosaur Haversian bone and endothermy, *Evolution 31,* 449–450.

Bowden, D. M., German, D. C., and Poynter, W. D. (1978). An autoradiographic, semistereotaxic mapping of major projections from locus coeruleus and adjacent nuclei in *Macaca mulatta, Brain Res. 145,* 257–276.

Bowker, R. M., Steinbusch, H. W. M., and Coulter, J. D. (1981). Serotonergic and peptidergic projections to the spinal cord demonstrated by a combined retrograde HRP histochemical and immunocytochemical staining method, *Brain Res. 211,* 412–417.

Bradbury, A. F., Smyth, D. G., Snell, C. R., Birdsall, N. J. M., and Hulme, E. C. (1976). C-fragment of lipotropin has a high affinity for brain opiate receptors, *Nature 260,* 793–795.

Brady, J. V. (1958). The paleocortex and behavioral motivation, in: *Biological and Biochemical Bases of Behavior* (H. F. Harlow and C. N. Woolsey, eds.), the University of Wisconsin Press, Madison, pp. 193–235.

Brady, J. V., and Nauta, W. J. H. (1953). Subcortical mechanisms in emotional behavior: Affective changes following septal forebrain lesions in the albino rat, *J. Comp. Physiol. Psychol. 46,* 339–346.

Brady, J. V., and Nauta, W. J. H. (1955). Subcortical mechanisms in emotional behavior: The duration of affective changes following septal and habenular lesions in the albino rat, *J. Comp. Physiol. Psychol. 48,* 412–420.

Brand, S., and Rakic, P. (1978). Time of origin of neurons in primate neostriatum: ^{3}H-thymidine autoradiographic analysis in the rhesus monkey, *Anat. Rec. 190,* 345–346.

Brattstrom, B. H. (1978). Learning studies in lizards, in: *The Behavior and Neurology of Lizards* (N. Greenberg

and P. D. MacLean, eds.), U.S. Government Printing Office, Washington, D.C., DHEW Publication No. (ADM) 77-491, pp. 173–181.
Bratz, E. (1899a). Ueber des Ammonshorn bei Epileptischen und Paralytkern, *Allg. Z. Psychiatr. 56,* 841–844.
Bratz, E. (1899b). Ammonshornbefunde bei Epileptischen, *Archiv. für Psychiatrie 31,* 820–836.
Brazier, M. A. B. (1964). Evoked responses recorded from the depths of the human brain, *Ann. N.Y. Acad. Sci. 112,* 33–59.
Brickner, R. M. (1934). An interpretation of frontal lobe function based upon the study of a case of partial bilateral frontal lobectomy, in: *Localization of Function in the Cerebral Cortex, Res. Publ. Assoc. Res. Nerv. Ment. Dis. 13,* 259–351.
Brickner, R. M. (1940). A human cortical area producing repetitive phenomena when stimulated, *J. Neurophysiol. 3,* 128–130.
Brickner, R. M. (1952). Brain of patient A. after bilateral frontal lobectomy; status of frontal lobe problem, *AMA Arch. Neurol. Psychiatry 68,* 293–313.
Bridges, R. S., and Grimm, C. T. (1982). Reversal of morphine disruption of maternal behavior by concurrent treatment with the opiate antagonist naloxone, *Science 218,* 166–168.
Bridgman, P. W. (1959). *The Way Things Are,* Harvard University Press, Cambridge, Mass.
Brierly, J. B., Corsellis, J. A. N., Hierons, C., and Nevin, S. (1960). Subacute encephalitis of later adult life, mainly affecting the limbic areas, *Brain 83,* 357.
Brillouan, L. (1956). *Science and Information,* Academic Press, New York.
Brink, A. S. (1955). Note on a very tiny specimen of *Thrinaxodon liorhinus, Palaeontol. Afr. 3,* 73–76.
Brink, A. S. (1956). Speculations on some advanced mammalian characteristics in the higher mammal-like reptiles, *Palaeontol. Afr. 4,* 77–95.
Brink, A. S. (1958). Note on a new skeleton of *Thrinaxodon liorhinus, Palaeontol. Afr. 6,* 15–22.
Brissaud, E. (1895). *Leçons sur les maladies nerveuses,* Masson, Paris.
Broadwell, R. D. (1975a). Olfactory relationships of the telencephalon and diencephalon in the rabbit. I. An autoradiographic study of the efferent connections of the main and accessory olfactory bulbs, *J. Comp. Neurol. 163,* 329–346.
Broadwell, R. D. (1975b). Olfactory relationships of the telencephalon and diencephalon in the rabbit. II. An autoradiographic and horseradish peroxidase study of the efferent connections of the anterior olfactory nucleus, *J. Comp. Neurol. 164,* 389–410.
Broca, P. (1861a). Remarques sur le siège de la faculté du langage articulé, suivies d'une observation d'aphonie (perte de la parole), *Bull. Soc. Anat. 36,* 330–357.
Broca, P. (1861b). Nouvelle observation d'aphonie produite par une lésion de la moitié postérieure des deuxième et troisième circonvolutions frontales, *Bull. Soc. Anat. 36,* 398–407.
Broca, P. (1865). Sur le siège de la faculté du langage articulé, *Bull. Soc. Anthropol. 6,* 377–393.
Broca, P. (1878). Anatomie comparée des circonvolutions cérébrales. Le grand lobe limbique et la scissure limbique dans la série des mammifères, *Rev. Anthropol. 1,* Ser. 2, 385–498.
Broca, P. (1879). Localisations cérébrales, Recherches sur les centres olfactifs, *Rev. Anthropol. 2,* 385–455.
Brockhaus, H. (1940). Die Cyto- und Myeloarchitektonik des Cortex claustralis und des Claustrum beim Menschen, *J. Psychol. Neurol. 49,* 249–348.
Brodal, A. (1947). The hippocampus and the sense of smell, *Brain 70,* 179–222.
Brodal, P. (1978). The corticopontine projection in the rhesus monkey. Origin and principles of organization, *Brain 101,* 251–283.
Brodmann, K. (1908). Beiträge zur histologischen Lokalisation der Grosshirnrinde. VI. Mitteilung: Die Cortexgliederung des Menschen, *J. Psychol. Neurol. 10,* 231–246.
Brodmann, K. (1909). Vergleichende Lokalisationslehre der Grosshirnrinde in ihren Prinzipien dargestellt auf Grund des Zellenbaues, *Leipzig Barth.*
Bronowski, J. (1966). The logic of the mind, *Am. Sci. 54,* 1–14.
Broom, R. (1932). *The Mammal-Like Reptiles of South Africa and the Origin of Mammals,* Witherby, London.
Broom, R., Robinson, J. T., and Schepers, G. W. H. (1949). *Sterkfontein Ape-Man: Plesianthropus,* Transvaal Museum, Memoir No. 4, Pretoria, South Africa.
Brown, F., Harris, J., Leakey, R., and Walker, A. (1985). Early *Homo erectus* skeleton from west Lake Turkana, Kenya, *Nature 316,* 788–792.
Brown, S., and Schäfer, E. A. (1888). An investigation into the functions of the occipital and temporal lobes of the monkey's brain, *Philos. Trans. R. Soc. London Ser. B 179,* 303–327.
Bruns, L. (1892). Über Störungen des Gleichgewichtes bei Stirnhirntumoren, *Deutsche. Med. Wchnschr. 18,* 138–140.

Bucher, K. L. (1970). Temporal lobe neocortex and maternal behavior in rhesus monkey. Doctoral thesis, School of Hygiene and Public Health, Johns Hopkins School of Medicine.

Buchwald, N. A., Wyers, E. J., Carlin, W. J., and Farley, R. E. (1961). Effects of caudate stimulation on visual discrimination, *Exp. Neurol. 4,* 23–36.

Buchwald, N. A., Price, D. D., Vernon, L., and Hull, C.D. (1973). Caudate intracellular response to thalamic and cortical inputs, *Exp. Neurol. 38,* 311–323.

Bucy, P. C. (1975). Heinrich Klüver, *Surg. Neurol. 3,* 229–231.

Bucy, P. C., and Klüver, H. (1940). Anatomic changes secondary to temporal lobectomy, *Arch. Neurol. Psychiatry 44,* 1142–1146.

Bucy, P. C., and Klüver, H. (1955). An anatomical investigation of the temporal lobe in the monkey (*Macaca mulatta*), *J. Comp. Neurol. 103,* 151–252.

Bucy, P. C., Keplinger, J. E., and Siqueira, E. B. (1964). Destruction of the "pyramidal tract" in man, *J. Neurosurg. 21,* 385–398.

Buechner, H. K. (1961). Territorial behavior in Uganda kob, *Science 133,* 698–699.

Buge, A., Escourolle, R., Rancurel, G., and Poisson, M. (1975). Mutisme akinetique et ramollissement bicingulaire. 3 observations anatomocliniques, *Rev. Neurol. 131,* 121–137.

Bunnell, B. N. (1966). Amygdaloid lesions and social dominance in the hooded rat, *Psychonom. Sci. 6,* 93–94.

Bunnell, B. N., Sodetz, F. J., and Shalloway, D. I. (1970). Amygdaloid lesions and social behavior in the golden hamster, *Physiol. Behav. 5,* 153–162.

Bunney, W. E., Jr., and Davis, J. M. (1965). Norepinephrine in depressive reactions, *Arch. Gen. Psychiatry 13,* 483–494.

Burden, W. D. (1928). Results of the Douglas Burden expedition to the island of Komodo. V. Observations on the habits and distribution of *Varanus komodoensis* Ouwens, American Museum Novitates, New York, No. 316, pp. 1–10.

Burns, R. S., LeWitt, P. A., Ebert, M. H., Pakkenberg, H., and Kopin, I. J. (1985). The clinical syndrome of striatal dopamine deficiency: Parkinsonism induced by 1-methyl-4-phenyl-1,2,3,6-tetrahydropyridine (MPTP), *N. Engl. J. Med. 312,* 1418–1421.

Buser, P., Bancaud, J., Talairach, J., and Szikla, G. (1968). Interconnexions amygdalo-hippocampiques chez l'Homme. Étude physiologique au cours d'explorations stéréotaxiques, *Rev. Neurol. 119,* 283–298.

Butcher, L. L., and Hodge, G. K. (1976). Postnatal development of acetylcholinesterase in the caudate–putamen nucleus and substantia nigra of rats, *Brain Res. 106,* 223–240.

Butler, A. B. (1978). Forebrain connections in lizards and the evolution of the sensory systems, in: *Behavior and Neurology of Lizards* (N. Greenberg and P. D. MacLean eds.), U.S. Government Printing Office, Washington, D.C., DHEW Publication No. (ADM) 77-491, pp. 65–78.

Butler, A. B., and Northcutt, R. G. (1971). Retinal projections in *Iguana iguana* and *Anolis carolinensis, Brain Res. 26,* 1–13.

Cadell, T. E. (1963). Effects of fornix section on learned and social behavior in rhesus monkeys, *Diss. Abstr. 24,* 2131, and University Microfilms, Inc., Ann Arbor.

Caine, E. D., Hunt, R. D., Weingartner, H., and Ebert, M. H. (1978). Huntington's dementia. Clinical and neuropsychological features, *Arch. Gen. Psychiatry 35,* 377–384.

Cairny, J. (1926). A general survey of the forebrain of *Sphenodon punctatum, J. Comp. Neurol. 42,* 255–348.

Calhoun, J. B. (1962). Population density and social pathology, *Sci. Am. 206,* 139–146.

Calhoun, J. B. (1964). The social use of space, in: *Physiological Mammalogy,* Volume 1 (W. Mayer and R. Van Gelder, eds.), Academic Press, New York, pp. 1–188.

Calhoun, J. B. (1971). Space and the strategy of life, in: *Behavior and Environment* (A. H. Esser, ed.), Plenum Press, New York, pp. 329–387.

Cameron, D. E., and Prados, M. D. (1948). Symposium on gyrectomy. Part 2—Bilateral frontal gyrectomy: Psychiatric results, in: *The Frontal Lobes, Res. Publ. Assoc. Res. Nerv. Ment. Dis., 27,* pp. 534–537.

Campbell, A. W. (1905). *Histological Studies on the Localization of Cerebral Functions,* Cambridge University Press, London.

Campbell, B. (1979). *Humankind Emerging,* 2nd Edition, Little, Brown, Boston.

Cannon, D. F. (1949). *Explorer of the Human Brain. The Life of Santiago Ramon y Cajal (1852–1934),* Schuman, New York.

Cannon, W. B. (1929). *Bodily Changes in Pain, Hunger, Fear, and Rage. An Account of Recent Researches into the Function of Emotional Excitement,* 2nd Edition, Appleton, New York.

Carlsson, A., and Lindqvist, M. (1963). Effect of chlorpromazine or haloperidol on formation of 3-methoxytyramine and normetanephrine in mouse brain, *Acta Pharmacol. Toxicol. 20,* 140–144.

Carlsson, A., Lindqvist, M., Magnusson, T., and Waldeck, B. (1958). On the presence of 3-hydroxytyramine in brain, *Science 127,* 471.

Carlsson, A., Falck, B., and Hillarp, N. A. (1962). Cellular localization of brain monoamines, *Acta Physiol. Scand. 56,* 1–28.

Carlyle, T. (1871/1884). Diary for June 15, 1871. Quoted by James Anthony Froude, in: *Thomas Carlyle,* Volume 2, Longmans, Green, New York, p. 407.

Carman, J. B., Cowan, W. M., and Powell, T. P. S. (1963). The organization of the cortico-striate connexions in the rabbit, *Brain 86,* 525–562.

Carman, J. B., Cowan, W. M., and Powell, T. P. S., and Webster, K. E. (1965). A bilateral cortico-striate projection, *J. Neurol. Neurosurg. Psychiatry 28,* 71–77.

Carmel, P. W. (1970). Efferent projections of the ventral anterior nucleus of the thalamus in the monkey, *Am. J. Anat. 128,* 159–183.

Carmichael, M., and MacLean, P. D. (1961). Use of squirrel monkey for brain research, with description of restraining chair, *Electroenceph. Clin. Neurophysiol. 13,* 128–129.

Carpenter, C. C. (1961a). Patterns of the social behavior of Merriam's Canyon lizard—Iguanidae, *Southwestern Naturalist 6,* 138–148.

Carpenter, C. C. (1961b). Patterns of social behavior in the desert iguana, *Dipsosaurus dorsalis, Copeia, 4,* 396–405.

Carpenter, C. C. (1962). Patterns of behavior in two Oklahoma lizards, *Am. Midl. Nat. 67,* 132–151.

Carpenter, C. C. (1967). Aggression and social structure of iguanid lizards, in: *Lizard Ecology; A Symposium* (W. W. Milstead, ed.), University of Missouri Press, Columbia, pp. 87–105.

Carpenter, M. B. (1976). Anatomical organization of the corpus striatum and related nuclei, in: *The Basal Ganglia, Res. Publ. Assoc. Res. Nerv. Ment. Dis. 55,* 1–36.

Carpenter, M. B. (1981). Anatomy of corpus striatum and brain stem integrating systems, in: *American Physiology Handbook of "Motor Control"* (V. Brooks, ed.), Williams and Wilkens, Baltimore, pp. 947–995.

Carpenter, M. B., and Peter, P. (1972). Nigrostriatal and nigrothalamic fibers in the rhesus monkey, *J. Comp. Neurol. 144,* 93–116.

Carpenter, M. B., and Strominger, N. L. (1967). Efferent fibers of the subthalamic nucleus in the monkey. A comparison of the efferent projections of the subthalamic nucleus, substantia nigra and globus pallidus, *Am. J. Anat. 121,* 41–71.

Carpenter, M. B., Fraser, R. A., and Shriver, J. E. (1968). The organization of pallidosubthalamic fibers in the monkey, *Brain Res. 11,* 522–559.

Carpenter, M. B., Nakano, K., and Kim, R. (1976). Nigrothalamic projections in the monkey demonstrated by autoradiographic technics, *J. Comp. Neurol. 165,* 401–416.

Carr, A. (1965). The navigation of the green turtle, *Sci. Am. 212,* 78–86.

Carroll, R. L. (1964). The earliest reptiles, *J. Linn. Soc. London Zool. 45,* 61–83.

Carter, D. A., and Fibiger, H. C. (1978). The projections of the entopeduncular nucleus and globus pallidus in rat as demonstrated by autoradiography and horseradish peroxidase histochemistry, *J. Comp. Neurol. 177,* 113–124.

Cartmill, M., Pilbeam, D., and Isaac, G. (1986). One hundred years of paleoanthropology, *Am. Sci. 74,* 410–420.

Casey, K. L. (1966). Unit analysis of nociceptive mechanisms in the thalamus of the awake squirrel monkey, *J. Neurophysiol. 29,* 727–750.

Casey, K. L., Cuenod, M., and MacLean, P. D. (1965). Unit analysis of visual input to posterior limbic cortex. II. Intracerebral stimuli, *J. Neurophysiol. 28,* 1118–1131.

Caveness, W. F. (1955). Emotional and psychological factors in epilepsy: General clinical and neurological considerations, *Am. J. Psychiatry 112,* 190–193.

Chamberlain, H. (1928). The inheritance of lefthandedness, *J. Hered. 19,* 557–559.

Chambers's Encyclopedia (1892a). *A Dictionary of Universal Knowledge,* Article on psychology, Lippincott, Philadelphia.

Chambers's Encyclopedia (1892b). *A Dictionary of Universal Knowledge,* Article on handedness, Chambers, Edinburgh, pp. 724–725.

Chance, M. (1969). Towards the biological definition of ethics, in: *Biology and Ethics* (J. Ebling, ed.), Academic Press, New York, pp. 3–13.

Chang, M. M., and Leeman, S. E. (1970). Isolation of a sialagogic peptide from bovine hypothalamus tissue and its characterization as substance P, *J. Biol. Chem. 245,* 4784–4790.

Chapman, L. F., Walter, R. D., Markham, C. H., Rand, R. W., and Crandall, P. H. (1967). Memory changes

induced by stimulation of hippocampus or amygdala in epilepsy patients with implanted electrodes, *Trans. Am. Neurol. Assoc. 92,* 50–56.
Chapman, W. P., Rose, A. S., and Solomon, H. C. (1948). Measurements of heat stimulus producing motor withdrawal reaction in patients following frontal lobotomy, in: *The Frontal Lobes, Res. Publ. Assoc. Res. Nerv. Ment. Dis. 27,* 754–768.
Chessick, R. D., and Bolin, R. R. (1962). Psychiatric study of patients with psychomotor seizures, *J. Nerv. Ment. Dis. 134,* 72–79.
Chiszar, D. (1978). Lateral displays in the lower vertebrates: Forms, functions and origins, in: *Contrasts in Behavior—Adaptations in the Aquatic and Terrestrial Environments* (E. S. Reese and F. J. Lighter, eds.), Wiley, New York, pp. 105–135.
Chitanondh, H. (1966). Stereotaxic amygdalotomy in the treatment of olfactory seizures and psychiatric disorders with olfactory hallucination, *Confin. Neurol. 27,* 181–196.
Chodos, A. (1986). Marginalia. String fever, *Am. Sci. 74,* 253–254.
Chomsky, N. (1968). *Language and Mind,* Harcourt, New York.
Christ, J. F. (1969). Derivation and boundaries of the hypothalamus, with atlas of hypothalamic grisea, in: *The Hypothalamus* (W. Haymaker, E. Anderson, and W. J. H. Nauta, eds.), Thomas, Springfield, Ill., pp. 13–60.
Chronister, R. B., and White, L. E., Jr. (1975). Fiberarchitecture of the hippocampal formation: Anatomy, projections, and structural significance, in: *The Hippocampus,* Volume 1 (R. L. Isaacson and K. H. Pribram, eds.), Plenum Press, New York, pp. 9–39.
Clark, G., Magoun, H. W., and Ranson, S. W. (1939). Hypothalamic regulation of body temperature, *J. Neurophysiol. 2,* 61–80.
Clark, J. D. (1984). The way we were. Speculating and accumulating: New approaches to the study of early human living, *Anthroquest 30,* 1, 18–19.
Clark, J. D., and Harris, J. W. K. (1985). Fire and its roles in early hominid lifeways, *Afr. Archaeol. Rev. 3,* 3–27.
Clark, R. W. (1976). *The Life of Bertrand Russell,* Knopf, New York.
Clark, W. E. L., and Meyer, M. (1947). The terminal connexions of the olfactory tract in the rabbit, *Brain 70,* 304–328.
Clark, W. E. L., and Meyer, M. (1950). Anatomical relationships between the cerebral cortex and the hypothalamus, *Br. Med. Bull. 6,* 341–345.
Coats, J. (1876). A study of two illustrative cases of epilepsy, *Br. Med. J. 2,* 647–649.
Cobb, S. (1943). *Borderlands of Psychiatry,* Harvard University Press, Cambridge, Mass.
Coggeshall, R. E., and MacLean, P. D. (1958). Hippocampal lesions following administration of 3-acetylpyridine, *Proc. Soc. Exp. Biol. Med. 98,* 687–689.
Colbern, D., Isaacson, R. L., Bohus, B., and Gispen, W. H. (1977). Limbic-midbrain lesions and ACTH-induced excessive grooming, *Life Sci. 21,* 393–402.
Colbert, E. H. (1945). The dinosaur book, *Am. Mus. Nat. Hist. 14.*
Colbert, E. H. (1966). *The Age of Reptiles,* Norton, New York.
Colbert, E. H. (1969). *Evolution of the Vertebrates,* Wiley, New York.
Colbert, E. H. (1972). Antarctic fossils and the reconstruction of Gondwanaland, *Nat. Hist, 81,* 66–73.
Collias, N. E. (1980). Basal telencephalon suffices for early socialization in chicks, *Physiol. Behav. 24,* 93–97.
Collins, E. T. (1932). The physiology of weeping, *Br. J. Ophthalmol. 16,* 1–20.
Comte, A. (1853/1955). Altruism, in: *The Oxford Universal Dictionary on Historical Principles* (C. T. Onions, ed.), Oxford University Press, London.
Connell, P. H. (1958). *Amphetamine Psychosis,* Chapman & Hall, London.
Conrad, C. D., and Stumpf, W. E. (1975). Direct visual input to the limbic system: Crossed retinal projections to the nucleus anterodorsalis thalami in the tree shrew, *Exp. Brain Res. 23,* 141–149.
Conrad, K., and Ule, G. (1951). Ein Fall von Korsakow-Psychose mit anatomischen Befund und klinischen Betrachtungen, *Dtsch. Z. Nervenheilkd. 165,* 430–455.
Conrad, L. C. A., Leonard, C. M., and Pfaff, D. W. (1974). Connections of the median and dorsal raphe nuclei in the rat: An autoradiographic and degeneration study, *J. Comp. Neurol. 156,* 179–206.
Cookson, J. H. (1962). Failure of lizards to learn a simple task, *Br. J. Herpetol. 3,* 40.
Cooper, I. S. (1956). *The Neurosurgical Alleviation of Parkinsonism,* Thomas, Springfield, Ill.
Cooper, I. S. (1961). *Parkinsonism. Its Medical and Surgical Therapy,* Thomas, Springfield, Ill.
Cope, E. D. (1892). On the homologies of the posterior cranial arches in the Reptilia, *Trans. Am. Philos. Soc. 27.*

Cordeau, J. P., and Mahut, H. (1964). Some long-term effects of temporal lobe resections on auditory and visual discrimination in monkeys, *Brain 87,* 177–190.

Coren, S., and Porac, C. (1977). Fifty centuries of right-handedness: The historical record, *Science 198,* 631–632.

Correll, R. E., and Scoville, W. B. (1967). Significance of delay in the performance of monkeys with medial temporal lobe resections, *Exp. Brain Res. 4,* 85–96.

Corsellis, J. A. N., Goldberg, G. J., and Norton, A. R. (1968). ''Limbic encephalitis'' and its association with carcinoma, *Brain 91,* 481–496.

Cotzias, G. C., Van Woert, M. H., and Schiffer, L. M. (1967). Aromatic amino acids and modification of parkinsonism, *N. Engl. J. Med. 276,* 374–379.

Courchesne, E., Hesselink, J. R., Jernigan, T. L., and Yeung-Courchesne, R. (1987). Abnormal neuroanatomy in a nonretarded person with autism. Unusual findings with magnetic resonance imaging, *Arch. Neurol. 44,* 335.

Courville, C. B. (1958). Traumatic lesions of the temporal lobe as the essential cause of psychomotor epilepsy, in: *Temporal Lobe Epilepsy* (M. Baldwin and P. Bailey, eds.), Thomas, Springfield, Ill., pp. 220–239.

Cowan, W. M., and Powell, T. P. S. (1955). The projection of the midline and intralaminar nuclei of the thalamus of the rabbit, *J. Neurol. Neurosurg. Psychiatry 18,* 266–279.

Cowan, W. M., Raisman, G., and Powell, T. P. S. (1965). The connexions of the amygdala, *J. Neurol. Neurosurg. Psychiatry 28,* 137–151.

Cowan, W. M., Gottlieb, D. I., Hendrickson, A. E., Price, J. L., and Woolsey, T. A. (1972). The autoradiographic demonstration of axonal connections in the central nervous system, *Brain Res. 37,* 21–51.

Cragg, B. G. (1961). Olfactory and other afferent connections of the hippocampus in the rabbit, rat, and cat, *Exp. Neurol. 3,* 588–600.

Crandall, P. H., Walter, R. D., and Rand, R. W. (1963). Clinical applications of studies on stereotactically implanted electrodes in temporal-lobe epilepsy, *J. Neurosurg. 20,* 827–840.

Creutzfeldt, O. D. (1956). Die Krampfausbreitung im Temporallappen der Katze. Die Krampfentladungen des Ammonshorns und ihre Beziehungen zum übrigen Rhinencephalon und Isocortex, *Schweizer Archiv. für Neurologie und Psychiatrie, 77,* 163–194.

Creutzfeldt, O. D., and Meyer-Mickeleit, R. W. (1953). Patterns of convulsive discharges of the hippocampus and their propagation, *Electroencephalogr. Clin. Neurophysiol. 3,* 43.

Crews, D. (1978). Integration of internal and external stimuli in the regulation of lizard reproduction, in: *The Behavior and Neurology of Lizards* (N. Greenberg and P. D. MacLean, eds), U.S. Government Printing Office, Washington, D. C., DHEW Publication No. (ADM) 77-491, pp. 149–171.

Crompton, A. W. (1963). On the lower jaw of *Diarthrognathus* and the origin of the mammalian lower jaw, *Proc. Zool. Soc. London 140,* 697–753.

Crompton, A. W., and Hylander, W. L. (1986). Changes in mandibular function following the acquisition of a dentary–squamosal jaw articulation, in: *The Ecology and Biology of Mammal-like Reptiles* (N. Hotton, P. D. MacLean, J. J. Roth, and E. C. Roth, eds), Smithsonian Institution Press, Washington, D.C., pp. 263–282.

Crompton, A. W., and Jenkins, F. A., Jr. (1973). Mammals from reptiles: A review of mammalian origins, *Annu. Rev. Earth Planet. Sci. 1,* 131–155.

Crompton, A. W., and Jenkins, F. A., Jr. (1979). Origin of mammals, in *Mesozoic Mammals* (J. A. Lillegraven, Z. Kielan-Jaworowska, and W. A. Clemens, eds.), University of California Press, Berkeley, pp. 59–73.

Crosby, E. C., and Humphrey, T. (1941). Studies of the vertebrate telencephalon. II. The nuclear pattern of the anterior olfactory nucleus, tuberculum olfactorium and the amygdaloid complex in adult man, *J. Comp. Neurol. 74,* 309–352.

Crosby, E. C., Humphrey, T., and Lauer, E. W. (1962). *Correlative Anatomy of the Nervous System,* Macmillan Co., New York.

Cuénod, M., Casey, K. L., and MacLean, P. D. (1965). Unit analysis of visual input to posterior limbic cortex. I. Photic stimulation, *J. Neurophysiol. 28,* 1101–1117.

Currie, S., Heathfield, K. W. G., Henson, R. A., and Scott, D. F. (1971). Clinical course and prognosis of temporal lobe epilepsy: A survey of 666 patients, *Brain 94,* 173–190.

Cushing, H. (1922). The field defects produced by temporal lobe lesions, *Brain 44,* 341–396.

Cushing, H., and Heuer, G. J. (1911). Distortions of the visual fields in cases of brain tumor. Statistical studies (first paper), *Bull. Johns Hopkins Hosp. 22,* 190–195.

Dafny, N., Bental, E., and Feldman, S. (1965). Effect of sensory stimuli on single unit activity in the posterior hypothalamus, *Electroencephalogr. Clin. Neurophysiol. 19,* 256–263.

Dahl, D., Ingram, W. R., and Knott, J. R. (1962). Diencephalic lesions and avoidance learning in cats, *Arch. Neurol. 7,* 314–319.

Dahlström, A., and Fuxe, K. (1964). Evidence for the existence of monoamine-containing neurons in the central nervous system. I. Demonstration of monoamines in the cell bodies of brain stem neurons, *Acta Physiol. Scand. 62,* 1–80.

Daitz, H. M. (1953). Note on the fibre content of the fornix system in man, *Brain 76,* 509–512.

Daitz, H. M., and Powell, T. P. S. (1954). Studies of the connexions of the fornix system, *J. Neurol. Neurosurg. Psychiatry 17,* 75–82.

Dale, H. H. (1914). The occurrence in ergot and action of acetyl-choline, *J. Physiol. (London) 48,* 3–4.

Daly, D. (1958). Ictal affect, *Am. J. Psychiatry 115,* 97–108.

Daly, D., and Mulder, D. W. (1957). Gelastic epilepsy, *Neurology 7,* 189–192.

Darley, F. L., Aronson, A. E., and Brown, J. R. (1975). *Motor Speech Disorders,* Saunders, Philadelphia.

Darlington, C. D. (1978). A diagram of evolution, *Nature 276,* 447–452.

Dart, R. A. (1925). *Australopithecus africanus:* The man-ape of South Africa, *Nature 115,* 195–199.

Dart, R. A. (1935). The dual structure of the neopallium: Its history and significance, *J. Anat. 69,* 3–19.

Dart, R. A. (1959). *Adventures with the Missing Link,* The Institutes Press, Philadelphia.

Davidoff, L. M., and Brenner, C. (1948). A new instrument for the performance of bifrontal lobotomy, in: *The Frontal Lobes, Res. Publ. Assoc. Res. Nerv. Ment. Dis. 27,* 638–641.

Davis, G. D. (1958). Caudate lesions and spontaneous locomotion in the monkey, *Neurology 8,* 135–139.

Davison, C., and Kelman, H. (1939). Pathologic laughing and crying, *Arch. Neurol. Psychiatry 42,* 595–643.

de Bruin, J. P. C. (1980). Telencephalon and behavior in teleost fish: A neuroethological approach, in: *Comparative Neurology of the Telencephalon* (S. O. E. Ebbesson, ed.), Plenum Press, New York, pp. 175–198.

Dejerine, J. (1895/1901). *Anatomie des centres nerveux,* two volumes, Rueff, Paris.

DeJong, R. N., Itabashi, H. H., and Olson, J. R. (1969). Memory loss due to hippocampal lesions. Report of a case, *Arch. Neurol. 20,* 339–348.

Delay, J., and Brion, S. (1954). Syndrome de Korsakoff et corps mamillaires, *Encéphale 43,* 193–200.

Delgado, J. M. R., Mark, V., Sweet, W., Revin, F., Weiss, G., Bach-y-Rita, G., and Hagiwara, R. (1968). Intracerebral radio stimulation and recording in completely free patients, *J. Nerv. Ment. Dis. 147,* 329–340.

Delgado, J. M. R., Delgado-Garcia, J. M., Amerigo, J. A., and Grau, C. (1975). Behavioral inhibition induced by pallidal stimulation in monkeys, *Exp. Neurol. 49,* 580–591.

Dell, P. (1952). Correlations entre le système végétatif et le système de la vie de relation. Mesencéphale, diencéphale et cortex cérébral, *J. Physiol. (London) 44,* 471–557.

Dell, P. (1972). Discussion, in: *Limbic System Mechanisms and Autonomic Function* (C. H. Hockman, ed.) Thomas, Springfield, Ill.

Dell, P., and Olson, R. (1951). Projections "secondaires" mesencéphaliques, diencéphaliques et amygdaliennes des afférences viscérales vagales, *C.R. Soc. Biol. 145,* 1088–1091.

DeLong, M. R. (1971). Activity of pallidal neurons during movement, *J. Neurophysiol. 34,* 414–427.

DeLong, M. R. (1973). Putamen: Activity of single units during slow and rapid arm movements, *Science 179,* 1240–1242.

Demos, R. (1937). Introduction, in: *The Dialogues of Plato,* two volumes (Trans. by B. Jowett), Random House, New York.

Dempsey, E. W., and Morison, R. S. (1943). The electrical activity of a thalamocortical relay system, *Am. J. Physiol. 138,* 283–296.

Dempsey, E. W., and Rioch, D. M. (1939). The localization in the brain stem of the oestrous responses of the female guinea pig, *J. Neurophysiol. 2,* 9–18.

Denniston, R. H., and MacLean, P. D. (1961). Erection display in male–male and male–female interaction in the squirrel monkey, *Saimiri sciureus, Am. Zool. 1,* 445.

Denny-Brown, D. (1946). Diseases of the basal ganglia and subthalamic nuclei, in: *Oxford Loose-leaf Medicine* (H. A. Christian, ed.), Oxford University Press, London.

Denny-Brown, D. (1951). The frontal lobes and their functions, in: *Modern Trends in Neurology,* 1st Series (A. Feiling, ed.), Harper & Row (Hoeber), New York, pp. 13–89.

Denny-Brown, D. (1960). Motor mechanisms—introduction: The general principles of motor integration, in: *Handbook of Physiology* (J. Field, H. W. Magoun, and V. E. Hall, eds.), American Physiological Society, Washington, D.C., pp. 781–796.

Denny-Brown, D. (1962). *The Basal Ganglia and their Relation to Disorders of Movement,* Oxford University Press, London.

de Olmos, J. S. (1969). A cupric-silver method for impregnation of terminal axon degeneration and its further use in staining granular argyrophilic neurons, *Brain Behav. Evol. 2,* 213–237.

de Olmos, J. S. (1972). The amygdaloid projection field in the rat as studied with the cupric-silver method, in: *The Neurobiology of the Amygdala* (B. E. Eleftheriou, ed.), Plenum Press, New York, p. 145–204.

de Olmos, J. S., and Heimer, L. (1980). Double and triple labeling of neurons with fluorescent substances; the study of collateral pathways in the ascending raphe system, *Neurosci. Lett. 19,* 7–12.

De Robertis, E. (1956). Submicroscopic changes of the synapse after nerve section in the acoustic ganglion of the guinea pig: An electron microscope study, *J. Biophys. Cytol. 2,* 503.

Descartes, R. (1637/1967). Discourse on the method of rightly conducting the reason and seeking for truth in the sciences, in: *Philosophical Works of Descartes,* Volume 1 (Trans. by E. S. Haldane and G. R. T. Ross), Dover, New York, pp. 81–130.

Descartes, R. (1644/1967). The principles of philosophy, in: *Philosophical Works of Descartes,* Volume 1 (Trans. by E. S. Haldane and G. R. T. Ross), Dover, New York, pp. 219–327.

Descartes, R. (1649/1967). The passions of the soul, in: *Philosophical Works of Descartes,* Volume 1 (Trans. by E. S. Haldane and G. R. T. Ross), Dover, New York, pp. 331–427.

Desmedt, J. E., and Mechelse, K. (1959). Mise en évidence d'une quatrième aire de projection acoustique dans l'écorce cérébrale du chat, *J. Physiol. (Paris) 51,* 448–449.

Desmond, A. J. (1976). *The Hot-Blooded Dinosaurs: A Revolution in Paleontology,* Dial Press/James Wade, New York.

de Spinoza, B. (1677/1955). The ethics, in: *The Chief Works of Benedict Spinoza* (Trans. by R. H. M. Elwes), Dover, New York, pp. 45–271.

DeVito, J. L., and White, L. E., Jr. (1966). Projections from the fornix to the hippocampal formation in the squirrel monkey, *J. Comp. Neurol. 127,* 389–398.

de Vries, E. (1910). Bemerkungen zur Ontogenie und vergleichenden Anatomie des Claustrums, *Folia Neurobiol. 4,* 481–513.

Diamond, I. T., Chow, K. L., and Neff, W. D. (1958). Degeneration of caudal medial geniculate body following cortical lesion ventral to auditory area II in cat, *J. Comp. Neurol. 109,* 349–362.

Dicks, D., Myers, R. E., and Kling, A. (1969). Uncus and amygdala lesions: Effects on social behavior in the free-ranging rhesus monkey, *Science 165,* 69–71.

Diepen, R., Janssen, P., Engelhardt, F., and Spatz, H. (1956). Recherches sur le çerveau de l'éléphant d'Afrique (Loxodonta africana Blum). II. Données sur l'hypothalamus, *Acta Neurol. Psychiatr. Belg. 11,* 759–788.

Dieterlen, F. (1959). Das Verhalten des Syrischen Goldhamsters (*Mesocricetus auratus* Waterhouse), *Z. Tierpsychol. 16,* 47–103.

Dietz, R. S., and Holden, J. C. (1970). The breakup of Pangaea, *Sci. Am. 223,* 30–41.

Diezel, P. B. (1955). Iron in the brain: A chemical and histochemical examination, in: *Biochemistry of the Developing Nervous System,* Academic Press, New York, pp. 145–152.

Dingwall, W. O. (1988). The evolution of human communicative behavior, in: *Linguistics: The Cambridge Survey,* Volume 3 (F. V. Newmeyer, ed.),Cambridge University Press, London, pp. 274–313.

Distel, H. (1978). Behavioral responses to the electrical stimulation of the brain in the green iguana, in: *The Behavior and Neurology of Lizards* (N. Greenberg and P. D. MacLean, eds.), U.S. Government Printing Office, Washington, D.C., DHEW Publication No. (ADM) 77-491, pp. 135–147.

Divac, I., and Oberg, R. G. E. (1979). Current conceptions of neostriatal functions. History and an evaluation, in: *The Neostriatum* (I. Divac and R. G. E. Oberg, eds.), Pergamon Press, New York, pp. 215–230.

Doane, B. K., and Livingston, K. E. (eds.) (1986). *The Limbic System: Functional Organization and Clinical Disorders,* Raven Press, New York.

Dollard, J., and Miller, N. E. (1950). *Personality and Psychotherapy: An Analysis in Terms of Learning, Thinking, and Culture,* McGraw–Hill, New York.

Domesick, V. B. (1969). Projections from the cingulate cortex in the rat, *Brain Res. 12,* 296–320.

Domesick, V. B. (1970). The fasciculus cinguli in the cat, *Brain Res. 20,* 19–32.

Dostoyevsky, F. (1962). *The Idiot* (Trans. by C. Garnett), Modern Library, New York.

Dott, N. M. (1938). Surgical aspects of the hypothalamus, in: *The Hypothalamus* (J. Beattie, G. Riddock, and N. M. Dott, eds.), Oliver & Boyd, Edinburgh, pp. 131–185.

Douglas, R. J. (1967). The hippocampus and behavior, *Psychol. Bull. 67,* 416–442.

Dow, R. S. (1942). Cerebellar action potentials in response to stimulation of the cerebral cortex in monkeys and cats, *J. Neurophysiol. 5,* 121.

Dow, R. S. (1974). Some novel concepts of cerebellar physiology, *Mt. Sinai J. Med. N.Y. 41,* 103–119.

Drachman, D. A., and Adams, R. D. (1962). Herpes simplex and acute inclusion-body encephalitis, *Arch. Neurol. 7,* 45–63.

Drachman, D. A., and Ommaya, A. K. (1964). Memory and the hippocampal complex, *Arch. Neurol. 10,* 411–425.

Droogleever-Fortuyn, J., and Stefens, R. (1951). On the anatomical relations of the intralaminar and midline cells of the thalamus, *Electroencephalogr. Clin. Neurophysiol. 3,* 393–400.

Droz, B., and Leblond, C. P. (1963). Axonal migration of proteins in the central and peripheral nerves as shown by radioautography, *J. Comp. Neurol. 121,* 325–346.

Druckman, R., and Chao, D. (1957). Laughter in epilepsy, *Neurology 7,* 26–36.

Dua, S., and MacLean, P. D. (1964). Localization for penile erection in medial frontal lobe, *Am. J. Physiol. 207,* 1425–1434.

Du Bois-Reymond, E. (1849). *Untersuchungen uber Thierische Elektricitat,* two volumes, Reiner, Berlin.

DuMond, F. V., and Hutchinson, T. C. (1967). Squirrel monkey reproduction: The "Fatted" male phenomenon and seasonal spermatogenesis, *Science 158,* 1067–1069.

Durward, A. (1934). Some observations on the development of the corpus striatum of birds, with special reference to certain stages in the common sparrow (*Passer domesticus*), *J. Anat. 68,* 492–499.

Duvall, D. (1986). A new question of pheromones: Aspects of possible chemical signaling and reception in the mammal-like reptiles, in: *The Ecology and Biology of Mammal-like Reptiles* (N. Hotton, P. D. MacLean, J. J. Roth, and E. C. Roth, eds.), Smithsonian Institution Press, Washington, D.C., pp. 219–238.

Earle, K. M. (1968). Studies on Parkinson's disease including X-ray fluorescent spectroscopy of formalin fixed brain tissue, *J. Neuropathol. Exp. Neurol. 27,* 1–14.

Earle, K. M., Baldwin, M., and Penfield, W. (1953). Incisural sclerosis and temporal lobe seizures produced by hippocampal herniation at birth, *Arch. Neurol. Psychiatry 69,* 27–42.

Ebbesson, S. O. E. (1980). The parcellation theory and its relation to interspecific variability in brain organization, evolutionary and ontogenetic development, and neuronal plasticity, *Cell Tissue Res. 213,* 179–212.

Economo, C. von (1929). *The Cytoarchitectonics of the Human Cerebral Cortex,* Oxford University Press, London.

Economo, C. von (1931). *Encephalitis lethargica. Its sequelae and treatment* (Trans. by K. O. Newman), Oxford University Press, London.

Economo, C. von, and Koskinas, G. N. (1925). *Die Cytoarchitektonik der Hirnrinde des erwachsenen Menschen,* Springer, Berlin.

Edinger, L. (1899). *The Anatomy of the Central Nervous System of Man and of Vertebrates in General* (Trans. by W. S. Hall), Davis, Philadelphia.

Edinger, L., and Wallenberg, A. (1902). Untersuchungen über den Fornix und das Corpus mamillare, *Arch. Psychiatr. 35,* 1–21.

Edinger, T. (1966). Brains from 40 million years of camelid history, in: *Evolution of the Forebrain* (R. Hassler and H. Stephan, eds.), Thieme, Stuttgart, pp. 153–261.

Ehret, R., and Schneider, E. (1961). Photogene Epilepsie mit suchtartiger Selbstauslösung kleiner Anfälle und wiederholten Sexualdelikten, *Arch. Psychiatr. Nervenkr. 202,* 75–94.

Ehringer, H., and Hornykiewicz, O. (1960). Verteilung von Noradrenalin und Dopamin (3-Hydroxytyramin) im Gehirn des Menschen und ihr Verhalten bei Erkrankungen des extrapyramidalen Systems, *Wien. Klin. Wochenschr. 38,* 1236–1239.

Eibl-Eibesfeldt, I. (1961). *Galapagos—The Noah's Ark of the Pacific* (Trans. by A. H. Brodrick), Doubleday, New York.

Eibl-Eibesfeldt, I. (1970). *Ethology: The Biology of Behavior,* Holt, Rinehart & Winston, New York.

Eibl-Eibesfeldt, I. (1971). !Ko-Buschleute (Kalahari)—Schamweisen und Spotten, *Homo 22,* 261–266.

Eibl-Eibesfeldt, I., and Wickler, W. (1968). Ethology, in: *International Encyclopedia of the Social Sciences,* Volume 5 (D. L. Sills, ed.), Macmillan Co., New York, pp. 186–193.

Eisenberg, L. (1958). The autistic child in adolescence, in: *Psychopathology: A Source Book* (C. F. Reed, I. E. Alexander, and S. S. Tomkins, eds.), Harvard University Press, Cambridge, Mass., pp. 15–24.

Eisenberg, L., and Kanner, L. (1958). Early infantile autism, 1943–1955, in: *Psychopathology: A Source Book* (C. F. Reed, I. E. Alexander, and S. S. Tomkins, eds.), Harvard University Press, Cambridge, Mass., pp. 3–14.

Eldredge, N., and Tattersall, I. (1982). *The Myths of Human Evolution,* Columbia University Press, New York.

Eleftheriou, B. E., and Zolovick, A. J. (1968). Effect of amygdaloid lesions on plasma and pituitary thyrotropin levels in the deermouse, *Proc. Soc. Exp. Biol. Med. 127,* 671–674.

Elliot, T. R. (1904). On the action of adrenalin, *J. Physiol. (London) 31,* 72, xx–xxi.
Elwers, M., and Critchlow, V. (1960). Precocious ovarian stimulation following hypothalamic and amygdaloid lesions in rats, *Am. J. Physiol. 198,* 381–385.
Elwers, M., and Critchlow, V. (1961). Precocious ovarian stimulation following interruption of stria terminalis, *Am. J. Physiol. 201,* 281–284.
Emmers, R., and Akert, K. (1963). *A stereotaxic atlas of the squirrel monkey (Saimiri sciureus),* University of Wisconsin Press, Madison.
Endröczi, E., and Lissak, K. (1959). The role of the mesencephalon and archicortex in the activation and inhibition of the pituitary–adrenocortical system, *Acta Physiol. Hung. 15,* 25.
Engel, J., Jr. (1984). The use of positron emission tomographic scanning in epilepsy, *Ann. Neurol. 15,* S180–S191.
Enlow, D. H., and Brown, S. O. (1957). A comparative study of fossil and recent bone tissue. Part II. Reptiles and birds, *Tex. J. Sci. 9,* 186–214.
Erickson, T. (1945). Erotomania (nymphomania) as an expression of cortical epileptiform discharge, *Arch. Neurol. Psychiatry 53,* 226–231.
Escueta, A. V. D., Bacsal, F. E., and Treiman, D. M. (1982). Complex partial seizures on closed-circuit television and EEG: A study of 691 attacks in 79 patients, *Ann. Neurol.11,* 292–300.
Esser, A. H. (1968). Dominance hierarchy and clinical course of psychiatrically hospitalized boys, *Child Dev. 39,* 147–157.
Esser, A. H. (1973). Cottage fourteen: Dominance and territoriality in a group of institutionalized boys, *Small Group Behav. 4,* 130–146.
Euler, C. S. von, and Gaddum, J. H. (1931). A unidentified depressor substance in certain tissue extracts, *J. Physiol. (London) 72,* 74–87.
Evans, L. T. (1936). A study of a social hierarchy in the lizard, *Anolis carolinensis, J. Genet. Psychol. 48,* 88–111.
Evans, L. T. (1938a). Cuban field studies on territoriality of the lizard, *Anolis sagrei, J. Comp. Psychol. 25,* 97–125.
Evans, L. T. (1938b). Courtship behavior and sexual selection of anolis, *J. Comp. Psychol. 26,* 475–497.
Evans, L. T. (1951). Field study of the social behavior of the black lizard, *Ctenosaura pectinata, Am. Mus. Novit. 1493,* 1–26.
Evans, L. T. (1958). Fighting in young and mature opossums, *Anat. Rec. 131,* 549.
Evans, L. T. (1959). A motion picture study of maternal behavior of the lizard, *Eumeces obsoletus* Baird and Girard, *Copeia 1959,* 103–110.
Evans, L. T. (1961). Structure as related to behavior in the organization of populations in reptiles, in: *Vertebrate Speciation,* A University of Texas Symposium, pp. 148–178.
Everett, J. W. (1965). Ovulation in rats from preoptic stimulation through platinum electrodes. Importance of duration and spread of stimulus, *Endocrinology 76,* 1195–1201.

Fabricius, E. (1971). Ethological evidence of genetically determined behaviour patterns, and conflicts between these patterns and changed environmental conditions, in: *Society, Stress and Disease,* Volume 1 (L. Levi, ed.), Oxford University Press, London, pp. 71–78.
Fagen, R. (1981). *Animal Play Behavior,* Oxford University Press, London.
Fager, C. A. (1968). Evaluation of thalamic and subthalamic surgical lesions in the alleviation of Parkinson's disease, *J. Neurosurg. 28,* 145–149.
Fahn, S., and Cote, L. J. (1968). Regional distribution of γ-aminobutyric acid (GABA) in brain of the rhesus monkey, *J. Neurochem. 15,* 209.
Falck, B. (1962). Observations on the possibilities for the cellular localization of monoamines with a fluorescence method, *Acta Physiol. Scand. 56,* 1–25.
Falck, B., Hillarp, N.-A., Thieme, G., and Torp, A. (1962). Fluorescence of catecholamines and related compounds condensed with formaldehyde, *J. Histochem. Cytochem. 10,* 348–354.
Falconer, M. A. (1970). Significance of surgery for temporal lobe epilepsy in childhood and adolescence, *J. Neurosurg. 33,* 233–252.
Falconer, M. A. (1971). Genetic and related aetiological factors in temporal lobe epilepsy, *Epilepsia 12,* 13–31.
Falconer, M. A. (1974). Mesial temporal (Ammon's horn) sclerosis as a common cause of epilepsy. Aetiology, treatment, and prevention, *Lancet 2,* 767–770.
Falconer, M. A., and Pond, D. A. (1953). Temporal lobe epilepsy with personality and behaviour disorders caused by an unusual calcifying lesion. Report of two cases in children relieved by temporal lobectomy, *J. Neurol. Neurosurg. Psychiatry 16,* 234–244.

Falconer, M. A., Serafetinides, E. A., and Corsellis, J. A. N. (1964). Etiology and pathogenesis of temporal lobe epilepsy, *Arch. Neurol. 10,* 233–248.

Falk, D. (1980). A reanalysis of the South African australopithecine natural endocasts, *Am. J. Phys. Anthropol. 53,* 525–539.

Falk, D. (1983). Cerebral cortices of East African early hominids, *Science 221,* 1072–1074.

Falk, D. (1987). Hominid paleoneurology, *Annu. Rev. Anthropol. 16,* 13–30.

Fallon, J. H., Koziell, D. A., and Moore, R. Y. (1978). Catecholamine innervation of the basal forebrain. II. Amygdala, suprarhinal cortex and entorhinal cortex, *J. Comp. Neurol. 180,* 509–532.

Faul, H. (1978). A history of geologic time, *Am. Sci. 66,* 159–175.

Faull, R. L. M., and Mehler, W. R. (1976). Studies of the fiber connections of the substantia nigra in the rat using the method of retrograde transport of horseradish peroxidase, *Soc. Neurosci. Abstr. 1,* 62.

Feindel, W., and Penfield, W. (1954). Localization of discharge in temporal lobe automatism, *AMA Arch. Neurol. Psychiatry 72,* 605–630.

Ferguson, S. M. (1962). Temporal lobe epilepsy: Psychiatric and behavioral aspects, *Bull. N.Y. Acad. Med. 38,* 668–679.

Ferguson, S. M., Rayport, M., Gardner, R., Kass, W., Weiner, H., and Reiser, M. (1969). Similarities in mental content of psychotic states, spontaneous seizures, dreams, and responses to electrical brain stimulation in patients with temporal lobe epilepsy, *Psychosom. Med. 31,* 479–498.

Fernandez de Molina, A., and Hunsperger, R. W. (1959). Central representation of affective reactions in forebrain and brain stem: Electrical stimulation of amygdala, stria terminalis, and adjacent structures, *J. Physiol. (London) 145,* 251–265.

Ferrari, W., Floris, E., and Paulesu, F. (1955). Sull'effetto eosinofilopenizzante dell' ACTH iniettato nella cisterna magna, *Boll. Soc. Ital. Biol. Sper. 31,* 859–862.

Ferrier, D. (1876/1966). *The Functions of the Brain,* Smith Elder, London, 1876; reprinted in 1966 by Dawsons of Pall Mall, London.

Field, J., Magoun, H. W., and Hall, V. E. (1959–60). *Handbook of Physiology,* Volumes I–III, Waverly Press, Baltimore.

Finch, G. (1941). Chimpanzee handedness, *Science 94,* 117–118.

Fink, R. P., and Heimer, L. (1967). Two methods for selective silver impregnation of degenerating axons and their synaptic endings in the central nervous system, *Brain Res. 4,* 369–374.

Fitch, H. S. (1970). Viviparity *versus* oviparity, in: *Reproductive Cycles in Lizards and Snakes* (F. B. Cross, P. S. Humphrey, R. M. Mengel, and E. H. Taylor, eds.), University of Kansas Printing Service, Lawrence, pp. 214–220.

Flanigan, S., Gabrieli, E. R., and MacLean, P. D. (1957). Cerebral changes revealed by radioautography with S^{35}-labeled l-methionine, *AMA Arch. Neurol. Psychiatry 77,* 588–594.

Flechsig, P. (1921). Die myelogenetische Gliederung der Leitungsbahnen des Linsenkernes beim Menschen, *Ber. Verh. Saechs. Akad. Wiss. Leipzig Math. Phys. Kl. 73,* 295–302.

Fleischer, S. F. (1973). Deficits in maternal behavior of rats with lesions of the septal area, Unpublished doctoral dissertation, Columbia University.

Fleischhauer, K. (1964). Fluorescenzmikroskopische Untersuchungen über den Stofftransport zwischen Ventrikelliquor und Gehirn, *Z. Zellforsch. Mikrosk. Anat. 62,* 639–654.

Flor-Henry, P. (1969). Psychosis and temporal lobe epilepsy: A controlled investigation, *Epilepsia, 10,* 363–395.

Flourens, P. (1824). *Recherches expérimentales sur les propriétés et les fonctions du système nerveux dans les animaux vertébrés,* Crevot, Paris.

Flynn, J. P., and Wasman, M. (1960). Learning and cortically evoked movement during propagated hippocampal afterdischarges, *Science 131,* 1607–1608.

Flynn, J. P., MacLean, P. D., and Kim, C. (1961). Effects of hippocampal afterdischarges on conditioned responses, in: *Electrical Stimulation of the Brain* (D. E. Sheer, ed.), University of Texas Press, Austin, pp. 382–386.

Foix, C., and Nicolesco, J. (1925). *Les noyaux gris centraux et la region mesencéphalo-sous-optique,* Masson, Paris.

Foley, J. O., and DuBois, F. S. (1937). Quantitative studies of the vagus nerve in the cat. I. The ratio of sensory to motor fibers, *J. Comp. Neurol. 67,* 49–67.

Foltz, E. L., and White, L. E., Jr. (1957). Experimental cingulumotomy and modification of morphine withdrawal, *J. Neurosurg. 14,* 655–673.

Foltz, E. L., and White, L. E., Jr. (1962). Pain "relief" by frontal cingulumotomy, *J. Neurosurg. 19,* 89–100.

Forel, A. (1872). Beiträge zur Kenntniss des Thalamus opticus und der ihn umgebenden Gebilde bei den Säugethieren, *Sitzungsber. Akad. Wiss. Wien 66,* 25–58.

Forel, A. (1937). *Out of My Life* (Trans. by B. Miall), Norton, New York.

Forman, D., and Ward, J. W. (1957). Responses to electrical stimulation of caudate nucleus in cats in chronic experiments, *J. Neurophysiol. 20,* 230–243.

Forster, F. M., and Liske, E. (1963). Role of environmental clues in temporal lobe epilepsy, *Neurology 13,* 301–305.

Forster-Nietzsche, E. (1954). Introduction: How Zarathustra came into being, in: *The Philosophy of Nietzsche,* Modern Library, New York, p. xix–xxxiii.

Fossey, D. (1971). More years with mountain gorillas, *National Geographic 140*(4), 574–584.

Fossey, D. (1976). The behavior of the mountain gorilla. Doctoral dissertation, University of Cambridge.

Fox, C. A. (1943). The stria terminalis, longitudinal association bundle and precommissural fornix fibers in the cat, *J. Comp. Neurol. 79,* 277–295.

Fox, C. A., and Rafols, J. A. (1975). The radial fibers in the globus pallidus, *J. Comp. Neurol. 159,* 177–200.

Fox, C. A., McKinley, W. A., and Magoun, H. W. (1944). An oscillographic study of olfactory system of cats, *J. Neurophysiol. 7,* 1–16.

Fox, C. A., Hillman, D. E., Siegesmund, K. A., and Sether, L. A. (1966). The primate globus pallidus and its feline and avian homologues: A Golgi and electron microscope study, in: *Evolution of the Forebrain, Phylogenesis and Ontogenesis of the Forebrain* (R. Hassler and H. Stephan, eds.), Thieme Verlag, Stuttgart, pp. 237–248.

Fox, C. A., Andrade, A. N., Hillman, D. E., and Schwyn, R. C. (1971). The spiny neurons in the primate striatum: A Golgi and electron microscopic study, *J. Hirnforsch. 13,* 181–201.

Fox, C. A., Andrade, A. N., Luqui, I. J., and Rafols, J. A. (1974). The primate globus pallidus: A Golgi and electron microscopic study, *J. Hirnforsch. 15,* 75–93.

Fox, C. A., Rafols, J. A., and Cowan, W. M. (1975). Computer measurements of axis cylinder diameters of radial fibers and "comb" bundle fibers, *J. Comp. Neurol. 159,* 201–224.

Frakes, L. A. (1979). *Climate Throughout Geologic Time,* Elsevier, Amsterdam.

Frantzen, E., Lennox-Buchthal, M., and Nygaard, A. (1968). Longitudinal EEG and clinical study of children with febrile convulsions, *Electroencephalogr. Clin. Neurophysiol. 24,* 197.

Franzen, E. A., and Myers, R. E. (1973a). Age effects of social behavior deficits following prefrontal lesions in monkeys, *Brain Res. 54,* 277–286.

Franzen, E. A., and Myers, R. E. (1973b). Neural control of social behavior: Prefrontal and anterior temporal cortex, *Neuropsychologia 11,* 141–157.

Freeman, W., and Watts, J. W. (1936). Prefrontal lobotomy in agitated depression. Report of a case, *Med. Ann. D.C. 5,* 326–329.

Freeman, W., and Watts, J. W. (1944). Physiological psychology, *Annu. Rev. Physiol. 6,* 517–542.

Freeman, W., and Watts, J. W. (1947). Psychosurgery during 1936–1946, *Arch. Neurol. Psychiatry 58,* 417.

Freeman, W., and Watts, J. W. (1950). *Psychosurgery, in the Treatment of Mental Disorders and Intractable Pain,* 2nd Edition, Thomas, Springfield, Ill.

Freemon, F. R., and Nevis, A. H. (1969). Temporal lobe sexual seizures, *Neurology 19,* 87–90.

Freud, S. (1900/1953). *The Interpretation of Dreams,* Hogarth Press, London.

Freud, S. (1938). *A General Introduction to Psychoanalysis,* Garden City Publishing, New York.

Freud, S. (1948). *Three Contributions to the Theory of Sex* (Trans. by A. A. Brill), Nervous and Mental Disease Monogr., New York.

Friedman, H. M., and Allen, N. (1969). Chronic effects of complete limbic lobe destruction in man, *Neurology 19,* 679–690.

Friedrichsen, C., and Melchior, J. C. (1954). Febrile convulsions in children, their frequency and prognosis, *Acta Paediatr. Scand. 43,* 307–317.

Frisch, K. von (1964). *Biology,* Bayerischer Schulbuch, Munich.

Fritsch, G., and Hitzig, E. (1870). Über die elektrische Erregbarkeit des Grosshirns, *Arch. Anat. Physiol. Wiss. Med. 37,* 300–332.

Frost, L. L., Baldwin, M., and Woods, C. D. (1958). Investigation of the primate amygdala: Movements of the face and jaws, *Neurology 8,* 543–546.

Fuller, J. L., Rosvold, H. E., and Pribram, K. H. (1957). The effect on affective and cognitive behavior in the dog of lesions of the pyriform–amygdala–hippocampal complex, *J. Comp. Physiol. Psychol. 50,* 89–96.

Fulton, J. F. (1949a). *Physiology of the Nervous System,* 3rd Edition, Oxford University Press, London.

Fulton, J. F. (1949b). *Functional localization in the frontal lobes and cerebellum,* Oxford University Press (Clarendon), London.

Fulton, J. F. (1951). *Frontal Lobotomy and Affective Behavior. A Neurophysiological Analysis,* Norton, New York.

Fulton, J. F. (1953). Discussion of a paper by H. Gastaut, *Epilepsia 2,* 77.

Fulton, J. F., and Jacobsen, C. F. (1935). The functions of the frontal lobes, a comparative study in monkeys, chimpanzees, and man, *Abstr. 2nd Int. Neurol. Congr. (London)* pp. 70–71.

Gaddum, J. H. (1953). Antagonism between lysergic acid diethylamide and 5-hydroxytryptamine, *J. Physiol. (London) 121,* 15 p.

Gaddum, J. H. (1954). Drugs antagonistic to 5-hydroxytryptamine, in: *Ciba Foundation Symposium on Hypertension,* Little, Brown, Boston, pp. 75–77.

Gajdusek, D. C. (1970). Physiological and psychological characteristics of Stone Age man, in: *Symposium on Biological Bases of Human Behavior, Eng. Sci. 33,* 26–33, 56–62.

Galaty, D. H. (1974). The philosophical basis of mid-nineteenth century German reductionism, *J. Hist. Med. 29,* 295–316.

Galdikas, B. M. F. (1978). Orangutan adaptation at Tanjung Puting Reserve, Central Borneo, Doctoral dissertation, University of California, Los Angeles.

Galvani, L. (1791/1953). *Commentary on the Effect of Electricity,* Waverly Press, Baltimore, p. 20.

Gamper, E. (1928). Zur Frage der Polioencephalitis haemorrhagica der chronischen Alkoholiker: Anatomische Befunde beim alkoholischen Korsakow und ihre Beziehungen zum klinischen Bild, *Dtsch. Z. Nervenheilkd. 102,* 122–129.

Gannicliffe, A., Saldanha, J. A., Itzhaki, R. F., and Sutton, R. N. P. (1985). Herpes simplex viral DNA in temporal lobe epilepsy, *Lancet i,* 214–215.

Ganser, S. (1882). Vergleichend-anatomische Studien über das Gehirn des Maulwurfs, *Morphol. Jahrb. 7,* 591–725.

Gans, A. (1924). Beitrag zur Kenntnis des Aufbas des nucleus dentatus auf zwei Teilen, namentlich auf Grund von Untersuchungen mit der Eisenreaktion, *Z. Gesamte Neurol. Psychiatr. 93,* 750–755.

Gardner, E. (1959). *Fundamentals of Neurology,* Saunders, Philadelphia.

Gascon, G. G., and Gilles, F. (1973). Limbic dementia, *J. Neurol. Neurosurg. Psychiatry 36,* 421–430.

Gaskell, W. H. (1886). On the structure, distribution, and function of the nerves which innervate the visceral and vascular systems, *J. Physiol. (London) 7,* 1–80.

Gastaut, H. (1954). Interpretation of the symptoms of "psychomotor" epilepsy in relation to physiologic data on rhinencephalic function, *Epilepsia 3,* 84–88.

Gastaut, H. (1970). Clinical and electroencephalographic classification of epileptic seizures, *Epilepsia 11,* 102–113.

Gastaut, H., Vigouroux, R., Corriol, J., and Badier, M. (1951). Effets de la stimulation électrique (par électrodes à demeure) du complexe amygdalien chez le chat non narcose, *J. Physiol. (Paris) 43,* 740–746.

Gastaut, H., Naquet, R., Vigouroux, R., and Corriol, J. (1952). Provocation de comportements émotionnels divers par stimulation rhinencéphalique chez le chat avec électrodes à demeure, *Rev. Neurol. 86,* 319–327.

Geier, S., Bancaud, J., Talairach, J., Bonis, A., Szikla, G., and Enjelvin, M. (1977). The seizures of frontal lobe epilepsy, *Neurology 27,* 951–958.

Gerard, R. W., Marshall, W. H., and Saul, L. J. (1936). Electrical activity of the cat's brain, *AMA Arch. Neurol. Psychiatry 36,* 675–738.

Gerebetzoff, M. A. (1942). Note anatomo-expérimentale sur le fornix, la corne d'Ammon et leur relations avec diverses structures encéphaliques, notamment l'épiphyse, *Acta Neurol. Psychiatr. Belg. 42,* 199–206.

Gergen, J. A., and MacLean, P. D. (1962). *A stereotaxic atlas of the squirrel monkey's brain (Saimiri sciureus),* U.S. Government Printing Office, Washington, D.C., PHS Publ. No. 933.

Gergen, J. A., and MacLean, P. D. (1964). The limbic system: Photic activation of limbic cortical areas in the squirrel monkey, *Ann. N.Y. Acad. Sci. 117,* 69–87.

von Gerlach, J. (1858). *Microskopische Studien aus dem Gebiete der menschlichen Morphologie,* Enke, Erlangen.

Gessa, G. L., Vargiu, L., and Ferrari, W. (1966). Stretchings and yawnings induced by adrenocorticotrophic hormone, *Nature 211,* 426.

Gessa, G. L., Pisano, M., Vargiu, L. Crabai, F., and Ferrari, W. (1967). Stretching and yawning movements after intracerebral injection of ACTH, *Rev. Can. Biol. 26,* 229–236.

Gibbs, E. L., Gibbs, F. A., and Fuster, B. (1948). Psychomotor epilepsy, *Arch. Neurol. Psychiatry 60,* 331–339.

Gibbs, F. A., and Gibbs, E. L. (1936). The convulsion threshold of various parts of the cat's brain, *Arch. Neurol. Psychiatry 35,* 109–116.
Gibbs, F. A., Gibbs, E. L., and Lennox, W. G. (1938). Cerebral dysrhythmias of epilepsy, *Arch. Neurol. Psychiatry 39,* 298–314.
Gill, G. N. (1985). The hypothalamic-pituitary control system, in: *Best and Taylor's Physiological Basis of Medical Practice* (J. B. West, ed.), Williams & Wilkins, Baltimore, pp. 856–871.
Gilles de la Tourette, G. (1885). Étude sur une affection nerveuse caractérisée par l'incoordination motrice accompagnée d'écholalie et de coprolalie (jumping, latah, myriachit), *Arch. Neurol. (Paris) 9,* 158–200.
Gingerich, P. D., Wells, N. A., Russell, D. E., and Shah, S. M. (1983). Origin of whales in epicontinental remnant seas: New evidence from the early Eocene of Pakistan, *Science 220,* 403–406.
Girgis, M., and Kiloh, L. G. (eds.) (1980). Limbic epilepsy and the dyscontrol syndrome, Elsevier/North-Holland, Amsterdam.
Gispen, W. H., and Isaacson, R. L. (1981). ACTH-induced excessive grooming in the rat, *Pharmacol. Ther. 12,* 209–246.
Glaser, G. H. (1957). Visceral manifestations of epilepsy, *Yale J. Biol. Med. 30,* 176–186.
Glaser, G. H. (1967). Limbic epilepsy in childhood, *J. Nerv. Ment. Dis. 144,* 391–397.
Glaser, G. H., and Pincus, J. H. (1969). Limbic encephalitis, *J. Nerv. Ment. Dis. 149,* 59–67.
Glees, P. (1944). The anatomical basis of cortico-striate connexions, *J. Anat. 78,* 47–51.
Glees, P. (1946). Terminal degeneration within the central nervous system as studied by a new silver method, *J. Neuropathol. Exp. Neurol. 5,* 54–59.
Glees, P., and Griffith, H. B. (1952). Bilateral destruction of the hippocampus (Cornu Ammonis) in a case of dementia, *Monatsschr. Psychiatr. Neurol. 123,* 193–204.
Glickenstein, M., May, J. G., and Mercier, B. E. (1985). Corticopontine projection in the macaque: The distribution of labelled cortical cells after large injections of horseradish peroxidase in the pontine nuclei, *J. Comp. Neurol. 235,* 343–359.
Gloor, P. (1972). Temporal lobe epilepsy: Its possible contribution to the understanding of the functional significance of the amygdala and of its interaction with neocortical–temporal mechanisms, in: *The Neurobiology of the Amygdala* (B. E. Eleftheriou, ed.), Plenum Press, New York, pp. 423–457.
Gloor, P., Olivier, A., and Ives, J. (1980). Prolonged seizure monitoring with stereotaxically implanted depth electrodes in patients with bilateral interictal temporal epileptic foci: How bilateral is bitemporal epilepsy?, in: *Advances in Epileptology: The Xth Epilepsy International Symposium* (J. A. Wada and J. K. Penry, eds.), Raven Press, New York, pp. 83–88.
Gloor, P., Olivier, A., Quesney, L. F., Andermann, F., and Horowitz, S. (1982). The role of the limbic system in experimental phenomena of temporal lobe epilepsy, *Ann. Neurol. 12,* 129–144.
Goin, C. J., and Goin, O. B. (1962). *Introduction to Herpetology,* Freeman, San Francisco.
Golby, F. (1937). An experimental investigation of the cerebral hemispheres of *Lacerta viridis, J. Anat. 71,* 332–355.
Goldby, F., and Gamble, H. J. (1957). The reptilian cerebral hemispheres, *Biol. Rev. 32,* 383–420.
Golden, G. S. (1972). Embryologic demonstration of a nigro-striatal projection in the mouse, *Brain Res. 44,* 278–282.
Goldman, P. S., and Nauta, W. J. H. (1977). An intricately patterned prefronto-caudate projection in the rhesus monkey, *J. Comp. Neurol. 171,* 369–386.
Goldman-Rakic, P. S., and Porrino, L. J. (1985). The primate mediodorsal (MD) nucleus and its projection to the frontal lobe, *J. Comp. Neurol. 242,* 535–560.
Golgi, C. (1873). Sulla struttura della sostanza grigia dell cervello, *Gazz. Med. Ital. Lombarda. 33,* 244–246.
Golub, L. M., Guhleman, H. V., and Merlin, J. K. (1951). Seizure patterns in psychomotor epilepsy, *Dis. Nerv. Syst. 12,* 1–4.
Gomez, J. A., Thompson, R. L., and Mettler, F. A. (1958). Effect of striatal damage on conditioned and unlearned behavior, *Trans. Am. Neurol. Assoc. 83,* 88–91.
Gonatas, N. K., Harper, C., Mitzutani, T., and Gonatas, J. O. (1979). Superior sensitivity of conjugates of horseradish peroxidase with wheat germ agglutinin for studies of retrograde axonal transport, *J. Histochem. Cytochem. 27,* 728–734.
Gonzalez, D., and Elvidge, A. R. (1962). On the occurrence of epilepsy caused by astrocytoma of the cerebral hemispheres, *J. Neurosurg. 19,* 470–482.
Goodall, J. (1981–82). Newsletter. The Jane Goodall Institute for Wildlife Research and Education, Winter 1981–1982.
Goodall, J., Bandora, A., Bergmann, E., Busse, C., Matama, H., Mpongo, E., Pierce, A., and Riss, D. (1979). Intercommunity interactions in the chimpanzee population of the Gombe National Park, in: *The*

Great Apes (D. A. Hamburg and E. R. McCown, eds.), Benjamin/Cummings, Menlo Park, Calif., pp. 13–54.

Gottlieb, D. I., and Cowan, W. M. (1973). Autoradiographic studies of the commissural and ipsilateral association connections of the hippocampus and dentate gyrus of the rat. I. The commissural connections, *J. Comp. Neurol. 149,* 393–422.

Gould, S. J. (1988). A novel notion of Neanderthal, *Nat. Hist. 97,* 16–21.

Gowers, W. R. (1881). *Epilepsy and Other Chronic Convulsion Diseases: Their Causes, Symptoms, and Treatment,* William Wood, New York.

Gowers, W. R. (1888). *A Manual of Diseases of the Nervous System,* Volume II, Churchill, London.

Gowers, W. R. (1901). *Epilepsy and Other Chronic Convulsive Diseases: Their Causes, Symptoms, and Treatment,* 2nd Edition, Churchill, London.

Graham, R. C., and Karnovsky, M. J. (1966). The early stages of absorption of injected horseradish peroxidase in the proximal tubule of mouse kidney: Ultrastructural correlates by a new technique, *J. Histochem. Cytochem. 14,* 291–299.

Grange, K. M. (1961). Pinel and eighteenth-century psychiatry, *Bull. Hist. Med. 35,* 442–453.

Grantham, E. C. (1951). Prefrontal lobotomy for the relief of pain with a report of a new operation technique, *J. Neurosurg. 8,* 405.

Grastyan, E. (1959). The hippocampus and higher nervous activity, in: *Second Conference on the Central Nervous System and Behavior,* Transactions, Josiah Macy, Jr. Foundation, New York, pp. 119–205.

Graybiel, A. M., and Hickey, T. L. (1982). Chemospecificity of ontogenetic units in the striatum: Demonstration by combining [^{3}H]thymidine neuronography and histochemical staining, *Proc. Natl. Acad. Sci. USA 79,* 198–202.

Graybiel, A. M., and Ragsdale, C. W., Jr. (1978). Histochemically distinct compartments in the striatum of human, monkey, and cat demonstrated by acetylthiocholinesterase staining, *Proc. Natl. Acad. Sci. USA 75,* 5723–5726.

Graybiel, A. M., and Ragsdale, C. W., Jr. (1980). Clumping of acetylcholinesterase activity in the developing striatum of the human fetus and young infant, *Proc. Natl. Acad. Sci. USA 77,* 1214–1218.

Graybiel, A. M., and Sciascia, T. R. (1975). Origin and distribution of nigrotectal fibers in the cat, *Soc. Neurosci. Abstr. 1,* 271.

Graybiel, A. M., Ragsdale, C. W., and Moon Edley, S. (1979). Compartments in the striatum of the cat observed by retrograde cell labeling, *Exp. Brain Res. 34,* 189–195.

Green, J. D., and Adey, W. (1956). Electrophysiological studies of hippocampal connections and excitability, *Electroencephalogr. Clin. Neurophysiol. 8,* 245–262.

Green, J. D., and Arduini, A. A. (1954). Hippocampal electrical activity in arousal, *J. Neurophysiol. 17,* 533–557.

Green, J. D., Maxwell, D. S., Schindler, W. J., and Stumpf, C. (1960). Rabbit EEG "theta" rhythm: Its anatomical source and relation to activity in single neurons, *J. Neurophysiol. 23,* 403–420.

Green, J. R., and Scheetz, D. G. (1964). Surgery of epileptogenic lesions of the temporal lobe, *Arch. Neurol. 10,* 135–148.

Greenberg, B., and Noble, G. K. (1944). Social behavior of the American chameleon (*Anolis carolinensis* Voigt), *Physiol. Zool. 14,* 392–439.

Greenberg, N. (1973). Behavior studies of the blue spiny lizard, Ph.D. thesis, Rutgers University, University Microfilms, Ann Arbor.

Greenberg, N. (1976). Thermoregulatory aspects of behavior in the blue spiny lizard *Sceloporus cyanogenys* (Sauria, Iguanidae), *Behaviour 59,* 1–21.

Greenberg, N. (1977a). An ethogram of the blue spiny lizard, *Sceloporus cyanogenys* (Reptilia, Lacertilia, Iguanidae), *J. Herpetol. 11,* 177–195.

Greenberg, N. B. (1977b). A neuroethological investigation of display behavior in the lizard *Anolis carolinensis* (Lacertilia, Iguanidae), *Am. Zool. 17,* 191–201.

Greenberg, N. (1978). Ethological considerations in the experimental study of lizard behavior, in: *The Behavior and Neurology of Lizards* (N. Greenberg and P. D. MacLean, eds.), U.S. Government Printing Office, Washington, D. C., DHEW Publication No. (ADM) 77-491, pp. 203–224.

Greenberg, N., Ferguson, J. L., and MacLean, P. D. (1976). A neuroethological study of display behavior in lizards, *Neuroscience 2,* 689.

Greenberg, N. B., MacLean, P. D., and Ferguson, J. L. (1979). Role of the paleostriatum in species-typical behavior of the lizard (*Anolis carolinensis*), *Brain Res. 172,* 229–241.

Greenberg, N., Scott, M., and Crews, D. (1984). Role of the amygdala in the reproductive and aggressive behavior of the lizard, *Anolis carolinensis, Physiol. Behav. 32,* 147–151.

Gregory, W. K. (1929/1967). *Our Face from Fish to Man,* Hafner, New York.

Gregory, W. K., and Simpson, G. G. (1926). Cretaceous mammal skulls from Mongolia, *Am. Mus. Novit 225,* 1–20.

Grinker, R. R. (1948). Discussion, in: *The Frontal Lobes, Res. Publ. Assoc. Res. Nerv. Ment. Dis. 27,* 502–503.

Groos, K. (1896/1915). *The Play of Animals,* Appleton-Century-Crofts, New York. (First edition in German.)

Grünthal, E. (1947). Über das klinische Bild nach umschriebenem beiderseitigem Ausfall der Ammonshornrinde, *Monatsschr. Psychiatr. Neurol. 113,* 1–16.

Gudden, B. von (1870). Experimental Untersuchungen über das peripherische und centrale Nervensystem, *Arch. Psychiatr. Nervenkr. 2,* 693–723.

Gudden, B. von (1881). Beitrag zur Kenntnis des Corpus mamillare und der sogenannten Schenkel des Fornix, *Arch. Psychiatr. Nervenkr. 11,* 428–452.

Gudden, B. von (1884). Über das Corpus mamillare und die sogenannten Schenkel des Fornix, *Versamm. Dtsch. Naturforsch. 57,* 126–127.

Gudden, H. (1896). Klinische und anatomische Beiträge zur Kenntnis der multiplen Alkoholneuritis nebst Bemerkungen über die Regenerationsvorgänge im peripheren Nervensystem, *Arch. Psychol. 28,* 643–741.

Guillemin, R., Ling, N., and Burgus, R. (1976). Endorphines, peptides d'origine hypothalaminque et neurohypophysaire à activité morphinomimétique, *C. R. Acad. Sci. Ser. D 282,* 783–785.

Guillery, R. W. (1955). A quantitative study of the mamillary bodies and their connexions, *J. Anat. 89,* 19–32.

Guillery, R. W. (1956). Degeneration in the post-commissural fornix and the mamillary peduncle of the rat, *J. Anat. 90,* 350–370.

Guillery, R. W. (1957). Degeneration in the hypothalamic connexions of the albino rat, *J. Anat. 91,* 91–115.

Guillery, R. W. (1959). Afferent fibres to the dorsomedial thalamic nucleus in the cat, *J. Anat. 93,* 403–419.

Guillette, L. J., Jr., and Hotton, N. (1986). The evolution of mammalian reproductive characteristics in therapsid reptiles, in; *The Ecology and Biology of Mammal-like Reptiles* (N. Hotton, P. D. MacLean, J. J. Roth, and E. C. Roth, eds.), Smithsonian Institution Press, Washington, D.C., pp. 239–250.

Gundy, G. C., Ralph, C. L., and Wurst, G. Z. (1975). Parietal eyes in lizards: Zoogeographical correlates, *Science 190,* 671–673.

Gunther, A. C. (1867). Contribution to the anatomy of *Hatteria* (*Rhynchocephalus* Owen), *Philos. Trans. R. Soc. London 157,* 595–629.

Gurdjian, E. S. (1927). The diencephalon of the albino rat, *J. Comp. Neurol. 43,* 1–114.

Guthrie, R. D., and Petocz, R. G. (1970). Weapon automimicry among mammals, *Am. Nat. 104,* 585–588.

Haber, S., and Elde, R. (1981). Correlation between met-enkephalin and substance P immunoreactivity in the primate globus pallidus, *Neuroscience 6,* 1291–1297.

Haber, S. N., Groenewegen, H. J., Grove, E. A., and Nauta, W. J. H. (1985). Efferent connections of the ventral pallidum. Evidence of a dual striato-pallidofugal pathway, *J. Comp. Neurol. 235,* 322–335.

Haeckel, E. (1876). *The History of Creation or the Development of the Earth and Its Inhabitants by the Action of Natural Causes* (Trans. by E. Ray Lankester), two volumes, King, London.

Hagamen, W. D., Zitzmann, E. K., and Reeves, A. G. (1963). Sexual mounting of diverse objects in a group of randomly selected, unoperated male cats, *J. Comp. Physiol. Psychol. 56,* 298–302.

Hallen, O. (1954). Das oral-petit mal: Beschreibung und Zergliederung der als uncinate-fit (Jackson) und psychomotor-fit (Lennox) bezeichneten epileptischen Äquivalente, *Dtsch. Z. Nervenheilkd. 171,* 236–260.

Hallgren, B., and Sourander, P. (1958). The effect of age on the non-haem iron in the human brain, *J. Neurochem. 3,* 41–51.

Hallowitz, R. A., and MacLean, P. D. (1977). Effects of vagal volleys on units of intralaminar and juxtalaminar thalamic nuclei in monkeys, *Brain Res. 130,* 271–286.

Halstead, W. C. (1947). *Brain and Intelligence. A Quantitative Study on the Frontal Lobes,* University of Chicago Press, Chicago.

Halstead, W. C. (1948). Specialization of behavioral functions and the frontal lobes, in: *The Frontal Lobes, Res. Publ. Assoc. Res. Nerv. Ment. Dis. 27,* 59–66.

Hamilton, A. M. (1882). On cortical sensory discharging lesions (sensory epilepsy), *N.Y. Med. J. 35,* 575–584.

Hanberry, J., and Jasper, H. (1953). Independence of diffuse thalamocortical projection system shown by specific nuclear destructions, *J. Neurophysiol. 16,* 252–271.

Hardison, O. B., Jr. (1986). A tree, a streamlined fish, and a self-squared dragon: Science as a form of culture, Ga. Rev. 40, 369–403.

Harlow, J. M. (1848). Passage of an iron rod through the head, *Boston Med. Surg. J. 39,* 389–393.

Harlow, J. M. (1868). Recovery from the passage of an iron bar through the head, *Mass. Med. Soc. Publ. 2,* 327–346.

Harman, P. J., and Berry, C. M. (1956). Neuroanatomical distribution of action potentials evoked by photic stimuli in cat fore- and midbrain, *J. Comp. Neurol. 105,* 395–416.

Harris, V. A. (1964). *The Life of the Rainbow Lizard,* Hutchinson, London.

Hassler, R. (1961). Motorische und sensible Effekte umschriebener Reizungen und Ausschaltungen im menschlichen Zwischenhirn, *Dtsch. Z. Nervenheilkd. 183,* 148–171.

Haymaker, W. (1956). *Bing's Local Diagnosis in Neurological Diseases,* Mosby, St. Louis.

Haymaker, W., and Schiller, F. (1970). *The Founders of Neurology,* Thomas, Springfield, Ill.

Haymaker, W., Anderson, E., and Nauta, W. J. H. (1969). *The Hypothalamus,* Thomas, Springfield, Ill.

Hayward, J. N. (1972). The amygdaloid nuclear complex and mechanisms of release of vasopressin from the neurohypophysis, in: *The Neurobiology of the Amygdala* (B. E. Eleftheriou, ed.), Plenum Press, New York, pp. 685–739.

Heath, J. E. (1962). Temperature-independent morning emergence in lizards of the genus Phrynosoma, *Science 138,* 891–892.

Heath, R. G. (1962). Common characteristics of epilepsy and schizophrenia: Clinical observation and depth electrode studies, *Am. J. Psychiatry 118,* 1013–1026.

Heath, R. G. (1963). Electrical self-stimulation of the brain in man, *Am. J. Psychiatry 120,* 571–577.

Heath, R. G., and Guerrero-Figueroa, R. (1965). Psychotic behavior with evoked septal dysrhythmia: Effects of intracerebral acetylcholine and gamma aminobutyric acid, *Am. J. Psychiatry 121,* 1080–1086.

Heath, R. G., and Monroe, R. R. (1954). Psychiatric observations, in: *Studies in Schizophrenia. A multidisciplinary approach to mind–brain relationships,* Harvard University Press, Cambridge, Mass., pp. 345–382.

Hebb, D. O. (1929/1963). Introduction, in: *Brain Mechanisms and Intelligence. A Quantitative Study of Injuries to the Brain,* Dover, New York, pp. v–xiii.

Hebb, D. O. (1949). *The Organization of Behavior,* Wiley, New York.

Hebb, D. O., and Penfield, W. (1940). Human behavior after extensive bilateral removal from the frontal lobes, *Arch. Neurol. Psychiatry 44,* 421–438.

Hediger, H. (1950). *Wild Animals in Captivity,* Butterworth, London.

Hediger, H. (1955). *The Psychology and Behaviour of Animals in Zoos and Circuses,* Butterworth, London.

Heimer, L. (1968). Synaptic distribution of centripetal and centrifugal nerve fibres in the olfactory system of the rat. An experimental anatomical study, *J. Anat. 103,* 413–432.

Heimer, L., and Wilson, R. D. (1975). The subcortical projections of the allocortex: Similarities in the neural associations of the hippocampus, the piriform cortex, and the neocortex, in: *Proceedings of the Golgi Centennial Symposium* (M. Santini, ed.), Raven Press, New York, pp. 177–193.

Henriksen, S. J., Bloom, F. E., McCoy, F., Ling, N., and Guillemin, R. (1978). β-Endorphin induces nonconvulsive limbic seizures, *Proc. Natl. Acad. Sci. USA 75,* 5221–5225.

Henry, C. E. (1963). Positive spike discharges in the EEG and behavior abnormality, in: *EEG and Behavior* (G. H. Glaser, ed.), Basic Books, New York, pp. 315–344.

Henschen, S. E. (1920/1922). *Klinische und pathologische Beitrage zur Pathologie des Gehirns,* Volumes V–VII, Nordiske Bokhandeln, Stockholm.

Herberg, L. J. (1963a). A hypothalamic mechanism causing seminal ejaculation, *Nature 198,* 219–220.

Herberg, L. J. (1963b). Seminal ejaculation following positively reinforcing electrical stimulation of the rat hypothalamus, *J. Comp. Physiol. Psychol. 56,* 679–685.

Herkenham, M. (1978). The connections of the nucleus reuniens thalami: Evidence for a direct thalamo-hippocampal pathway in the rat, *J. Comp. Neurol. 177,* 589–610.

Herkenham, M., and Nauta, W. J. H. (1977). Afferent connections of the habenular nuclei in the rat: A horseradish peroxidase study, with a note on the fiber-of-passage problem, *J. Comp. Neurol. 173,* 123–146.

Herkenham, M., and Pert, C. B. (1980). In vitro autoradiography of opiate receptors in rat brain suggests loci of "opiatergic" pathways, *Proc. Natl. Acad. Sci. USA 77,* 5532–5536.

Herman, L. M. (ed.) (1980). *Cetacean Behavior: Mechanisms and Functions,* Wiley, New York.

Hermann, L. (1868). *Untersuchungen zur Physiologie der Muskeln und Nerven,* August Hirschwald, Berlin.

Herrick, C. J. (1910). The morphology of the forebrain in Amphibia and Reptilia, *J. Comp. Neurol. 20,* 413–547.

Herrick, C. J. (1914). The medulla oblongata of larval amblystoma, *J. Comp. Neurol. 28,* 343–427.

Herrick, C. J. (1921). A sketch of the origin of the cerebral hemispheres, *J. Comp. Neurol. 32,* 429–454.

Herrick, C. J. (1926). *Brains of Rats and Men: A Survey of the Origin and Biological Significance of the Cerebral Cortex,* University of Chicago Press, Chicago.

Herrick, C. J. (1933). The functions of the olfactory parts of the cerebral cortex, *Proc. Natl. Acad. Sci. USA 19,* 7–14.

Herrick, C. J. (1948). *The Brain of the Tiger Salamander,* University of Chicago Press, Chicago.

Hershkovitz, P. (1984). Taxonomy of squirrel monkeys genus *Saimiri* (Cebidae, Platyrrhini): A preliminary report with description of a hitherto unnamed form, *Am. J. Primatol. 7,* 155–210.

Hess, W. R. (1932). *Die Methodik der lokalisierten Reizung und Ausschaltung subkortikaler Hirnabschnitte,* Thieme, Stuttgart.

Hess, W. R. (1954). *Diencephalon: Autonomic and Extrapyramidal Functions,* Grune & Stratton, New York.

Hess, W. R., and Brugger, M. (1943). Das subkortikale Zentrum der affektiven Abwehrreaktion, *Helv. Physiol. Pharmacol. Acta 1,* 33–52.

Hess, W. R., and Meyer, A. E. (1956). Triebhafte Fellreinigung der Katze als Symptom diencephaler Reizung, *Helv. Physiol. Pharmacol. Acta 14,* 397–410.

Hess, W. R., Brugger, M., and Bucher, V. (1945–46). Zur Physiologie von Hypothalamus, Area praeoptica und Septum, sowie angrenzender Balken- und Stirnhirnbereiche, *Monatsschr. Psychiatr. Neurol. 3,* 17–59.

Hewitt, W. (1958). The development of the human caudate and amygdaloid nuclei, *J. Anat. 92,* 377–382.

Hierons, R., and Saunders, M. (1966). Impotence in patients with temporal-lobe lesions, *Lancet 2,* 761–763.

Hill, D. (1949). The electroencephalographic concept of psychomotor epilepsy: A summary, *4th Int. Neurol. Congr. Rep.* pp. 27–33.

Hill, J. M. (1980). Sex difference in brain iron, *Soc. Neurosci. Abstr. 6,* 131.

Hill, J. M. (1981). Changes in brain iron during the estrous cycle, *Soc. Neurosci. Abstr. 7,* 219.

Hill, J. M. (1982). Brain iron: Sex difference and changes during the estrous cycle and pregnancy, in: *The Biochemistry and Physiology of Iron* (P. Saltman and J. Hegenauer, eds.), Elsevier/North-Holland, Amsterdam, pp. 599–601.

Hill, R. G., Mitchell, J. F., and Pepper, C. M. (1977). The excitation and depression of hippocampal neurones by iontophoretically applied enkephalins, *J. Physiol. (London) 272,* 50–51.

Hilton, S. M., and Zbrozyna, A. W. (1963). Amygdaloid region for defence reaction and its efferent pathway to the brain stem, *J. Physiol. (London) 165,* 160–173.

Hinde, R. A. (1972). *Non-Verbal Communication,* Cambridge University Press, London.

Hines, M. (1923). The development of the telencephalon in *Sphenodon punctatum, J. Comp. Neurol. 35,* 483–537.

Hingston, R. W. G. (1933). *The Meaning of Animal Colour and Adornment; Being a New Explanation of the Colours, Adornments and Courtships of Animals, their Songs, Moults, Extravagant Weapons, the Differences between their Sexes, the Manner of Formation of their Geographical Varieties and other Allied Problems,* Arnold, London.

Hirsch, K. R. (1979). The oldest vertebrate egg? *J. Paleon. 53,* 1068–1084.

His, W. von (1904). *Die Entwickelung des menschlichen Gehirns,* Hirzel, Leipzig.

Hjorth-Simonsen, A. (1972). Projection of the lateral part of the entorhinal area to the hippocampus and fascia dentata, *J. Comp. Neurol. 146,* 219–232.

Hjorth-Simonsen, A., and Jeune, B. (1972). Origin and termination of the hippocampal perforant path in the rat studied by silver impregnation, *J. Comp. Neurol. 144,* 215–232.

Hockman, C. H., Talesnik, J., and Livingston, K. E. (1969). Central nervous system modulation of baroceptor reflexes, *Am. J. Physiol. 217,* 1681–1689.

Hodes, R. S., Peacock, S. M., and Heath, R. G. (1951). Influence of the forebrain on somato-motor activity. I. Inhibition, *J. Comp. Neurol. 94,* 381–408.

Hoffman, H. H., and Kuntz, A. (1957). Vagus nerve components, *Anat. Rec. 127,* 551–567.

Hofstadter, D. R. (1979). *Gödel, Escher, Bach: An Eternal Golden Braid,* Basic Books, New York.

Hogben, L. (1937/1946). *Mathematics for the Million,* Norton, New York.

Holloway, R. L. (1981). Volumetric and asymmetry determination on recent hominid endocasts: Spy I and II, Djebel Ihroud I, and the Sale *Homo erectus* specimens, with some notes on Neandertal brain size, *Am. J. Phys. Anthropol. 55,* 385–393.

Holmes, G. (1907). A form of familial degeneration of the cerebellum, *Brain 30,* 466–489.

Holmes, G. (1922a). The Croonian Lectures on the clinical symptoms of cerebellar disease and their interpretation. Lecture I, *Lancet 2,* 1177–1182.

Holmes, G. (1922b). The Croonian Lectures on the clinical symptoms of cerebellar disease and their interpretation. Lecture III, *Lancet 1,* 59–65.

Hong, J. S., Yang, H.-Y. T., Fratta, W., and Costa, E. (1977). Determination of methionine enkephalin in discrete regions of rat brain, *Brain Res. 134,* 383–386.

Hoogland, P. V. (1977). Efferent connections of the striatum in *Tupinambis nigropunctatus, J. Morphol. 152,* 229–246.

Hooshmand, H., and Brawley, B. W. (1969). Temporal lobe seizures and exhibitionism, *Neurology 19,* 1119–1124.

Hopkins, D. A., and Lawrence, D. G. (1975). On the absence of a rubrothalamic projection in the monkey with observations on some ascending mesencephalic projections, *J. Comp. Neurol. 161,* 269–294.

Horel, J. A. (1978). The neuroanatomy of amnesia. A critique of the hippocampal memory hypothesis, *Brain 101,* 403–445.

Horn, A. S., and Snyder, S. H. (1971). Chlorpromazine and dopamine: Conformational similarities that correlate with the antischizophrenic activity of phenothiazine drugs, *Proc. Natl. Acad. Sci. USA 68,* 2325–2328.

Hornykiewicz, O. (1963). Die topische Lokalisation und das Verhalten von Noradrenalin und Dopamin (3-Hydroxytyramine) in der Substantia nigra des normalen and Parkinson-kranken Menschen, *Wien. Klin. Wochenschr. 75,* 309–312.

Hornykiewicz, O. (1966). Dopamine (3-hydroxytyramine) and brain function, *Pharmacol. Rev. 18,* 925–964.

Horowitz, M. J., Adams, J. E., and Rutkin, B. B. (1968). Visual imagery on brain stimulation, *Arch. Gen. Psychiatry 19,* 469–486.

Hotton, N. (1959). The pelycosaur tympanum and early evolution of the middle ear, *Evolution 13,* 99–121.

Hover, E. L., and Jenssen, T. A. (1976). Descriptive analysis and social correlates of agonistic displays of *Anolis limifrons* (Sauria, Iguanidae), *Behaviour 58,* 173–191.

Howard, H. E. (1920). *Territory in Bird Life,* John Murray, London. Reprinted Atheneum, New York, 1964.

Howard, H. E. (1929). *An Introduction to the Study of Bird Behaviour,* Cambridge University Press, London.

Hughes, J. (1975). Isolation of an endogenous compound from the brain with pharmacological properties similar to morphine, *Brain Res. 88,* 295–308.

Hughes, J., Smith, T. W., Kosterlitz, H. W., Fothergill, L. A., Morgan, B. A., and Morris, H. R. (1975). Identification of two related pentapeptides from the brain with potent opiate agonist activity, *Nature 258,* 577–579.

Hughes, W. L., Bond, V. P., Brecher, G., Cronkite, E. P., Painter, R. B., Quastler, H., and Sherman, F. G. (1958). Cellular proliferation in the mouse as revealed by autoradiography with tritiated thymidine, *Proc. Natl. Acad. Sci. USA 44,* 476–483.

Hume, D. (1739/1888). *A Treatise of Human Nature,* Volumes 1–3 (L. A. Selby-Bigge, ed.), Oxford University Press, London.

Hunsperger, R. W. von (1956). Affektreaktionen auf elektrische Reizung im Hirnstamm der Katze, *Helv. Physiol. Acta 14,* 70–92.

Hunsperger, R. W., and Bucher, V. M. (1967). Affective behavior produced by electrical stimulation in the forebrain and brain stem of the cat, *Prog. Brain Res. 27,* 103–127.

Hunter, R., Logue, V., and McMenemy, W. H. (1963). Temporal lobe epilepsy supervening on longstanding transvestism and fetishism: A case report, *Epilepsia 4,* 60–65.

Huntington, G. (1872). On chorea, *Med. Surg. Rep. 26,* 317–321.

Hutton, E. L. (1943). Early results of prefrontal leukotomy, *Lancet 2,* 362.

Ironside, R. (1956). Disorders of laughter due to brain lesions, *Brain 79,* 589–609.

Irving, E. (1977). Drift of the major continental blocks since the Devonian, *Nature 270,* 304–309.

Irving, W. N. (1987). New dates from old bones, *Nat. Hist. 96,* 8–14.

Isaacson, R. L. (1974). *The Limbic System,* Plenum Press, New York. See also 2nd Edition, 1982.

Isaacson, R. L., and Wickelgren, W. O. (1962). Hippocampal ablation and passive avoidance, *Science 138,* 1104–1106.

Itaya, S. K., Van Hoesen, G. W., and Jenq, C. B. (1981). Direct retinal input to the limbic system of the rat, *Brain Res. 226,* 33–42.

Jackson, J. H. (1874). A case of the right hemiplegia and loss of speech from local softening of the brain, *Br. Med. J. 1,* 804–805.

Jackson, J. H. (1875/1958). On the anatomical and physiological localization of movements in the brain, in:

Selected Writings of John Hughlings Jackson, Volume 1 (J. Taylor, ed.), Basic Books, New York, pp. 37–76.

Jackson, J. H. (1879–80/1958). On affections of speech from disease of the brain, in: *Selected Writings of John Hughlings Jackson,* Volume 2 (J. Taylor, ed.), Basic Books, New York, pp. 171–204.

Jackson, J. H. (1880/1958). On right- or left-sided spasm at the onset of epileptic paroxysms, and on crude sensation warnings and elaborate mental states, in: *Selected Writings of John Hughlings Jackson,* Volume 1 (J. Taylor, ed.), Basic Books, New York, pp. 308–317.

Jackson, J. H. (1882/1958). On some implications of dissolution of the nervous system, in: *Selected Writings of John Hughlings Jackson,* Volume 2 (J. Taylor, ed.), Basic Books, New York, pp. 29–44.

Jackson, J. H. (1884/1958). Evolution and dissolution of the nervous system, in: *Selected Writings of John Hughlings Jackson,* Volume 2 (J. Taylor, ed.), Basic Books, New York, pp. 45–75.

Jackson, J. H. (1888/1958). On a particular variety of epilepsy (''intellectual aura''), one case with symptoms of organic brain disease, in: *Selected Writings of John Hughlings Jackson,* Volume 1 (J. Taylor, ed.), Basic Books, New York, pp. 385–405.

Jackson, J. H. (1890/1958). On convulsive seizures. Lumleian lectures delivered at Royal College of Physicians, in: *Selected Writings of John Hughlings Jackson,* Volume 1 (J. Taylor, ed.), Basic Books, New York, pp. 412–457.

Jackson, J. H. (1894/1958). The factors of insanities, in: *Selected Writings of John Hughlings Jackson,* Volume 2 (J. Taylor, ed.), Basic Books, New York, pp. 411–421.

Jackson, J. H. (1958). Lectures on the diagnosis of epilepsy, in: *Selected Writings of John Hughlings Jackson,* Volume 1 (J. Taylor, ed.), Basic Books, New York, pp. 276–307.

Jackson, J. H., and Beevor, C. E. (1889). Epilepsy with olfactory aura. (3rd hand report of case presented before Medical Soc of London.) *Lancet 1,* 381.

Jackson, J. H., and Colman, W. S. (1898/1958). Case of epilepsy with tasting movements and ''dreamy state''—very small patch of softening in the left uncinate gyrus, *Selected Writings of John Hughlings Jackson,* Volume 1 (J. Taylor, ed.), Basic Books, New York, pp. 458–463.

Jackson, J. H., and Stewart, J. P. (1899/1958). Epileptic attacks with a warning of a crude sensation of smell and with the intellectual aura (dreamy state) in a patient who had symptoms pointing to gross organic disease of the right temporo-sphenoidal lobe, *Selected Writings of John Hughlings Jackson,* Volume 1 (J. Taylor, ed.), Basic Books, New York, pp. 464–473.

Jacob, F. (1977). Evolution and tinkering, *Science 196,* 1161–1166.

Jacobowitz, D. M. (1979). Hypothesis for the local control of norepinephrine release, in: *Catecholamines: Basic and Clinical Frontiers,* Volume 2 (E. Usdin, I. J. Kopin, and J. Barchas, eds.), Pergamon Press, Elmsford, N.Y., pp. 1792–1794.

Jacobowitz, D. M., and MacLean, P. D. (1978). A brainstem atlas of catecholaminergic neurons and serotonergic perikarya in pygmy primate (*Cebuella pygmaea*), *J. Comp. Neurol. 177,* 397–416.

Jacobson, M. (1811). Descriptionis anatomique d'un orgran observé dans les Mammifères, *Ann. Mus. Hist. Nat. (Paris) 18,* 412–424.

Jakob, A. (1923). Die Extrapyramidalin Erkrankungen, *Monogr. Neurol. Psychiatr. 37.*

Jakob, A. (1925). The anatomy, clinical syndromes and physiology of the extrapyramidal system, *Arch. Neurol. Psychiatry 13,* 596–620.

James, W. (1890). *The Principles of Psychology,* Volume II, Holt, New York.

Jandolo, B., Gessini, L., Occhipinti, E., and Pompili, A. (1977). Laughing and running fits as manifestation of early traumatic epilepsy, *Eur. Neurol. 15,* 177–182.

Janz, D. (1955). Anfallsbild und Verlaufsform epileptischer Erkrankungen, *Nervenarzt 26,* 20–28.

Jasper, H. H. (1962). Mechanisms of epileptic automatism, *Epilepsia 3,* 281–390.

Jasper, H. H. (1964). Some physiological mechanisms involved in epileptic automatisms, *Epilepsia 5,* 1–20.

Jasper, H. D., and Kershman, J. (1941). Electroencephalographic classification of the epilepsies, *Arch. Neurol. Psychiatry 45,* 903–943.

Jaspers, K. (1913/1963). *General Psychopathology* (Trans. by J. Hoenig and M. W. Hamilton), University of Chicago Press, Chicago.

Jayaraman, A., Batton, R. R., III, and Carpenter, M. B. (1977). Nigrotectal projections in the monkey: An autoradiographic study, *Brain Res. 135,* 147–152.

Jeans, J. (1943). *Physics and Philosophy,* Cambridge University Press, London.

Jenssen, T. A. (1978). Display diversity in anoline lizards and problems of interpretation, in: *The Behavior and Neurology of Lizards* (N. Greenberg and P. D. MacLean, eds.), U.S. Government Printing Office, Washington, D.C., DHEW Publication No. (ADM) 77-491, pp. 269–286.

Jenssen, T. A., and Hover, E. L. (1976). Display analysis of the signature display of *Anolis limifrons* (Sauria: Iguanidae), *Behaviour 57*, 227–240.

Jenssen, T. A., and Rothblum, L. A. (1977). Display repertoire analysis of *Anolis townsendi* (Sauria: Iguanidae), from Cocos Island, *Copeia* 103–109.

Jerison, H. J. (1973). *Evolution of the Brain and Intelligence*, Academic Press, New York.

Johannsen, W. (1911). The genotype conception of heredity, *Am. Nat. 45*, 129–159.

Johanson, D. C., and Taieb, M. (1976). Plio-Pleistocene hominid discoveries in Hadar, Ethiopia, *Nature 260*, 293–297.

Johanson, D. C., and White, T. D. (1979). A systematic assessment of early African hominids, *Science 203*, 321–330.

Johnson, J. L., and Aprison, M. H. (1971). The distribution of glutamate and total free amino acids in thirteen specific regions of the cat central nervous system, *Brain Res. 26*, 141–148.

Johnson, R. T. (1982). *Viral Infections of the Nervous System*, Raven Press, New York.

Johnston, J. B. (1916). The development of the dorsal ventricular ridge in turtles, *J. Comp. Neurol. 26*, 481–505.

Johnston, J. B. (1923). Further contributions to the study of the evolution of the forebrain, *J. Comp. Neurol. 35*, 337–481.

Johnston, M. V., and Coyle, J. T. (1979). Histological and neurochemical effects of fetal treatment with methylazoxymethanol on rat neocortex in adulthood, *Brain Res. 170*, 135–155.

Jolicoeur, P., Pirlot, P., Baron, G., and Stephan, H. (1984). Brain structure and correlation patterns in Insectivora, Chiroptera, and primates, *Syst. Zool. 33*, 14–29.

Jolliffe, N., Colbert, C. N., and Joffee, P. M. (1936). Observations on etiologic relationship of vitamin B (B_1) to polyneuritis in alcohol addict, *Am. J. Med. Sci. 191*, 515–526.

Jones, E. G. (1985). *The Thalamus*, Plenum Press, New York.

Jones, E. G., and Leavitt, R. Y. (1974). Reetrograde axonal transport and the demonstration of non-specific projections to the cerebral cortex and striatum from thalamic intralaminar nuclei in the rat, cat, and monkey, *J. Comp. Neurol. 154*, 349–378.

Jones, E. G., Burton, H., Saper, C. B., and Swanson, L. W. (1976). Midbrain, diencephalic and cortical relationships of the basal nucleus of Meynert and associated structures in primates, *J. Comp. Neurol. 167*, 385–420.

Jones, E. G., Coulter, J. D., Burton, H., and Porter, R. (1977). Cells of origin and terminal distribution of corticostriatal fibers arising in the sensory-motor cortex of monkeys, *J. Comp. Neurol. 173*, 53–80.

Jones, E. G., Wise, S. P., and Coulter, J. D. (1979). Differential thalamic relationships of sensory-motor and parietal cortical fields in monkeys, *J. Comp. Neurol. 183*, 833–882.

Jones, T. C., Thorington, R. W., Hu, M. M., Adams, E., and Cooper, R. W. (1973). Karyotypes of squirrel monkeys (*Saimiri sciureus*) from different geographic regions, *Am. J. Phys. Anthropol. 38*, 269–278.

Jouvet, M. (1967). Neurophysiology of the states of sleep, *Physiol. Rev. 47*, 117–177.

Jouvet, M. (1969). Biogenic amines and the states of sleep, *Science 163*, 32–39.

Jouvet, M., Michel, F., and Mounier, D. (1960). Analyse électroencéphalographique comparée due sommeil physiologique chez le chat et chez l'homme, *Rev. Neurol. 103*, 189–205.

Jubb, K. V., Saunders, L. Z., and Coates, H. V. (1956). Thiamine deficiency encephalopathy in cats, *J. Comp. Pathol. 66*, 217–227.

Jung, R. (1949). Hirnelektrische Untersuchungen über den Elektrokrampf: die Erregungsablaufe in corticalen und subcorticalen Hirnregionen bei Katze und Hund, *Arch. Psychiatr. Nervenkr. 183*, 206–244.

Jung, R., and Hassler, R. (1960). The extrapyramidal motor system, in: *Handbook of Physiology* (J. Field, H. W. Magoun, and V. E. Hall, eds.), American Physiological Society, Washington, D.C., pp. 863–927.

Jung, R., and Kornmüller, A. E. (1938). Eine Methodik der Ableitung lokalisierter Potentialschwankungen aus subcorticalen Hirngebieten, *Arch. Psychiatr. Nervenkr. 109*, 1–30.

Juorio, A. V., and Vogt, M. (1967). Monoamines and their metabolites in the avian brain, *J. Physiol.* (*London*) *189*, 489–518.

Jürgens, U. (1976). Projections from the cortical larynx area in the squirrel monkey, *Exp. Brain Res. 25*, 401–411.

Jürgens, U. (1983). Afferent fibers to the cingular vocalization region in the squirrel monkey, *Exp. Neurol. 80*, 395–409.

Jürgens, U. (1984). The efferent and afferent connections of the supplementary motor area, *Brain Res. 300*, 63–81.

Jürgens, U., and Ploog, D. (1970). Cerebral representations of vocalization in the squirrel monkey, *Exp. Brain Res. 10*, 532–554.

Jürgens, U., and Pratt, R. (1979a). Role of the periaqueductal grey in vocal expression of emotion, *Brain Res. 167,* 367–378.

Jürgens, U., and Pratt, R. (1979b). The cingular vocalization pathway in the squirrel monkey, *Exp. Brain Res. 34,* 499–510.

Kaada, B. R. (1951). Somato-motor, autonomic and electrocorticographic responses to electrical stimulation of 'rhinencephalic' and other structures in primates, cat and dog. A study of responses from the limbic, subcallosal, orbito-insular, piriform and temporal cortex, hippocampus–fornix and amygdala, *Acta Physiol. Scand. 23* (Suppl. 83), 1–285.

Kaitz, S. S., and Robertson, R. T. (1981). Thalamic connections with limbic cortex. II. Corticothalamic projections, *J. Comp. Neurol. 195,* 527–545.

Kalil, K. (1978). Patch-like termination of thalamic fibers in the putamen of the rhesus moneky: An autoradiographic study, *Brain Res. 140,* 333–339.

Kallen, B. (1951). On the ontogeny of the reptilian forebrain. Nuclear structures and ventricular sulci, *J. Comp. Neurol. 95,* 307–347.

Kamrin, R. P. (1966). Temporal lobe epilepsy caused by unruptured middle cerebral artery aneurysms, *Arch. Neurol. 14,* 421–427.

Kandel, E. R., Spencer, W. A., and Brinley, F. J., Jr. (1961). Electrophysiology of hippocampal neurons. I. Sequential invasion and synaptic organization, *J. Neurophysiol. 24,* 225–242.

Kanner, L. (1943). Autistic disturbances of affective contact, *Nerv. Child 3,* 217–250.

Kant, I. (1781/1899). *Critique of Pure Reason* (Trans. by J. M. D. Meiklejohn), Colonial Press, New York.

Karamian, A. I., Zagorulko, T. M., and Bilyan, R. N. (1984). New electrophysiological evidence on location of visual representation in limbic area of the neocortex, *Sechenov Physiol. J. USSR 70,* 1256–1264.

Karplus, J. P., and Kreidl, A. (1909). Gehirn und Sympathicus. I. Mitteilung: Zwischenhirnbasis und Halssympathicus, *Pfluegers Arch. Gesamte Physiol. 129,* 138–144.

Karplus, J. P. and Kreidl, A. (1910). Gehirn und Sympathicus. II. Mitteilung: Ein Sympathicuszentrum im Zwischenhirn, *Pfluegers Arch. Gesamte Physiol. 135,* 401–416.

Karten, H. J. (1969). The organization of the avian telencephalon and some speculations on the phylogeny of the amniote telencephalon, *Ann. N.Y. Acad. Sci. 167,* 164–179.

Kaufmann, W. (1968). *Nietzsche: Philosopher, Psychologist, Antichrist,* 3rd revised edition, Random House, New York.

Kawakami, M., Seto, K., Terasawa, E., Yoshida, K., Miyamoto, T., Sekiguchi, M., and Hattori, Y. (1968). Influence of electrical stimulation and lesion in limbic structure upon biosynthesis of adrenocorticoid in the rabbit, *Neuroendocrinology 3,* 337–348.

Kebabian, J. W., and Calne, D. B. (1979). Multiple receptors for dopamine, *Nature 277,* 93–96.

Keefer, D. A., and Stumpf, W. E. (1975). Atlas of estrogen-concentrating cells in the central nervous sytem of the squirrel monkey, *J. Comp. Neurol. 160,* 419–442.

Keith, A. (1932). Review of *The mammal-like reptiles of South Africa and the origin of mammals, J. Anat. 66,* 669–671.

Kelly, A. H., Beaton, L. E., and Magoun, H. W. (1946). A midbrain mechanism for facio-vocal activity, *J. Neurophysiol. 9,* 181–189.

Kemp, J. M. (1970). The termination of strio-pallidal strio-nigral fibres, *Brain Res. 17,* 125–128.

Kemp, J. M., and Powell, T. P. S. (1971). The site of termination of afferent fibres in the caudate nucleus, *Philos. Trans. R. Soc. (London) 262,* 413–427.

Kennard, M. A., and Fulton, J. F. (1942). Corticostriatal interrelations in monkey and chimpanzee, in: *The Diseases of the Basal Ganglia, Res. Publ. Assoc. Ment. Dis. 21,* 228–245.

Kennedy, F. (1932). The symptomatology of frontal and temporosphenoidal tumors, *AMA Arch. Neurol. 98,* 864–866.

Kermack, D. M., Kermack, K. A., and Mussett, F. (1968). The Welsh pantothere Kuehneotherium praecursoris, *J. Linn. Soc. London Zool. 47,* 407–423.

Kershman, J. (1949). "The borderland of epilepsy": A reconsideration, *Arch. Neurol. Psychiatry 62,* 551–559.

Kesarev, V. S., Malofeyeva, L. I., and Trykova, O. V. (1977). Ecological specificity of cetacean neocortex, *J. Hirnforsch. 18,* 447–460.

Kety, S. S. (1959). Biochemical theories of schizophrenia, *Science 129,* 1528–1532, 1590–1596.

Kevetter, G. A., and Winans, S. S. (1981a). Connections of the corticomedial amygdala in the golden hamster. I. Efferents of the "vomeronasal amygdala," *J. Comp. Neurol. 197,* 81–98.

Kevetter, G. A., and Winans, S. S. (1981b). Connections of the corticomedial amygdala in the golden hamster. II. Efferents of the "olfactory amygdala," *J. Comp. Neurol. 197,* 99–111.

Khachaturian, H., Lewis, M. E., Hollt, V., and Watson, S. J. (1983). Telencephalic enkephalinergic systems in the rat brain, *J. Neurosci. 3,* 844–855.

Kim, C. (1960). Nest building, general activity, and salt preference of rats following hippocampal ablation, *J. Comp. Physiol. Psychol. 53,* 11–16.

Kim, J. S., Hassler, R., Haug, P., and Paik, K. S. (1977). Effect of frontal cortex ablation on striatal glutamic acid level in rat, *Brain Res. 132,* 370–374.

Kimble, D. P., and Pribram, K. H. (1963). Hippocampectomy and behavior sequences, *Science 139,* 824–825.

Kimble, D. P., Rogers, L., and Hendrickson, C. W. (1967). Hippocampal lesions disrupt maternal, not sexual, behavior in the albino rat, *J. Comp. Physiol. Psychol. 63,* 401–407.

King, D. W., and Ajmone Marsan, C. (1977). Clinical features and ictal patterns in epileptic patients with EEG temporal lobe foci, *Ann. Neurol. 2,* 138–147.

King, J. C., Tobet, S. A., Snavely, F. L., and Arimura, A. A. (1982). LHRH immunopositive cells and their projections to the median eminence and organum vasculosum of the lamina terminalis, *J. Comp. Neurol. 209,* 287–300.

Kinnard, M. A., and MacLean, P. D. (1967). A platinum micro-electrode for intracerebral exploration with a chronically fixed stereotaxic device, *Electroencephalogr. Clin. Neurophysiol. 22,* 183–186.

Kirby, R. J., and Kimble, D. P. (1968). Avoidance and escape behavior following striatal lesions in the rat, *Exp. Neurol. 20,* 215–227.

Kirzinger, A., and Jürgens, U. (1982). Cortical lesion effects and vocalization in the squirrel monkey, *Brain Res. 233,* 299–315.

Kitching, J. W. (1963). *Bone, Tooth & Horn Tools of Paleolithic Man,* Manchester University Press, Manchester.

Kleist, K. (1931). Gehirnpathologische und lokalisatorische Ergebnisse. Die Störungen der Ichleistungen und ihre Lokalisation im Orbital-, Innen- und Zwischernhirn, *Monatsschr. Psychiatr. Neurol. 79,* 338–350.

Kling, A. (1968). Effects of amygdalectomy and testosterone on sexual behavior of male juvenile macaques, *J. Physiol. Psychol. 65,* 466.

Kling, A., and Mass, R. (1974). Alterations of social behavior with neural lesions in nonhuman primates, in: *Primate Aggression, Territoriality, and Xenophobia: A Comparative Perspective* (R. L. Holloway, ed.), Academic Press, New York, pp. 361–386.

Kling, A., Lancaster, J., and Benitone, J. (1970). Amygdalectomy in the free-ranging vervet (*Cercopithecus aethiops*), *J. Psychiatr. Res. 1,* 191–199.

Klüver, H. (1951). Functional differences between the occiptal and temporal lobes with special reference to the interrelations of behavior and extracerebral mechanisms, in: *Cerebral Mechanisms in Behavior* (L. A. Jeffress, ed.), Wiley, New York, pp. 147–182.

Klüver, H. (1958). "The temporal lobe syndrome" produced by bilateral ablations. Reprinted from Ciba Foundation Symposium on the *Neurological Basis of Behavior,* pp. 175–182.

Klüver, H., and Bucy, P. C. (1937). "Psychic blindness" and other symptoms following bilateral temporal lobectomy in rhesus monkeys, *Am. J. Physiol. 119,* 352–353.

Klüver, H., and Bucy, P. C. (1938). An analysis of certain effects of bilateral temporal lobectomy in the rhesus monkey, with special reference to "psychic blindness," *J. Psychol. 5,* 33–54.

Klüver, H., and Bucy, P. C. (1939a). A preliminary analysis of functions of the temporal lobes in monkeys, *Trans. Am. Neurol. Assoc.* 170–175.

Klüver, H., and Bucy, P. C. (1939b). Preliminary analysis of functions of the temporal lobes in monkeys, *Arch. Neurol. Psychiatry 42,* 979–1000.

Knight, R. P. A. (1865). *A Discourse on the Worship of Priapus and its Connexion with Mystic Theology of the Ancients,* London, privately printed.

Knott, J. R., Ingram, W. R., and Correll, R. E. (1960). Effects of certain subcortical lesions on learning and performance in the cat, *AMA Arch. Neurol. 2,* 247–259.

Koehler, O. (1954). Vom Erbgut der Sprache, *Homo 5,* 97–104.

Koelle, G. B. (1954). The histochemical localization of cholinesterases in the central nervous system of the rat, *J. Comp. Neurol. 100,* 211–228.

Koestler, A. (1963). *The Sleepwalkers,* Grosset & Dunlap, New York.

Koestler, A. (1964). *The Act of Creation,* Macmillan Co., New York.

Koestler, A. (1967). *The Ghost in the Machine,* Hutchinson, London. (Reprinted 1968 by Macmillan, New York.)

Koikegami, H., Fuse, S., Yokohama, T., Watanabe, T., and Watanabe, H. (1955). Contributions to the comparative anatomy of the amygdaloid nuclei of mammals with some experiments of their destruction or stimulation, *Folia Psychiatr. Neurol. Jpn. 8,* 336–370.

Kolb, L. (1949). An evaluation of lobotomy and its potentialities for future research in psychiatry and the basic sciences, *J. Nerv. Ment. Dis. 110,* 112–148.

Korsakoff, S. S. (1887). Disturbance of psychic function in alcoholic paralysis and its relation to the disturbance of the psychic sphere in multiple neuritis of nonalcoholic origin, *Vestn. Psichiatr. 4*(2).

Kortlandt, A. (1940). Wechselwirkung zwischen Instinkten, *Arch. neerl. Zool. 4,* 442–520.

Koskoff, Y. D., Dennis, W., Lazovik, D., and Wheeler, E. T. (1948). The psychological effects of frontal lobotomy performed for the alleviation of pain, in: *The Frontal Lobes, Res. Publ. Assoc. Res. Nerv. Ment. Dis. 27,* 723–753.

Kovacs, S., Sandor, A., Vertes, A., and Vertes, M. (1965). The effect of lesions and stimulation of the amygdala on pituitary–thyroid function, *Acta Physiol. Acad. Sci. Hung. 27,* 221–227.

Kramer, H. C. (1954). Laughing spells in patients, after lobotomy, *J. Nerv. Ment. Dis. 119,* 517–522.

Krauthamer, G. M. (1979). Sensory functions of the neostriatum, in: *The Neostriatum* (I. Divac and R. G. E. Oberg, eds.), Pergamon Press, Elmsford, N.Y., pp. 263–289.

Krayenbühl, H., Wyss, O. A. M., and Yasargil, M. G. (1961). Bilateral thalamotomy and pallidotomy as treatment for bilateral parkinsonism, *J. Neurosurg. 18,* 429–444.

Krettek, J. E., and Price, J. L. (1977a). The cortical projections of the mediodorsal nucleus and adjacent thalamic nuclei in the rat, *J. Comp. Neurol. 171,* 157–192.

Krettek, J. E., and Price, J. L. (1977b). Projections from the amygdaloid complex to the cerebral cortex and thalamus in the rat and cat, *J. Comp. Neurol. 172,* 687–722.

Krettek, J. E., and Price, J. L. (1977c). Projections from the amygdaloid complex and adjacent olfactory structures to the entorhinal cortex and to the subiculum in the rat and cat, *J. Comp. Neurol. 172,* 723–752.

Krettek, J. E., and Price, J. L. (1978). Amygdaloid projections to subcortical structures within the basal forebrain and brainstem in the rat and cat, J. Comp. Neurol. 178, 225–254.

Krieg, W. J. S. (1932). The hypothalamus of the albino rat, *J. Comp. Neurol. 55,* 19–89.

Krieg, W. J. S. (1946). Connections of the cerebral cortex. I. The albino rat. B. Structure of the cortical areas, *J. Comp. Neurol. 84,* 277–323.

Krieg, W. J. S. (1947). Connections of the cerebral cortex. I. The albino rat. C. Extrinsic connections, *J. Comp. Neurol. 86,* 267–394.

Krieg, W. J. S. (1963). *Connections of the Cerebral Cortex,* Brain Books, Evanston, Ill.

Krieger, M. S., Conrad, L. C., and Pfaff, D. W. (19779). An autoradiographic study of the efferent connections of the ventromedial nucleus of the hypothalamus, *J. Comp. Neurol. 183,* 785–816.

Kubie, J. L., and Halpern, M. (1975). Laboratory observations of trailing behavior in garter snakes, *J. Comp. Physiol. Psychol. 89,* 667–674.

Kubie, J. L., and Halpern, M. (1979). Chemical senses involved in garter snake prey trailing, *J. Comp. Physiol. Psychol. 93,* 648–667.

Kubie, J. L., Vagvolgyi, A., and Halpern, M. (1978). The role of the vomeronasal and olfactory systems in the courtship behavior of male garter snakes, *J. Comp. Physiol. Psychol. 92,* 627–641.

Kubie, L. S. (1953). Some implications for psychoanalysis of modern concepts of the organization of the brain, *Psychoanal. Q. 22,* 21–68.

Kuhar, M. J., Pert, C. B., and Snyder, S. H. (1973). Regional distribution of opiate receptor binding in monkey and human brain, *Nature 245,* 447–450.

Kühlenbeck, H. (1938). The ontogenetic development and phylogenetic significance of the cortex telencephali in the chick, *J. Comp. Neurol. 69,* 273–301.

Kunzle, H. (1975). Bilateral projections from precentral motor cortex to the putamen and other parts of the basal ganglia. An autoradiographic study in *Macaca fascicularis, Brain Res. 88,* 195–209.

Kuo, J.-S., and Carpenter, M. B. (1973). Organization of pallidothalamic projections in the rhesus monkey, *J. Comp. Neurol. 151,* 201–228.

Kupffer, C. (1859). *De cornu ammonis textura disquisitionis, praecipuae in cuniculis institutae,* Dorpat.

Kuru, M. (1956). The spino-bulbar tracts and the pelvic sensory vagus. Further contributions to the theory of the sensory dual innervation of the viscera, *J. Comp. Neurol. 104,* 207–231.

Kuypers, H., Kievit, J., and Groen-Klevant, A. (1974). Retrograde axonal transport of horseradish peroxidase in rat's forebrain, *Brain Res. 67,* 211–218.

Kuypers, H. G. J. M., Catsman-Berrevoets, C. E., and Padt, R. E. (1977). Retrograde axonal transport of fluorescent substances in the rat's forebrain, *Neurosci. Lett. 6,* 127–135.

Laborit, H., Huguenard, P., and Alluaume, R. (1952). Un nouveau stabilisateur vegetatif, le 4560 RP, *Presse Med. 60,* 206–208.

Lammers, H. J. (1971). The neural connections of the amygdaloid complex in mammals, in: *The Neurobiology of the Amygdala* (B. E. Eleftheriou, ed.), Plenum Press, New York.

Landau, E. (1919). The comparative anatomy of the nucleus amygdali, the claustrum and the insular cortex, *J. Anat. 53,* 351–360.

Langley, J. N. (1893). Preliminary account of the arrangement of the sympathetic nervous system based chiefly on observations upon pilomotor nerves, *Proc. R. Soc. London 52,* 547–556.

Langley, J. N. (1900). The sympathetic and other related systems of nerves, in: *Textbook of Physiology,* Volume 2 (E. A. Schäfer, ed.), Pentland, Edinburgh, pp. 616–696.

Langley, J. N. (1903). The autonomic nervous system, *Brain 26,* 1–26.

Langley, J. N. (1905). On the reaction of cells and of nerve-endings to certain poisons, chiefly as regards the reaction of striated muscle to nicotine and to curari, *J. Physiol. (London) 33,* 374–413.

Langley, J. N. (1921). *The Autonomic Nervous System,* Heffer, Cambridge.

Langley, J. N., and Dickinson, W. L. (1889). On the local paralysis of peripheral ganglia, and on the connection of different classes of nerve fibres with them, *Proc. R. Soc. London 46,* 423–431.

Langston, J. W., Ballard, P., Tetrud, J. W., and Irwin, I. (1983). Chronic parkinsonism in humans due to a product of meperidine-analog synthesis, *Science 219,* 979–980.

Lasek, R., Joseph, B. S., and Whitlock, D. G. (1968). Evaluation of radioautographic neuroanatomical tracing method, *Brain Res. 8,* 319–336.

Lashley, K. S. (1929/1963). *Brain Mechanisms and Intelligence: A Quantitative Study of Injuries to the Brain,* Dover, New York.

Lassek, A. M. (1957). *The Human Brain: From Primitive to Modern,* Thomas, Springfield, Ill.

Laursen, A. M. (1962). Movements evoked from the region of the caudate nucleus in cats, *Acta Physiol. Scand. 54,* 175–184.

Laursen, A. M. (1963). Corpus striatum, *Acta Physiol. Scand. 59,* 1–106.

LaVail, J. H., and LaVail, M. M. (1972). Retrograde axonal transport in the central nervous system, *Science 176,* 1416.

Lawicka, W., and Konorski, J. (1959). Physiological mechanisms of delayed reactions. III. The effects of prefrontal ablations on delayed reactions in dogs, *Acta Biol. Exp. (Warsaw) 19,* 221–231.

Lawick-Goodall, J. V. (1968). The behavior of free-living chimpanzees in the Gombe Stream, *Anim. Behav. 1,* 161–311.

Lawick-Goodall, J. V. (1969). Mother–offspring relationships in free-ranging chimpanzees, in: *Primate Ethology* (D. Morris, ed.), Anchor Books, Doubleday, New York, pp. 365–436.

Lawick-Goodall, J. V. (1971). *In the Shadow of Man,* Houghton Mifflin, Boston.

Leakey, L. S. B. (1959). A new fossil skull from Olduvai, *Nature 184,* 491–493.

Leakey, L. (1961). New finds at Olduvai Gorge, *Nature (Lond.) 189,* 649–650.

Leakey, L. S. B., Tobias, P. V., and Napier, J. R. (1964). A new species of the genus *Homo* from Olduvai Gorge, *Nature 202,* 7–9.

Leakey, R. (1973). Further evidence of lower Pleistocene hominids from East Rudolf, North Kenya, *Nature 242,* 170–173.

Leakey, R. E. F., and Walker, A. C. (1976). *Australopithecus, Homo erectus* and the single species hypothesis, *Nature 261,* 572–574.

Le Beau, J. (1952). The cingular and precingular areas in psychosurgery (agitated behaviour, obsessive compulsive states, epilepsy), *Acta Psychiatr. Neurol. Scand. 27,* 305–316.

Lee, K., Diaz, M., and Melchior, J. C. (1981). Temporal lobe epilepsy—Not a consequence of childhood febrile convulsions in Denmark, *Acta Neurol. Scand. 63,* 231–236.

Lehtinen, L., and Kivalo, A. (1965). Laughter epilepsy, *Acta Neurol. Scand. 41,* 255–261.

Leiner, H. C., Leiner, A. L., and Dow, R. S. (1986). Does the cerebellum contribute to mental skills? *Behav. Neurosci. 100,* 443–454.

Lenneberg, E. H. (1967). *Biological Foundations of Language,* Wiley, New York.

Lennox, W. G. (1951). Phenomena and correlates of psychomotor triad, *Neurology 1,* 357–371.

Lennox, W. G. (1960). *Epilepsy and Related Disorders,* two volumes (with the collaboration of M. Lennox), Little, Brown, Boston.

Lennox, W. G., and Cobb, S. (1933). Epilepsy: Aura in epilepsy; A statistical review of 1,359 cases, *Arch. Neurol. Psychiatry 30,* 374–387.

Leonard, C. M. (1969). The prefrontal cortex of the rat. I. Cortical projection of the mediodorsal nucleus. II. Efferent connections, *Brain Res. 12,* 321–343.

Leonard, C. M. and Scott, J. W. (1971). Origin and distribution of the amygdalofugal pathways in the rat: An experimental neuroanatomical study, *J. Comp. Neurol. 141,* 313–330.

Lewis, F. T. (1923–24). The significance of the term hippocampus, *J. Comp. Neurol. 35,* 213–230.

Lewis, M. E., Pert, C. B., and Herkenham, M. (1983). Opiate receptor localization in rat cerebral cortex, *J. Comp. Neurol. 216,* 339–358.

Lewis, P. R., and Shute, C. C. D. (1967). The cholinergic limbic system: Projections to the hippocampal formation, medial cortex, nuclei of the ascending cholinergic reticular system, and the subfornical organ and supra-optic crest, *Brain 90,* 521–540.

Lewis, P. R., Shute, C. C. D., and Silver, A. (1967). Confirmation from choline acetylase analyses of a massive cholinergic innervation to the rat hippocampus, *J. Physiol. (London) 191,* 215–224.

Lewy, F. H. (1942/1966). Historical introduction: The basal ganglia and their diseases, in: *The Diseases of the Basal Ganglia, Res. Publ. Assoc. Res. Nerv. Ment. Dis. 21,* 1–20.

Leyton, A. S. F., and Sherrington, C. S. (1917). Observations on the excitable cortex of the chimpanzee, orangutan, and gorilla, *Q. J. Exp. Physiol. 11,* 135–222.

Li, C. H., and Chung, D. (1976). Isolation and structure of an untriakonta peptide with opiate activity from camel pituitary glands, *Proc. Natl. Acad. Sci. USA 73,* 1145–1148.

Liberson, W. T., and Akert, K. (1955). Hippocampal seizure states in guinea pig, *Electroencephalogr. Clin. Neurophysiol. 7,* 211–222.

Lieberman, P. (1985). On the evolution of human syntactic ability. Its pre-adaptive bases—motor control and speech, *J. Hum. Evol. 14,* 657–668.

Lieberman, P. (in press). The origins of some aspects of human language and cognition, in: *The Human Revolution: Behavioral and Biological Perspectives on the Origins of Modern Humans* (P. Mellars and C. B. Stringer, eds.), Edinburgh Univ. Press, Edinburgh.

Lieberman, P., and Crelin, E. S. (1971). On the speech of Neanderthal man, *Linguistic Inquiry 2,* 203–222.

Lieberman, P., Crelin, E. S., and Klatt, D. H. (1972). Phonetic ability and related anatomy of the newborn and adult human, Neanderthal Man, and the chimpanzee, *Am. Anthropol. 74,* 287–307.

Liles, S. L., and Davis, G. D. (1969). Interrelation of caudate nucleus and thalamus in alteration of cortically induced movement, *J. Neurophysiol. 32,* 564–573.

Lilly, J. C. (1958). Learning motivated by subcortical stimulation: The ''start'' and the ''stop'' patterns of behavior, in: *Reticular Formation of the Brain,* Little, Brown, Boston, pp. 705–727.

Lin, N. (1963). Territorial behavior in the cicada-killer wasp, *Behaviour 20,* 15–33.

Lindegren, C. C. (1966). *The Cold War in Biology,* Planarian Press, Ann Arbor.

Lindenberg, R. (1955). Compression of brain arteries as pathogenetic factor for tissue necroses and their areas of predilection, *J. Neuropathol. Exp. Neurol. 14,* 223–243.

Lindsay, J., Ounsted, C., and Richards, P. (1980). Long-term outcome in children with temporal lobe seizures. IV: Genetic factors, febrile convulsions and the remission of seizures, *Dev. Med. Child Neurol. 22,* 429–439.

Lindvall, O. (1975). Mesencephalic dopaminergic afferents to the lateral septal nucleus of the rat, *Brain Res. 87,* 89–95.

Lindvall, O., and Bjorklund, A. (1979). Dopaminergic innervation of the globus pallidus by collaterals from the nigrostriatal pathway, *Brain Res. 172,* 169–173.

Lipps, T. (1960). Empathy, inner imitation, and sense-feelings (from *Arch. Gesamte Psychol. 1,* 1903), in: *A Modern Book of Esthetics: An Anthology* (M. Rader, ed.), 3rd Edition, Holt, Rinehart & Winston, New York, pp. 374–382.

Lisk, R. D. (1962). Diencephalic placement of estradiol and sexual receptivity in the female rat, *Am. J. Physiol. 203,* 493–496.

List, C. F., Dowman, C. E., Bagchi, B. K., and Bebin, J. (1958). Posterior hypothalamic hamartomas and gangliogliomas causing precocious puberty, *Neurology 8,* 164–174.

Livingston, K. E. (1953). Cingulate cortex isolation for the treatment of psychoses and psychoneuroses, *Proc. Assoc. Res. Nerv. Ment. Dis. 21,* 374–378.

Livingston, R. B. (1978). *Sensory Processing, Perception, and Behavior,* Raven Press, New York.

Lloyd, K. G. (1977). CNS compensation to dopamine neurone loss in Parkinson's disease, in: *Parkinson's Disease: Neurophysiological, Clinical, and Related Aspects* (F. S. Messiha and A. D. Kenny, ed.), Plenum Press, New York, pp. 255–265.

Locke, J. (1690/1894). *An essay concerning human understanding,* two volumes (A. C. Fraser, ed.), Oxford Press, London.

Loe, P. R., and Benevento, L. A. (1969). Auditory–visual interaction in single units in the orbito-insular cortex of the cat, *Electroencephalogr. Clin. Neurophysiol. 26,* 395–398.

Loewy, A. D., and Burton, H. (1978). Nuclei of the solitary tract: Efferent projections to the lower brain stem and spinal cord of the cat, *J. Comp. Neurol. 181,* 421–450.

Lohman, A. H. M. (1963). The anterior olfactory lobe of the guinea pig, *Acta Anat. 53,* 109.

Lohman, A. H. M., and Lammers, H. J. (1963). On the connections of the olfactory bulb and the anterior olfactory nucleus in some mammals, *Prog. Brain Res. 3,* 149–162.

Loiseau, P., Cohadon, F., and Cohadon, S. (1971). Gelastic epilepsy. A review and report of five cases, *Epilepsia 12,* 313–323.

Loizou, L. A. (1972). The postnatal ontogeny of monoamine-containing neurones in the central nervous system of the albino rat, *Brain Res. 40,* 395–418.

Long, C. A. (1969). The origin and evolution of mammary glands, *Bio. Sci. 19,* 519–523.

Loo, Y. T. (1931). The forebrain of the opossum, *Didelphis virginiana.* Part II. Histology, *J. Comp. Neurol. 52,* 1–148.

Lorente de Nó, R. (1933). Studies on the structure of the cerebral cortex: I. The area entorhinalis, *J. Psychol. Neurol. 45,* 381–438.

Lorente de Nó, R. (1934). Studies on the structure of the cerebral cortex. II. Continuation of the study of the ammonic system, *J. Psychol. Neurol. 46,* 113–177.

Lorente de Nó, R. (1949). Architectonics and structure of the cerebral cortex, in: *Physiology of the Nervous System* (J. F. Fulton, ed.), Oxford University Press, London.

Lorenz, K. (1935). Der Kumpan in der Umwelt des Vogels, *J. Ornithol. 83,* 137–213.

Lorenz, K. (1937). The companion in the bird's world, *Auk 54,* 245–273.

Lorenz, K. (1966). *On Aggression,* Harcourt, Brace & World, New York.

Ludwig, B. I., and Ajmone Marsan, C. (1975). Clinical ictal patterns in epileptic patients with occipital electroencephalographic foci, *Neurology 25,* 463–471.

Ludwig, B. I., Ajmone Marsan, C., and Van Buren, J. (1975). Cerebral seizures of probable orbitofrontal origin, *Epilepsia 16,* 141–158.

Lundberg, P. O. (1960). Cortico-hypothalamic connexions in the rabbit, *Acta Physiol. Scand. 49,* 72.

Lundberg, P. O. (1962). The nuclei gemini. Two hitherto undescribed nerve cell collections in the hypothalamus of the rabbit, *J. Comp. Neurol. 119,* 311–316.

Luria, A. R. (1969). Frontal lobe syndromes, in: *Handbook of Clinical Neurology,* Volume 2 (P. V. Vinken and G. W. Bruyn, eds.), North-Holland, Amsterdam, pp. 725–757.

Luria, A. R., and Homskaya, E. D. (1964). Disturbance in the regulative role of speech with frontal lobe lesions, in: *The Frontal Granular Cortex and Behavior* (J. M. Warren and K. Akert, eds.), McGraw-Hill, New York, pp. 353–371.

Ma, N. S. F., Jones, T. C., Thorington, R. W., and Cooper, R. W. (1974). Chromosome banding patterns in squirrel monkeys (*Saimiri sciureus*), *J. Med. Primatol. 3,* 120–137.

Macchi, G., Bentivoglio, M., Minciacchi, D., and Molinari, M. (1981). The organization of the claustroneocortical projections in the cat studied by means of the HRP retrograde axonal transport, *J. Comp. Neurol. 195,* 681–695.

MacLean, P. D. (1949a). A new nasopharyngeal lead, *Electroencephalogr. Clin. Neurophysiol. 1,* 110–112.

MacLean, P. D. (1949b). Psychosomatic disease and the "visceral brain." Recent developments bearing on the Papez theory of emotion, *Psychosom. Med. 11,* 338–353.

MacLean, P. D. (1950). Developments in electroencephalography: The basal and temporal regions, *Yale J. Biol. Med. 22,* 437–451.

MacLean, P. D. (1952). Some psychiatric implications of physiological studies on frontotemporal portion of limbic system (visceral brain), *Electroencephalogr. Clin. Neurophysiol. 4,* 407–418.

MacLean, P. D. (1954). The limbic system and its hippocampal formation. Studies in animals and their possible application to man, *J. Neurosurg. 11,* 29–44.

MacLean, P. D. (1955). The limbic system ("visceral brain") in relation to central gray and reticulum of the brain stem. Evidence of interdependence in emotional processes, *Psychosom. Med. 17,* 355–366.

MacLean, P. D. (1957a). Chemical and electrical stimulation of hippocampus in unrestrained animals. I. Methods and electroencephalographic findings, *AMA Arch. Neurol. Psychiatry 78,* 113–127.

MacLean, P. D. (1957b). Chemical and electrical stimulation of hippocampus in unrestrained animals. II. Behavioral findings, *AMA Arch. Neurol. Psychiatry 78,* 128–142.

MacLean, P. D. (1957c). Visceral functions of the nervous system, *Annu. Rev. Physiol. 19,* 397–416.

MacLean, P. D. (1958). Contrasting functions of limbic and neocortical systems of the brain and their relevance to psychophysiological aspects of medicine, *Am. J. Med. 25,* 611–626.

MacLean, P. D. (1959). The limbic system with respect to two basic life principles, in: *Transactions of Second*

Conference on the Central Nervous System and Behavior, Josiah Macy, Jr. Foundation, New York, pp. 31–118.

MacLean, P. D. (1960a). Psychosomatics, in: *Handbook of Physiology, Neurophysiology III,* American Physiological Society, Washington, D.C., 1723–1744.

MacLean, P. D. (1960b). John F. Fulton (1899–1960). A midsummer reminiscence, *Yale J. Biol. Med. 33,* 85–93.

MacLean, P. D. (1962). New findings relevant to the evolution of psychosexual functions of the brain, *J. Nerv. Ment. Dis. 135,* 289–301.

MacLean, P. D. (1964a). Mirror display in the squirrel monkey, *Saimiri sciureus, Science 146,* 950–952.

MacLean, P. D. (1964b). Man and his animal brains, *Mod. Med. (Chicago) 32,* 95–106.

MacLean, P. D. (1966a). The limbic and visual cortex in phylogeny: Further insights from anatomic and microelectrode studies, in: *Evolution of the Forebrain* (R. Hassler and H. Stephan, eds.), Thieme, Stuttgart, pp. 443–453.

MacLean, P. D. (1966b). Studies on the cerebral representation of certain basic sexual functions, in: *Brain and Behavior, III. Brain and Gonadal Function* (R. A. Gorski and R. E. Whalen, eds.), University of California Press, Berkeley/Los Angeles, pp. 35–79.

MacLean, P. D. (1967). The brain in relation to empathy and medical education, *J. Nerv. Ment. Dis. 144,* 374–382.

MacLean, P. D. (1968). Alternative neural pathways to violence, in: *Alternatives to Violence* (L. Ng, ed.), Time-Life Books, New York, pp. 24–34.

MacLean, P. D. (1969a). The hypothalamus and emotional behavior, in: *The Hypothalamus* (W. Haymaker, E. Anderson, and W. J. H. Nauta, eds.), Thomas, Springfield, Ill., pp. 659–678.

MacLean, P. D. (1969b). The internal–external bonds of the memory process, *J. Nerv. Ment. Dis. 149,* 40–47.

MacLean, P. D. (1969c). The paranoid streak in man, in: *Beyond Reductionism* (A. Koestler and J. R. Smythies, eds.), Hutchinson, London, pp. 258–278.

MacLean, P. D. (1970a). The triune brain, emotion, and scientific bias, in: *The Neurosciences. Second Study Program* (F. O. Schmitt, ed.), Rockefeller University Press, New York, pp. 336–349.

MacLean, P. D. (1970b). The limbic brain in relation to the psychoses, in: *Physiological Correlates of Emotion* (P. Black, ed.), Academic Press, New York, pp. 129–146.

MacLean, P. D. (1972a). Cerebral evolution and emotional processes: New findings on the striatal complex, *Ann. N.Y. Acad. Sci. 193,* 137–149.

MacLean, P. D. (1972b). Implications of microelectrode findings on exteroceptive inputs to the limbic cortex, in: *Limbic System Mechanisms and Autonomic Function* (C. H. Hockman, ed.), Thomas, Springfield, Ill., pp. 115–136.

MacLean, P. D. (1973a). A triune concept of the brain and behavior. Lecture I. Man's reptilian and limbic inheritance; Lecture II. Man's limbic brain and the psychoses; Lecture III. New trends in man's evolution, in: *The Hincks Memorial Lectures* (T. Boag and D. Campbell, eds.), University of Toronto Press, Toronto, pp. 6–66.

MacLean, P. D. (1973b). The brain's generation gap: Some human implications, *Zygon J. Relig. Sci. 8,* 113–127.

MacLean, P. D. (1973c). New findings on brain function and sociosexual behavior, in: *Contemporary Sexual Behavior: Critical Issues in the 1970s* (J. Zubin and J. Money, eds.), Johns Hopkins University Press, Baltimore, pp. 53–74.

MacLean, P. D. (1973d). An evolutionary approach to the investigation of psychoneuroendocrine functions, in: *Hormones and Brain Function* (K. Lissák, ed.), Plenum Press, New York, pp. 379–389.

MacLean, P. D. (1974). Discussion, *Res. Publ. Assoc. Res. Nerv. Ment. Dis. 52,* 148.

MacLean, P. D. (1975a). On the evolution of three mentalities, *Man–Environment–Systems 5,* 213–224. Reprinted 1977, in: *New Dimensions in Psychiatry: A World View,* Volume 2 (S. Arieti and G. Chrzanowski, eds.), Wiley, New York, pp. 305–382. Reprinted 1978, in: *Human Evolution, Biosocial Perspectives* (S. L. Washburn and E. R. McCown, eds.), Benjamin/Cummings, Menlo Park, Calif. pp. 32–57. Reprinted 1982, in: *General Systems Theory and the Psychological Sciences,* Volume 1 (W. Gray, J. Fidler, and J. Battista, eds.), Intersystems Publications, Seaside, pp. 43–54.

MacLean, P. D. (1975b). The imitative–creative interplay of our three mentalities, in: *Astride the Two Cultures: Arthur Koestler at 70* (H. Harris, ed.), Hutchinson, London, pp. 187–211.

MacLean, P. D. (1975c). Role of pallidal projections in species-typical display behavior of squirrel monkey, *Trans. Am. Neurol. Assoc. 100,* 110–113.

MacLean, P. D. (1975d). Sensory and perceptive factors in emotional functions of the triune brain, in: *Emotions–Their Parameters and Measurement* (L. Levi, ed.), Raven Press, New York, pp. 71–92.
MacLean, P. D. (1975e). An ongoing analysis of hippocampal inputs and outputs: Microelectrode and anatomic findings in squirrel monkeys, in: *The Hippocampus,* Volume 1 (R. L. Isaacson and K. H. Pribram, eds.), Plenum Press, New York, pp. 177–211.
MacLean, P. D. (1977). An evolutionary approach to brain research on prosematic (nonverbal) behavior, in: *Reproductive Behavior and Evolution* (J. S. Rosenblatt and B. R. Komisaruk, eds.), Plenum Press, New York, pp. 137–164.
MacLean, P. D. (1978a). Challenges of the Papez heritage, in: *Limbic Mechanisms* (K. Livingston and O. Hornikiewicz, eds.), Plenum Press, New York, pp. 1–15.
MacLean, P. D. (1978b). Effects of lesions of globus pallidus on species-typical display behavior of squirrel monkeys, *Brain Res. 149,* 175–196.
MacLean, P. D. (1978c). *A Mind of Three Minds: Educating the Triune Brain,* University of Chicago Press, Chicago, pp. 308–342.
MacLean, P. D. (1978d). Why brain research on lizards?, in: *The Behavior and Neurology of Lizards* (N. Greenberg and P. D. MacLean, eds.), U.S. Government Printing Office, Washington, D. C., DHEW Publication, No. (ADM) 77-491, pp. 1–10.
MacLean, P. D. (1981). Role of transhypothalamic pathways in social communication, in: *Handbook of the Hypothalamus,* Volume 3 (P. Morgane and J. Pankseep, eds.), Dekker, New York, pp. 259–287.
MacLean, P. D. (1982). On the origin and progressive evolution of the triune brain, in: *Primate Brain Evolution* (E. Armstrong and D. Falk, eds.), Plenum Press, New York, pp. 291–316.
MacLean, P. D. (1984). Commentary on "Disorders of the Limbic System," *Integr. Psychiatry 2,* 102–103.
MacLean, P. D. (1985a). Brain evolution relating to family, play, and the separation call, *Arch. Gen. Psychiatry 42,* 405–417.
MacLean, P. D. (1985b). Editorial: Evolutionary psychiatry and the triune brain, *Psychol. Med. 15,* 219–221.
MacLean, P. D. (1986a). Culminating developments in the evolution of the limbic system: The thalamocingulate division, in: *The Limbic System: Functional Organization and Clinical Disorders* (B. K. Doane and K. F. Livingston, eds.), Raven Press, New York, pp. 1–28.
MacLean, P. D. (1986b). Ictal symptoms relating to the nature of affects and their cerebral substrate, in: *Emotion: Theory, Research, and Experience,* Volume 3, (R. Plutchik and H. Kellerman, eds.), Academic Press, New York, pp. 61–90.
MacLean, P. D. (1986c). Neurobehavioral significance of the mammal-like reptiles, in: *The Ecology and Biology of Mammal-like Reptiles* (N. Hotton III, P. D. MacLean, J. J. Roth, and E. C. Roth, eds.), Smithsonian Institution Press, Washington, D. C., pp. 1–21.
MacLean, P. D. (1987a). The midline frontolimbic cortex and the evolution of crying and laughter, in: *The Frontal Lobes Revisited* (E. Perecman, ed.), IRBN Press, New York, pp. 121–140.
MacLean, P. D. (1987b). Triune brain, in: *Encyclopedia of Neuroscience* (G. Adelman, ed.), Birkhäuser Boston, Cambridge, pp. 1235–1237.
MacLean, P. D. (1988). Cytochemical tracing of cerebral connections of midline frontal cortex in Saimiri monkeys, *Soc. Neurosci. 14,* 692.
MacLean, P. D., and Arellano, Z. A. P. (1950). Basal lead studies in epileptic automatisms, *Electroencephalogr. Clin. Neurophysiol. 2,* 1–16.
MacLean, P. D., and Creswell, G. (1970). Anatomical connections of visual system with limbic cortex of monkey, *J. Comp. Neurol. 138,* 265–278.
MacLean, P. D., and Delgado, J. M. R. (1953). Electrical and chemical stimulation of frontotemporal portion of limbic system in the waking animal, *Electroencephalogr. Clin. Neurophysiol. 5,* 91–100.
MacLean, P. D., and Newman, J. D. (1988). Role of midline frontolimbic cortex in production of the isolation call of squirrel monkeys, *Brain Res. 450,* 111–123.
MacLean, P. D., and Ploog, D. W. (1962). Cerebral representation of penile erection, *J. Neurophysiol. 25,* 29–55.
MacLean, P. D., and Pribram, K. H. (1953). Neuronographic analysis of medial and basal cerebral cortex. I. Cat, *J. Neurophysiol. 16,* 312–323.
MacLean, P. D., Horwitz, N. H., and Robinson, F. (1952). Olfactory-like responses in pyriform area to non-olfactory stimulation, *Yale J. Biol. Med. 25,* 159–172.
MacLean, P. D., Flanigan, S., Flynn, J. P., Kim, C., and Stevens, J. R. (1955-56). Hippocampal function: Tentative correlations of conditioning, EEG, drug, and radioautographic studies, *Yale J. Biol. Med. 28,* 380–395.

MacLean, P. D., Denniston, R. H., Dua, S., and Ploog, D. W. (1962). Hippocampal changes with brain stimulation eliciting penile erection, in: *Physiologie de L'hippocampe,* Colloques Internationaux du Centre National de la Recherche Scientifique, *107,* 491–510.

MacLean, P. D., Denniston, R. H., and Dua, S. (1963a). Further studies on cerebral representation of penile erection: Caudal thalamus, midbrain, and pons, *J. Neurophysiol. 26,* 273–293.

MacLean, P. D., Dua, S., and Denniston, R. H. (1963b). Cerebral localization for scratching and seminal discharge, *Arch. Neurol. 9,* 485–497.

MacLean, P. D., Yokota, T., and Kinnard, M. D. (1968). Photically sustained on-responses of units in posterior hippocampal gyrus of awake monkey, *J. Neurophysiol. 31,* 870–883.

MacNeilage, P. F. (1987). The evolution of hemispheric specialization for manual function and language, in: *Higher Brain Functions: Recent Explorations of the Brain's Emergent Properties* (S. P. Wise, ed.), Wiley, New York, pp. 285–309.

MacPherson, J. M., Rasmusson, D. D., and Murphy, J. (1980). Activities of neurons in "motor" thalamus during control of limb movement in the primate, *J. Neurophysiol. 44,* 11–28.

Macrae, D. (1954). On the nature of fear, with reference to its occurrence in epilepsy, *J. Nerv. Ment. Dis. 120,* 385–393.

Magendie, F. (1823). On the functions of the corpora striata and corpora quadrigemina, *Lancet 1,* 439–441.

Magendie, F. (1841). *Leçons sur les fonctions et les maladies du système nerveux,* two volumes, Lecaplain, Paris.

Magistretti, P., Uren, R., Blume, H., Schomer, D., and Royal, H. (1982). Delineation of epileptic focus by single photon emission tomography, *Eur. J. Nucl. Med. 7,* 484–485.

Magnus, O., and Lammers, H. J. (1956). The amygdaloid–nuclear complex, *Folia Psychiatr. Neurol. Neurochir. Neerl. 59,* 1–28.

Magoun, H. W., Atlas, D., Ingersoll, E. H., and Ranson, S. W. (1937). Associated facial, vocal and respiratory components of emotional expression: An experimental study, *J. Neurol. Psychopathol. 17,* 241–255.

Mahut, H., Zola-Morgan, S., and Moss, M. (1982). Hippocampal resections impair associative learning and recognition memory in the monkey, *J. Neurosci. 2,* 1214–1229.

Malamud, N. (1966). The epileptogenic focus in temporal lobe epilepsy from a pathological standpoint, *Arch. Neurol. 14,* 190–195.

Malamud, N., and Skillikorn, S. A. (1956). Relationship between the Wernicke and the Korsakoff syndrome, *Arch. Neurol. Psychiatry 76,* 585–596.

Malmo, R. B. (1948). Symposium on gyrectomy. Part 3—Psychological aspects of frontal gyrectomy and frontal lobotomy in mental patients, in: *The Frontal Lobes, Res. Publ. Assoc. Res. Nerv. Ment. Dis. 27,* 537–564.

Mandelbrot, B. B. (1977). *Fractals: Form, Chance, and Dimension,* W. H. Freeman, San Francisco.

Mandelbrot, B. B. (1983). *The Fractal Geometry of Nature,* W. H. Freeman, New York.

Mantyh, P. W. (1983). Connections of midbrain periaqueductal gray in the monkey. I. Ascending efferent projections, *J. Neurophysiol. 49,* 567–581.

Marcellini, D. L. (1970). Ethoecology of *Hemidactylus frenatus* (Sauria, Gekkonidae) with emphasis on acoustic behavior, An unpublished dissertation submitted to the University of Oklahoma.

Marchi, V., and Algeri, G. (1885). Sulle degenerazioni discendenti consecutive a lesioni sperimentale in diverse zone della corteccia cerebrale, *Riv. Sper. Freniatr. 11,* 492–494.

Margerison, J. H., and Corsellis, J. A. N. (1966). Epilepsy and the temporal lobes. A clinical, electroencephalographic and neuropathological study of the brain in epilepsy, with particular reference to the temporal lobes, *Brain 89,* 499–530.

Marshak, A. (1976). Implications of the paleolithic symbolic evidence for the origin of language, *Am. Sci. 64,* 136–145.

Marshall, W. H., Talbot, S. A., and Ades, H. W. (1943). Cortical response of the anesthetized cat to gross photic and electrical afferent stimulation, *J. Neurophysiol. 6,* 1–15.

Martin, G. F., Jr., and Hamel, E. G., Jr. (1967). The striatum of the opossum, *Didelphis virginiana.* Description and experimental studies, *J. Comp. Neurol. 131,* 491–516.

Martin, J. P. (1927). Hemichorea resulting from a local lesion of the brain (the syndrome of the body of Luys), *Brain 50,* 637–651.

Martin, J. P. (1950). Fits of laughter (sham mirth) in organic cerebral disease, *Brain 73,* 453–464.

Martin, J. P. (1967). *The Basal Ganglia and Posture,* Pitman Medical, London.

Maske, H. (1955). Über den topochemischen Nachweis von Zink im Ammonshorn verschiedener Säugetiere, *Naturwissenschaften 42,* 424.

Mason, J. W. (1958). The central nervous system regulation of ACTH secretion, in: *Reticular Formation of the Brain,* (H. H. Jasper, ed.), Little, Brown, Boston, pp. 645–662.

Mason, J. W. (1959). Plasma 17-hydroxycorticosteroid levels during electrical stimulation of the amygdaloid complex in conscious monkeys, *Am. J. Physiol. 196,* 44–48.

Maspes, P. E., and Pagni, C. A. (1964). Stereoelectroencephalographic study of three cases of psychomotor epilepsy, *Confin. Neurol. 24,* 321–335.

Masters, R. D. (1981). Linking ethology and political science: Photographs, political attention, and presidential elections, in: *Biopolitics: Ethnological and physiological approaches. New directions for methodology of social and behavioral sciences, No. 7* (M. Watts, ed.), Jossey–Bass, San Francisco, pp. 61–89.

Maurus, M., Mitra, J., and Ploog, D. (1965). Cerebral representation of the clitoris in ovariectomized squirrel monkeys, *Exp. Neurol. 13,* 283–288.

Mayhew, W. W. (1963). Observations on captive *Amphibolurus pictus,* an Australian agamid lizard, *Herpetologica 19,* 81–88.

Mayr, E. (1977). Concepts in the study of animal behavior, in: *Reproductive Behavior and Evolution* (J. S. Rosenblatt and B. R. Komisaruk, eds.), Plenum Press, New York, pp. 1–16.

McBride, R. L., and Sutin, J. (1976). Projections of the locus coeruleus and adjacent pontine tegmentum in the cat, *J. Comp. Neurol. 165,* 265–284.

McGeer, P. L., Eccles, J. C., and McGeer, E. G. (1978). *Molecular Neurobiology of the Mammalian Brain,* Plenum Press, New York.

McGuinness, C., Dalsass, M., Proshansky, E., and Krauthamer, G. (1976). Afferent connections of the nucleus centrum medianum in the cat, *Soc. Neurosci. Abstr. 2,* 67.

McHugh, P. R., and Smith, G. P. (1967). Plasma 17-OHCS response to amygdaloid stimulation with and without afterdischarges, *Am. J. Physiol. 212,* 619–622.

McIntyre, H. D., Mayfield, F. H., and McIntyre, A. P. (1954). Ventromedial quadrant coagulation in the treatment of the psychoses and neuroses, *Am. J. Psychiatry 3,* 112–119.

McLardy, T. (1969). Ammonshorn pathology and epileptic dyscontrol, *Nature 221,* 877–878.

McLennan, H., and York, D. H. (1967). The action of dopamine on neurones of the caudate nucleus, *J. Physiol. (London) 189,* 393–402.

Meadows, D. H., Meadows, D. L., Randers, J., and Behrens, W. W., III (1972). *The Limits to Growth,* Universe Books, New York.

Mehler, W. R. (1966). Further notes on the centre médian nucleus of Luys, in: *The Thalamus* (D. P. Purpura and M. D. Yahr, eds.), Columbia University Press, New York, pp. 109–122.

Mehler, W. R. (1980). Subcortical afferent connections of the amygdala in the monkey, *J. Comp. Neurol. 190,* 733–762.

Mehler, W. R., Feferman, M. E., and Nauta, W. J. H. (1960). Ascending axon degeneration following anterolateral cordotomy, *Brain 83,* 718–750.

Meibach, R. C., and Siegel, A. (1977a). Efferent connections of the hippocampal formation in the rat, *Brain Res. 124,* 197–224.

Meibach, R. C., and Siegel, A. (1977b). Thalamic projection of the hippocampal formation: Evidence for an alternate pathway involving the internal capsule, *Brain Res. 134,* 1–12.

Meldrum, B. S., and Brierly, J. B. (1973). Prolonged epileptic seizures in primates, *Arch. Neurol. 28,* 10–17.

Mellgren, S. I., and Srebro, B. (1973). Changes in acetylcholinesterase and distribution of degeneration fibres in the hippocampal region after septal lesions in the rat, *Brain Res. 52,* 19–36.

Meltzer, S. J. (1906–07). The factors of safety in animal structure and animal economy, *Harvey Lect.* 139–169.

Merlis, J. K. (1970). Proposals for an international classification of the epilepsies, *Epilepsia 11,* 114–119.

Mesulam, M.-M. (1976). The blue reaction product in horseradish peroxidase neurohistochemistry: Incubation parameters and visibility, *J. Histochem. Cytochem. 24,* 1273.

Mesulam, M.-M. (1978). Tetramethyl benzidine for horseradish peroxidase neurohistochemistry: A non-carcinogenic blue reaction-product with superior sensitivity for visualizing neural afferents and efferents, *J. Histochem. Cytochem. 26,* 106–117.

Mesulam, M.-M., and Mufson, E. J. (1982). Insula of the old world monkey. III: Efferent cortical output and comments on function, *J. Comp. Neurol. 212,* 38–52.

Mesulam, M.-M., and Van Hoesen, G. (1976). Acetylcholinesterase-rich projections from the basal forebrain of the rhesus monkey to neocortex, *Brain Res. 109,* 152–157.

Mettler, F. A. (1942). Relation between pyramidal and extrapyramidal functions, *Res. Publ. Assoc. Res. Nerv. Ment. Dis. 21,* 150–227.

Mettler, F. A. (1948). *Neuroanatomy,* Mosby, St. Louis.
Mettler, F. A., Ades, H. W., Lipman, E., and Culler, E. A. (1939). The extrapyramidal system. An experimental demonstration of function, *Arch. Neurol. Psychiatry 41,* 984–995.
Meyer, A., Beck, E., and McLardy, T. (1947). Prefrontal leucotomy: A neuro-anatomical report, *Brain 70,* 18–49.
Meyer, J. S., and Barron, D. W. (1960). Apraxia of gait: A clinico-pathological study, *Brain 83,* 261–284.
Meyer, M., and Allison, A. C. (1949). Experimental investigation of the connexions of the olfactory tracts in the monkey, *J. Neurol. Neurosurg. Psychiatry 12,* 274–286.
Meyers, R. (1942). The modification of alternating tremors, rigidity, and festination by surgery of the basal ganglia, *Res. Publ. Assoc. Res. Nerv. Ment. Dis. 21,* 602–665.
Meyerson, B. J. (1964). Central nervous monoamines and hormone induced estrus behavior in the spayed cat, *Acta Physiol. Scand. 63,* 1–32.
Meynert, T. (1868). Studien über das pathologisch-anatomische Material der Wiener Irrenanstalt, *Vjschr. Psychiatr. 1.*
Meynert, T. (1872). The brain of mammals, in: *A Manual of Histology* (S. Stricker, ed.), Wood, New York, pp. 650–760.
Michael, C. R. (1969). Retinal processing of visual images, *Sci. Am. 220,* 104–114.
Michael, R. P. (1965). Oestrogens in the central nervous system, *Br. Med. Bull. 21,* 87–90.
Miller, N., and Dollard, J. (1941). *Social Learning and Imitation,* Yale University Press, New Haven, Conn.
Milner, B. (1964). Some effects of frontal lobectomy in man, in: *The Frontal Granular Cortex and Behavior* (J. M. Warren and K. Akert, eds.), McGraw–Hill, New York, pp. 313–334.
Milner, B., Corkin, S., and Teuber, H.-L. (1968). Further analysis of the hippocampal amnesic syndrome: 14-year follow-up study of H.M., *Neuropsychologia 6,* 215–234.
Mishkin, M. (1972). Cortical visual areas and their interactions, in: *Brain and Human Behavior* (A. G. Karczmar and J. C. Eccles, eds.), Springer-Verlag, Berlin, pp. 187–208.
Mishkin, M. (1978). Memory in monkeys severely impaired by combined but not by separate removal of amygdala and hippocampus, *Nature 273,* 297–298.
Mishkin, M. (1982). A memory system in the monkey, *Philos. Trans. R. Soc. London Ser. B 298,* 85–95.
Miyadi, D. (1964). Social life of Japanese monkeys, *Science 143,* 783–786.
Miyata, M., and Sasaki, M. (1983). HRP studies on thalamocortical neurons related to the cerebellocerebral projection in the monkey, *Brain Res. 274,* 213–224.
von Monakow, C. (1895). Experimentelle und pathologische-anatomische Untersuchungen über die Haubenregion, den Sehhügel und die Regio subthalamica, nebst Beiträgen zur Kenntnis früherworbener Gross- und Kleinhirndefekte, *Arch. Psychiatr. Nervenkr. 27,* 1–128.
Moniz, E. (1936). A cirurgia ao serviço da psiquiatria, *A Méd. contemp. 54,* 159.
Monod, J. (1971). *Chance and Necessity,* Knopf, New York.
Monroe, R. R. (1986). Episodic behavioral disorders and limbic ictus, in: *The Limbic System: Functional Organization and Clinical Disorders* (B. K. Doane and K. E. Livingston, ed.), Raven Press, New York, pp. 251–266.
Montagu, A. (1956). *The Biosocial Nature of Man,* Grove Press, New York.
Montagu, A. (1976). *The Nature of Human Aggression,* Oxford University Press, New York.
Montagu, K. A. (1957). Catechol compounds in rat tissues and in brains of different animals, *Nature 180,* 244–245.
Moore, R. Y. (1978). Catecholamine innervation of the basal forebrain. I. The septal area, *J. Comp. Neurol. 177,* 665–684.
Moore, R. Y., and Halaris, A. E. (1975). Hippocampal innervation by serotonin neurons of the midbrain raphe in the rat, *J. Comp. Neurol. 164,* 171–184.
Moore, R. Y., Bhatnagar, R. K., and Keller, A. (1971). Anatomical and chemical studies of a nigro-striatal projection of the cat, *Brain Res. 30,* 119–135.
Morest, D. K. (1961). Connexions of dorsal tegmental nucleus in rat and rabbit, *J. Anat. 95,* 229–246.
Morest, D. K. (1967). Experimental study of the projections of the nucleus of the tractus solitarius and the area postrema in the cat, *J. Comp. Neurol. 130,* 277–299.
Morest, D. K. (1970). A study of neurogenesis in the forebrain of opossum pouch young, *Z. Anat. Entwicklungsgesch. 130,* 265–305.
Morgane, P. J., McFarland, W. L., and Jacobs, M. S. (1982). The limbic lobe of the dolphin brain: A quantitative cytoarchitectonic study, *J. Hirnforsch. 23,* 465–552.

Morgane, P. J., Jacobs, M. S., and Galaburda, A. (1986). Evolutionary morphology of the dolphin brain, in: *Dolphin Behavior and Cognition: Comparative and Ecological Aspects* (R. Buhr, R. Schusterman, J. Thomas, and F. Wood, eds.), Erlbaum, Hillsdale, NJ, pp. 5–29.

Morison, R. S., and Dempsey, E. W. (1942). A study of thalamo-cortical relations, *Am. J. Physiol. 135,* 281–300.

Morris, D. (1957). "Typical intensity" and its relations to the problem of ritualisation, *Behaviour 11,* 1–12.

Morris, D. (1967). *The Naked Ape,* McGraw–Hill, New York.

Morris, D. (1979). Gestures. The hands may be as eloquent as the tongue, *Nat. His. 88,* 114–121.

Morris, W. (1858/1933). The haystack in the flood, in: *The Defence of Guenevere and Other Poems,* Oxford University Press, London, pp. 131–136.

Morrison, F., and Poletti, C. E. (1980). Hippocampal influence on amygdala unit activity in awake squirrel monkeys, *Brain Res. 192,* 353–369.

Mosko, S., Lynch, G., and Cotman, C. W. (1973). The distribution of septal projections to the hippocampus of the rat, *J. Comp. Neurol. 152,* 163–174.

Mufson, E. J., Mesulam, M.-M., and Pandya, D. N. (1979). Insular cortex and amygdala have reciprocal connections in the rhesus monkey, *Soc. Neurosci. Abstr. 5,* 280.

Mulder, D. W., and Daly, D. (1952). Psychiatric symptoms associated with lesions of temporal lobe, *J. Am. Med. Assoc. 150,* 173–176.

Mulder, D. W., Daly, D., and Bailey, A. A. (1954). Visceral epilepsy, *AMA Arch. Intern. Med. 93,* 481–493.

Mullan, S., and Penfield, W. (1959). Illusions of comparative interpretation and emotion, *AMA Arch. Neurol. Psychiatry 81,* 269–284.

Müller-Preuss, P., and Jürgens, U. (1976). Projections from the 'cingular' vocalization area in the squirrel monkey, *Brain Res. 103,* 29–43.

Murphy, M. R. (1970). Territorial behavior of the caged male golden hamster, *Proc. Am. Psychol. Assoc. 78,* 237–238.

Murphy, M. R. (1971). Natural history of the Syrian golden hamster—a reconnaissance expedition, *Abstr. Am. Zool. 11,* 632.

Murphy, M. R. (1973). Effects of female hamster vaginal discharge on the behavior of male hamsters, *Behav. Biol. 9,* 367–375.

Murphy, M. R. (1976). Olfactory stimulation and olfactory bulb removal: Effects on territorial aggression in male Syrian golden hamsters, *Brain Res. 113,* 95–110.

Murphy, M. R. (1977). Interspecific sexual preferences of female hamsters, *J. Comp. Physiol. Psychol. 91,* 1337–1346.

Murphy, M. R., MacLean, P. D., and Hamilton, S. C. (1981). Species-typical behavior of hamsters deprived from birth of the neocortex, *Science 213,* 459–461.

Murphy, M. R., MacLean, P. D., and Hamilton, S. C. (1981). Species-typical behavior of hamsters deprived from birth of neocortex, *Science 213,* 459–461.

Murray, E. A., and Mishkin, M. (1983). Severe tactual memory deficits in monkeys after combined removal of the amygdala and hippocampus, *Brain Res. 270,* 340–344.

Myers, K., Hale, C. S., Mykytowycz, R., and Hughes, R. L. (1971). The effects of varying density and space on sociality and health in animals, in: *Behavior and Environment* (A. H. Esser, ed.), Plenum Press, New York, pp. 148–187.

Myers, R. D., and Beleshin, D. B. (1971). Changes in serotonin release in hypothalamus during cooling or warming of the monkey, *Am. J. Physiol. 220,* 1746–1753.

Myers, R. E. (1963). Projections of the superior colliculus in monkey, *Anat. Rec. 145,* 264.

Myers, R. E., and Swett, C., Jr. (1970). Social behavior deficits of free-ranging monkeys after anterior temporal cortex removal: A preliminary report, *Brain Res. 18,* 551–556.

Nadler, R. D. (1975). Sexual cyclicity in captive lowland gorillas, *Science 189,* 813–814.

Namba, M. (1957). Cytoarchitektonische Untersuchungen am Striatum, *J. Hirnforsch. 3,* 24–48.

Naquet, R. (1954). Effects of stimulation of the rhinencephalon in the waking cat, *Electroencephalogr. Clin. Neurophysiol. 6,* 711–712.

Narabayashi, H., Nagao, T., Saito, Y., Yoshida, M., and Nagahata, M. (1963). Stereotaxic amygdalotomy for behavior disorders, *Arch. Neurol. 9,* 1–16.

Narkiewicz, O. (1964). Degenerations in the claustrum after regional neocortical ablations in the cat, *J. Comp. Neurol. 123,* 335–356.

Nauta, H. J. W., Pritz, M. B., and Lasek, R. J. (1974). Afferents to the rat caudoputamen studied with horseradish peroxidase. An evaluation of a retrograde neuroanatomical research method, *Brain Res. 67,* 219–238.

Nauta, W. J. H. (1953). Some projections of the medial wall of the hemisphere in the rat's brain (cortical areas 32 and 25, 24 and 29), *Anat. Rec. 115,* 352.

Nauta, W. J. H. (1956). An experimental study of the fornix system in the rat, *J. Comp. Neurol. 104,* 247–271.

Nauta, W. J. H. (1958). Hippocampal projections and related neural pathways to the mid-brain in the cat, *Brain 81,* 319–340.

Nauta, W. J. H. (1962). Neural associations of the amygdaloid complex in the monkey, *Brain 85,* 505–520.

Nauta, W. J. H. (1964). Some efferent projections of the prefrontal cortex in the monkey, in: *The Frontal Granular Cortex and Behavior* (J. M. Warren and K. Akert, eds.), McGraw-Hill, New York, pp. 394–409.

Nauta, W. J. H., and Domesick, V. B. (1978). Crossroads of limbic and striatal circuitry: Hypothalamo-nigral connections, in: *Limbic Mechanisms: The Continuing Evolution of the Limbic System Concept* (K. E. Livingston and O. Hornykiewicz, eds.), Plenum Press, New York, pp. 75–93.

Nauta, W. J. H., and Gygax, P. A. (1954). Silver impregnation of degenerating axons in the central nervous system: A modified technic, *Stain Technol. 29,* 91–93.

Nauta, W. J. H., and Karten, H. J. (1970). A general profile of the vertebrate brain, with sidelights on the ancestry of cerebral cortex, *The Neurosciences. Second Study Program* (F. O. Schmitt, ed.), Rockefeller University Press, New York, pp. 7–26.

Nauta, W. J. H., and Kuypers, H. G. J. M. (1958). Some ascending pathways in the brain stem reticular formation of the cat, in: *Reticular Formation of the Brain* (H. H. Jasper, ed.), Little, Brown, Boston.

Nauta, W. J. H., and Mehler, W. R. (1966). Projections of the lentiform nucleus in the monkey, *Brain Res. 1,* 3–42.

Nauta, W. J. H., and Whitlock, D. G. (1954). An anatomical analysis of the non-specific thalamic projection system, in: *Brain Mechanisms and Consciousness* (J. Delafresnaye, ed.), Blackwell, Oxford, pp. 81–116.

Nelson, K. B., and Ellenberg, J. H. (1978). Prognosis in children with febrile seizures, *Pediatrics 61,* 720–727.

Nettles, W. C., Jr., Xie, Z. N., Ball, D., Shenkir, C. A., and Vinson, S. (1982). Reversal of morphine disruption of maternal behavior by concurrent treatment with the opiate antagonist naloxone, *Science 218,* 166–168.

New Larousse Encyclopedia of Mythology (1968). Hamlyn, London.

Newman, J. D. (1979). Central nervous system processing of sounds in primates, in: *Neurobiology of Social Communication in Primates: An Evolutionary Perspective* (H. Steklis and M. Raleigh, eds.), Academic Press, New York, pp. 69–109.

Newman, J. D. (1985a). The infant cry of primates. An evolutionary perspective, In: *Infant Crying: Theoretical and Research Perspectives* (B. M. Lester and C. F. Z. Boukydis, eds.), Plenum Press, New York, pp. 307–323.

Newman, J. D. (1985b). Squirrel monkey communication, in: *Handbook of Squirrel Monkey Research* (L. A. Rosenblum and C. L. Coe, eds.), Plenum Press, New York, pp. 99–126.

Newman, J. D., and MacLean, P. D. (1982). Effects of tegmental lesions on the isolation call of squirrel monkeys, *Brain Res. 232,* 317–330.

Newman, J. D., and MacLean, P. D. (1985). Importance of medial frontolimbic cortex in production of the isolation call of squirrel monkeys, *Soc. Neurosci. Abstr. 16,* 495.

Newman, J. D., Murphy, M. R., and Harbaugh, C. R. (1982). Naloxone-reversible suppression of isolation call production after morphine injections in squirrel monkeys, *Soc. Neurosci. Abstr. 8,* 940.

Newman, J. D., and Symmes, D. (1982). Inheritance and experience in the acquisition of primate acoustic behavior, in: *Primate Communication* (C. T. Snowden, C. H. Brown, and M. R. Peterson, eds.), Cambridge University Press, Cambridge, pp. 259–278.

Newton, I. (1729/1962). *Principia Mathematica* (Trans. by A. Motte), University of California Press, Berkeley.

Nice, M. M. (1933). The theory of territorialism and its development, in: *Fifty Years Progress of American Ornithology,* American Ornithologists' Union, Lancaster, pp. 89–100.

Nicolis, G. (1975). Dissipative instabilities, structure, and evolution, *Adv. Chem. Phys. 29,* 29–47.

Nicolis, J. S., and Protonotarios, E. N. (1979). Bifurcation in nonnegotiable games: A paradigm for self-organisation in cognitive systems, *Int. J. Bio-Med. Comput. 10,* 417–447.

Nicoll, R. A., Siggins, G. R., Ling, N., Bloom, F. E., and Guillemin, R. (1977). Neuronal actions of endorphins and enkephalins among brain regions: A comparative microiontophoretic study, *Proc. Natl. Acad. Sci. USA 74,* 2584–2588.

Nielsen, J. M. (1946). *Agnosia, Apraxia, Aphasia,* 2nd Edition, Harper & Row, New York.

Nielsen, J. M., and Jacobs, L. L. (1951). Bilateral lesions of the anterior cingulate gyri, *Bull. Los Angeles Neurol. Soc. 16,* 231–234.

Nietzsche, F. (1888/1968). Thus spoke Zarathustra, in: *The Portable Nietzsche* (Trans. by W. Kaufmann), Viking Press, New York.

Nietzsche, F. (1908/1969). *Ecce Homo* (Trans. by W. Kaufmann), Vintage Books, New York, pp. 215–335.

Niki, H., Sakai, M., and Kubota, K. (1972). Delayed alternation performance and unit activity of the caudate head and medial orbitofrontal gyrus in the monkey, *Brain Res. 38,* 343–353.

Nilges, R. G. (1944). The arteries of the mammalian cornu ammonis, *J. Comp. Neurol. 80,* 177–190.

Nissl, F. von (1892a). Über die Veranderungen der Ganglienzellen am Facialiskern des Kaninchens nach Ausreissung der Nerven, *Allg. Z. Psychiatr. Ihre Grenzgeb. 48,* 192–198.

Nissl, F. von (1892b). Über die sogenannten Granula der Nervenzellen, *Neurol. Zentralbl. 13,* 676–685.

Nissl, F. von (1913). Die Grosshirnanteile des Kaninchens, *Arch. Psychiatr. Nervenkr. 52,* 866–953.

Noble, G. K. (1936). Function of the corpus striatum in the social behavior of fishes, *Anat. Rec. 64,* 34.

Noble, G. K., and Bradley, H. T. (1933). The mating behavior of lizards; its bearing on the theory of sexual selection, *Ann. N.Y. Acad. Sci. 35,* 25–100.

Noirot, E. (1968). Ultrasounds in young rodents. II: Changes with age in albino rats, *Anim. Behav. 16,* 129–134.

Noirot, E. (1972). Ultrasounds and maternal behavior in small rodents, *Dev. Psychobiol. 5,* 371–387.

Norgren, R. (1976). Taste pathways to hypothalamus and amygdala, *J. Comp. Neurol. 166,* 17–30.

Norgren, R., and Leonard, C. M. (1973). Ascending central gustatory pathways, *J. Comp. Neurol. 150,* 217–238.

Norris, F. H., and Lanauze, H. E. (1960). Generalized automatism evoked in cat from temporal depth electrodes, *Electroencephalogr. Clin. Neurophysiol. 12,* 887.

Northrop, F. S. C. (1963). *Man, Nature, and God. A Quest for Life's Meaning,* Pocket Books, New York.

Obersteiner, H. (1887/1890). *The Anatomy of the Central Nervous Organs in Health and in Disease* (Trans. by A. Hill), Griffin, London.

Offen, M. L., Davidoff, R. A., Troost, B. T., and Richey, E. T. (1976). Dacrystic epilepsy, *J. Neurol. Neurosurg. Psychiatry 39,* 829–834.

Okon, E. E. (1972). Factors affecting ultrasound production in infant rodents, *J. Zool. 168,* 139–148.

Olds, J. (1958). Selective effects of drives and drugs on "reward" systems of the brain, in: *Neurological Basis of Behavior* (G. E. W. Wolstenholme and C. M. O'Connor, eds.), Churchill, London, pp. 124–141.

Olds, J., and Milner, P. (1954). Positive reinforcement produced by electrical stimulation of septal areas and other regions of the rat brain, *J. Comp. Physiol. Psychol. 47,* 419–427.

O'Leary, J. L., and Bishop, G. H. (1938). Margins of the optically excitable cortex in the rabbit, *Arch. Neurol. Psychiatry 40,* 482–499.

Oliver, G., and Schäfer, E. A. (1895). The physiological effects of extracts of the suprarenal capsules, *J. Physiol. (London) 18,* 230–276.

Olivier, A., Parent, A., Simard, H., and Poirier, L. J. (1970). Cholinesterasic striatopallidal and striantonigral efferents in the cat and the monkey, *Brain Res. 18,* 273–282.

Olson, C. R., and Graybiel, A. M. (1980). Sensory maps in the claustrum of the cat, *Nature 288,* 479–481.

Olson, E. C. (1944). Origin of mammals based upon cranial morphology of the therapsid suborders, *Geol. Soc. Am. 55,* 1–136.

Olson, E. C. (1959). The evolution of mammalian characters, *Evolution 13,* 344–353.

Olson, L., Seiger, A., and Fuxe, K. (1972). Heterogeneity of striatal and limbic dopamine innervation: Highly fluorescent islands in developing and adult rats, *Brain Res. 44,* 282–288.

Olton, D. S. (1978). The function of septo-hippocampal connections in spatially organized behaviour, in: *Functions of the Septo-Hippocampal System* (K. Elliott and J. Whelan, eds.), Elsevier—Excerpta Medica, North-Holland, Amsterdam, pp. 327–349.

Ommaya, A. K. (1968). Mechanical properties of tissues of the nervous system, *J. Biomech. 1,* 127–138.

Orbach, J., Milner, B., and Rasmussen, T. (1960). Learning and retention in monkeys after amygdala–hippocampus resection, *Arch. Neurol. 3,* 230–251.

Orr, A. H., MacLean, P. D., Creecy, R. H., Eckardt, M. J., and Vaughn, W. J. (1981). An approach to the study of respiration-related units in the forebrain, *Electroencephalogr. Clin. Neurophysiol. 51,* 31.

Orthner, H., and Roeder, F. (1962). Erfahrungen mit stereotaktischen Eingriffen, *Acta Neurochir. 10,* 572–629.

Osborn, H. F. (1903). The reptilian subclasses Diapsida and Synapsida and the early history of the Diaptosauria, *Mem. Am. Mus. Nat. Hist. 1,* 451–519.

Osler, W. (1894). *On Chorea and Choreiform Affections,* Blakiston, Philadelphia.

Ottersen, O. P. (1980). Afferent connections to the amygdaloid complex of the rat and cat: II. Afferents from the hypothalamus and the basal telencephalon, *J. Comp. Neurol. 194,* 267–289.

Ottersen, O. P., and Ben-Ari, Y. (1979). Afferent connections to the amygdaloid complex of the rat and cat, *J. Comp. Neurol. 187,* 401–424.

Paasonen, M. K., and Vogt, M. (1956). The effects of drugs on the amounts of substance P and 5-hydroxytryptamine in mammalian brain, *J. Physiol. (London) 131,* 617–626.

Paasonen, M. K., MacLean, P. D., and Giarman, N. J. (1957). 5-Hydroxytryptamine (serotonin, enteramine) content of structures of the limbic system, *J. Neurochem. 1,* 326–333.

Pagels, H. R. (1982). *The Cosmic Code: Quantum Physics as the Language of Nature.* Simon & Schuster, New York.

Pagni, C. A. (1963). Étude électro-clinique des post-décharges amydalo-hippocampiques chez l'homme par moyen d'electrodes de profondeur placées avec méthode stéréotaxique, *Confin. Neurol. 23,* 477–499.

Pagni, C. A., and Marossero, F. (1965). Some observations on the human rhinencephalon: A stereoelectroencephalographic study, *Electroencephalogr. Clin. Neurophysiol. 18,* 260–271.

Pakkenberg, H., and Brody, H. (1965). The number of cells in the substantia nigra in paralysis agitans, *Acta Neuropathol. 5,* 320–324.

Palmer, A. R. (1974). Search for the Cambrian world, *Am. Sci. 62,* 216–224.

Pandya, D. N., and Barnes, C. L. (1987). Architecture and connections of the frontal lobe, in: *The Frontal Lobes Revisited* (E. Perecman, ed.), IRBN Press, New York, pp. 41–72.

Panksepp, J. (1981). The ontogeny of play in rats, *Dev. Psychobiol. 14,* 327–332.

Papez, J. W. (1929). *Comparative Neurology* (A manual and text for the study of the nervous system of vertebrates), Hafner, New York.

Papez, J. W. (1937). A proposed mechanism of emotion, *Arch. Neurol. Psychiatry 38,* 725–743.

Papez, J. W. (1942). A summary of fiber connections of the basal ganglia with each other and with other portions of the brain, *Res. Publ. Assoc. Res. Nerv. Ment. Dis. 21,* 21–68.

Parent, A. (1979). Monoaminergic systems of the brain, in: *Biology of the Reptilia,* Volume 10 (C. Gans, G. Northcutt, and P. S. Ulinski, eds.), Academic Press, New York, pp. 247–285.

Parent, A. (1986). *Comparative Neurobiology of the Basal Ganglia,* Wiley, New York.

Parent, A., and Olivier, A. (1970). Comparative histochemical study of the corpus striatum, *J. Hirnforsch. 12,* 73–81.

Parent, A., Boucher, R., and O'Reilly-Fromentin, J. (1981). Acetylcholinesterase-containing neurons in cat pallidal complex: Morphological characteristics and projection towards the neocortex, *Brain Res. 230,* 356–361.

Parkinson, J. (1817). *An Essay on the Shaking Palsy,* Whittingham & Rowland, London.

Parrington, F. R. (1941). On two mammalian teeth from the lower Rhaetic of Somerset, *Ann. Mag. Nat. Hist. 8,* 140–144.

Parrington, F. R. (1971). On the upper Triassic mammals, *Philos Trans. R. Soc. London Ser. B 261,* 231–272.

Parson, B. S. (1924). *Left-handedness,* Macmillan Co., New York.

Pasik, P., Pasik, T., and DiFiglia, M. (1979). The internal organization of the neostriatum in mammals, in: *The Neostriatum* (I. Divac and R. G. E. Oberg, eds.), Pergamon Press, Elmsford, N.Y., pp. 5–36.

Pasquier, D. A., and Reinoso-Suarez, F. (1976). Direct projections from hypothalamus to hippocampus in the rat demonstrated by retrograde transport of horseradish peroxidase, *Brain Res. 108,* 165–169.

Passouant, P., and Cadilhac, J. (1962). Les rythmes théta hippocampiques au cours du sommeil, *Physiologie de l'Hippocampique* Colloques Internationaux du Centre National de la Recherche Scientifique, Montpellier, *107,* 331–347.

Patten, B. M. (1953). *Human Embryology,* 2nd Edition, McGraw-Hill, New York.

Pavlov, I. P. (1928). *Lectures on Conditioned Reflexes* (Trans. by W. H. Gantt), International Publishers, New York.

Penfield, W. (1948). Symposium on gyrectomy. Part 1—Bilateral frontal gyrectomy and postoperative intelligence, in: *The Frontal Lobes, Res. Publ. Assoc. Res. Nerv. Ment. Dis. 27,* 529–534.

Penfield, W. (1952). Memory mechanisms, *Arch. Neurol. Psychiatry 67,* 178–191.

Penfield, W. (1956). Epileptogenic lesions, *Acta Neurol. Psychiatr. Belg. 56,* 75–88.

Penfield, W., and Erickson, T. C. (1941). *Epilepsy and Cerebral Localization,* Thomas, Springfield, Ill.

Penfield, W., and Faulk, M. E., Jr. (1955). The insula. Further observations on its function, *Brain 78,* 445–470.

Penfield, W., and Jasper, H. (1954). *Epilepsy and the Functional Anatomy of the Human Brain,* Little, Brown, Boston.

Penfield, W., and Milner, B. (1958). Memory deficit produced by bilateral lesions in the hippocampal zone, *AMA Arch. Neurol. Psychiatry 79,* 475–497.

Penfield, W., and Perot, P. (1963). The brain's record of auditory and visual experience. A final summary and discussion, *Brain 86,* 596–696.

Penfield, W., and Rasmussen, T. (1952). *The Cerebral Cortex of Man,* Macmillan Co., New York.

Penfield, W., and Roberts, L. (1959). *Speech and Brain-Mechanisms,* Princeton University Press, Princeton, N.J.

Penfield, W., and Welch, K. (1951). The supplementary motor area of the cerebral cortex, *Arch. Neurol. Psychiatry 66,* 289–317.

Penry, J. K. (1975). Perspectives in complex partial seizures, in: *Advances in Neurology,* Volume 11 (J. K. Penry and D. D. Daly, eds.), Raven Press, New York, pp. 1–14.

Perachio, A. A., Marr, L. D., and Alexander, M. (1979). Sexual behavior in male rhesus monkeys elicited by electrical stimulation of preoptic and hypothalamic areas, *Brain Res. 177,* 127–144.

Perry, T. L., Berry, K., Hansen, S., Diamond, S., and Mok, C. (1971). Regional distribution of amino acids in human brain obtained at autopsy, *J. Neurochem. 18,* 513.

Pert, C. B., and Snyder, S. H. (1973a). Opiate receptor: Demonstration in nervous tissue, *Science 179,* 1011–1014.

Pert, C. B., and Snyder, S. H. (1973b). Properties of opiate receptor binding in rat brain, *Proc. Natl. Acad. Sci USA 70,* 2243–2247.

Peters, R. P., and Mech, L. D. (1975). Scent-marking in wolves, *Am. Sci. 63,* 628–637.

Peterson, G. M. (1931). A preliminary report on the right and left handedness of the rat, *J. Comp. Psychol. 12,* 243.

Petrovicky, P. (1971). Structure and incidence of Gudden's tegmental nuclei in some mammals, *Acta Anat. 80,* 273–286.

Pfaff, D. W. (1968a). Autoradiographic localization of testosterone-^{3}H in the female rat brain and estradiol-^{3}H in the male rat brain, *Experientia 24,* 958–959.

Pfaff, D. W. (1968b). Uptake of ^{3}H-testosterone by the female rat brain. An autoradiographic study, *Endocrinology 82,* 1149–1155.

Pfaff, D. W., and Keiner, M. (1973). Atlas of estradiol-concentrating cells in the central nervous system of the female rat, *J. Comp. Neurol. 151,* 121–158.

Pfaff, D. W., Gerlach, J. L., McEwen, B. S., Ferin, M., Carmel, P., and Zimmerman, E. A. (1976). Autoradiographic localization of hormone-concentrating cells in the brain of the female rhesus monkey, *J. Comp. Neurol. 170,* 279–294.

Pfleger, L. (1880). Beobachtungen über Schrumpfung und Sclerose des Ammonshornes bei Epilepsie, *Allg. Z. Psychiatr. 36,* 359–365.

Phillips, G. B., Victor, M., Adams, R. D., and Davidson, C. S. (1952). A study of the nutritional defect in Wernicke's syndrome: The effect of a purified diet, thiamine, and other vitamins on the clinical manifestations, *J. Clin. Invest. 31,* 859–871.

Phillips, R. E. (1964). "Wildness" in the mallard duck: Effects of brain lesions and stimulation on "escape behavior" and reproduction, *J. Comp. Neurol. 122,* 139–156.

Phillips, R. E. (1968). Approach–withdrawal behavior of peach-faced lovebirds, *Agapornis roseicollis,* and its modification by brain lesions, *Behaviour 3,* 163–184.

Piaget, J. (1967). *Six Psychological Studies* (Trans. by A. Tenzer), Random House, New York.

Pickard, G. E., and Silverman, A.-J. (1981). Direct retinal projections to the hypothalamus, piriform cortex, and accessory optic nuclei in the golden hamster as demonstrated by a sensitive anterograde horseradish peroxidase technique, *J. Comp. Neurol. 196,* 155–172.

Pickel, V. M., Segal, M., and Bloom, F. E. (1974). A radioautographic study of the efferent pathways of the nucleus locus coeruleus, *J. Comp. Neurol. 155,* 15–42.

Plato (1937). Protagoras, in: *The Dialogues of Plato,* Volume 1 (Trans. by B. Jowett), Random House, New York.

Ploog, D. (1979). Phonation, emotion, cognition, with reference to the brain mechanisms involved, in: *Brain and Mind, Ciba Foundation Series 69,* Excerpta Medica, Amsterdam, pp. 79–98.

Ploog, D. W. (1981). Neurobiology of primate audio-vocal behavior, *Brain Res. Rev. 3,* 35–61.

Ploog, D. W., and MacLean, P. D. (1963a). On functions of the mammillary bodies in the squirrel monkey, *Exp. Neurol. 7,* 76–85.

Ploog, D. W., and MacLean, P. D. (1963b). Display of penile erection in squirrel monkey (*Saimiri sciureus*), *Anim. Behav. 11,* 32–39.

Ploog, D. W., Hopf, S., and Winter, P. (1967). Ontogenese des Verhaltens von Totenkopf-Affen (*Saimiri sciureus*), *Psychol. Forsch. 31,* 1–41.

Plotnik, R. (1968). Changes in social behavior of squirrel monkeys after anterior temporal lobectomy, *J. Comp. Physiol. Psychol. 66,* 369–377.

Plutarch (1962). Moralia: *A Letter to Apollonius,* Volume II (Trans. by F. C. Babbitt), Harvard University Press, Cambridge, Mass., pp. 180–185.

Plutchik, R. (1980). *Emotion: A Psychoevolutionary Synthesis,* Harper & Row, New York.

Poeck, K. (1969). Pathologic laughing and crying, in: *Handbook of Clinical Neurology,* Volume 2 (P. J. Vinken and G. W. Bruyn, eds.), Wiley–Interscience, New York, pp. 356–367.

Poeck, K., and Pilleri, G. (1963). Pathologisches Lachen und Weinen, *Schweiz. Arch. Neurol. Psychiatr. 92,* 323–370.

Poirier, L. J. (1960). Experimental and histological study of midbrain dyskinesias, *J. Neurophysiol. 23,* 534–551.

Poirier, L. J., and Sourkes, T. L. (1965). Influence of the substantia nigra on the catecholamine content of the striatum, *Brain 88,* 181–192.

Poletti, C. E., and Creswell, G. (1977). Fornix system efferent projections in the squirrel monkey: An experimental degeneration study, *J. Comp. Neurol. 175,* 101–127.

Poletti, C. E., and Sujatanond, M. A. (1980). Evidence for a second hippocampal efferent pathway to hypothalamus and basal forebrain comparable to fornix system: A unit study in the awake monkey, *J. Neurophysiol. 44,* 514–531.

Poletti, C. E., Kinnard, M. A., and MacLean, P. D. (1973). Hippocampal influence on unit activity of hypothalamus, preoptic region, and basal forebrain in awake, sitting squirrel monkeys, *J. Neurophysiol. 36,* 308–324.

Polyak, S. (1957). *The Vertebrate Visual System,* University of Chicago Press, Chicago.

Pool, J. L. (1954). The visceral brain of man, *J. Neurosurg. 11,* 45–63.

Porrino, L. J., and Goldman-Rakic, P. S. (1982). Brainstem innervation of prefrontal and anterior cingulate cortex in the rhesus monkey revealed by retrograde transport of HRP, *J. Comp. Neurol. 205,* 63–76.

Porter, R. W. (1954). The central nervous system and stress-induced eosinopenia, *Recent Prog. Horm. Res. 10,* 1–27.

Powell, E. W. (1966). Septal efferents in the cat, *Exp. Neurol. 14,* 328–337.

Powell, E. W. (1973). Limbic projections to the thalamus, *Exp. Brain Res. 17,* 394–401.

Powell, T. P. S., and Cowan, W. M. (1954). The connexions of the midline and intralaminar nuclei of the thalamus of the rat, *J. Anat. 88,* 307–319.

Powell, T. P. S., and Cowan, W. M. (1955). An experimental study of the efferent connexions of the hippocampus, *Brain 78,* 115–132.

Powell, T. P. S., and Cowan, W. M. (1956). A study of thalamo-striate relations in the monkey, *Brain 79,* 364–390.

Powell, T. P. S., Guillery, R. W., and Cowan, W. M. (1957). A quantitative study of the fornix-mamillo-thalamic system, *J. Anat. 91,* 419–437.

Powell, T. P. S., Cowan, W. M., and Raisman, G. (1965). The central olfactory connexions, *J. Anat. 99,* 791–813.

Powers, J. B., Fields, R. B., and Winans, S. S. (1979). Olfactory and vomeronasal system participation in male hamsters' attraction to female vaginal secretions, *Physiol. Behav. 22,* 77–84.

Pretorius, J. K., Phelan, K. D., and Mehler, W. R. (1979). Afferent connections of the amygdala in rat, *Anat. Rec. 193,* 657.

Pribram, K. H. (1967). The limbic systems, efferent control of neural inhibition and behavior, *Prog. Brain Res. 27,* 318–336.

Pribram, K. H. (1971). *Languages of the Brain: Experimental Paradoxes and Principles in Neuropsychology,* Prentice-Hall, Englewood Cliffs, N.J.

Pribram, K. H., and Bagshaw, M. (1953). Further analysis of the temporal lobe syndrome utilizing frontotemporal ablation, *J. Comp. Neurol. 99,* 347–375.

Pribram, K. H., and Fulton, J. F. (1954). An experimental critique of the effects of anterior cingulate ablations in monkey, *Brain 77,* 33–44.

Pribram, K. H., and MacLean, P. D. (1953). Neuronographic analysis of medial and basal cerebral cortex. II. Monkey, *J. Neurophysiol. 16,* 324–340.

Pribram, K. H., Chow, K. L., and Semmes, J. (1953). Limit and organization of the cortical projection from the medial thalamic nucleus in monkey, *J. Comp. Neurol. 98,* 433–448.

Price, J. L. (1973). An autoradiographic study of complementary laminar patterns of termination of afferent fibers to the olfactory cortex, *J. Comp. Neurol. 150,* 87–108.

Prigogine, I., and Lefever, R. (1975). Stability and self-organization in open systems, *Adv. Chem. Phys. 29,* 1–28.

Prus, J. (1898). Die Leitungsbahnen und Pathogenese der Rindenepilepsie, *Wien. Klin. Wochenschr. 11,* 857–863.

Putnam, T. J. (1926). Studies on the central visual system. II. A comparative study of the form of the geniculostriate visual system of mammals, *Arch. Neurol. Psychiatry 16,* 285–300.

Pye-Smith, P. H. (1871). On left-handedness, *Guy's Hosp. Rep. 16,* 141–146.

Quiroga, J. C. (1979). The brain of two mammal-like reptiles (Cynodontia–Therapsida), *J. Hirnforsch. 20,* 341–350.

Quiroga, J. C. (1980). The brain of the mammal-like reptile *Probainognathus jenseni* (Therapsida, Cynodontia). A correlative paleo-neoneurological approach to the neocortex at the reptile–mammal transition, *J. Hirnforsch. 21,* 299–336.

Quock, R. M., and Horita, A. (1974). Apomorphine: Modification of its hyperthermic effect in rabbits by p-chlorophenylalanine, *Science 183,* 539–540.

Radna, R. J., and MacLean, P. D. (1981a). Vagal elicitation of respiratory-type and other unit responses in striopallidum of squirrel monkeys, *Brain Res. 213,* 29–44.

Radna, R. J., and MacLean, P. D. (1981b). Vagal elicitation of respiratory-type and other unit responses in basal limbic structures of squirrel monkeys, *Brain Res. 213,* 45–61.

Radna, R. J., and Vaughn, W. J. (1977). Computer Assisted Unit Data Acquisition/Reduction. Natl. Tech. Inf. Syst., Springfield, Report No. NIH/DO-77/003, Accession No. PB 270, 745.

Radna, R. J., and Vaughn, W. J. (1978). Computer assisted unit data acquisition/reduction, *EEG J. 44,* 239–242.

Radna, R. J., MacLean, P. D., Orr, A., and Eckhardt, M. (1980). Forebrain units discharging in phase with respiration, *Electroencephalogr. Clin. Neurophysiol. 49,* 23P–27P.

Raisman, G. (1966). The connexions of the septum, *Brain 89,* 317–348.

Raisman, G., Cowan, W. M., and Powell, T. P. S. (1965). The extrinsic afferent, commissural and association fibres of the hippocampus, *Brain 88,* 963–996.

Raisman, G., Cowan, W. M., and Powell, T. P. S. (1966). An experimental analysis of the efferent projection of the hippocampus, *Brain 89,* 83–108.

Ramón y Cajal, S. (1891). Significacion fisiologica de las expaniones protoplasmaticas y nerviosas de las celulas de la sustanica gris por el, *Rev. Cienc. Med. Barcelona 17,* 637–679, 715–723.

Ramón y Cajal, S. (1893). *Nuevo Concepto de la Histologia de los Centros Nerviosos,* Imprenta de Henrich y C.ª en Comandita, Barcelona.

Ramón y Cajal, S. (1893/1968). *The Structure of Ammon's Horn* (Trans. by L. M. Kraft), Thomas, Springfield, Ill.

Ramón y Cajal, S. (1901–02/1955). *Studies on the Cerebral Cortex (Limbic Structures)* (Trans. by L. M. Kraft), Lloyd-Luke, London/Year Book, Chicago.

Ramón y Cajal, S. (1909–11). *Histologie duè système nerveux de l'homme et des vertébrés,* two volumes (Trans. by L. Azoulay), Maloine, Paris.

Ramón y Cajal, S. (1904/1954). *Neuron Theory or Reticular Theory?* (Trans. by M. U. Purkiss and C. A. Fox), Consejo Superior de Investigaciones Cientificas, Madrid.

Ramón y Cajal, S. (1928/1959). *Degeneration and Regeneration of the Nervous System* (Trans. and ed. by R. M. May), Hafner, New York.

Ramón y Cajal, S. (1933). ¿Neuronismo o Reticularismo? Las pruebas objetivas de la unidad anatómica de las células nerviosas, *Arch. Neurobiol. Madrid 13,* 144.

Rand, A. S. (1968). A nesting aggregation of iguanas, *Copeia 3,* 552–561.

Rand, A. S., and Rand, W. M. (1978). Display and dispute settlement in nesting iguanas, in: *The Behavior and Neurology of Lizards* (N. Greenberg and P. D. MacLean, eds.), U.S. Government Printing Office, Washington, D. C., DHEW Publication No. (ADM) 77-491, pp. 245–251.

Randrup, A., and Munkvad, I. (1974). Pharmacology and physiology of stereotyped behavior, *J. Psychiatr. Res. 11,* 1–10.

Ranson, S. W. (1939). Somnolence caused by hypothalamic lesions in the monkey, *Arch. Neurol. Psychiatry 41,* 1–23.

Ranson, S. W., and Berry, C. (1941). Observations on monkeys with bilateral lesions of the globus pallidus, *Arch. Neurol. Psychiatry 46,* 504–508.

Ranson, S. W., and Ranson, M. (1939). Pallidofugal fibers in the monkey, *Arch. Neurol. Psychiatry 42,* 1059–1067.

Ranson, S. W., and Ranson, S. W., Jr. (1942). Efferent fibers of the corpus striatum, in: *The Diseases of the Basal Ganglia, Res. Publ. Assoc. Res. Nerv. Ment. Dis. 21,* 69–76.

Ranson, S. W., Ranson, S. W., Jr., and Ranson, M. (1941). Corpus striatum and thalamus of a partially decorticate monkey, *Arch. Neurol. Psychiatry 46,* 402–415.

Rees, H. D., Switz, G. M., and Michael, R. P. (1980). The estrogen-sensitive neural system in the brain of female cats, *J. Comp. Neurol. 193,* 789–804.

Reis, D. J., and Oliphant, M. C. (1964). Bradycardia and tachycardia following electrical stimulation of the amygdaloid region in monkey, *J. Neurophysiol. 27,* 893–912.

Remillard, G. M., Ethier, R., and Andermann, F. (1974). Temporal lobe epilepsy and perinatal occlusion of the posterior cerebral artery: A syndrome analogous to infantile hemiplegia and a demonstrable etiology in some patients with temporal lobe epilepsy, *Neurology 24,* 1001–1009.

Renshaw, B., Forbes, A., and Morison, B. R. (1940). Activity of isocortex and hippocampus: Electrical studies with micro-electrodes, *J. Neurophysiol. 3,* 74–105.

Reymond, M.-C., and Battig, K. (1964). Investigation sur le ''syndrome septal'' du Rat consécutif à des lésions électrolytiques, *J. Physiol. (Paris) 56,* 807–828.

Ricardo, J. A., and Koh, E. T. (1978). Anatomical evidence of direct projections from the nucleus of the solitary tract to the hypothalamus, amygdala, and other forebrain structures in the rat, *Brain Res. 153,* 1–26.

Richardson, T. F., and Winokur, G. (1967). Déjà vu in psychiatric and neurosurgical patients, *Arch. Gen. Psychiatry 17,* 622–625.

Richter, E. (1966).Über die Entwicklung des Globus pallidus und des Corpus subthalamicum beim Menschen, in: *Evolution of the Forebrain* (R. Hassler and H. Stephan, eds.), Thieme, Stuttgart, pp. 285–295.

Rife, D. C. (1940). Handedness with special reference to twins, *Genetics 25,* 178–186.

Riley, H. A. (1960). *An Atlas of the Basal Ganglia, Brain Stem, and Spinal Cord,* Hafner, New York.

Rimland, B. (1964). *Infantile Autism; the Syndrome and its Implications for a Neural Theory of Behavior,* Appleton-Century-Crofts, New York.

Rinvik, E., Grofova, I., and Ottersen, P. (1976). Demonstration of nigrotectal and nigroreticular projections in the cat by axonal transport of proteins, *Brain Res. 112,* 388–394.

Rioch, D. M., Wislocki, G. B., and O'Leary, J. L. (1940). A precis of preoptic, hypothalamic, and hypophysial terminology, with atlas, in: *Hypothalamus and Central Levels of Autonomic Function, Res. Publ. Assoc. Res. Nerv. Ment. Dis. 20,* 3–30.

Riss, W., Burstein, S. D., and Johnson, R. W. (1963). Hippocampal or pyriform lobe damage in infancy and endocrine development of rats, *Am. J. Physiol. 204,* 861–866.

Ritvo, E. R., Freeman, B. J., Scheibel, A. B., Duong, T., Robinson, H., Guthrie, D., and Ritvo, A. (1986). Lower Purkinje cell counts in the cerebella of four autistic subjects: Initial findings of the UCLA-NSAC Autopsy Reseach Report, *Am. J. Psychiatry 143,* 7.

Roberts, L. H. (1975). Evidence for the laryngeal source of ultrasonic and audible cries of rodents, *J. Zool. 175,* 243–257.

Roberts, T. S., and Akert, K. (1963). Insular and opercular cortex and its thalamic projection in Macaca mulatta, *Schweiz. Arch. Neurol. Psychiatr. 92,* 1–43.

Robertson, R. T., and Kaitz, S. S. (1981). Thalamic connections with limbic cortex. I. Thalamocortical projections, *J. Comp. Neurol. 195,* 501–525.

Robertson, R. T., Kaitz, S. S., and Robards, M. J. (1980). A subcortical pathway links sensory and limbic systems of the forebrain, *Neurosci. Lett. 17,* 161–165.

Robinson, B. W. (1967). Vocalization evoked from forebrain in *Macaca mulatta, Physiol. Behav. 2,* 345–354.

Robinson, B. W., and Mishkin, M. (1966). Ejaculation evoked by stimulation of the preoptic area in monkey, *Physiol. Behav. 1,* 269–272.

Robinson, B. W., and Mishkin, M. (1968). Penile erection evoked from forebrain structures in *Macaca mulatta, Arch. Neurol. 19,* 184–198.

Robinson, F., and Lennox, M. A. (1951). Sensory mechanisms in hippocampus, cingulate gyrus, and cerebellum of the cat, *Fed. Proc. 10,* 110–111.

Rodin, E. A., Mulder, D. W., Faucett, R. L., and Bickford, R. G. (1955). Psychologic factors in convulsive disorders of focal origin, *AMA Arch. Neurol. Psychiatry 74,* 365–374.

Rogers, F. T. (1922). Studies of the brain stem. VI. An experimental study of the corpus striatum of the pigeon as related to various instinctive types of behavior, *J. Comp. Neurol. 35,* 21–59.

Roland, P. E., Meyer, E., Shibasaki, T., Yamamoto, Y. L., and Thompson, C. J. (1982). Regional cerebral blood flow changes in cortex and basal ganglia during voluntary movements in normal human volunteers, *J. Neurophysiol. 48,* 467–480.

Rolls, E. T., Thorpe, S. J., Maddison, S., Roper-Hall, A., Puerto, A., and Perret, D. (1979). Activity of neurones in the neostriatum and related structures in the alert animal, in: *The Neostriatum* (I. Divac and R. G. E. Oberg, eds.), Pergamon Press, Elmsford, N.Y., pp. 163–182.

Romer, A. S. (1958). Phylogeny and behavior with special reference to vertebrate evolution, in: *Behavior and Evolution* (A. Roe and G. C. Simpson, eds.), Yale University Press, New Haven, Conn., pp. 48–75.

Romer, A. S. (1966). *Vertebrate Paleontology,* University of Chicago Press, Chicago.

Romer, A. S. (1967). Major steps in vertebrate evolution, *Science 158,* 1629–1637.

Rosabal, F. (1967). Cytoarchitecture of the frontal lobe of the squirrel monkey, *J. Comp. Neurol. 130,* 87–108.

Rose, J. E., and Malis, L. I. (1965). Geniculo-striate connections in the rabbit. II. Cytoarchitectonic structure of the striate region and of the dorsal lateral geniculate body; organization of the geniculo-striate projections, *J. Comp. Neurol. 125,* 121–139.

Rose, J. E., and Woolsey, C. N. (1948). Structure and relations of limbic cortex and anterior thalamic nuclei in rabbit and cat, *J. Comp. Neurol. 89,* 279–348.

Rose, J. E., and Woolsey, C. N. (1958). Cortical connections and functional organization of the thalamic auditory system of the cat, in: *Biological and Biochemical Basis of Behavior* (H. F. Harlow and C. N. Woolsey, eds.), University of Wisconsin Press, Madison, pp. 127–150.

Rose, M. (1912). Histologische Lokalisation der Grosshirnrinde der kleinen Säugetiere, *J. Psychol. Neurol. 19,* 391–479.

Rose, M. (1926). Über das histogenetische Prinzip der Einteilung der Grosshirnrinde, *J. Psychol. Neurol. 32,* 97–160.

Rose, M. (1927). Gyrus limbicus anterior and Regio retrosplenialis (Cortex holoprotoptychos quinquestratificatus). Vergleichende Architektonik bei Tier und Mensch, *J. Psychol. Neurol. 35,* 65–173.

Rose, M. (1931). Cytoarchitektonischer Atlas der Grosshirnrinde des Kaninchens, *J. Psychol. Neurol. 43,* 353–440.

Rosenblueth, A., Wiener, N., and Bigelow, J. (1943). Behavior, purpose, and teleology, *Philos. Sci. 10,* 18–24.

Rosenblum, L. A., and Cooper, R. W. (1968). *The Squirrel Monkey,* Academic Press, New York.

Rosene, D. L., and Van Hoesen, G. W. (1977). Hippocampal efferents reach widespread areas of cerebral cortex and amygdala in the rhesus monkey, *Science 198,* 315–317.

Rosner, B. S., Blankfein, R. J., and Davis, R. A. (1966). Depth electrographic studies of the caudate nucleus in man, *Electroencephalogr. Clin. Neurophysiol. 20,* 391–396.

Rosvold, H. E. (1968). The prefrontal cortex and caudate nucleus: A system for effecting correction in response mechanisms, in: *Mind as a Tissue* (C. Rupp, ed.), Harper & Row (Hoeber), New York, pp. 21–38.

Rosvold, H. E., Mirsky, A. F., and Pribram, K. H. (1954). Influence of amygdalectomy on social behavior in monkeys, *J. Comp. Physiol. Psychol. 47,* 173–178.

Roth, J. J., and Ralph, C. L. (1976). Body temperature of the lizard (*Anolis carolinensis*): Effect of parietalectomy, *J. Exp. Zool. 198,* 17–27.

Roth, J. J., and Roth, E. C. (1980). The parietal–pineal complex among paleovertebrates: Evidence for temperature regulation, in: *A Cold Look at the Warm-Blooded Dinosaurs* (R. D. K. Thomas and E. C. Olson, eds.), Westview Press, Boulder, Colo., pp. 189–231.

Royce, G. J. (1978). Autoradiographic evidence for a discontinuous projection to the caudate nucleus from the centromedian nucleus in the cat, *Brain Res. 146,* 145–150.

Royce, G. J. (1982). Laminar origin of cortical neurons which project upon the caudate nucleus: A horseradish peroxidase investigation in the cat, *J. Comp. Neurol. 205,* 8–29.

Rubens, A. B. (1975). Aphasia with infarction in the territory of the anterior cerebral artery, *Cortex 11,* 239–250.

Rubenstein, E. H., and Delgado, J. M. R. (1963). Inhibition induced by forebrain stimulation in the monkey, *Am. J. Physiol. 205,* 941–948.

Rubinstein, H. S., and Freeman, W. (1940). Cerebellar agenesis, *J. Nerv. Ment. Dis. 92,* 489–502.

Rundles, R. W., and Papez, J. W. (1937). Connections between the striatum and the substantia nigra in a human brain, *Arch. Neurol. Psychiatry 38,* 550–563.

Russell, B. (1921). *The Analysis of Mind,* Allen & Unwin, London.

Rust, A. (1974). Handwerkliches Konnen und Lebensweise des Steinzeitmenschen, in: *Mannheimer forum 73/74. Ein Panorama der Naturwissenschaften* (zurammengestellt und redigiert von H.v. Ditfurth), Boehringer Mannheim GmbH, pp. 193–247.

Rylander, G. (1948). Personality analysis before and after frontal lobotomy, in: *The Frontal Lobes, Res. Publ. Assoc. Res. Nerv. Ment. Dis. 27,* 691–705.

Sackeim, H. A., Greenberg, M. S., Weiman, A. L., Gur, R. C., Hungerbuhler, J. P., and Geschwind, N. (1982). Hemispheric asymmetry in the expression of positive and negative emotions. Neurologic evidence, *Arch. Neurol. 39,* 210–218.

Salk, L. (1960). The effects of the normal heartbeat sound on the behavior of the new-born infant: Implications for mental health, *World Ment. Health 12,* 168–175.

Sanides, F. (1969). Comparative architectonics of the neocortex of mammals and their evolutionary interpretation, *Ann. N.Y. Acad. Sci. 16,* 404–423.

Sano, K., and Malamud, N. (1953). Clinical significance of sclerosis of the cornu ammonis. Ictal "psychic phenomena," *Arch. Neurol. Psychiatry 70,* 40–53.

Saper, C. B., and Loewy, A. D. (1980). Efferent connections of the parabrachial nucleus in the rat, *Brain Res. 197,* 291–317.

Saper, C. B., Swanson, L. W., and Cowan, W. M. (1976). The efferent connections of the ventromedial nucleus of the hypothalamus of the rat, *J. Comp. Neurol. 169,* 409–442.

Saper, C. B., Swanson, L. W., and Cowan, W. M. (1979). An autoradiographic study of the efferent connections of the lateral hypothalamic area in the rat, *J. Comp. Neurol. 183,* 689–706.

Sar, M., Stumpf, W. E., Miller, R. J., Chang, K.-J., and Cuatrecasas, P. (1978). Immunohistochemical localization of enkephalin in rat brain and spinal cord, *J. Comp. Neurol. 182,* 17–38.

Sarich, V. M., and Wilson, A. C. (1967). Immunological time scale for hominid evolution, *Science 158,* 1200–1203.

Sawa, M., Ueki, Y., Aritam, M., and Harada, T. (1954). Preliminary report on the amygdaloidectomy on the psychotic patients, with interpretation of oral–emotional manifestation in schizophrenics, *Folia Psychiatr. Neurol. Jpn. 7,* 309–329.

Sawyer, C. H. (1972). Functions of the amygdala related to the feedback actions of gonadal steroid hormones, in: *The Neurobiology of the Amygdala* (B. E. Eleftheriou, ed.), Plenum Press, New York, pp. 745–762.

Scalia, F. (1966). Some olfactory pathways in the rabbit brain, *J. Comp. Neurol. 126,* 285–310.

Scalia, F., and Winans, S. S. (1975). The differential projections of the olfactory bulb and accessory olfactory bulb in mammals, *J. Comp. Neurol. 161,* 31–56.

Scarff, J. E. (1948). Unilateral lobotomy with relief of ipsilateral, contralateral, and bilateral pain, *J. Neurosurg. 5,* 288–293.

Schäfer, E. A. (1895). *Quain's Elements of Anatomy,* Volume III, Part I, Longmans, Green, New York, pp. 139–203.

Schäfer, E. A. (1898/1900). The cerebral cortex, in: *Text-Book of Physiology* (E. A. Schäfer, ed.), Pentland, Edinburgh.

Schaffer, K. (1890). Pathologie und pathologische Anatomie der Lyssa, Eine klinisch-histologische Studie, *Beitr. Pathol. Anat. 7,* 191–244.

Schaffer, K. (1892). Beitrag zur Histologie der Ammonshornformation, *Arch. Mikrosk. Anat. 39,* 611–632.

Schaller, G. B. (1963). *The Mountain Gorilla. Ecology and Behavior,* University of Chicago Press, Chicago.

Scheibel, M. E., and Scheibel, A. B. (1970). The rapid Golgi method. Indian summer or renaissance?, in: *Contemporary Research Methods in Neuroanatomy* (W. J. H. Nauta and S. O. E. Ebbesson, eds.), Springer-Verlag, Berlin.

Schell, G. R., and Strick, P. L. (1984). The origin of thalamic inputs to the arcuate premotor and supplementary motor areas, *J. Neurosci. 4,* 539–560.

Schenk, V. W. D. (1959). Unilateral atrophy of the fornix, in: *Recent Neurological Research* (A. Biemond, ed.), Elsevier, Amsterdam, pp. 168–179.

Schildkraut, J. J. (1965). The catecholamine hypothesis of affective disorders: A review of supporting evidence, *Am. J. Psychiatry 122,* 509–522.

Schiller, F. (1967). The vicissitudes of the basal ganglia (Further landmarks in cerebral nomenclature), *Bull. Hist. Med. 41,* 515–538.

Schneider, R. (1949). Ein Beitrag zur Ontogenese der Basalganglien des Menschen, *Anat. Nachr. 1,* 115–137.

Scholz, W. (1959). The contribution of patho-anatomical research to the problem of epilepsy, *Epilepsia 1,* 36–55.

Schön Ybarra, M. A. (1984). Locomotion and postures of red howlers in a deciduous forest–savanna interface, *Am. J. Phys. Anthropol. 63,* 65–76.

Schreiner, L., and Kling, A. (1953). Behavioral changes following rhinencephalic injury in cat, *J. Neurophysiol. 16,* 643–659.

Schrodinger, E. (1951). *Science and Humanism: Physics in our Time,* Cambridge University Press, London.

Scollo-Lavizzari, G., and Akert, K. (1963). Cortical area 8 and its thalamic projection in *Macaca mulatta, J. Comp. Neurol. 121,* 259–269.

Scott, J. P. (1962). Introduction to animal behaviour, in: *The Behaviour of Domestic Animals* (E. S. E. Hafez, ed.), Williams & Wilkins, Baltimore, pp. 1–20.

Scoville, W. B. (1954). The limbic lobe in man, *J. Neurosurg. 11,* 64–66.

Scoville, W. B., and Milner, B. (1957). Loss of recent memory after bilateral hippocampal lesions, *J. Neurol. Neurosurg. Psychiatry 20,* 11–21.

Scoville, W. B., Dunsmore, R. H., Liberson, W. T., Henry, C. E., and Pepe, A. (1953). Observations on medial temporal lobotomy and uncotomy in the treatment of psychotic states, in: *Psychiatric Treatment, Res. Publ. Assoc. Res. Nerv. Ment. Dis. 31,* 347–369.

Segal, M., and Landis, S. (1974). Afferents to the hippocampus of the rat studied with the method of retrograde transport of horseradish peroxidase, *Brain Res. 78,* 1–15.

Segundo, J. P., and Machne, X. (1956). Unitary responses to afferent volleys in lenticular nucleus and claustrum, *J. Neurophysiol. 19,* 325–339.

Sequiera, L. W., Carrasco, L. H., Curry, A., Jennings, L. C., Lord, M. A., and Sutton, R. N. P. (1979). Detection of herpes-simplex viral genome in brain tissue, *Lancet* 1, 609–612.

Serafetinides, E. A., and Falconer, M. A. (1962). The effects of temporal lobectomy in epileptic patients with psychosis, *J. Ment. Sci. 108,* 584–593.

Shakow, D., and Rapaport, D. (1964). The influence of Freud on American psychology, *Psychological Issues,* Volume 13, International Universities Press, New York.

Shands, H. C. (1977). Science, linguistic science, and the invention of the future, *Semiotica 19,* 85–102.

Shapiro, A. K. (1970). Symposium. Gilles de la Tourette's syndrome, *N.Y. J. Med. 70,* 2193–2214.

Sheehan, D. (1936). Discovery of the autonomic nervous system, *Arch. Neurol. Psychiatry 35,* 1081–1115.

Shellshear, J. L. (1920/21). The basal arteries of the forebrain and their functional significance, *J. Anat. 55,* 27–35.

Shenkin, H. A., and Lewey, F. H. (1943). Aura of taste preceding convulsions associated with a lesion of the parietal operculum: Report of case, *Arch. Neurol. 50,* 375–378.

Sheridan, P. J. (1979). The nucleus interstitialis striae terminalis and the nucleus amygdaloideus medialis: Prime targets for androgen in the rat forebrain, *Endocrinology 104,* 130–136.

Sherk, H., and LeVay, S. (1981). The visual claustrum of the cat. III. Receptive field properties, *J. Neurosci. 1,* 993–1002.

Sherrington, C. S. (1900). Cutaneous sensations, in: *Text-Book of Physiology,* Volume 2 (E. A. Schäfer, ed.), Pentland, Edinburgh, pp. 920–1001.

Sherrington, C. S. (1906/1947). *The Integrative Action of the Nervous System,* 2nd Edition, Yale University Press, New Haven, Conn.

Shizume, K., Fukashi, M., Liro, S., Matsuda, K., Nagataki, S., and Okinaka, S. (1962). Effect of electrical stimulation of the limbic system on pituitary thyroidal function, *Endocrinology 71,* 456–463.

Showers, M. J. C. (1959). The cingulate gyrus: Additional motor area and cortical autonomic regulator, *J. Comp. Neurol. 112,* 231–287.

Sidman, R. L., and Angevine, J. B., Jr. (1962). Autoradiographic analysis of time of origin of nuclear versus cortical components of mouse telencephalon, *Anat. Rec. 142,* 326–327.

Sidman, R. L., and Rakic, P. (1982). Development of the human central nervous system, in: *Histology and Histopathology of the Nervous System,* Volume I (W. Haymaker and R. D. Adams, eds.), Thomas, Springfield, Ill., pp. 3–145.

Siegel, A., and Tassoni, J. P. (1971a). Differential efferent projections from the ventral and dorsal hippocampus of the cat, *Brain Behav. Evol. 4,* 185–200.

Siegel, A., and Tassoni, J. P. (1971b). Differential efferent projections of the lateral and medial septal nuclei to the hippocampus in the cat, *Brain Behav. Evol. 4,* 210–219.

Siegel, A., Edinger, H., and Ohgami, S. (1974). The topographical organization of the hippocampal projection to the septal area: A comparative neuroanatomical analysis in the gerbil, rat, rabbit, and cat, *J. Comp. Neurol. 157,* 359–378.

Sigrist, F. (1945). Zur Physiologie des Vicq d'Azyrschen Bündels und seiner umittelbaren Umgebung, *Helv. Physiol. Acta 3,* 361–372.

Simon, E. T., Hiller, J. M., and Edelman, I. (1973). Stereospecific binding of potent narcotic analgesic (^{3}H) etorphine to rat brain homogenates, *Proc. Natl. Acad. Sci. USA 70,* 1947–1949.

Simons, E. L. (1965). New fossil apes from Egypt and the initial differentiation of the Hominoidea, *Nature 205,* 135–139.

Simpson, D. A. (1952). The efferent fibres of the hippocampus in the monkey, *J. Neurol. Neurosurg. Psychiatry 15,* 79–92.

Simpson, G. G. (1927). Mesozoic Mammalia. IX. The brain of Jurassic mammals, *Am. J. Sci. 214,* 259–268.

Singer, M. (1962). *The Brain of the Dog in Section,* Saunders, Philadelphia.

Singer, M., and Yakovlev, P. I. (1954). *The Human Brain in Sagittal Section,* Thomas, Springfield, Ill.

Skeen, L. C., and Hall, W. C. (1977). Efferent projection of the main and assessory olfactory bulb in the tree shrew (*Tupaia glis*), *J. Comp. Neurol. 172,* 1–35.

Slater, E., Beard, H. W., and Glithero, E. (1963). The schizophrenia-like psychoses of epilepsy. I. Psychiatric aspects, *Br. J. Psychiatry 109,* 95–150.

Slotnick, B. M. (1967). Disturbances of maternal behavior in the rat following lesions of the cingulate cortex, *Behaviour 24,* 204–236.

Slotnick, B. M. (1969). Maternal behavior deficits following forebrain lesions in the rat, *Am. Soc. Zool. 9,* 1068–1069.

Slotnick, B. M. (1975). Neural and hormonal basis of maternal behavior in the rat, in: *Hormonal Correlates of Behavior,* Volume 2 (B. Eleftheriou and R. Sprott, eds.), Plenum Press, New York, pp. 585–656.

Slotnick, B. M., and Katz, H. M. (1974). Olfactory learning-set formation in rats, *Science 185,* 796–798.

Slotnick, B. M., and McMullen, M. F. (1972). Intraspecific fighting in albino mice with septal forebrain lesions, *Physiol. Behav. 8,* 333–337.

Slotnick, B. M., and Nigrosh, B. J. (1975). Maternal behavior in mice with cingulate cortical, amygdala, or septal lesions, *J. Comp. Physiol. Psychol. 88,* 118–127.

Slusher, M. A. (1966). Effects of cortisol implants in the brainstem and ventral hippocampus on diurnal corticosteroid levels, *Exp. Brain Res. 1,* 184–194.

Small, J. G., Milstein, V., and Stevens, J. R. (1962). Are psychomotor epileptics different? A controlled study, *Arch. Neurol. 7,* 187–194.

Smart, I. H. M., and Sturrock, R. R. (1979). Ontogeny of the neostriatum, in: *The Neostriatum* (I. Divac and R. G. E. Oberg, eds.), Pergamon Press, Elmsford, N.Y., pp. 127–146.

Smith, G. E. (1895). The connection between the olfactory bulb and the hippocampus, *Anat. Anz. 10,* 470–474.

Smith, G. E. (1901). Notes upon the natural subdivision of the cerebral hemisphere, *J. Anat. Physiol. 35,* 431–454.

Smith, G. E. (1902). On the homologies of the cerebral sulci, *J. Anat. 36,* 309–319.

Smith, G. E. (1903–04). The morphology of the occipital region of the cerebral hemisphere in man and ape, *Anat. Anz. 24,* 436–451.

Smith, G. E. (1910). The term "archipallium," a disclaimer, *Anat. Anz. Jena 35,* 429.

Smith, G. E. (1918–19). A preliminary note on the morphology of the corpus striatum and the origin of the neopallium, *J. Anat. 53,* 271–291.

Smith, G. E. (1919). The significance of the cerebral cortex [Croonian Lectures], *Br. Med. J. 1,* 796–797; *2,* 11–12.

Smith, J. C., and Sales, G. D. (1980). Ultrasonic behavior and mother–infant interactions in rodents, in: *Maternal Influences and Early Behavior* (R. W. Bell and W. P. Smotherman, eds.), SP Medical & Scientific Books, New York, pp. 105–133.

Smith, J. W. (1969). Messages of vertebrate communication, *Science 165,* 145–150.

Smith, O. C. (1930). The corpus striatum, amygdala, and stria terminalis of Tamandua tetradactyla, *J. Comp. Neurol. 51,* 65–127.

Smith, W. J., Chase, J., and Katz, A. (1974). Tongue showing: A facial display of humans and other primate species, *Semiotica 11,* 201–246.

Smith, W. K. (1945). The functional significance of the rostral cingular cortex as revealed by its responses to electrical excitation, *J. Neurophysiol. 8,* 241–255.

Snow, C. P. (1967). *Variety of Men,* Scribner's, New York.

Snyder, M., and Diamond, I. T. (1968). The organization and function of the visual cortex in the tree shrew, *Brain Behav. Evol. 1,* 244–288.

Snyder, S. H. (1972). Catecholamines in the brain as mediators of amphetamine psychosis, *Arch. Gen. Psychiatry 27,* 169–179.

Sodetz, F. J., and Bunnell, B. N. (1970). Septal ablation and the social behavior of the golden hamster, *Physiol. Behav. 5,* 79–88.

Sommer, W. (1880). Erkrankung des Ammonshorns als aetiologisches Moment der Epilepsie, *Arch. Psychiatr. 10,* 631–675.

Sorenson, E. R., and Gajdusek, D. C. (1969). Nutrition in the Kuru region, *Acta Trop. 26,* 281–330.

Spatz, H. (1922a). Über Beziehungen zwischen der Substantia nigra des Mittelhirnfusses und dem Globus pallidus des Linsenkernes, *Verh. Anat. Ges. 31,* 159–180.

Spatz, H. (1922b). Über den Eisennachweis im Gehirn, besondes in Zentren des Extrapyramidal-motorischen Systems, *Z. Gesamte Neurol. Psychiatr. 77,* 261–390.

Spatz, H. (1924). Zur Ontogenese des Striatum und des Pallidum, *Dtsch. Z. Nervenheilkd. 81,* 185–190.

Spatz, H. (1937). Über die Bedeutung der basalen Rinde, auf Grund von Beobachtungen bei Pickscher Krankheit und bei gedeckten Hirnverletzungen, *Z. Neurol. Psychiatr. 158,* 208–232.

Spatz, M., and Laqueur, G. L. (1968). Transplacental chemical induction of microencephaly in two strains of rats. I. (33404), *Proc. Soc. Exptl. Biol. Med. 129,* 705–710.

Spector, I., Hassmannova, Y., and Albe-Fessard, D. (1970). A macrophysiological study of functional organization of the claustrum, *Exp. Neurol. 29,* 31–51.

Spencer, H. (1896). *Principles of Psychology,* two volumes, Appleton, New York.

Spencer, H. J. (1976). Antagonism of cortical excitation of striatal neurons by glutamic acid diethyl ester: Evidence for glutamic acid as a excitatory transmitter in the rat striatum, *Brain Res. 102,* 91–101.

Spencer, S. S. (1981). Depth electroencephalography in selection of refractory epilepsy for surgery, *Ann. Neurol. 9,* 207–214.

Spencer, S. S., Spencer, D. D., Williamson, P. D., and Mattson, R. H. (1983). Sexual automatisms in complex partial seizures, *Neurology 33,* 527–533.

Spielmeyer, W. (1927). Die pathogenese des epileptischen Krampfes. Histopathologischer Teil, *Z. Gesamte Neurol. Psychiatr. 109,* 501–520.

Spielmeyer, W. (1930). The anatomic substratum of the convulsive state, *Arch. Neurol. Psychiatry 23,* 869–875.

Sprague, J. M., and Meyer, M. (1950). An experimental study of the fornix in the rabbit, *J. Anat. 84,* 354–368.

Squire, L. R. (1986). Mechanisms of memory, *Science 232,* 1612–1619.

Squire, L. R., and Zola-Morgan, S. (1983). The neurology of memory: The case for correspondence between the findings for human and nonhuman primate, in: *The Physiological Basis of Memory,* (J. A. Deutsch, ed.) Academic Press, New York, pp. 199–268.

Stamm, J. S. (1954). Control of hoarding activity in rats by the median cerebral cortex, *J. Comp. Physiol. Psychol. 47,* 21–27.

Stamm, J. S. (1955). The function of the median cerebral cortex in maternal behavior of rats, *J. Comp. Physiol. Psychol. 48,* 347–356.

Stamps, J. A., and Barlow, G. W. (1972). Variation and stereotype in the displays of *Anolis aeneus* (Sauria: Iguanidae), *Behav. 47,* 67–94.

Starzl, T. E., and Magoun, H. W. (1951). Organization of the diffuse thalamic projection system, *J. Neurophysiol. 14,* 133–146.

Starzl, T. E., and Whitlock, D. G. (1952). Diffuse thalamic projection system in monkey, *J. Neurophysiol. 15,* 449–468.

Stauder, K. H. (1935). Epilepsie und Schlafenlappen, *Arch. Psychiatr. Nervenkr. 104,* 181–212.

Steen, L. A. (1988). The science of patterns, *Science 240,* 611–616.

Stephan, H. (1964). Die kortikalen Anteile des limbischen Systems, *Nervenarzt 9,* 396–401.

Stephan, H. (1979). Comparative volumetric studies on striatum in insectivores and primates, *Appl. Neurophysiol. 42,* 78–80.

Stephan, H., and Andy, O. J. (1964). Quantitative comparisons of brain structures from insectivores to primates, *Am. Zool. 4,* 59–74.

Stepien, L. S., Cordeau, J. P., and Rasmussen, T. (1960). The effect of temporal lobe and hippocampal lesions on auditory and visual recent memory in monkeys, *Brain 83,* 470–489.

Sternberger, L. A. (1974). *Immunocytochemistry,* Prentice–Hall, Englewood Cliffs, N.J.

Stevens, J. R. (1957). The "march" of temporal lobe epilepsy, *AMA Arch. Neurol. Psychiatry 77,* 227–236.

Stevens, J. R. (1975). Interictal clinical manifestations of complex partial seizures, *Adv. Neurol. 11,* 85–112.

Stevens, J. R., Glaser, G. H., and MacLean, P. D. (1955). The influence of sodium amytal on the recollection of seizure states, *Trans. Am. Neurol. Assoc. 79,* 40–45.

Stevens, J. R., Kim, C., and MacLean, P. D. (1961). Stimulation of caudate nucleus: Behavioral effects of chemical and electrical stimulation, *Arch. Neurol. 4,* 47–54.

Stevens, J. R., Mark, V. H., Erwin, F., Pacheco, P., and Suematsu, K. (1969). Deep temporal stimulation in man, *Arch. Neurol. 21,* 157–169.

Stevens, S. S. (1951). *Handbook of Experimental Psychology,* Wiley, New York.

Steward, O., and Scoville, S. A. (1976). Cells of origin of entorhinal cortical afferents to the hippocampus and fascia dentata of the rat, *J. Comp. Neurol. 169,* 347–370.

Stier, E. (1911). *Untersuchungen über Linkshändigkeit und die functionellen Differenzen der Hirnhälften,* Fischer, Jena.

Stierlin, H. (1974). Karl Jaspers' psychiatry in the light of his basic philosophic position, *J. Hist. Behav. Sci. 10,* 213–223.

Stonorov, D. (1972). Protocol at the annual brown bear fish feast, *Nat. Hist. 81,* 66–94.

Straus, W. (1964). Factors affecting the cytochemical reaction of peroxidase with benzidine and the stability of the blue reaction product, *J. Histochem. Cytochem. 12,* 462.

Strick, P. L. (1976). Anatomical analysis of ventrolateral thalamic input to primate motor cortex, *J. Neurophysiol. 39,* 1020–1031.

Strong, O. S. (1895). The cranial nerves of Amphibia, *J. Morphol. 10,* 101–217.

Stuart, D. G., Porter, R. W., Adey, W. R., and Kamikawa, Y. (1964). Hypothalamic unit activity. I. Visceral and somatic influences, *Electroencephalogr. Clin. Neurophysiol. 16,* 237–247.

Stumpf, W. E. (1968). Estradiol-concentrating neurons: Topography in the hypothalamus by dry-mount autoradiography, *Science 162,* 1001–1003.

Subirana, A. (1953). Discussion of a paper by H. Gastaut, *Epilepsia 2,* 95–96.

Subirana, A., and Oller-Daurelia, L. (1953). The seizures with a feeling of paradisiacal happiness as the onset manifestation of certain temporal symptomatic epilepsies, *C. R. du Ve Congres Neurologique International,* Lisbonne, *IV,* 246–250.

Sudakov, K., MacLean, P. D., Reeves, A. G., and Marino, R. (1971). Unit study of exteroceptive inputs to claustrocortex in awake, sitting, squirrel monkey, *Brain Res. 28,* 19–34.

Suess, E. (1904). *The face of the earth* (Trans. by H. B. C. Sollas), Oxford University Press (Clarendon), London, pp. 417–420.

Sutton, D., Larson, C., and Lindeman, R. C. (1974). Neocortical and limbic lesion effects on primate phonation, *Brain Res. 71,* 61–75.

Swanson, L. W. (1976). An autoradiographic study of the efferent connections of the preoptic region in the rat, *J. Comp. Neurol. 167,* 227–256.

Swanson, L. W., and Cowan, W. M. (1975). Hippocampo-hypothalamic connections: Origin in subicular cortex, not Ammon's horn, *Science 189,* 303–304.

Swanson, L. W., and Cowan, W. M. (1979). The connections of the septal region in the rat, *J. Comp. Neurol. 186,* 621–656.

Swash, M. (1972). Released involuntary laughter after temporal lobe infarction, *J. Neurol. Neurosurg. Psychiatry 35,* 108–113.

Switzer, R., and Hill, J. (1979). Globus pallidus component in olfactory tubercle: Evidence based on iron distribution, *Soc. Neurosci. Abstr. 5,* 79.

Switzer, R., Hill, J., and Heimer, L. (1982). The globus pallidus and its rostroventral extension into the olfactory tubercle of the rat: A cyto- and chemoarchitectural study, *Neuroscience 7,* 1891–1904.

Symmes, D., Newman, J. D., Talmage-Riggs, G., and Lieblich, A. K. (1979). Individuality and stability of isolation peeps in squirrel monkeys, *Anim. Behav. 27,* 1142–1152.

Szabo, J. (1962). Topical distribution of the striatal efferents in the monkey, *Exp. Neurol. 5,* 21–36.

Szabo, J. (1967). The efferent projections of the putamen in the monkey, *Exp. Neurol. 19,* 463–476.

Szabo, J. (1970). Projections from the body of the caudate nucleus in the rhesus monkey, *Exp. Neurol. 27,* 1–15.

Tagliamonte, A., Tagliamonte, P., Gessa, G. L., and Brodie, B. B. (1969). Compulsive sexual activity induced by p-chlorophenylalanine in normal and pinealectomized male rats, *Science 166,* 1433–1435.

Talairach, J., Bancaud, J., Geier, S., Bordas-Ferrer, M., Szikla, G., and Rusu, M. (1973). The cingulate gyrus and human behaviour, *Electroencephalogr. Clin. Neurophysiol. 34,* 45–52.

Talland, G. A. (1965). An amnesic patient's disavowal of his own recall performance and its attribution to the interviewer, *Psychiatr. Neurol. 149,* 67–76.

Tarr, R. S. (1977). Role of the amygdala in the intraspecies aggressive behavior of the iguanid lizard, *Sceloporus occidentalis, Physiol. Behav. 18,* 1153–1158.

Tennyson, V. M., Barrett, R. E., Cohen, G., Cote, L., Heikkila, R., and Mytilineou, C. (1972). The developing neostriatum of the rabbit: Correlation of fluorescence histochemistry, electron microscopy, endogenous dopamine levels, and [^{3}H]dopamine uptake, *Brain Res. 46,* 251–285.

Terenius, L. (1973). Stereospecific interaction between narcotic analgesics and synaptic plasma membrane fraction of rat cerebral cortex, *Acta Pharmacol. Toxicol. 33,* 317–320.

Terkel, J., and Rosenblatt, J. S. (1972). Humoral factors underlying maternal behavior at parturition: Cross transfusion between freely moving rats, *J. Comp. Physiol. Psychol. 80,* 365.

Terreberry, R. R., and Neafsey, E. J. (1983). Rat medial frontal cortex: A visceral motor region with a direct projection to the solitary nucleus, *Brain Res. 278,* 245–249.

Terzian, H., and Ore, G. D. (1955). Syndrome of Klüver and Bucy. Reproduced in man by bilateral removal of the temporal lobes, *Neurology 5,* 373–380.

Teschemacher, O. K. E., Cox, B. M., and Goldstein, A. (1975). A peptide-like substance from pituitary that acts like morphine, *Life Sci. 16,* 1771–1776.

Theodore, W. H., Brooks, R., Sato, S., Patronas, N., Margolin, R., Di Chiro, G., and Porter, R. J. (1984). The role of positron emission tomography in the evaluation of seizure disorders, *Ann. Neurol. 15,* S176–S179.

Thoa, N. B., Tizabi, Y., and Jacobowitz, D. M. (1977). The effect of isolation on catecholamine concentration and turnover in discrete areas of the rat brain, *Brain Res. 131,* 259–269.

Thomas, R. D. K., and Olson, E. C. (1980). *A Cold Look at the Warm-Blooded Dinosaurs,* West View Press, Boulder.

Thompson, D. W. (1917/1952). *On Growth and Form,* two volumes, Cambridge University Press, London.

Thompson, J. M., Woolsey, C. N., and Talbot, S. A. (1950). Visual areas I and II of cerebral cortex of rabbit, *J. Neurophysiol. 13,* 277–288.

Thompson, R. (1964). A note on cortical and subcortical injuries and avoidance learning by rats, in: *The Frontal Granular Cortex and Behavior* (J. M. Warren and W. Akert, eds.), McGraw–Hill, New York, pp. 16–27.

Thomson, J. A. (1893). Embryology, in: *Chamber's Encyclopedia: A Dictionary of Universal Knowledge,* Volume 4, Lippincott, Philadelphia, pp. 317–322.

Thomson, K. S. (1988). Marginalia. Ontogeny and phylogeny recapitulated, *Am. Sci. 76,* 273–275.

Thorington, R. W. (1968). Observation of squirrel monkeys in a Colombian forest, in: *The Squirrel Monkey* (L. A. Rosenblum and R. W. Cooper, eds.), Academic Press, New York, pp. 69–85.

Tiger, L. (1970). *Men in Groups,* Vintage Books, New York.

Tilney, F. (1936). The development and constituents of the human hypophysis, *Bull. Neurol. Inst. N.Y. 5,* 387–436.

Tilney, F., and Morrison, J. F. (1912). Pseudo-bulbar palsy, clinically and pathologically considered, with the clinical report of five cases, *J. Nerv. Ment. Dis. 39,* 500–535.

Tilney, F., and Riley, H. A. (1930). *The Form and Functions of the Central Nervous System,* Harper & Row (Hoeber), New York.

Tinbergen, N. (1940). Die Übersprungbewegung, *Z. Tierpsychol. 4,* 1–40.

Tinbergen, N. (1951). *The Study of Instinct,* Oxford University Press (Claredon), London.

Tobias, P. V. (1964). The Olduvai Bed I hominine with special reference to its cranial capacity, *Nature 202,* 3–4.

Tobias, T. J. (1975). Afferents to prefrontal cortex from the thalamic mediodorsal nucleus in the rhesus monkey, *Brain Res. 83,* 191–212.

Todd, R. B. (1861). *Clinical Lectures* (L. S. Beale, ed.), 2nd Edition, Churchill, London, pp. 267, 275.

Torch, W. C., Hirano, A., and Solomon, S. (1977). Anterograde transneuronal degeneration in the limbic system: Clinical–anatomic correlation, *Neurology 27,* 1157–1163.

Toth, N. (1985). Archeological evidence for the preferential right-handedness in the lower and middle Pleistocene, and its possible implications, *J. Hum. Evol. 14,* 607–614.

Tow, M. P., and Whitty, C. W. M. (1953). Personality changes after operations on the cingulate gyrus in man, *J. Neurol. Neurosurg. Psychiatry 16,* 186–193.

Trachy, R. E., Sutton, D., and Lindeman, R. C. (1981). Vocalization: Differential effects of anterior cingulate damage on response rate and acoustical features in the rhesus monkey, *Am. J. Primatol. 1,* 43–55.

Trankell, A. (1950). *Vansterhanthet hos barn i skolaldern* (with a summary in English), Forum, Helsingfors.

Travis, R. P., Jr., and Sparks, D. L. (1968). Unitary responses and discrimination learning in the squirrel monkey: The globus pallidus, *Physiol. Behav. 3,* 187–196.

Travis, R. P., Jr., Hooten, T. F., and Sparks, D. L. (1968). Single unit activity related to behavior motivated by food reward, *Physiol. Behav. 3,* 309–318.

Tredgold, R. F., and Soddy, K. (1963). *Tredgold's Textbook of Mental Deficiency (Subnormality),* 10th Edition, Williams & Wilkins, Baltimore.

Trembly, B. (1956). A study of the functions of the septal nuclei, Thesis, Yale University School of Medicine.

Trembly, B., and Sutin, J. (1961). Septal projections to the dorsomedial thalamic nucleus in the cat, *Electroencephalogr. Clin. Neurophysiol. 13,* 880–888.

Tretiakoff, C. (1919). Contribution à l'étude de l'anatomie pathologique du locus niger de Soemmering, Thèse de Paris, Paris.

Trimble, M. R. (1986). Radiological studies in epileptic psychosis, in: *The Limbic System: Functional Organization and Clinical Disorders* (B. K. Doane and K. E. Livingston, eds.), Raven Press, New York, pp. 195–200.

Trinkaus, E., and LeMay, M. (1982). Occipital bunning among later Pleistocene hominids, *Am. J. Phys. Anthropol. 57,* 27–35.

Tsai, M. H., Garber, B. B., and Larramendi, L. M. H. (1981a). ^{3}H-thymidine autoradiographic analysis of telencephalic histogenesis in the chick embryo: I. Neuronal birthdates of telencephalic compartments, *in situ, J. Comp. Neurol. 198,* 275–292.

Tsai, M. H., Garber, B. B., and Larramendi, L. M. H. (1981b). ^{3}H-thymidine autoradiographic analysis of telencephalic histogenesis in the chick embryo: II. Dynamics of neuronal migration, displacement, and aggregation, *J. Comp. Neurol. 198,* 293–306.

Turner, W. (1890). The convolutions of the brain: A study in comparative anatomy, *J. Anat. Physiol. 25,* 105–153.

Twarog, B. M., and Page, I. H. (1953). Serotonin content of some mammalian tissues and urine and a method for its determination, *Am. J. Physiol. 175,* 157–161.

Uchimura, T. (1928). Zur Pathogenese der ortlich elektiven Ammonshornerkrankung, *Z. Gesamte Neurol. Psychiatr. 114,* 567–601.

Ule, G. (1951). Korsakow-Psychose nach doppelseitiger Ammonshornzerstörung mit transneuronaler Degeneration der Corpora mamillaria, *Dtsch. Z. Nervenheilkd. 165,* 446–456.

Ulinski, P. S. (1978). A working concept of the organization of the anterior dorsal ventricular ridge, in: *The Behavior and Neurology of Lizards* (N. Greenberg and P. D. MacLean, eds.), U.S. Government Printing Office, Washington, D.C., DHEW Publication No. (ADM) 77-491, pp. 121–132.

Ungerstedt, U. (1971). Stereotaxic mapping of the monoamine pathways in the rat brain, *Acta Physiol. Scand. 367,* 1–48.

Valenstein, E. S., and Nauta, W. J. H. (1959). A comparison of the distribution of the fornix system in the rat, guinea pig, cat, and monkey, *J. Comp. Neurol. 113,* 337–363.

Valladas, H., Reyss, J. L., Joron, J. L., Valladas, G., Bar-Yosef, O., and Vandermeersch, B. (1988). Thermoluminescence dating of Mousterian 'Proto-Cro-Magnon' remains from Israel and the origin of modern man, *Nature 331,* 614–616.

Valverde, F. (1965). *Studies on the Piriform Lobe,* Harvard University Press, Cambridge, Mass.

van Bogaert, L., Radermecker, J., and Devos, J. (1955). Sur une Observation mortelle d'encéphalite aiguë nécrosante: (Sa situation vis-à-vis du groupe des encéphalites transmises par arthropodes et de l'encéphalite herpétique), *Rev. Neurol. 92,* 329–356.

Van Buren, J. M. (1961). Sensory, motor and autonomic effects of mesial temporal stimulation in man, *J. Neurosurg. 28,* 273–288.

Van Buren, J. M., and Ajmone Marsan, C. (1960). A correlation of autonomic and EEG components in temporal lobe epilepsy, *Arch. Neurol. 3,* 683–703.

Van Buren, J. M., and Borke, R. C. (1972). The mesial temporal substratum of memory. Anatomical studies in three individuals, *Brain 95,* 599–632.

Van Buren, J. M., Li, C. L., and Ojemann, G. A. (1966). The fronto-striatal arrest response in man, *Electroencephalogr. Clin. Neurophysiol. 21,* 114–130.

Van der Kooy, D., and Kuypers, H. G. J. M. (1979). Fluorescent retrograde double labeling: Axonal branching in the ascending raphe and nigral projections, *Science 204,* 873–875.

Van der Kooy, D., McGinty, J. F., Koda, L. Y., Gerfen, C. R., and Bloom, F. E. (1982). Visceral cortex: A direct connection from prefrontal cortex to the solitary nucleus in the rat, *Neurosci. Lett. 33,* 123–127.

Vanegas, H., and Flynn, J. P. (1968). Inhibition of cortically-elicited movement by electrical stimulation of the hippocampus, *Brain Res. 11,* 489–506.

Van Hoesen, G. W., and Pandya, D. N. (1975). Some connections of the entorhinal (area 28) and perirhinal (area 35) cortices of the rhesus monkey. I. Temporal lobe afferents, *Brain Res. 95,* 1–24.

Van Hoesen, G. W., Yeterian, E. H., and La Vizzo-Mourey, R. (1981). Widespread corticostriate projections from temporal cortex of the rhesus monkey, *J. Comp. Neurol. 199,* 205–219.

Van Reeth, P. C. (1959). Un cas d'épilepsie temporale autoprovoqué et le problème de l'autostimulation cérébrale hédonique, *Acta Neurol. Belg. 4,* 490–495.

Van Reeth, P., Dierkens, J., and Luminet, D. (1958). L'hypersexualité dans l'épilepsie et les tumeurs du lobe temporal, *Acta Neurol. Psychiatr. Belg. 2,* 194–218.

Van Valen, L. (1960). Therapsids as mammals, *Evolution 14,* 304–313.

Van Valkenburg, C. T. (1912). Caudal connections of the corpus mamillare, *Proc. Acad. Sci. Amst. 14,* 1118–1121.

Van Woert, M. H., Jutkowitz, R., Rosenbaum, D., and Bowers, M. B., Jr. (1976). Gilles de la Tourette's syndrome: Biochemical approaches, in: *The Basal Ganglia, Res. Publ. Assoc. Res. Nerv. Ment. Dis. 55,* 459–465.

Vaz Ferreira, A. (1951). The cortical areas of the albino rat studied by silver impregnation, *J. Comp. Neurol. 95,* 177–244.

Verhaart, W. J. C. (1950). Fiber analysis of the basal ganglia, *J. Comp. Neurol. 93,* 425–440.

Vessie, P. R. (1932). On the transmission of Huntington chorea for 300 years—the Bures family group, *J. Nerv. Ment. Dis. 76,* 553–573.

Victor, M. (1981). Diseases of the nervous system due to nutritional deficiency, in: *Clinical Medicine* (J. Spittell, Jr., ed.), Harper & Row, New York, pp. 1–30.

Victor, M., and Yakovlev, P. I. (1955). S. S. Korsakoff's psychic disorder in conjunction with peripheral neuritis. A translation of Korsakoff's original article with brief comments on the author and his contribution to clinical medicine, *Neurology 5,* 394–406.

Victor, M., Adams, R. S., and Collins, G. H. (1971). *The Wernicke–Korsakoff Syndrome. A Clinical and Pathological Study of 245 Patients, 82 with Post-Mortem Examinations,* Davis, Philadelphia.

Vilensky, J. A., and Van Hoesen, G. W. (1981). Corticopontine projections from the cingulate cortex in the rhesus monkey, *Brain Res. 205,* 391–395.

Villablanca, J. R., and Marcus, R. J. (1975). Effects of caudate nuclei removal in cats. Comparison with effects of frontal cortex ablation, in: *Brain Mechanisms in Mental Retardation* (N. A. Buchwald and M. A. B. Brazier, eds.), Academic Press, New York, pp. 273–311.

Villiger, E. (1931). *Brain and Spinal Cord* (W. H. F. Addison, ed.), Lippincott, Philadelphia.

Vogt, B. A. (1976). Retrosplenial cortex in the rhesus monkey: A cytoarchitectonic and Golgi study, *J. Comp. Neurol. 169,* 63–98.

Vogt, B. A., and Peters, A. (1981). Form and distribution of neurons in rat cingulate cortex: Areas 32, 24, and 29, *J. Comp. Neurol. 195,* 603–625.

Vogt, B. A., Rosene, D. L., and Pandya, D. N. (1979). Thalamic and cortical afferents differentiate anterior from posterior cingulate cortex in the monkey, *Science 204,* 205–207.

Vogt, B. A., Rosene, D. L., and Peters, A. (1981). Synaptic termination of thalamic and callosal afferents in cingulate cortex of the rat, *J. Comp. Neurol. 201,* 265–283.

Vogt, B. A., Pandya, D. N., and Rosene, D. L. (1987). Cingulate cortex of rhesus monkey. I. Cytoarchitecture and thalamic afferents, *J. Comp. Neurol. 262,* 256–270.

Vogt, C. (1909). La myéloarchitecture du thalamus du cercopitheque, *J. Psychol. Neurol. 12,* 285–324.

Vogt, C., and Vogt, O. (1919a). Zur Kenntnis der pathologischen Veranderungen des Striatum und zur Pathophysiologie der dabei auftretenden Krankheitsersuchungen, *S. B. Heidelberger Akad. Wiss. Math. Nat. Kl. Abt. B Biol. Wiss. Abh. 14,* 3–56.

Vogt, C., and Vogt, O. (1919b). Allgemeine Ergebnisse unserer Hirnforschung, *J. Psychol. Neurol. 25,* Erg. Hoft 3.

Vogt, C., and Vogt, O. (1920). Lehre der Erkrankungen des striaren Systems, *J. Psychol. Neurol. 25,* 628–846.

Vogt, C., and Vogt, O. (1922). Erkrankungen der Grosshirnrinde im Lichte der Topistik, Pathoklise und Pathoarchitektonik, *J. Psychol. Neurol. 28,* 1.

Vogt, C., and Vogt, O. (1953). Gestaltung der topistischen Hirnforschung und ihre Forderung durch den Hirnbau und seine Anomalien, *J. Hirnforsch. 1,* 1–46.

Vogt, O. (1898). Sur un faisceau septo-thalamique, *C. R. Soc. Biol. 50,* 206–207.

Vogt, O. von (1910). Die myeloarchitektonische Felderung des menschlichen Stirnhirns, *J. Psychol. Neurol. 15,* 221–232.

Vonderahe, A. R. (1944). The anatomic substratum of emotion, *The New Scholasticism 18,* 76–95.

Voneida, T. J., and Trevarthen, C. B. (1969). An experimental study of transcallosal connections between the proreus gyri of the cat, *Brain Res. 12,* 384–395.

von Foerster, H., Mora, P. M., and Amiot, L. W. (1960). Doomsday: Friday 13 November, A. D. 2026, *Science 132,* 1291–1295.

Votaw, C. L. (1960). Certain functional and anatomical relations of the cornu ammonis of the macaque monkey. II. Anatomical relations, *J. Comp. Neurol. 114,* 283–293.

Wada, J., and Rasmussen, T. (1960). Intra-carotid injection of sodium amytal for the lateralization of cerebral speech dominance, *J. Neurosurg. 17,* 266–282.

Wadeson, R. W., Jr. (1965). Ego and central nervous system function—A frame of reference, *Perspect. Biol. Med. 8,* 520–532.

Waldeyer, W. (1891). Über einige neuere Forschungen im Gebiete der Anatomie des Centralnervensystems, *Dtsch. Med. Wochenschr. 17,* 1213–1218, 1244–1246, 1267–1269, 1287–1289, 1331–1332, 1352–1356.

Walker, A., Leakey, R. E., Harris, J. M., and Brown, F. H. (1986). 2.5-Myr *Australopithecus boisei* from west of Lake Turkana, Kenya, *Nature 322,* 517–522.

Walker, A. E. (1938). *The Primate Thalamus,* University of Chicago Press, Chicago.

Walker, A. E. (1940). A cytoarchitectural study of the prefrontal area of the macaque monkey, *J. Comp. Neurol. 73,* 59–86.

Wall, P. D., Glees, P., and Fulton, J. F. (1951). Corticofugal connexions of posterior orbital surface in rhesus monkey, *Brain 74,* 66–71.

Waller, W. H. (1850). Experiments on the section of the glossopharyngeal and observations of the alterations produced thereby in the structure of their primitive fibres, *Philos. Trans. 140,* 423–429.

Walsh, F. B. (1971). On the neurologist Dr. Frank Ford, *Johns Hopkins Med. J. 128,* 105–107.

Ward, A. A. (1948). The cingulate gyrus: Area 24, *J. Neurophysiol. 11,* 13–23.

Warren, J. M. (1953). Handedness in the rhesus monkey, *Science 118,* 622–623.

Warren, J. M. (1980). Handedness and laterality in humans and other animals, *Physiol. Psychol. 8,* 351–359.

Warren, J. M. Abplanalp, J. M., and Warren, H. B. (1967). The development of handedness in cats and monkeys, in: *Early Behavior* (E. H. Stevenson, E. H. Hess, and H. Rheingold, eds.), Wiley, New York, pp. 73–101.

Washburn, S. L., and DeVore, I (1961). Social life of baboons, *Sci. Am. 204,* 62–71.

Wasman, M., and Flynn, J. P. (1962). Directed attack elicited from hypothalamus, *Arch. Neurol. 6,* 220–227.

Watson, D. M. S. (1913). Further notes on the skull, brain, and organs of special sense of Diademodon, *Annu. Mag. Nat. Hist. 12,* 217–228.

Watson, J. B. (1924). *Behaviorism,* The People's Institute Publishing Co., New York.

Watts, C. R., and Stokes, A. W. (1971). The social order of turkeys, *Sci. Am. 224,* 112–118.

Waxler, M., and Rosvold, H. E. (1970). Delayed alternation in monkeys after removal of the hippocampus, *Neuropsychologia 8,* 137–146.

Webster, K. E. (1961). Cortico-striate interrelations in the albino rat, *J. Anat. 95,* 532–544.

Webster, K. E. (1965). The cortico-striatal projection in the cat, *J. Anat. 99,* 329–337.

Webster, K. E. (1979). Some aspects of the comparative study of the corpus striatum, in: *The Neostriatum* (I. Divac and R. G. E. Oberg, eds.), Pergamon Press, Elmsford, N.Y., pp. 107–126.

Webster's New World Dictionary (1968).

Wegener, A. L. (1915). *Die Entstehung der Kontinente und Ozeane,* Vieweg, Leipzig.

Weil, A. A. (1959). Ictal emotions occurring in temporal lobe dysfunction, *AMA Arch. Neurol. 1,* 87–97.

Weil, A. A. , Nosik, W. A., and Demmy, N. (1958). Electroencephalographic correlation of laughing fits, *Am. J. Med. Sci. 235,* 301–308.

Weisberger, B. A. (November 19, 1972). Book review of "*Black Mountain,*" The Washington Post, Washington, D.C.

Weiskrantz, L., and Warrington, E. K. (1975). The problem of the amnesic syndrome in man and animals, in: *The Hippocampus,* Volume 2 (R. L. Isaacson and K. H. Pribram, eds.), Plenum Press, New York, pp. 411–428.

Weiss, P., and Holland, Y. (1967). Neuronal dynamics and axonal flow: II. The olfactory nerve as model test object, *Proc. Natl. Acad. Sci. USA 57,* 258–264.

Weisskopf, V. F. (1983). The origin of the universe, *Am. Sci. 71,* 473–480.

Wernicke, C. (1881). *Lehrbuch der Gehirnkrankheiten für Arzte und Studierende,* Volume 2, Fischer, Kassel, pp. 229–242.

West, B. J., and Goldberger, A. L. (1987). Physiology in fractal dimensions, *Am. Sci. 75,* 354–365.

Westoll, T. S. (1943). The hyomandibular of *Eusthenopteron* and the tetrapod middle ear, *Proc. R. Soc. London Ser. B 131,* 393–414.

Westoll, T. S. (1945). The mammalian middle ear, *Nature 155,* 114–115.

Wever, E. G. (1965). Structure and function of the lizard ear, *J. Aud. Res. 5,* 331–371.

Wever, E. G. (1974). The lizard ear: Gekkonidae, *J. Morphol. 143,* 121–166.

Wheatley, M. D. (1944). Hypothalamus and affective behavior in cats: Study of effects of experimental lesions, with anatomic correlations, *Arch. Neurol. Psychiatry 52,* 296–316.

White, J. C., and Sweet, W. H. (1969). *Pain and the Neurosurgeon. A Forty-year Experience,* Thomas, Springfield, Ill.

White, L. E., Jr. (1965). Olfactory bulb projections of the rat, *Anat. Rec. 152,* 465–480.

Whitlock, D. G., and Nauta, W. J. H. (1956). Subcortical projections from the temporal neocortex in *Macaca mulatta, J. Comp. Neurol. 106,* 183–212.

Whitman, C. O. (1919). *The Behavior of Pigeons,* Posthumous work (H. Carr, ed.), Carnegie Institute Publication No. 257.

Wickler, W. von (1966). Ursprung und biologische Deutung des Genitalpräsentierens männlicher Primaten, *Z. Tierpsychol. 23,* 422–437.

Wiener, N. (1948). *Cybernetics, or Control and Communication in the Animal and the Machine,* Wiley, New York.

Wiesendanger, R., and Wiesendanger, M. (1985). The thalamic connections with medial area 6 (supplementary motor cortex) in the monkey (*Macaca fascicularis*), *Exp. Brain Res. 59,* 91–104.

Williams, D. (1956). The structure of emotions reflected in epileptic experiences, *Brain 79,* 29–67.

Willis, T. (1664). *Cerebri anatome, cui accessit nervorum descriptio et usus,* Martyn & Allestry, London.

Williston, S. W. (1925). *The Osteology of the Reptiles,* Harvard University Press, Cambridge, Mass.

Wilson, D. (1891). *The Right Hand: Left-Handedness,* Macmillan & Co., London.

Wilson, M. E., and Cragg, B. G. (1969). Projections from the medial geniculate body to the cerebral cortex in the cat, *Brain Res. 13,* 462–475.

Wilson, S. A. K. (1912). Progressive lenticular degeneration: A familial nervous disease associated with cirrhosis of the liver, *Brain 34,* 295–509.

Wilson, S. A, K. (1914). An experimental research into the anatomy and physiology of the corpus striatum, *Brain 36,* 427–492.

Wilson, S. A. K. (1924). Some problems in neurology. II. Pathological laughing and crying, *J. Neurol. Psychopathol. 4,* 299–333.

Wilson, S. A. K. (1925a). Disorders of motility and of muscle tone, with special reference to the corpus striatum. Lecture I, *Lancet 2,* 1–10.

Wilson, S. A. K. (1925b). Disorders of motility and of muscle tone, with special reference to the corpus striatum. Lecture II, *Lancet 2,* 53–62.

Wilson, S. A. K. (1925c). Disorders of motility and of muscle tone, with special reference to the corpus striatum. Lecture III, *Lancet 2,* 169–178.

Wilson, S. A. K. (1925d). Disorders of motility and of muscle tone, with special reference to the corpus striatum. Lecture IV, *Lancet 2,* 215–276.

Wilson, S. A. K. (1930). Nervous semeiology, with special reference to epilepsy, *Br. Med. J. 2,* 50–54.

Winans, S. S., and Scalia, F. (1970). Amygdaloid nucleus: New afferent input from the vomeronasal organ, *Science 170,* 330–332.

Winslow, J. B. (1732). *Exposition anatomique de la structure due corps humain,* Volume IV, Aux Depens de la Campagne, Amsterdam, Section 361.

Winter, P. (1969). Dialects in squirrel monkeys: Vocalization of the roman arch type, *Folia Primatol. 10,* 216–229.

Winter, P., Ploog, D., and Latta, J. (1966). Vocal repertoire of the squirrel monkey (*Saimiri sciureus*), its analysis and significance, *Exp. Brain Res. 1,* 359–384.

Wise, S. P., and Herkenham, M. (1982). Opiate receptor distribution in the cerebral cortex of the rhesus monkey, *Science 218,* 387–389.

Witt, E. D., and Goldman-Rakic, P. S. (1983). Intermittent thiamine deficiency in the rhesus monkey. I. Progression of neurological signs and neuroanatomical lesions, *Ann. Neurol. 13,* 376–395.

Wolff, P. H. (1969). The natural history of crying and other vocalizations in early infancy, in: *Determinants of Infant Behvior,* Volume 4 (B. M. Foss, ed.), Methuen, London, pp. 81–109.

Woods, J. W. (1956). Loss of aggressiveness in wild rats following lesions in the rhinencephalon, *XXth Int. Physiol. Congr. Abstr. Commun.* pp. 978–979.

Woollard, H. H., and Harpman, A. (1939). The cortical projection of the medial geniculate body, *J. Neurol. Psychiatry 2,* 35–44.

Woolley, D. W., and Shaw, E. (1954). A biochemical and pharmacological suggestion about certain mental disorders, *Proc. Natl. Acad. Sci. USA 20,* 228–231.

Woolsey, C. N., and Walzl, E. M. (1942). Topical projection of nerve fibers from local regions of the cochlea to the cerebral cortex of the cat, *Bull. Johns Hopkins Hosp. 71,* 315–344.

Woolsey, C. N., Marshall, W. H., and Bard, P. (1942). Representation of cutaneous tactile sensibility in the cerebral cortex of the monkey as indicated by evoked potentials, *Bull. Johns Hopkins Hosp. 70,* 399–441.

Wyler, A. R., and Bolender, N.-F. (1983). Preoperative CT diagnosis of mesial temporal sclerosis for surgical treatment of epilepsy, *Ann. Neurol. 13,* 59–64.

Wysoki, C. J. (1979). Neurobehavioral evidence for the involvement of the vomeronasal system in mammalian reproduction, *Neurosci. Biobehav. Rev. 3,* 301–341.

Wyss, J. M., Swanson, L. W., and Cowan, W. M. (1979). A study of subcortical afferents to the hippocampal formation in the rat, *Neuroscience 4,* 463–476.

Yakovlev, P. I. (1948). Motility, behavior, and the brain. Stereodynamic organization and neural coordinates of behavior, *J. Nerv. Ment. Dis. 107,* 313–335.

Yakovlev, P. I. (1966). The central "paradox" of Parkinson's disease. The Second Symposium on Parkinson's Disease, *J. Neurosurg. 2* (Suppl.), 292–301.

Yakovlev, P. I., Locke, S., Koskoff, D. Y., and Patton, R. A. (1960). Limbic nuclei of thalamus and connections of limbic cortex: I. Organization of projections of anterior group of nuclei and of midline nuclei of thalamus to anterior cingulate gyrus and hippocampal rudiment in monkey, *Arch. Neurol. 3,* 620–641.

Yap, P. M. (1952). The Latah reaction: Its pathodynamics and nosological position, *J. Ment. Sci. 98,* 505–564.

Yeterian, E. H., and Van Hoesen, G. W. (1978). Cortico-striate projections in the rhesus monkey: The organization of certain cortico-caudate connections, *Brain Res. 139,* 43–63.

Yokota, T., and MacLean, P. D. (1968). Fornix and fifth-nerve interaction on thalamic units in awake, sitting squirrel monkeys, *J. Neurophysiol. 31,* 358–370.

Yokota, T., Reeves, A. G., and MacLean, P. D. (1967). Intracellular olfactory response of hippocampal neurons in awake, sitting squirrel monkeys, *Science 157,* 1072–1073.

Yokota, T., Reeves, A. G., and MacLean, P. D. (1970). Differential effects of septal and olfactory volleys on intracellular responses of hippocampal neurons in awake, sitting monkeys, *J. Neurophysiol. 33,* 96–107.

Zahl, P. A. (1973). One strange night on turtle beach, *Natl. Geogr. 144,* 570–581.

Zieglgansberger, W., French, E. D., Siggins, G. R., and Bloom, F. E. (1979). Opioid peptides may excite hippocampal pyramidal neurons by inhibiting adjacent inhibitory interneurons, *Science 205,* 415–417.

Zola-Morgan, S., Squire, L. R., and Mishkin, M. (1982). The neuroanatomy of amnesia: Amygdala–hippocampus versus temporal stem, *Science 218,* 1337–1339.

Zuckerman, S. (1932). *The Social Life of Monkeys and Apes,* Kegan Paul, London.

Indexes

Author Index

Subject Index